P9-BJT-251

Slope Stability

Geotechnical Engineering and Geomorphology

Edited by

M. G. Anderson

Department of Geography, University of Bristol

and

K. S. Richards

Department of Geography, University of Cambridge

JOHN WILEY & SONS

Chichester · New York · Brisbane · Toronto · Singapore

Library of Congress Cataloging-in-Publication Data:

Slope stability — geotechnical engineering and
geomorphology.

Includes index.
1. Slopes (Soil mechanics) — Addresses, essays,
lectures. 2. Engineering geology — Addresses, essays,
lectures. 3. Geomorphology — Addresses, essays,
lectures. I. Anderson, M. G. II. Richards, K. S.
TA710.S553 1986 624.1′51 86-4063

ISBN 0 471 91021 X

British Library Cataloguing in Publication Data:

Slope stability: geotechnical engineering
and geomorphology.
1. Mass-wasting
I. Anderson, M. G. II. Richards, K. S.
551.4′36 QE598.2

ISBN 0 471 91021 X

Phototypeset for Dobbie Typesetting Service, Plymouth, Devon
Printed and bound in Great Britain

List of Contributors

M. G. Anderson	*Department of Geography, University of Bristol, University Road, Bristol BS8 1SS, England*
V. Cotecchia	*Instituto di Geologia Applicata e Geotechnica, Universita di Bari, Via Re David, 200, Bari, Italy*
J. C. Cripps	*Department of Geology, University of Sheffield, Mappin Street, Sheffield S1 3JD, England*
S. C. Francis	*Merlin Profilers (Velpro) Ltd, Duke House, Duke Street, Woking GU21 5BA, England*
D. Fredlund	*Department of Civil Engineering, University of Saskatchewan, Saskatoon, Saskatchewan, Canada S7N 0W0*
R. A. Freeze	*Department of Geological Sciences, University of British Columbia, Vancouver, British Columbia, Canada V6T 1W5*
D. R. Greenway	*Geotechnical Control Office, 6th Floor, Empire Centre, 68 Mody Road, Tsim Sha Tsui East, Kowloon, Hong Kong*
C. Harris	*Geography Section, Department of Geology, University College, Cardiff, PO Box 78, Cardiff CF1 1XL, U.K.*
S. R. Hencher	*Department of Earth Sciences, University of Leeds, Leeds LS2 9JT, England*
R. Le B. Hooke	*Department of Geology and Geophysics, University of Minnesota, Minneapolis, Minnesota 55455, USA*
R. H. Johnson	*Formerly, School of Geography, University of Manchester, Manchester M13 9PL, England*
M. J. Kirkby	*School of Geography, University of Leeds, Leeds LS2 9JT, England*
P. E. Kneale	*School of Geography, University of Leeds, Leeds LS2 9JT, England*
N. R. Lorriman	*Department of Geography, The University, Hull HU6 7RX, England*
D. F. T. Nash	*Department of Civil Engineering, Queen's Building, University of Bristol, University Walk, Bristol BS8 1TR, England*
T. Okimura	*Faculty of Engineering, Kobe University, Rokkodai Nada, Kobe 657, Japan*

K. OKUNISHI — *Disaster Prevention Research Institute, Kyoto University, Gokanosho, Uji-City 611, Japan*

K. S. RICHARDS — *Department of Geography, University of Cambridge, Downing Place, Cambridge CB2 3EN, England*

M. J. SELBY — *Department of Earth Sciences, University of Waikato, Private Bag, Hamilton, New Zealand*

J. M. SHEN — *Geotechnical Control Office, 6th Floor, Empire Centre, 68 Mody Road, Tsim Sha Tsui East, Kowloon, Hong Kong*

R. K. TAYLOR — *School of Engineering and Applied Science, University of Durham, Old Shire Hall, Durham DH1 3LE, England*

R. K. TORRANCE — *Department of Geography, Carleton University, Ottawa, Canada K1S 5B6*

Contents

Slope Stability
Edited by M. G. Anderson and K. S. Richards

Chapter 1

Modelling slope stability: the complimentary nature of geotechnical and geomorphological approaches

M. G. Anderson
Department of Geography
University of Bristol, Bristol BS8 1SS

and

K. S. Richards
Department of Geography
University of Cambridge, Cambridge CB2 3EN

Problems of slope stability, instability, and the associated mass movement processes of slope failure, represent research themes common to both geotechnical engineers and geomorphologists, although their perspectives clearly differ. While the engineering concern is generally site specific and limited by project design-life considerations to time periods of less than 100 years, for the geomorphologist longer term slope stability and slope evolution, involving a mixture of slope processes operating over tens of thousands of years, constitute a major area of enquiry (e.g. Kirkby, 1984). This simple distinction in terms of time-scale is, however, complicated by additional concerns. For engineers, the practical problems arising when fossil shears are reactivated by natural or artificial disturbances place a premium on understanding the history of slope development over geological (late Quaternary) time-scales, as the emphasis of the collection of studies introduced by Skempton (1976) illustrates. On the other hand, geomorphologists are necessarily interested in the process mechanisms preceding and during failure, in the types of mass movement process involved and the translation of debris. The apparent dichotomy in terms of time-scale is thus largely an illusion, and the true distinction lies in the objectives of analysis of slope stability: a statement of the stability of a specific slope for the engineer, and an appraisal of the role of slope failure processes in slope evolution for the geomorphologist. Given, therefore, that it is the objective rather than the methodology that distinguishes engineering and geomorphological approaches, there should be much to gain from a recognition of the common ground and potential for collaboration between engineers and geomorphologists.

It would appear that there are three major elements involved in the analysis of slope stability. The first is the essentially static consideration, characterized by limit-equilibrium models, of the stability of specific slopes. These are often calibrated by using minimal pore pressure data based on standpipe piezometers having slow response times, and are employed in back-analysis to estimate shear strength parameters at failure. The second element involves application of these models over short time-scales (seasonal, annual, or a few years), during which dynamic variability of external 'forcing' variables must be allowed for in the analysis—factors such as the slope hydrology, which causes transient peaks in soil moisture and pore pressure in relation to infiltration processes during rainstorms, or basal erosion which increases slope height and angle. Thirdly, there is the application of stability analyses to the explanation of long-term slope evolution, over periods when it is necessary to make assumptions about the response of pore-water pressures to climatically induced variations in average and extreme slope hydrology conditions. A final, separate issue concerns the role of weathering, which constitutes a common concern for both the engineer and the geomorphologist. The 'long-term' stability (i.e. over about 100 years) of cut slopes involves weathering-related processes of rebound, dilation and the ingress of water, with geotechnical consequences originally identified by Skempton (1964). In the 'long-term' of slope evolution, weathering processes result in progressive comminution of slope regolith materials (Carson and Petley, 1970), which changes the median grain size and size-sorting characteristics which influence the angle of internal friction and regolith stability. Furthermore, pedogenesis changes the vertical distribution of particle size, bulk density, and hydraulic conductivity within the regolith, with attendant influences on the slope hydrology.

Although these three elements involve the application of essentially similar models of slope stability, the problems of their calibration differ markedly. In the site-specific case, it is normal to assume fixed soil strength properties, worst-case hydrological and therefore pore-water pressure conditions, and two-dimensional representation of the failure surface. The two overriding concerns are with the selection of an appropriate stability analysis, and with data reduction—that is, with the choice of relevant parameters from a wide range of potentially available data. The engineering objective is to obtain parsimonious solutions in terms of both data and computational requirements and it is unnecessary for the model to represent the *process* of failure accurately and in detail as long as it successfully predicts the *stability* of the slope within the limits specified by the chosen factor of safety. The second level of sophistication in modelling slope stability is coincident with relaxation of the temporal constraint and introduction of dynamic hydrological conditions and time-variant material properties which reflect the geotechnical response to weathering. It is at this level that the interests of the geomorphologist and the engineer are most clearly equivalent, as regards model structure, data requirements, and application of model output. Accordingly, the emphasis of this volume is on the instrumentation and calibration of the key hydrological inputs to stability models, and the subtleties of calibration presented by the influences of vegetation loading, inhomogeneity of material, basal conditions, and earthquake stressing—all dynamic influences operating over a range of time-scales, which equally inform modelling applications in the site-specific engineering context and the more generalized study of slope evolution. Geomorphological applications of slope stability analysis to the interpretation of long-term development of hillslopes suffer the problem not of a potential embarrassment of data, but of a paucity which necessitates assumptions about both the appropriate model and its parameterization (Anderson *et al.*, 1980).

Recognition of the subtleties of dynamic calibration of stability analyses is therefore essential.

In the engineering context, there is a considerable *potential* data base which is nevertheless rarely exploited in full, and simplifying assumptions are generally acceptable. This is because in site investigation work, dynamic process conditions can safely be ignored as long as probable worse-case parameter values can be estimated. Engineering solutions are given flexibility by the imposition of conservative factors of safety, and it is therefore possible to make *assumptions* about input data which could in fact be *measured* (in the absence of economic constraints). The geomorphologist concerned with long-term stability problems is, however, concerned with predicting conditions at failure, and therefore for a factor of safety of unity. The model output requirement is therefore more rigorous, although the consequences of error are less dramatic. It would appear that there is convergence between short-term site investigation and long-term slope evolution applications in that reductionist approaches to data characterize both, albeit respectively, by choice and by necessity. It is possible that this similarity has encouraged geomorphologists to employ the simpler stability models and calibration assumptions (e.g. infinite slope planar slide with water table at the surface; Skempton and de Lory, 1957) in studying slope development, perhaps unaware of the generous factor of safety assumptions permissible in an engineering analysis. Thus, in long-term geomorphological analysis, there is comparatively little evidence to assist in narrowing the choices of model type and appropriate parameter values. Alternative model formulations can therefore be constructed, each of which is capable of predicting observed slope form, although their implications regarding process conditions and stability mechanisms may be very different. The essays in this collection emphasize the dynamic elements that complicate the calibration of short-term engineering analysis and the interpretation of natural slope development in both the short and long term, and focus on areas of common interest between geotechnical engineers and geomorphologists.

The limit equilibrium methods forming the framework for analysis of slope stability are reviewed by Nash in Chapter 2. The range of moment and force equilibrium methods available illustrates the problem faced by the geomorphologist lacking detailed information on shear surface geometry and pore-water pressure distribution, and back-analyses for failures (i.e. where the factor of safety = 1) indicate the variation of shear strength parameters, especially cohesion, implied at failure (Skempton and Hutchinson, 1969; Fredlund and Krahn, 1977). Calibration of these stability models requires data on shear strength properties and pore pressure conditions. The former are derived from a familiar range of field and laboratory techniques (Petley, 1984), while the latter demand improved techniques capable of instrumenting rapid groundwater and soil suction responses to rainfall without damping the transient peak conditions: Anderson and Kneale review some of the methods in Chapter 3. Sweeney (1982) has observed that if the influence of soil suction on shear strength is ignored, certain tropical slopes have factors of safety less than unity and yet show no signs of distress. A complication in this regard is the status of current techniques used for the estimation of basic shear strength. Without extremely accurate methods for shear strength estimation, the influence of suction on shear strength cannot be reliably inferred from back-analysis. Sweeney and Robertson (1979) have pointed out the problems of making laboratory measurements of effective stress shear strength parameters which represent the field behaviour of residual soils. They emphasize particularly the difficulties of testing at low stress levels, the incomplete knowledge of factors affecting effective stress shear strength parameters, such as structure, mineralogy and grading, and the problems associated with sample disturbance. Hencher *et al.* (1984), in a review of the back-analysis of landslide case studies in Hong Kong, are able to add specific support to this complexity by showing that even for a single landslide (with *no* groundwater) there are numerous possible solutions by back-analysis; a situation which may be attributable to complex hydrological conditions which are not always capable of incorporation in stability analysis.

In many respects, therefore, a key element in the interpretation of slope stability is rigorous incorporation of the hydrological element. This is explicitly considered in Chapter 4 by Fredlund, who outlines the development of a modified Coulomb criterion capable of incorporating suction effects such as those increasingly seen to play a significant role in maintaining stability of tropical slopes at abnormally steep angles (Rouse and Reading, 1985). Finite-difference models of soil water response to design storms form the basis for predicting minima of factors of safety achieved during the peaks of pore-water pressure. However, the hydrological characteristics of slope soils depend on other conditions such as the vegetation cover and soil structure. In Chapter 6, Greenway demonstrates the complex interplay between vegetation loading of a slope, strength augmentation by roots, and modification of the slope hydrology as a consequence of evapotranspiration and altered infiltration characteristics. Anderson and Shen, using a similar modelling framework in Chapter 7 to define the dynamic soil-water regime, deal with structural properties of the soil. In Hong Kong—a sub-tropical location with frequent cyclone rainstorms, steep slopes mantled by residual soils, and considerable development pressure—analysis of slope stability has been forced to develop rapidly to a state-of-the-art condition. Steep slopes there are often protected by 'chunam', a soil–cement mix which reduces infiltration during rain and increases soil suction and slope stability. Chapter 7 outlines a modelling framework for assessing the soil hydrological effects of intact and degraded chunam, having implications for temporal variation of slope stability; the strategy could, of course, be used in other situations where natural soil surface crusting, or subsurface impeded drainage, occur.

Concern with soils which can be assumed homogeneous or structured in a simple layered manner is evident in most of the chapters in the collection, but Hencher discusses the evaluation of joints and structures in strength-anisotropic rocks in Chapter 5. This provides a necessary background for the consideration in Chapter 8 of groundwater effects in heavily fractured bedrock. Japanese geotechnical engineers have been forced, by the need to predict landslide hazards in a heavily populated and mountainous country, to develop simple but reliable methods of modelling slope stability in rocks of such structural complexity as to preclude analytical approaches, and Okunishi and Okimura review several applications of one such model. Tectonic processes constitute a major concern for geotechnical engineers in earthquake-prone regions, such as Japan and southern Italy, and for long earthquakes have been recognized by geomorphologists as a significant factor in hillslope denudation in susceptible locations. Simonett (1967), for example, estimated landslide erosion depths following an earthquake in Papua-New Guinea, and demonstrated a decline in erosion with distance from the epicentre. In Chapter 9, Cotecchia reviews the experience of engineering geology in earthquake-prone southern Italy, where again the emphasis of the engineering profession reflects the importance of the factors triggering instability and a settlement history which has left a legacy of hilltop towns surrounded by unstable slopes.

Slope stability is controlled in the long term by basal erosion conditions and weathering of slope materials. The pattern of slope evolution caused by different scales of mass movement process depends on the relative rates of erosion and weathering, and the emphasis in Chapter 10 (Richards and Lorriman) is on slope failure in locations such as marine cliffs and river-banks where the active control is the rate of basal erosion. Under these conditions, it is necessary to augment the stability analyses presented in Chapter 2 by additional models in which tensile and compressive strengths are more relevant material properties than the shear strength. Taylor and Cripps, in discussing weathering processes and geotechnical profiles in over-consolidated clays occupy a pivotal position in the book (Chapter 13) since the consequences of these issues for both the engineering and evolutionary time-scales of analysis are evident in, respectively, the time-dependent models of geotechnical stability developed by Skempton (1964) and Carson and Petley (1970) and noted above.

Two distinctive approaches to the modelling of long-term slope evolution are presented by Kirkby and Freeze in Chapters 11 and 12. Kirkby adopts a framework in which the slope geometry reflects the mechanical properties of materials and the processes of failure only in a generalized way. Explicit reference to stability analyses is absent but assumptions are necessarily made about the limiting angles of both stability and of deposition of failed masses, in order to solve the continuity equation for sediment transfer on the slope and predict the evolution of slope geometry. Freeze, by contrast, makes a number of explicit assumptions about the hydrogeology of slopes in different climatic environments, themselves characterized very simply by average storm conditions, and develops a sensitivity analysis of the response of limiting slope angle to the climatic and hydrogeological parameters by assuming a circular-arc failure model. These two approaches are not judged by their quantitative predictions, but by the inherent logic of their model structures, and illustrate the calibration problems arising in the study of slope evolution.

The last six chapters present geomorphological studies of slope development and processes, classified either by material properties or climatic environment. In each case the understanding of natural slope forms and their associated mass movement processes is improved by reference to the dynamic influences which also allow complex modelling of site problems. Quick clays (Torrance, Chapter 14) are characterized by catastrophic loss of strength on remoulding and by rapid retrogressive failure (flow-slides). The former can only be understood in the context of the depositional and weathering history of the material, and the latter require a more dynamic modelling strategy of post-failure mechanics than is provided by standard limit-equilibrium methods. In Chapter 15, Selby, by contrast, considers rock slopes in which stress-release joint development during landform evolution causes a gradual mutual adjustment of rock slope morphology and rock mass strength, in which the slope failure processes are conditioned by the structural influences discussed in Chapter 5 but in turn influence the development of these structures. An important contrast also exists between Chapter 16 and the final group of three chapters. Hooke reviews the role of mass movement processes—specifically, debris flows—in creating constructional fan landforms in semi-arid environments. This chapter demonstrates that there is a considerable danger in interpreting past mass movement in the context of secular climatic change; in a dry environment, rare but normally extreme storms provide the trigger for slope failure, and these may occur at random at any time. Chapter 18 then provides a review, and a detailed case study, of the role of past mass movements in controlling Quaternary slope evolution in a presently temperate environment. Johnson stresses the interaction of a range of failure processes and the problems of simplistically associating these with specific periods of time without adequate dating control. Weathering, pedogenesis, and accumulation of material in potentially unstable locations, all represent long-term processes which may be truncated by short-term random storm effects causing failure, and this implies that periods of slope failure may not necessarily be explained by climatic change alone. Nevertheless, periglacial conditions in the history of slopes in present temperate environments will have encouraged mobilization of the regolith over permanently frozen substrates, and Harris (Chapter 17) explains the thaw-consolidation process that allows this to occur on extremely gentle slopes. The modern arctic environment provides an analogue for interpretation of low-angle shear surfaces in Britain, and allows direct examination of diagnostic sedimentological fabric properties that can be used to reconstruct past processes when observed in fossil form in areas formerly subject to periglacial conditions. Finally, in Chapter 19, Francis demonstrates that even various aspects of the simple infinite slope planar slide model create problems when geomorphologists attempt to exploit it as a means of interpreting slope development by mass movement processes, invoking the concept of threshold slopes.

Notwithstanding the research thrusts relating to individual elements that have been introduced (such as shear strength and suction, strength testing procedures, pore-water pressure, and rock structure) there are three more general (and thus more complex) themes which would benefit from further research. It is important that these themes are simultaneously explored and the research implications assessed, since they provide frameworks for model development and inference with which the individual research elements must be compatible. The three themes are:

1. *Post-failure mechanisms* Stability analysis is essentially static even when dynamic variation of external conditions is incorporated: the analysis assesses whether a slope might fail, but provides no insight into the behaviour of mass movement debris after failure. This is necessary in the modelling of long-term slope evolution, since the distance of translation downslope controls the change of slope form. It is also relevant to the engineer in the contexts of hazard assessment and land use zoning. Post-failure mechanics including loss of strength on remoulding, dilation and velocity of the moving mass, and the dispersal of pore water are dynamic processes which control the distance of translation and the conditions for cessation of motion. Models of the threshold for cessation of motion have been developed for debris flows, in terms of flow thickness and slope gradient (Takahashi and Yoshida, 1979), but are lacking for other mass movement processes. General information on the distance of displacement may derive from consideration of the morphometry of mass movements (Brunsden, 1973; Hansen, 1984). For example, the ratio of the length of the depositional zone to the length of shear surface reflects the type of failure, and the water content at failure relative to the liquid limit of the material (Crozier, 1973). However, the processes operating after failure need to be understood before displacement can be predicted.

   Thus, detailed mechanisms of slope failure constitute an important theme for collaborative research. For example, van Asch (1984) and Iverson (1986) have developed models to predict the progress of failure. The former deals with progressive creep of failed blocks on a pre-existing shear surface, while the latter presents a detailed sediment budget for a landslide complex undergoing progressive failure as downslope translation of failed blocks on the lower slope removes support for the material upslope. This model has been calibrated by several years of detailed monitoring of a single landslide site, and illustrates the methodology necessary for site appraisal of the interaction of mass movement processes which affect different magnitudes and frequencies of material transfer on an unstable slope. Slope degradation at sites experiencing both deep-seated rotational failure and shallow translational mudslides, such as the coastal slopes investigated by Brunsden (1974) and Bromhead (1978), can only be understood by proper consideration of the processes operating to diminish material strength, adjust the moisture regime in failed debris, and transfer material downslope. In complex failure zones the importance of dynamic consideration of the landslide process is apparent, and taken in conjunction with the evaluation of dynamic external influences such as rain-induced pore-water pressure fluctuations, rapid basal erosion, weathering, vegetation effects and earthquake stress, suggest a range of joint concerns for geotechnical engineers and geomorphologists.

2. *Two- and three-dimensional slope geometry* A significant limitation of conventional stability analysis in the context of interpretation of natural slope development is its emphasis on two-dimensional models; extension to three-dimensional slide geometry either concludes that side-effects are of negligible importance, or relies on predetermined shear surface geometry to maintain analytical simplicity. However, geomorphological slope studies have increasingly abandoned emphasis on the two-dimensional profile form both in morphological analysis (Parsons, 1979) and in assessment of feedbacks between hydrological processes and landform development (Crabtree and Burt, 1983). Convergence of surface and subsurface flow into hollows increases chemical weathering at these sites, and also favours accumulation of colluvium. Reneau *et al.* (1984) have shown that colluvium-filled hollows in California store the products of slope weathering and transport processes, until the combination of a critical depositional thickness and a rainfall event capable of generating critical pore-water pressures results in failure of the colluvium store by a rapid mass movement process. The three-dimensional geometry of the slope encourages the convergence of sediment transport to the storage site, but also increases the local pore-water pressure peaks by establishing convergence of soil-water flow. Thus although the engineer may find side-effects of marginal significance in a limit equilibrium stability analysis, the calibration of a coupled hydrological model predicting soil moisture and pore-pressure response to rainfall may be improved by consideration of lateral subsurface inflow (Anderson and Pope, 1984). This is therefore a second area for further research.
3. *Weathering effects* The combination of long- and short-term processes is an important element of geomorphological interpretation of slope development, and progressive accumulation of weathering products until failure is triggered by heavy rain, is one example. Carson and Petley (1970) also argue that a discontinuous pattern of slope angle decline occurs as the limiting angle of stability of the regolith changes when weathering alters its particle-size distribution until a period of instability can be triggered. A further secular process which has been little researched in this context, but which is likely to be a significant element in long-term slope stability because of its relevance for the soil moisture conditions which control stability, is pedogenesis. Progressive illuviation of clay into a soil B-horizon increases the differential of hydraulic conductivity between the upper and lower soil horizons, resulting in an increased probability that a perched

water table can be generated during a storm whose intensity is less than the surface infiltration capacity but whose duration is such that A-horizon soil-water storage capacity is filled. Ellis and Richards (1985) draw attention to the possible role of this process in encouraging catastrophic slope failure of post-glacial podzolized soils on the slopes of Ulvådalen in west-central Norway, during an extreme storm in 1962. Although the *temporal* pedogenic influence may not be relevant to the engineer, the *spatial* pattern of soil profile structure and its effect on soil moisture storage and transmission may be a matter for joint concern with geomorphologists.

REFERENCES

Anderson, M. G., and Pope, R. G. (1984). 'The incorporation of soil water physics models into geotechnical studies of landslide behaviour.' *Proc. 4th Int. Symp. Landslides*, Toronto, **1**, 349–53.

Anderson, M. G., Richards, K. S., and Kneale, P. E. (1980). 'The role of stability analysis in the interpretation of the evolution of threshold slopes.' *Trans. Inst. Brit. Geogrs.* N.S., **5**, 100–12.

Bromhead, E. N. (1978). 'Large landslides in London Clay at Herne Bay, Kent.' *Q. J. Eng. Geol.*, **11**, 291–304.

Brunsden, D. (1973). 'The application of systems theory to the study of mass movement.' *Geologica Applicata e Idrogeologia*, Univ. of Bari, **8**, 185–207.

Brunsden, D. (1974). 'The degradation of a coastal slope, Dorset, England.' *Inst. Brit. Geogrs.* Spec. Publ., 7, 79–98.

Carson, M. A., and Petley, D. J. (1970). 'The existence of threshold hillslopes in the denudation of the landscape.' *Trans. Inst. Brit. Geog.*, **49**, 71–95.

Crabtree, R. W., and Burt, T. P. (1983). 'Spatial variation in solutional denudation and soil moisture over a hillslope hollow.' *Earth Surf. Proc. Landforms*, **8**, 151–60.

Crozier, M. J. (1973). 'Techniques for the morphometric analysis of landslips.' *Zeitschrift fur Geomorphologie*, **17**, 78–101.

Ellis, S., and Richards, K. S. (1985). 'Pedogenic and geotechnical aspects of Late Flandrian slope instability in Ulvådalen, west-central Norway.' In *Geomorphology and Soils*. Richards K. S., Arnett, R. R., and Ellis, S. (eds.), George Allen & Unwin, London, pp. 328–47.

Fredlund, D., and Krahn, J. (1977). 'Comparison of slope stability methods of analysis.' *Can. Geotech. J.*, **14**, 429–39.

Hansen, M. J. (1984). 'Strategies for classification of landslides.' In *Slope Instability*. Brunsden, D., and Prior D. B. (eds.), John Wiley, Chichester, pp. 1–25.

Hencher, S. R., Massey, J. B., and Brand, E. W. (1984). 'Application of back analysis to Hong Kong landslides.' *Proc. 4th Int. Symp. Landslides*, Toronto, 631–8.

Iverson, R. M. (1986). 'Dynamics of slow landslides: a theory for time-dependent behaviour.' In *Hillslope Processes*. Abrahams, A. (ed.), George Allen & Unwin, London, 297–317.

Kirkby, M. J. (1984). 'Modelling cliff development in South Wales: Savigear revisited.' *Zeitschrift fur Geomorphologie*, **28**, 405–26.

Parsons, A. J. (1979). 'Plan form and profile form of hillslopes.' *Earth Surf. Proc. Landforms*, **4**, 395–402.

Petley, D. J. (1984). 'Ground investigation, sampling and testing for studies of slope stability.' In *Slope Instability*. Brunsden, D. and Prior, D. B. (eds.), John Wiley, Chichester, pp. 67–101.

Reneau, S. L., Dietrich, W. E., Wilson, C. J., and Rogers, J. D. (1984). 'Colluvial deposits and associated landslides in the northern San Francisco Bay area, California, USA.' *Proc. 4th Int. Symp. Landslides*, Toronto, **1**, 425–30.

Rouse, C. and Reading A. (1985). 'Soil mechanics and natural slope stability.' In *Geomorphology and Soils*. Richards, K. S., Arnett, R. R. and Ellis, S. (eds.), George Allen & Unwin, London, pp. 159–79.

Simonett, D. S. (1967). 'Landslide distribution and earthquakes in the Bewani and Torricelli Mountains, New Guinea.' In *Landform Studies from Australia and New Guinea*. Jennings, J. N., and Mabbutt, J. A. (eds.), Methuen, London, 64–84.

Skempton, A. W. (1964). 'Long-term stability of clay slopes.' *Geotechnique*, **14**, 75–102.
Skempton, A. W. (1976). 'Introduction. A discussion on "Valley slopes and cliffs in Southern England: morphology, mechanics, and Quaternary history".' *Phil. Trans. Roy. Soc., London*, **283A**, 423–6.
Skempton, A. W., and de Lory, F. A. (1957). 'Stability of natural slopes in the London Clay.' *Proc. 4th Int. Conf. Soil Mech. Fndtn. Eng.*, London, **2**, 378–81.
Skempton, A. W., and Hutchinson, J. N. (1969). 'Stability of natural slopes and embankment foundations.' State-of-the-Art Report *7th Int. Conf. Soil Mech. Fndtn. Eng.*, Mexico, 291–355.
Sweeney, D. J. (1982). 'Some *in situ* soil suction measurements in Hong Kong's residual soil slopes.' *Proc. 7th SE Asia Geotech. Conf.* McFeat-Smith, I. and Lumb, P. (eds.), Hong Kong Inst. Eng., 91–105.
Sweeney, D. J., and Robertson, P. K. (1979). 'A fundamental approach to slope stability problems in Hong Kong.' *Hong Kong Engineer*, **7**, 35–44.
Takahashi, T., and Yoshida, H. (1979). 'Study on the deposition of debris flows (1)—deposition due to abrupt change of bed slope.' *Annals of the Disaster Prevention Research Institute*, Kyoto University, **22B-2**, 315–28.
van Asch, Th. W. J. (1984). 'Creep processes in landslides.' *Earth Surf. Proc. Landforms*, **9**, 573–83.

Slope Stability
Edited by M. G. Anderson and K. S. Richards

Chapter 2

A Comparative Review of Limit Equilibrium Methods of Stability Analysis

DAVID NASH
Department of Civil Engineering
University of Bristol, Bristol BS8 1TR

2.1 INTRODUCTION

A quantitative assessment of the stability of a slope is clearly important when a judgement is needed about whether the slope is stable or not, and decisions are to be made as a consequence. There are a number of different methods of stability analysis available, but the procedures are broadly similar in concept. The slope under consideration and the soil forming it are modelled theoretically, the loadings on the slope are included and a failure criterion for the soil is introduced. The analysis then indicates whether the failure criterion is reached, and a comparison may then be made between these conditions and those under which the modelled slope would just fail. It is important to realize that the results of such an analysis are of limited value in themselves, as they are dependent on the theoretical models adopted for the slope and the soil. However, when combined with experience of their application in similar conditions, the results are a useful input to the decision-making process.

In this chapter several methods of analysis are described which are commonly used for assessing the stability of slopes. Most of these are limit equilibrium methods and later in this chapter the theoretical basis for this class of method is examined, and the different methods are compared. But first we must define our terms and introduce the fundamentals of soil mechanics. The treatment of the shear strength of soils and seepage presented here is necessarily rather brief. For a fuller treatment the reader is referred to the standard textbooks on soil mechanics such as those by Terzaghi and Peck (1967), Lambe and Whitman (1969), and Cedergren (1967).

2.2 FUNDAMENTALS

2.2.1 Nature of Soils

Soils consist of an assemblage of particles with water and air filling the void spaces between. The particles are generally either rock fragments (gravel size and larger) or rock-forming minerals (sand size and smaller). The rock-forming minerals are either massive minerals such as quartz and feldspar or clay minerals such as kaolinite, illite, chlorite, montmorillonite, and halloysite. In general the massive minerals form angular rotund particles of silt and sand size (0.002–2 mm), while the clay minerals form flake-shaped particles which are generally smaller than 0.002 mm in dimensions.

The presence of water strongly influences the physical interaction of soil particles. Massive minerals do not absorb much water as their surfaces are usually inactive. However, clay minerals are formed from sheets of silicates which are frequently charged electrically and tend to absorb water readily. This ability to absorb water results in the characteristic plasticity of clays and in their swelling and shrinkage behaviour. The ions present in the pore water can influence the electrochemical interaction between clay particles, and the geochemistry may be completely altered by weathering (see Chapters 13 and 14). Thus the water may influence the shear strength parameters of a soil, and an appreciation of the significance of geochemistry can be important for an understanding of long-term morphological processes.

2.2.2 The Principle of Effective Stress

Stresses are transmitted through a soil (or rock) both by the soil skeleton and by the pore fluid. The soil skeleton can transmit normal stresses and shear stresses through the interparticle contacts, but the pore fluid can exert only all-round pressure. It is the stresses transmitted by the soil skeleton through the interparticle contacts that control the strength and deformation of the soil. Where stresses applied to the soil are wholly supported by the pore fluid pressure, they are not felt by the contacts between particles and hence the soil behaviour is not affected. This is the basis of the *principle of effective stress.*

The *effective stress* (σ') acting on any plane is defined by the equation:

$$\sigma' = \sigma - u \tag{2.1}$$

in which σ is the *total stress* acting on the plane and u is the *pore pressure*. The total stress is equal to the total force per unit area acting normal to the plane, and the pore pressure may vary independently as is discussed in section 2.2.5. The effective stress is approximately equal to the average intergranular force per unit area, and the equation shows that it cannot be measured directly but that total stress and pore pressure must first be determined separately.

The principle of effective stress is fundamental to soil mechanics and is thus central to understanding the behaviour of slopes.

2.2.3 Influence of Stress History

The compressibility and shear behaviour of a soil deposit are strongly influenced by its stress history. If a soil has not been subjected to a greater vertical effective stress in the past than it is carrying at present it is said to be *normally-consolidated.* Conversely, if in the past the soil has carried a larger vertical effective stress than at present, it is said to be *over-consolidated.* In general, normally-consolidated soils are weaker and more compressible than over-consolidated soils under the same state of stress.

2.2.4 Shear Strength

Soils derive their strength from the contacts between particles which can transmit the normal and shear forces. In general these interparticle contacts are primarily frictional, and so shear strength is directly governed by the effective stresses. The shear strength of a soil is fully mobilized when a soil element can only just support the stresses imposed on it and large plastic deformations are occurring.

Figure 2.1(a) shows an element of soil whose shear strength has been fully mobilized under principal stresses σ_{1_f} and σ_{3_f}. The Mohr's circle for this condition is shown in Figures 2.1(b) together with the circles for other similar tests on the same soil in which the strength has been fully mobilized under different combinations of stresses. The intermediate principal stress is not shown as it has only a minor influence on the shear strength. The lines drawn on the diagram which just touch all the circles are termed the *failure envelope.* By definition the soil cannot sustain a state of stress given by point X as this lies outside the failure envelope.

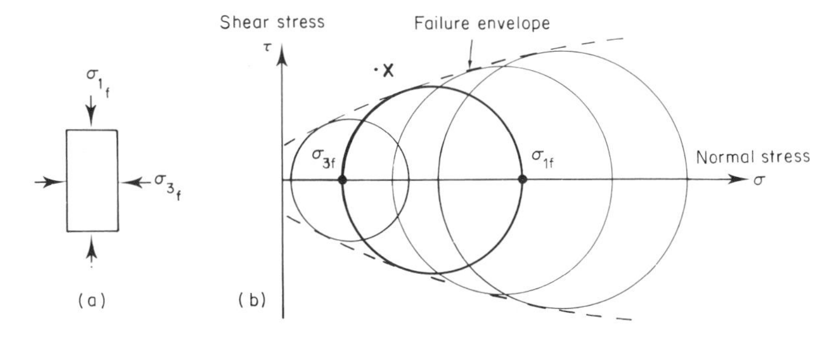

Figure 2.1 (a) Element of soil at failure. (b) Mohr's circles

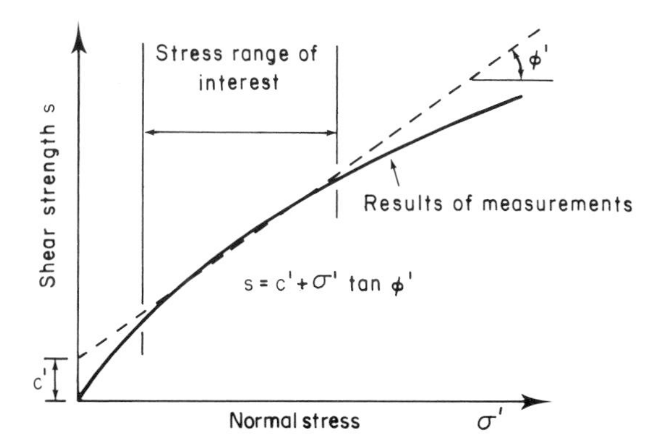

Figure 2.2 Straight line approximation of curved failure envelope

For most soils the failure envelope is curved (see Figure 2.2) but it is usual to approximate it over a limited stress range by the Mohr–Coulomb linear relationship:

$$s = c' + \sigma' \tan\phi' \tag{2.2}$$

where c' is the *cohesion* and ϕ' is the *angle of shearing resistance* determined using effective stresses. Where the shear strength parameters are determined under particular conditions they are given a subscript. Thus c_r', ϕ_r' denote the residual strength parameters (effective stress), and c_u, ϕ_u denote the undrained strength parameters (total stress).

2.2.4.1 Shear Strength of Cohesionless Soils

Sands and gravels are known as cohesionless soils as they evidently have no shear strength when they are unconfined. Their strength when confined arises from interparticle friction and the interlocking of the particles. Figure 2.3 illustrates the effect of interlock. In Figure 2.3(a) interlocked rows of cylinders are being sheared relative to one another. For significant

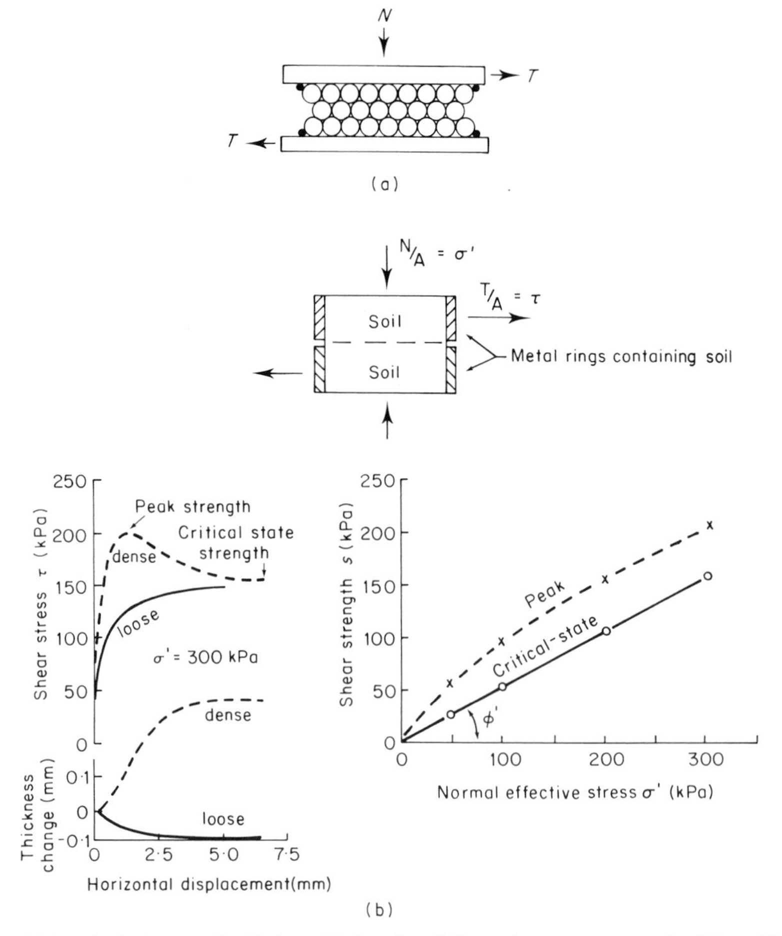

Figure 2.3 (a) Interlocked rows of cylinders. (b) Results of direct shear tests on sands. (*From Kenney, 1984.*)

movement to occur the rows must separate initially, and this involves work against the applied normal stress as well as work in overcoming the friction between pairs of cylinders. After some initial movement the cylinders can slide directly over each other.

A similar process occurs in sands (see Figure 2.3(b)). When dense sands are sheared they dilate initially (i.e. increase in overall volume), and this gives a high strength (peak strength). If the shearing is continued the strength drops and eventually the sand may be sheared at constant volume. In contrast, loose sands contract during shearing and the strength gradually increases with further deformation until the sand is shearing at constant volume. Careful measurements have shown that under the same normal stress the constant volume condition is the same irrespective of the initial density and this is known as the *critical state* of the sand.

2.2.4.2 Shear Strength of Cohesive Soils

Clays are known as cohesive soils as they generally possess a significant shear strength when unconfined. Although there is sometimes some bonding between clay particles, often this strength is due to suction (pore pressure less than atmospheric pressure) within a clay specimen. The suction results in positive effective stresses and hence in some shear strength of the soil.

There are two important differences between clays and sands which distinguish their behaviour in shear. Firstly, the permeability of clays is very much less than that of sands, and this inhibits the movement of water if there is a tendency to change volume. As a result it may take years after a change of surface loading on a deposit of clay for excess pore pressures to dissipate and for the effective stresses to reach equilibrium. We must, therefore, distinguish between *drained* and *undrained* shear behaviour of clays. The other important difference between clays and sands concerns the shape of the particles; clay particles are plate-like. When a shear plane has formed in a clay and there has been substantial shear deformation across it, the adjacent clay particles tend to align themselves parallel to the plane. The shear strength can then be significantly smaller than that of the adjacent clay mass. This is known as the *residual strength*.

Although clays and sands have different mineralogy and thus different interparticle friction, in many respects they behave similarly when sheared. The initial behaviour of clays in shear when they are fully drained is similar to that of sands. Like dense sands, heavily over-consolidated clays tend to dilate on shearing and the shear strength reaches a peak at small strains. If shearing is then continued the moisture content will increase until shearing takes place at constant volume (i.e. *critical state* conditions). Further straining results in the formation of a shear plane and a drop of shear strength to the residual value.

Normally-consolidated clays behave similarly to loose sands, tending to contract on shearing. The shear strength is fully mobilized when critical state conditions have been reached, before it drops to the residual value after large deformations have occurred. Failure envelopes from some tests on clays are shown in Figure 2.4.

2.2.4.3 Drained and Undrained Strength

The above discussion focused on the behaviour of clays and sands when full drainage is allowed. The shear strength of a soil is dependent on the effective stresses whatever the

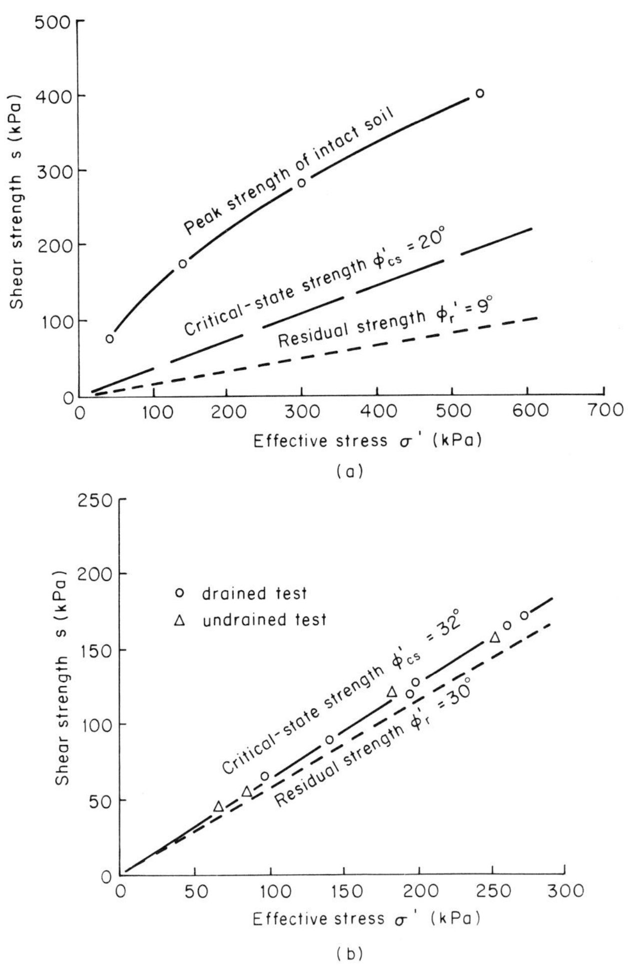

Figure 2.4 Results of triaxial compression tests: (a) on heavily over-consolidated London clay; (b) on normally consolidated Oslo clay. (*From Kenney, 1984.*)

conditions of drainage. However, when movement of the pore water is restricted, the pore pressure increases in a soil which is trying to contract and decreases in one trying to dilate. The change of pore pressure directly affects the effective stresses and hence the shear strength.

It has been found empirically that the strength of a saturated soil is constant if its volume remains unchanged. This is illustrated in Figure 2.5 which shows the result of testing several identical specimens of saturated clay in a triaxial apparatus with different confining pressures. If no consolidation (i.e. no drainage) is allowed, the specimens have the same *undrained* shear strength and it appears that the clay is purely cohesive (i.e. $\phi_u = 0$). The explanation for this behaviour is that the pore pressure developed in each specimen is different by an amount equal to the difference in confining pressures, and hence the effective stresses are the same. This behaviour is in contrast to what happens if the drainage is not restricted; the specimens would have different *drained* strengths.

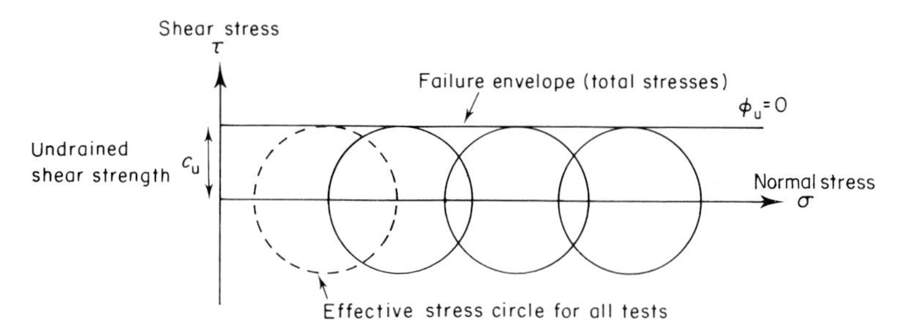

Figure 2.5 Results of undrained triaxial tests on saturated clay

When considering field problems in which the loading or unloading occurs sufficiently rapidly that drainage does not occur, the undrained shear strength may be applied in the stability analysis when a total stress analysis is used (see section 2.4). The undrained strength of a clay may be determined in the laboratory, or *in-situ* in the field (e.g. by vane tests) if the test is quick enough that the affected soil is effectively undrained. In practice the measured undrained strength is affected by the type of test and it should be used with caution.

2.2.5 Groundwater and Seepage

The mechanical properties of soils and rocks are strongly influenced by the presence of water in the void spaces. The water has two main effects. Firstly, a change of water pressure in the void spaces can directly affect the effective stresses which control the shear strength and compressibility of soils and rocks. Secondly, the water can affect the interparticle contacts as was discussed in section 2.2.1. While an appreciation of the latter effects can be important for an understanding of long-term morphological processes, the *pore-water pressure* effects influence the stability of a slope directly. In this section we shall consider the determination of pore pressures in groundwater bodies generated by rainfall, infiltration, and seepage. Evaluation of the pore pressures is of course essential if an effective stress stability analysis is being carried out (see section 2.4).

2.2.5.1 Static Groundwater

If a hole is excavated into wet ground, the standing water level in the hole defines the water table. The water table can be measured on site by installation of a standpipe in a borehole (see Chapter 3), but this will not distinguish whether or not the water table is perched. The void spaces in a soil are interconnected even in a clay. In a saturated soil the voids are by definition full of water, and in the absence of flow the pore-water pressure increases linearly with depth below the water table (hydrostatic pressure distribution, see Figure 2.6). In fine-grained soils, capillary action may cause *suction* which can lift the pore water above the water table. In the saturated zone above the water table the pore-water pressure decreases linearly with height. In partially saturated soils the suction cannot be calculated directly and must be determined by on-site measurement (see Chapter 3). Suction in a soil mass increases the effective stresses and the shear strength. Thus it has a direct influence on slope stability and this subject is considered in detail in Chapter 4.

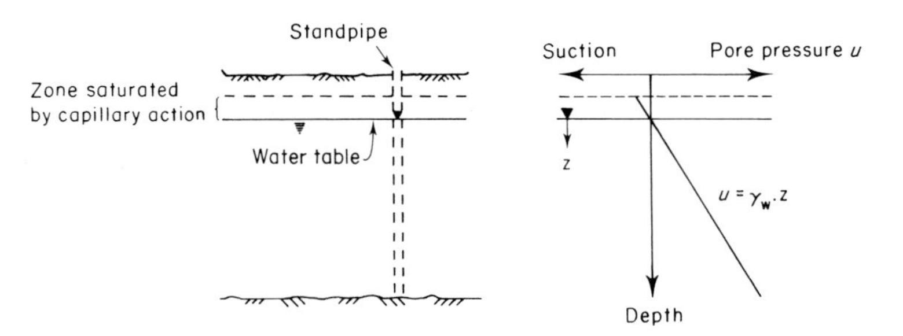

Figure 2.6 Hydrostatic pore pressure distribution

2.2.5.2 Quasi-static Groundwater

Where groundwater is present in a confined aquifer, for example in a permeable rock stratum beneath a covering of clay, it is often important to establish the distribution of water pressure with depth. Piezometers (see Figure 2.8) are used for this as each one is sealed into the ground to measure the pore-water pressure at a particular point (see Chapter 3). If the groundwater is static the pore-water pressure distribution is linear with depth within the aquifer, but in general it is not hydrostatic below the water table in the overburden. Such a situation is common where rock slopes are covered with a mantle of drift or head deposits and the clarification of the hydrogeology is crucial to an understanding of the slope processes.

2.2.5.3 Seepage

In general, and particularly in slopes, the groundwater is not static. A drop of water, flowing from one point in a soil mass to another, follows a flow path which is locally rather tortuous as the droplet passes individual soil particles, and its velocity of flow will vary considerably. However the overall flow path is reasonably smooth and the water progresses steadily through the soil. This process is called seepage.

In the 1850s H. Darcy carried out experiments on the steady flow of water through soils and showed empirically that the quantity of flow Q per unit time is given by:

$$Q = kiA \tag{2.3}$$

where k is the *coefficient of permeability* of the soil, i is the *hydraulic gradient*, and A is the cross-sectional area of the soil conducting the flow. This can best be understood by reference to Figure 2.7 which shows an apparatus for measuring soil permeability. The hydraulic gradient i is the rate of decrease of *total head* with distance in the direction of flow, so here

$$i = (h_1 - h_2)/L$$

Total head is defined by Bernoulli's equation as equal to the sum of *pressure head*, *elevation head*, and *velocity head*. As seepage velocities are usually small, the last term is generally

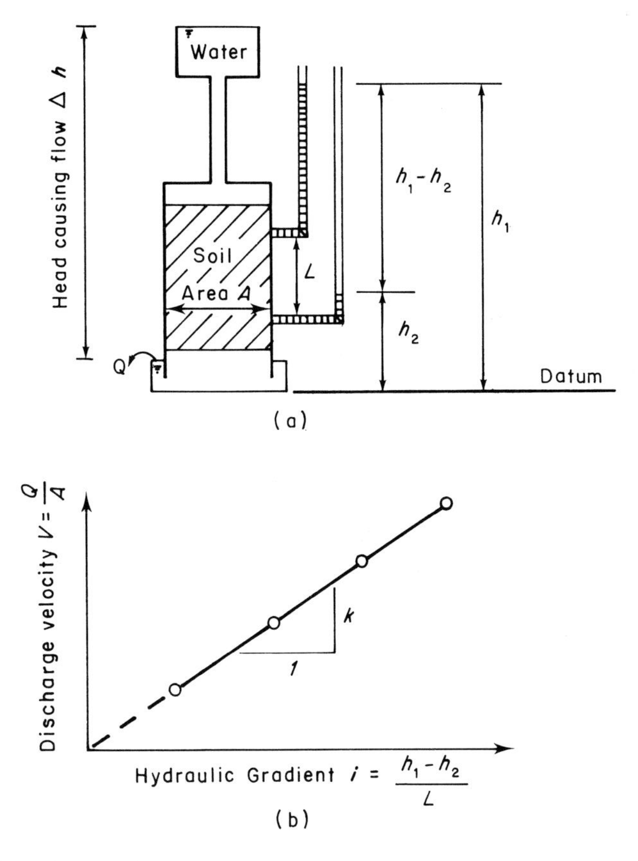

Figure 2.7 (a) Seepage apparatus. (b) Test results illustrating Darcy's law

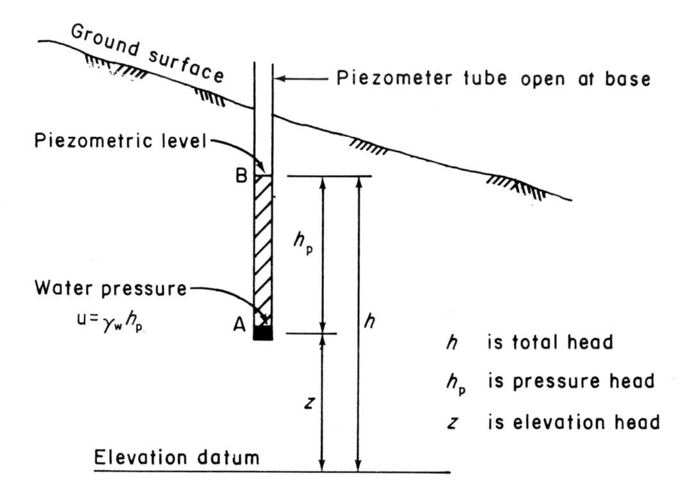

Figure 2.8 Total head, pressure head, and elevation head

negligible. Total head h can be determined in the field with a piezometer (see Figure 2.8), and is given by:

$$\text{Total head } h = \text{Pressure head } h_p + \text{Elevation head } z \tag{2.4}$$

Pore pressure u is given by:

$$u = \text{pressure head} \times \text{unit weight of water} = h_p \gamma_w \tag{2.5}$$

For one-dimensional flow through homogeneous soils there is a linear variation of total head with distance. Two- and three-dimensional flow through incompressible saturated soils is governed by Laplace's equation

$$k_x \frac{\partial^2 h}{\partial x^2} + k_y \frac{\partial^2 h}{\partial y^2} + k_z \frac{\partial^2 h}{\partial z^2} = 0 \tag{2.6}$$

where k_x, k_y, k_z are the permeabilities in the x, y and z directions.

Where the problem can be simplified to two-dimensional flow through a homogeneous soil the graphical technique of flow net sketching can be used. Figure 2.9 shows a flow net for a slope in an isotropic soil. The net consists of orthogonal *flow lines* (which indicate the seepage path) and *equipotentials* (lines joining points of equal total head). Since the flow lines and equipotentials are drawn so that they form curved 'squares', the head drop between any two adjacent equipotentials is a constant. It is then easy to determine the total head at any point N, and hence the pore pressure can be evaluated using equations (2.4) and (2.5). If the permeability is known the seepage quantity in a unit time can be determined from:

$$Q = k \cdot \frac{N_f}{N_d} \cdot H \tag{2.7}$$

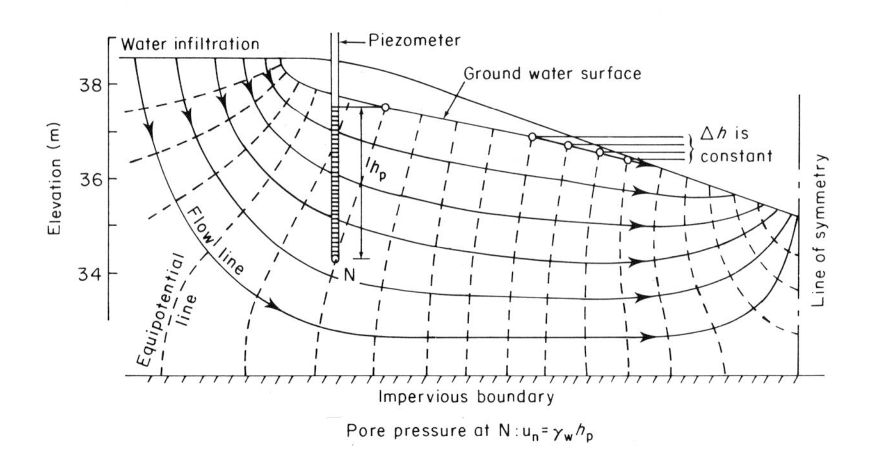

Figure 2.9 Sketched flow net for gully slope in isotropic soil. (*From Kenney, 1984.*)

where N_f is the number of flow channels, N_d is the number of equal head drops on the flow net, and H is the total head drop. Flow nets can be adapted for flow through soils with anisotropic permeability, but they are difficult to draw for layered soils. For three-dimensional flow problems, flow through layered soils and transient flow problems it is often necessary to use a numerical analysis to evaluate the distribution of total head.

In conclusion, it is obviously necessary for anyone involved in assessing slope stability to have a good understanding of soil mechanics. Hopefully the brief presentation here will serve as a basis for understanding the various methods of analysis discussed in this book, but reference should be made to the standard textbooks for a fuller discussion.

2.3 SLOPE CHARACTERIZATION

At the start of a stability analysis, the slope geometry, loadings, soil and groundwater conditions must be defined. A good appreciation of the geology and hydrogeology is obviously essential, and often it is useful to classify the instability mechanism. There are three major classes of slope movement: falls, slides, and flows. A number of schemes of classification have been proposed, and Figure 2.10 shows that proposed by Skempton and Hutchinson (1969).

In general a two-dimensional analysis will be made, and the geometry must be simplified so that representative cross-sections may be drawn (e.g. Figure 2.11). The loadings, soil, and groundwater conditions can be shown on each cross-section which can then be used as the basis for the stability analysis.

The principal loading on a slope is usually the self-weight of the soil, but surface loadings such as those from buildings and prestressed anchors can be included. Where a slope is partially submerged, the water pressure normal to the slope should be included. An alternative

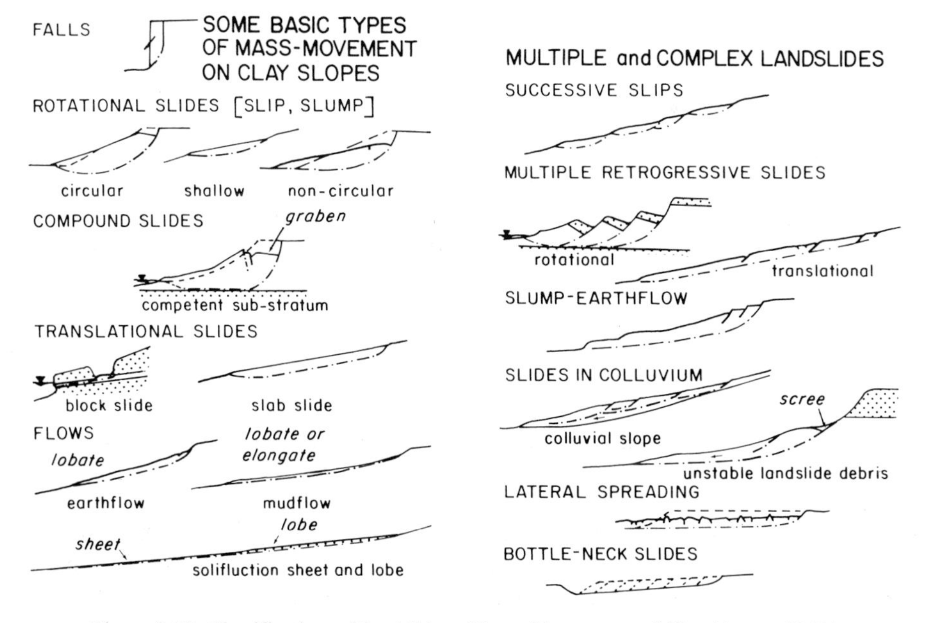

Figure 2.10. Classification of landslides. (*From Skempton and Hutchinson, 1969.*)

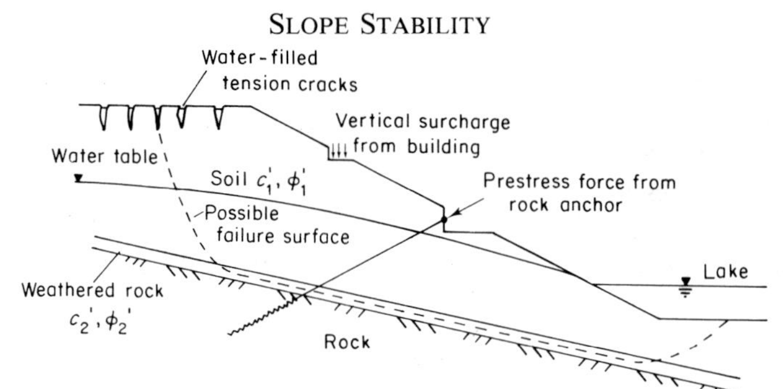

Figure 2.11 Representative cross-section

procedure using the submerged unit weight of the soil was given by Bishop (1955). When considering the stability of a slope during earthquakes, the inertia forces resulting from ground accelerations may be included (a pseudo-static analysis), although such an analysis is not always reliable (see Chapter 9).

An early decision must be made whether a total or an effective stress analysis is to be made (see section 2.4). Then representative shear strength parameters must be selected for each soil type, usually on the basis of laboratory tests (e.g. triaxial, shear box) on representative samples. Sometimes the strength may be determined in the field by *in-situ* tests.

The selection of shear strength parameters is obviously crucial to the analysis and requires experience. For the analysis of *first-time* slides it might be expected that the peak strength would be operative. However, the shear strains within a slope are generally not uniform, and for strain-softening materials such as over-consolidated clays this implies *progressive failure* of the soil (see section 2.8.4). The critical state shear strength represents a useful lower bound strength for first-time slides, and this has been confirmed by back-analysis of failures in cuttings (Skempton, 1977).

Many slopes contain pre-existing shear surfaces as a result of previous shear movements. There are a number of situations in which this may have occurred, including previous landsliding, tectonic folding, and periglacial phenomena and it is possible that movements on these shear surfaces could be reactivated. Under these circumstances it is appropriate to use the residual shear strength parameters as a lower bound value.

Tension cracks may form at the crest of a slope, particularly in cohesive or partially saturated soils. The theoretical maximum depth of a tension crack in a cohesive soil is given by:

$$z = \frac{2c_u}{\gamma} \tag{2.8}$$

where c_u is the undrained shear strength and γ is the total unit weight of the soil. The possibility that tension cracks could fill with water in periods of heavy rain should be considered carefully.

The distribution of pore pressures within the slope is required if an effective stress analysis is being carried out. Where possible this is obtained directly from instrumentation (see Chapter 3), but often a model of the groundwater is needed as a framework for interpreting observations and for interpolation. In homogeneous ground conditions in the steady state it is generally conservative to consider that the pore pressures are effectively hydrostatic below the water table. A more accurate pore-pressure distribution may be obtained from a flow net (see Figure 2.9), or an equivalent numerical analysis. With any of the above, the pore pressures along a potential failure surface are input into the analysis explicitly.

Alternatively, the pore pressure may be expressed implicitly by the *pore pressure ratio* r_u given by:

$$r_u = \frac{u}{\gamma z} \qquad (2.9)$$

where u is the pore pressure at a point being considered and γz is the weight of soil above. This is often useful when the precise distribution of pore pressures is unknown, perhaps because of lack of instrumentation or because the pore pressures are altering under a change of loading conditions. If pore pressures are specified in this way use can be made of *stability charts* (see section 2.7.3) for the analysis of slopes in homogeneous soils.

2.4 CHOICE BETWEEN TOTAL AND EFFECTIVE STRESS

When analysing the stability of a slope, a decision must be made whether to use a total or an effective stress analysis. The choice generally follows from the classification of a stability problem as *short* or *long term*. Slope failures generally result from a change of loading on the soil and if this occurs quickly (e.g. in engineering problems such as excavation of a cutting), the stability during and immediately after the change (i.e. in the short term) may need to be assessed. This will be particularly important if the change of loading results in a change of pore-water pressure in the soil mass and the change is rapid compared to the consolidation time for the soil. If the change of loading is slow compared to the consolidation time of the soil or if the loading is a natural fluctuation of groundwater levels as occurs in natural slopes the problem is considered to be long term.

In principle, a total or an effective stress approach could be used to analyse any slope, although since soils are predominantly frictional materials an effective stress analysis seems inherently more logical especially for the analysis of long-term problems. In practice for short-term stability problems a total stress analysis is often simpler and more convenient as there is usually difficulty in predicting pore-pressure changes.

In specifying the shear strength parameters for a total stress analysis it is assumed that for saturated soils $\phi_u = 0$ and c_u is the undrained shear strength, i.e. the soil behaves as if it were purely cohesive. In an effective stress analysis the effective strength parameters c', ϕ' are used and the pore pressure must be specified as an independent variable.

These distinctions between short and long term can best be made with the aid of an example. Figure 2.12 shows the changes in pore pressure and factor of safety when a cutting is excavated in clay, a problem involving unloading. During excavation the average shear stress on a failure surface increases while the total normal stresses decrease. If the clay is saturated the reduction of normal stresses is accompanied by a short-term *reduction* of the pore pressures. The magnitude of the reduction depends on the nature of the clay and would generally be greater for over-consolidated clays than for normally-consolidated clays. In the long term the pore pressures will *increase* to their equilibrium values, thus resulting

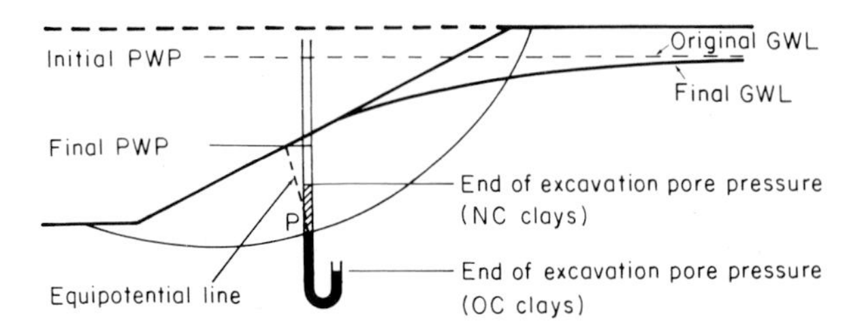

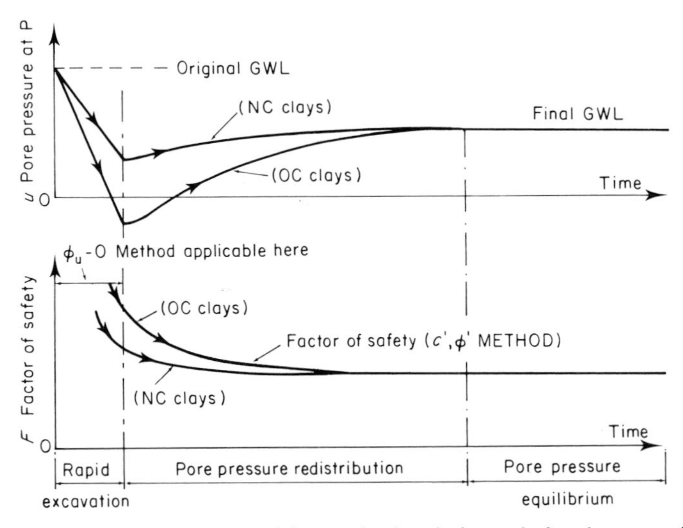

Figure 2.12 The changes in pore pressure and factor of safety during and after the excavation of a cut in clay. (*Reproduced from Bishop and Bjerrum, 1960, by permission of the American Society of Civil Engineers.*)

in a further reduction in the effective stresses in the clay, and hence a reduction in its strength and thus in the stability of the slope.

Several conclusions can be reached about the stability of such an excavated slope:

1. The long-term stability is likely to be the most critical and can best be analysed with an effective stress analysis using equilibrium groundwater conditions.
2. It is generally not necessary to assess the stability in the short term nor to predict the changes of pore pressure during excavation.
3. If an assessment of the stability in the short term is required (e.g. for a temporary excavation) a total stress analysis may be convenient using the undrained shear strengths of the clay. However, since excavation is accompanied by a reduction of pore pressures and these can increase again quickly, the results of a short-term analysis should be treated with caution especially where the clay is fissured (Bishop and Bjerrum, 1960). The question 'How long is short term?' is considered in detail by Chandler (1984).
4. If pore pressures are monitored during and after excavation an assessment of the stability can be made with an effective stress analysis at any time.

The opposite conclusions can be reached about problems involving loading (e.g. foundations, embankments on soft ground); generally the short term stability is critical.

Most problems involving natural slopes may be classified as long term, although failures generally result from small changes of loading. For example, the groundwater level may rise in a period of heavy rainfall resulting in a decrease of the effective stress and hence in the shear strength along a potential slip surface. Alternatively erosion at the toe of the slope may increase the shear stresses and decrease the resistance. Usually an effective stress analysis is the most appropriate for these conditions, but in soils such as loose sand or quick clay (Aas, 1981) in which small shear strains can cause a sudden build-up of pore pressure, great care is needed in deciding on appropriate shear strength parameters and pore pressures. In these circumstances a total stress analysis may be considered, although there is then the real difficulty of determining the appropriate undrained shear strength.

2.5 EARLY DEVELOPMENT OF STABILITY ANALYSIS

The analysis of slopes has its origin in the work of Coulomb in 1776. In his work on the stability of retaining walls, Coulomb introduced the concept of shear resistance of the soil as being the sum of cohesive and frictional components (Heyman, 1972). Coulomb developed a wedge analysis (see Figure 2.13) and by searching for the most critical failure surface he evaluated the active thrust on a retaining wall. He extended the analysis to find the critical height H_c of an unsupported vertical cut in a soil possessing both cohesion and friction, which using modern terminology can be stated as

$$H_c = \frac{4c}{\gamma} \cdot \frac{\cos\phi}{1-\sin\phi} \tag{2.10}$$

During the first half of the nineteenth century, some careful field observations were made of slides in cuttings and embankments, mainly associated with construction of the railways and canals. In the 1840s Gregory and Colthurst reported notable failures in Britain, while in France, Collin made a major contribution (Skempton, 1949, 1979). Collin (1846) studied a number of failures in clays and concluded that slip surfaces are generally curved and that slips occur when the cohesion is just exceeded by the gravitational forces. He made measurements of (undrained) cohesion, and introduced it into a stability analysis of failure on a cycloidal slip surface, an analysis which was similar to our modern $\phi_u = 0$ analysis.

In the early part of this century the more modern methods of stability analysis were developed in Sweden (Petterson, 1955; Bjerrum and Flodin, 1960). During construction of

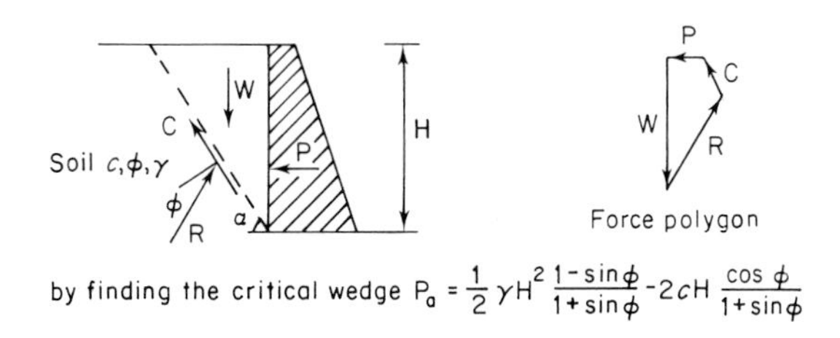

Figure 2.13 Coulomb's analysis of the thrust on a smooth retaining wall

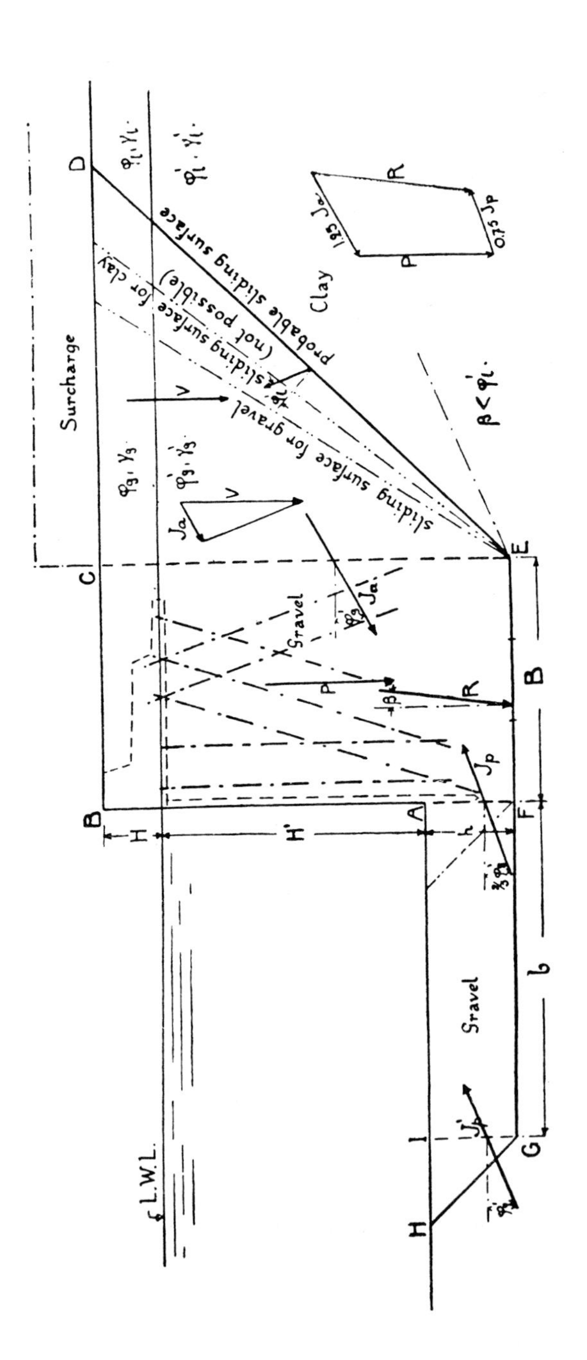

Figure 2.14 (a) Method proposed by Fellenius in 1911 for computing the stability of quay walls

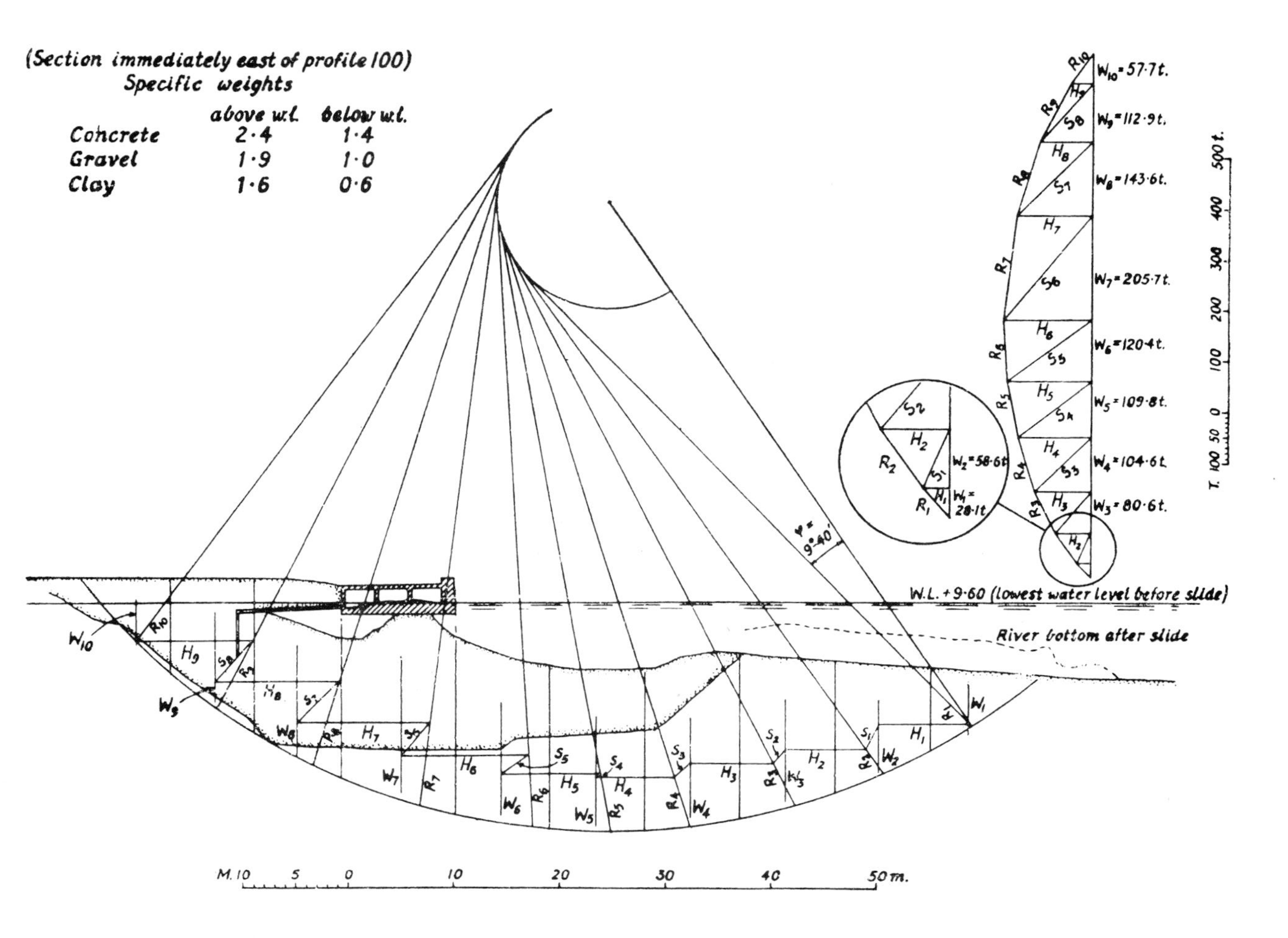

Figure 2.14 (b) Cross-section of the Stigberg Quay, Gothenberg, with a circular-cylindrical slip surface, presented by Petterson in 1916

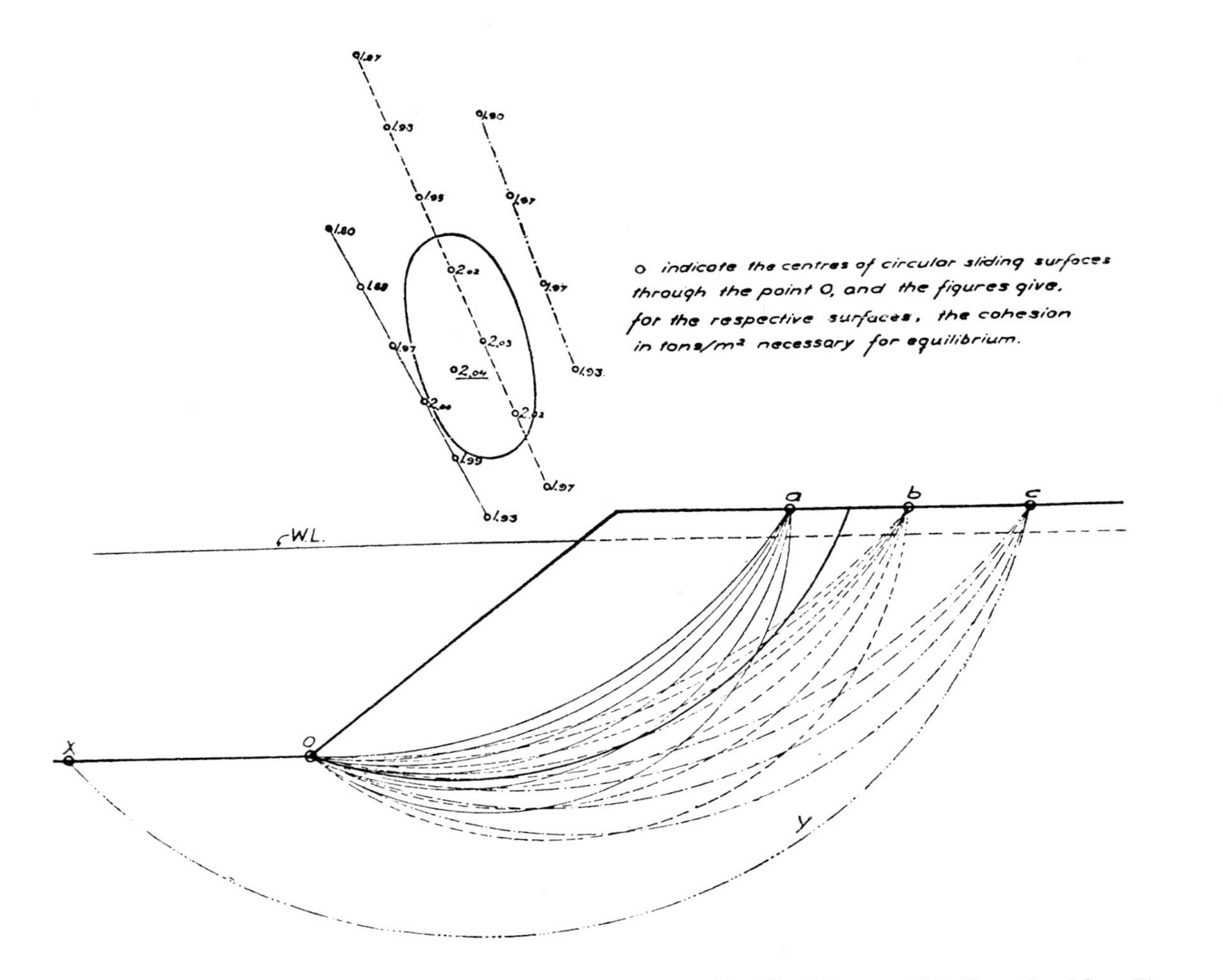

Figure 2.14 (c) Fellenius's presentation of the principle of the $\phi = 0$ analysis in Teknisk Tidsskrift 12 January 1918. (*Reproduced from Bjerrum and Flodin, 1960, by permission of the Institution of Civil Engineers.*)

Gothenberg harbour there were several failures of quay walls and in 1910 Fellenius developed the *wedge analysis* shown in Figure 2.14(a). This analysis assumed frictional behaviour of the soil and it was followed by the *friction circle* method presented by Hultin and Peterson in 1916. In this analysis (see Figure 2.14(b)) for the first time, the slope was divided into a number of slices, and a graphical technique was used to evaluate the equilibrium of each slice and the overall slope. The interslice forces were assumed horizontal and the submerged weight was used for soil below the water table. This method was later developed by Taylor (1948) but like the *logarithmic spiral* method (Rendulic, 1935), it has been superseded by analytical methods. In 1918 Fellenius extended the circular arc analysis to a cohesive soil ($\phi_u = 0$ analysis—see section 2.6.2) and showed how it could be used to search for the most critical slip circle (see Figure 2.14(c)). Later Fellenius (1927, 1936) extended the analysis to soils with both friction and cohesion (the ordinary method of slices—see section 2.6.2.4).

Some of the early methods of analysis implicitly acknowledged that the friction developed on the sliding surfaces depended on the effective weight of the soil above. However, it was the discovery of the principle of effective stress by Terzaghi in the early 1920s (e.g. Terzaghi, 1925) that led to its incorporation into stability analysis with pore pressure specified as an independent variable. Concurrently effective stress testing of soils was developed so that shear strength parameters could then be determined and specified in the analysis.

In 1954 Bishop presented his method of slices for circular arc analysis, a method which is still widely used today. Similar methods were developed for the analysis of slips on non-circular slip surfaces (e.g. Janbu *et al.*, 1956). All these methods are limit equilibrium methods of analysis, and some of the commonly used ones are described in the next section. As will be discussed in section 2.8.2 this class of method is approximate, but experience shows that it can be extremely useful in practice.

2.6 MODERN LIMIT EQUILIBRIUM METHODS OF ANALYSIS

2.6.1. Introduction

In limit equilibrium methods we postulate that the slope might fail by a mass of soil sliding on a failure surface. At the moment of failure, the shear strength is fully mobilized all the way along the failure surface, and the overall slope and each part of it are in static equilibrium. The shear strength of the soil is normally given by the Mohr–Coulomb failure criterion:

$$s = c_u \quad \text{(for total stress analyses)} \tag{2.11a}$$

$$s = c' + \sigma' \tan\phi' \quad \text{(for effective stress analyses)} \tag{2.11b}$$

In presenting the methods here, no distinction is made between peak, critical state, and residual shear strength parameters, but of course in practice the parameters must be selected with care.

In the analysis of stable slopes the shear strength mobilized under equilibrium conditions is less than the available shear strength, and it is conventional to introduce a factor of safety F defined by:

$$F = \frac{\text{Shear strength available}}{\text{Shear strength required for stability}} \tag{2.12}$$

A number of slip surfaces are considered and the critical one is identified; the corresponding (smallest) factor of safety is then taken to be the factor of safety of the slope.

When using limit equilibrium methods there is in principle little restriction on the shape of the slip surface. As was noted by Collin (Skempton, 1949) slip surfaces are generally curved, but they can have plane sections especially when influenced by topography, stratigraphic horizons or discontinuities. When choosing a method of analysis for a particular slope it is important to consider the likely shape of the failure surface, and thus whether a circular or a non-circular method of analysis is the more appropriate. The methods of analysis described here are classified by shape of the failure surface in Table 2.1. In all of them a two-dimensional failure is assumed, i.e. the slip is assumed to be infinitely wide; this assumption is discussed later in section 2.8.3. These methods of analysis are not suited to slopes whose failure entails significant changes of geometry, for example rock falls involving toppling failure.

Table 2.1 Methods of Analysis

Method	Circular	Non-circular	Overall Moment Equilibrium	Overall Force Equilibrium	Assumptions about interslice forces
Infinite slope		*		*	Parallel to slope
Wedge analysis		*		*	Define inclination
$\phi_u = 0$	*		*		
Ordinary	*		*		Resultant parallel to base of each slice
Bishop	*	(*)	*		Horizontal
Janbu simplified	(*)	*		*	Horizontal
Lowe and Karafiath	*	(*)		*	Define inclination
Spencer	*	(*)	*	*	Constant inclination
Morgenstern and Price	*	*	*	*	$X/E = \lambda.f(x)$
Janbu rigorous	*	*	*	*	Define thrust line
Frelund and Krahn GLE	*	*	*	*	$X/E = \lambda.f(x)$

Note E and X are horizontal and vertical components of interslice forces respectively.

In the sections which follow a number of limit equilibrium methods of analysis are presented which are widely used today. First the linear methods are given from which the factor of safety may be determined directly. Then the more complex methods of slices are described. For ease of comparison a common nomenclature has been adopted so that the differences can be clarified. For simplicity the effects of surface loadings have been omitted but these can readily be included in the equations.

2.6.2 Linear Methods

The methods of analysis which are most amenable to hand calculation are the infinite slope analysis, $\phi_u = 0$ analysis, ordinary method of slices, and the wedge or sliding block analysis. These methods are simple to use since in each there is a linear equation for the

INFINITE SLOPE ANALYSIS

Failure is assumed to occur by sliding of a slab of soil on a plane slip surface which is parallel to the ground surface.

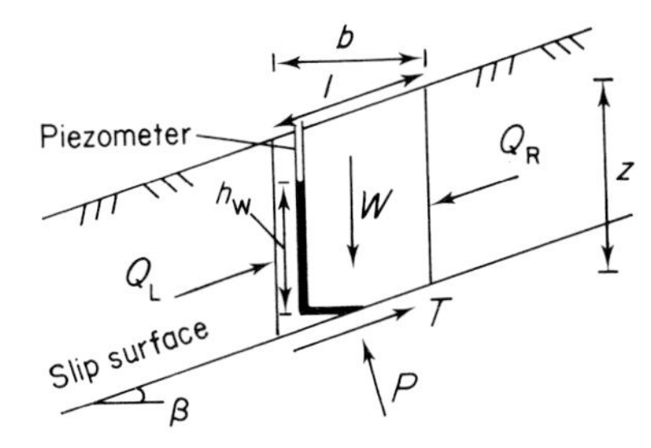

Soil properties: c' ϕ' γ
Water unit weight γ_w

Slice weight $W = \gamma z b$
Pore pressure at base $u = \gamma_w h_w$
and $r_u = u/\gamma z$

For slice shown: at base total normal stress σ, shear stress τ, pore pressure u

Since the slope is infinite: $Q_L = Q_R$

Resolving perpendicular to slope: $P = W\cos\beta = \sigma l$ so $\sigma = \dfrac{W}{b}\cos^2\beta$

parallel to slope $T = W\sin\beta = \tau l$ so $\tau = \dfrac{W}{b}\sin\beta\cos\beta$

Failure criterion $s = c' + (\sigma - u)\tan\phi'$

Mobilized shear strength $\tau = s/F$ where F is factor of safety

$$\text{Hence } \frac{W}{b}\sin\beta\cos\beta = \frac{1}{F}\left(c' + \left[\frac{W}{b}\cos^2\beta - u\right]\tan\phi'\right)$$

$$\text{so } F = \frac{c' + [\gamma z\cos^2\beta - u]\tan\phi'}{\gamma z\sin\beta\cos\beta}$$

$$\text{or } F = \frac{c'/\gamma z + [\cos^2\beta - r_u]\tan\phi'}{\sin\beta\cos\beta}$$

If there is steady seepage parallel to the slope:

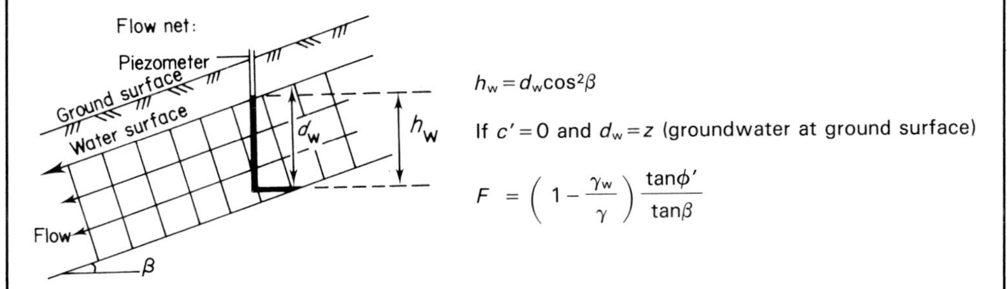

$h_w = d_w\cos^2\beta$

If $c' = 0$ and $d_w = z$ (groundwater at ground surface)

$$F = \left(1 - \frac{\gamma_w}{\gamma}\right)\frac{\tan\phi'}{\tan\beta}$$

Note Haefili (1948) examines the case where flow is emerging horizontally through the face of the slope.

Figure 2.15 Infinite slope analysis. (*From Haefili, 1948; Skempton and Delory, 1957.*)

factor of safety; for this reason they are extremely useful in practice. However, as will be discussed later in section 2.6.2.4, the assumptions made in the ordinary method of slices result in factors of safety which are conservative.

2.6.2.1 Infinite Slope Analysis

Landslides in which a planar mass of soil slides on a slip surface which is approximately parallel to the ground surface (see Figure 2.15) can be analysed effectively using the infinite slope analysis (Skempton and Delory, 1957). In this analysis the soil is assumed to slide on a plane slip surface which is parallel to the ground surface and the slope is assumed to be infinite in extent at an inclination β to the horizontal. The ground and groundwater conditions are assumed not to vary along the slip.

Landslides in natural slopes in clays frequently occur during periods of heavy rainfall. The groundwater level may then rise to the ground surface and steady seepage down the slope results. In these circumstances if $c'=0$ the factor of safety is given by:

$$F = \left(1 - \frac{\gamma_w}{\gamma}\right)\frac{\tan\phi'}{\tan\beta} \qquad (2.13)$$

It can be seen that $F=1$ when $\beta \doteqdot \phi'/2$ (2.14)

Where there has been previous landslipping, the angle of shearing resistance may already be reduced to its residual value ϕ'_r, which for clays with a high clay friction may be as low as 8°–12° (e.g. Skempton, 1964; Lupini *et al.*, 1981). This explains why even very shallow natural slopes in clay soils may be unstable.

2.6.2.2 Wedge Analysis

There are situations in which the slip surface can be approximated by two or three straight lines. This may occur when the slope is underlain by a strong stratum such as rock or there is a weak stratum included within or beneath the slope. In these circumstances an accurate assessment of the stability may be made by splitting the slope into several blocks of soil and examining the equilibrium of each block. Figure 2.16 shows the wedge method applied to the analysis of a sloping core dam (Seed and Sultan, 1967). It will be seen that there are more unknowns than there are equations of equilibrium. The problem may be made determinate by making assumptions about the inclination of the forces between the blocks, and by assuming that the same proportion of the available shear strength is mobilized all along the slip surfaces. Several trials are required to find a value of factor of safety for which the force polygon closes.

2.6.2.3 $\phi_u=0$ Method

This is the simplest of the circular arc methods of analysis (see Figure 2.17). Failure is assumed to occur by rotation of a rigid block of soil on a cylindrical failure surface along which the undrained shear strength of the soil c_u is mobilized. In the early analyses (e.g. Fellenius, 1918) the shear strength was assumed to be

WEDGE METHOD

The slope is divided into several blocks which are assumed to slide on plane failure surfaces. By examining the horizontal and vertical equilibrium of each block, the factor of safety may be obtained. An assumption must be made about the interwedge forces.

Example: Stability of a Sloping Core Dam (Seed and Sultan, 1967).

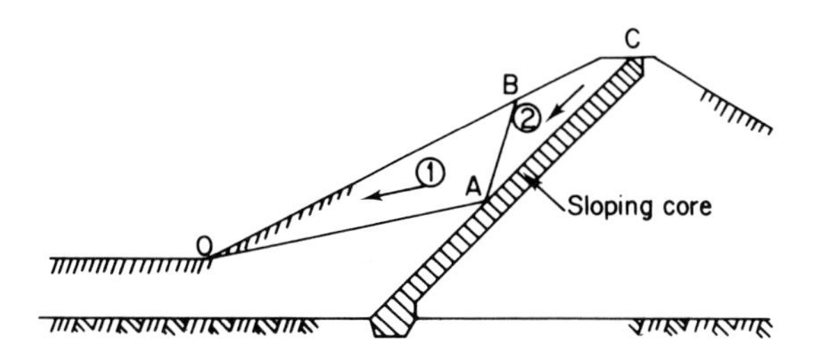

Force polygon:

Wedges and forces:

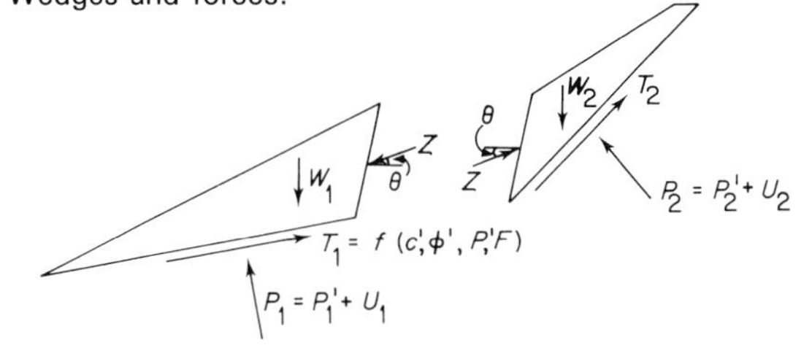

$\tan\phi_m' = \dfrac{\tan\phi'}{F}$

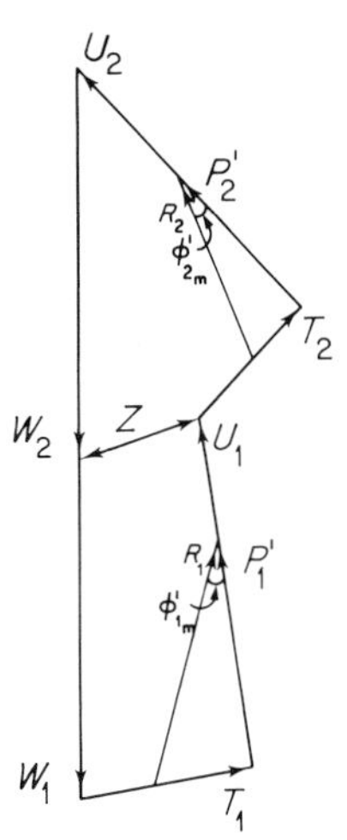

There are eight unknowns: P_1 P_2 Z θ T_1 T_2 F_1 F_2

For each block there are three equations : Horizontal equilibrium
Vertical equilibrium
Mohr–Coulomb $T = (C + (P - U)\tan\phi)/F$

As there are two more unknowns than equations, two assumptions are required. Normally the angle of the interwedge forces θ is assumed, and F_1 is taken equal to F_2. A trial and error procedure is adopted, and the force polygons are constructed assuming various values of F until they close.

Figure 2.16 Wedge method of analysis

ϕ_u = O METHOD

Failure is assumed to occur by rotation of a block of soil on a cylindrical slip surface on which the undrained strength may be mobilized.

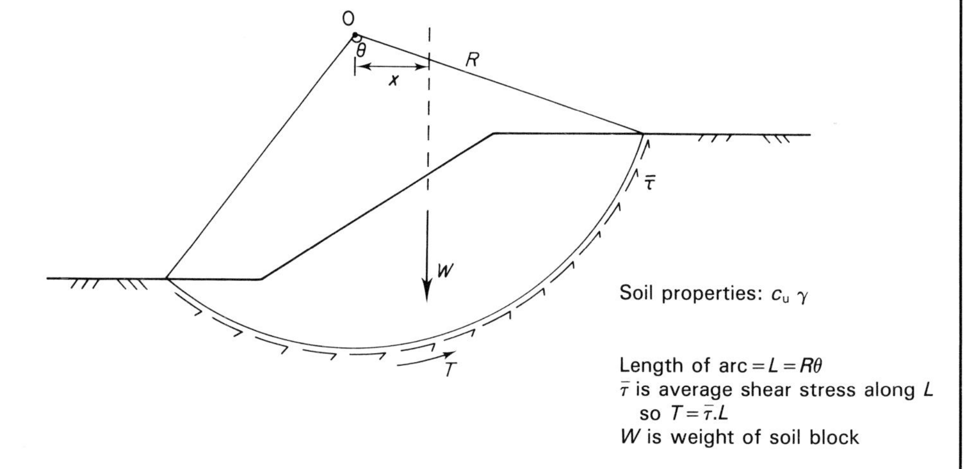

Moments about O: overturning moment $= Wx$
restoring moment $= TR$

Failure criterion: $s = c_u$

Mobilized shear strength $\tau = s/F$ so $\bar{\tau} = c_u/F$ where F is factor of safety

In equilibrium $Wx = TR$

$$\text{hence } Wx = \frac{c_u LR}{F}$$

$$F = \frac{c_u LR}{Wx}$$

The analysis may easily be adapted to take account of varying shear strength, and the presence of surcharges and water at the toe.

Figure 2.17 $\phi_u = 0$ method—circular arc analysis. (*From Fellenius, 1918.*)

constant along the failure surface, but with the introduction of the fall-cone test and other methods of determining the shear strength the variation of shear strength around the circle could be included (e.g. Skempton and Golder, 1948). The overall stability of the soil mass is examined by taking moments about the centre of the circle.

The use of undrained shear strength in this analysis implies that the pore pressures and effective stresses in the soil have not had time to reach equilibrium under an applied loading. Thus it can be applied appropriately for end of construction conditions—for example, the short-term stability of an embankment on soft clay. In general it is not appropriate to apply this method to the analysis of natural slopes in clays except when they are subjected to a sudden change of loading.

ORDINARY METHOD OF SLICES

Failure is assumed to occur by rotation of a block of soil on a cylindrical slip surface centred on O. By examining moment equilibrium about O an expression for the factor of safety is obtained. It is assumed that the resultant of the interslice forces on each slice is parallel to its base.

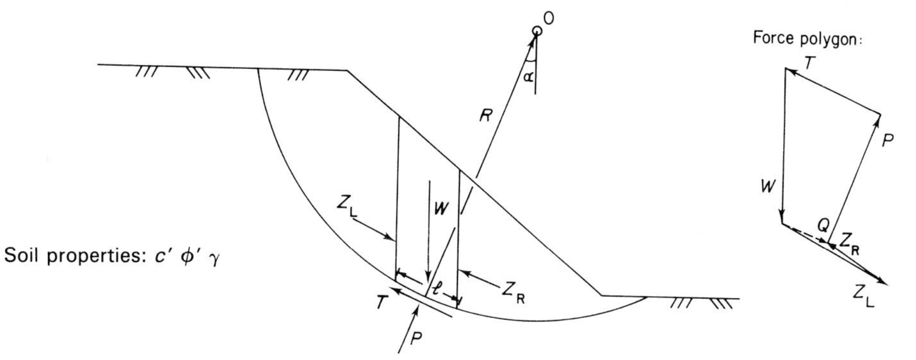

For slice shown: at base – total normal stress σ, shear stress τ, pore pressure u

Failure criterion: $s = c' + (\sigma - u)\tan\phi'$

Mobilized shear strength $\tau = s/F$ where F is factor of safety

$$\text{Now } P = \sigma l \quad T = \tau l \quad \text{so } T = \frac{l}{F}(c'l + (P - ul)\tan\phi') \tag{1}$$

Assume that the resultant of the interslice forces Q is parallel to base of slice.

$$\text{Resolving normal to base of slice } P = W\cos\alpha \tag{2}$$

$$\text{Overall MOMENT Equilibrium about O: } \Sigma\, WR\sin\alpha = \Sigma\, TR \tag{3}$$

(note that interslice forces are internal and their net moment is zero).

$$\text{so } \Sigma\, W\sin\alpha = \Sigma \frac{l}{F}(c'l + (P - ul)\tan\phi')$$

$$\text{hence } F_m = \frac{\Sigma(c'l + (P - ul)\tan\phi')}{\Sigma\, W\sin\alpha} \tag{4}$$

$$\text{substitute for } P\text{: } F_m = \frac{\Sigma(c'l + (W\cos\alpha - ul)\tan\phi')}{\Sigma W\sin\alpha} \tag{5}$$

This equation does not contain F on the right-hand side and so is easily solved by hand calculation. However the false assumption about the interslice forces results in errors which may be as large as 60% (Whitman and Bailey, 1967).

Figure 2.18 Ordinary method of slices (or Swedish method of analysis). (*From Fellenius, 1927, 1936.*)

2.6.2.4 Ordinary or Swedish Method

In the $\phi_u = 0$ analysis the undrained shear strength mobilized around the slip surface was assumed to be independent of the stress level. In an effective stress analysis, the shear strength on the slip surface is related to the effective normal stress by the Mohr–Coulomb failure criterion and thus the variation of the normal stress around the failure surface must be determined. This may be achieved by dividing the failure mass into a number of slices.

The Ordinary or Swedish method (Fellenius, 1927, 1936) is the simplest method of slices to use, and is amenable to hand calculations. In this method (Figure 2.18) the normal force acting on the base of any slice is determined by resolving forces normal to the base of the slice. To make the problem determinate, the assumption is made that the resultant of the interslice forces acting on any slice is parallel to its base. Thus the normal force is given by:

$$P = W\cos\alpha \tag{2.15}$$

By taking moments about the centre of the circle the overall stability is examined, and a value of the factor of safety is obtained.

Unfortunately the assumptions about the interslice forces do not satisfy statics, and this may lead to an underestimate of the factor of safety by as much as 60% (Turnbull and Hvorslev, 1967; Whitman and Bailey, 1967). For this reason this method is not used much nowadays.

2.6.3 Non-linear Methods—Methods of Slices

2.6.3.1 Introduction

If an effective stress analysis of a slope is to be made, the effective stresses must be determined around the failure surface. In practice this is achieved if the failure mass of soil is divided into a number of slices. Figure 2.19 shows a typical slice with the forces which act on it. Many of these forces and their lines of action are unknown at the start of the analysis. In Table 2.2 it is shown that for a slope divided into n slices there are in general $5n-2$ unknowns while there are only $3n$ equations of statics available. Thus $2n-2$ assumptions must be made for the problem to be rendered statically determinate.

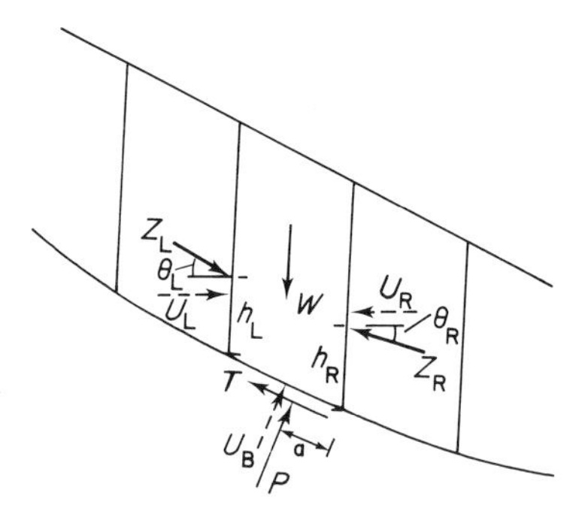

Figure 2.19 Forces on a typical slice

In the next section the equations of equilibrium are formulated quite generally, and non-linear equations are obtained for the factor of safety. Their solution necessitates an iterative procedure, but first the necessary assumptions must be made. There are several classes of assumption which may be made:

1. Assumptions about the distribution of normal stress round the slip surface.
2. Assumptions about the position of the line of thrust of the interslice forces.
3. Assumptions about the inclination of the interslice forces.

Table 2.2 Equations and Unknowns in Methods of Slices

For slope divided into n slices (see Figure 2.19 for typical slice):

Equations available: total $3n$ (vertical, horizontal, and moment equilibrium)

Unknowns:	1	Safety factor F used to relate shear forces T to normal forces P
	n	Normal total forces P on base of slice (pore-water forces U_B are known)
	n	Positions a of forces P
	$n-1$	Interslice total forces Z (pore-water forces U_L, U_R are known)
	$n-1$	Inclinations θ of interslice forces
	$n-1$	Heights h of interslice forces
total	$5n-2$	

Thus $2n-2$ assumptions are required for the problem to be statically determinate.

Common assumptions:	n	Position of P taken as centre of slice
	$n-1$	Inclinations θ of interslice forces *or* heights h of line of thrust
total	$2n-1$	This implies that the problem is overspecified.

In most methods of analysis the normal force P is assumed to act at the centre of the base of each slice; this is reasonable providing the slices are thin, and reduces the number of assumptions required to $n-2$.

In many methods of analysis, an assumption is made about the inclinations of the interslice forces (see Table 2.1). As this involves another $n-1$ assumptions the problem is now overspecified. The analysis may then be carried out either satisfying moment equilibrium or horizontal force equilibrium, yielding two factors of safety, F_m and F_f. In general these are different although if a constant value of interslice force inclination θ is assumed there is one value of θ for which F_m and F_f are identical. This is the basis of Spencer's method (see section 2.6.3.6).

Since all the methods involve assumptions, none will yield the 'correct' value of factor of safety. However, it will be shown in section 2.8.2.2 that most of the methods yield an acceptable result.

2.6.3.2 General Formulation (GLE)

Fredlund and Krahn (1977) have shown that the equations of equilibrium can be formulated quite generally as is shown in Figure 2.20. The formulation is the same for circular and non-circular slip surfaces, although for the latter a fictional centre of rotation is adopted. By considering overall moment equilibrium and overall force equilibrium separately, two different expressions are obtained for the factor of safety, F_m and F_f.

Each expression contains the total normal force acting on the base of each slice which is given by the expression:

$$P = [W - (X_R - X_L) - \frac{1}{F}(c'l\sin\alpha - ul\tan\phi'\sin\alpha)]/m_\alpha \tag{2.16}$$

$$\text{where } m_\alpha = \cos\alpha\ (1 + \tan\alpha\ \frac{\tan\phi'}{F}) \tag{2.17}$$

GENERAL METHOD OF SLICES

Failure is assumed to occur by sliding of a block of soil on a non-circular (or circular) slip surface. By examining overall moment equilibrium about an assumed centre of rotation or overall force equilibrium, two expressions are obtained for factor of safety. An assumption must be made about the interslice forces.

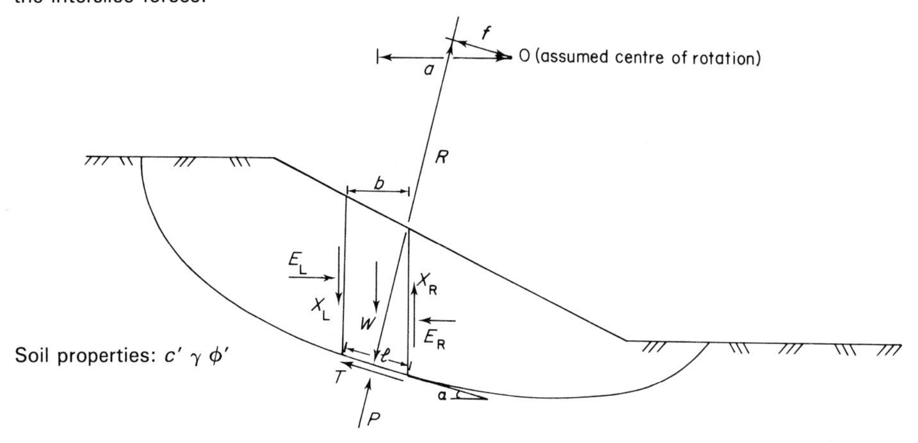

For slice shown: at base – total normal stress σ, shear stress τ, pore pressure u

Failure criterion: $s = c' + (\sigma - u)\tan\phi'$

Mobilized shear strength $\tau = s/F$ where F is factor of safety

$$\text{Now } P = \sigma l \qquad T = \tau l \qquad \text{so } T = \frac{1}{F}(c'l + (P - ul)\tan\phi') \tag{1}$$

$$\text{Resolve vertically: } P\cos\alpha + T\sin\alpha = W - (X_R - X_L) \tag{2}$$

Rearranging and substituting for T gives

$$P = \left[W - (X_R - X_L) - \frac{1}{F}(c'l\sin\alpha - ul\tan\phi'\sin\alpha)\right]/m_\alpha \tag{3}$$

$$\text{where } m_\alpha = \cos\alpha\left(1 + \tan\alpha\frac{\tan\phi'}{F}\right)$$

Resolve horizontally: $T\cos\alpha - P\sin\alpha + E_R - E_L = 0$

Rearranging and substituting for T gives

$$E_R - E_L = P\sin\alpha - \frac{1}{F}[c'l + (P - ul)\tan\phi']\cos\alpha \tag{4}$$

$$\text{Overall MOMENT equilibrium (about O)}: \Sigma Wd = \Sigma TR + \Sigma Pf \tag{5}$$

Rearranging and substituting for T gives

$$F_m = \frac{\Sigma[c'l + (P - ul)\tan\phi']R}{\Sigma(Wd - Pf)} \tag{6}$$

For circular slip surfaces $f = 0$ $\quad d = R\sin\alpha \quad$ $R = \text{constant}$

$$\text{so } F_m = \frac{\Sigma[c'l + (P - ul)\tan\phi']}{\Sigma W\sin\alpha} \tag{6a}$$

Figure 2.20 continued on next page

continued

Overall FORCE equilibrium

In the absence of surface loading $\Sigma(E_R - E_L) = 0$ (7a)

$\Sigma(X_R - X_L) = 0$ (7b)

so from (4) $$\Sigma(E_R - E_L) = \Sigma P\sin\alpha - \Sigma \frac{1}{F_f}(c'l + (P - ul)\tan\phi')\cos\alpha = 0 \quad (8)$$

$$\text{so } F_f = \frac{\Sigma\,(c'l + (P - ul)\tan\phi')\cos\alpha}{\Sigma\,P\sin\alpha} \quad (9)$$

In order to solve for F_m and F_f, P must be evaluated, and this requires evaluation of X_R, P_L the interslice shear forces. As the problem is indeterminate an assumption must be made. Some common assumptions are:

$X_R - X_L = 0$	Bishop (1955)
$\frac{X}{E}$ = constant	Spencer (1967)
$\frac{X}{E} = \lambda f(x)$	Morgenstern and Price (1965)

In general $F_m = F_f$ and Bishop (1955) showed that F_m is much less sensitive to the assumption about interslice forces than F_f.

Figure 2.20 General method of slices. (*From Fredlund and Krahn, 1977.*)

Values of the function m_α are plotted in Figure 2.21. As the expression for P contains the interslice shear forces X_R and X_L, assumptions are required to render the problem determinate as was discussed above. It is these assumptions and whether overall force equilibrium, moment equilibrium or both are considered that distinguish the different methods which follow.

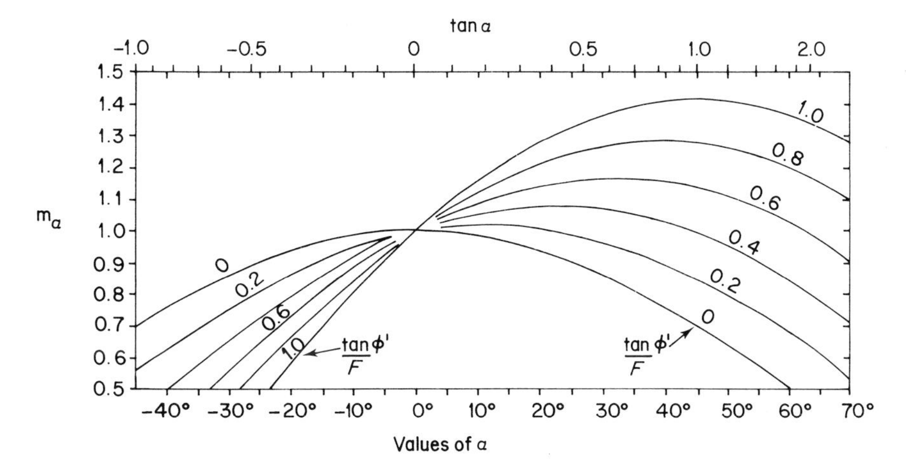

Figure 2.21 Plot of factor m_α against inclination of base of slice α. (*Reproduced by permission of the Norwegian Geotechnical Institute.*)

BISHOP'S SIMPLIFIED METHOD OF SLICES

Failure is assumed to occur by rotation of a block of soil on a cylindrical slip surface centred on O. By examining overall moment equilibrium about O an expression for the factor of safety is obtained. It is assumed that the interslice forces are horizontal.

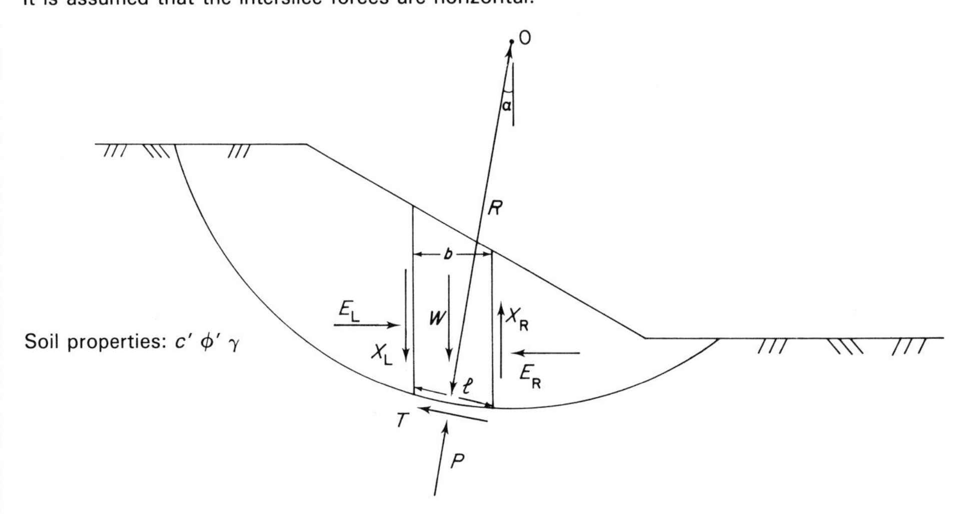

For slice shown: at base – normal stress σ, shear stress τ, pore pressure u

Mohr–Coulomb failure criterion: $s = c' + (\sigma - u) \tan\phi'$
Mobilized shear strength $\tau = s/F$ where F is factor of safety

$$\text{Now } P = \sigma l \qquad T = \tau l \qquad \text{so } T = \frac{1}{F}(c'l + (P - ul)\tan\phi') \tag{1}$$

Resolve vertically: $P\cos\alpha + T\sin\alpha = W - (X_R - X_L)$

Assuming $X_R = X_L = 0$ (i.e. interslice forces horizontal)

$$P = [W - \frac{1}{F}(c'l\sin\alpha - ul\tan\phi' \sin\alpha)]/m_\alpha \tag{2}$$

$$\text{where } m_\alpha = \cos\alpha(1 + \tan\alpha \frac{\tan\phi'}{F})$$

Overall MOMENT equilibrium (about O): $\Sigma\, WR\sin\alpha = \Sigma\, TR$ (3)

Rearranging and substituting for T gives

$$F_m = \frac{\Sigma\,(c'l + (P - ul)\tan\phi')}{\Sigma\, W\sin\alpha} \tag{4}$$

As this equation contains F on both sides it has to be solved iteratively. Convergence is usually quick and so the method is suitable for hand calculation, although it is time consuming.

Figure 2.22 Bishop's simplified method of slices. (*From Bishop, 1955.*)

2.6.3.3 Bishop's Routine Method

Bishop (1955) originally presented his method for analysis of circular slip surfaces (see Figure 2.22), but it can be applied to non-circular slip surfaces by adopting a fictional centre of rotation. In this method it is assumed that the interslice shear forces may be neglected. The total normal force is assumed to act at the centre of the base of each slice, and is determined by resolving the forces on each slice vertically.

The normal force is then given by the expression:

$$P = [W - \frac{1}{F}(c'l\sin\alpha - ul\tan\phi'\sin\alpha)]/m_\alpha \quad (2.18)$$

where m_α is given by equation (2.17) and is shown in Figure 2.21. By taking moments about the centre of the circle the overall stability is examined and a value of the factor of safety F_m is obtained. This method is moderately suitable for hand calculations and convergence is normally rapid.

This method of analysis involves a total of $2n-1$ assumptions. Thus the problem is overspecified, and in general overall horizontal equilibrium is not satisfied. Bishop (1955) discussed the significance of this and showed that the factor of safety is not particularly sensitive to the value of the interslice shear forces providing overall moment equilibrium is satisfied. This point is further discussed in section 2.8.2.2.

2.6.3.4 Janbu's Simplified Method

Janbu *et al.* (1956) published one of the first routine methods for the analysis of non-circular slip surfaces (see Figure 2.23). In this method the assumption is made that the interslice shear forces are zero and thus the expression obtained for the total normal force on the base of each slice is the same as that obtained by Bishop. By examining overall horizontal force equilibrium a value of the factor of safety F_o is obtained. At first sight this appears to be different from that obtained in the general formulation of Fredlund and Krahn, but in fact the two expressions are effectively the same.

To allow for the effect of the interslice shear forces, a correction factor f_o is applied; thus the factor of safety of the slope is given by:

$$F_f = f_o . F_o \quad (2.19)$$

To obtain suitable values of the correction factor f_o, Janbu *et al.* calibrated this analysis against Janbu's rigorous method for a variety of slopes. They found that f_o depends on the geometry of the problem as well as the soil conditions, and they prepared a design chart (Figure 2.24) from which f_o may be obtained.

Again in this method of analysis the problem is overspecified so in general overall moment equilibrium is not satisfied. However, like Bishop's method it is amenable to hand calculation and so is useful in practice. As is discussed in section 2.8.2.2 the factor of safety F_f determined using a force equilibrium procedure is much more sensitive to the assumption about interslice forces than the factor of safety F_m determined by satisfying moment equilibrium. For this reason it seems preferable to use a method of analysis in which moment equilibrium is satisfied.

JANBU'S SIMPLIFIED METHOD

Failure is assumed to occur by sliding of a block of soil on a non-circular slip surface. By examining overall force equilibrium an expression for factor of safety is obtained. It is assumed that the interslice shear forces are zero, but a correction factor is introduced to allow for them.

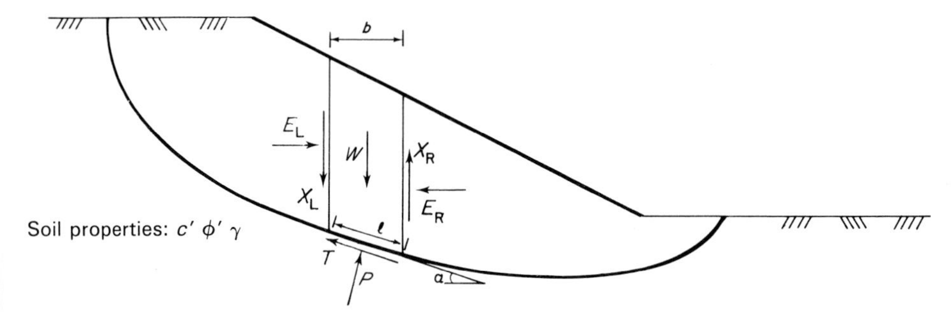

For slice shown: at base – total normal stress σ, shear stress τ, pore pressure u.

Failure criterion: $s = c' + (\sigma - u)\tan\phi'$

Mobilized shear strength $\tau = s/F$ where F is factor of safety.

$$\text{Now } P = \sigma l \qquad T = \tau l \qquad \text{so } T = \frac{1}{F}(c'l + (P - ul)\tan\phi') \tag{1}$$

Resolve vertically: $P\cos\alpha + T\sin\alpha = W - (X_R - X_L)$

Assume $X_L = X_R = 0$ (i.e. interslice forces horizontal)

Rearranging and substituting for T gives

$$P = [W - \frac{1}{F}(c'l\sin\alpha - ul\tan\phi'\sin\alpha)]/m_\alpha \tag{2}$$

$$\text{where } m_\alpha = \cos\alpha(1 + \tan\alpha\frac{\tan\phi'}{F})$$

Resolve parallel to base of slice: $T + (E_R - E_L)\cos\alpha = (W - (X_R - X_L))\sin\alpha$

again assume $X_L = X_R = 0$; rearrange, and substitute for T

$$\text{so } E_R - E_L = W\tan\alpha - \frac{1}{F}(c'l + (P - ul)\tan\phi')\sec\alpha \tag{3}$$

Overall FORCE equilibrium:

In the absence of surface loading $\Sigma(E_R - E_L) = 0$ (4)

$$\text{so } \Sigma(E_R - E_L) = \Sigma W\tan\alpha - \frac{1}{F}\Sigma(c'l + (P - ul)\tan\phi')\sec\alpha = 0 \tag{5}$$

$$\text{whence } F_o = \frac{\Sigma(c'l + (P - ul)\tan\phi')\sec\alpha}{\Sigma W\tan\alpha} \tag{6}$$

To take account of the interslice shear forces. Janbu *et al.* applied a correction factor f_o (see Figure 2.24)

$$\text{where } F_f = f_o . F_o \tag{7}$$

Note In their original formulation, Janbu *et al.* eliminated P and obtained the expression

$$F_o = \frac{\Sigma[c'b + (W - ub)\tan\phi']/n_\alpha}{\Sigma W\tan\alpha}$$

in which $n_\alpha = \cos\alpha . m_\alpha$

Both these expressions are equivalent to the expression for F_f (Figure 2.20 equation (9)) obtained by Fredlund and Krahn by resolving vertically and horizontally for each slice.

Figure 2.23 Janbu's simplified method. (*From Janbu* et al., *1956*.)

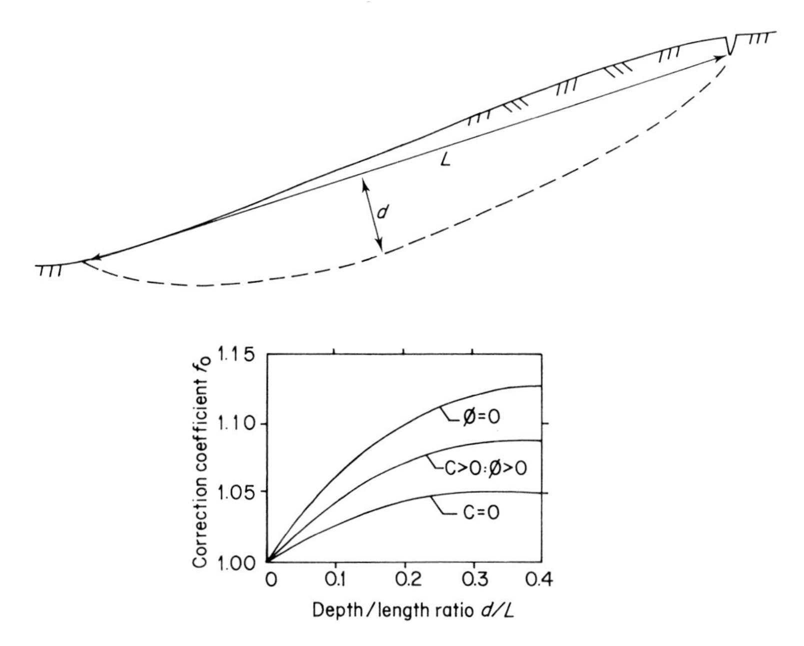

Figure 2.24 Correction factor f_0 for use in Janbu's simplified method. (*Reproduced from Janbu* et al., *1956, by permission of Norwegian Geotechnical Institute.*)

FORCE EQUILIBRIUM METHOD

Failure is assumed to occur by sliding of a block of soil on a circular (or non-circular) slip surface. By examining the force equilibrium of each slice graphically, the factor of safety is obtained. An assumption must be made about the inclination of the interslice forces.

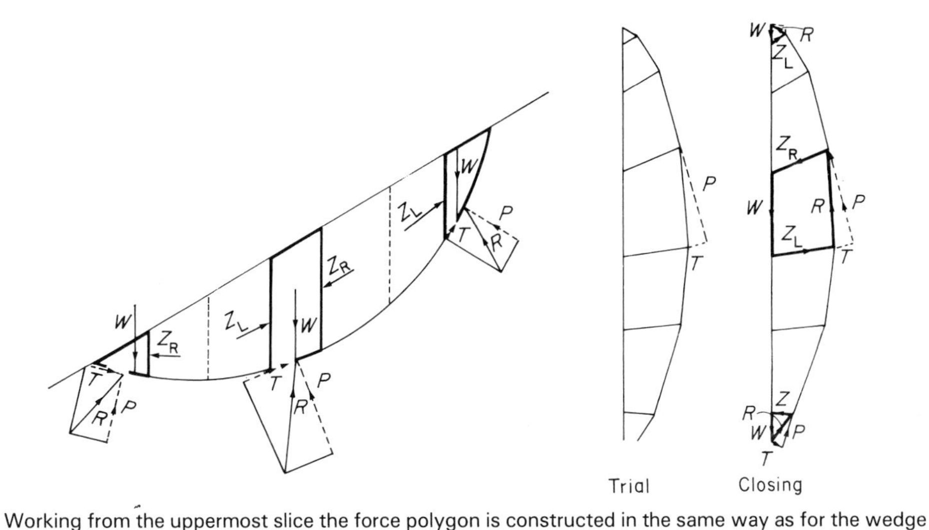

Working from the uppermost slice the force polygon is constructed in the same way as for the wedge analysis (see Figure 2.16), assuming various values of the factor of safety F, until it closes.

Figure 2.25 Force equilibrium method

SPENCER'S METHOD

Failure is assumed to occur by rotation of a block of soil on a cylindrical slip surface centred on O. By examining overall moment equilibrium and overall force equilibrium two expressions are obtained for the factor of safety. The interslice forces are assumed to be at a constant inclination, and the inclination is found at which the two expressions give the same factor of safety.

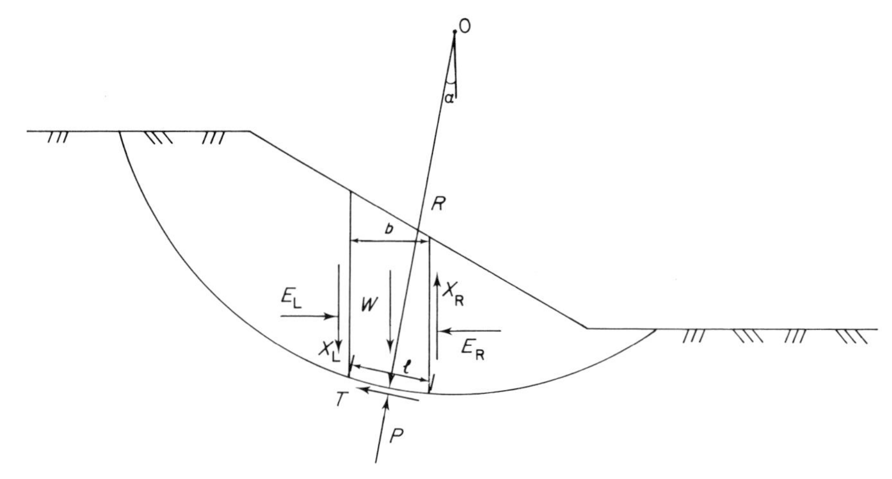

For slice shown: at base — normal stress σ, shear stress τ, pore pressure u

Mohr-Coulomb failure criteria: $s = c' + (\sigma - u)\tan\phi'$

Mobilized shear strength $\tau = s/F$ where F is factor of safety.

$$\text{Now } P=\sigma l \quad T=\tau l \quad \text{so } T = \frac{l}{F}(c'l + (P-ul)\tan\phi') \tag{1}$$

$$\text{Resolve vertically: } P\cos\alpha + T\sin\alpha = W - (X_R - X_L) \tag{2}$$

Rearranging, and substituting for T gives

$$P = [W - (X_R - X_L) - \frac{l}{F}(c'l\sin\alpha - ul\tan\phi'\sin\alpha)]/m_\alpha \tag{3}$$

$$\text{where } m_\alpha = \cos\alpha(1 + \tan\alpha \frac{\tan\phi'}{F})$$

$$\text{Resolve horizontally: } T\cos\alpha - P\sin\alpha + E_R - E_L = 0 \tag{4}$$

Rearranging, and substituting for T gives

$$E_R - E_L = P\sin\alpha - \frac{l}{F}\left[c'l + (P-ul)\tan\phi'\right]\cos\alpha \tag{5}$$

$$\text{Assume that } \frac{X}{E} = \tan\theta = \text{constant throughout the slope.} \tag{6}$$

$$\text{Overall MOMENT equilibrium (about O): } \Sigma WR\sin\alpha = \Sigma TR \tag{7}$$

Rearranging and substituting for T gives

$$F_m = \frac{\Sigma(c'l + (P-ul)\tan\phi')}{\Sigma W\sin\alpha} \tag{8}$$

Figure 2.26 continued on next page

continued

Overall FORCE equilibrium:

In the absence of surface loading $\Sigma\ (E_R - E_L) = 0$ (9a)
$\Sigma\ (X_R - X_L) = 0$ (9b)

So from (5) $\Sigma\ (E_R - E_L) = \Sigma P \sin\alpha - \frac{1}{F_f}\ \Sigma(c'l + (P - ul)\ \tan\phi')\ \cos\alpha$

$$\text{whence } F_f = \frac{\Sigma(c'l + (P - ul)\ \tan\phi')\ \sec\alpha}{\Sigma(W - (X_R - X_L))\ \tan\alpha} \quad (10)$$

The solution is reached iteratively. First it is assumed that $X_R - X_L = 0$. Then values of E and X are calculated using equations (5) and (6); the values of shear force X lag by one iteration. The inclination θ of the interslice forces is adjusted so that $F_m = F_f$.

Note In Spencer's original formulation the resultant of the interslice forces Q is used so that $Q\cos\theta = E_L - E_R$, $Q\sin\theta = X_L - X_R$. P and T are eliminated from equations (2) and (4) above to find an expression for Q. Overall moment and force equilibrium are satisfied by the conditions $\Sigma QR\cos(\alpha - \theta) = 0$, $\Sigma Q = 0$ and a value of θ is found at which both expressions yield the same factor of safety.

Figure 2.26 Spencer's method. (*Adapted from Spencer, 1967, by permission of the Institution of Civil Engineers.*)

2.6.3.5 Other Force Equilibrium Methods

In addition to Janbu's simplified method there are several methods of analysis in which only the equations of force equilibrium are satisfied (e.g. Lowe and Karafiath 1960; Sherrard *et al.*, 1963; Seed and Sultan, 1967; US Army Corps of Engineers, 1970). These methods differ in the assumptions which are made about the inclination of the interslice forces. For example Lowe and Karafiath propose that the interslice forces are assumed to be inclined at an angle equal to the mean of the inclinations of the ground surface and the slip surface at the top and bottom of the interslice boundary. Sherrard *et al.* suggest that the inclination of interslice force should be taken as equal to the inclination of the slope.

Originally, a graphical procedure was adopted (see Figure 2.25) for the analysis in which a value of factor of safety F_f was assumed and a trial force polygon was drawn. If the last slice is in equilibrium the assumed value of F_f is correct. This procedure is an extension of the wedge analysis (see section 2.6.2.2).

Nowadays the analysis may be accomplished by a numerical equivalent of the graphical procedure using the general formulation to obtain F_f. An iterative procedure is used in which the interslice shear forces are set to zero on the first iteration and are then calculated using

$$X/E = f(x) \quad (2.20)$$

for subsequent iterations. The side force function $f(x)$ defines the inclination of the interslice forces.

Again in these methods of analysis overall moment equilibrium is not satisfied as the problem is overspecified, and thus it seems preferable to use a method of analysis in which moment equilibrium is satisfied (see section 2.8.2.2).

2.6.3.6 Spencer's Method

Spencer (1967) originally presented his method for the analysis of circular slip surfaces (see Figure 2.26), but it can be applied to non-circular slip surfaces by adopting a fictional centre of rotation. In this method it is assumed that the interslice forces are at a constant inclination θ throughout the slope, so that

$$X/E = \tan\theta \tag{2.21}$$

The normal force on the base of the slice is thus:

$$P = [W-(E_R-E_L)\tan\theta - \frac{1}{F}(c'l\sin\alpha - ul\tan\phi'\sin\alpha)]/m_\alpha \tag{2.22}$$

By considering overall force equilibrium and overall moment equilibrium, two values of factor of safety F_f and F_m are obtained. This is because a total of $2n-1$ assumptions have been made and the problem is overspecified. A value of θ can then be found for which the two values coincide, and this is taken as the factor of safety of the slope.

Spencer examined the relationship between F_f and F_m for a typical problem (see Figure 2.27). This shows that the factor of safety derived from moment equilibrium F_m is comparatively insensitive to the interslice shear force. This is in line with Bishop's opinion, and justifies the widespread application of Bishop's method.

2.6.3.7 Morgenstern and Price's Method

Morgenstern and Price (1965) describe a method of analysis which may be applied to circular and non-circular slip surfaces. They assumed that the stresses and forces vary continuously across the slip, and by resolving normal and parallel to the base of each slice formulated the equations of equilibrium quite generally. The assumption is then made that the interslice shear forces X are related to the interslice normal forces E by:

$$X/E = \lambda.f(x) \tag{2.23}$$

where $f(x)$ is a function which varies continuously across the slip, and λ is a scaling factor. For a given function $f(x)$, values of λ and F are found for which overall force equilibrium and overall moment equilibrium are satisfied. Thus $F=F_m=F_f$.

Some functions $f(x)$ which may be used are shown in Figure 2.28. Morgenstern and Price state that the factor of safety is relatively insensitive to the choice of $f(x)$, providing this choice is made sensibly. This is confirmed by a study made by Fredlund and Krahn (1977) which is discussed in section 2.8.2.2. The choice of $f(x)$ can be assessed by considering the distribution of normal stress on interslice boundaries. In general there should be no effective tensile stresses, and the shear stresses must be less than those required for local critical equilibrium.

Morgenstern and Price's original analysis is slightly involved and is not presented in detail here, but Fredlund and Krahn (1977) have shown that almost identical results may be obtained using their general formulation of the equations of equilibrium together with Morgenstern and Price's assumption about the interslice shear forces (equation (2.23) above). The normal force P on the base of each slice is then given by inserting values

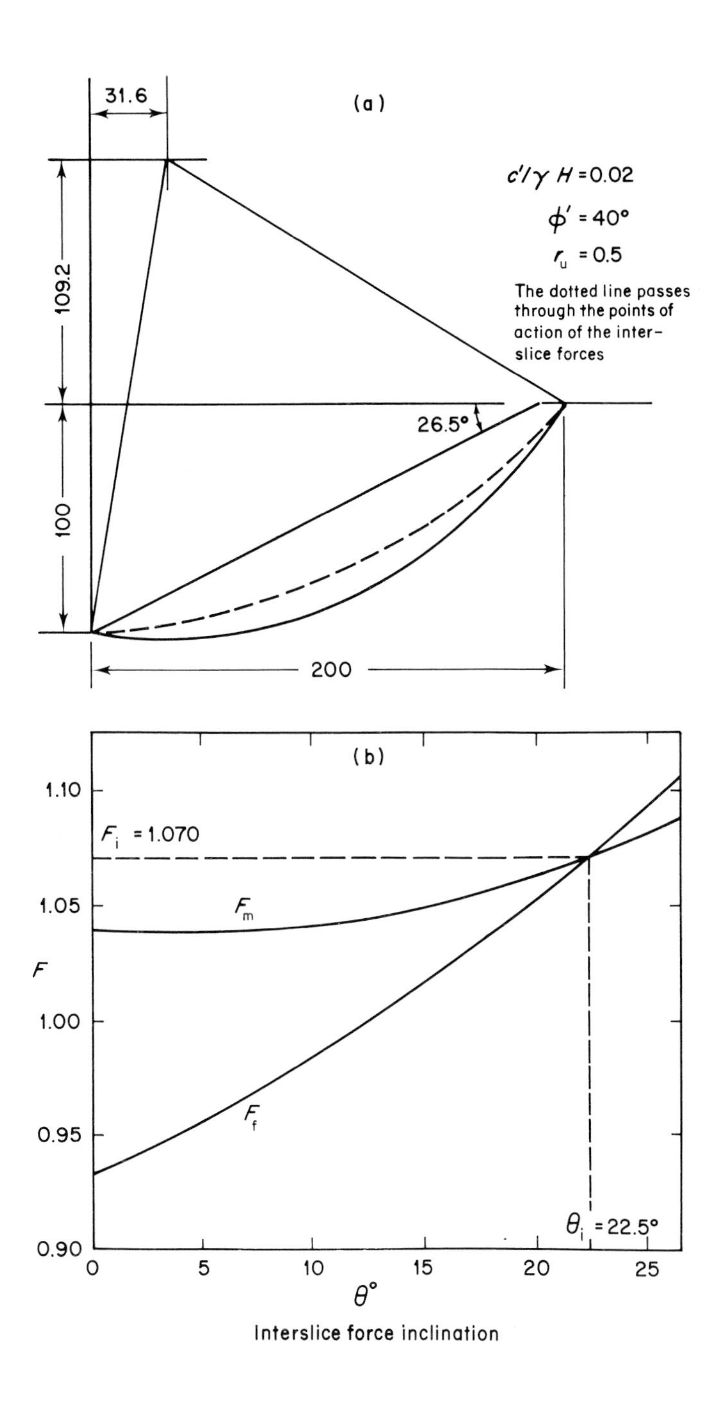

Figure 2.27 Variation of F_m and F_f with inclination of interslice forces. (*Reproduced from Spencer, 1967, by permission of the Institution of Civil Engineers.*)

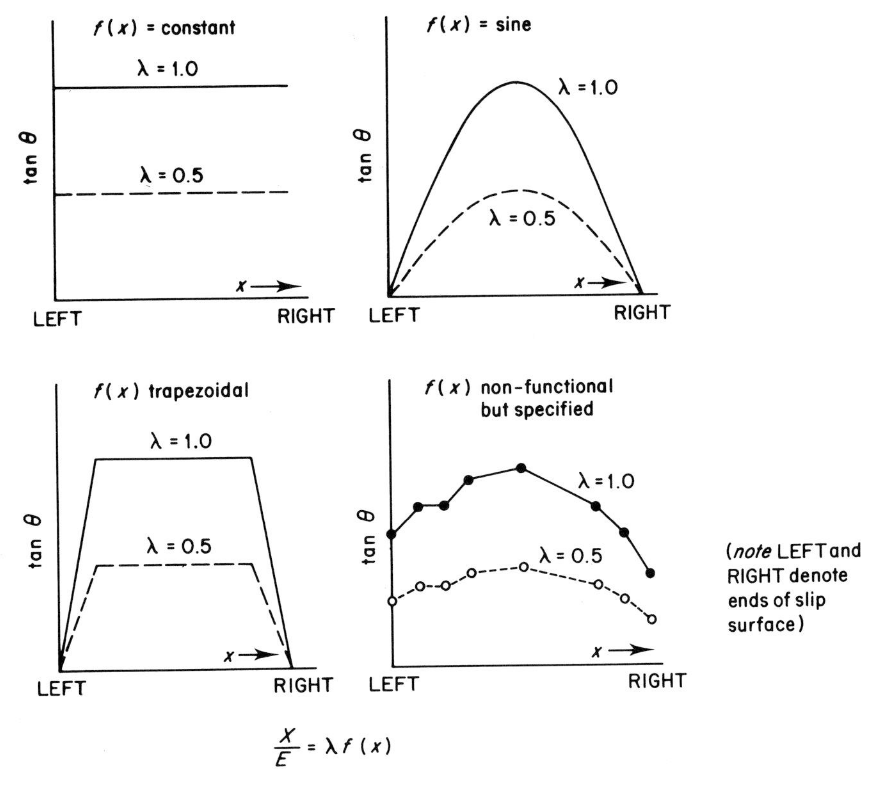

Figure 2.28 Typical interslice force inclination functions used in Morgenstern and Price analaysis. (*Reproduced from Fredlund and Krahn, 1977, by permission of the National Research Council of Canada.*)

for X_L and X_R determined from equation (2.23) into the general expression (equation 2.16) for P. The value of λ for which $F_m = F_f$ can then be found iteratively.

2.6.3.8 Janbu's Rigorous Analysis

Janbu (1954a) described a method of analysis which may be applied to both circular and non-circular slip surfaces (see Figure 2.29). The original formulation was extended to cover the analysis of bearing capacity and earth pressure problems by Janbu (1957). This was the first method of slices in which overall force equilibrium and overall moment equilibrium are satisfied.

Janbu formulated the general equations of equilibrium by resolving vertically and parallel to the base of each slice. By considering overall force equilibrium an expression for the factor of safety F_f was obtained. As with Janbu's simplified method, this appears to be different from that obtained in the general formulation of Fredlund and Krahn but again the two expressions are effectively the same. To render the problem statically determinate the position of the line of thrust of the interslice forces is assumed. By taking moments about the centre of the base of each slice overall moment equilibrium is implicitly satisfied, and the interslice shear forces can be calculated. These are then inserted into the expression for the factor of safety, and so both overall force equilibrium and overall moment equilibrium are satisfied.

JANBU'S RIGOROUS ANALYSIS

Failure is assumed to occur by sliding of a block of soil on a non-circular slip surface. By examining overall force equilibrium an expression for factor of safety is obtained. The interslice forces are evaluated by considering the moment equilibrium of each slice. For this it is necessary to assume a position of the line of thrust of the interslice forces. Overall moment equilibrium is satisfied implicitly.

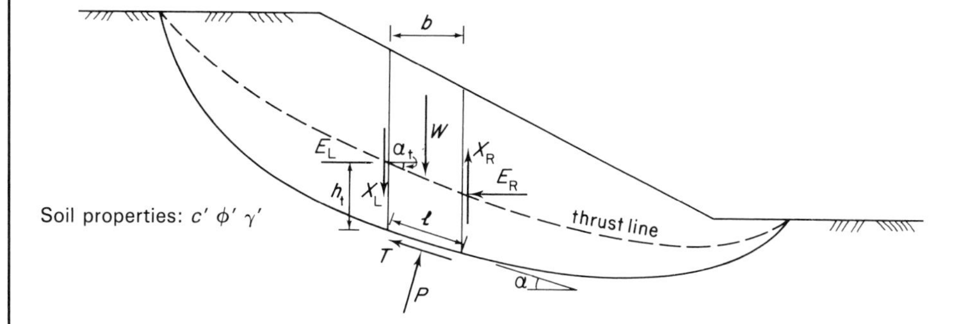

For slice shown: at base — total normal stress σ, shear stress τ, pore pressure u.

Failure criterion: $s = c' + (\sigma - u)\tan\phi'$

Mobilized shear strength $\tau = s/F$ where F is factor of safety

Now $P = \sigma l \quad T = \tau l$ so $T = \dfrac{l}{F}(c'l + (P - ul)\tan\phi')$ (1)

Resolve vertically: $P\cos\alpha + T\sin\alpha = W - (X_R - X_L)$

Rearranging, and substituting for T gives

$$P = \left[W - (X_R - X_L) - \frac{l}{F}(c'l\sin\alpha - ul\tan\phi'\sin\alpha)\right]/m_\alpha \tag{2}$$

$$\text{where} \quad m_\alpha = \cos\alpha\left(1 + \tan\alpha\,\frac{\tan\phi'}{F}\right)$$

Resolve parallel to base of slice: $T + (E_R - E_L)\cos\alpha = (W - (X_R - X_L))\sin\alpha$

Rearranging, and substituting for T gives

$$E_R - E_L = (W - (X_R - X_L))\tan\alpha - \frac{l}{F}(c'l + (P - ul)\tan\phi')\sec\alpha \tag{3}$$

Take moments about centre of base of slice (for thin slice):

$$E_R\, b\tan\alpha_t - X_R b - (E_R - E_L)h_t = 0$$

$$\text{or} \quad X_R = E_R\tan\alpha_t - (E_R - E_L)\frac{h_t}{b} \tag{4}$$

Overall FORCE equilibrium:

In the absence of surface loading $\Sigma(E_R - E_L) = 0$ (5a)

$$\Sigma(X_R - X_L) = 0 \tag{5b}$$

so from (3) $\Sigma(E_R - E_L) = \Sigma(W - (X_R - X_L))\tan\alpha - \dfrac{l}{F_f}\Sigma(c'l + (P - ul)\tan\phi')\sec\alpha = 0$

(6)

Figure 2.29 continued on next page

continued

whence $$F_f = \frac{\Sigma(c'l + (P - ul)\tan\phi')\sec\alpha}{\Sigma(W - (X_R - X_L))\tan\alpha} \quad (7)$$

The solution is reached iteratively. First it is assumed that $X_R - X_L = 0$. Then values of E and X are calculated using (3) and (4) above; the values of shear force X lag by one iteration. As moment equilibrium is satisfied by (4), $F_f = F_m$.

Note in Janbu's original formulation P is eliminated and the following expression for F is obtained:

$$F = \frac{\Sigma[c'b + (W - (X_R - X_L) - ub)\tan\phi']/n_\alpha}{\Sigma(W - (X_R - X_L))\tan\alpha}$$

in which $n_\alpha = \cos\alpha . m_\alpha$

This expression is equivalent to the expression for F_f (Equation (19) in Figure 2.20) obtained by Fredlund and Krahn by resolving vertically and horizontally for each slice.

Figure 2.29 Janbu's rigorous analysis. (*From Janbu 1954a, 1957.*)

2.7 RESULTS OF ANALYSES

The methods of analysis presented above have a wide application. They may be used for the analysis of slopes with complicated geometry, non-homogenous soil conditions, seepage, and for circular or non-circular failure surfaces. With the advent of computers their use has become routine.

The results of a stability analysis are usually expressed by a factor of safety which is applied to the shear strength of the soils as in equation (2.12). Alternatively, the analysis could be adapted to give the slope angle at which failure would occur, or the highest groundwater level or the ultimate surface loading. In general different failure surfaces are examined and the one yielding the smallest factor of safety found. This is then the factor of safety of the slope.

Where circular failure surfaces are being analysed it is often convenient to show the results graphically. By marking the factor of safety against the position of the centre of each circle drawn on a cross-section of the slope, contours of factor of safety can be plotted (see Figure 2.30). When using this procedure the circle radius must be defined implicitly (e.g. all circles pass through the toe).

2.7.1 Resistance Envelope

Where the shear strength parameters are not known the results of an analysis can be presented as a resistance envelope (Kenney, 1967). First some shear strength parameters c', ϕ' are assumed and a number of failure surfaces are analysed. The average shear stress $\bar{\tau}$ along each failure surface is determined (e.g. in Bishop's analysis $\bar{\tau}$ = overturning moment/(radius × arc length), as well as the factor of safety F. By substituting into the Mohr–Coulomb equation

$$\bar{\tau} = (c' + \bar{\sigma}'\tan\phi')/F$$

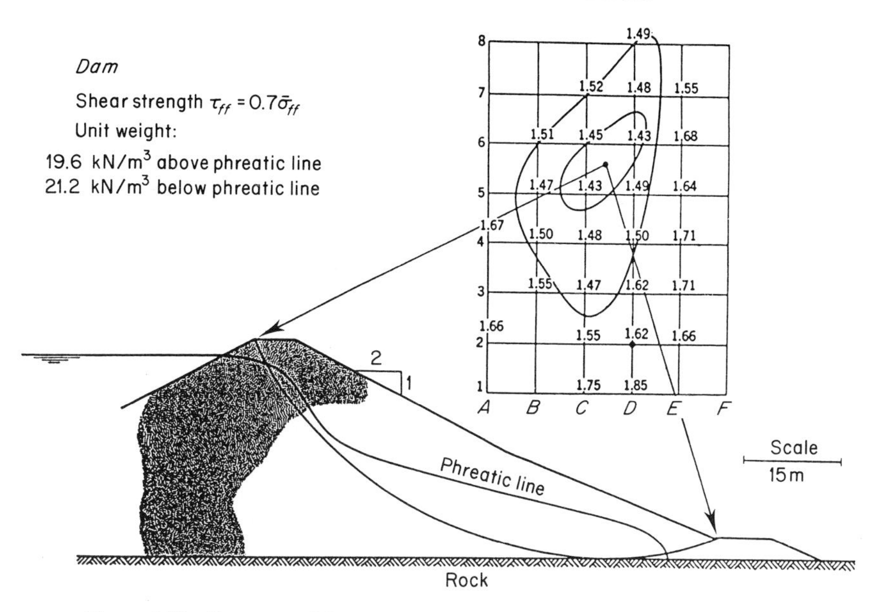

Figure 2.30 Contours of factor of safety. (*From Lambe and Whitman, 1969.*)

the average normal effective stress $\bar{\sigma}'$ for that failure surface may be found. The values $\bar{\tau}$, $\bar{\sigma}'$ are independent of the assumed shear strength parameters and may be plotted on a diagram (see Figure 2.31). A curved resistance envelope may be drawn to enclose the stress points. Lines drawn tangent to this envelope define mobilized shear strength parameters c'/F, $\tan\phi'/F$ and the critical failure surface on which they are mobilized. Thus if a slope has failed and the failure surface is known, the shear strength parameters may be determined.

2.7.2 Stability Charts

In some circumstances it may be useful to make a rapid assessment of stability without resorting to a detailed analysis. A number of stability charts are available in which the results of the analysis of many slopes of simple cross-section with homogeneous soil conditions are presented. The variables considered usually include the slope height H and the angle β, the shear strength parameters c and ϕ, the unit weight γ of the soil, and the factor of safety F. The analyses have made use of the fact that these variables can be grouped into dimensionless parameters as follows:

Stability number: $N = \dfrac{c}{\gamma H}$ (Taylor, 1948)

or $\lambda_{c\phi} = \dfrac{\gamma H}{c} \tan\phi$ (Janbu, 1954b, Spencer, 1967)

Pore pressure ratio $r_u = u/\gamma z$ (see section 2.3)

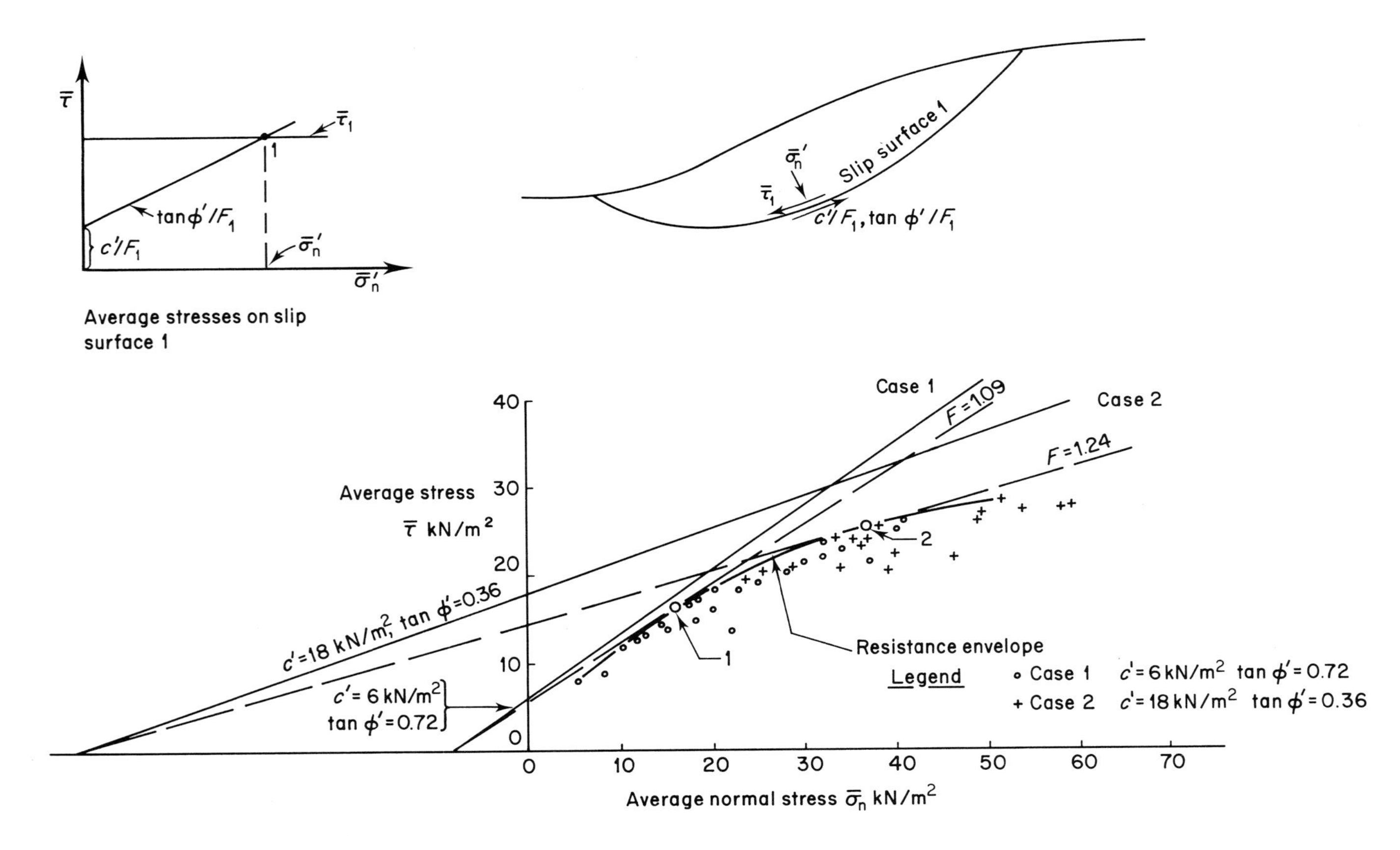

Figure 2.31 Resistance envelope. *(Reproduced from Kenney, 1967, by permission of the Norwegian Geotechnical Institute.)*

The use of these dimensionless parameters reduces the number of variables to be investigated and for given conditions the most critical circular slip surface has then been found. For homogeneous soil conditions where $\lambda_{c\phi} < 1$, Duncan and Wright (1980) found that the critical slip surface passes beneath the toe of the slope provided that the slope angle is less than about 50° and that the depth of the slip circle is not limited by the presence of a firm soil stratum or rock. Where $\lambda_{c\phi} > 2$ the critical slip circle passes through or above the toe of the slope.

In the following sections, two of the most commonly used stability charts are described. Table 2.3 lists many of the other charts which are available in the literature.

2.7.2.1 Taylor's Charts

Taylor (1948) used the friction circle method to examine the stability of homogeneous slopes without seepage and introduced the stability number $N = c/\gamma H$. While his results for soils for which $\phi > 0$ are of limited use, those for purely cohesive soils ($\phi_u = 0$) are useful for the short-term analysis (total stress) of slopes in clay. A stability chart derived from Taylor's is shown in Figure 2.32.

For a given slope of height H underlain by a hard stratum at depth DH, the stability number can be read off directly and hence the undrained shear strength $c_{u_{req}}$ required for stability calculated. The factor of safety against undrained failure is given by:

$$F = \frac{c_u}{c_{u_{req}}} \tag{2.24}$$

The solution assumes no tension cracking or surcharges are present and that the slope is not partially submerged. The stability chart can be used for the analysis of fully submerged slopes by the use of the submerged unit weight γ_{sub}.

Example

A wide excavation 5 metres deep with side slopes at 20° to the horizontal is to be made in a uniform deposit of soft clay. The clay has an undrained shear strength of 20 kN m^{-2} and unit weight 16 kN m^{-3} and is underlain by rock at a depth of 10 metres. The factor of safety may be estimated using the chart as follows:

Depth factor $D = 10/5 = 2$

For $\beta = 20°$ from Figure 2.32 $\dfrac{\gamma H}{c} = 6.2$

The undrained shear strength required for stability
$c_{u_{req}} = 16 \times 5/6.2 = 12.9$ kN/m^2

Hence factor of safety $F = \dfrac{c_u}{c_{u_{req}}} = \dfrac{20}{12.9} = 1 \cdot 55$

Table 2.3 Summary of Some Stability Charts

Author	Parameters	Slope angles β	Methods	Remarks
Taylor (1948)	c_u	0–90°	$\phi=0$	
	$c\ \phi$	0–90°	Friction Circle	Dry slopes only
Janbu (1954b)	$c\ \phi\ r_u$	11–90°	Swedish	Toe circles only; uses $\lambda_{c\phi}=\frac{\gamma H \tan\phi'}{c'}$
Bishop and Morgenstern (1960)	$c\ \phi\ r_u$	11–26.5°	Bishop	
Gibson and Morgenstern (1962)	c_u	0–90°	$\phi=0$	c_u increasing linearly with depth
Morgenstern (1963)	$c\ \phi$	11–26.5°	Bishop	Rapid drawdown
Spencer (1967)	$c\ \phi\ r_u$	0–34°	Spencer	Toe circles only
Hunter and Schuster (1968)	c_u	0–90°	$\phi=0$	Extended Gibson and Morgenstern to include finite strength at ground level
O'Connor and Mitchell (1977)	$c\ \phi\ r_u$	11–26.5°	Bishop	Extended Bishop and Morgenstern to $c/\gamma H=0.1$
Hoek and Bray (1977)	$c\ \phi$	0–90°	Friction Circle	Extended Taylor to include groundwater and tension cracks.
	$c\ \phi$	0–90°	Wedge	3-dimensional wedge
Cousins (1978)	$c\ \phi\ r_u$	0–45°	Friction Circle	Extended Taylor using $\lambda_{c\phi}$
Charles and Soares (1984)	ϕ	26.5–63.4°	Bishop	Non-linear failure envelope $\tau=A(\sigma')^b$

Note In general, partial submergence of slope is not considered.

2.7.2.2 Bishop and Morgenstern's Charts

Bishop and Morgenstern (1960) used Bishop's methods of slices to examine the stability of slopes in soils with effective shear strength parameters c' and ϕ'. Pore pressure was introduced as an independent variable using the pore pressure ratio r_u. As one would expect the factor of safety decreases with increasing pore pressure, and Bishop and Morgenstern expressed the factor of safety by the equation:

$$F = m - n.r_u \tag{2.25}$$

where m and n are dimensionless parameters which depend on slope angle β, $c'/\gamma H$, ϕ' and the depth factor D. Figure 2.33 shows one of the charts prepared by Bishop and Morgenstern. The range of slopes covered has been extended by O'Connor and Mitchell (1977).

Example

Estimate the factor of safety of a slope 30 metres high with inclination 3:1. The soil is of unit weight 20 kN m^{-3} and has shear strength parameters $c' = 15$ kN m^{-2}, $\phi' = 27.5°$.

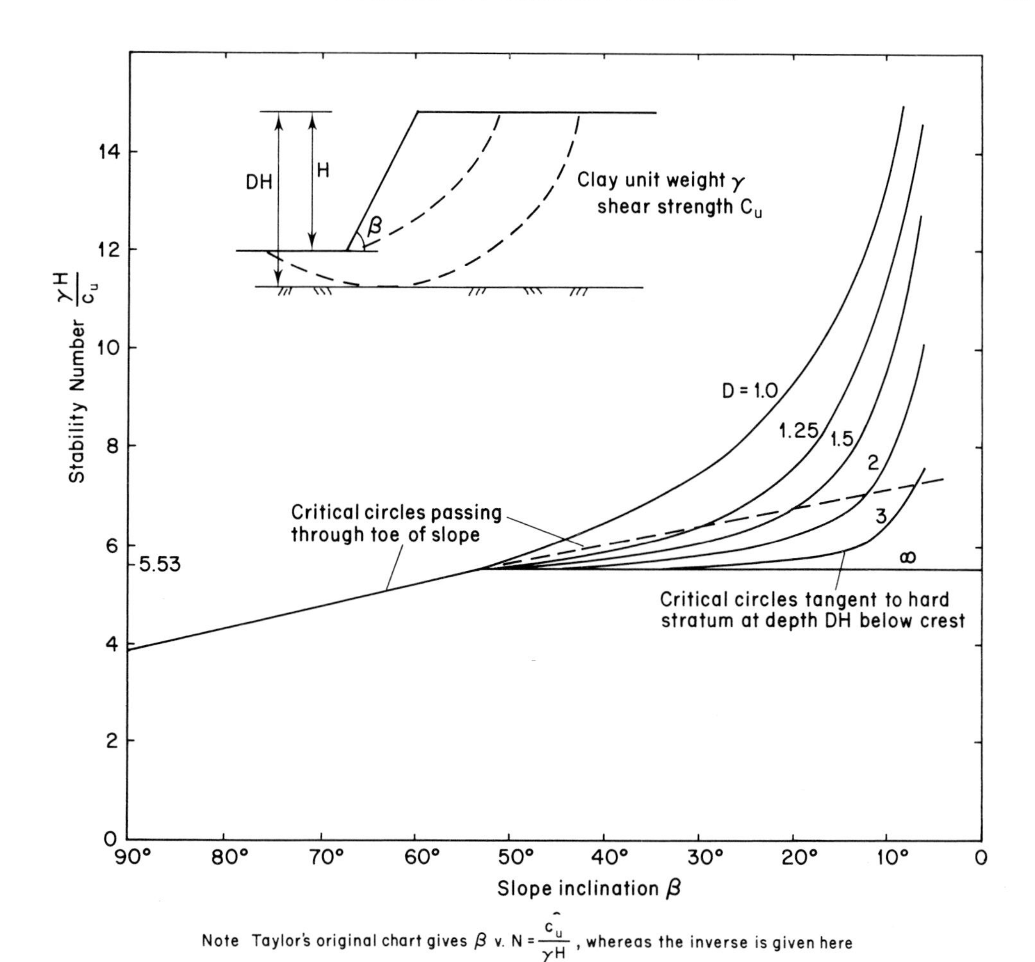

Figure 2.32 Stability chart for cohesive soil ($\phi_u = 0$). (*After Taylor, 1948.*)

The base of the slope is underlain by a hard stratum at 7.5 metres depth and the average pore pressure ratio $\bar{r}_u$ is 0.3.

$$\text{Depth factor } D = 37.5/30 = 1.25$$

$$\frac{c'}{\gamma H} = 15/20 \times 30 = 0.025$$

From charts, for 3:1 slope $m = 2.21$, $n = 2.10$

so Factor of Safety $F = m - n.r_u = 2.21 - 2.10 \times 0.3 = 1.58$

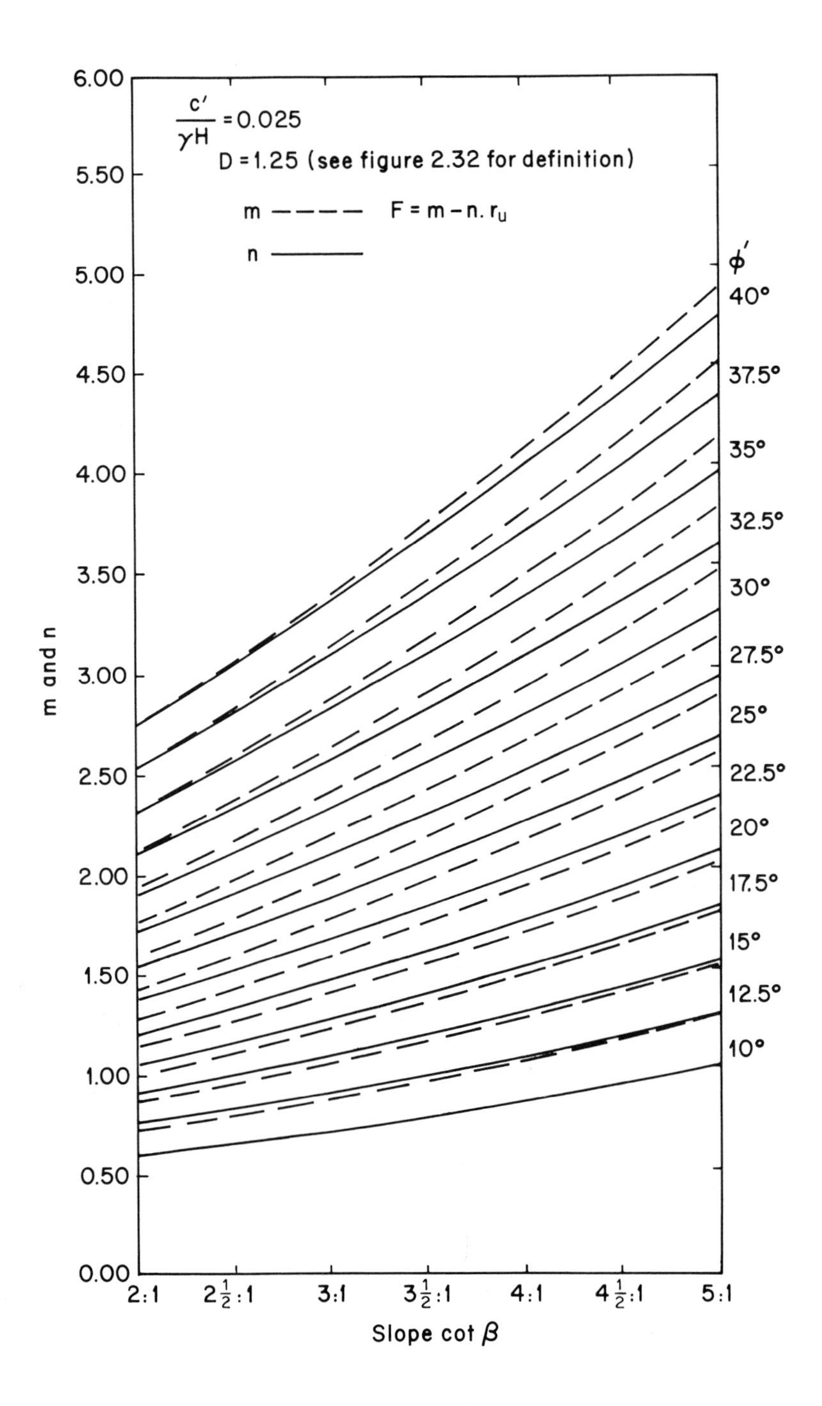

Figure 2.33 Stability coefficients m and n for $\frac{c'}{\gamma H} = 0{\cdot}025$ and $D = 1{\cdot}25$. (*Reproduced from Bishop and Morgenstern, 1960, by permission of the Institution of Civil Engineers.*)

2.8 DISCUSSION

2.8.1. Factor of Safety

In the design of slopes, the factor of safety on shear strength traditionally has several functions:

1. To take into account uncertainty of shear strength parameters due to soil variability, and the relationship between the strength measured in the laboratory and the operational field strength.
2. To take into account uncertainties in the loading on the slope (e.g. surface loading, unit weight, pore pressures).
3. To take into account the uncertainties in the way the model represents the actual conditions in the slope, which include:
   (a) the possibility that the critical failure mechanism is slightly different from the one which has been identified, and
   (b) that the model is not conservative.
4. To ensure deformations within the slope are acceptable.

The factor of safety does not allow for the possibility of gross errors, for example a bad choice of failure mechanism, such as ignoring the presence of existing shear surfaces in a slope.

It has been pointed out by De Mello (1977) that a factor of safety of 1.0 does not indicate that failure of a slope is necessarily imminent. The real factor of safety is strongly influenced by minor geological details, stress–strain characteristics of the soil, actual pore-pressure distribution, initial stresses, progressive failure, and numerous other factors. Although it is usual for a deterministic analysis to be made, it is good practice for the analysis to include a study of the sensitivity of the factor of safety to the choice of each parameter over which there is uncertainty. The major design decisions are then taken on the basis of a chosen factor of safety (or factor of ignorance!).

Recently research has focused on this design process and several procedures are being developed. If the factors affecting the stability can be identified and their variability defined it is possible to introduce this into the analysis. At its simplest, different factors of safety are applied to c' and to $\tan\phi'$ to reflect the confidence with which they are known, as well as the different mobilization of the cohesive and frictional components of strength with strain.

If the variability of the material parameters (c', ϕ', γ) and pore pressures can be expressed statistically (Harr, 1977), a probabilistic analysis can then yield a probability of failure. The difficulty with this approach is that some of the most important areas of uncertainty (e.g., bad choice of failure mechanism due to minor geological detail) necessitate the application of extreme value statistics if they are to be included in the analysis, and this is fraught with difficulty (De Mello, 1977). These procedures are still based on limit equilibrium analysis, but offer the possibility of a more rational approach to design—a development which is overdue in geotechnical engineering.

2.8.2 Accuracy of Limit Equilibrium Methods

In applying a limit equilibrium method of analysis it is important to appreciate the limitations of this class of method of stability analysis, as well as the significance of choosing one method rather than another. In the next section the theoretical basis of the limit equilibrium method is examined. Then the accuracy of a number of the methods and some of their principal limitations are considered.

2.8.2.1 Theoretical Background

An exact stability analysis would involve solving simultaneously the conditions of equilibrium and compatibility throughout the slope. The conditions which must be satisfied are as follows:

1. Each point within the soil mass must be in equilibrium.
2. The stresses within the soil must be in equilibrium with the stresses applied to the soil.
3. The strains occurring at a point must be compatible with the strains at all surrounding points.
4. The strains at every point must be related to the stresses by an appropriate stress–strain relationship for the soil.
5. The failure criterion for the soil (e.g. Mohr–Coulomb) should not be violated at any point in the slope.

Clearly a complete knowledge of the stress–strain behaviour of the soil would be required and the calculations would be very complicated. In general this approach is impractible for routine stability analysis.

Numerical techniques such as finite element analysis can be used to obtain an approximate distribution of the stresses and strains throughout a slope. With modern computers these techniques are extremely powerful and they are particularly useful for analysing the conditions in a stable slope or embankment when it is subjected to changes of loading or geometry. However their use for analysing slopes which are on the point of failure is less satisfactory and in general their use is limited by the difficulty of modelling the stress–strain behaviour of the soil.

Several classes of methods of analysis are available for examining the conditions under which failure of a slope might occur. These include:

1. Methods based on the upper and lower bound theorems of plasticity theory.
2. Limit equilibrium methods.

In these methods the calculations are simplified by ignoring some of the conditions of equilibrium and compatibility, but as a consequence of the simplification they are only approximate. A full treatment of these methods as applied to soil mechanics is given by Chen (1975) and Atkinson (1981), but it is useful to consider them here briefly.

Upper and lower bound theorems

In the theory of plasticity there are two important theorems which can be used to determine limits to the collapse load of a structure. If the conditions of equilibrium are ignored an upper bound to the collapse load may be calculated. If the conditions of compatibility are ignored a lower bound to the collapse load may be calculated. The theorems can only be proved for perfectly plastic materials and this is done by invoking the principle of virtual work. They may be applied to bars, frameworks and continua like soil.

Stated fully:

> *Upper bound theorem.* If a set of external loads acts on a plastic collapse mechanism and the work done by the external loads in an increment of displacement equals the work done by the internal stresses, collapse must occur and the external loads are an upper bound to the true collapse loads.

Notice that the external loads are not necessarily in equilibrium with the internal stresses, and that the mechanism of collapse is not necessarily the actual collapse mechanism. By examining different mechanisms, the upper bound value may be improved (i.e. reduced).

> *Lower bound theorem.* If a set of external loads acting on a structure are in equilibrium with the internal stresses and the internal stresses nowhere exceed the failure criterion of the material, collapse cannot occur and the external loads are a lower bound to the true collapse loads.

Notice that the strains and displacements are not considered and that the state of stress is not necessarily the actual state of stress at collapse. By examining different admissible states of stress the lower bound value may be improved (i.e. increased).

Of course, soils are not perfectly plastic materials. However at failure under undrained conditions they behave as if they are perfectly plastic, so the bound theorems apply. At failure under drained conditions their behaviour is not perfectly plastic, but it can be shown that the upper bound theorem applies. The significance of these theorems in geotechnical engineering can best be appreciated by applying them to an example.

Example—Stability of vertical cut in cohesive soil

A long vertical cut of height H is to be made in a homogeneous clay soil of undrained shear strength c_u and unit weight γ overlying rock, as shown in Figure 2.34. Excavation will be sufficiently rapid that the soil remains undrained. What is the critical height of the cut H_c at which collapse occurs?

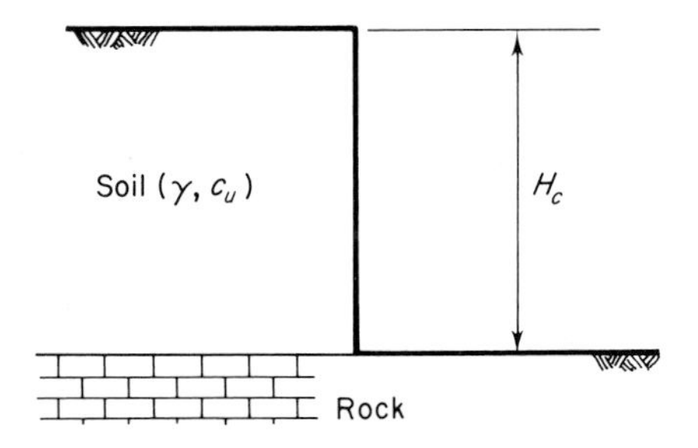

Figure 2.34 Critical height of vertical cut in cohesive soil

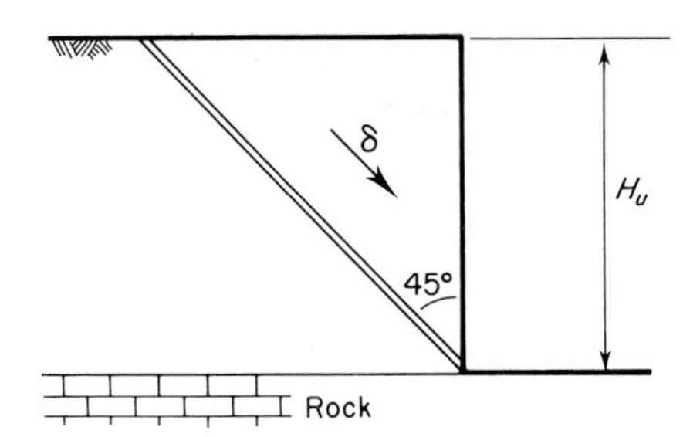

Figure 2.35 Mechanism of plastic collapse for upper bound solution for vertical cut in cohesive soil

Upper bound solution

To formulate an upper bound solution we must postulate a mechanism by which failure can take place. One such mechanism is shown in Figure 2.35. Here a wedge of soil is underlain by a single straight slip plane. Failure takes place as a rigid body movement under the action of the applied gravitational stress, and is resisted by shearing on the slip plane.

For a small displacement δ parallel to the slip plane:

Work done by gravity = Weight × vertical displacement

$$= \tfrac{1}{2}\,\gamma H^2 \frac{\delta}{\sqrt{2}}$$

Work dissipated along slip plane = length × shear stress × displacement

$$= H\sqrt{2}c_u\delta$$

On equating internal and external work we obtain an upper bound for the height of the cut at collapse H_u:

$$H_u = \frac{4c_u}{\gamma}$$

This chosen mechanism is not necessarily the true collapse mechanism and a smaller upper bound height may be obtained by investigating other mechanisms.

Lower bound solution

The lower bound solution to this problem requires an examination of the state of stress in the soil in the vicinity of the cut. Suitable states of stress may vary smoothly throughout the soil or there may be sudden discontinuities, but in either case the conditions of equilibrium must be satisfied and nowhere may the stresses exceed the failure criteria for the soil.

One admissible state of stress is shown in Figure 2.36(a). In this case the shear stresses on horizontal and vertical planes are zero which is clearly correct at the ground surface and at the cut face. The stresses on two soil elements A and B are shown, and are examined on a Mohr diagram in Figure 2.36(b). It can be seen that the circle for element A just touches the failure envelope when $\gamma H = 2c_u$, and that the circle for element B is smaller. Thus the lower bound value H_l for the height of the cut at collapse is given by:

$$H_l = \frac{2c_u}{\gamma}$$

This lower bound value may be improved by considering other admissible states of stress.

Thus from upper and lower bound theorems we have found that the critical height at collapse H_c lies in the range:

$$\frac{2c_u}{\gamma} < H_c < \frac{4c_u}{\gamma}$$

By examining other collapse mechanisms and other states of stress improved values of the upper and lower bounds to the height at collapse can be found. For this problem Heyman (1973) has shown that the range can be narrowed to

$$\frac{2 \cdot 83c_u}{\gamma} < H_c < \frac{3 \cdot 83c_u}{\gamma}$$

Limit equilibrium method

In the limit equilibrium method, we first postulate a collapse mechanism with blocks of soil sliding on an arbitrary arrangement of slip surfaces. Then without exceeding the failure criterion on the slip surfaces, we ensure that the complete mechanism and each component of it are in statical equilibrium. By examining a number of different mechanisms the critical

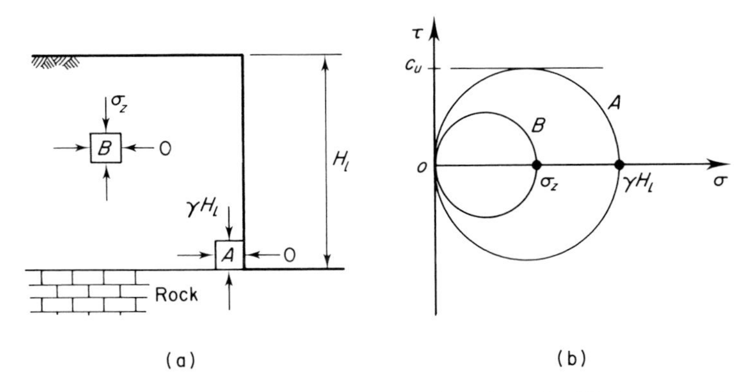

Figure 2.36 Lower bound solution for vertical cut in cohesive soil. (a) Equilibrium state of stress. (b) Mohr's circle of stress

one is found, and the external loading on this is taken to be approximately equal to the true collapse load.

Thus the limit equilibrium method combines some features of the upper and lower bound methods. Like the upper bound method, a collapse mechanism is postulated but as there are no restrictions on the shape of the slip surface it is assumed that the surrounding soil can deform sufficiently for the mechanism to become compatible. Like the lower bound method, equilibrium is examined but no check is made on the stresses in the soil surrounding the slip surface. Thus in general the limit equilibrium method gives neither a lower nor an upper bound value of the collapse loading, nor is the critical failure mechanism found from the analysis necessarily the actual mechanism at collapse. However, if the soil is homogeneous and the shape of the slip surface is limited to those permitted for an upper bound solution (plane or logarithmic spiral) the limit equilibrium solution may be allowable as an upper bound.

Example—Stability of vertical cut in cohesive soil

By way of illustration, the limit equilibrium method is applied to the problem considered above. Figure 2.37 shows a possible mode of failure for the soil behind the cut. Here a wedge of soil slides as a rigid block on a plane slip surface inclined at an angle θ to the vertical. At collapse the shear strength is fully mobilized on the shear plane and the forces are in equilibrium. Resolving the forces parallel to the slip plane:

$$T = W\cos\theta \qquad \text{i.e. } H\sec\theta . c_u = \frac{1}{2}\gamma H^2 \tan\theta . \cos\theta$$

$$\text{so } H = \frac{4c_u}{\gamma \sin 2\theta}$$

By considering different inclinations of the slip plane, the minimum value of H can be found by inspection or by differentiation. This occurs when $\theta = 45°$, and gives as the critical height of the cut from the limit equilibrium analysis:

$$H_{le} = \frac{4c_u}{\gamma}$$

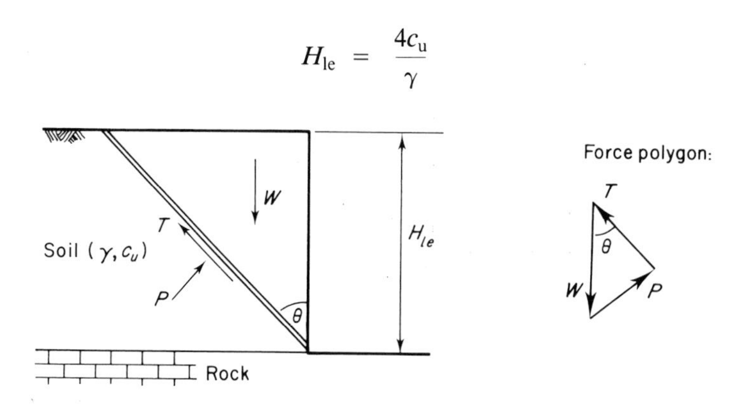

Figure 2.37 Limit equilibrium wedge analysis of vertical cut in cohesive soil

A smaller value may be obtained by considering other failure mechanisms, and Fellenius (1927) found that with a circular failure surface the critical height is given by:

$$H_{le} = \frac{3 \cdot 85 c_u}{\gamma}$$

This value is the same as the critical height found from applying the upper bound theorem and is thus an upper bound value itself. However, this agreement between the two methods is not general as was noted above.

Chen (1975) has compared the results of limit equilibrium analysis for dry slopes in homogeneous soil with an upper bound analysis. In applying the upper bound theorem he considered a slip surface defined by a logarithmic spiral (see Figure 2.38). He expressed the results of the analyses by a stability number $N = \gamma H/c$, and compared them with results given by Taylor (1948) obtained using the ordinary method of slices, and the friction circle and log spiral methods of analysis. The results of these analyses are shown in Table 2.4, and comparison suggests that the limit equilibrium method gives results which may be close to those obtained by applying the upper bound theorem. Unfortunately, comparable results from a lower bound analysis are not available.

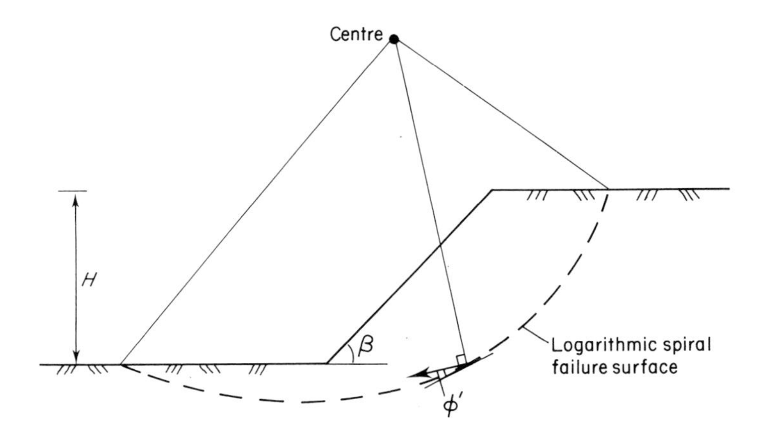

Figure 2.38 Upper bound solution for stability of cut considered by Chen, 1975

In conclusion, although the limit equilibrium method of analysis gives neither a true upper nor lower bound value of the collapse loading, experience has shown that when used with care a good estimate of the collapse loading may be obtained. It has numerous advantages over the more rigorous plasticity methods, as account can readily be taken of non-homogeneity of soil, seepage, and surface loadings. Because of this the method is widely used in engineering practice.

2.8.2.2 Comparison of Limit Equilibrium Methods of Analysis

The limit equilibrium methods of analysis described in section 2.6 above are widely used for the analysis of slopes, embankments, and excavations. With the exception of the

Table 2.4 Comparison of Stability Number $N = \gamma H/c$ by Methods of Limit Equilibrium and Limit Analysis (*Reproduced from Taylor, 1948; Chen, 1975, by permission of Elsevier Science Publishers*)

		Limit equilibrium			limit analysis
Slope angle β (°)	Friction angle ϕ (°)	slices	ϕ circle	logspiral	logspiral
90	0	3.83	3.83	3.83	3.83
	5	4.19	4.19	4.19	4.19
	15	5.02	5.02		5.02
	25	6.06	6.06	6.06	6.06
75	0	4.57	4.57	4.57	4.56
	5	5.13	5.13		5.14
	15	6.49	6.52		6.57
	25	8.48	8.54		8.58
60	0	5.24	5.24	5.24	5.25
	5	6.06	6.18	6.18	6.16
	15	8.33	8.63	8.63	8.63
	25	12.20	12.65	12.82	12.74
45	0	5.88	5.88*	5.88*	5.53*
	5	7.09	7.36		7.35
	15	11.77	12.04		12.05
	25	20.83	22.73		22.90
30	0	6.41*	6.41*	6.41*	5.53*
	5	8.77*	9.09*		9.13*
	15	20.84	21.74		21.69
	25	83.34	111.1	125.0	119.93
15	0	6.90*	6.90*	6.90*	5.53*
	5	13.89*	14.71*	14.71*	14.38*
	10		43.62		45.49

*Critical failure surface passes below toe.

ordinary method of slices, extensive experience suggests that the methods are useful and reliable if they are applied appropriately. The accuracy of the analysis of a particular slope depends on the accuracy with which the geometry of the slope, the groundwater conditions, and the soil properties can be defined, the accuracy with which the analysis models the actual conditions in the slope, as well as the inherent accuracy of the method of analysis. In general the greatest inaccuracies are connected with the groundwater conditions, the soil properties, and the stability model of the slope.

As was shown in the previous section, none of the limit equilibrium methods of analysis is exact, nor is any an upper or lower bound solution. In section 2.6 it was shown that each of the methods satisfies different conditions of equilibrium and uses different assumptions about the interslice forces (see also Table 2.1). A number of studies have been made in which several of the methods are used to analyse the same problem (e.g. Brinch Hansen, 1966; Whitman and Bailey, 1967; Fredlund and Krahn, 1977; Fredlund *et al.*, 1981; and Duncan and Wright, 1980).

Brinch Hansen's study

Brinch Hansen (1966) used several limit equilibrium methods to determine the bearing capacity of a strip foundation on weightless soil ($c' = 0$, $\phi' = 30°$) with a uniform unit surcharge. This problem is interesting as it is one for which an exact solution has been obtained by Prandtl using plasticity theory. The bearing capacity q_f may be expressed by

$$q_f = N_q . s$$

where s is the surface load and N_q is the bearing capacity factor. Brinch Hansen's results are shown in Figure 2.39, and their range is disturbingly large with results both greater and smaller than the correct value.

There are two reasons for this variation. Firstly, most of the methods use a circular failure surface which is a poor approximation to the true rupture line. Secondly, the interslice

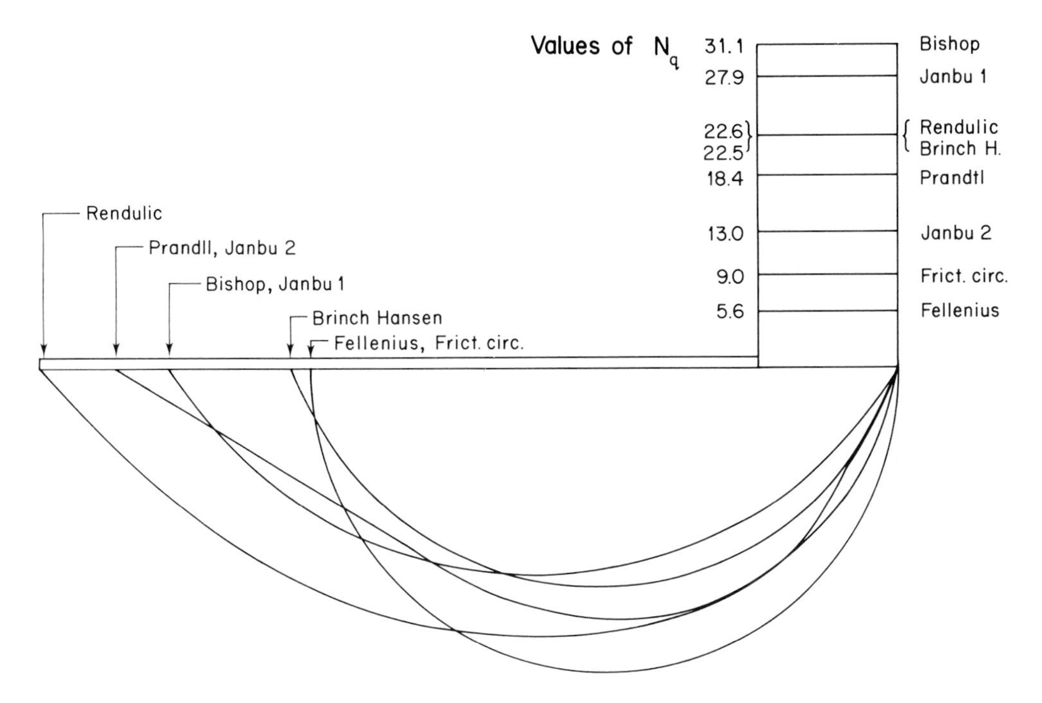

Author	Method	Rupture Line	N_q	% $N_{q\text{Prandtl}}$	$F = \dfrac{\tan 30°}{\tan\phi'_m}$
Fellenius	Slices	Circle	5.6	31	0.58
Krey	Friction Circle	Circle	9.0	49	0.75
Janbu *et al.*	Slices	Prandtl	13.0	71	0.88 × 1.05 = 0.92
Prandtl	Plasticity	Prandtl	18.4	100	1.0
Brinch Hansen	Equilibrium	Circle	22.5	113	1.07
Rendulic	Extreme	Spiral	22.6	114	1.07
Janbu *et al.*	Slices	Circle	27.9	151	1.15 × 1.05 = 1.21
Bishop	Slices	Circle	31.1	169	1.19

Figure 2.39 Comparison of bearing capacity factors determined by different methods. (*Adapted from Brinch Hansen, 1966, by permission of the Danish Geotechnical Institute.*)

Table 2.5 Comparison of Factors of Safety for Example Problem (*Reproduced from Fredlund and Krahn, 1977, by permission of the National Research Council of Canada.*)

Case no.	Example problem*	Ordinary method	Simplified Bishop method	Spencer's method			Janbu's simplified method	Janbu's rigorous method†	Morgenstern–Price method $f(x)$ = constant	
				F	θ	λ			F	λ
1	Simple 2:1 slope, 12 m high, $\phi' = 20°$, $c' = 28.75$ kPa	1.928	2.080	2.073	14.81	0.237	2.041	2.008	2.076	0.254
2	Same as 1 with a thin, weak layer with $\phi' = 10°$, $c' = 0$	1.288	1.377	1.373	10.49	0.185	1.448	1.432	1.378	0.159
3	Same as 1 except with $r_u = 0.25$	1.607	1.766	1.761	14.33	0.255	1.735	1.708	1.765	0.244
4	Same as 2 except with $r_u = 0.25$ for both materials	1.029	1.124	1.118	7.93	0.139	1.191	1.162	1.124	0.116
5	Same as 1 except with a piezometric line	1.693	1.834	1.830	13.87	0.247	1.827	1.776	1.833	0.234
6	Same as 2 except with a piezometric line for both materials	1.171	1.248	1.245	6.88	0.121	1.333	1.298	1.250	0.097

*Width of slice is 0.3 m and the tolerance on the non-linear solutions is 0.001

†The line of thrust is assumed at 0.333.

forces are generally not included satisfactorily. Turnbull and Hvorslev (1967) have shown that where the failure surface is inclined steeply at the toe the ordinary method underestimates the shear resistance there and Bishop's method overestimates it. Brinch Hansen's results illustrate this well, and if the interslice forces were included by using one of the more rigorous methods the differences from the theoretical value would be quite small.

This example is a very extreme one and for normal slope stability analyses the differences are generally much smaller. This is because the vertical loading on the failure surface is usually more evenly distributed in slope stability problems and the angles of inclination of the failure surface at the toe are smaller.

Fredlund and Krahn's study

Fredlund and Krahn (1977) compared the results of a number of methods of analysis when applied to an example slope stability problem (see Figure 2.40). Various combinations of geometry, soil and groundwater conditions were considered, and some of their results are given in Table 2.5. These results show a much smaller variation than those for the bearing capacity problem, and with the exception of the ordinary method the range of factors of safety is less than 4% for all six cases.

The sensitivity of the factor of safety to the assumptions about interslice forces and the conditions of equilibrium which are satisfied was examined (see Figure 2.41). Here the factors of safety F_m, and F_f, determined using constant side force function $f(x)$, are plotted against the scaling factor λ for the interslice forces.

This shows again that the factor of safety F_m, determined from satisfying moment equilibrium, is relatively insensitive to the interslice force assumption. In the cases considered, the differences between factors of safety obtained from Bishop's method (satisfying moment equilibrium) and Spencer's and Morgenstern and Price's methods (with λ chosen to satisfy force and moment equilibrium) are no more than 0.4%. In contrast the

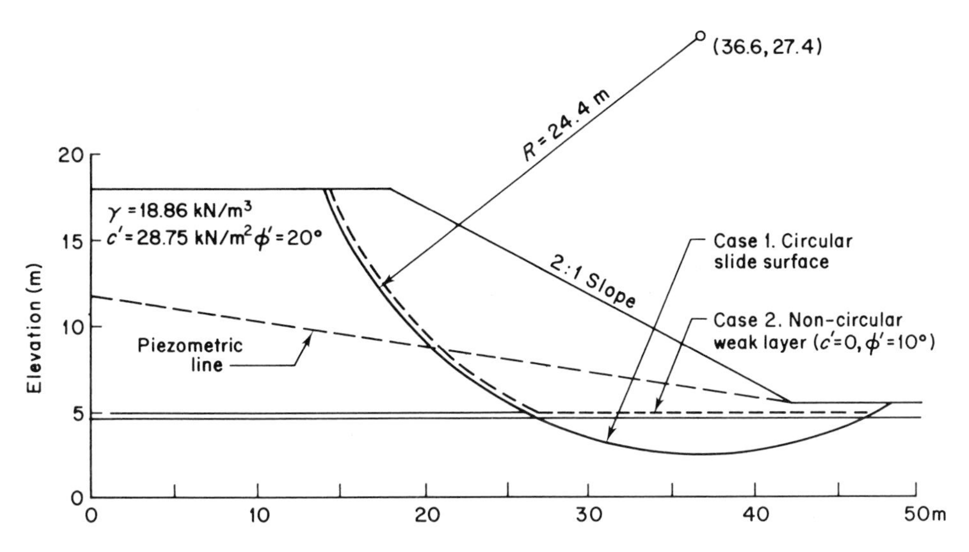

Figure 2.40 Example problem using circular and non-circular sliding surfaces. (*Adapted from Fredlund and Krahn, 1979, by permission of the National Research Council of Canada.*)

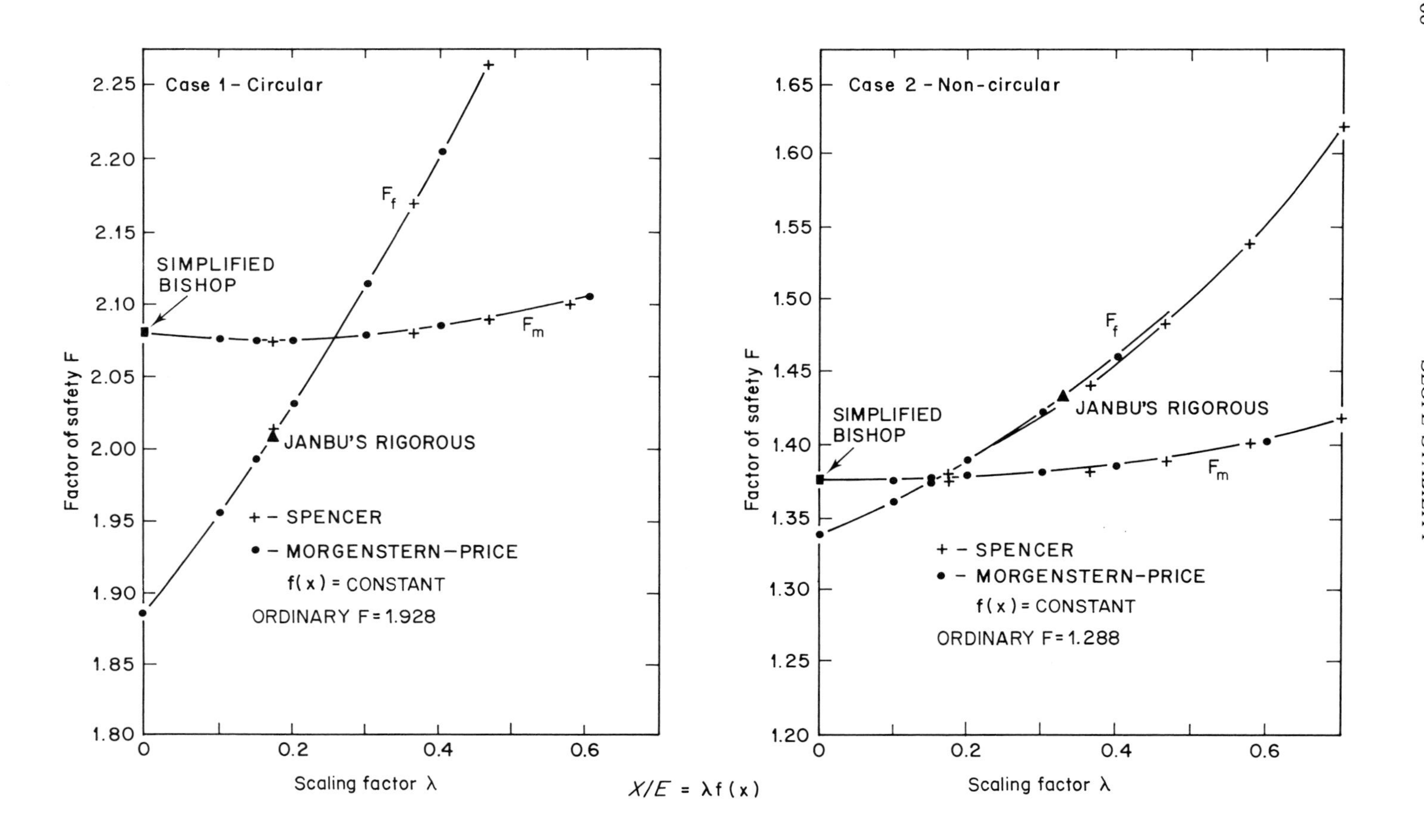

Figure 2.41 Influence of interslice forces on factors of safety. (*Adapted from Fredlund and Krahn, 1977, by permission of the National Research Council of Canada.*)

Table 2.6 Influence of Side Force Function in Morgenstern–Price Method. (*Reproduced from Fredlund and Krahn, 1977, by permission of the National Research Council of Canada.*)

Case no.	Example problem	University of Alberta programme side force function				University of Saskatchewan SLOPE programme side force function					
		Constant		Half sine		Constant		Half sine		Clipped sine*	
		F	λ	F	λ	F	λ	F	λ	F	λ
1	Simple 2:1 slope, 12 m high, $\phi' = 20°$, $c' = 28.75$ kPa	2.085	0.257	2.085	0.314	2.076	0.254	2.076	0.318	2.083	0.390
2	Same as 1 with a thin, weak layer with $\phi' = 10°$, $c' = 0$	1.394	0.182	1.386	0.218	1.378	0.159	1.370	0.187	1.364	0.203
3	Same as 1 except with $r_u = 0.25$	1.772	0.351	1.770	0.432	1.765	0.244	1.764	0.304	1.779	0.417
4	Same as 2 except with $r_u = 0.25$ for both materials	1.137	0.334	1.117	0.441	1.124	0.116	1.118	0.130	1.113	0.138
5	Same as 1 except with a piezometric line	1.838	0.270	1.837	0.331	1.833	0.234	1.832	0.290	1.832	0.300
6	Same as 2 except with a piezometric line for both materials	1.265	0.159	Not converging		1.250	0.097	1.245	0.101	1.242	0.104

*Cordinates $x = 0$, $y = 0.5$, and $x = 1.0$, $y = 0.25$.

Note: University of Alberta programme based on original Morgenstern–Price formulation; University of Saskatchewan programme based on Fredlund and Krahn's GLE formulation; Tolerance on both solutions is 0.001.

factor of safety F_f determined by satisfying force equilibrium is very sensitive to λ, and by implication those methods which satisfy force equilibrium alone (e.g. Janbu's simplified method before correction, Lowe and Karafiath, etc.) are less accurate than Bishop's method which satisfies moment equilibrium alone.

Fredlund and Krahn also investigated the influence on the factor of safety of varying the side force function $f(x)$ in the Morgenstern and Price's analysis. Table 2.6 shows their results for the example problem, and indicates that the factor of safety is fairly insensitive to the choice of $f(x)$.

Similar conclusions about the accuracy of the limit equilibrium methods of analysis were reached by Duncan and Wright (1980) who carried out a parametric study of homogeneous slopes. They found that Bishop's method and the methods satisfying all conditions of equilibrium all give values of factor of safety which are within 5% of the value obtained by using a logarithmic spiral analysis (Rendulic, 1935) which is an upper bound value. As a result of these comparative studies we may conclude that:

1. The methods which satisfy all conditions of equilibrium (Janbu's rigorous, Spencer's and Morgenstern and Price's methods) all give accurate results ($\pm 5\%$) for the analysis of the slopes.
2. Bishop's method which only satisfies moment equilibrium gives similarly accurate results except where the slip surface is steeply inclined at the toe.
3. Other methods which do not satisfy all conditions of equilibrium (ordinary method, force–equilibrium methods) may be highly inaccurate.
4. Where the slip surface is steeply inclined at the toe a method should be chosen which gives a sensible distribution of interslice forces.

2.8.3 Three-dimensional Effects

Although the limit equilibrium analyses described in section 2.6 are normally applied to a single vertical cross-section, landslips are of course three-dimensional and are frequently bowl-shaped (see Figure 2.42). The end-effects can be very significant in practice (e.g. Azzouz *et al.*, 1981) and would generally enhance the stability. There are no rigorous techniques available for the analysis of a generalized three-dimensional slip mass, but some solutions are available where the geometry can be approximated to regular shapes (e.g. Azzouz and Baligh, 1978; Chen and Chameau, 1983). An indication of the influence of the end-effects can be obtained by analysing a number of cross-sections and using an averaging technique (Lambe and Whitman, 1969).

For rock slopes the three-dimensional effects are very important as the stability is generally controlled by the presence of pre-existing discontinuities. If the slip can be approximated to a block of material sliding on two planar surfaces, then a wedge analysis can be used (Hoek and Bray, 1977).

2.8.4 Progressive Failure

The use of a limit equilibrium method for the analysis of slopes necessitates the assumption that the shear strength is mobilized to the same degree all around a potential surface. The factor of safety (on shear strength) is thus an average value for the assumed critical

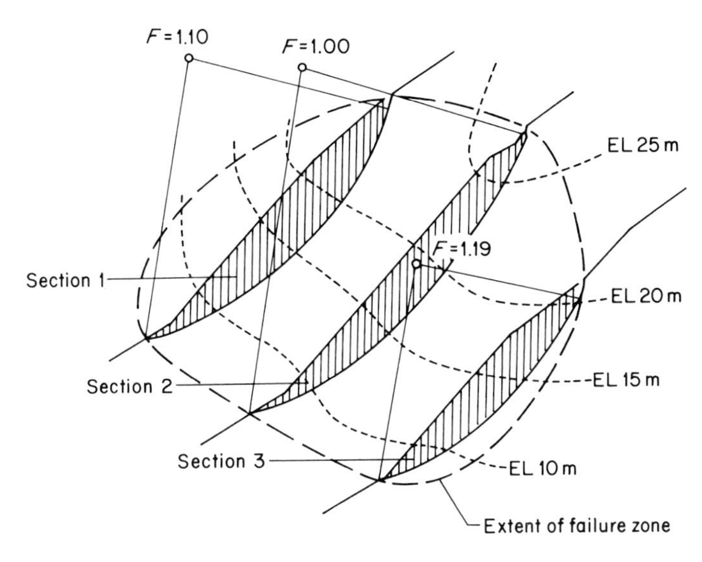

Figure 2.42 Approximate three-dimensional reconstruction of the failure surface at Lodalen. (*After Sevaldson, 1956.*)

failure surface. No information is obtained about the actual distribution of stresses within the slope.

A number of analyses have been made to examine the stress distributions within stable slopes, and the development of failure. Bishop (1952) used an elastic stress analysis of a stable embankment to show that shear stresses vary markedly around a circular surface (see Figure 2.43). Over nearly half the total length of the surface the shear stresses exceed the average value by up to 70%. Although the analysis is only approximate, it suggests that for a homogeneous embankment, a safety factor of even 1.5 from a total stress analysis implies some overstressing. Similar results have been obtained from finite element analyses of slopes and excavations.

The finding that the peak strength is reached at some points on a slip surface before others, is particularly significant for the analysis of first-time slides. This is the problem of progressive failure. Most clays are not ideal plastic non-brittle materials, and when their peak strength has been reached the strength decreases with increasing strain. Thus if failure occurs, the average mobilized strength will be somewhat less than the peak strength around the failure surface and it may approach the critical state strength (Skempton, 1977). This problem is particularly important when considering the stability of slopes in heavily over-consolidated clays where there is a large amount of recoverable strain energy (Bjerrum, 1967; Burland *et al.*, 1977). A similar problem arises in slopes in highly structured sensitive or quick clays found in Scandinavia and Canada in which the breakdown of the clay structure is accompanied by a build-up of pore pressure (Bjerrum and Kenney, 1967; Aas, 1981).

With the development of finite element methods of analysis it is now possible to examine such problems numerically, but this necessitates very careful soil testing so that the stress–strain behaviour can be properly defined.

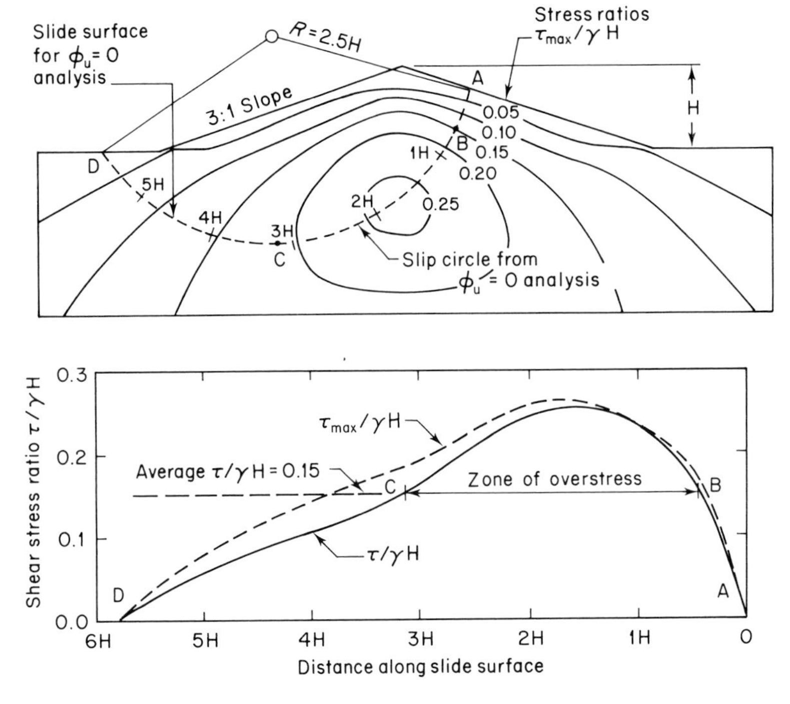

Figure 2.43 Distribution of shear stresses in a uniform embankment and foundation using elastic analysis. (*After Bishop, 1952; Turnbull and Hvorslev, 1967.*)

2.8.5 Dynamic Stability

The limit equilibrium methods discussed above are applicable to the analysis of slopes in static equilibrium. They are not well suited to the analysis of dynamic stability of slopes, for example debris flows (see Chapter 16), avalanches, and slopes under earthquake loading (see Chapter 9). Although horizontal loading of the soil mass can be included (pseudo-static analysis) the dynamic effects on the pore pressures and shear strength parameters are difficult to model. These effects are often of major importance, and much more research is needed before reliable routine methods of analysis are available for these situations. A full discussion of the dynamic analysis of slopes during earthquakes is beyond the scope of this book, and the reader is referred to the work of Seed (1968, 1979), and Ishihara (1985) for a fuller discussion.

2.9 CONCLUSIONS

The limit equilibrium methods of analysis described here are widely used for the analysis of slopes, embankments, and excavations. Although at first sight each method may appear to be quite different, we have seen that the only significant difference is the assumption made about the position and inclination of the interslice forces. It is important to understand the theoretical basis of a particular method before applying it in practice.

With the exception of the ordinary method of slices, extensive experience suggests that the methods are useful and reliable if they are applied appropriately. This experience has been gained from the analysis of slopes which have failed. In using these methods to analyse slopes which have not failed it is important to appreciate that the principal areas of uncertainty are the soil properties, the groundwater conditions, and the mechanism of a possible failure.

A thorough site investigation is essential to establish the soil and groundwater conditions, followed by careful soil testing in a reputable laboratory. The geotechnical specialist responsible for the project should be closely involved at all stages of the investigation so that minor details are not overlooked.

Once the soil and groundwater conditions have been established, a method of analysis may be chosen which is appropriate to the anticipated mechanism of failure. It is often useful to ascertain the sensitivity of the result of the analysis to small changes in the assumed parameters so that engineering decisions may be based on a full understanding of the problem. In conclusion it must be reiterated that a sophisticated analysis is no substitute for experience and engineering judgement.

REFERENCES

Aas, G. (1981). 'Stability of natural slopes in quick clays.' *Proc. 10th Int. Conf. SMFE*, Stockholm, **3**, 333–8.

Atkinson, J. H. (1981). *Foundations and Slopes*. McGraw-Hill, London.

Azzouz, A. S., and Baligh, M. M. (1978). 'Three dimensional stability of slopes.' *Research Report R78-8, No. 595*. Dept. of Civ. Eng., MIT, Cambridge, Mass. 349pp.

Azzouz, A. S., Baligh, M. M., and Ladd, C. C. (1981). Three dimensional stability analyses of four embankment failures. *Proc. 10th Int. Conf. SMFE*, Stockholm, **3**, 343–6.

Bishop, A. W. (1952). *The stability of earth dams*. Unpublished Ph.D. Thesis. London University, London, England.

Bishop, A. W. (1955). 'The use of the slip circle in the stability analysis of slopes.' *Geotechnique*, **5**, 7–17.

Bishop, A. W., and Bjerrum, L. (1960). 'The relevance of the triaxial test to the solution of stability problems.' *ASCE Research Conf. on Shear Strength of Cohesive Soils*, Boulder, Colorado, 437–501.

Bishop, A. W., and Morgenstern, N. R. (1960). 'Stability coefficients for earth slopes.' *Geotechnique*, **10**, 129–50.

Bjerrum, L. (1967). 'Progressive failure in slopes of overconsolidated plastic clay and clay shales.' *ASCE, J. Soil Mech. Fndtn. Div.*, **93** (SM5), 1–49.

Bjerrum, L., and Flodin, N. (1960). 'The development of soil mechanics in Sweden, 1900–1925.' *Geotechnique*, **10**, 1–18.

Bjerrum, L., and Kenney, T. C. (1967). 'Effect of structure on the shear behaviour of normally consolidated quick clays.' *Proc. Geotech. Conf.*, Oslo, Norway, **2**, 19–27.

Brinch Hansen, J. (1966). 'Comparison of methods for stability analysis'. Bulletin No 21, Danish Geotechnical Institute.

Burland, J. B., Longworth, T. I., and Moore, J. F. A. (1977). 'A study of ground movement and progressive failure caused by a deep excavation in Oxford Clay.' *Geotechnique*, **27**, 557–91.

Cedergren, H. R. (1967). *Seepage, Drainage and Flow Nets*. Wiley, New York.

Chandler, R. J. (1984). 'Recent European experience of landslides in over-consolidated clays and soft rocks. State-of-the-art lecture.' *Proc. 4th Int. Symp. Landslides*, Toronto, **1**, 61–81.

Charles, J. A., and Soares, M. M. (1984). 'Stability of compacted rockfill slopes.' *Geotechnique*, **34**, 61–70.

Chen, R. H., and Chameau, J.-L. (1983). 'Three dimensional limit equilibrium analysis of slopes.' *Geotechnique*, **32**, 31–40.

Chen, W. F. (1975). *Limit Analysis and Soil Plasticity*. Elsevier, Amsterdam.
Collin, A. (1846) (trans. W. Schriever). *Landslides in Clay*. University of Toronto Press, 1956.
Cousins, B. F. (1978). 'Stability charts for simple earth slopes.' *ASCE, J. Geotech. Eng. Div.*, **104** (GT2), 267–79.
De Mello, V. F. B. (1977). 'Reflections on design decisions of practical significance to embankment dams.' *Geotechnique*, **27**, 279–355.
Duncan, J. M., and Wright, S. G. (1980). 'The accuracy of equilibrium methods of slope stability analysis.' *Proc. Int. Symp. Landslides*, New Delhi.
Fellenius, W. (1918). 'Kaj-och jordrasen i Göteborg'. *Teknisk Tidsskrift V.U.*, **48**, 17–19.
Fellenius, W. (1927). Erdstatische Berechnungen mit Reibung und Kohäsion (Adhäsion) und unter Annahme kreis-zylinderischer Gleitflächen. Ernst, Berlin.
Fellenius, W. (1936). 'Calculation of stability of earth dams.' *Trans. 2nd Int. Congr. Large Dams*, **4**, 445.
Fredlund, D. G., and Krahn, J. (1977). 'Comparison of slope stability methods of analysis.' *Can. Geotech. J.* **14**, 429–439.
Fredlund, D. G., Krahn, J., and Pufahl, D. E. (1981). 'The relation between limit equilibrium slope stability methods.' *Proc. Int. Conf. SMFE*, Stockholm, **3**, 409–16.
Gibson, R. E., and Morgenstern, N. R. (1962). 'A note on the stability of cuttings in normally-consolidated clays.' *Geotechnique*, **12**, 212–216.
Haefeli, R. (1948). 'The stability of slopes acted upon by parallel seepage.' *Proc. 2nd Int. Conf. SMFE*, Rotterdam, **1**, 57–62.
Harr, M. E. (1977). *Mechanics of Particulate Media—A Probabilistic Approach*. McGraw-Hill, New York, 543pp.
Heyman, J. (1972). *Coulomb's Memoir in Statics*. Cambridge University Press.
Heyman, J. (1973). 'The stability of a vertical cut.' *Int. J. Mech. Sci.*, **15**, 845–54.
Hoek, E., and Bray, J. W. (1977). *Rock Slope Engineering*. Inst. of Min. and Met., London, 402pp.
Hunter, J. H., and Schuster, R. L. (1968). 'Stability of simple cuttings in normally consolidated clays.' *Geotechnique*, **18**, 372–8.
Ishihara, K. (1985). 'Stability of natural deposits during earthquakes.' *Proc. 11th Int. Conf. SMFE*, San Francisco, **1**, 321–76.
Janbu, N. (1954a). 'Application of composite slip surfaces for stability analysis.' *Eur. Conf. Stability Earth Slopes*, Stockholm, **3**, 43–9.
Janbu, N. (1954b). 'Stability analysis of slopes with dimensionless parameters.' *Harvard Soil Mechanics Series* No 46, 811pp.
Janbu, N. (1957). 'Earth pressure and bearing capacity calculations by generalised procedure of slices.' *Proc. 4th Int. Conf. SMFE*, London, **2**, 207–12.
Janbu, N., Bjerrum, L., and Kjaernsli, B. (1956). 'Veiledning ved løsning av fundamenterings oppgaver' (in Norwegian with English summary: Soil mechanics applied to some engineering problems). *Norwegian Geotechnical Institute*, Publ. No. 16.
Kenney, T. C. (1967). 'Shear strength of soft clay.' *Proc. Geotech. Conf.*, Oslo, **2**, 49–55.
Kenney, T. C. (1984). 'Properties and behaviour of soils relevant to slope instability.' Chapter 2 in *Slope Instability*. Brunsden, D., and Prior, D. B. (eds.), Wiley, Chichester.
Lambe, T. W., and Whitman, R. V. (1969). *Soil Mechanics*. Wiley, New York.
Lowe, J., and Karafiath, L. (1960). 'Stability of earth dams upon drawdown.' *Proc. 1st Panamerican Conf. SMFE*, Mexico, **2**, 537–52.
Lupini, J. F., Skinner, A. E., and Vaughan, P. R. (1981). 'The drained residual strength of soils.' *Geotechnique*, **31**, 181–213.
Morgenstern, N. R. (1963). 'Stability charts for earth slopes during rapid drawdown.' *Geotechnique*, **13**, 121–32.
Morgenstern, N. R., and Price, V. E. (1965). 'The analysis of the stability of generalised slip surfaces.' *Geotechnique*, **15**, 79–93.
O'Connor, M. J. and Mitchell, R. J. (1977). 'An extension of the Bishop and Morgenstern slope stability charts.' *Can. Geotech. J.*, **14**, 144–51.
Petterson, K. E. (1955). 'The early history of circular sliding surfaces.' *Geotechnique*, **5**, 275–96.
Rendulic, L. (1935). 'Ein Beitrag zur Bestimmung der Gleitsicherheit.' *Der Bauingenieur*, No 19/20.

Seed, H. B. (1968). 'Landslides during earthquakes due to soil liquefaction.' Terzaghi Lecture. *ASCE, J. Soil. Mech. Fndtn Div.*, **94** (SM5), 193–261.
Seed, H. B. (1979). 'Considerations in the earthquake-resistant design of earth and rockfill dams.' *Geotechnique*, **29**, 215–63.
Seed, H. B., and Sultan, H. A. (1967). 'Stability analyses for a sloping core embankment.' *ASCE, J. Soil Mech. Fndtn. Div.*, **93**, (SM5), 69–83.
Sevaldson, R. A. (1956). 'The slide in Lodalen, 6 October 1954.' *Geotechnique*, **6**, 1–16.
Sherrard, J. L., Woodward, R. J., Gizienski, S. G., and Clevenger, W. A. (1963). *Earth and Earth-rock Dams*. Wiley, New York.
Skempton, A. W. (1949). 'Alexandre Collin. A note on his pioneer work in soil mechanics.' *Geotechnique*, **1**, 216–21.
Skempton, A. W. (1964). 'Long-term stability of clay slopes.' *Geotechnique*, **14**, 77–102.
Skempton, A. W. (1977). 'Slope stability of cuttings in brown London clay.' *Proc. 9th Int. Conf. SMFE*, Tokyo, **3**, 261–70.
Skempton, A. W. (1979). 'Landmarks in early soil mechanics.' *Proc. 7th Eur. Conf. SMFE*, Brighton, **5**, 1–26.
Skempton, A. W., and Delory, F. A. (1957). 'Stability of natural slopes in London clay.' *Proc. 4th Int. Conf. SMFE*, London, **2**, 378–81.
Skempton, A. W. and Golder, H. Q. (1948). 'Practical examples of the $\phi = 0$ analysis of stability of clays.' *Proc. 2nd Int. Conf. SMFE*, Rotterdam, **2**, 63–70.
Skempton, A. W., and Hutchinson, J. N. (1969). 'Stability of natural slopes and embankment foundations'. State-of-the-art report.' *Proc. 7th Int. Conf. SMFE*, Mexico City, **2**, 291–335.
Spencer, E. (1967). 'A method of analysis of the stability of embankments assuming parallel inter-slice forces.' *Geotechnique*, **17**, 11–26.
Sultan, H. A., and Seed, H. B. (1967). 'Stability of sloping core earth dams.' *ASCE, J. Soil Mech. Fndtn. Div.*, **93** (SM5), 45–68.
Taylor, D. W. (1948). *Fundamentals of Soil Mechanics*. Wiley, New York.
Terzaghi, K. (1925). *Erdbaumechanik*. Deuticke, Vienna.
Terzaghi, K., and Peck, R. B. (1967). *Soil Mechanics in Engineering Practice*. Wiley, New York.
Turnbull, W. J., and Hvorslev, M. L. (1967). 'Special problems in slope stability.' *ASCE, J. Soil Mech. Fdtn. Div.*, **93** (SM4), 499–528.
US Army Corps of Engineers (1970). 'Engineering and design, stability of earth and rockfill dams.' *Dept. of the Army, Corps of Engineers, Engineer Manual EM1110-2-1902.*
Whitman, R. V., and Bailey, W. A. (1967). 'Use of computers for slope stability analysis.' *ASCE, J. Soil. Mech. Fndtn Div.*, **93**, (SM4), 475–98.

Slope Stability
Edited by M. G. Anderson and K. S. Richards

Chapter 3

Instrumentation of Pore Pressure and Soil Water Suction

M. G. ANDERSON
Department of Geography
University of Bristol, Bristol BS8 1SS

and

P. E. KNEALE
School of Geography
University of Leeds, Leeds LS2 9JT

3.1 INTRODUCTION

Slope hydrology is an element of slope instability that is reflected explicitly in slope stability models (e.g. Chapters 2, 4, and 8), or is implicitly required in the examination of longer term hillslope development (e.g. Chapters 12 and 19). In many slope stability problems and in many materials (especially residual soils) uncertainties relating to the strength parameters of a soil are perhaps similar to the difficulties of predicting pore pressure or suction values for stability analysis. While this factor is of critical, and frequently underrated importance, it must be appreciated that pore pressure/suction monitoring is undertaken for perhaps four primary reasons:

1. To calibrate stability analysis models.
2. To provide a means of estimating the worst groundwater condition a site may experience.
3. To provide initial results as a prelude to full site instrumentation.
4. To identify the dominant groundwater/soil water processes.

Instrumentation selection, installation, and monitoring is thus a major element in site investigation work (Table 3.1). While this chapter discusses these components in the context of pore pressure monitoring systems, it must be noted that both detailed project definition

Table 3.1 Schematic Flow Chart for Planning a Monitoring Operation. (*After Geotechnical Control Office, 1984. Reproduced by permission of the Hong Kong Government.*)

1 Ground Behaviour Warning Levels and Contingency Action	2 General Monitoring Plan	3 Detailed Monitoring Plan
PROJECT DEFINITION Geometry; geology; groundwater; stress; construction programme	TERMS OF REFERENCE Monitoring objectives; budget	PERSONNEL No. of persons; allocation of responsibilities; liaison and reporting channels
↓ GROUND BEHAVIOUR Mechanism; critical locations; magnitudes;	↓ WHAT TO MEASURE Displacement; water; pressure; load	↓ This chapter INSTRUMENTS Selection; calibration; detailed layout
↓ CONTINGENCY PLANNING Decisions on hazard warning levels; action plans if warning levels exceeded	↓ WHERE TO MEASURE Identify key locations and depths; establish priorities	↓ INSTALLATION Define installation locations, times and procedures
	↓ WHEN TO MEASURE Project duration; frequency of readings; frequency of reports	↓ MONITORING Define detailed monitoring programme
		↓ DATA PROCESSING Draft and print data sheets and graphs; set up computation procedures
		↓ REPORTING Define reporting requirements; timing; contents; responsibilities

and terms of reference will be additional ingredients which should be used as necessary to modify configurations for *specific* project demands. Of course, in many investigations it is necessary to undertake more than one of the above objectives of pore pressure monitoring simultaneously, due to cost constraints. These objectives may potentially interact and, bearing in mind the particular significance of pore-water conditions, it is important that a sound review is provided of available instrumentation. This chapter seeks to undertake that task.

It must not be forgotten that there is also an increasing interaction between model requirement, instrumentation development, and subsequent model refinement. Thus, while instrumentation may, in the simplest case, facilitate the articulation of a stability model

Table 3.2 Selected Advances in Water Monitoring Equipment

Date		Author
	Hydraulic/mechanical based systems	
1922	Tensiometer developed	Gardner *et al.*
1949	Casagrande piezometer	Casagrande
1951	Theory of piezometer response times (later extended by Gibson, 1963; Brand and Premchidt, 1982)	Hvorslev
	Electronic based systems	
1962	Tensiometer–pressure transducer system	Bianchi
1969	Psychrometers developed	Richards
1975	Automatic multiple vibrating wire	Massarsch *et al.*
1977	Automatic transducer system for positive and negative pressures	Anderson and Burt
1980	Radio frequency soil moisture sensor	McKim *et al.*
1980	Theory of tensiometer response times	Towner
1982	Automatic bubbler system for standpipe piezometer monitoring	Pope *et al.*
1984	Heat dissipation sensor for high suctions (>100 kPa)	Lee and Fredlund
1984	Development of modem RS232 links between field data logger and office computer system	Geotechnical instruments
1986	Acoustic depth monitoring for standpipe piezometers	Anderson

at a given site, instrumentation can equally be the *source* of model refinement (especially with the recent introduction of automatic systems over the complete pressure/suction range, as Table 3.2 illustrates). At this point, costs may intervene in the subsequent degree of application and exposure that such an enhanced model may experience. Recent advances in combined models of soil water, groundwater, and slope stability, for example, provide a good example of such interaction (Anderson, 1983; Anderson *et al.*, 1983; Anderson and Pope, 1984).

The following sections review instrumentation in three classes; piezometer systems, tensiometers, and sensors for suction measurement in excess of 90 kPa; Chapters 2 and 4 demonstrate the need for accurate monitoring systems throughout the pressure–suction range.

3.2 PIEZOMETER SYSTEMS

3.2.1 Standpipes

Wells (diameter >0.5 m) are not normally used in engineering studies, but well records may be utilized during survey to give a general indication of the water-table position. The well level is unlikely ever to be in true equilibrium with the soil water conditions, even where soils are highly permeable, due to the large volume of water that must move into the well for a given increase in pore-water pressure. Standpipes are slotted or perforated tubes placed in boreholes and surrounded by sand or gravel (Figure 3.1). The narrower diameter (20–50 mm) allows self-de-airing and the response time is reduced compared with a well. However, a filter is not usually incorporated in this type of installation, water entering

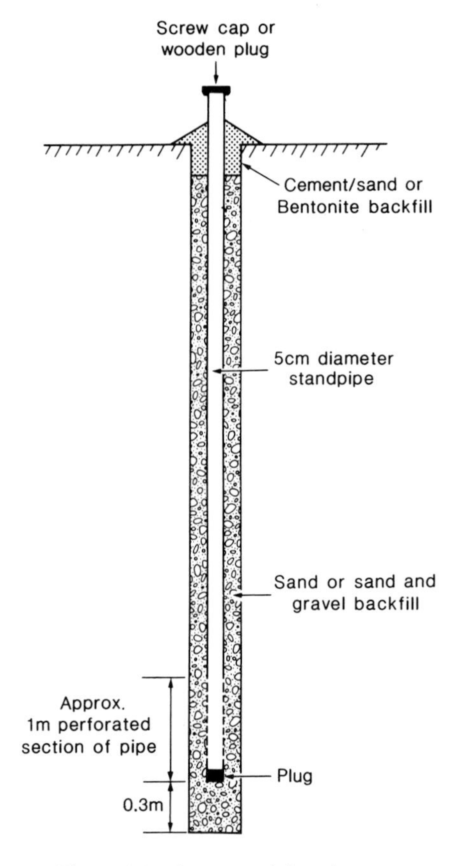

Figure 3.1 Open standpipe piezometer

or leaving the tube via slots or perforations in the tube wall (Figure 3.1). In time this may cause silting in the standpipe and reduce the accuracy of the readings.

3.2.2 Standpipe Piezometers

Hydraulic piezometers more usually comprise a porous ceramic piezometer element sealed into the ground but connected to a remote readout station by a water-filled tube. This acts as a filter to sediment movement and improves the contact between the water in the soil and water in the piezometer. Open hydraulic piezometers are connected to a vertical rigid tube from the probe to the surface and are available in a range of diameters and tip probe types. This tube is open to the atmosphere and monitoring is undertaken using some type of dip-meter. This instrumentation is simple, light, and battery operated. The principal problem is in accurately measuring the depth of the water level using such a cable based method. Operator care in straightening the cable before measurement should minimize inaccuracies, as will using piezometer tubing that is twice the diameter of the probe and cable. In addition, droplets of water on the side of the tube can give false readings unless the probe is designed to avoid this.

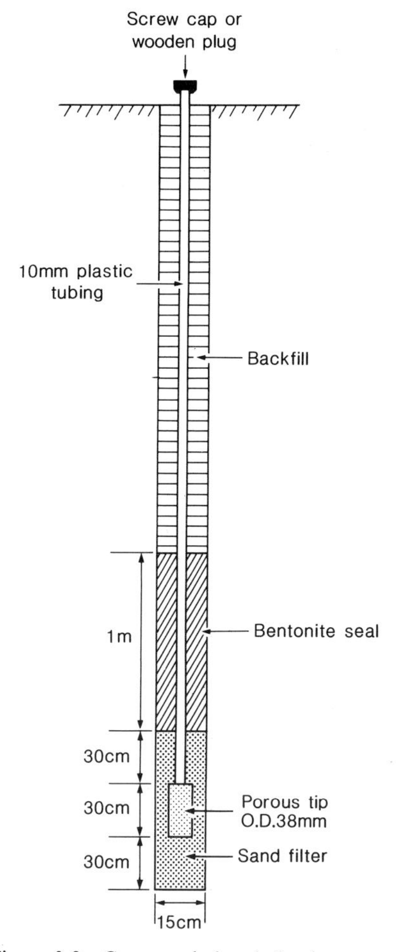

Figure 3.2 Casagrande borehole piezometer

Figure 3.2 shows the standard Casagrande type piezometer (Casagrande, 1949). The installation method for Casagrande piezometers is somewhat time consuming but if carefully followed should yield very precise results. A borehole is drilled to the required depth of the measuring probe. A casing is then driven to the final depth and the material in the casing removed. The standpipe is located in position and the lower end backfilled with sand as the casing is partially withdrawn. A bentonite, or cement bentonite, seal is made above the sand filter and the remainder of the tube is grouted with an impermeable grout. A grout mix that is suitable for soils is a 1:1:3 ratio of cement, bentonite, and water. It is important that the grout is less permeable than the soil. Grout volumes should be checked and compared with the volume of the hole over the length to be grouted, thereby attempting to ensure sealing of the piezometer is as good as possible (Geotechnical Control Office, 1984). The advantages of the Casagrande piezometer are that it is simple, inexpensive, has no metallic elements so has a long field life, and should be self de-airing if the standpipe is larger than 12 mm in diameter (otherwise it can be installed with a riser pipe for de-airing).

3.2.3 Closed Hydraulic Piezometers

Closed piezometers replace the open standpipe and dip-meter monitoring method with a design in which the pipe is fully water (or oil) filled such that changes in pressure occurring in the soil are more rapidly reflected in changes in pressure within the piezometer. The monitoring device may be a Bourdon gauge or mercury manometer (where the fluid filled tubes can be brought to a surface gauge housing), or by a diaphragm and electrical monitoring device placed in the tip of the piezometer itself. In these cases, a single wire is routed to a recorder house.

The closed piezometer design is suitable for installation at sites where tips can be buried and construction allowed to continue, and in remote or inaccessible sites. The elimination of the air-filled standpipe serves to decrease the response time of the piezometer system. However, if the soil around the piezometer tip is not saturated then air may pass across the tip and the presence of air bubbles in the piezometric lines reduces the accuracy of the measurements. A range of piezometer tips is available to suit the soil conditions. The selection of a tip must be made to balance the importance of a fast response with the necessity for regular de-airing.

In general terms, closed piezometer systems have three advantages over open types:

1. Depending on the type of pressure gauge and probe volume concerned, they have a more rapid response time and so are suitable for work in impermeable soils.
2. They can be used to measure small negative pore water pressures.
3. They may be installed with a central reading system so operator time at a site is reduced as is disturbance to the site around individual piezometers.

They are, however, generally less straightforward to install and, in a number of cases, require regular de-airing if accurate results are to be obtained.

Closed hydraulic piezometers are generally installed in a similar manner to standpipe piezometers, although the flexibility of the reading lines is a major advantage over the standpipe type. In cases where this type of installation is used, very particular attention should be paid to grouting to effect a complete seal. The hydraulic lines must be made of a rigid material, so that expansion in the tube does not affect either the recorded pressure or the response time.

The United States Bureau of Reclamation tip design (Figure 3.3) is appropriate particularly for emplacement during embankment and dam construction. The two tubes

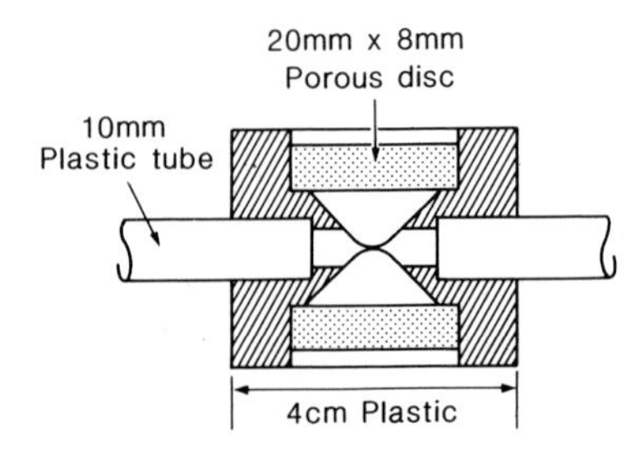

Figure 3.3 United States Bureau of Reclamation tip design

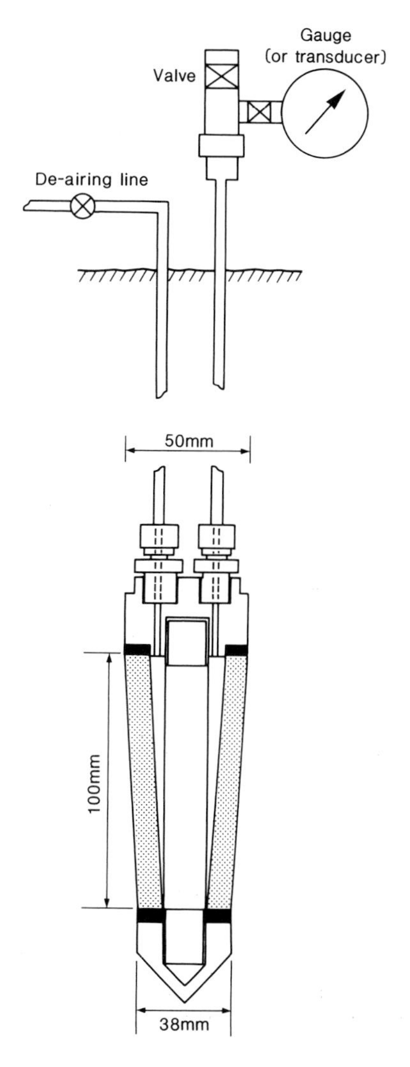

Figure 3.4 Bishop high air entry tip

allow for de-airing of the tip at regular intervals, and re-equilibration after de-airing should be very fast. Like the USBR tip, the Bishop high air entry tip (Figure 3.4) has a long record of reliability in the field. Theoretically, it should require less frequent de-airing than the USBR tip and is more flexible in terms of installation. Usually a borehole is drilled or augered to within 50 cm of the required depth and the Bishop tip pushed or driven into place depending upon soil type. The hole is then backfilled ensuring the hydraulic lines are not obstructed.

3.2.4 Pneumatic Piezometers

The pneumatic piezometer is generally used for the measurement of water pressures in the range 0 to 200 kPa. The system essentially comprises a piezometer tip in which water pressure

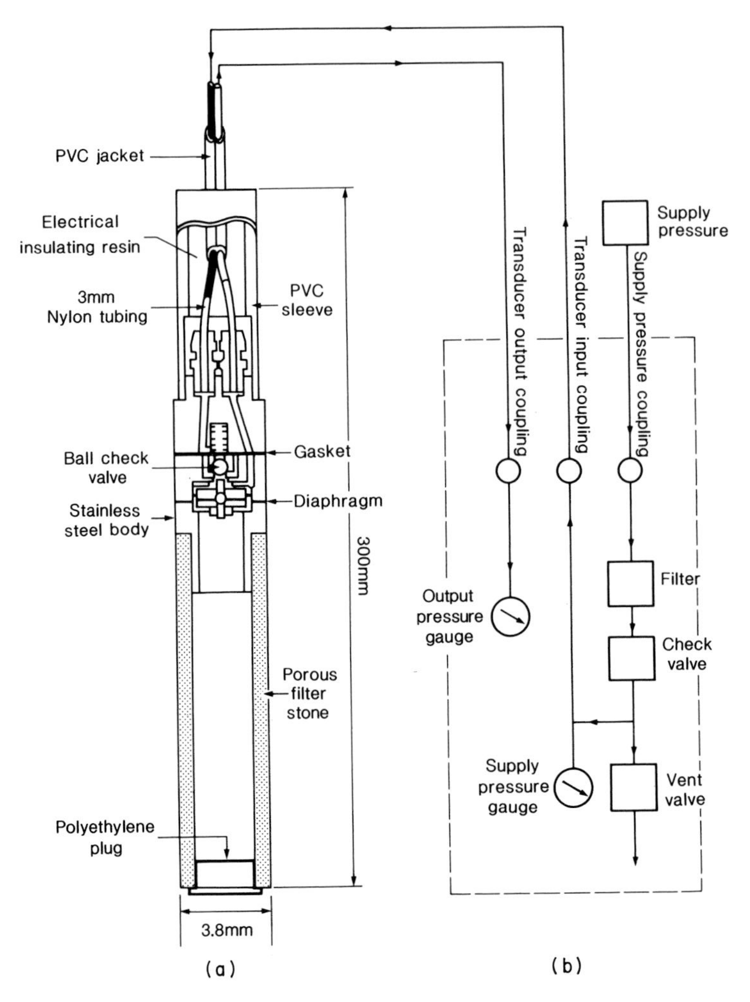

Figure 3.5 Pneumatic piezometer system showing (a) pore-pressure transducer and (b) output and control unit

is balanced by pneumatic pressure. In the type of configuration illustrated in Figure 3.5, the pneumatic pressure is applied through flexible tubing which may be as far as 500 m from the piezometer. Mikkelsen and Wilson (1983) report that automatically recording pneumatic systems are becoming more readily available, with output being converted to an electrical readout at a central surface station. A problem common to all diaphragm piezometers is the inability to resaturate the ceramic *in situ*. In some instances it has been reported that twin tubes for circulating water through the ceramic have been added, representing a rather cumbersome addition. Thus, effectively the utility of these piezometers is limited to areas in which it is thought unlikely that negative water pressures are likely to occur. Pope *et al.* (1982) make the additional points that, notwithstanding the disadvantage of restricting installations to continuously saturated areas, a pneumatic piezometer system is complicated to install and would be more expensive than the bubbler system which is described below.

The piezometer tips used in pneumatic piezometers are constructed of brass and PVC. They contain a flexible nitrite diaphragm (see Figure 3.5) which has a volumetric displacement of the order of 0.04 cm^3.

3.2.5 Vibrating Wire Piezometers

Figure 3.6 illustrates the basic components of a vibrating wire piezometer. An electromagnetic coil, located close to a central wire, is used both to pluck the wire and to convert the wire vibrations so produced to an electrical output current whose frequency is identical to the natural resonant frequency of the wire. Changing pressure causes deflection

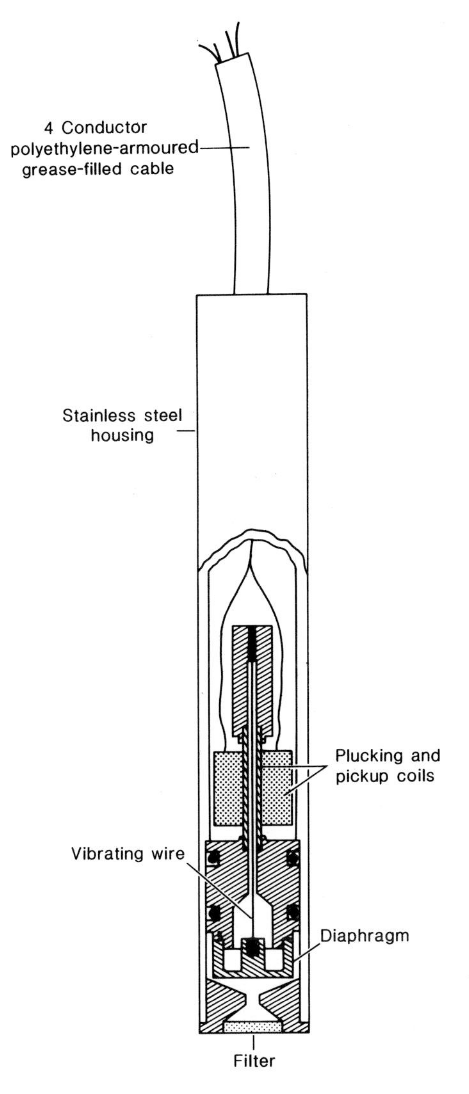

Figure 3.6 Vibrating wire piezometer

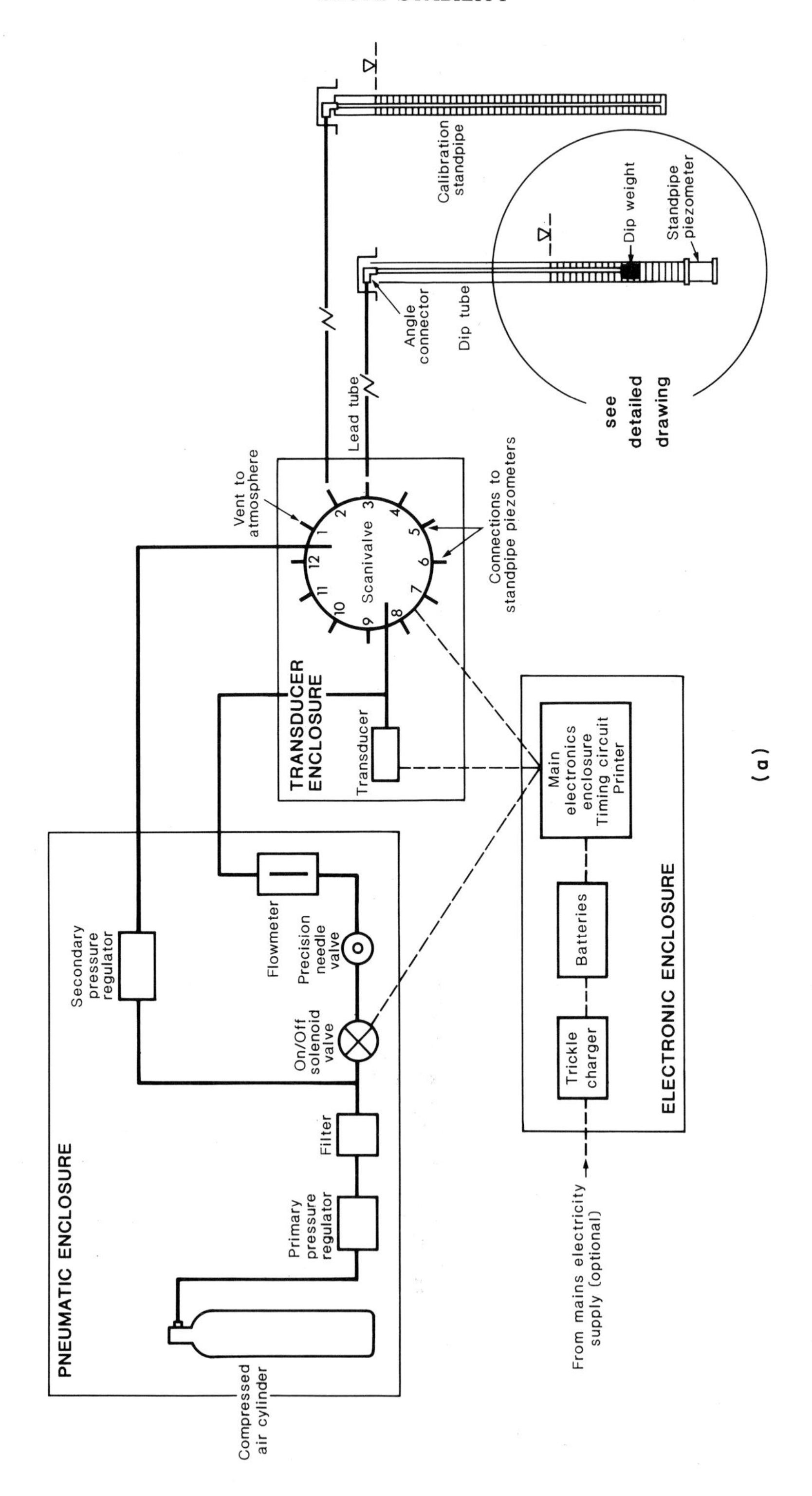
PNEUMATIC ENCLOSURE
Compressed air cylinder
Primary pressure regulator
Filter
On/Off solenoid valve
Precision needle valve
Flowmeter
Secondary pressure regulator
TRANSDUCER ENCLOSURE
Transducer
Scanivalve
1 2 3 4 5 6 7 8 9 10 11 12
Vent to atmosphere
Connections to standpipe piezometers
Lead tube
Calibration standpipe
Angle connector
Dip tube
Dip weight
Standpipe piezometer
see detailed drawing
ELECTRONIC ENCLOSURE
Main electronics enclosure Timing circuit Printer
Batteries
Trickle charger
From mains electricity supply (optional)

(a)

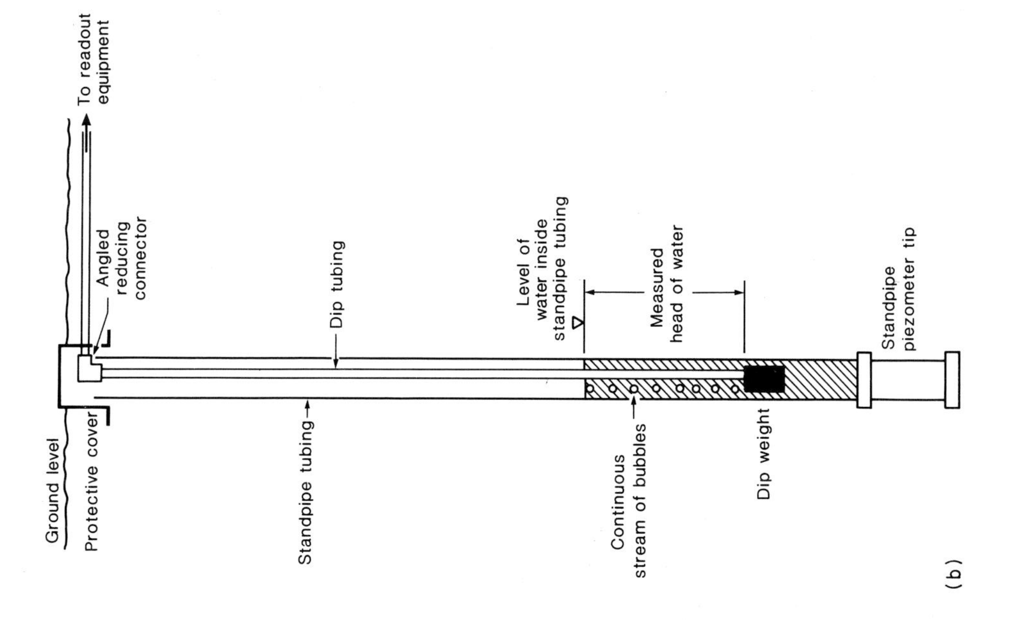

Figure 3.7 Automatic bubbler piezometer system showing: (a) system components; (b) detail of dip tube. (After Geotechnical Instruments Ltd.)

of the diaphragm, thereby altering the tension of the wire and its resonant frequency. There is therefore for each pressure a corresponding frequency output. The cable used to transmit the frequency signals is designed for direct burial in the ground. It has two thick outer wraps of tough high-density polythene separated by a metallic armoured sheath. The cable is grease-filled to discourage water ingress in the event that it is cut. Both cable and coil are protected from lightning damage. It should be noted that lead wire resistance changes brought about by water penetration, temperature variations, and contact resistance do not affect the frequency of the output signal. It is necessary to use a filter stone to prevent soil particles from impinging directly on the diaphragm. The standard filter stone is made from sintered stainless steel with a 50 micron pore size. In partially saturated soils, where air is to be excluded from the piezometer cavity, a high air entry filter stone may be specified. Negative pressures (suction) can be measured in theory, although most applications of such instruments are restricted to positive pressures. A typical operating temperature range is -29 °C to 65 °C, with a thermal zero shift of 0.05% full scale deflection per °C. The diaphragm volumetric displacement at full scale is small (0.01 cm^3) and vibrating wire piezometers thereby achieve a very rapid response, with the possibility of the lead cables being installed over several kilometres.

3.2.6 The Automatic Bubbler System

The basic design of the bubbler system is shown in Figure 3.7. The principle of the system is that air is blown out of a small tube within the standpipe piezometer at a controlled rate. The back-pressure within the tube is then approximately equal to the hydrostatic head of water in the piezometer above the base of the outlet tube. Clearly, as the water level changes then the back-pressure varies in direct response. An integral element in the system design is a rotary wafer switch or 'Scanivalve' which allows air to be switched in sequence to a number of piezometers in turn (see Burt, 1978; Anderson and Burt, 1977). It is therefore possible for just one transducer to service a number of piezometers. Pope *et al.* (1982) report a series of tests that were undertaken to determine the most suitable air flow rate for various lengths of tubing connecting the Scanivalve to the piezometer. Higher air flow rates result in a faster response, but friction losses in the tubing also increase. For a fixed flow rate of 40 cm^3 min^{-1}, Figure 3.8 shows the time to develop a back-pressure in the system when the lead tube to the piezometer is increased from 0 m to 120 m (with the dip-tube length fixed at 20 m). In practice, therefore, a restriction of the system is the placement of the recording enclosures within a distance not exceeding 150 m from the piezometers (Insley and McNicholl, 1982).

An examination of air flow rates for this sytem has been undertaken by Pope *et al.* (1982). The results show (Figure 3.9) that a flow rate of 50 cm^3 min^{-1} was appropriate, since at this value the column of water within the standpipe was not lifted, the air friction loss in the tubing was only 0.035 m head of water per 100 m of lead tube, and the system response time was 2 minutes per metre head of water (also for 100 m lead tube). In addition, this was a flow rate that could be maintained by the air cylinder for many weeks.

Following the design scheme adopted by Anderson and Burt (1977), two reference pressures are included in the Scanivalve switching unit. The first is a vent to atmosphere, the reading of which is stored and subtracted from all subsequent readings by the microprocessor system. The second reference is a water filled standpipe that subjects the

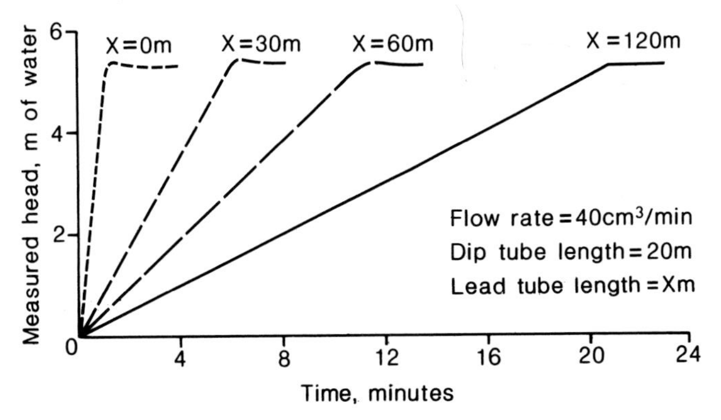

Figure 3.8 Effect of lead tube length in bubbler system (Figure 3.7) on the time to develop a back pressure

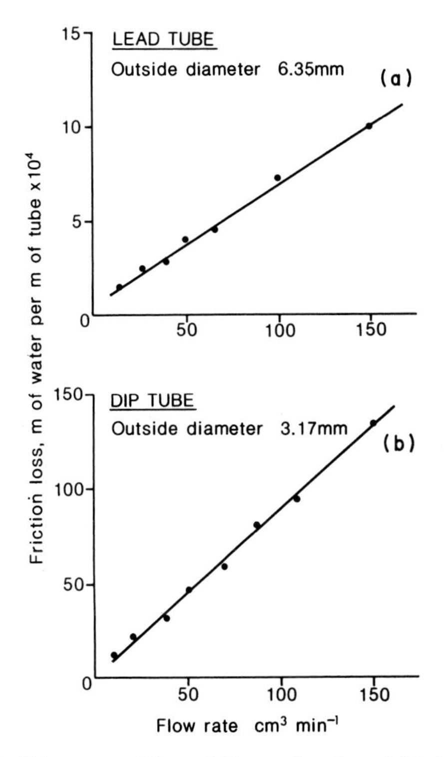

Figure 3.9 Flow rate in bubbler system (Figure 3.7) as a function of friction loss in (a) the lead tube and (b) the dip tube. (*After Pope* et al., *1982*, Proc. 7th SE Asian Geotechnical Conference. *Reproduced by permission of the Hong Kong Institution of Engineers.*)

transducer to the maximum pressure range likely to be encountered in the monitoring programme. In this manner, therefore, the calibration of the transducer is automatically checked on each cycle of readings. A dwell time of approximately five minutes per piezometer is generally set, allowing a cycle of 10 piezometers and two references to be completed in an hour. An enhanced configuration of this type can accommodate up to 55 piezometers.

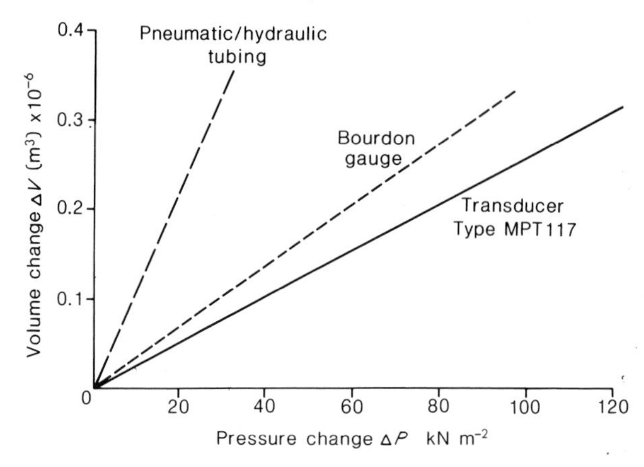

Figure 3.10 Pressure change and induced volume changes for a transducer, Bourdon gauge, and hydraulic tubing

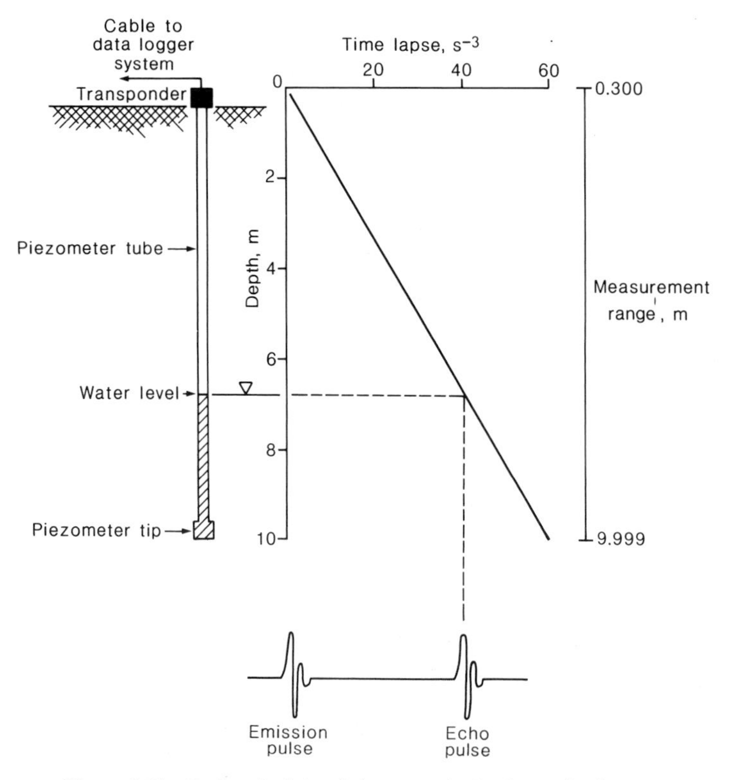

Figure 3.11 Basic principle of the acoustic depth monitoring system

System stiffness is an important aspect of monitoring systems used for both positive and negative pore pressure monitoring. In particular, the stiffness of the transducer and associated hydraulic/or pneumatic lead lines is of greater significance in this context. The volume change associated with pressure changes can be readily determined in the laboratory and Figure 3.10 shows the results of testing an MPT 117 transducer with the pressure and volume changes measured by using dead load test apparatus (of course, other transducers will have different stiffness characteristics).

Stiffness in the positive range (defined as V/P) of this transducer is thus $2.56 \times 10^{-9}\,m^5/kN$. Undertaking a similar test on a Bourdon gauge yields a stiffness of $3.4 \times 10^{-9}\,m^5/kN$. The standard hydraulic or pneumatic tubing used for piezometer installation (and the lead tubes used in the bubbler system, Figure 3.7) can be subjected to the same stiffness testing. The tubing, being polythene covered nylon tubing (2.7 mm ID 4.8 mm OD) has a stiffness of $1.20 \times 10^{-8}\,m^5/kN$ per metre length.

3.2.7 Acoustic Depth Monitoring System

Acoustic depth monitoring provides an alternative system to that of the bubbler method described above. It is a similar technique to echo sounding, but in the medium of air rather than water. The monitoring system comprises a transponder located at the top of a piezometer tube and linked by a cable to the data logger. The transponder emits pulses of high frequency sound which travel down the piezometer tube. Milliseconds later, it detects the echoes returning from the water surface. Since the sound pulses travel at known velocity, the time lapse between their emission and the arrival of the echo signals is a linear fraction of the distance between the transponder and water surface. Figure 3.11 illustrates this basic principle. There are restrictions upon the range of water surface depths that can be determined however. If the water surface is very close to the transponder (less than 0.3 m) the leading edge of the echo pulse returns before the trailing edge of the emission pulse has departed, and thus

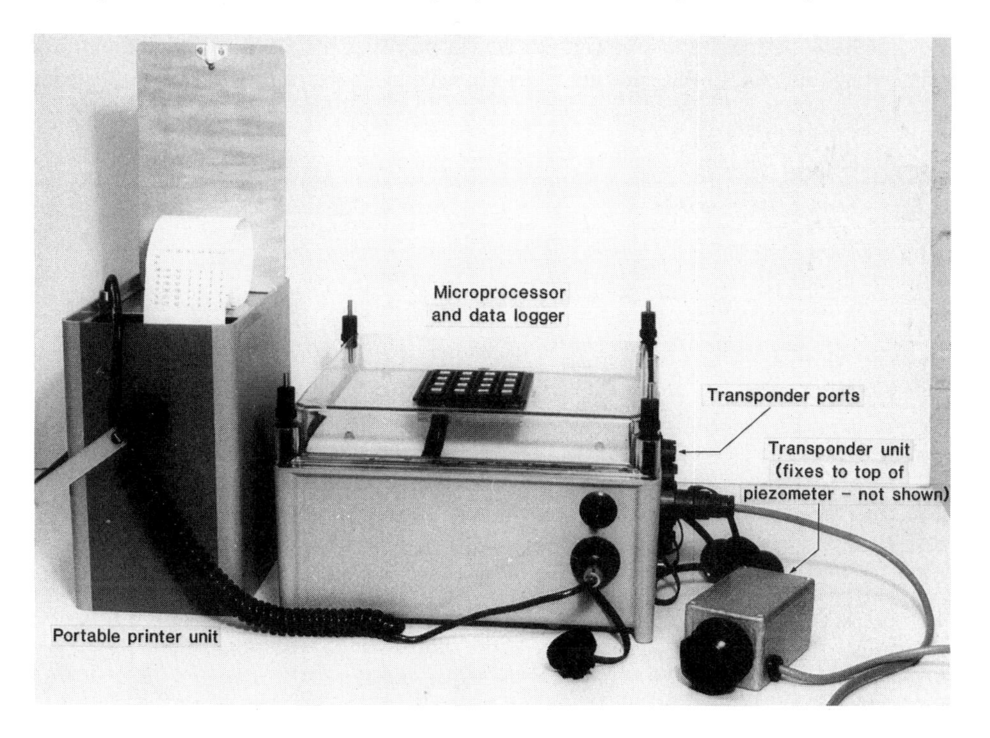

Figure 3.12 The main elements of the acoustic depth monitoring system. (After Anderson, 1986.)

measurements are not possible in the top 0.3 m, while the maximum measurement depth is 9.9 m. The velocity of sound in air varies with temperature, but a thermal sensor in the transponder automatically maintains accurate conversion of time lapse to distance.

The system utilizes the standard piezometer tube as an acoustic wave-guide. The tubing may be vertical, inclined, curved, or even knotted, but must retain an approximately constant cross-sectional area. The distance is measured along the tube axis irrespective of the orientation of the tube and thus a correction should be applied if true depth is required for non-vertical installations. The low current consumption of a six piezometer system with data logger means that in excess of 2,000 readings may be taken (on the system shown in Figure 3.12 the time base can be pre-set from 15 minutes to 24 hours).

3.3 TENSIOMETER SYSTEMS

3.3.1 Introduction

Soil suction instrumentation can be divided into two types according to the methodology of the monitoring system: tensiometric methods which are suitable for recording suctions up to approximately 80–90 kPa, and thermal conductivity sensors which are capable of providing estimates of soil suction up to 1,500 kPa. Both systems may be of a manual reading type or automatic. Systems for the measurement of soil suctions in excess of 90 kPa are considered in section 3.4. Table 3.3 presents a selected list of installations and reviews that have recently been undertaken in this field.

A tensiometer measures soil suction by obtaining equilibrium across a high air entry porous medium between the soil suction and a confined reservoir of water within the tensiometer system. As soluble salts are free to pass through the porous medium, the osmotic

Table 3.3 Reports detailing tensiometer and thermal conductivity systems

Tensiometer Systems	Thermal Conductivity Sensors
Field installations (> 0 °C)	*Field installations*
Anderson and Burt (1977)	Lee (1983)
Anderson and Kneale (1980)	Department of Main Roads, NSW, Australia (1977)
Rice (1969)	Phene *et al.* (1973)
Williams (1978)	
Watson (1965)	
Harr (1977)	
Sweeney (1982)	
Chipp *et al.* (1982)	
Field installation (< 0 °C)	
Colbeck (1976)	
Ingersoll (1980), (1981)	
McKim *et al.* (1976)	
Reviews of tensiometric techniques	*Review of thermal conductivity methods*
Ingersoll (1981)	Lee (1983)
McKim *et al.* (1980)	
Schmugge *et al.* (1980)	

component of suction is eliminated and the tensiometer measures only the matric component of suction. A useful treatment of the components of suction is given by Krahn and Fredlund (1972). At a temperature of 20 °C at sea level water boils or cavitates at a pressure of about 97 kPa, i.e. at approximately 1 atmosphere (or bar) of pressure below atmospheric pressure. This places an upper limit on the tensiometer range for measuring matric suction as the water in the tensiometer system cavitates as 1 bar of suction is approached. For sites at significant elevations above sea level, the range of the tensiometer is further reduced. It is of interest to note that fine grained soils can contain water at matric suctions of much greater than 1 bar without cavitation of the water. Apparently this is because water in the very fine pores of the soil takes on a molecular structure that can sustain very high tensions or suctions (Croney *et al.*, 1952). A further problem is that even though the tensiometer is filled with de-aired water, air in solution in the soil water diffuses through the high air entry medium, during the equalization process. This air starts to come out of solution at about 85 kPa suction at sea level and can reduce the accuracy of soil suction measurements. This figure is reduced by approximately 3.5 kPa for every 300 m increase in elevation (Sweeney, 1982).

3.3.2 Manual Tensiometers

Jetfill tensiometers have been available from Soil Moisture Corporation for several years, and have proved to be robust and accurate for the measurement of soil suction in Hong Kong and elsewhere. Ingersoll (1981) details the mode of operation and specification of these units, and McFarlane (1981) has discussed installation procedures (Figure 3.13). A smaller unit (2100 series—Figure 3.14) is available with a ceramic cup approximately 8 mm diameter (see Ingersoll, 1981).

The 'quick draw' tensiometer (Figure 3.15) is an adaptation of the jetfill tensiometer, designed to give very rapid response times and is intended for portable use. The principal restriction is the available probe depth—a maximum of approximately 0.5 m. Notwithstanding this limitation, response times are very rapid. Figure 3.16 illustrates results given by Sweeney (1982) in an application in Hong Kong. Similar results have been obtained by the author in both the Caribbean and the UK.

As can be seen from Figure 3.15, the high air entry ceramic tip is connected to a vacuum dial gauge by means of a very small bore capillary tube, which runs inside the tensiometer rod. As Sweeney (1982) observes, the internal water volume of the unit is very small, so that the detrimental effect of the large thermal expansion and contraction of water is reduced to an insignificant level. This allows soil suction measurements to be made quickly and accurately, even when the temperature of the tensiometer is greatly different from that of the soil.

3.3.3 Automatic Tensiometer Systems

Transducer–tensiometer systems for the measurement of soil suction (and pressure) have been available for a number of years. Burt (1978) describes such a system in detail with a single transducer serving 22 tensiometers through a Scanivalve switch. This system is analogous to the bubbler system, described earlier, in this latter respect. Figure 3.17 shows the basic layout of the system, while Figure 3.18 shows the Scanivalve and electronic

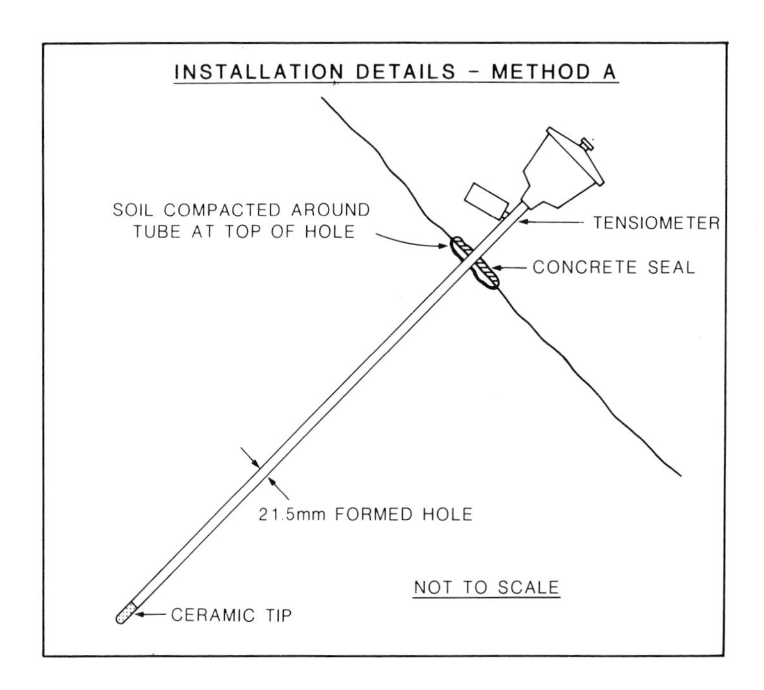

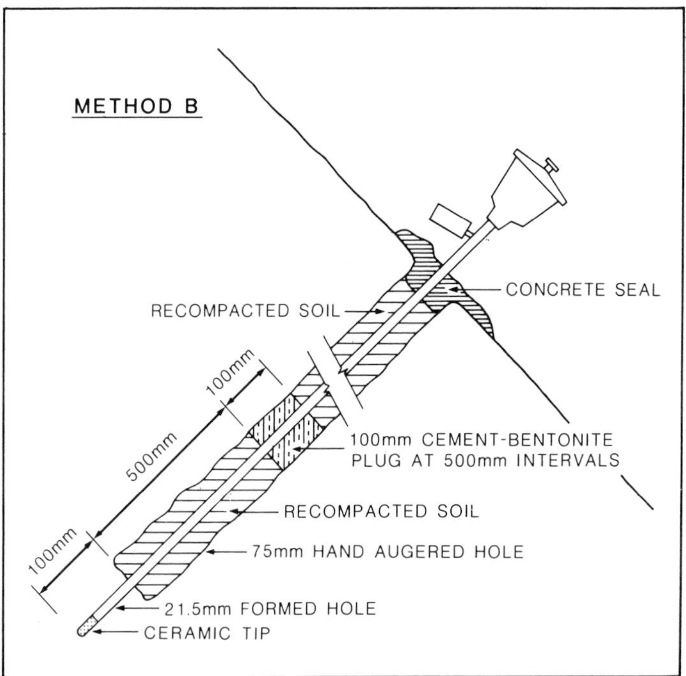

Figure 3.13 Installation details for manual tensiometers. (*After McFarlane, 1981.*)

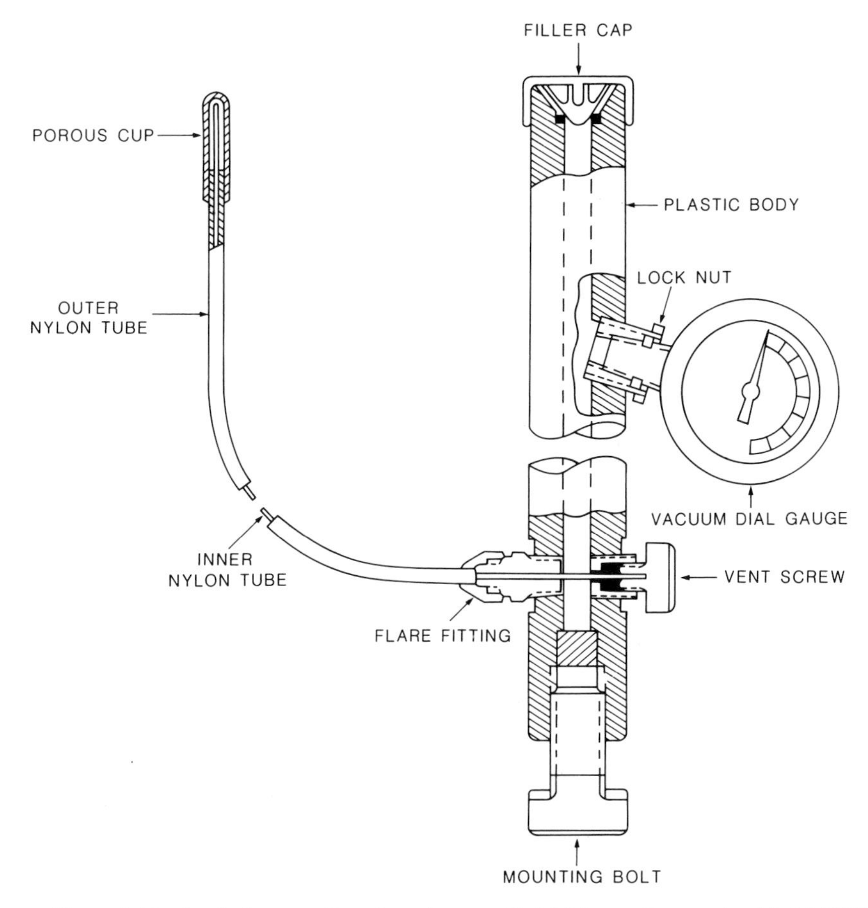

Figure 3.14 Small diameter (8 mm) tensiometer (Soil Moisture Corporation Model 2100)

enclosures (the thermal printer unit here can be replaced by a solid state cartridge data logging system). Just as for the bubbler system described above, two references are used in the Scanivalve cycle (one to atmosphere, the other to maximum suction range expected in the field). The hydraulic lines connecting the bulkhead Scanivalve fittings (Figure 3.18) to the tensiometer are identical in specification to those used in the bubbler system to ensure system stiffness. In this respect, it is important in both cases to maintain as short a lead as possible between the Scanivalve and transducer. The tensiometer design for such a system is of specific interest to maintain stiffness. Figure 3.18 shows the design adopted by Anderson and Burt (1977). The tensiometer consists of a 22 mm external diameter, one bar air entry, porous ceramic cup glued to a cylindrical perspex body of a similar diameter. A rubber bung through which a logger reading line and a shorter de-airing return line pass, forms a seal at the upper end. The de-airing line can either be terminated at the soil surface above the tensiometer, or returned to a centrally located de-airing board housed with the Scanivalve enclosure.

Employing a differential transducer (say with an operating range of -100 kPa to $+200$ kPa) allows such a system to record pressure as well as suction, thereby not only

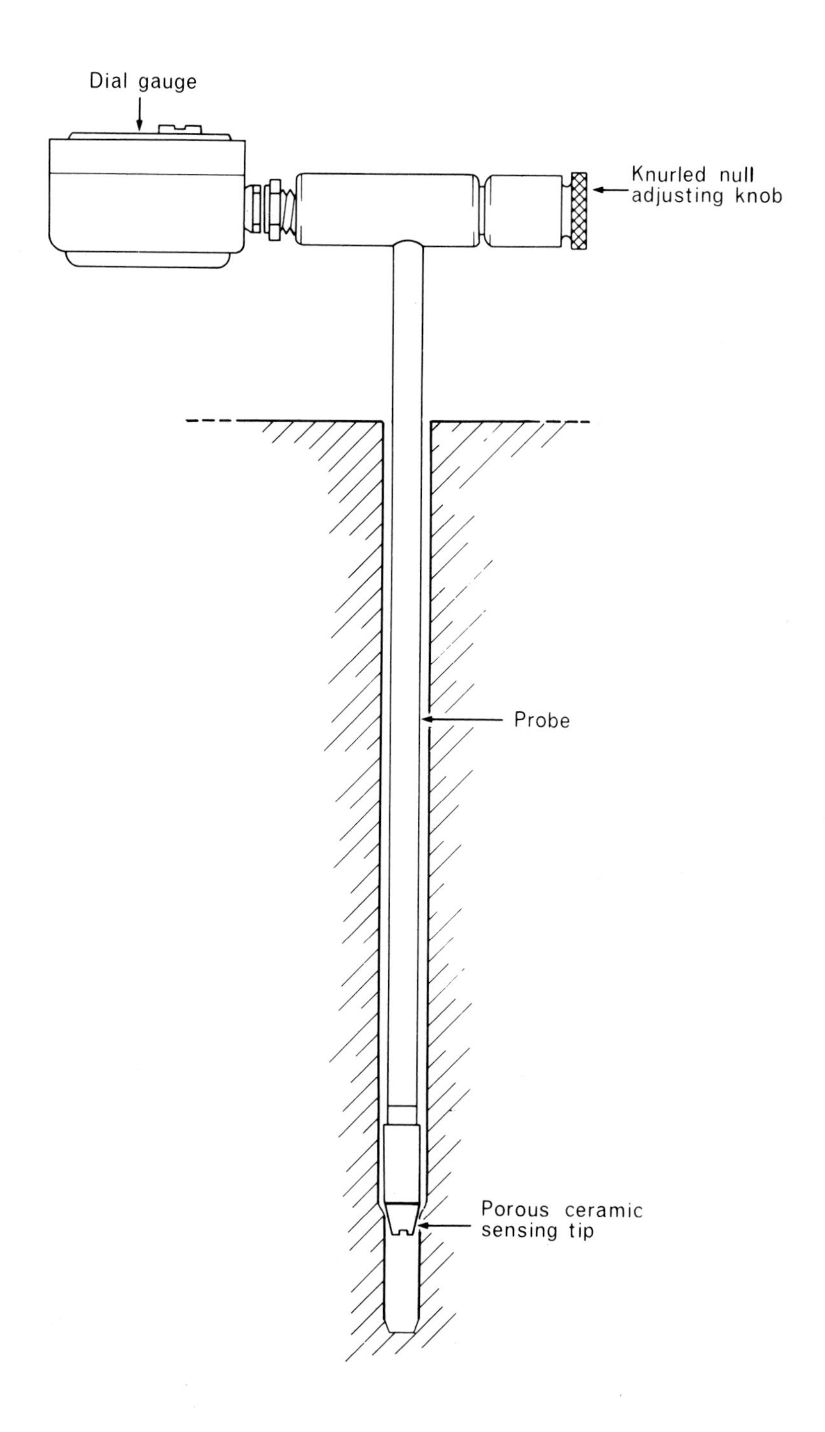

Figure 3.15 'Quick draw' tensiometer

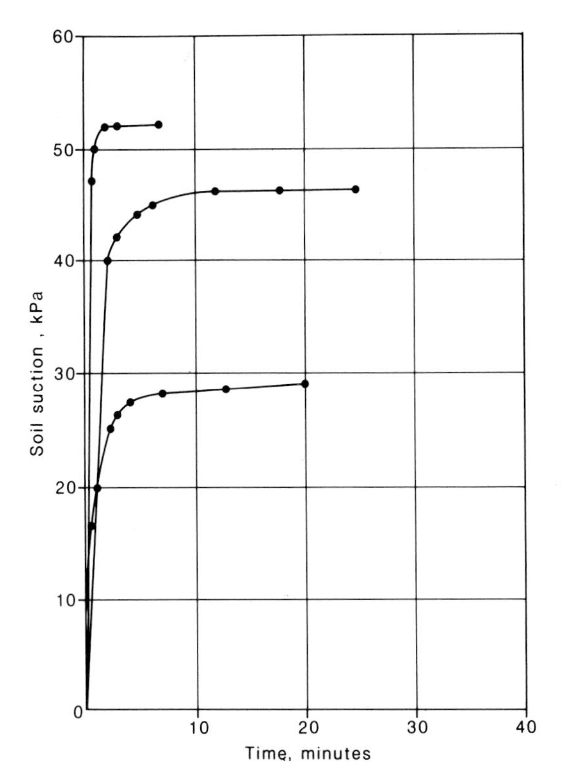

Figure 3.16 Response times of the 'quick draw' tensiometer. (*After Sweeney, 1982*, Proc. 7th SE Asian Geotechnical Conference. *Reproduced by permission of the Hong Kong Institution of Engineers.*)

increasing the flexibility of the system in terms of its disposition in the context of the groundwater level, but also allowing a more enhanced elevation range to be accommodated.

Response time information is critical to the employment of transducer tensiometer systems. Notwithstanding this obvious requirement to maximize the full potential of such systems in terms of tensiometer reading frequency using automatic systems, there is only a restricted number of studies that have examined field response times. Table 3.4 summarizes certain of the information available. In general, response times are of the order of 60 seconds, and in consequence a 3-minute connection time to each tensiometer on a fully automatic system may be considered appropriate. Anderson and Kneale (1980) showed that recovery to 99% of the equilibrium value could be achieved for step changes of 10 kPa within 120 seconds, with tensiometers installed in Oxford Clay. The use of tubing as specified in Anderson and Kemp (1985) for the tensiometer connections to the Scanivalve enclosure, a very short (30 cm or less) 'reading' line from the Scanivalve to the transducer, and pre-installation pressure testing of all the components and fittings, are the principal design guides which will ensure response times are of the order of 1–2 minutes. It is, however, important to note that transducer stiffness may not be identical in the positive and negative directions, and that, especially in the context of less expensive transducers, calibration close to 0 kPa may require particular care. The general installation procedures for the tensiometer emplacement follow those described in the above section.

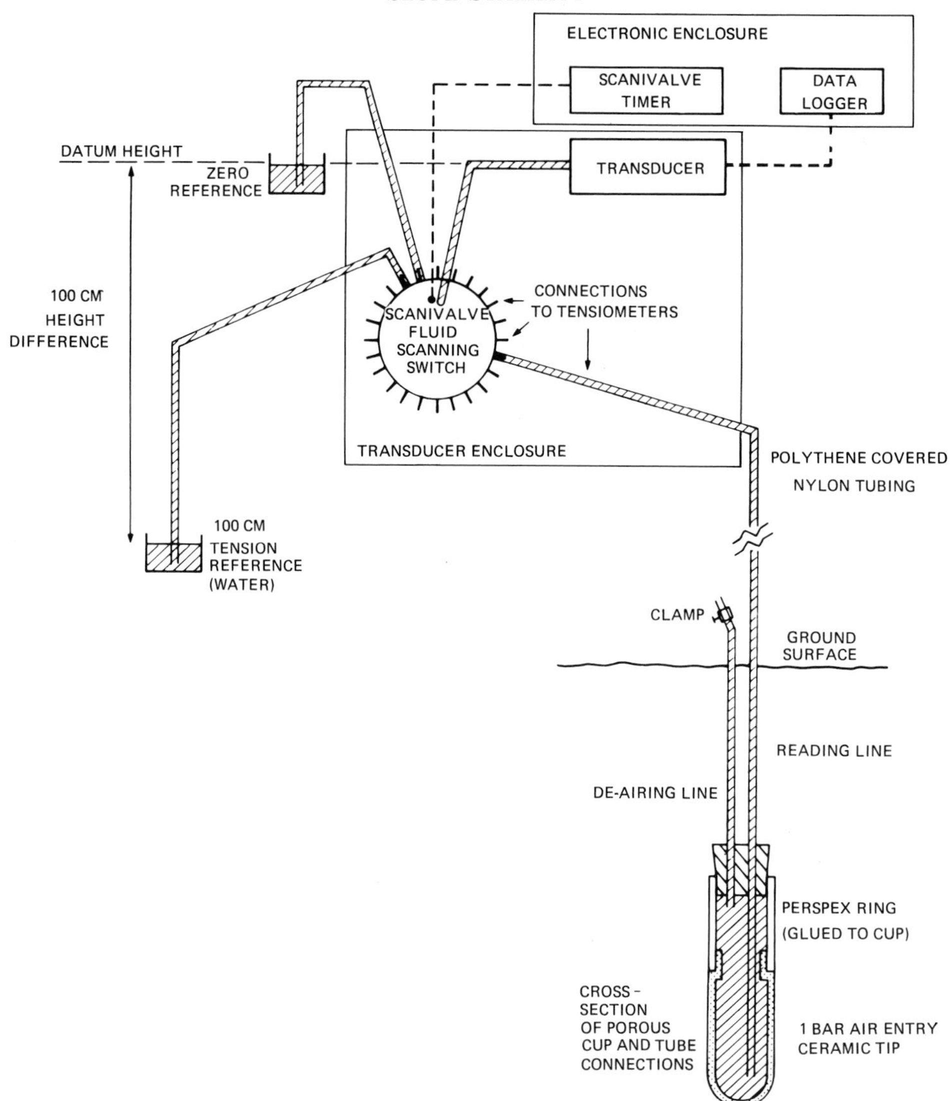

Figure 3.17 Transducer-tensiometer system components

Individual pressure transducer–tensiometer (PTT) systems have been developed. One such system is shown in Figure 3.19. This unit differs from the Scanivalve based system reported above, only in that *each* ceramic cup has its own transducer. While this dispenses with the need for a Scanivalve fluid switch system, the configuration is significantly more expensive, since it requires one transducer per ceramic cup. Three variants have been developed: basic non-purgeable, purgeable, and borehole. The resolution of soil suction measurement claimed by the manufacturers is 0.1 kPa.

The non-purgeable PTT is simply a standard 50 mm × 25 mm porous pot, cemented to a collar which mates via an O-ring seal to a pressure transducer. The porous pot is filled with de-aerated water before installation and must be removed from the soil periodically

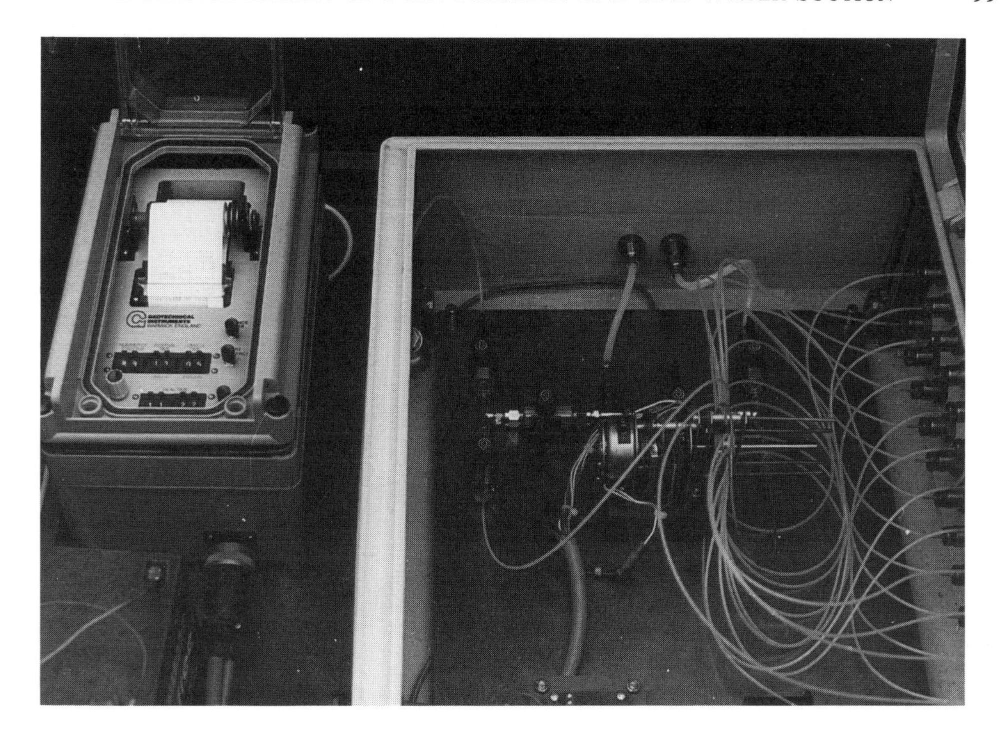

Figure 3.18 Transducer and electronic components of the Scanivalve tensiometer system (Figure 3.17)

for refilling. The purgeable PTT employs the same type of porous pot and transducer but is also fitted with purging tubes and valves. Figure 3.19 illustrates the purgeable unit. The borehole PTT uses the non-purgeable units with the porous cup shrouded and potted with plaster of Paris into an open-sided mounting.

3.4 SYSTEMS FOR MEASURING SUCTIONS IN EXCESS OF 90 kPa: THERMAL CONDUCTIVITY SENSORS

The rate of heat dissipation in a porous material of low heat conductivity is sensitive to water content, therefore the water content of a porous material can be measured by supplying a heat source at a point centred within this material, and by measuring the temperature rise at that point. Lee and Fredlund (1984) outline the theory of the sensor and the temperature–suction calibration that is required for each probe. Figure 3.20 shows the basic circuit design for the Moisture Control System (MCS) equipment—the sensor is used as a heating element and a temperature measuring device. Thus the sensor consists of a diode which is surrounded by a heating coil and embedded in a porous medium. In a study of the MCS 6000 sensor, Lee (1983) reports the following principal findings. Firstly, the sensitivity in the range 0–100 kPa is ±6 kPa. Above 200 kPa the results from the sensor must be very open to question—see Figure 3.21. However, a principal constraint on its use is the time required for the sensor to give a stable reading. On the desorption cycle, that time increases with a decrease in the water content, while for the absorption cycles,

Table 3.4 Response Times of Transducer–Tensiometer Systems

Authors	Date	Cup Material	Cup Air Entry Value	Transducer Sensitivity	Empirical/ Theoretical	Soil Material	Response Time	Level of Accuracy
Klute, A. and Peters, D. B.	1962	fritted glass	—	3×10^3 mbar cm^{-3}	E	—	1 s	—
Watson, K. K.	1965	ceramic	86 cm water	1×10^2 psi cm^{-3}	E	sand	0.1 s	—
			1,300 cm water	1×10^2 psi cm^{-3}	E	sand	29.6 s	
Watson, K. K. and Jackson, R. J.	1967	ceramic	1,300 cm water	5×10^{-5} cm^3 mbar^{-1}	T	—	9.1 s	—
Young, N. C. (36)	1968							
Fitzsimmons, D. W. and Young, N. C.	1972	porvic (polyvinal chloride)	—	3×10^4 mbar cm^{-3}	E	depending on material and saturation	1 s to 60 s	—
Williams, T. H. L.	1978	ceramic	1,000 cm water	1×10^{-4} cm^3 mbar^{-1}	T	—	1 s	—
Boels, D. *et al.*	1978	ceramic	—		T	heavy-clay–wet	60 s	1%
						–medium	98 s	1%
						–dry		
						sandy clay–wet	60 s	1%
						loam–medium	125 s	1%
						–dry	450 s	1%
Towner, G. D.	1980				T			
Anderson, M. G. and Kneale, P. E.	1980	ceramic	1,000 cm water		E	clay	100 s	—

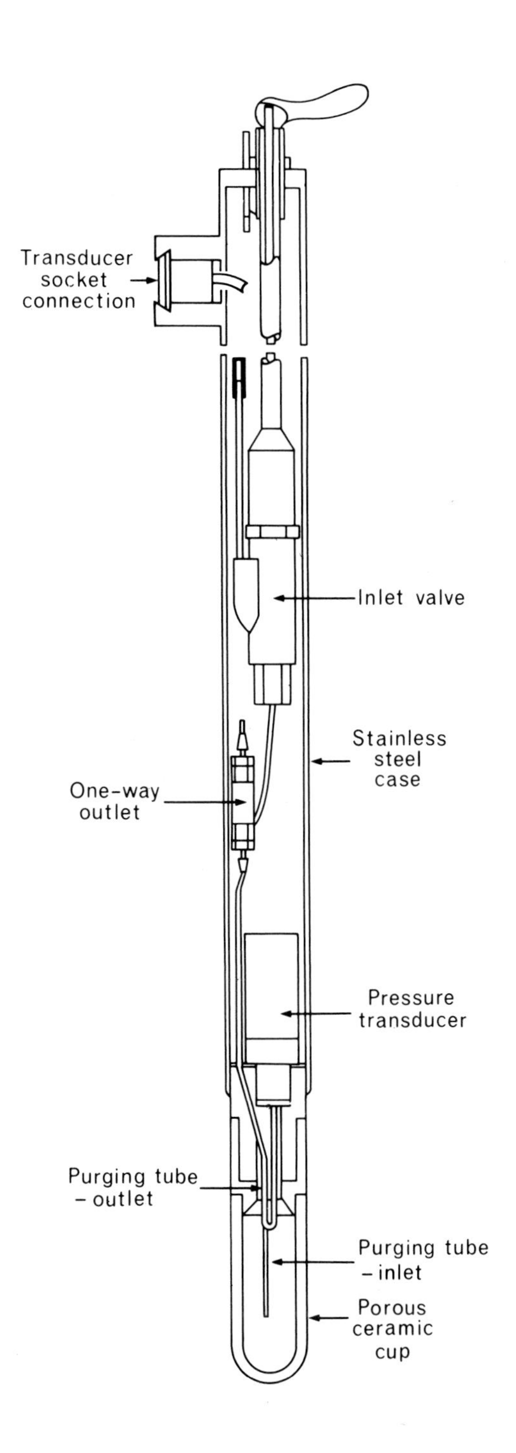

Figure 3.19 Individual pressure transducer system

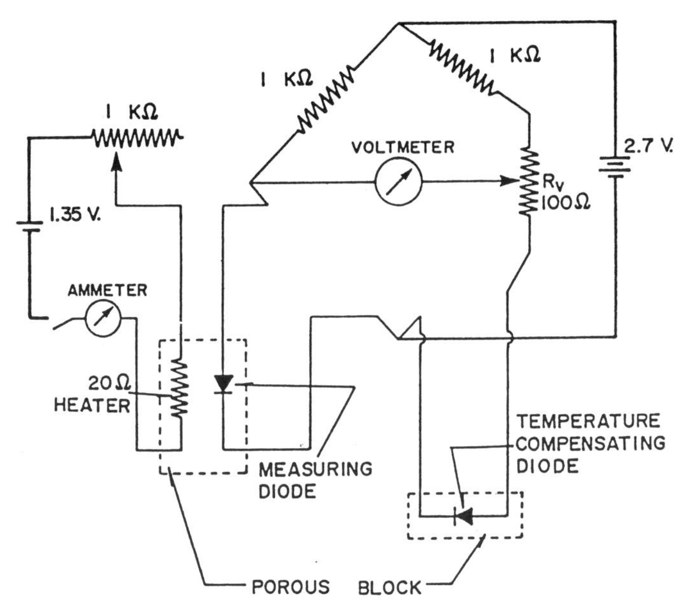

Figure 3.20 Circuit design for MCS suction system. (*After Lee, 1983.*)

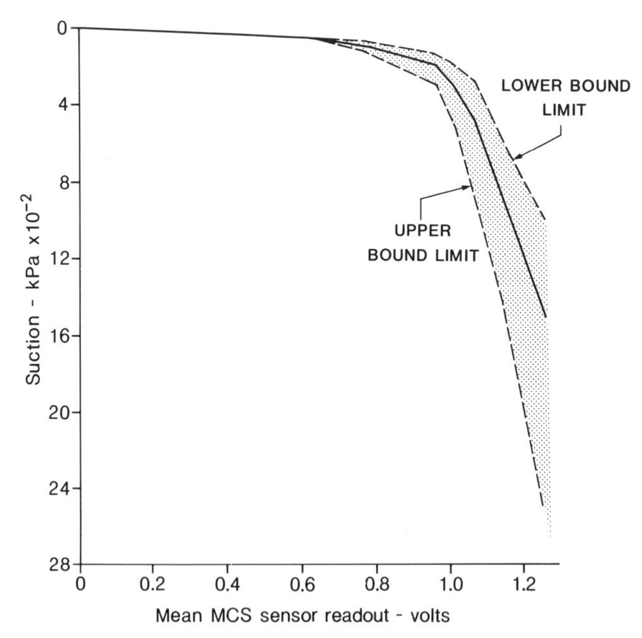

Figure 3.21 MCS sensor suction—output response. (*After Lee, 1983.*)

the response time exhibits a greater degree of non-linearity. Figure 3.22 shows that in either case response times are of the order of 160 hours, illustrating the severe constraints on such systems for transient soil water conditions.

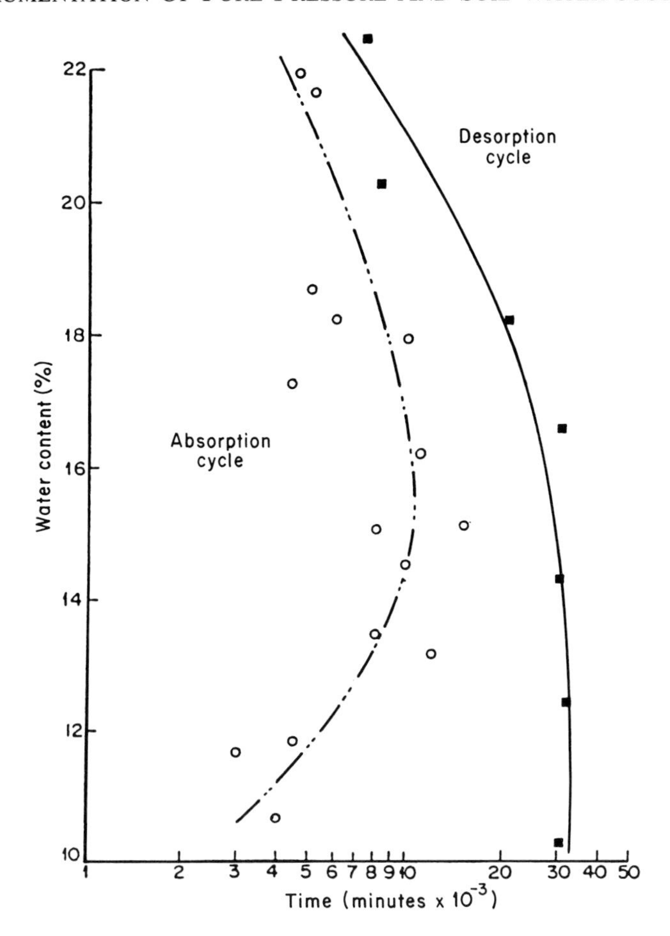

Figure 3.22 Response times for the MCS sensor. (*After Lee, 1983.*)

The sensor output is a voltage measurement which has a 1:1 linear relationship with soil suction between 10 and 60 kPa. Above suction values of 60 kPa the sensor must be calibrated; the calibration results will vary from sensor to sensor (Lee, 1983). In a comprehensive review of the MCS system tested in plastic and non-plastic soils, Lee and Fredlund (1984) report that the system is suitable for geotechnical engineering purposes for the suction range 0–200 kPa. Errors in this suction range (of approximately 10%) can be attributed to entrapped air during installation (Nagpal and Boersma, 1973), temperature (Gardner, 1955), and hysteretic effects.

3.5 DISCUSSION

The basic question concerning instrumentation, following the above review, is that of the selection of instrumentation for given circumstances and site investigation requirements. Four primary objectives were raised in the introductory section, and Figure 3.23 illustrates one of these scenarios. Instrumentation choice in research design requires initial decisions of response time and cost to be made as this figure indicates. Response times have been the

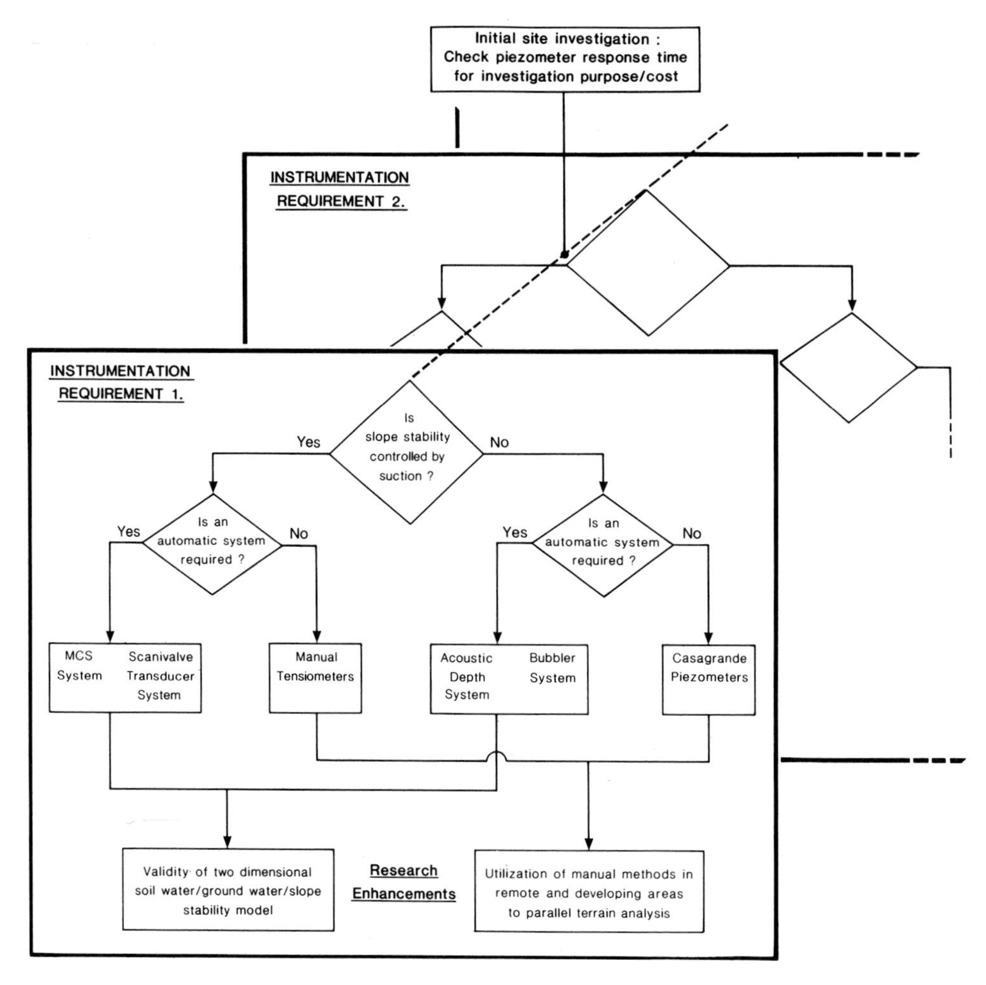

Figure 3.23 Illustration of the interrelationships between site investigation objectives, instrumentation requirements and options and subsequent model research and development

subject of much research, both in theory and in the field (see Table 3.4, and Figures 3.22 and 3.24).

There are three aspects to response time considerations:

1. The theoretical system response.
2. Change of response time with time (due to air incursion into the ceramic, for example).
3. The generalization of (1) and (2) above to produce a summary response guide (see Figure 3.24).

The simplest system for which to determine the response time is that of the hydrostatic time lag for open piezometers (Figures 3.1 and 3.2). Hvorslev (1951) presented the analysis for this system, shown in Figure 3.24(a). Under recovery, then, the total volume of flow (V) needed for full equalization of the pressure difference 'H' is $V=AH$, where A is the

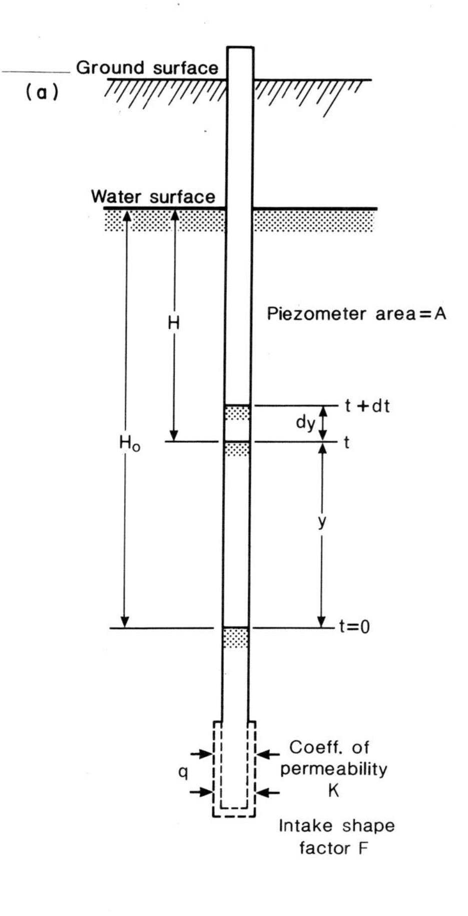

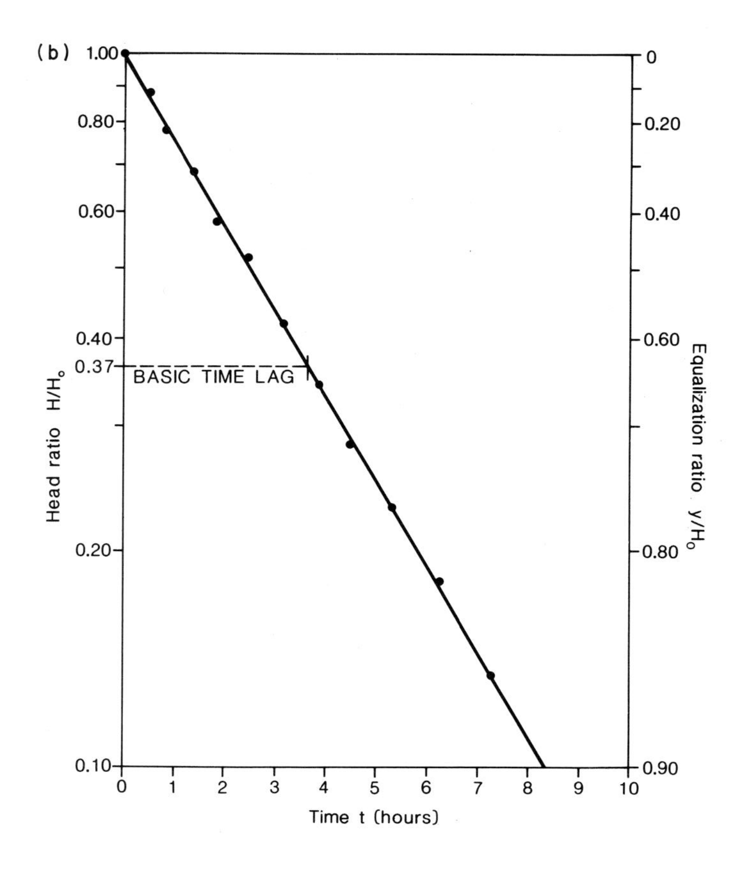

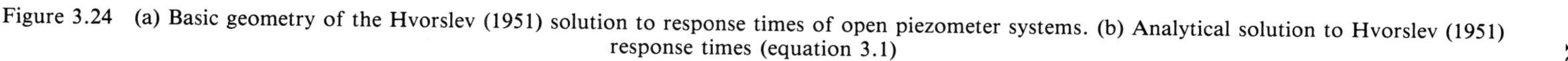
Figure 3.24 (a) Basic geometry of the Hvorslev (1951) solution to response times of open piezometer systems. (b) Analytical solution to Hvorslev (1951) response times (equation 3.1)

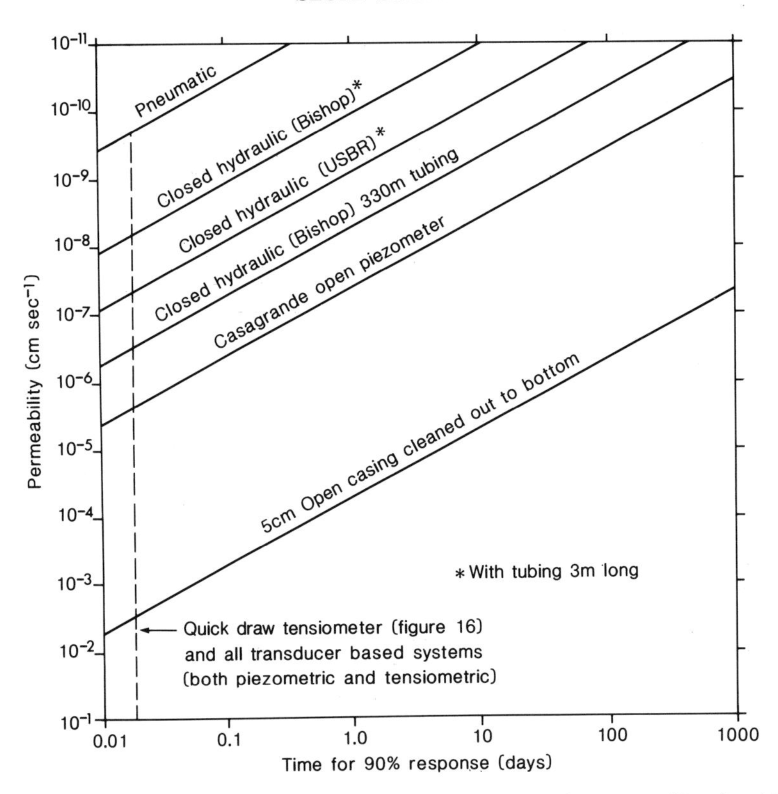

Figure 3.25 Selected response times for hydraulic pneumatic and electronic systems. (*Based on Mitchell, 1983,* Earth Structures Engineering. *Reproduced by permission of George Allen & Unwin.*)

cross-sectional area of the standpipe. Hvorslev then defined the basic time lag (T) as the time required for pressure equalization when the *original* flow rate $q = FKH$ is maintained (where F is a shape factor of the intake ceramic and K is the soil hydraulic conductivity). Thus:

$$T = \frac{V}{q} = \frac{AH}{FKH} = \frac{A}{FK} \tag{3.1}$$

Figure 3.24(b) shows the solution to extensions of equation (3.1) in which the equalization ratio (y/H_0), is plotted against time. The basic time lag (equation (3.1)) occurs at an equalization of 63%. Shape factors are defined by Hvorslev (1951) for different ceramics; for a spherical shaped intake ceramic $F = 2\pi D$, and for an hemispherical ceramic $F = \pi D$. This then is the calculation basis for the response times of the open systems shown in Figure 3.25.

By contrast, for systems that are closed and employ transducers, the response times are, for most practical purposes, instantaneous (Figure 3.25), although response times are of course lengthened by tube distances, as we have shown in the case of bubbler piezometer systems (Figure 3.8). While we have reviewed the response times of tensiometer–transducer systems (Table 3.4), Towner (1980) has provided a set of criteria to determine whether the

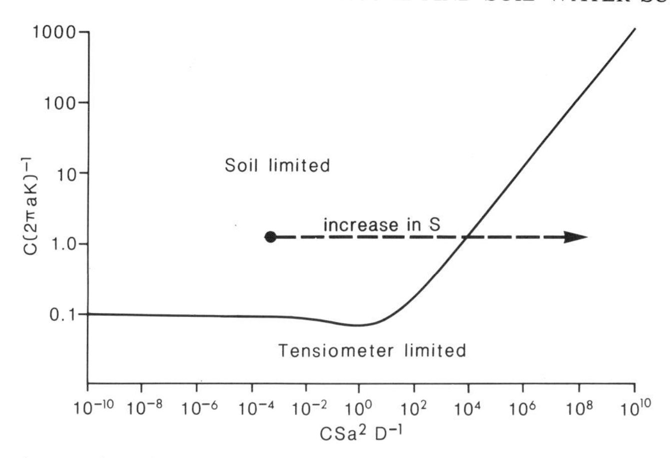

Figure 3.26 The demarcation of 'soil limited' and 'tensiometer limited' conditions for tensiometer response times. See text for complete explanation. (*Based on Towner, 1980,* J. Soil Science, **31**. *Reproduced by permission of Blackwell Scientific Publications.*)

response time of a tensiometer system will be governed by the soil or tensiometer. If C is the cup conductance (ml s^{-1} cm^{-1} water—defined as the volume of liquid crossing the ceramic per unit time per unit pressure difference across the ceramic), S the tensiometer sensitivity (cm cm^{-3}—defined as the ratio of tensiometer fluid pressure to flux of water across the ceramic), a is the radius of the tensiometer tip (cm), K the soil hydraulic conductivity (cm s^{-1}) and D is the diffusion coefficient (K times the suction moisture gradient—cm^2 s^{-1}), then Figure 3.26 provides a summary of those conditions in the context of tensiometer response. If $C/(2\pi aK)$ for a given tensiometer is less than 0.069 (Figure 3.26), the tensiometer in the field will behave as predicted. If, however, this value is exceeded, the soil properties may influence the response. Figure 3.26 shows the interesting phenomenon that an increase in sensitivity of a system that is soil limited can eventually lead to the system becoming tensiometer limited. Towner (1980) argues that there could be a case for deliberately *increasing* the response time such that the tensiometer characteristics are independent of the unknown soil properties which are dynamic—thus remaining within the tensiometer limited zone of Figure 3.26. If this were done, then the dynamic system behaviour would always be a function of the known tensiometer characteristics.

Comparative recording of different systems at the same location are rare. However, Kemp (1986) illustrates the situation where a standpipe piezometer and an automatic Scanivalve system were employed to measure pore pressures at approximately 2 metres depth in a residual volcanic soil in St Lucia, West Indies. *Both* piezometer systems were continously recorded by transducers, ensuring that accurate comparisons could be made. Results showed that even for low intensity tropical storms, implied differences in groundwater levels could exceed 0.5 m, with significant dampening occurring in the case of the standpipe.

It is not feasible to comment upon the implications of such differences in terms of the factor of safety calculation in *general* terms, due to the variability in pore pressures and other variables relevant to stability. Anderson and Howes (1985) have shown that variability, however induced, may be expected to be most significant in relatively shallow slopes, and yet such slopes may not be critical in terms of stability. The clear implication is that, in any particular investigation, optimization of instrumentation cannot be undertaken to yield

Table 3.5 Summary of selected piezometer and tensiometer characteristics. (*After Geotechnical Control Office, 1984. Reproduced by permission of the Hong Kong Government.*)

	Type	Range	Response	De-airing	Remote Reading	Long Term Reliability	Other: Advantages	Other: Disadvantages	Recommendation
POSITIVE PRESSURE	Open Hydraulic (Casagrande)	Atmospheric to top of standpipe level	Slow	Self de-airing	Not normally, but possible with Halcrow buckets or bubbler system	Very good	Cheap, simple to read and maintain. *In situ* permeability measurement possible	Vandal damage often irreparable	First choice for measurement within positive pressure range unless very rapid response or remote reading required
	Closed Hydraulic (Low air entry pressure)	Any positive pressure	Moderate	Can be deaired	Yes	Depends on pressure measuring system 1) Mercury manometer– very good 2) Bourdon gauge— poor in humid atmosphere 3) Pressure transducer— moderate but easily replaced	Fairly cheap. *In situ* permeability measurement possible	Gauge house usually required. Regular de-airing necessary. Uncovered tubing liable to rodent attack	Useful when remote reading required and for artesian pressures
	Closed Hydraulic (High air entry pressure)	−1 atmosphere to any positive pressure	Moderate	Can be deaired		As above	Fairly cheap. *In situ* permeability measurements in low permeability soil are possible	As above. Very regular de-airing required when measuring suctions	Useful for measuring small suctions
	Pneumatic	Any positive pressure	Rapid	Cannot be de-aired. Only partially self de-airing	Yes Some head loss over long distance	Moderate to poor but very little long term experience available	Fairly cheap. No gauge house required	No method of checking if pore water or pore air pressure is measured	Only suitable when tip almost always below groundwater level and no large suctions occur
	Electric vibrating-wire type	Any positive pressure	Rapid	As above	Yes but special cable required	Signal quality degenerates with time. Instrument life about ten years but reliability of instrument that cannot be checked is always suspect		As above. Expensive. Zero reading liable to drift and cannot be checked	Not generally recommended
	Electric resistance type	Any positive pressure	Rapid	As above	Yes but with care because of transmission losses	Poor		As above	Not recommended
SUCTION	Tensiometer	−1 to positive pressure	Moderate to rapid	Can be deaired	Yes	Good	Cheap, simple to read and maintain	Vandal damage often irreparable. Regular de-airing required	First choice for measuring pore suction
	Psychro-meter	Below −1 atmosphere	Variable	Not relevant	Short distances only	Instrument life one to two years. Little long term experience available		Not accurate between 0 and −1 atmosphere	Research stage at the moment

recommendations beyond those summarized in Figure 3.25 and Table 3.5 without reference to appropriate site stability analyses (see Chapter 2). Response times, with certain of the systems described above, do not remain constant with time (as Figure 3.25 might imply). Diaphragm piezometers are, as we have noted, a particular problem in which there exist conflicting levels of optimism on the ability of high air entry ceramics to prevent air penetration. Once air penetration has occurred over time, a high air entry ceramic will retard resaturation by the surrounding pore water. If positive pore-water pressures are of greatest interest, then clearly a low air entry ceramic would be the better choice. While twin tube hydraulic systems (e.g. Figure 3.4) suffer from the same air entry problems over time (Vaughan, 1974), air can be removed by re-circulating water under a carefully controlled back-pressure to minimize pore pressure changes in the soil surrounding the piezometer tip. However, should the ceramic have become desaturated, then air can flow through at low pressure (the air entry value being reduced); the ceramic is only 100% effective when fully saturated (Mikkelsen and Wilson, 1984).

However, response time is only one factor in equipment selection. Cost represents a further factor. These two elements in many investigations may serve to complicate the scenario given in Figure 3.23, as the simple choice/cost pattern in Figure 3.27 illustrates. Costs shown here are intended to be illustrative only, but they nevertheless firmly show the significance of the duration of the project and the manner of assumed capital cost depreciation. In the examples shown here for the bubbler system, in the first instance the capital costs are apportioned over only six months, while in the second instance, the capital cost is shown apportioned over a three-year period with an assumed project usage of 50% over that period.

Clearly, costs depend upon location, supplier, the frequency of manual readings and labour, and no firm guidance can be given in general terms. The only recommendation that is given here is that since cost is a general *constraint* on both the number and scope of objectives that can be realized in any project, then it is paramount that alternative instrumentation strategies are costed (as shown in Figure 3.27) and matched to key investigative questions (as shown in Figure 3.23). Only in this manner can inference be maximized objectively within such cost constraints. The increased availability of alternative systems makes this all the more important.

Of course, while the discussion has emphasized the significance of comparative response time/cost aspects, these are not the only elements in instrumentation selection. The Hong Kong Geotechnical Control Office (1984) has provided a useful summary table of other factors, in the context of both positive and negative pressure measurement (Table 3.5). While the review above has extended this framework to include other suction recording equipment (e.g. MCS sensor and transducer systems), it is clear that documentation is now needed on a *reliability*/cost basis to enhance the existing selection criteria.

Field instrumentation of soil water conditions has received inputs from three principal groups: engineers, manufacturers, and geomorphologists. Until relatively recently, geomorphologists have not given this type of instrumentation sufficiently rigorous attention in either slope investigations or hillslope hydrology, in spite of the process-based utilization of models in both fields that require closely specified soil water conditions. This review has thus focused, with one or two exceptions (e.g. Anderson, 1986; Burt, 1978; Kemp, 1986) primarily upon 'engineering' based instrumentation developments which are generally available at only relatively high cost levels (Figure 3.27). There is little doubt that

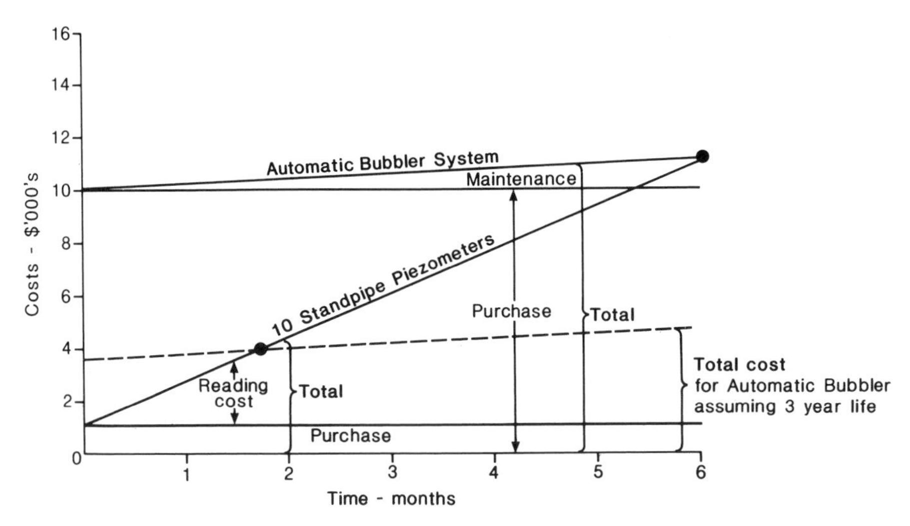

Figure 3.27 Comparative illustrative costs of bubbler piezometer system and manual piezometers (read daily)

geomorphologists must become more attuned to specific instrumentation characteristics (e.g. system stiffness, Figure 3.10; and response times, Figure 3.25) becoming an integral and *specified* element in process experimental design. There is no doubt that this requires a significant resource base, but there is no alternative if research enhancement in the context of hillslope studies (Figure 3.23) is to be developed.

ACKNOWLEDGEMENT

The senior author acknowledges with gratitude the helpful advice and discussions with the late Richard Pope, which contributed so much not only to the structure of this chapter but also to Dr Anderson's geotechnical research experience in Hong Kong.

REFERENCES

Anderson, M. G. (1983). 'The prediction of soil suction for slopes in Hong Kong.' Final Technical Report on PWD CE 3/81. *Geotechnical Control Office*, *Hong Kong*, 242pp.

Anderson, M. G. (1986). 'Hydrological and stability analysis of a complex soil slope, Tomo, Java.' Final Technical Report on R3743. *Overseas Development Administration*, London, 132pp.

Anderson, M. G., and Burt, T. P. (1977). 'Automatic monitoring of soil moisture conditions in a hillslope spur and hollow.' *J. Hydrology*, **33**, 27–36.

Anderson, M. G., and Howes, S. (1985). 'Development and application of a combined soil water—slope stability model.' *Q.J. Eng. Geol.*, **18**, 225–36.

Anderson, M. G., and Kemp, M. J. (1985). 'The prediction of pore water pressure conditions in road cut slopes, St Lucia, West Indies.' Final Technical Report on R3426, *Overseas Development Administration*, London, 208pp.

Anderson, M. G., and Kneale, P. E. (1980). 'Pore water changes in a road embankment.' *Highway Engineer*, **25**, 193–5.

Anderson, M. G., McNicholl, D. P., and Shen, J. M. (1983). 'On the effect of topography in controlling soil water conditions, with specific regard to cut slope piezometric levels.' *Hong Kong Engineer*, **11**, 35–41.

Anderson, M. G., and Pope, R. G. (1984). 'The incorporation of soil water physics models into geotechnical studies of landslide behaviour.' *Int. Soc. Soil Mech. Fndtn Eng.*, 349–53.

Bianchi, W. B. L. (1962). 'Measuring soil moisture tension changes.' *Agric. Eng.*, **43**, 398–9.

Boels, D., Van Gils, J. B. H. M., Veerman, G. J., and Wit, K. E. (1978). 'Theory and system of automatic determination of soil moisture characteristics and unsaturated hydraulic conductivities.' *Soil Sci.*, **126**, 191–9.

Brand, E. W., and Premchitt, J. (1982). 'Response times of cylindrical piezometers.' *Geotechnique*, **32**, 203–16.

Burt, T. P. (1978). 'An automatic fluid scanning switch tensiometer system.' *British Geomorphological Research Group*, Technical Bulletin 21.

Casagrande, A. (1949). 'Soil mechanics in the design and construction of the Logan International Airport.' *J. Boston Soc. Civ. Eng.*, **36**, 192–221.

Chipp, P. N., Clare, D. G., Henkel, D. J., and Pope, R. G. (1982). 'Field measurement of suction in colluvium covered slopes in Hong Kong.' In *Proc. 7th SE Asian Geotech. Conf.* McFeat-Smith, I., and Lumb, P. (eds.), Hong Kong Institute Engineers, 49–62.

Colbeck, S. C. (1976). 'On the use of tensiometers in snow hydrology,' *J. Glaciology*, **17**, 135–40.

Croney, D., Coleman, J. D., and Bridge, P. M. (1952). 'The suction of moisture held in soil and other porous materials.' Road Research Technical Paper, No. 24, HMSO, London.

Department of Main Roads, NSW, Australia (1977). 'Investigation of soil moisture conditions and seasonal moisture variations in Illawarra Division.' Materials and Research Laboratory. Report 1R.11.

Fitzsimmons, D. W., and Young, N. C. (1972). 'Tensiometer-pressure transducer system for studying unsteady flow through soils.' *Trans. Am. Soc. Agric. Eng.*, **15**, 272–5.

Gardner, R. (1955). 'Relation of temperature to moisture tension of soil.' *Soil Sci.*, **79**, 257–65.

Gardner, W. O., Israelsen, W., Edelfsen, N. E., and Conrad, H. (1922). 'The capillary potential function and its relation to irrigation practice.' *Phys. Rev. Ser.*, **2**, 20, 196.

Geotechnical Control Office (1984). *Geotechnical Manual for Slopes* (2nd edn). Public Works Department, Hong Kong, 295pp.

Gibson, R. E. (1963). 'An analysis of system flexibility and its effects on time lags in pore pressure measurements.' *Geotechnique*, **13**, 1–9.

Harr, R. D. (1977). 'Water flux in soil and subsoil on a steep forested slope.' *J. Hydrology*, **33**, 37–58.

Hvorslev, M. J. (1951). 'Time lag and soil permeability in groundwater observations.' US Army Corps of Engineers, Waterways Experiment Station, Bulletin 38, 48pp.

Ingersoll, J. E. (1980). *Soil tensiometers for use at temperatures below freezing* (unpublished report).

Ingersoll, J. E. (1981). 'Laboratory and field use of soil tensiometers above and below 0°C.' US Army Corps of Engineers, Cold Regions Research and Engineering Laboratory, Hanover, N.H., Special Report, 81–7.

Insley, H., and McNicholl, D. (1982). 'Groundwater monitoring of a soil slope in Hong Kong.' *Proc. 7th SE Asian Geotech. Conf.* McFeat-Smith, I., and Lumb, P. (eds.), Hong Kong Institute of Engineers, 63–75.

Kemp, M. J. (1986). *Stability processes in tropical soils*. Unpublished Ph.D. Thesis. University of Bristol, 383pp.

Klute, A., and Peters, D. B. (1962). 'A recording tensiometer with a short response time.' *Proc. Soil Sci. Soc. Am.*, **26**, 87–8.

Krahn, J., and Fredlund, D. G. (1972). 'On total, matric and osmotic suction.' *Soil Sci.*, **114**, 339–48.

Lee, R. K. C. (1983). *Measurement of soil suction using the MCS 6000 sensor*. Unpublished M.S. Thesis, University of Saskatchewan, 162pp.

Lee, R. K. C., and Fredlund, D. G. (1984). 'Measurement of soil suction using the MCS 6000 sensor.' *Proc. Int. Conf. Expansive Soils*, Adelaide, 50–54.

McFarlane, J. (1981). 'Suction–moisture relationships in residual and remoulded Hong Kong soils.' Public Works Department, Hong Kong, Materials Division, Report 24.

McKim, H. L., Berg, R. L., McGraw, R. W., Atkins, R. T., and Ingersoll, J. (1976). 'Development of a remote-reading tensiometer/transducer system for use in subfreezing temperatures.' *Conference on Soil Water Problems in Cold Regions, Edmonton*, 31–45.

McKim, H. L., Walsh, J. E., and Arion, D. N. (1980). Review of techniques for measuring soil moisture *in situ*.' *US Army Corps of Engineers*, Cold Regions Research and Engineering Laboratory, Hanover, N.H., Special Report 80-31.

Massarsch, K. R., Broms, B. B., and Sundquist, O. (1975). 'Pore pressure determination with multiple piezometer.' Proc. Conf. *In Situ* Measurement of Soil Properties, *Am. Soc. Civ. Eng.*, North Carolina, **1**, 260–5.

Mikkelsen, P. E. and Wilson, S. D. (1983). 'Field instrumentation: accuracy, performance, automation and procurement.' In *Field Measurements in Geomechanics*. Kovari, K. (ed.), Balkema, Rotterdam, pp. 251–72.

Mitchell, R. J. (1983). *Earth Structures Engineering*. George Allen & Unwin, Boston, 265pp.

Nagpal, N. K., and Boersma, L. (1973). 'Air entrapment as a possible source of error in the use of a cylindrical heat probe.' *Soil Sci. Soc. Am. Proc.*, **37**, 828–32.

Phene, C. J., Hoffman, G. J., and Austin, R. S. (1973). 'Controlling automated irrigation with soil matric potential sensor.' *Trans. Am. Soc. Agric. Eng.*, **16**, 773–6.

Pope, R. G., Weeks, R. C., and Chipp, P. N. (1982). 'Automatic recording of standpipe piezometers.' *Proc. 7th SE Asian Geotech. Conf.* McFeat-Smith, I., and Lumb, P. (eds.), Hong Kong Institute of Engineers, 77–89.

Rice, R. (1969). 'A fast response field tensiometer system.' *Trans. Am. Soc. Agric. Eng.* **12**, 48–50.

Richards, B. G. (1969). 'Pyschrometric technique for measuring soil-water potential.' *CSIRO Div. Soil Tech. Report*, **9**, 1–32.

Schmugge, T. J., Jackson, T. J., and McKim, H. L. (1980). 'Survey of methods for soil moisture determination.' *Water Resources Res.*, **16**, 961–79.

Sweeney, D. J. (1982). 'Some *in situ* soil suction measurements in Hong Kong's residual soil.' In *Proc. SE Asian Geotechnical Conference*. McFeat-Smith, I., and Lumb, P. (eds.), Hong Kong Institute of Engineers, 91–105.

Towner, G. D. (1980). 'Theory of time response of tensiometers.' *J. Soil Sci.*, **31**, 607–21.

Vaughan, P. R. (1974). 'The measurement of pore pressures with piezometers.' In *Field Instrumentation in Geotechnical Engineering*. British Geotechnical Society, Wiley, Chichester, pp. 411–22.

Watson, K. K. (1965). 'Some operating characteristics of a rapid response tensiometer system.' *Water Resources Res.*, **1**, 577–86.

Watson, K. K., and Jackson, R. J. (1967). 'Temperature effects on a tensiometer–pressure transducer system.' *Proc. Soil Sci. Soc. Am.*, **31**, 156–60.

Williams, T. H. Lee (1978). 'An automatic scanning and recording tensiometer system.' *J. Hydrology*, **39**, 175–83.

Young, N. C. (1968). *Using a tensiometer–pressure transducer apparatus to study one and two-dimensional imbibition*. M.Sc. Thesis, Univ. Idaho, Moscow, Idaho.

Slope Stability
Edited by M. G. Anderson and K. S. Richards

Chapter 4

Slope Stability Analysis Incorporating the Effect of Soil Suction

D. G. FREDLUND
Department of Civil Engineering
University of Saskatchewan, Saskatchewan, Canada S7N 0WO

4.1 INTRODUCTION

Slope stability analyses have become a common analytical tool to assess the factor of safety of natural and man-made slopes. Saturated shear strength parameters are generally used in the analysis. The portion of the soil profile above the groundwater table where the pore-water pressures are negative is usually ignored. This is a reasonable assumption for many situations where the major portion of the slip surface passes through saturated soil. However, for situations where the groundwater table is deep or where the concern is over the possibility of a shallow failure, there is need to understand how to perform slope stability analyses where the soil is unsaturated. The main objective of this paper is to discuss the developments in slope stability analysis to incorporate the effect on strength of suction in the unsaturated zone.

In order to maintain an analytical procedure for unsaturated soil mechanics, several aspects must be addressed (Fredlund, 1979). First, it must be possible to measure appropriate shear strength parameters for the unsaturated soils. Preferably these should provide a smooth transition to the saturated shear strength parameters. Second, techniques must be available to measure or estimate the *in situ* negative pore-water pressure conditions. If an assessment of the factor of safety is required for a particular instant in the history of the slope, it is possible either to measure the suction *in situ* or on laboratory samples. The problem becomes more demanding when it is necessary to predict possible changes in the soil suction profile throughout the seasons. Third, the conventional, limit equilibrium, slope stability analysis must be extended to incorporate the shear strength equation for unsaturated soils. Once verified through a series of case histories, the method of slope stability analysis for unsaturated soils becomes an extention of the procedure used for saturated soils.

The analysis of mass movements in unsaturated soils is a relatively new area of research. There are few complete case histories in which all the above factors have been assimilated. While it is recognized that further research is required, examples will be given to demonstrate how the factor of safety can be computed for unsaturated soil slopes. The present aspect most in need of further research is the measurement and prediction of soil suction profiles with the seasons.

4.2 SHEAR STRENGTH OF UNSATURATED SOILS

Shear strength tests on unsaturated soils commenced at Imperial College in the 1950s under the supervision of Bishop (Bishop *et al.* 1960). These were the first tests in which the pore-air and pore-water pressures were independently monitored as the soil was deformed to failure. The shear strength at failure was presented in terms of the total normal stress, σ_n, the pore-air pressure, u_a, and the pore-water pressure, u_w.

A suggested shear strength equation included an empirical soil parameter, χ, which proved difficult to quantify experimentally. As a result little attempt was made to incorporate the shear strength equation into practical problems.

In 1977, Fredlund and Morgenstern described the stress state variables for an unsaturated soil and explained their usage in an independent manner. This made it unnecessary to evaluate the empirical χ parameter. The proposed independent stress state variables for analysing practical problems were $(\sigma - u_a)$ and $(u_a - u_w)$. This led to the formulation of an extended Mohr–Coulomb failure criterion for unsaturated soils (Fredlund *et al.*, 1978).

4.2.1 Shear Strength Theory and Verification

The effective shear strength of a *saturated* soil is controlled by one stress state variable, $(\sigma_n - u_w)$. The corresponding Mohr–Coulomb failure envelope is a straight line defined by an intercept, c' (i.e. effective cohesion) and a slope, tan ϕ' (i.e. tangent of the effective angle of internal friction). An *unsaturated* soil requires two independent stress state variables that must be given consideration in defining the failure envelope. Therefore, it is most logical to extend the two-dimensional failure line for a saturated soil to a three-dimensional failure surface for an unsaturated soil (Figure 4.1). If the surface is assumed to be planar, the unsaturated shear strength, τ, can be written as follows:

$$\tau = c' + (\sigma_n - u_a) \tan \phi' + (u_a - u_w) \tan \phi^b \tag{4.1}$$

where: σ_n = total normal stress.

The friction angle, ϕ^b, is equal to the slope of a plot of matric suction $(u_a - u_w)$ versus shear strength when $(\sigma_n - u_a)$ is held constant. The effective angle of internal friction is the same for all suction values since the plane is assumed to have no 'warp'. The stress circles corresponding to failure conditions can be plotted on a three-dimensional diagram with the two stress state variables plotted on the horizontal axes and the shear strength as the ordinate (Figure 4.2). Test data can be interpreted by either a mathematical (Fredlund *et al.*, 1978) or a graphical method (Fredlund, 1981a). The interpretation of the data is easier to visualize using the graphical method and for this reason will be used throughout this chapter.

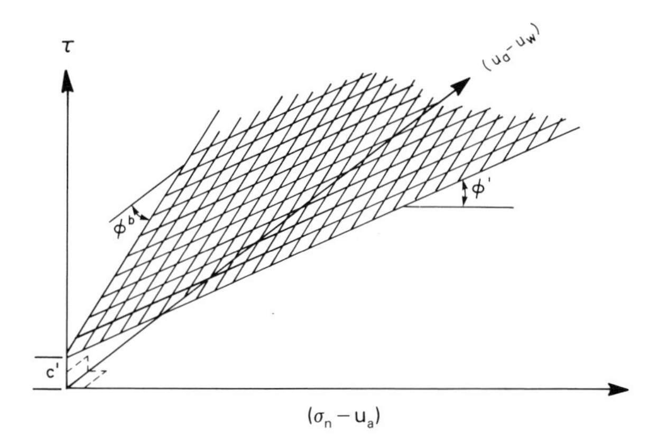

Figure 4.1 Planar surface representing the shear strength equation for an unsaturated soil

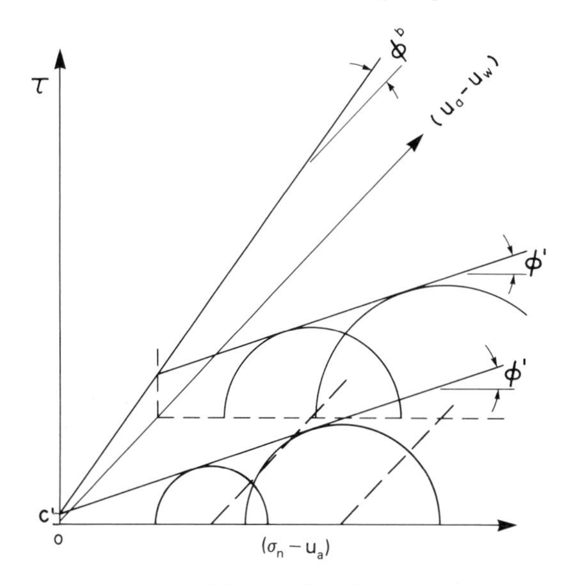

Figure 4.2 Mohr circles at failure for an unsaturated soil

Using the graphical method, the equation can be visualized as a two-dimensional graph with matric suction contoured as the third variable (Figure 4.3). Consequently, the ordinate intercepts (i.e. when $\sigma_n - u_a$ is equal to zero) of the various matric suction contours can be plotted versus matric suction to give the friction angle, ϕ^b (Figure 4.4).

The angle, ϕ^b, can either be visualized as a friction angle or as a component of cohesion. If it is considered as a component of cohesion, the total cohesion of the soil, c, has two components.

$$c = c' + (u_a - u_w) \tan \phi^b \tag{4.2}$$

Figure 4.5 shows the increase in cohesion due to matric suction for various ϕ^b values.

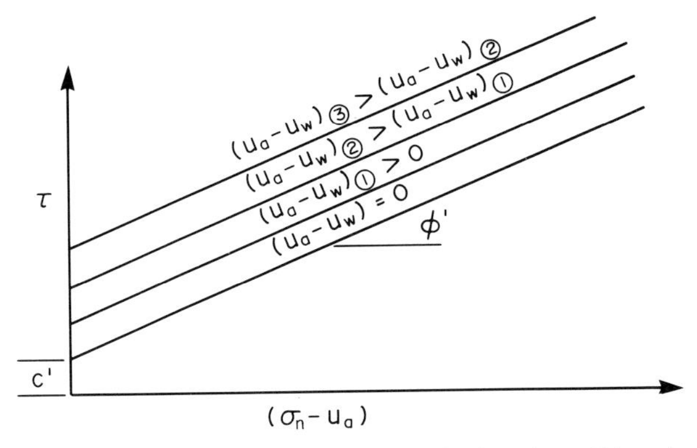

Figure 4.3 Failure surfaces for an unsaturated soil viewed parallel to the $(u_a - u_w)$ axis

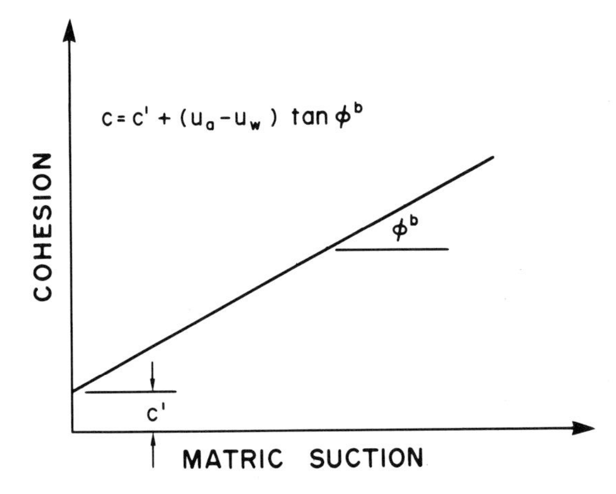

Figure 4.4 The increase in shear strength (cohesion) with matric suction when $(\sigma - u_a)$ is zero

The shear strength equation for an unsaturated soil can then be reduced to the same form of equation as that used for a saturated soil.

$$\tau = c + (\sigma_n - u_a) \tan \phi' \tag{4.3}$$

For most practical problems the pore-air pressure can be assumed to be atmospheric or equal to zero gauge.

There is a smooth transition from the unsaturated soil shear strength equation to the conventional shear strength equation for saturated soils. As saturation is approached, the pore-air pressure, u_a, becomes equal to the pore-water pressure, u_w. At this point, the proposed shear strength equation reverts to the conventional shear strength equation for a saturated soil.

The justification for the linear form of the unsaturated soil shear strength equation must be based on experimental laboratory test data. Bishop *et al.* (1960), presented several sets of triaxial test data on unsaturated soils. Two sets of data were from constant water

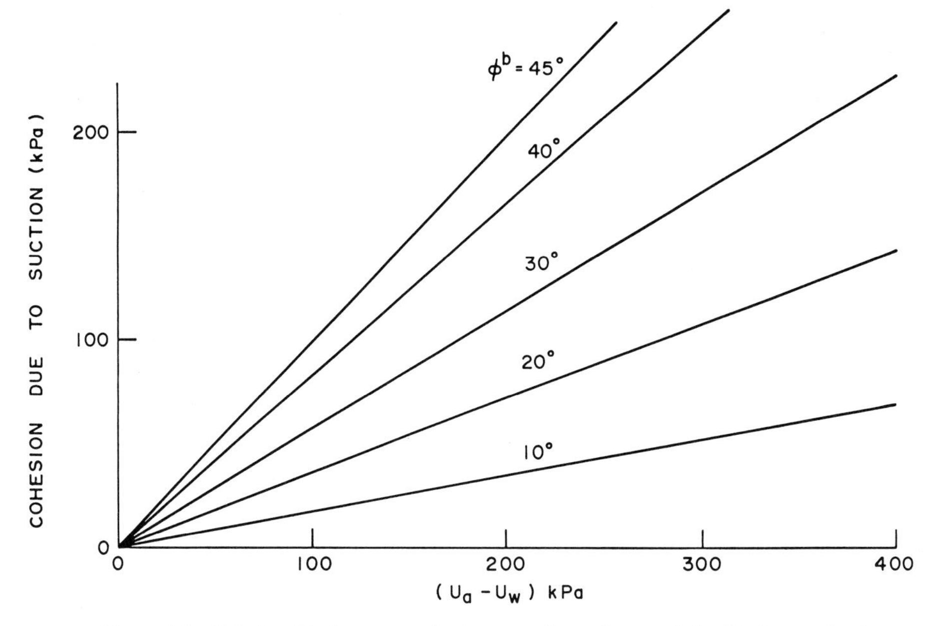

Figure 4.5 Relationship between cohesion, matric suction, and the friction angle, ϕ^b

Table 4.1 Triaxial Tests on Unsaturated Soils (Bishop *et al.*, 1960)

Soil Type	c' (kPa)	ϕ' (degrees)	ϕ^b (degrees)	Correlation*
Compacted shale $w = 18.6\%$	15.8	24.8	18.1	0.970
Boulder clay $w = 11.6\%$	9.6	27.3	21.7	0.974

*Correlation coefficient is for the computation of ϕ^b.

content tests with pore-air and pore-water pressure measurements. The soils tested were compacted shale and a boulder clay. The data have been re-analysed using the graphical method and the results are summarized in Table 4.1.

It can be seen that the correlation coefficients for the plot of ordinate intercepts versus matric suction are high (i.e. between 0.97 and 0.98). If the surface describing stress combinations at failure were ideally planar, the ordinate intercepts versus matric suction plot should be a straight line having a correlation coefficient of 1.

Satija (1978) initiated an extensive laboratory testing programme on statically compacted Dhanauri Clay. Four series of constant water content ($\overline{CW}$) and consolidated drained (CD) tests were conducted with pore-air and pore-water pressure measurements. The soil was compacted at two densities while maintaining the same initial water contents. Both the constant water content and consolidated drained test were performed at each of the compaction conditions. The samples were allowed to equalize to various initial stress states to form an equally spaced grid of tests. The original data have been re-analysed using the graphical method. Figure 4.6 shows the failure stress circles for one series of the constant

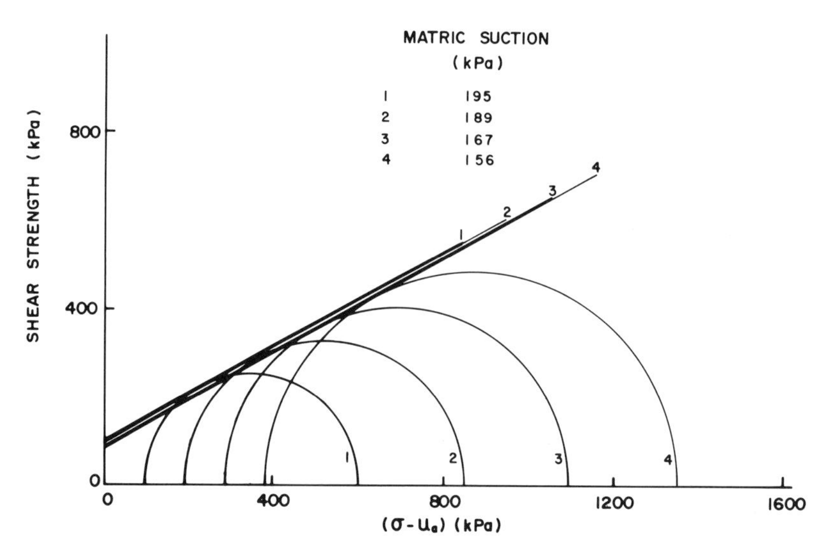

Figure 4.6 Mohr failure circles for Dhanauri Clay from $\overline{\text{CW}}$ tests, series III. (*From Satija, 1978.*)

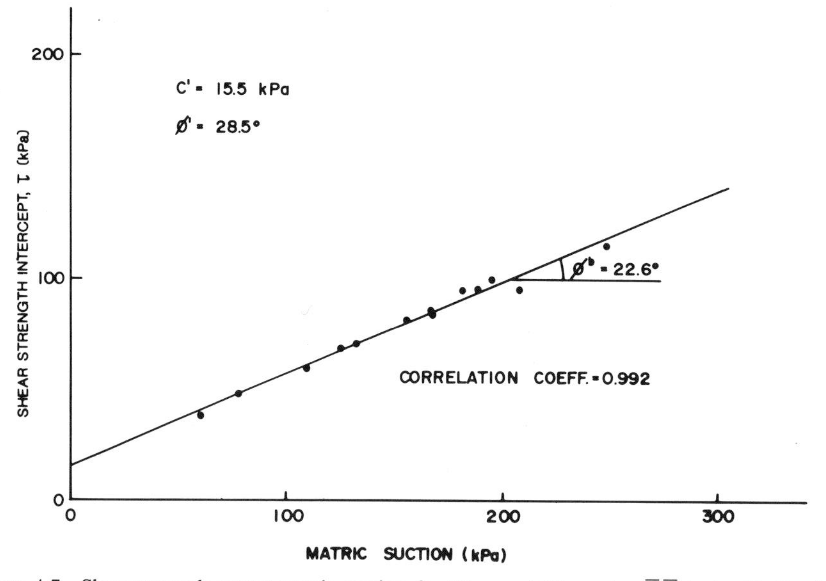

Figure 4.7 Shear strength versus matric suction for Dhanauri Clay using $\overline{\text{CW}}$ tests. (*Satija, 1978*)

water content tests performed on samples with a high initial density. The shear strength versus matric suction plot for all the constant water content tests on the high initial density samples is shown in Figure 4.7. The results are summarized in Table 4.2. The correlation coefficients for all the plots of ordinate intercepts versus matric suction are high and reveal no tendency towards a 'warped' surface. The results show that the ϕ^b values obtained from the CD tests were lower than the ϕ^b values obtained from the $\overline{\text{CW}}$ tests. The reason for this difference has not been fully pursued.

Table 4.2 Triaxial Tests on Unsaturated Dhanauri Clay. (*Satija, 1978*)

Soil Type	c' (kPa)	ϕ' (degrees)	ϕ^b (degrees)	Correlation Coefficients*
CD Test $\gamma_d = 15.5$ kN/m^3 $w = 22.2\%$	37.3	28.5	16.2	0.974
CD Test $\gamma_d = 14.5$ kN/m^3 $w = 22.2\%$	20.3	29.0	12.6	0.963
$\overline{\text{CW}}$ Test $\gamma_d = 15.5$ kN/m^3 $w = 22.2\%$	15.5	28.5	22.6	0.992
$\overline{\text{CW}}$ Test $\gamma_d = 14.5$ kN/m^3 $w = 22.2\%$	11.3	29.0	16.5	0.971

*Refers to the best-fit line for ϕ^b.

Escario (1980) performed direct shear tests with controlled pore pressures on compacted Madrid Grey Clay. The tests results have been re-analysed using the graphical method and are shown in Figure 4.8. The ϕ^b angle is 16.1 degrees (Figure 4.9). The correlation coefficient is high and there is no indication of 'warping' of the surface describing the stress conditions at failure.

To date, the published test data demonstrate that the surface describing the stress conditions at failure for an unsaturated soil, is essentially planar. There could possibly be some curvature, particularly at low suctions. However, the unsaturated soil shear strength (equation (4.1)) is sufficiently close to being planar to be satisfactory for engineering purposes.

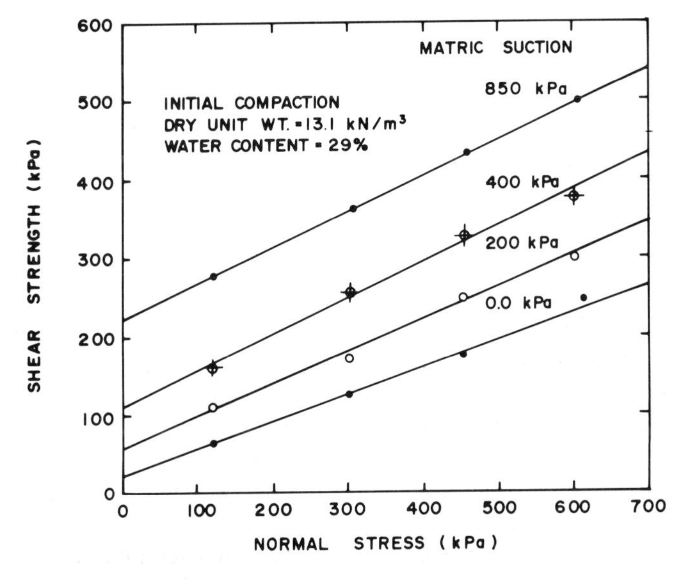

Figure 4.8 Shear strength of Madrid Grey Clay at various suction levels. (*Escario, 1980*)

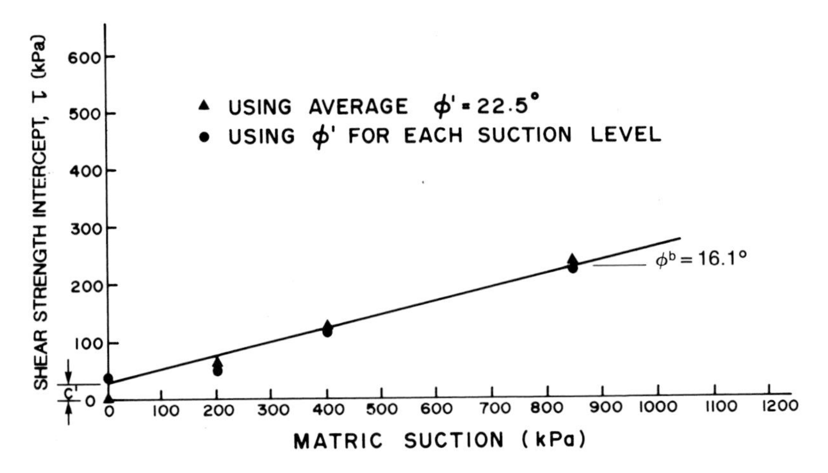

Figure 4.9 Shear strength versus matric suction for Madrid Grey Clay. (*Escario, 1980*)

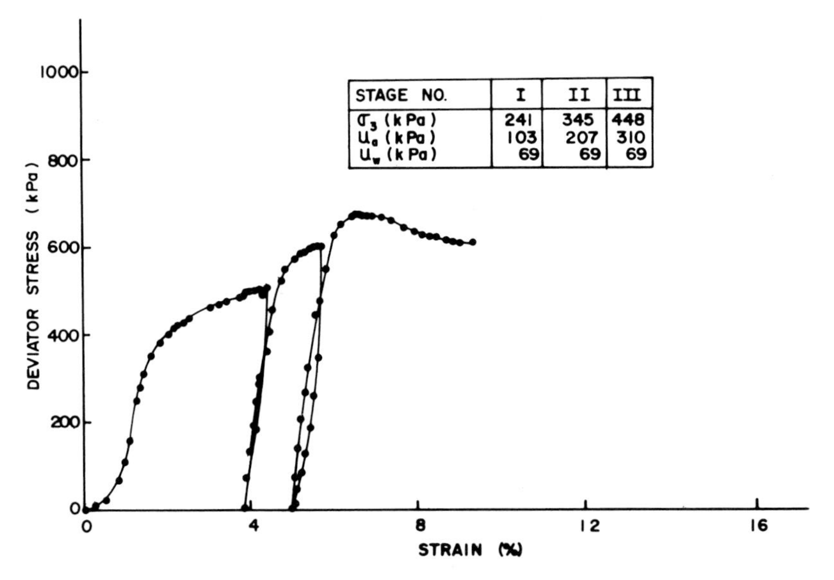

Figure 4.10 Stress versus strain curve for decomposed granite sample no. 10

Based on the assumption that the failure surface for an unsaturated soil was planar, Ho and Fredlund (1982a) proposed a single specimen, multistage triaxial test procedure to obtain the ϕ^b angle. A total of 17 samples of decomposed granite and rhyolite from Hong Kong were tested. Figure 4.10 shows a typical stress loading of the sample. The matric suction was controlled at differing values during each of the three stages (i.e. CD test). Figure 4.11 shows a corresponding plot for the interpretation of the shear strengths and the determination of ϕ^b.

The average measured angle, ϕ^b, was 15.3 degrees for decomposed granite and 13.8 degrees for the decomposed rhyolite (Ho and Fredlund, 1982b). Generally, the third stage

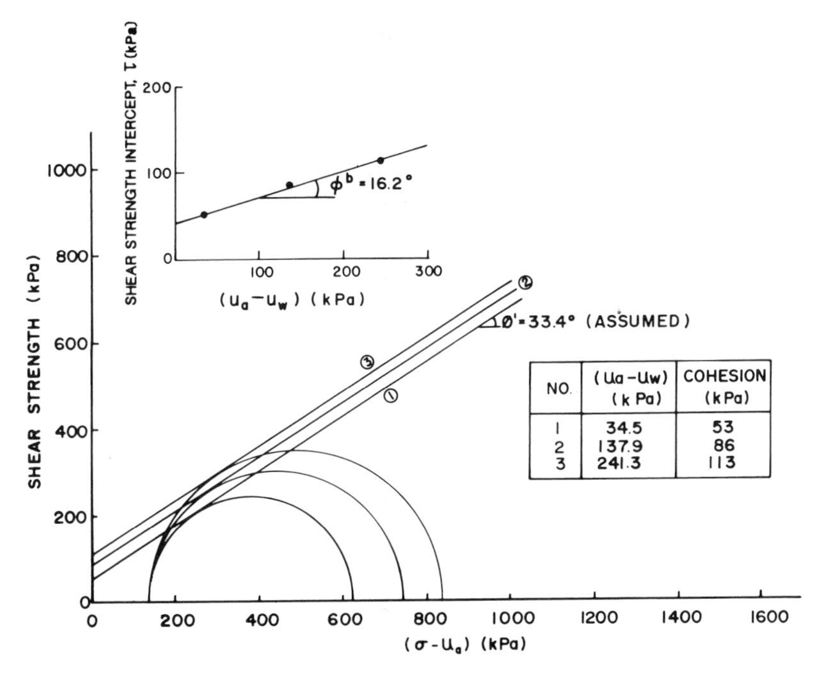

Figure 4.11 Mohr circles and the determination of ϕ^b for decomposed granite sample no. 10

of loading tended to reduce the measured ϕ^b angle. This was recognized as being related to the accumulation of excessive strains in the specimens. Using the first two stages of loading, the standard deviation of the measured ϕ^b angles was approximately ± 5 degrees.

4.2.2 Testing Apparatus and Procedure

A conventional triaxial cell for testing saturated soils (Bishop and Henkel, 1962) can be modified to accommodate the testing of unsaturated soils. The test procedure can be conducted either on one specimen (i.e. multistage) or on several specimens at similar initial (i.e. pre-testing) stress and volume–weight conditions. Details will be given for the multistage procedure (Ho and Fredlund, 1982a).

The major modification to a conventional triaxial cell involves the sealing of a high air entry disc on to the base pedestal of the triaxial cell. The high air entry disc allows the measurement of the pore-water pressure independent of the pore-air pressure. The disc must be sealed around the edge of the pedestal using a water resistant epoxy. It should then be checked for leaks by applying an air pressure to the saturated disc. The air entry value of the disc must be higher than the maximum matric suction that will be applied to the sample. Generally, a ⅜ inch thick, 5 bar (505 kPa) ceramic disc is most satisfactory. A coarse or low air entry disc can be placed on the top of the specimen to control the air pressure.

Figure 4.12 shows a modified triaxial cell. The plumbing layout for the triaxial apparatus control board is shown in Figure 4.13. Although the ceramic disc does not allow the passage of free air, dissolved air can diffuse through the water in the disc and collect as free air

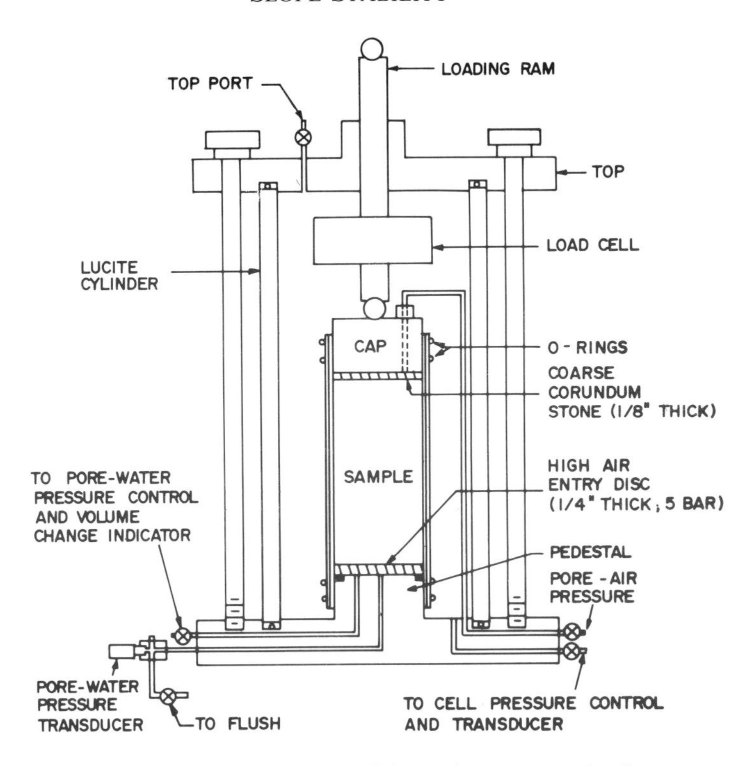

Figure 4.12 Modified triaxial cell for testing unsaturated soils

bubbles at the base of the disc. The free air bubbles can block the passage of water into the sample if the sample dilates or swells and can cause misleading measurements of the total water volume change.

A flushing system can be used to cope with the diffused air problem. An extra pore-water drainage line with a control valve must be connected to the base of the triaxial apparatus. By closing the valve on the pore-water drainage line leading to the pore-water pressure control and opening the valve on the added drainage line, diffused air accumulating below the disc can be periodically flushed.

The testing procedure involves the control of the air and water pressures during the entire test rather than their measurements in a closed system. Suction in the sample is maintained constant during the application of the deviator stress. Maintaining the pore-air and pore-water pressure is similar to performing a 'slow' or drained test on a saturated soil. The procedure used to impose suctions greater than 100 kPa (1 atm) is known as the axis-translation technique (Hilf, 1948; Bocking and Fredlund, 1980).

A multistage testing procedure is used to obtain maximum information from a limited number of tests and to eliminate the effect of soil variability. The test procedure is as follows:

A trimmed specimen is mounted in the triaxial cell and two rubber membranes are placed around the specimen. O-rings are placed over the membrane on the bottom pedestal. A spacer (two pieces of 3.2 mm (⅛ in) plastic tubing) can be placed between the membranes and the upper loading cap so that air within the sample can escape while water is added to the sample to reduce its suction.

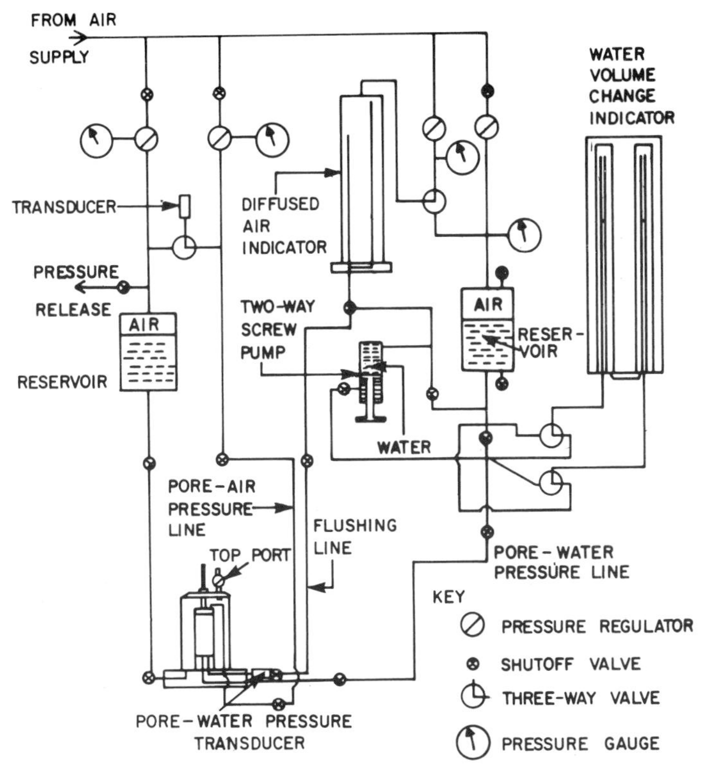

Figure 4.13 Schematic diagram of plumbing layout for triaxial apparatus control board

The object of this part of the testing procedure is to relax the suction to a low value before Stage I of loading. The following procedure is suggested for reducing the initial suction of the soil. First, the Lucite cylinder is put in place around the specimen and water is added up to a level about 0.5 cm (0.2 in) below the top of the specimen. The water provides lateral support to the specimen while additional water is slowly added to the sample through the air pressure line connected to the top loading cap. During this stage, the air pressure line is connected to a water reservoir to pass water into the top of the specimen. Water cannot readily be added to the bottom of the specimen because of the low permeability of the ceramic disc and the danger of cracking the disc by upward pressure. The specimen is then left for several hours to allow the distribution of water. The saturation process is continued until air can no longer be seen escaping from around the top of the specimen (i.e. from between the membrane and the top loading cap).

It is now possible to remove the spacer from between the membranes and the upper loading cap. At this point, the line connected to the top loading cap is disconnected from the water reservoir and connected directly to the air pressure line.

Once the sample has imbibed water, the top O-rings are placed around the loading cap. The stresses associated with the first stage of testing are applied and the sample is allowed to consolidate. A typical set of stresses for Stages I, II, III are given in Table 4.3. The associated stress state variables are given in Table 4.4. Once no further water volume change can be detected from the sample, the sample is in equilibrium with the applied stresses.

Table 4.3 Typical set of stresses for Stages I, II, and III

Stage	σ_3* (kPa)	u_a (kPa)	u_w (kPa)
I	250	100	50
II	350	200	50
III	500	350	50

*where σ_3 is a minor principal stress.

Table 4.4 Associated stress state variables

Stage	$(\sigma_3 - u_a)$ (kPa)	$(u_a - u_w)$ (kPa)
I	150	50
II	150	150
III	150	300

After consolidation is complete, the stresses are maintained while the sample is loaded at a constant strain rate. The choice of strain rate is based primarily on the coefficient of consolidation observed during the consolidation of the sample and the permeability properties of the high air entry disc (Ho and Fredlund, 1982c). For many soils, the high air entry disc permeability will control the rate of strain. Suggested typical strain rates for low plasticity soils are in the order of 0.001 to 0.004% strain per minute. For high plasticity soils the rates could be considerably slower.

The procedure for applying the deviator stress is as follows.

The deviator stress is applied until it is apparent that the stress has reached a peak value. At this point, the vertical load is 'backed-off' the sample (Figure 4.14). A new set of stresses for Stage II are applied to the specimen, consolidation is again allowed, and the loading process is repeated as before. The procedure is further repeated for Stage III.

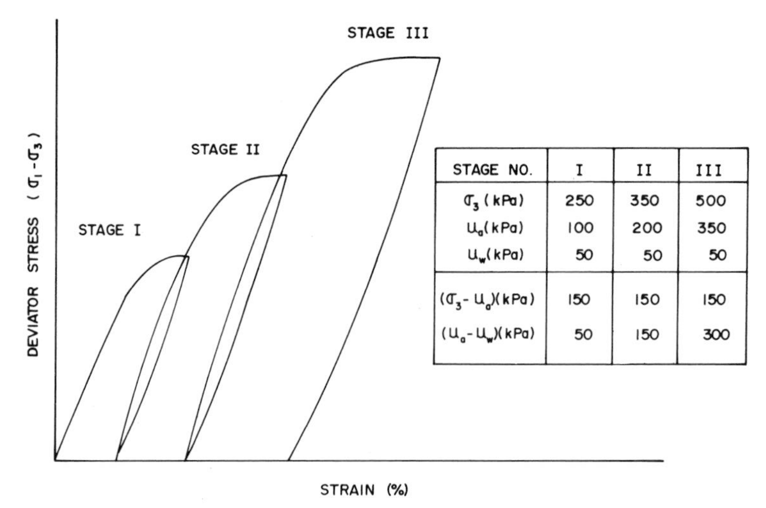

Figure 4.14 Idealized stress versus strain curve for a multistage triaxial test

4.2.3 Typical Test Data and their Interpretation

The test data can be readily interpreted using the graphical technique. The stress circles corresponding to the failure conditions are plotted on a two-dimensional graph with matric suction contoured as the third variable.

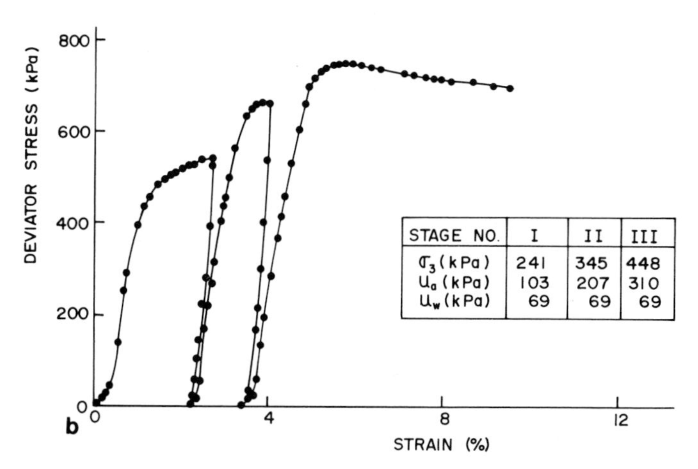

Figure 4.15 Stress versus strain curves for decomposed rhyolite sample 11C from Hong Kong

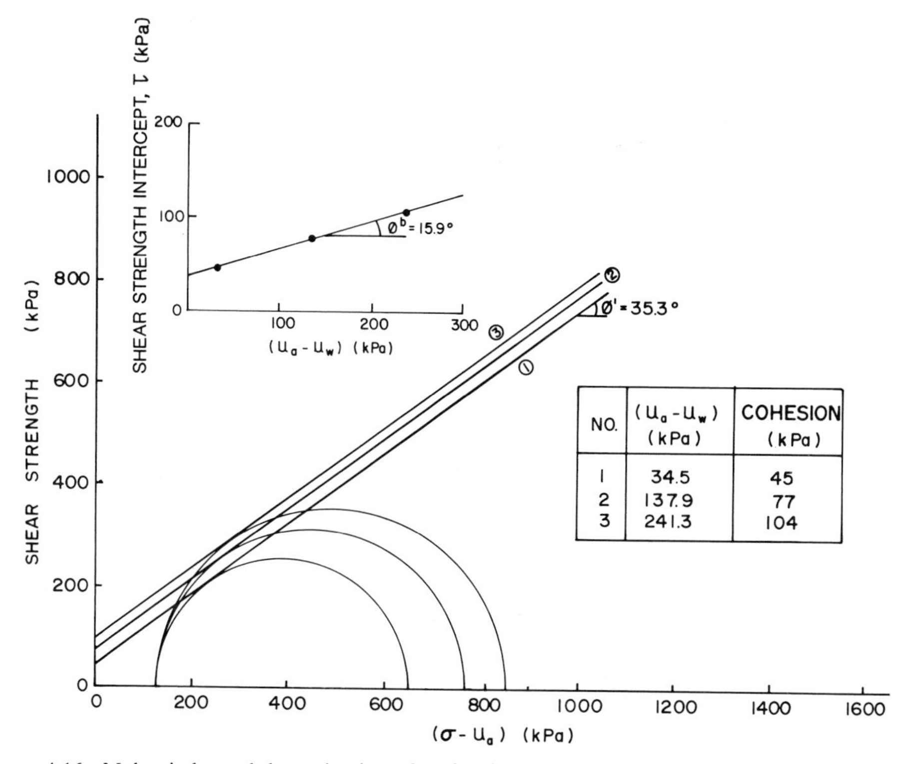

Figure 4.16 Mohr circles and determination of ϕ^b for decomposed rhyolite sample 11C from Hong Kong

To interpret the data it is necessary to know the effective angle of friction ϕ'. The ordinate intercepts (i.e. when $\sigma_n - u_a$ is equal to zero) of the various matric suction contours is plotted versus matric suction to give the angle, ϕ^b (Figure 4.4).

Figure 4.15 shows a typical stress versus strain curve for a decomposed rhyolite from Hong Kong. The manner for obtaining ϕ^b is shown in Figure 4.16. A line at a slope, ϕ', is drawn tangent to each Mohr circle. The shear strength intercepts on the ordinate are then plotted versus the matric suction for each stage and the ϕ^b angle is measured.

The *in situ* negative pore-water pressures must be either measured or predicted using an analytical method. This is an integral part in the computation of the factor of safety of a slope in unsaturated soils. However, these topics are dealt with in detail in other chapters, for example Chapter 3 describes the instrumentation which can be used to measure negative pore-water pressures.

4.3 MODIFICATIONS TO SLOPE STABILITY ANALYSIS

Any one of a number of commonly used limit equilibrium methods of slices could be used to demonstrate the effect on the analysis of the soil being unsaturated. The General Limit Equilibrium (GLE) method proposed by Fredlund *et al.* (1981) will be used since it satisfies both force and moment equilibrium and allows the visualization of other methods as special cases of the more general formulation (see Chapter 2).

The aspect of the factor of safety equations which changes for an unsaturated soil is the description of the shear strength. One of two approaches can be used for the factor of safety derivations (Fredlund, 1981a).

Firstly, it is possible to consider the matric suction term as part of the cohesion of the soil. In other words, the matric suction increases the cohesion of the soil. Therefore, the factor of safety equations do not need to be re-derived. Rather, the cohesion of the soil, c, simply has two components in accordance with equation (4.2). The mobilized shear force at the base of a slice, S_m can be written,

$$S_m = \frac{\beta}{F}[c + (\sigma_n - u_w)\tan\phi'] \quad (4.4)$$

where F = factor of safety. More specifically, the factor by which the shear strength of the soil must be reduced to bring the soil mass into a state of limiting equilibrium.

β = length across the base of a slice.

This approach has the advantage that the shear strength equation retains its conventional form. Therefore, it is possible to utilize a computer program written for saturated soils to solve unsaturated soils problems (Fredlund, 1981b). The stability analysis can be performed with the following considerations (Ching *et al.*, 1984): (a) the soil in the negative pore-water region must be sub-divided into several discrete layers; (b) each layer must embrace a soil layer of constant total cohesion (i.e. a specific matric suction range); and (c) the pore-air and pore-water pressures must be set to zero. This approach has the disadvantage that the cohesion is not a continuous function and the cohesions must be manually computed.

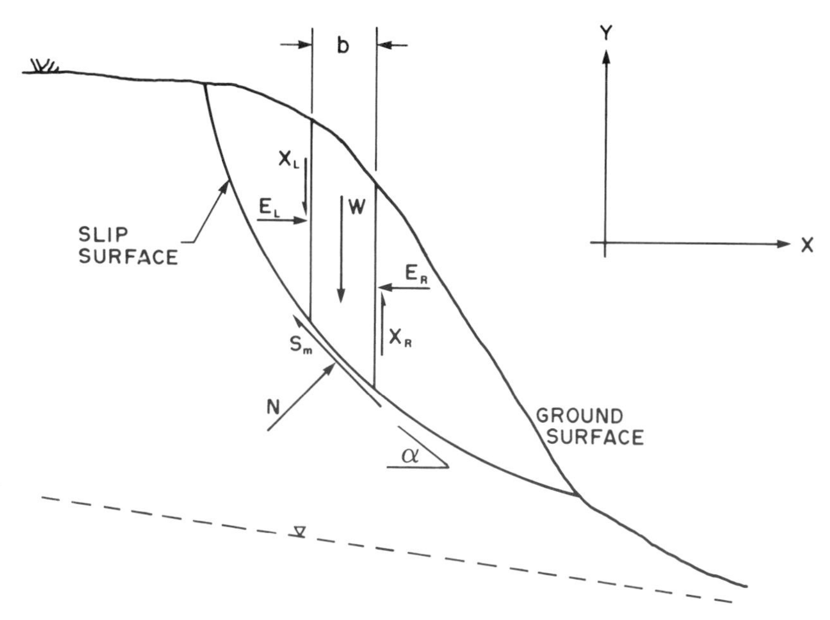

Figure 4.17 Forces acting on each slice for a slope stability analysis

Secondly, it is possible to re-derive the factor of safety equations using the shear strength equation for the unsaturated soil (i.e. equation (4.1)). The mobilized shear force at the base of a slice can be written,

$$S_m = \frac{\beta}{F}\,[c' + (\sigma_n - u_a)\tan\phi' + (u_a - u_w)\tan\phi^b] \tag{4.5}$$

The forces acting on a slice within a sliding soil mass are shown in Figure 4.17 and can be defined as follows:

W = total weight of a slice of width 'b' and height 'h'.
N = total normal force acting on the base of a slice. It is equal to $\sigma_n\beta$ where σ_n is the normal stress acting over the sloping distance, β.
E_L, E_R = horizontal interslice normal force on the left and right sides of a slice, respectively.
X_L, X_R = vertical interslice shear force on the left and right sides of a slice, respectively.
R = radius or moment arm associated with the mobilized shear resistance, S_m.
x = horizontal distance from the centre of each slice to the centre of moments.
f = offset distance from the normal force to the centre of moments.
α = angle between the base of the slice and the horizontal.

The normal force at the base of a slice, N, is derived by summing forces in the vertical direction:

$$N = \frac{W - (X_R - X_L) - \dfrac{c'\beta\sin\alpha}{F} + u_a\dfrac{\beta\sin\alpha}{F}(\tan\phi' - \tan\phi^b) + u_w\dfrac{\beta\sin\alpha}{F}\tan\phi^b}{\cos\alpha + \dfrac{\sin\alpha\tan\phi'}{F}} \tag{4.6}$$

For most analyses the pore-air pressure can be set to zero and equation (4.6) becomes:

$$N = \frac{W - (X_R - X_L) - \dfrac{c' \beta \sin\alpha}{F} + u_w \dfrac{\beta \sin\alpha \tan\phi^b}{F}}{\cos\alpha + \dfrac{\sin\alpha \tan\phi'}{F}} \tag{4.7}$$

When the soil becomes saturated, ϕ^b can be set to ϕ' and therefore the same equation can be used for both saturated and unsaturated soils. Computer coding can be written such that ϕ^b is used whenever the pore-water pressure is negative and ϕ' is used when the pore-water pressure is positive.

Two independent factor of safety equations can be derived; one with respect to moment equilibrium and the other with respect to horizontal force equilibrium. Moment equilibrium can be satisfied with respect to an arbitrary point above the central portion of the slip surface. For a circular slip surface, the centre of rotation is an obvious centre of moments. The centre of moments is immaterial when both force and moment equilibria are satisfied. When only moment equilibrium is satisfied, the computed factor of safety varies slightly with the point selected for the summation of moments:

$$F_m = \frac{\Sigma\{c' \beta R + [N - u_w \beta \dfrac{\tan\phi^b}{\tan\phi'} - u_a \beta (1 - \dfrac{\tan\phi^b}{\tan\phi'})] R \tan\phi'\}}{\Sigma Wx - \Sigma Nf} \tag{4.8}$$

where: F_m = factor of safety with respect to moment equilibrium.

The factor of safety with respect to force equilibrium is derived by summing the forces in the horizontal direction for all slices:

$$F_f = \frac{\Sigma\{c' \beta \cos\alpha + [N - u_w \beta \dfrac{\tan\phi^b}{\tan\phi'} - u_a \beta (1 - \dfrac{\tan\phi^b}{\tan\phi'})] \tan\phi' \cos\alpha\}}{\Sigma N \sin\alpha} \tag{4.9}$$

where: F_f = factor of safety with respect to force equilibrium.

When the pore-air pressure is zero or atmospheric, the entire pore-air pressure term can be dropped. The above formulations apply for both saturated and unsaturated soils. When the soil is saturated, the ϕ^b term must be set equal to ϕ'. The computer coding PC-SLOPE, distributed by Geo-Slope Programming Ltd, performs the stability analysis of unsaturated soils in accordance with equations (4.7), (4.8), and (4.9).

The summation of horizontal forces on each slice can be used to compute the total interslice normal force, E:

$$(E_R - E_L) = [W - (X_R - X_L)] \tan\alpha - \frac{S_m}{\cos\alpha} \tag{4.10}$$

The assumption is made that the interslice shear force, X, is related to the interslice normal force, E, by a mathematical function (Morgenstern and Price, 1965):

$$X = \lambda f(x) E \tag{4.11}$$

where: $f(x)$ = a functional relationship which describes the manner in which the magnitude of X/E varies across the slip surface.

λ = a scaling constant which represents the percentage of the function, $f(x)$, used for solving the factor of safety equations.

Until recently, the $f(x)$ was arbitrarily selected. Wilson and Fredlund (1983) and Fan (1983) performed finite element analyses on slopes of varying inclinations and studied a wide variety of possible slip surfaces. The results yielded an empirical interslice force function of the form:

$$f(x) = Ke^{\frac{-C^n\omega^n}{2}} \tag{4.12}$$

where:

e = base of the natural logarithm.

K = magnitude of the interslice force function at midslope (i.e. maximum value).

C = variable to define the inflection points.

n = variable to specify the flatness or sharpness of curvature.

ω = dimensionless x-position relative to the midpoint of the slope.

The variable, K, is a function of the slope inclination and the depth factor. The constants 'C' and 'n' are related to the slope inclination. A summary of the magnitudes of all constants is presented by Fredlund (1984).

The above slope stability analysis can be simplified if desired, by (a) electing to consider only circular slip surfaces; (b) electing to satisfy only the moment or the force equilibrium factor of safety equation; or (c) assuming the interslice shear forces are negligible.

4.4 APPLICATIONS OF SLOPE STABILITY ANALYSIS

The increase in the factor of safety due to negative pore-water pressures (or matric suction) can readily be demonstrated by studying the influence of increasing the cohesion. Particularly on shallow slip surfaces, the cohesion component significantly affects the computed factor of safety. Fredlund (1981a) selected a typical cross-section and soil properties for a slope in Hong Kong (Figure 4.18) and demonstrated the effect of increasing the cohesion on a selected slip surface (Figure 4.19). The factor of safety increased two-fold for an increase in cohesion of 60 kPa. Conversely, it can readily be appreciated that the factor of safety of a slope can decrease significantly when the cohesion due to matric suction is decreased during a prolonged wet period.

The additional factors which require engineering judgement when performing slope stability analyses on unsaturated soil slopes, are the assessment of ϕ^b and the matric suction profile. Test results to date indicate that the ϕ^b angle is always less than ϕ'. Typical values for ϕ^b range from 13 to 20 degrees.

The assessment of the relevant negative pore-water (or matric suction) profile is difficult and depends upon the problem being addressed. Typical situations can be outlined as follows:

1. There may be interest in performing a back-analysis on a slope which has just failed. Measurements of actual negative pore-water pressures just prior to failure or just

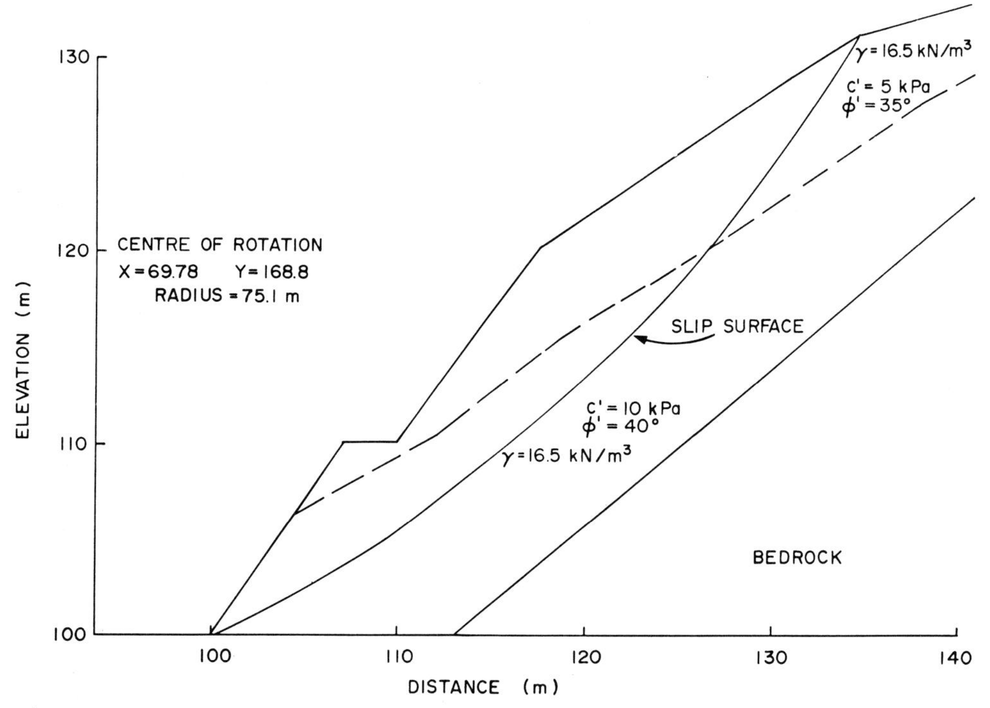

Figure 4.18 Example of a typical slope in Hong Kong (the material properties relate to two zones in granitic colluvium above the bedrock)

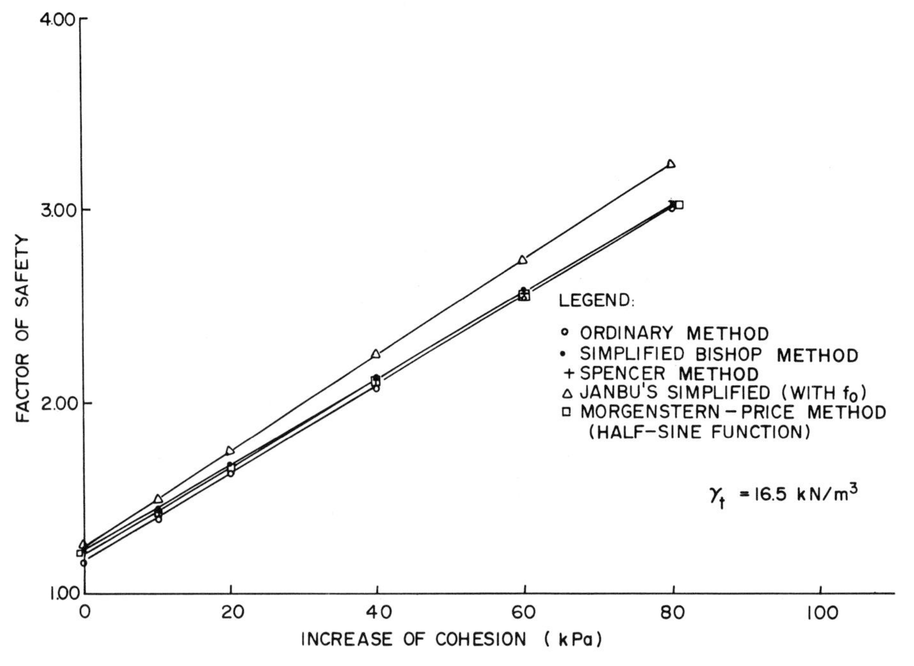

Figure 4.19 Increase in factor of safety for an increase in cohesion for the example slope

following the failure would be most relevant for analytical purposes (Fontura *et al.*, 1984).

2. When assessing the short-term stability of a cut slope, an ongoing monitoring of the negative pore-water pressures should provide the most valuable input into the analysis. The computed factor of safety must be maintained well in excess of 1.0 to ensure stability.
3. In the case of the assessment of the stability of natural or man-made slopes, the negative pore-water pressure profile to be used in an analysis is more difficult to ascertain (Lumb, 1975). Most natural soil deposits are desiccated near ground surface and contain cracks and fissures forming a secondary soil structure of varying depths. This provides easy access of water to the soil and can play a major role in rapidly decreasing the matric suction during periods of prolonged rainfall. Some soil deposits, although intact, may be highly porous and allow a rapid reduction in matric suction during rainy periods.

Other soil deposits are relatively intact and appear to maintain their negative pore-water profiles even during prolonged rainy periods (Sweeney, 1982). In these situations, the engineer must select the lower limit of the pore-water pressure profile for design purposes. In some situations this may be an unwise solution to the problem whereas in other situations it may be the only realistic solution (Widger and Fredlund, 1979). Certainly there is need for more detailed case histories in order to ensure a higher level of confidence in design.

In the early 1980s two sites were selected in Hong Kong where the relevant variables would be measured to study the influence of matric suction on the factor of safety (Sweeney and Robertson, 1979). The two sites were Fung Fai Terrace and Thorpe Manor (Ching *et al.*, 1984). In each case the soil stratigraphy was determined from numerous borings. Undisturbed soil samples were used to measure the shear strength parameters in the laboratory. Negative pore-water pressures were measured *in situ*. Stability analyses were then performed to assess the effect of soil suction. Parametric type analyses were also performed using varying percentages of the negative hydrostatic condition in the analysis. The negative pore-water pressures were converted to equivalent cohesion values using a ϕ^b angle of 15° (Figure 4.20).

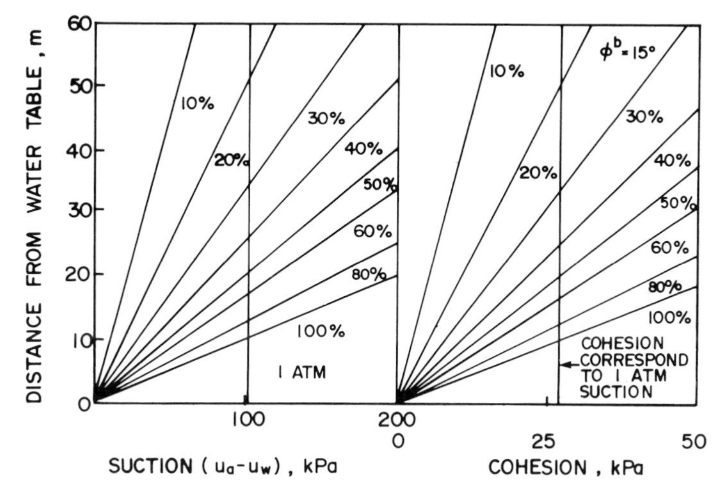

Figure 4.20 Equivalent increase in cohesion for various soil suction profiles

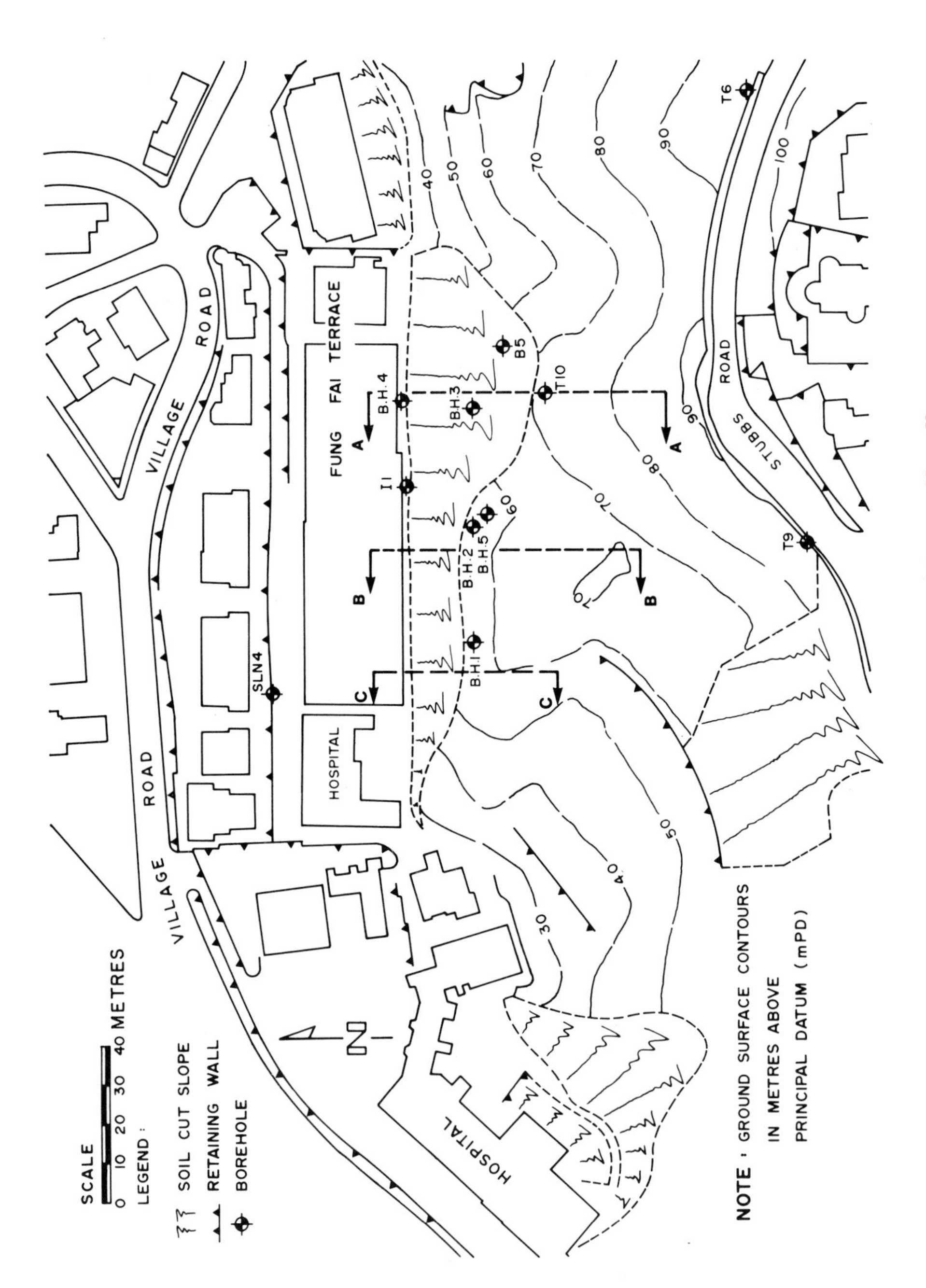

Figure 4.21 Site plan for Fung Fai Terrace, Hong Kong

4.4.1 Study Site 1: Fung Fai Terrace

Fung Fai Terrace is located in the north-central part of Hong Kong Island. The site consists of a row of residential buildings as shown in Figure 4.21. At the back of these buildings is a steep cut slope with an average inclination of 60 degrees to the horizontal and a maximum height of 35 metres. The cutting has been protected from infiltration of surface water by a layer of soil cement and lime plaster (i.e. locally referred to as chunam plaster; see Chapter 7) and has been in place for more than 40 years. Small but dangerous failures have occurred periodically at the crest of the cut slope and the low calculated factor of safety causes some concern. These circumstances prompted a detailed investigation.

Three cross-sections A–A, B–B and C–C are shown in Figures 4.22 to 4.24. The stratigraphy consists primarily of weathered granite. There is a layer of granitic colluvium, 4 to 5 metres thick, present at the top of the slope. Beneath the colluvium is a layer of completely to highly weathered granite of about 10 metres thickness. Bedrock is situated 20 to 30 metres below the surface. The water table is located well into the bedrock. It is estimated that the water table may rise by 5 and 8 metres under the influence of heavy rains with return periods of 10 and 1,000 years, respectively. The groundwater level does not directly affect the stability analyses.

Undisturbed core samples were tested to establish the pertinent strength parameters. Results are given in Table 4.5. The average ϕ^b angle for the soils was taken as 15 degrees (Ho and Fredlund, 1982b).

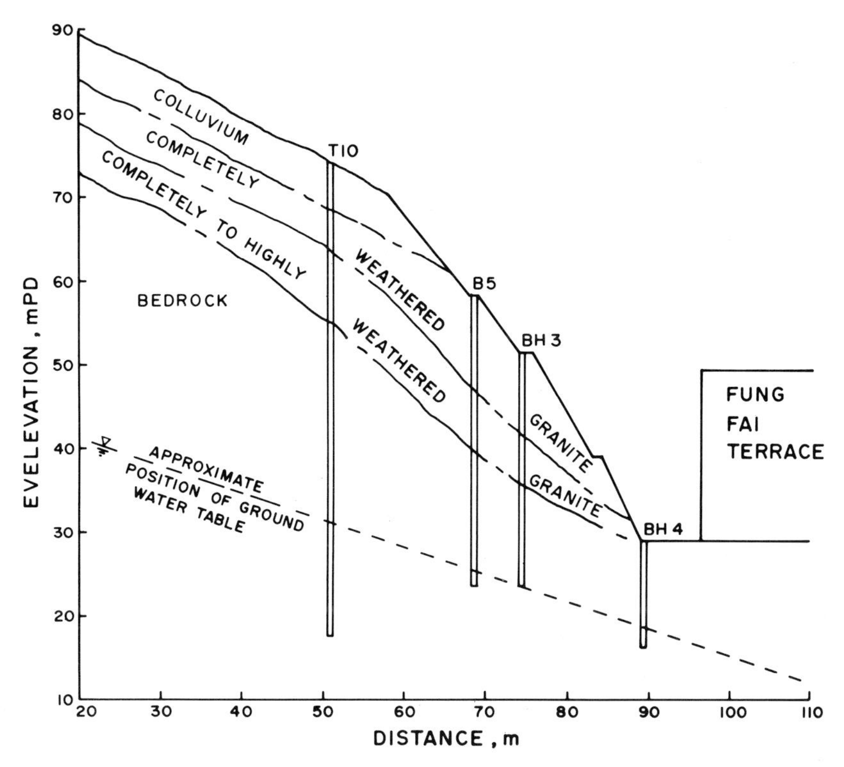

Figure 4.22 Section A–A for Fung Fai Terrace, Hong Kong

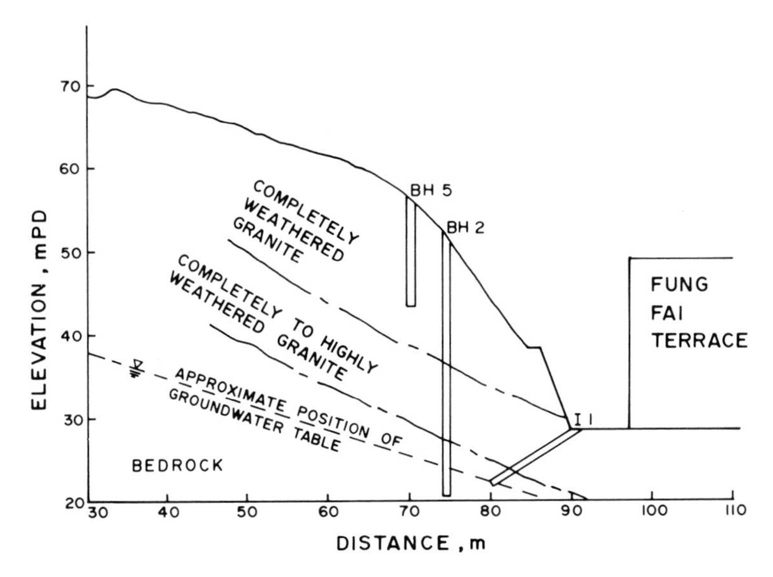

Figure 4.23 Section B–B for Fung Fai Terrace

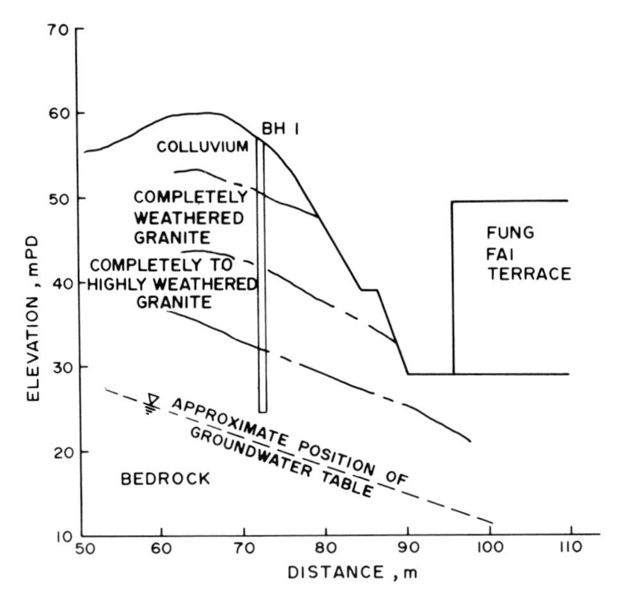

Figure 4.24 Section C–C for Fung Fai Terrace

Table 4.5 Strength Properties for Soils at Fung Fai Terrace

Soil Type	Unit Weight (kN/m^3)	c' (kPa)	ϕ' (degree)
Colluvium	19.6	10.0	35.0
Completely weathered granite	19.6	15.1	35.2
Completely to highly weathered granite	19.6	23.5	41.5

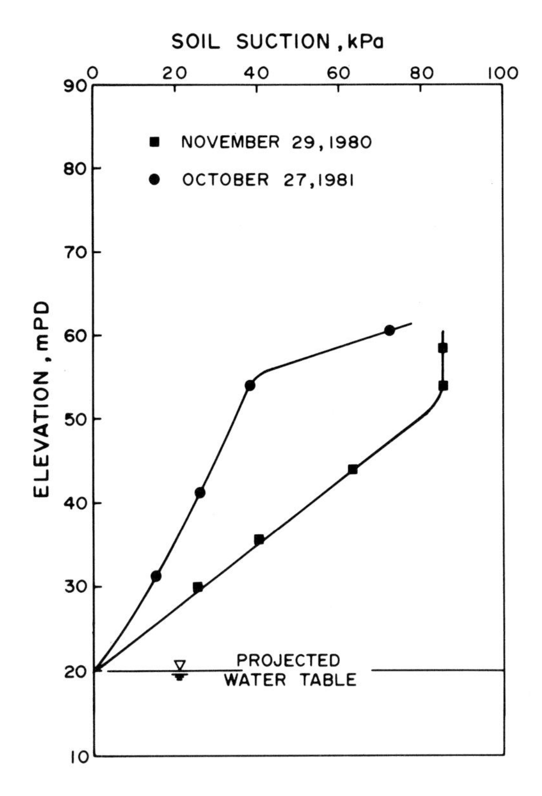

Figure 4.25 Suction measurements at Fung Fai Terrace

Soil suctions were measured at this site using a tensiometer inserted through small openings made into the face of the slope. Figure 4.25 shows two typical suction profiles obtained from near section A–A. The suctions varied considerably since the measurements were influenced by the proximity of the slope face. Suctions on the upper part of the profile could not be accurately measured because the capacity of the tensiometer was exceeded.

4.4.1.1 Stability Analysis at Fung Fai Terrace

Limit equilibrium stability analyses were performed on the three cross-sections shown in Figures 4.22 to 4.24. The assumption was made that the resultant interslice forces were horizontal. The computations were performed using the SLOPE-II computer program (Fredlund, 1981b). Circular surfaces were analysed to determine factor of safety. All critical surfaces passed through the toe of the slope. Table 4.6 summarizes the stability results without the effect of soil suction.

The most critical factor of safety is 0.86. The results indicate that the slope would be unstable if all conditions were representative. The fact that this slope has remained stable implies that the analysis is not completely representative of the field conditions. An additional strength is available, possibly due to soil suction.

Table 4.6 Stability Results for Fung Fai Terrace without the Effect of Suction

Section	Centre of Rotation* (metres) X-Coordinate	Y-Coordinate	Radius	Factor of Safety
A–A	232.5	190.0	216.0	0.864
B–B	143.8	120.0	89.5	0.910
C–C	171.6	118.1	120.8	0.881

*Critical centre of rotation.

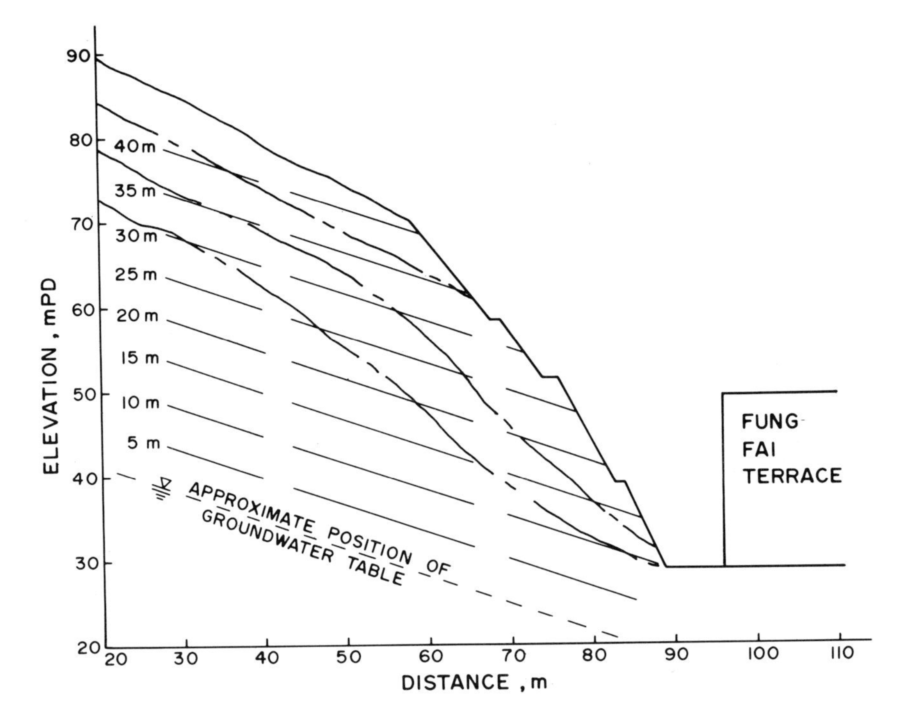

Figure 4.26 Subdivision of section A–A at Fung Fai Terrace for slope stability analysis

The cross-sections were re-analysed including the effect of soil suction. Each of the cross-sections was further divided into sub-strata drawn parallel to the water table in order to account for matric suction. Each sub-stratum was 5 metres thick. Figure 4.26 shows the sub-division for cross-section A–A. Each of the sub-strata was assumed to have a different total cohesion, c, as described by equation (4.2).

A parametric study was conducted to demonstrate changes in the factor of safety in response to variations in the matric suction. Suction profiles as shown in Figure 4.20 were assumed. Results for the parametric study are summarized in Table 4.7 and plotted in Figure 4.27. Figure 4.27 shows that a suction profile of 10 to 20% negative hydrostatic pressure is required to render a factor of safety of 1. The factor of safety for various sections is increased by 10 to 40% for matric suction profiles corresponding to 10 to 100% of negative

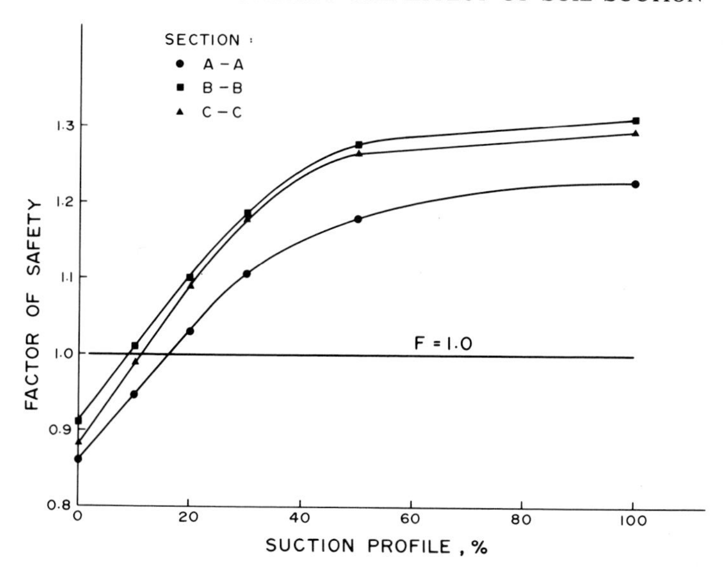

Figure 4.27 Results of a parametric slope stability study for Fung Fai Terrace

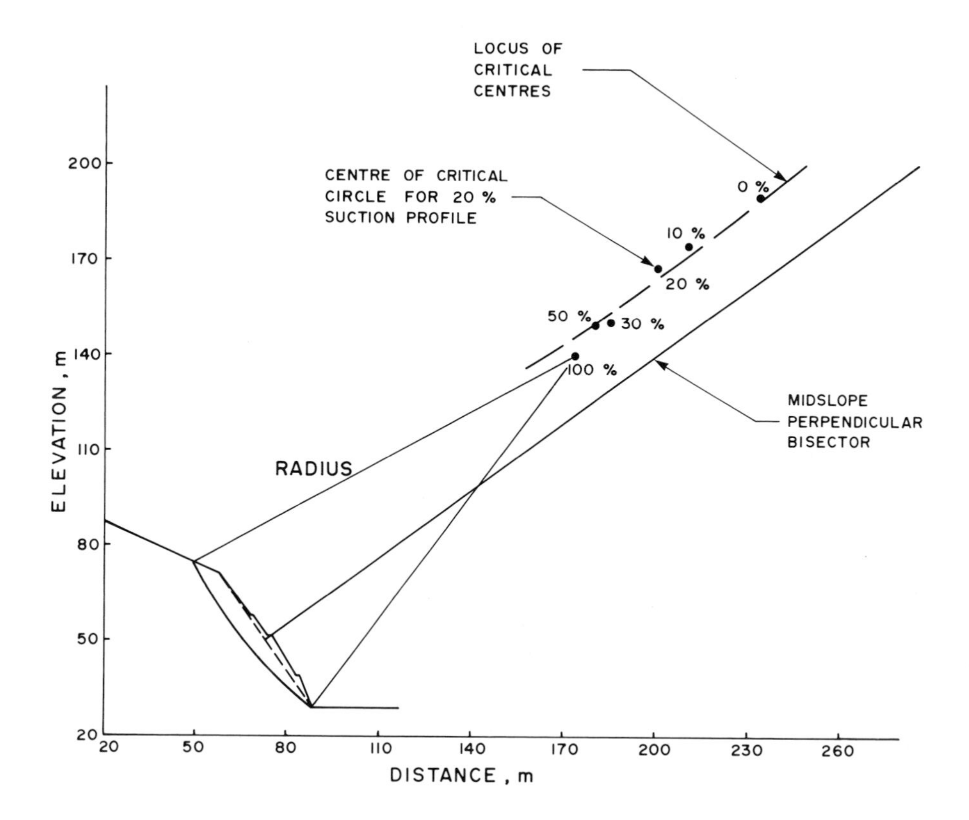

Figure 4.28 Critical centres for various suction profiles on section A–A

Table 4.7 Stability Results for Parametric Study for Fung Fai Terrace

A. Section A–A

Per cent of hydrostatic condtions	Centre of Rotation (m) X-Coordinate	Y-Coordinate	Radius	Factor of Safety
10	210.0	175.0	190.0	0.948
20	200.0	167.0	178.0	1.030
30	185.0	150.0	155.0	1.108
50	180.0	150.0	151.0	1.179
100	173.3	139.0	139.0	1.226

B. Section B–B

Per cent of hydrostatic conditions	Centre of Rotation (metres) X-Coordinate	Y-Coordinate	Radius	Factor of Safety
10	130.6	112.5	78.8	1.011
20	130.6	112.5	78.8	1.097
30	133.1	117.5	81.4	1.184
50	130.0	117.5	79.8	1.274
100	143.8	132.5	99.7	1.308

C. Section C–C

Per cent of hydrostatic conditions	Centre of Rotation (metres) X-Coordinate	Y-Coordinate	Radius	Factor of Safety
10	151.3	102.5	95.7	0.991
20	138.8	96.3	83.1	1.088
30	134.1	93.1	77.8	1.179
50	138.8	96.3	83.1	1.267
100	138.8	96.3	83.1	1.296

hydrostatic pressure, respectively. Figure 4.28 shows the variation in the position of the critical centre for section A–A. The critical slip surface tends to penetrate deeper into the slope as the cohesion increases.

Stability calculations were also performed using the actual matric suction values obtained from the field (Figure 4.25). The average increase in cohesion for each soil sub-stratum was calculated from the actual matric suction profile up to a maximum suction value of one atmosphere (Figure 4.25). The results are presented in Table 4.8. The overall factor of safety is approximately 1.10 based on the suction profile measured on 29 November 1980 whereas it is about 1.01 based on the suction profile measured on 27 October 1981.

4.4.2 Study Site 2: Thorpe Manor

Thorpe Manor is a site located in the Mid Levels district of Hong Kong Island. It has been proposed for a high-rise residential building. An unusually steep and high cut slope exists below the site, accommodating an existing residential building. This led to a detailed

Table 4.8 Stability Results with the Effect of Suction for Fung Fai Terrace

A. Suction Profile (29 November 1980)

Section	Centre of Rotation (metres) X-Coordinate	Y-Coordinate	Radius	Factor of Safety
A–A	176.3	141.9	143.0	1.072
B–B	133.1	117.5	81.4	1.143
C–C	138.8	96.3	83.1	1.132

B. Suction Profile (27 October 1981)

Section	Centre of Rotation (metres) X-Coordinate	Y-Coordinate	Radius	Factor of Safety
A–A	201.3	167.5	178.6	0.984
B–B	165.0	125.0	122.2	1.046
C–C	156.9	108.8	104.1	1.014

investigation to access the long-term stability of the site taking into account the imposed loads from the new building and induced changes in surface and subsurface drainage.

Figure 4.29 shows the site plan of Thorpe Manor, which is topographically situated at the front of a spur protruding from the main hillside. The cut slope under consideration is below a major access road and its critical cross-section, A–A, is shown in Figure 4.30. The slope is inclined at 60 degrees to the horizontal and has an average height of 30 metres. The stratigraphy consists entirely of weathered rhyolite. The surficial material is a completely weathered rhyolite of 5 to 10 metres in thickness. The second stratum is a layer of completely to highly weathered rhyolite varying from 5 to 10 metres in thickness. Underlying is another layer of slightly weathered rhyolite. Bedrock is located approximately 20 to 30 metres below the surface. The water table lies well below the ground surface. It is estimated that the water table will rise less than 5 to 8 metres under the influence of heavy rain with return periods of 10 and 1,000 years, respectively. Therefore, the water table does not directly influence the stability analysis.

Undisturbed core samples were tested to obtain the saturated shear strength parameters. The ϕ^b angle for the soils was independently evaluated. Table 4.9 gives a summary for the soil properties. *In situ* soil suction measurements were made from an exploratory caisson shaft installed near the cut slope (Figure 4.29). Suction profiles obtained during the rainy season of 1980 are plotted in Figure 4.31. These profiles are relatively uniform, except for variations which occurred as a result of infiltration and fluctuation in the position of the water table.

4.4.2.1 Stability Analysis for Thorpe Manor

Stability analyses using circular slip surfaces were performed on section A–A. The computations were first made assuming saturated conditions. A parametric study was included in order to evaluate the effect of the changes in the soil suction profile and the water table on the computed factor of safety. Various suction profiles were

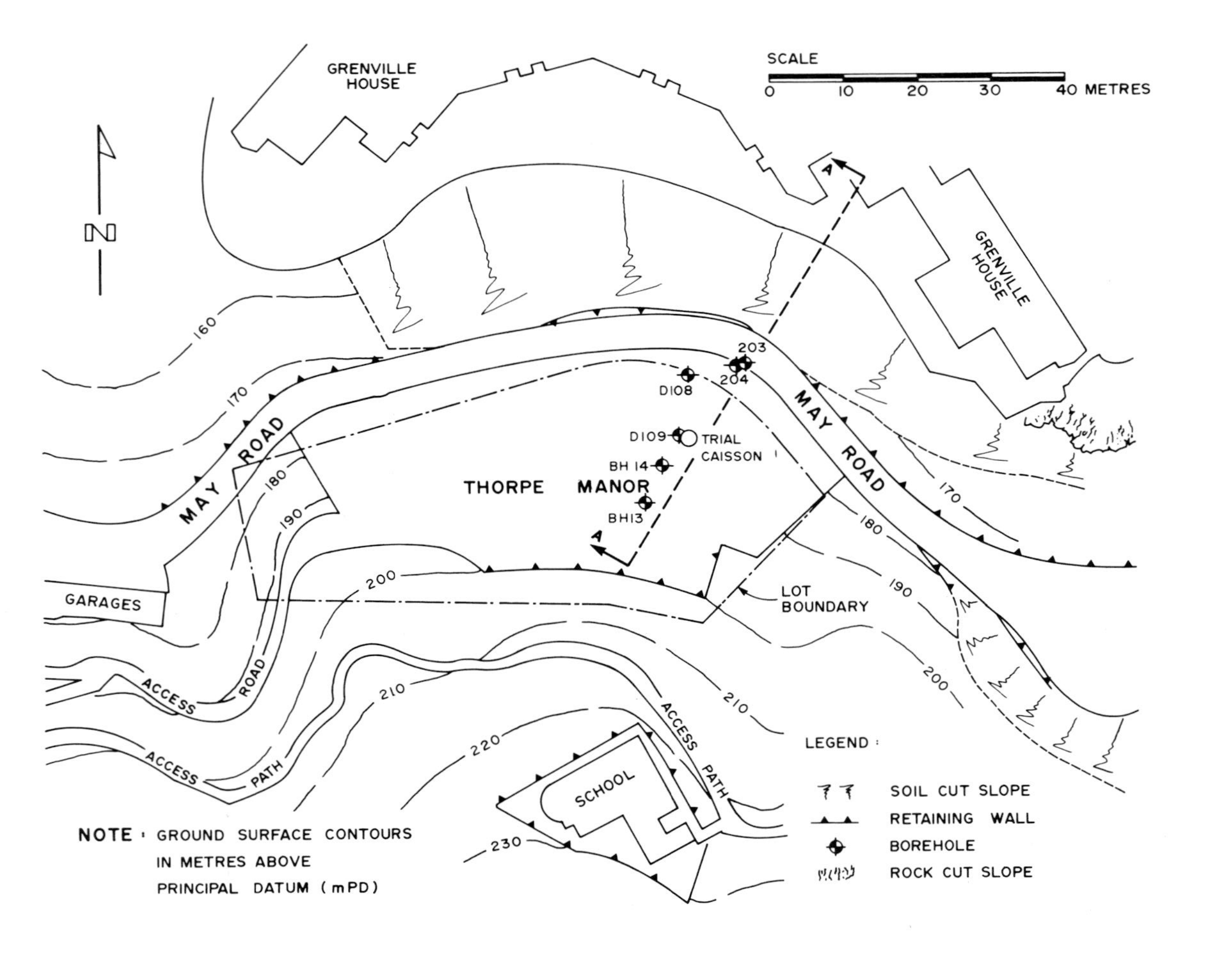

Figure 4.29 Site plan for Thorpe Manor, Hong Kong

Table 4.9 Properties for Soils at Thorpe Manor

Soil Type	Unit Weight (kN/m^3)	c' (kPa)	ϕ' (degree)	ϕ^b (degree)
Completely weathered rhyolite	18.4	10.1	42.6	12.0
Completely to highly weathered rhyolite	21.4	12.0	43.9	12.0

Table 4.10 Stability Results for Thorpe Manor

A. Approximate Water Table

Suction Profile (%)	Centre of Rotation (metres) X-Coordinate	Y-Coordinate	Radius	Factor of Safety
0	148.8	205.0	76.1	1.046
10	141.3	202.5	69.6	1.114
20	139.7	202.5	68.6	1.181
30	138.1	202.5	67.7	1.242
50	135.0	202.5	66.0	1.342
100	135.0	202.5	66.0	1.428
Actual*	126.9	192.5	53.3	1.254

*Suction profile of 2 September 1980

B. Water Table Corresponding to 1:10 Year Rain

Suction Profile (%)	Centre of Rotation (metres) X-Coordinate	Y-Coordinate	Radius	Factor of Safety
10	145.0	205.0	73.8	1.091
20	141.3	202.5	69.6	1.139
30	150.0	212.5	82.8	1.191
50	141.9	207.5	74.0	1.270
100	136.9	202.5	67.1	1.370

C. Water Table Corresponding to 1:1,000 Year Rain

Suction Profile (%)	Centre of Rotation (metres) X-Coordinate	Y-Coordinate	Radius	Factor of Safety
10	145.0	205.0	73.8	1.078
20	141.3	202.5	69.6	1.114
30	160.0	220.0	94.9	1.159
50	141.9	207.5	74.0	1.216
100	148.1	212.5	81.7	1.320

assumed. Water tables corresponding to heavy rains with return periods of 10 and 1,000 years, respectively, were used. Results from the stability analyses are summarized in Table 4.10.

The critical factor of safety for the cut slope without the effect of soil suction is approximately 1.05, suggesting that this slope is in a nearly unstable condition although

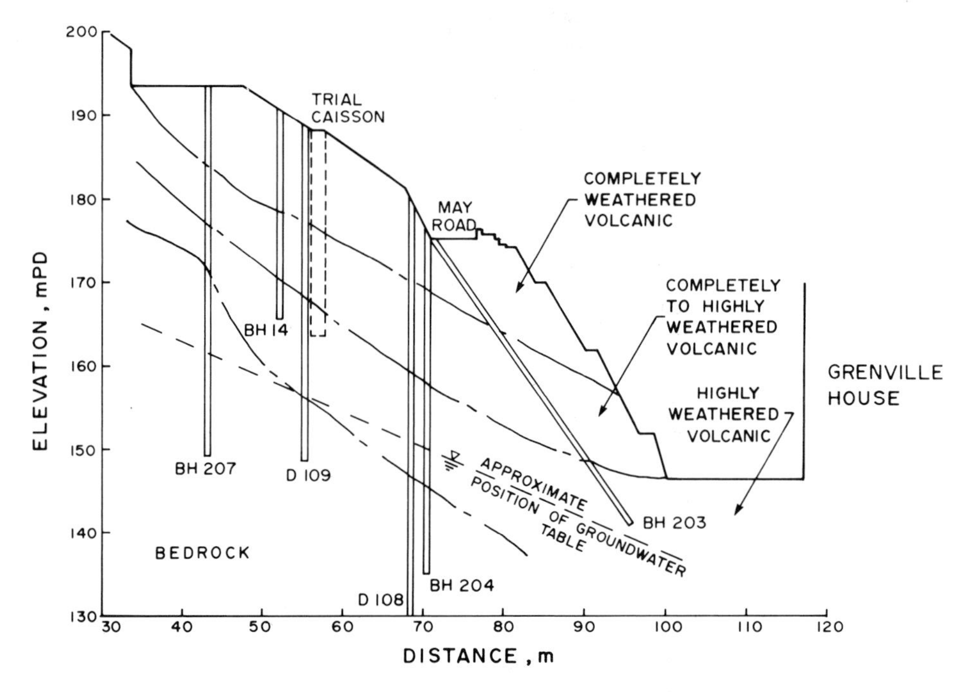

Figure 4.30 Section A-A for Thorpe Manor, Hong Kong

no distress is observed. Its computed factor of safety is increased to 1.25 when including the actual soil suctions. In other words, matric suction contributes approximately 20% towards an increased factor of safety.

Results from the parametric study show that factor of safety computations are sensitive to changes in the suction profile but less sensitive to the position of the water table. The computed critical factor of safety is 1.43 when using a matric suction profile equivalent to 100% of negative hydrostatic pressure.

4.5 CONCLUSIONS

During the past few years, a much clearer appreciation has emerged regarding the influence of soil suction on the stability of slopes. The shear strength equation for unsaturated soils has gained wide acceptance, and testing procedures have been proposed for measuring the shear strength parameters for unsaturated soils. The new strength parameter required is the angle, ϕ^b. This angle appears to be commonly of the order of 15 degrees; however, further testing and research are required for a better understanding of this soil parameter. The theoretical formulations for conventional limit equilibrium methods of slices have been extended to embrace unsaturated soils. These equations can readily be solved using computers. The greatest need relates to the measurement of *in situ* negative pore-water pressures. More studies are needed of the changes in pore-water pressure from season to season and further case studies are needed of

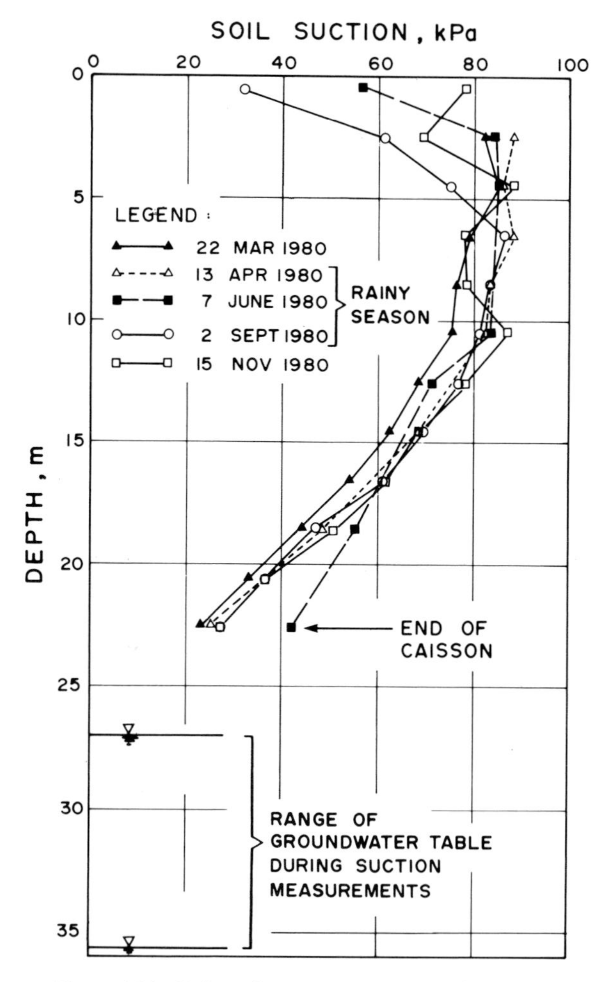

Figure 4.31 Soil suction measurements at Thorpe Manor

stability problems in unsaturated soils to promote confidence in the analysis outlined in this chapter.

REFERENCES

Bishop, A. W., Alpan, I., Blight, G. E., and Donald, I. B. (1960). 'Factors controlling the shear strength of partly saturated cohesive soils'. ASCE Research Conference on Shear Strength of Cohesive Soils, University of Colorado, Boulder, Colorado, pp. 503–32.

Bishop, A. W., and Henkel, D. J. (1962). *The Measurement of Soil Properties in the Triaxial Test* (2nd edn). Edward Arnold, London.

Bocking, K. and Fredlund, D. G. (1980). 'Limitations of the axis-translation technique. *Proc. 4th Conf. Expansive Soils*, Denver, Colorado, pp. 117–35.

Ching, R. K. H., Sweeney, D. J., and Fredlund, D. G. (1984). 'Increase in factor of safety due to soil suction for two Hong Kong slopes.' *Proc. 4th Int. Symp. Landslides*, Toronto, Canada, **1**, 617–24.

Escario, V. (1980). 'Suction controlled penetration and shear tests.' *Proc. 4th Int. Conf. Expansive Soils*, Denver, Colorado, **II**, 781–8.

Fan, K. (1983). *'Evaluation of interslice side forces for lateral earth force and slope stability problems'*. M.Sc. Thesis, University of Saskatchewan, Saskatoon, Canada.

Fontura, S. A. B., de Campos, L. E. P., and Costa Filno, L. M. (1984). 'A reanalysis of some slides in gneissic residual soils.' *Proc. 4th Int. Symp. Landslides*, Toronto, Canada, **1**, 625–30.

Fredlund, D. G. (1979). 'Appropriate concepts and technology for unsaturated soils.' 2nd Canadian Geotechnical Colloquium. *Can. Geotech. J.*, **16**, 1, 121–39.

Fredlund, D. G. (1981a). 'The shear strength of unsaturated soils and its relationship to slope stability problems in Hong Kong.' *Hong Kong Engineer, J. Hong Kong Inst. Engrs.*, April, 37–45.

Fredlund, D. G. (1981b). *'SLOPE-II Computer Program', User's Manual S-10.* Geo-Slope Programming Ltd, Calgary, Canada.

Fredlund, D. G. (1984). 'Analytical methods for slope stability analysis', State-of-the-Art. *Proc. 4th Int. Symp. Landslides*, Toronto, Canada, **1**, 229–50.

Fredlund, D. G., and Morgenstern, N. R. (1977). 'Stress state variables for unsaturated soils'. *ASCE, J. Geotech. Eng. Div.*, **103** (GT5), 447–466.

Fredlund, D. G., Krahn, J., and Pufahl, D. E. (1981). 'The relationship between limit equilibrium slope stability methods'. *Proc. 10th Int. Conf. Soil Mech. Fndtn. Eng.*, Stockholm. **3**, 409–16.

Fredlund, D. G., Morgenstern, N. R., and Widger, R. A. (1978). 'Shear strength of unsaturated soils.' *Can. Geotech. J.*, **15**, 313–21.

Hilf, J. (1948). 'Estimating construction pore pressures in rolled earth dams.' *Proc. 2nd Int. Conf. Soil Mech. Fndtn. Eng.*, Rotterdam, **III**, 234–40.

Ho, D. Y. F., and Fredlund, D. G. (1982a). 'A multistage triaxial test for unsaturated soils.' *Geotech. Test. J.*, ASTM, June, 18–25.

Ho, D. Y. F., and Fredlund, D. G. (1982b). 'Increase in strength due to suction for two Hong Kong soils.' *Proc. Conf. Eng. Const. Tropical Residual Soils*, ASCE, Honolulu, Hawaii, 263–95.

Ho, D. Y. F., and Fredlund, D. G. (1982c). 'Strain rates for unsaturated soil shear strength testing.' *Proc. 7th SE Asian Geotech. Conf.*, Nov., 787–803.

Lumb, P. (1975). 'Slope failures in Hong Kong'. *Q. J. Eng. Geol.*, **8**, 31–65.

Morgenstern, N. R., and Price, V. E., (1965). 'The analysis of the stability of general slip surfaces.' *Geotechnique*, **15**, no. 1, 79–93.

Satija, D. J. (1978). *Shear behaviour of partially saturated soils*, Unpublished Ph.D. thesis, Indian Institute of Technology, Delhi, India.

Sweeney, D. J. (1982). 'Some *in situ* soil suction measurements in Hong Kong's residual soil slopes.' *Proc. 7th SE Asian Regional Conf.* **1**, 91–106.

Sweeney, D., and Robertson, P. (1979). 'A fundamental approach to slope stability in Hong Kong.' *Hong Kong Engineer*. **7**, no. 10, 35–44.

Widger, R. A., and Fredlund, D. G. (1979). 'Stability of swelling clay embankments.' *Can. Geotech. J.*, no. 16, 140–51.

Wilson, G. W., and Fredlund, D. G. (1983). 'The evaluation of the interslice side forces for slope stability analysis by the finite element method.' *9th Can. Congr. Appl. Mechan.*, Saskatoon, Canada, 37 pp.

Slope Stability
Edited by M. G. Anderson and K. S. Richards

Chapter 5

The Implications of Joints and Structures for Slope Stability

S. R. HENCHER
Department of Earth Sciences
University of Leeds, Leeds LS2 9JT

5.1 INTRODUCTION

Rocks and soils typically contain many structural weaknesses which significantly reduce the shear strength of the mass below that of the intact material. Such discontinuities also have a controlling influence on the movement of groundwater through the mass. Hoek (1971) and other authors have pointed out that were it not for the presence of such discontinuities, even weak rock could stand vertically to heights of several hundred metres. Discontinuities result in markedly anisotropic engineering properties as in the case illustrated in Figure 5.1 where sliding has occurred along a planar joint through otherwise strong rock. The significance of discontinuities is sometimes less readily apparent as in the case reported by Douglas and Voight (1969) who noted a preferred orientation of microfractures through granite resulting in a directional variation in compressive strength. Where joints or structures in the rock mass are adversely orientated, site investigation for assessing slope stability must be aimed at identifying those blocks of rock that might move, taking into consideration the existing or proposed geometry of the slope. Methods of analysis must be capable of dealing with the sliding of such blocks possibly on several surfaces at once. The main difference between such 'rock' slope analytical methods and the generalized 'soil' methods discussed in chapter 2 is that for the former the geometry of failure is controlled by pre-existing planes of weakness. Generalized methods involve the search for a critical failure surface through material that is taken to be isotropic in strength. The labelling of methods as 'rock' and 'soil' is not entirely satisfactory because soil masses may contain adverse joints (particularly soils formed through weathering of rock) and will require 'rock' methods of investigation and analysis (Deere and Patton, 1971; Koo, 1982) and conversely, closely jointed rock through which irregular failure surfaces may develop must be analysed using generalized 'soil' methods (Hoek and Bray, 1981).

Figure 5.1 Single adverse joint controlling stability through strong rock

This chapter is arranged in five sections. The first section deals with the nature and origin of discontinuities in soil and rock masses and the second with methods for investigating and describing the geotechnically significant features of discontinuities. Shear strength of individual discontinuities and of closely jointed rock or soil masses in terms of effective stress is covered in the third section and methods for determining the hydrogeological conditions in the fourth. The final section discusses the analysis of the 'geotechnical model' determined by the methods described in the earlier sections.

5.2 ORIGIN AND CLASSIFICATION OF DISCONTINUITIES

An investigation into the extent, orientation, and distribution of discontinuities for a particular slope will only be truly effective when the geological nature of the structures is taken into account. Recognition of discontinuity type will almost certainly allow properties to be predicted and extrapolated with more confidence than could otherwise be justified. A few authors, notably Deere and Patton (1971) and especially Piteau (1970, 1973), have emphasized the importance of geology in extrapolating and assessing the importance of particular discontinuities. Piteau (1973) in presenting a well-documented case where a careful appraisal of jointing allowed patterns to be predicted in the rock mass away from the sampling point comments 'the analyst must assess whether the location at which the joint

Table 5.1 Geotechnical Classification of Discontinuities Common to all Rock and Soil Types

Discontinuity Type	Physical Characteristics	Geotechnical Aspects	Comments
Tectonic joints	Persistent fractures resulting from tectonic stresses. Joints often occur as related groups or 'sets'. Joint systems of conjugate sets may be explained in terms of regional stress fields.	Tectonic joints are classified as 'shear' or 'tensile' according to probable origin. Shear joints are often less rough than tensile joints. Joints may die out laterally resulting in impersistence and high strength.	May only be extrapolated confidently where systematic and where the geological origin is understood.
Faults	Fractures along which displacement has occurred. Any scale from millimetres to hundreds of kilometres. Often associated with zones of sheared rock.	Often low shear strength particularly where slickensided or containing gouge. May be associated with high groundwater flow or act as barriers to flow. Deep zones of weathering may occur along faults. Recent faults may be seismically active.	Mappable, especially where rocks either side can be matched. Major faults often recognized as photo lineations due to localized erosion.
Sheeting joints	Rough, often widely spaced fractures; parallel to the ground surface; formed under tension as a result of unloading.	May be persistent over tens of metres. Commonly adverse (parallel to slopes). Weathering concentrated along them in otherwise good quality rock.	Readily identified due to individuality and relationship with topography.
Lithological boundaries	Boundaries between different rock types. May be of any angle, shape, and complexity according to geological history.	Often mark distinct changes in engineering properties such as strength, permeability and degree and style of jointing. Commonly form barriers to groundwater flow.	Mappable allowing interpolation and extrapolation providing the geological history is understood.

Table 5.2 Geotechnical Classification of Discontinuities Characteristic of Particular Rock and Soil Types

Rock or Soil type	Discontinuity type	Physical characteristics	Geotechnical aspects	Comments
Sedimentary	Bedding planes/ bedding plane joints	Parallel to original deposition surface and marking a hiatus in deposition. Usually almost horizontal in unfolded rocks.	Often flat and persistent over tens or hundreds of metres. May mark changes in lithology, strength and permeability. Commonly close, tight, and with considerable cohesion. May become open due to weathering and unloading.	Geologically mappable and therefore, may be extrapolated providing structure understood. Other sedimentary features such as ripple marks and mud-cracks may aid interpretation and affect shear strength.
	Shaley cleavage	Close parallel discontinuities formed in mudstones during diagenesis and resulting in fissility.		
	Random fissures	Common in recent sediments probably due to shrinkage and minor shearing during consolidation. Not extensive but important mass feature.	Controlling influence for strength and permeability for many clays.	Best described in terms of frequency.
Igneous	Cooling joints	Systematic sets of hexagonal joints perpendicular to cooling surfaces are common in lavas and sills. Larger intrusions typified by doming joints and cross joints.	Columnar joints have regular pattern so easily dealt with. Other joints often widely spaced with variable orientation and nature.	Either entirely predictable or fairly random.
Metamorphic	Slaty cleavage	Closely spaced, parallel and persistent planar integral discontinuities in fine grained strong rock.	High cohesion where intact but readily opened due to weathering or unloading. Low roughness.	Formed by regional stresses and therefore mappable over wide areas.
	Schistosity	Crenulate or wavy foliation with parallel alignment of minerals in coarser grained rocks.	Often foliations coated with minerals such as talc, and chlorite giving a low shear strength.	Less mappable than slaty cleavage but general trends recognizable.

data are collected has been subjected to the same geological history of deformation as the location where extrapolation is to be made. If their histories are found to differ, extrapolation is not valid.' Too often the approach to the collection and processing of data is almost purely statistical with only scant regard to the geological history of the site. The emphasis in the rock mechanics literature, certainly in recent years, has been much more concerned with such aspects as statistical sampling errors and idealized joint shapes than with the true characteristics and properties of the various types of discontinuity.

For geotechnical purposes, a discontinuity may be defined as a boundary or break within the soil or rock mass which marks a change in engineering characteristics or which itself results in a marked change in the mass properties. This definition clearly includes such features as lithological boundaries, faults, bedding planes, and tectonic joints but also can be stretched to include microstructures such as a preferred orientation of microfractures. In all these cases the orientation and extent of the discontinuity or set of discontinuities could be measured, at least in theory, and would lead to anisotropic behaviour of the rock or soil mass when loaded. Tables 5.1 and 5.2 provide a simple classification of the most common types of discontinuity and list their typical characteristics and geotechnical significance. Table 5.1 deals with discontinuities which are common to all rock types while Table 5.2 lists discontinuities which are restricted to rocks of particular types. It should be noted that the terms discontinuity and joint are often used synonymously, joints being the most common type of discontinuity.

5.2.1 Tectonic Joints

Tectonic joints which are formed as the result of orogenic stress in the earth's crust are common to all rock types and may even be found in recent sediments (Burford and Dixon, 1978). They often occur in distinct 'sets', a term which is sometimes defined for rock mechanics purposes as a series of parallel joints (Herget, 1977) such as those illustrated in Figure 5.2. Structural geologists may also use the term 'set' to describe a group of joints of common origin which need not be parallel to one another (Hobbs *et al.*, 1976). The geometrical relationship between sets may sometimes be interpreted with respect to a regional stress pattern or a local geological structure such as a fault or a fold (see Price, 1966). Other joints may be non-systematic, often forming cross-joints between the systematic joints. In some cases well defined sets of joints are recognized that are apparently independent of tectonic stress fields (de Sitter, 1964; Fookes and Denness, 1969). Similarly, joints which have the appearance of tectonic joints are found in rocks which have been neither folded nor faulted. De Sitter (1964) comments that in such cases the joints themselves are indicative of a certain level of tectonic activity. Price (1959, 1966) attributes all tectonic jointing to the retention and subsequent release of strain energy in the rock.

Where the geometrical pattern of joints may be explained by reference to overall geological structure, then extrapolation and interpolation of joint data may be carried out with reasonable confidence. In many cases, however, it is very difficult to relate the joint pattern to a known cause and then a more cautious approach must be adopted. As noted in Table 5.1, joints formed as the result of shear stresses are commonly less rough than joints formed under tension and might therefore be expected to exhibit lower shear strengths.

Figure 5.2 Parallel set of tectonic joints

Figure 5.3 Faults through bedded sedimentary rocks

5.2.2 Faults

Faults such as those illustrated in Figure 5.3 are fractures along which displacement has taken place. They may occur singly or in groups forming shear zones. Faults are often discordant to other structures, particularly lithological boundaries and bedding, and are much rarer discontinuities than are joints. The distinction between joints and faults is strictly defined by some authors (see Fecker, 1978, for example), but in practice it is less easy to separate them. For example, discontinuities which exhibit slickensides, clearly indicating that displacement has taken place, are often regarded as joints rather than faults where they occur as sets throughout the mass rather than as isolated shear planes. Slickensides on such joints are probably indicative of minor internal movements within the mass (Skempton *et al.*, 1969). Similarly, where a zone of intense and discordant jointing is encountered, the term shear zone or fault might be employed even though there is no clear evidence of displacement along the zone.

Faults often cause geotechnical problems, not only in slopes but also in foundations and especially tunnels due to their association with shattered and sheared rocks and the fact that they often have relatively high permeabilities and may carry a lot of groundwater (see discussion by Sharp *et al.*, 1972). Weathering is often concentrated along faults and the rock may show signs of hydrothermal alteration. Earthquakes are associated with recent faulting and if it is suspected that faults may be seismically active, this may be checked by instrumentation, literature review or field observation (see Clark *et al.*, 1972; Sherard *et al.*, 1975; Ben Menahen, 1976; Donovan, 1978, for example). Observations of offset rivers, truncated spurs, and other geomorphological features may all indicate that faults are active. Faults are recognized by many features, particularly the displacement of recognizable beds and by crushed zones of rock (fault breccia and mylonite). Major faults are often preferentially eroded and trends can be recognized from air photographs.

5.2.3 Lithological Boundaries

Geological boundaries between different soils and rocks often mark sharp changes in engineering properties and are, therefore, significant for stability analysis. This is reflected by the often close correspondence between geomorphology and underlying rock type, a fact which geologists exploit to interpolate boundaries between exposures. Geological boundaries, as distinct from many other discontinuities, are readily identified in drill core and can be interpolated with reasonable confidence. The same cannot be said for weathering zones through the rock which may vary laterally in an unpredictable manner. Dykes or sills commonly form barriers to groundwater as in the case illustrated in Figure 5.4, where a perched water table developed above a dolerite dyke through decomposed granite and caused a landslide (Hencher and Martin, 1984). Where the lithological boundary marks a change from 'soil' to 'rock', then different methods of analysis will be required for the two component materials of the slope.

5.2.4 Sheeting Joints

Sheeting joints in hard rocks are typically rough and extensive, running parallel to the topography of the present-day, or that of the recent geological past. An example of topography controlled by extensive sheet jointing is given in Figure 5.5. There is some debate (Twidale,

Figure 5.4 Landslide caused by perching of water above dyke

Figure 5.5 Sheeting joints through granite

Figure 5.6 Parallel microfractures through weathered granite

1973) as to which came first, the jointing pattern or the geomorphological form. The majority of workers, however, attribute such joints to tensile stresses following unloading due to erosion or perhaps the removal of ice load. A discussion of the mechanism envisaged is given by Nichols (1980).

Sheeting joints occur in all lithologies and have been reported in quite recent sediments (Fookes, 1965; Skempton *et al.*, 1969). Characteristically they decrease in frequency with depth below ground surface and may cross geological boundaries. Rather surprisingly, they may be found in rocks that are otherwise quite highly jointed. It might have been expected in such cases that tensile loads could have been accommodated by movement along existing joints without need for further fracturing of the intact rock. Such occurrences might indicate that the 'tectonic' joints only fully developed *after* the sheeting joints as a result of weathering and stress relief.

A common feature in weathered rocks in Hong Kong is a mass fabric of microfracturing parallel to topography as illustrated in Figure 5.6, and this has similarly been attributed to unloading (Hencher and Martin, 1982). Examples are known of failures occurring along such orientated microfractures in the same way as along other discontinuities.

5.2.5 Bedding Planes and Bedding Plane Joints

Bedding planes in sediments mark either a change in sediment type or a hiatus of deposition. While imposing on the rock or soil mass a marked anisotropy, bedding planes are often closed and retain a strong cohesion. Because of unloading or weathering, however, they may open up to form bedding plane joints. In drill core it is sometimes difficult to tell whether the degree of jointing observed is representative of the rock in the ground or due to drilling disturbance. Most bedding planes are fairly flat and smooth but sedimentary features such as ripple marks or load casts associated with bedding surfaces may lend surfaces a rough texture. Planes commonly extend over wide distances although beds may pinch out due to the coalescence of different planes.

5.2.6 Fissures

Fissures are a common feature of unlithified clays and silts, are generally of limited length, and often appear essentially random in orientation although some authors have managed to relate them to specific stress fields (Skempton *et al.*, 1969; Fookes and Denness, 1969; McGown *et al.*, 1974). They are generally thought to develop during diagenesis and are best described as a mass feature of the soil. The presence of such fissures has been shown to influence the strength and permeability of soils considerably (Skempton and La Rochelle, 1965).

5.2.7 Cooling Joints

In igneous rocks, joints form as the magma cools. The most striking examples are the columnar joints formed perpendicular to the cooling surface in lavas (see Figure 5.7). The regular pattern of such jointing allows them to be readily accounted for in stability analysis.

Figure 5.7 Columnar jointing through rhyolite

Cooling joints in larger intrusions often form a doming pattern together with cross-joints (see discussion in Price, 1966; Gamon and Finn, 1984). Layering of igneous rocks due to density segregation of minerals on cooling can also be found as well as flow banding and such features will result in anisotropic engineering behaviour.

5.2.8 Metamorphic Fabrics

Metamorphic rocks formed under pressure commonly contain well-defined sets of discontinuities. Slaty cleavage, a close pattern of parallel planar discontinuities, is imposed on rocks by regional stresses and results in markedly anisotropic properties (Brown *et al.*, 1977). Phyllites and schists result from higher grades of metamorphism and typically exhibit foliation, the surfaces of which are often coated with minerals such as chlorite, talc and mica which may have low frictional properties. Such foliation is generally more wavy than slaty cleavage and this will contribute to shear strength. Metamorphic fabrics which owe their origins to regional stresses are often mappable over wide areas.

5.3 INVESTIGATION

5.3.1 Introduction

Stability analysis of slopes requires a knowledge of the distribution, geometry, and engineering properties of the discontinuities in the mass. The quantity of data collected, the methods employed, the quality of information obtained and its eventual usefulness will depend on many factors, notably the nature and seriousness of the problem, the accessibility of exposure, the time and cost justifiable for the task, and the experience and local knowledge of the investigator.

There is general agreement (ISRM, 1978; Herget, 1977) regarding the scope of description required to characterize the nature of discontinuities and the main attributes are listed below and illustrated schematically in Figure 5.8.

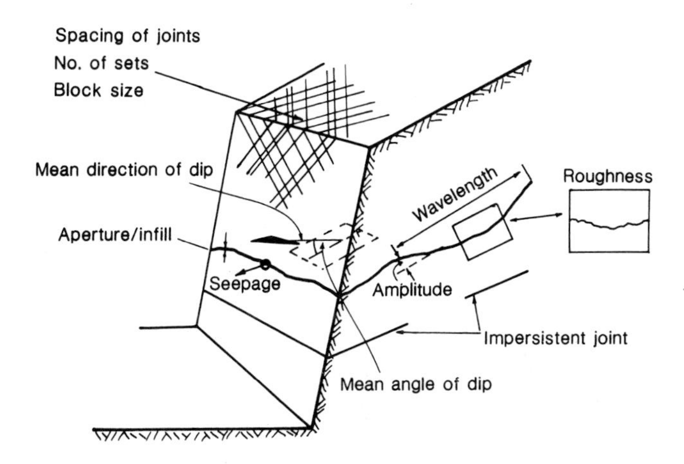

Figure 5.8 Description for the characterization of discontinuities

- Orientation (mean dip and dip direction).
- Spacing
- Persistence
- Roughness and Waviness
- Wall Strength
- Aperture
- Filling
- Seepage
- Number of sets
- Block size

Three main methods are used for obtaining data and will be discussed here, viz:

1. Photographic interpretation and measurement.
2. Mapping exposed faces.
3. Drilling.

Methods for assessing or measuring shear strength and water pressures will be discussed in later sections.

5.3.2 Photographic Methods

Photographic methods are particularly important for collecting data on the stability of natural hillsides but can also provide useful information for more detailed local studies. Trends of major joint sets and faults can often be identified from aerial photographs (Norman and Huntingdon, 1974) and angles of dip may sometimes be measured or estimated using stereographic pairs of photographs or by considering the traces of lineaments crossing variable topography. Allum (1966) points out that the significance of major faults or joints is often more readily appreciated from aerial photographs than in the field. Terrestial photogrammetry may be used for tracing joints in exposures and excavations (Moore, 1974; ISRM, 1978) and good quality photographs are commonly used as a basis for locating discontinuities and recording data in the field (see Starr *et al.*, 1981, for example). Useful though remote methods are for providing data at a relatively low cost particularly where access is difficult, detailed slope stability analysis requires a knowledge of characteristics such as roughness and degree of weathering of particular joints which can only be obtained by inspection. Furthermore, the usefulness of photographic methods is limited by the amount of exposure.

5.3.3 Mapping Exposed Surfaces

Provided that suitable exposures are available, mapping and inspection are the best ways for assessing the characteristics of discontinuities. Qualities such as roughness and persistence can only be determined by observing the lateral traces and surface features of discontinuities and therefore cannot be determined from drill core. Often the best approach to surface mapping is a combination of statistical sampling and concentration on areas of particular importance. Piteau (1973) reports a case where despite good exposure and experienced personnel, structural domains were not recognised in the field and only became apparent following the statistical processing of data.

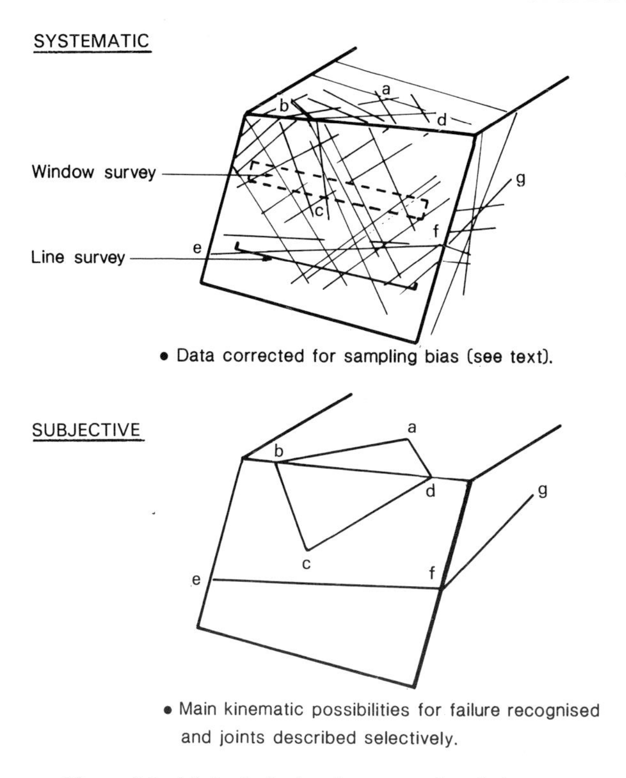

Figure 5.9 Methods for logging exposed rock faces

Systematic methods for collecting data involve logging all discontinuities intersecting selected lines or occurring in 'windows' set out on the exposed face (see Herget, 1977 and Figure 5.9). The use of proformas such as that presented in Figure 5.10 encourages objective and full description. Data may be treated statistically and corrected for sampling bias due to the orientations of the sampling lines (Terzhagi, 1965; Priest and Hudson, 1981; Hudson and Priest, 1983).

A danger of using a totally statistical, 'objective' approach is that the need to make engineering judgements in the field is under-emphasized. Data may be collected with little thought as to their implications and critical but rare data may be overshadowed in analysis by vast quantities of statistically correct but irrelevant data. Many authors caution against an over-statistical approach to collecting geological data (Whitten, 1966; Hoek and Bray, 1981), and at the other extreme a survey may be carried out in a totally subjective manner with only those discontinuities considered of importance being described (see Figure 5.9). Such an approach requires suitably experienced personnel to do the work and most importantly good, relevant exposure. Clearly the most suitable method for collecting data for a particular project will depend upon the nature of the problem, the quality of exposure, the resources available for investigation and the experience and expertise of the personnel involved.

There are certain characteristics of joints that are very difficult to measure even where the exposure is excellent. Impersistence in particular is one of the main factors contributing to slope stability and yet is extremely difficult to investigate. Joints can suddenly die out

GENERAL INFORMATION

Site | Date (Day, Month, Year) | Operator | Discontinuity data sheet No. ___ of ___

NATURE AND ORIENTATION OF DISCONTINUITIES

Chainage or No.	Type	Dip	Dip direction	Persistence	Aperture	Infilling	Consistency	Roughness	Differentially weathered zone width mm	Waviness wavelength	Waviness amplitude	Water	Remarks

Type
0. Fault zone
1. Fault
2. Joint
3. Cleavage
4. Schistosity
5. Shear
6. Fissure
7. Tension crack
8. Foliation
9. Bedding

Dip, Dip direction
(Expressed in degrees)

Persistence
(Expressed in metres)

Aperture
1. Wide (>200mm)
2. Mod. wide (60–200mm)
3. Mod. narrow (20–60mm)
4. Narrow (6–20 mm)
5. Very narrow (2–6mm)
6. Ext. narrow (<2 mm)
7. Tight

Nature of infilling
1. Clean
2. Surface staining
3. Non-cohesive
4. Cohesive
5. Cemented
6. Calcite
7. Chlorite, talc
8. Others-specify

Consistency of infilling

Soil strength	Rock strength
1. Soft	5. Weak
2. Firm	6. Mod. strong
3. Stiff	7. Strong
4. Hard	8. Very strong

Roughness
1. Polished
2. Slickensided
3. Smooth
4. Rough
5. Defined ridges
6. Small slope
7. Very rough

Waviness
Express wavelength and amplitude in metres

Water
1. Dry
2. Seepage
3. Slight flow <0.1 1/sec
4. Mod. flow 0.1–1 1/sec
5. High flow >0.1 1/sec

Figure 5.10 Proforma for the collection of discontinuity data. Record characteristics of each discontinuity encountered at chainages along survey line.

Figure 5.11 Impersistent joint in volcanic rock

as in the case illustrated in Figure 5.11, or may contain rock bridges that can provide considerable cohesional strength. Persistence can be estimated realistically only by observing the general characteristics of sets of joints in exposed faces. In practice, joints are normally regarded as persistent unless proved otherwise and it is accepted that design might err on the conservative side. Some authors have proposed mathematical methods to deal with impersistence (Jennings, 1970; Einstein *et al.*, 1983) and the matter is discussed at length by ISRM (1978).

Roughness and waviness (the distinction between which is not always clear) are essentially small-scale textural surface variations and larger scale undulations of the discontinuity surface respectively, and can only be measured on actual exposed joint surfaces. Both are important for defining the attitude of joints and for their contribution to shear resistance as discussed later. If the exposure is good enough, plates of various diameters may be used to express the variation in attitude of the joint over different base lengths (Fecker and Rengers, 1971; Hoek and Bray, 1981). In some studies, if the exposure at the actual site is not good enough for roughness to be measured, then exposures of similar joints in the vicinity perhaps in quarries or at the coast may be studied and the results interpreted intelligently (Richards and Cowland, 1982). This type of survey is imperative if rational decisions based on estimate of shear strength are to be taken.

Seepages in exposures together with evidence such as lush vegetation, localized weathering, and stains on joints in exposed faces can all be important for understanding the hydrogeological conditions. A detailed knowledge of groundwater

Methods		Comments
Borehole Periscope	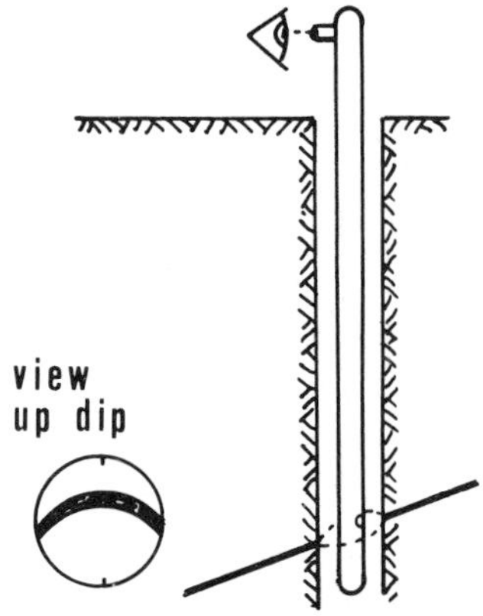 view up dip	• Simple device that works successfully to about 30 m. • Orientation, dip and depth can be measured. • Aperture/infill can be observed/photographed. • Water ingress seen if hole pumped dry.
Impression Packer	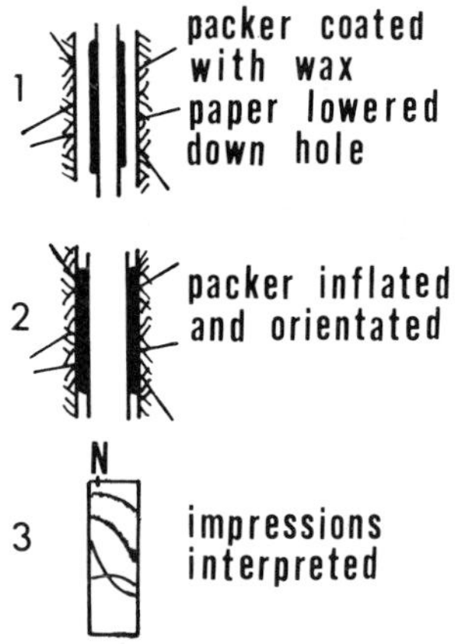 1 packer coated with wax paper lowered down hole 2 packer inflated and orientated N 3 impressions interpreted	• Excellent impressions achieved. • Joint characteristics other than orientation are unknown unless reference is made to core. • Care must be taken that compass is fully set before withdrawal.
Television Camera	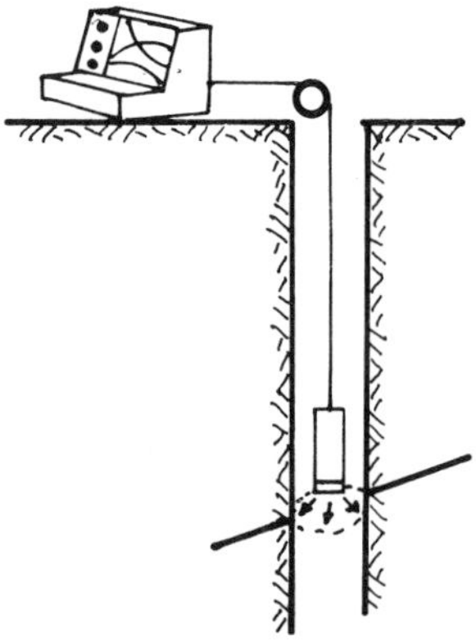	• Video record can be made of hole • Some systems only suitable for general viewing of quality rather than detailed measurement of joint data.

Figure 5.12 Downhole methods for measuring the orientation of discontinuities

Figure 5.13 Borehole periscope in inclined hole being used to take a photograph of a discontinuity

conditions, however, can only come in the long-term from instrumentation which can cope with short-term variations in pressure as discussed later (Patton, 1983; Cowland and Richards, 1985).

5.3.4 Drilling

Discontinuities within the rock mass can be sampled only by drilling or by excavation. In rock, core drilling is generally carried out using either double- or preferably triple-tube barrels to reduce disturbance. The core obtained can be logged to record the frequency of joints and samples may be selected for testing. The sampling size is, however, generally much too small to allow a representative roughness to be assessed.

Orientation of joints in rock core is not easy to measure. As drilling proceeds core may be rotated and only by very carefully piecing together of core (with 100% recovery) or by reference to some feature of known orientation such as regional bedding can the dip direction of joints be estimated. Various methods such as scratching a side during drilling are used for orientating core and these are discussed by Hoek and Bray (1981). Several devices can be used to measure the orientation of discontinuities in the wall of the drill hole after the core has been removed and are illustrated in Figures 5.12 to 5.14. Even if the orientation could be measured accurately, the problem of how representative is the core of the whole rock mass would remain.

Figure 5.14 Impression packer showing traces of discontinuities. (Photograph by A. Cipullo)

5.4 SHEAR STRENGTH

Slope stability analysis requires knowledge of the shear resistance along potential failure planes. If a failure can occur along discrete joints or joint sets then the shear strength of the individual joints must be assessed. If a more complex failure surface passing in part along joints and in part through intact material is possible then parameters are required that take into account that mode of failure.

It must be emphasized that shear resistance must always be expressed in terms of effective stress to take account of water pressures as discussed in the next section on groundwater.

Major advances have been made over the last two decades (Hoek, 1984) in our understanding of the factors controlling the shear strength of jointed masses and some acceptable methods for measuring or estimating strength have been established and are discussed below.

Table 5.3 Factors Contributing to the Shear Strength of Rock Discontinuities

Increasing normal load A_1 A_2	ADHESION (BASIC FRICTION) —Adhesion over true area of contact (A_1, A_2) • Proportional to normal load • Does not cause dilation • No reduction with displacement • Same for different roughnesses
Increasing normal load	INTERLOCKING AND PLOUGHING (ADDITIONAL FRICTION COMPONENT) —Surface texture component • Proportional to normal load • Does not cause dilation • Reduces with displacement with production of debris • Increases with rougher surface texture
$i°$	OVERRIDING • Work done due to dilation • Purely geometrical effect • Uphill movement results in increased shear strength (as measured in the horizontal)
	COHESION – Shearing of major asperities • Distinct from interlocking (see above) in that it is not proportional to normal load • Does not cause dilation • Lost after peak strength • Relative contributions of overriding and cohesion depend on stress level

5.4.1 Shear Strength of Individual Joints

The term shear strength is used to describe the total resistance against shearing developed along a surface and the main contributing factors for persistent joints are illustrated in Table 5.3. These factors can be separated mathematically although in most cases several factors will operate at the same time—for example shearing of asperities will often be accompanied by dilation. The term friction is restricted to describe the resistance proportional to normal load (a combination of factors 1 and 2 in Table 5.3). Patton (1966) demonstrated that friction and the effects of roughness on shear strength of rock joints could be separated in a practical way and Barton (1971, 1973) similarly distinguishes between these two components in his empirical equation for rock shear strength based on carefully scaled model tests:

$$\tau_p = \sigma_n \tan \left[JRC \log_{10} \left(\frac{JCS}{\sigma_n} \right) + \Phi_b \right]$$

where τ_p = peak shear strength
σ_n = effective normal stress
JRC = joint roughness coefficient
JCS = joint wall compressive strength
Φ_b = "basic" friction angle

Barton discusses the measurement of JRC and JCS in the papers referenced above and in later papers (Barton and Choubey, 1977; Barton and Bandis, 1980) and claims that his equation can predict shear strength extremely closely. Recent work by Bandis (Bandis *et al.*, 1981) has demonstrated the importance of scale effects in determining JRC and JCS values. The Φ_b value is the frictional resistance for 'flat or residual surfaces' and is a convenient reference point obtained from direct shear tests on planar rough-sawn or sand blasted surfaces of rock.

The preferred method of the author for deriving realistic shear strength parameters involves direct shear testing of representative samples of natural joint surfaces and interpreting the results obtained with respect to field-scale roughness. An advantage of testing natural surfaces is that the influence of surface coating and natural surface texture may be investigated, the significance of which might not be apparent were saw-cut surfaces of rock to be used. For example, during the investigation of a rock slope failure on a shallowly dipping plane in 1982, careful shear testing demonstrated the importance of a thin coating of chlorite which, at low stresses gave an effective friction angle of only 19 degrees but at higher stresses was less significant due to increasing contribution of the underlying monzonite rock surface. These detailed observations allowed the failure mechanism to be understood.

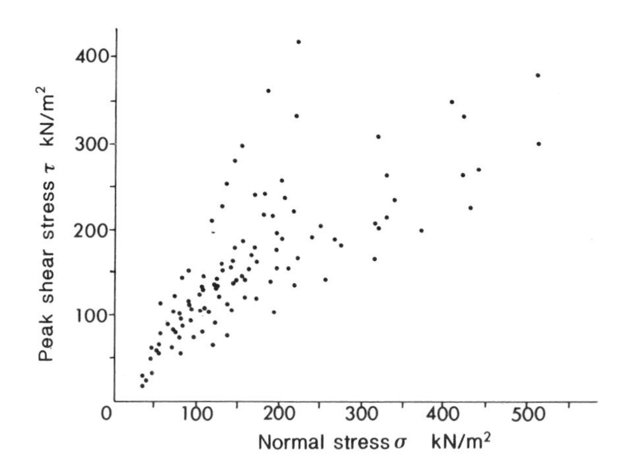

Figure 5.15 Peak shear strengths from multistage tests on rough discontinuities through granite

In carrying out shear tests on small rock samples it must be recognised that individual tests will give different results according to the roughness of each sample. As an example, the results of a series of tests on natural granite joint samples of various roughnesses are presented in Figure 5.15. In the early 1970s, the tendency was to test larger and larger samples, either *in situ* or in the laboratory with the aim of obtaining parameters that would incorporate the effect of larger scale joint roughness. In fact such tests often just give even wider scatter and at much greater expense than do small scale tests of the type reported by Ross-Brown and Walton (1975).

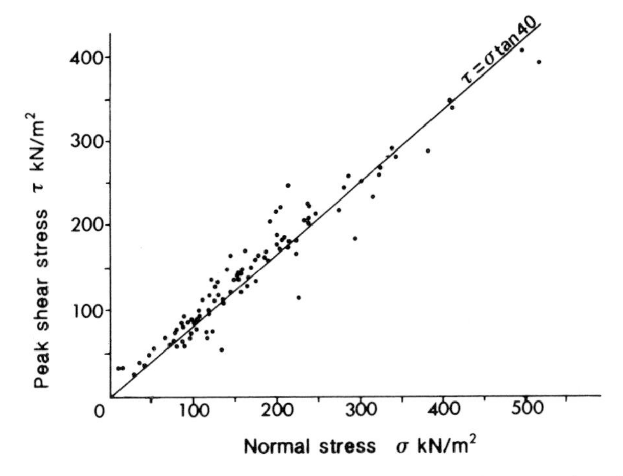

Figure 5.16 Peak shear strengths corrected for dilation and compression during tests

Figure 5.17 Planar tectonic joint through volcanic rock

Direct shear tests on small samples of rock joints can only be expected to provide the following data:

1. The basic frictional resistance for groups of like joints once the effects of sample variability have been removed.
2. Information on the mechanics of shearing of the natural joint that can aid in the selection of a roughness coefficient for the joint *in situ*.

Figure 5.18 Rough sheeting joint through volcanic rock

Interpretation requires the correction of data for the dilation or compression angle at peak strength which allows friction coefficients for effectively planar, 'natural' surfaces to be derived (Hencher and Richards, 1982). Figure 5.16 shows the data of Figure 5.15 corrected for the effects of individual sample roughness and the reduction in scatter is encouraging. The basic friction angle thus obtained was about 6 degrees higher than that for saw-cut surfaces of the same rock.

Once parameters describing the basic frictional strength have been obtained then the effects of field roughness must be taken into account and this is where engineering judgement is required. Examples of a planar tectonic joint and a rough sheeting joint through the same volcanic rock are given in Figure 5.17 and 5.18. A practical method for measuring the roughness of representative joints has been proposed by Fecker and Rengers (1971) and is described well in Hoek and Bray (1981). The method is explained schematically in Figure 5.19.

Different 'orders' of roughness are measured using different diameter base plates, measurements being taken on a grid pattern over the joint surface. It is generally found that the smaller the measuring plate, the greater the deviation from the average angle of inclination of the plane. Reference to the damage caused to asperities during shear tests at the correct stress levels, will aid in deciding the allowable roughness angle that will cause dilation of the plane during sliding (Richards and Cowland, 1982). This angle, $i°$, can be added to the basic friction angle so that the shear strength of joints becomes:

$$\tau = \sigma' \tan(\Phi + i) + c$$

where σ' is the effective normal stress

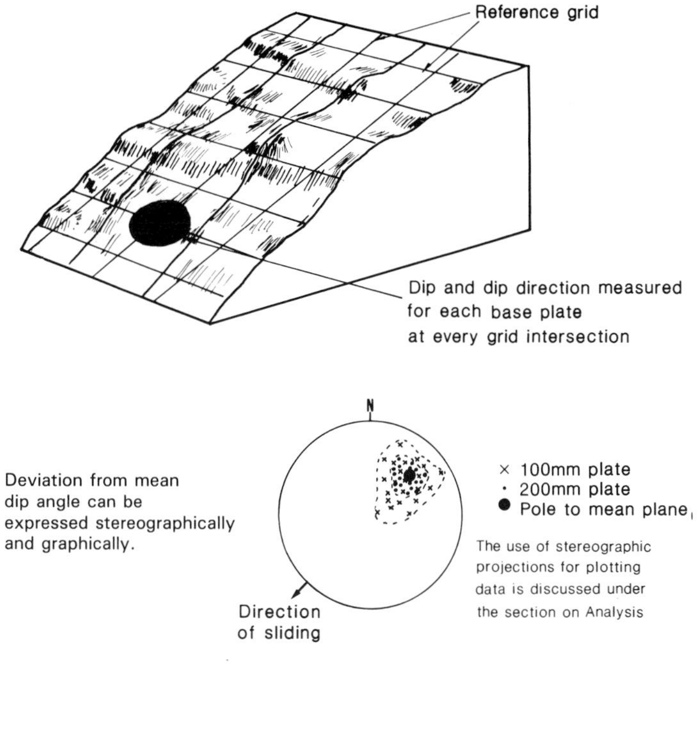

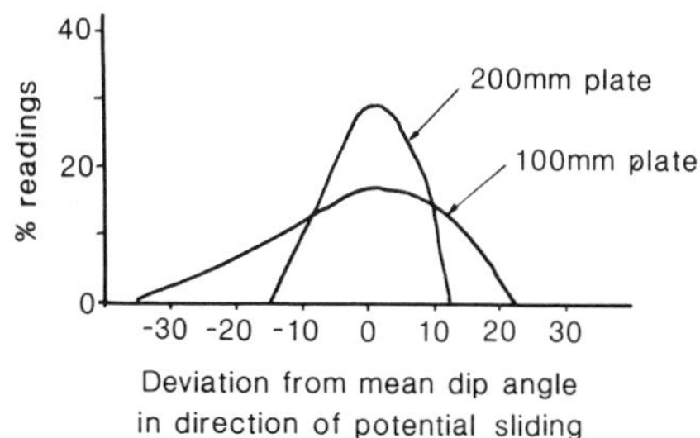

Figure 5.19 Method for quantifying the roughness of a joint

$\Phi°$ is the corrected friction angle from shear tests
$i°$ is the maximum allowable roughness angle
c is cohesion (see discussion below).

For the granite joints discussed earlier, field measurements together with observations of surface damage caused during shear tests suggested that a roughness angle of 15° could be allowed above the corrected friction angle of 40°. The effects of weathering (or infill) could be accommodated by reducing the roughness angle. The procedure for this practical and rational assessment of shear strength by measurement is illustrated in Figure 5.20.

Apparent cohesion due to shearing through of asperities is seldom measured at the low stresses typical of small rock slopes (say up to 50 metres in height) and in fact it is probably

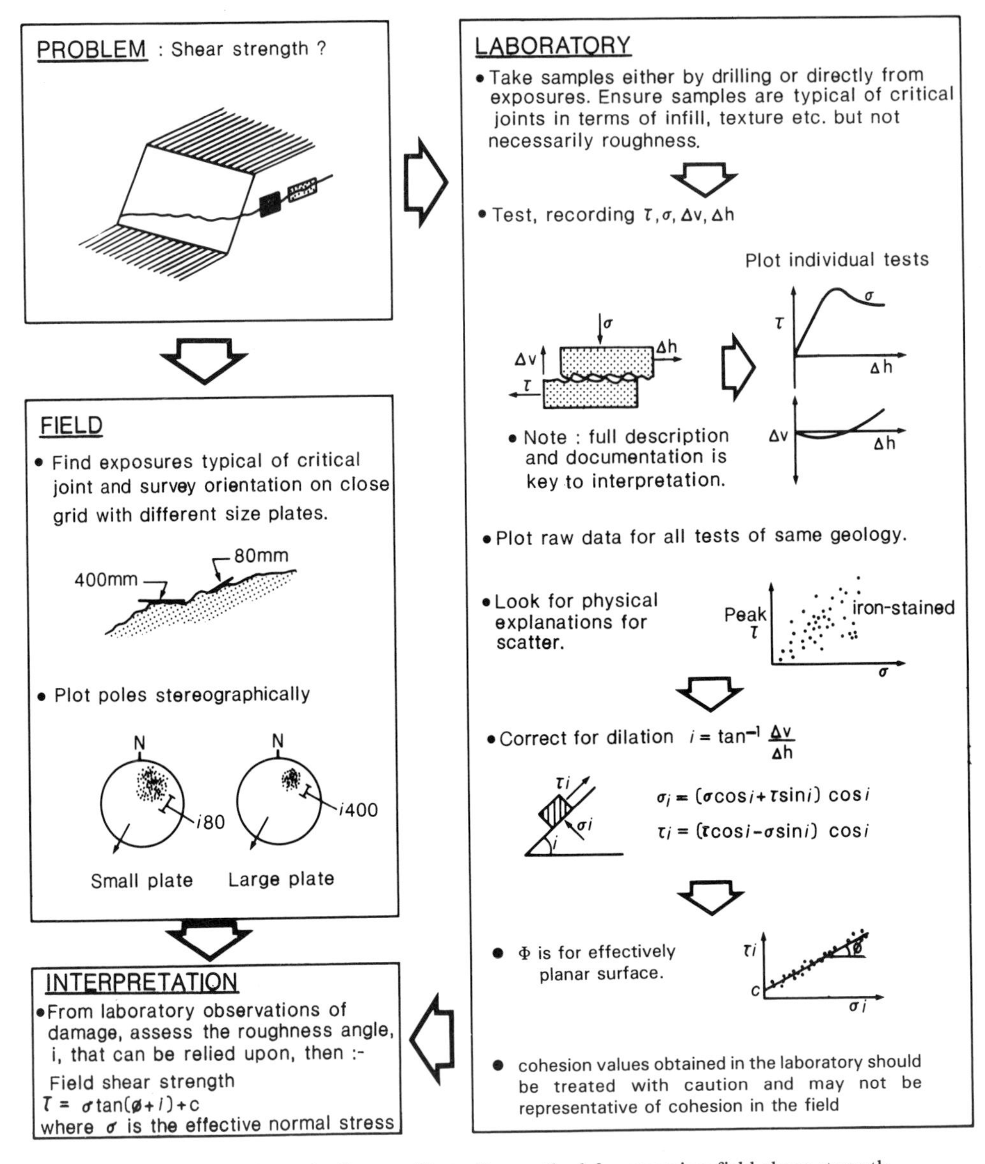

Figure 5.20 Schematic diagram illustrating method for assessing field shear strength

good practice to force test results through the origin in defining the basic friction angle unless there is an obvious cohesional factor. A true cohesion will result from impersistence of joints and several authors (see Jennings, 1970 and Einstein *et al.*, 1983 for example) have addressed the problem theoretically but at present, in the author's opinion, there is no better method available than to use engineering judgement following careful field description. Field values for cohesion might be calculated by back-analysis of failed slopes or by assuming a Factor of Safety of 1.0 for unfailed slopes but often other parameters are so poorly known that this proves impossible (Leroueil and Tavernas, 1981; Hencher *et al.*, 1984).

5.4.2 Shear Strength of Closely Jointed Rock and Soil Masses

Many slope failures occur through closely jointed soil and rock where the joints contribute to a general weakening effect of the mass rather than providing a simple joint-controlled shear surface (Skempton and La Rochelle, 1965; Koo, 1982). Representative shear parameters are very difficult to measure for such modes of failure although several authors have tested large-scale jointed models or large samples which have provided an insight into the factors involved (Hoek, 1984). Hoek and Brown (1980) have reviewed available data and have suggested several empirical strength equations for which the unknowns are rock type, quality of rock mass and compressive strength of the intact material. The authors experience of back-analysis of slope failure through such closely jointed materials is that the equations of Hoek and Brown seem of the right order and in fact there seems little alternative to their use. It is hoped that future published case histories will allow the equations to be tested further. One area for improvement might be to make some allowance for the *degree* of adversity of the jointing and Koo (1982) and Harris (1984) have noted the importance of this factor.

5.5 GROUNDWATER

Groundwater has a considerable influence on slope stability with rainfall being the most common triggering cause of landslides in many parts of the world. Brand *et al.* (1984) for example estimate that more than 90% of landslides in Hong Kong are the direct result of rainstorms (1,500 failures occurring during two rainstorms in 1982).

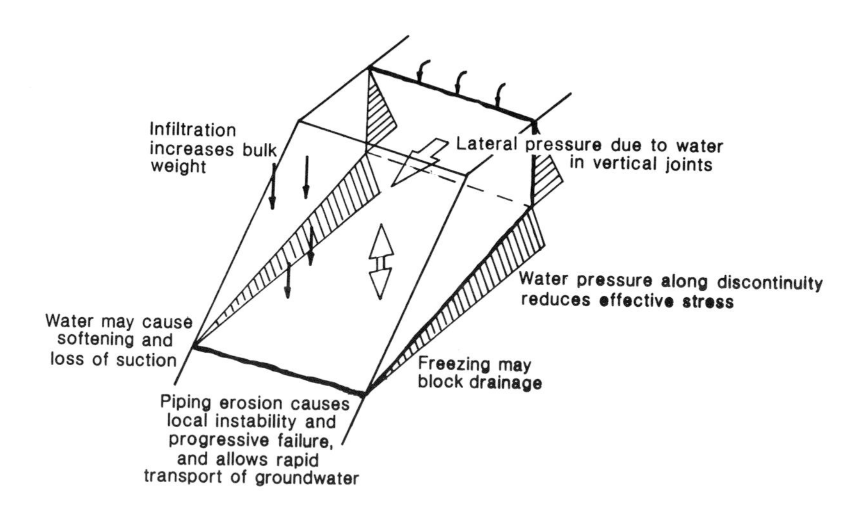

Figure 5.21 Influence of water on slope stability

The main ways in which groundwater affects the stability of slopes in soils and rock masses are illustrated schematically in Figure 5.21 and listed below:

1. Water pressures along the sliding surface reduce the effective normal stress transmitted across the surface thereby limiting the shear resistance developed.

2. Water pressures in vertical joints or fissures apply a force which may initiate sliding.
3. Water infiltration increases the unit weight of the material above the slip plane. This leads to an increased shear load without proportional increase in shear resistance due to (1) above.
4. Internal erosion due to piping.
5. Freezing of groundwater can temporarily restrict drainage and in addition may cause loosening of the mass because of the expansive force produced.
6. Water can cause softening of material as well as loss of suction; both reduce strength.

Drainage measures must be taken into account in the design of remedial or preventive works and the potential corrosion of anchoring systems or reinforcement must also be considered. The complexity of groundwater flow through discontinuous rock or soil will depend upon the geological structure. Certain features will have a controlling influence, notably the distribution of relatively permeable and impermeable lithologies, the degree, distribution and nature of jointing, the presence of faults which may allow rapid ingress

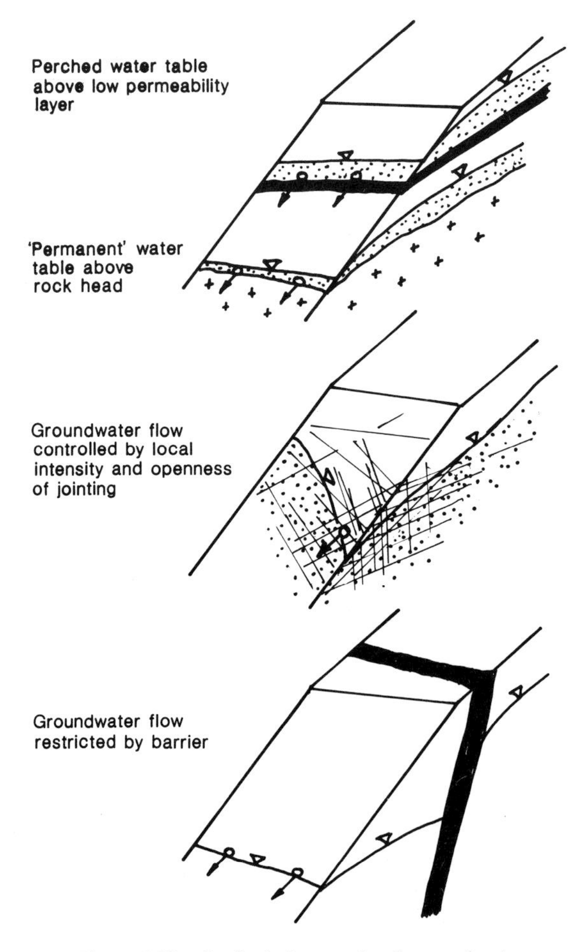

Figure 5.22 Geological controls of groundwater

and distribution of water and the position of rock head in areas of weathered materials. These are illustrated diagrammatically in Figure 5.22.

Detailed information on the hydraulic conductivity or permeability of sections of the mass can be gained by carrying out field tests or by theoretical consideration of the degree and openness of jointing (Hoek and Bray, 1981). The use of dye tracers can be helpful in determining drainage paths. Relevant data gained during site investigation will include levels where water is encountered and depths at which drilling water is lost, particularly if the loss of water can be traced to seepage in an existing face. The nature of the core itself can also be revealing. Joints which carry a lot of water may be particularly weathered and coated with deposits of oxides and clay.

From such investigations it is possible to establish the factors controlling the hydrogeological pattern but to gain an accurate assessment of water pressures to use in design it is generally necessary to install piezometers. The selection of piezometer type and location within the ground should take the following factors into account:

1. The geological and probable hydrogeological structure.
2. The conceivable modes of failure.
3. The expected rate of change of groundwater pressure.

Piezometers are sometimes placed at the base of boreholes with little reference to the site conditions and nature of the problem, the intention being to try to get a general view of the groundwater system. This is fine if the situation is not complex geologically but where, for example, there is a possibility that a perched water table may develop then it is vital that this be investigated specifically. Similarly, where a particularly adverse joint has been identified then piezometers must be installed to measure pressures on that joint alone. Patton (1983) comments that usually too few piezometers are used in slope stability studies to get a true picture and recommends the use of a 'modular' piezometer system which can measure pressures at different levels in a single drillhole. If piezometers are placed at the wrong location then misleading or insufficient data will be obtained. For example, of eleven failures investigated by the author in 1982, six had piezometers either through or close to the failure scar (Hencher *et al.*, 1984) but none of those piezometers provided data representative of conditions on the failure scar at failure (as proved by field observation or back-analysis). This was in some cases due to poor choice in location of piezometers relative to the critical geological structures, and in others due to an inadequate monitoring system.

The sophistication of the monitoring system must match the expected fluctuations in water pressure (see Chapter 3). If these are slow changing, then standpipes which are dipped on an occasional basis may be sufficient. A simple device comprising a string of plastic containers (buckets) suspended in the pipe can indicate maximum levels which occur between manual readings and is described by Brand *et al.* (1983). Conditions where transient response is expected to be important or where there is a high degree of complexity may demand that an automatic monitoring system be set up (Pope *et al.*, 1982). Cowland and Richards (1985) describe an automatic system for measuring water pressures on specific rock joints and show how such a system can reveal behaviour that would have been very difficult to predict. They demonstrate the transient nature of water pressure through a slope following intense rainfall with pulses of groundwater pressure migrating rapidly along the surfaces of large joints. This produces less severe conditions than might be suspected using results from a less sophisticated system.

Once a reasonable understanding of the groundwater pattern and the factors contributing to changes in groundwater pressure are established, modelling techniques may be used to predict the effect of changes such as increased infiltration, or excavation. Commonly used methods include flow nets (Wittke and Louis, 1966), electrical analogue methods (Hoek and Bray, 1981) and, more rarely, computer techniques (Leach and Herbert, 1982). The incorporation of 'design' water pressures in analysis is discussed in the next section.

5.6 ANALYSIS OF SLOPE STABILITY IN JOINTED ROCKS AND SOILS

5.6.1 Introduction

This section deals primarily with the analysis of slopes through which well-defined adverse discontinuities have been recognized. Two other special cases will be discussed briefly:

1. Closely jointed rock masses where soil mechanics methods of analysis are applicable.
2. Natural slopes where only very limited geotechnical information is available.

The degree of confidence in the results from any analysis will depend upon the quantity and quality of available data which will generally reflect: (a) the seriousness of the problem, i.e. risk versus consequence; (b) the resources available; (c) the difficulties of investigation.

It is sometimes not appreciated how difficult it is to obtain adequate information for the analysis of jointed rock masses using drillholes. The results of analysis based on such an investigation must, therefore, be treated with caution and in the case of the design for a new cutting, assumptions must be checked during construction.

The importance of any uncertainties in the input data can be quantified by carrying out a series of analyses in which each parameter is varied in turn within its probable range of values. Such a study is called a sensitivity analysis and is particularly useful when trying to assess the cost effectiveness of different options for preventive or remedial works (e.g. drainage or anchoring). Several authors have taken this type of analysis further and have expressed the likelihood of failure as a probability based on the statistical variation of the input parameters (McMahon, 1975; Piteau and Martin, 1977; Priest and Brown, 1983).

5.6.2 Discontinuity-controlled Failures

The analysis of discontinuity-controlled failures generally involves two interrelated stages:

1. Firstly, the available joint data must be sorted to identify those along which failures might occur. This may sometimes be done for an existing slope by eye providing the exposure is good.
2. Secondly, each joint-defined block must be analysed to determine whether the available shear resistance exceeds the forces tending to cause failure.

A combination of various techniques is used to carry out the analysis:

1. Stereographic projection is a geometrical construction which allows individual joints and other planes to be represented and their angular relationships to be measured; it

is used primarily for sorting data into a manageable form and for preliminary assessment of the potential for a block of rock to move. Detailed analysis can also be carried out using stereographic projection but that requires all data to be expressed as angles which involves some difficulty.

2. Graphical methods in which force vectors are drawn as a polygon are used to assess the stability of particular blocks. The geometry of each block must be pre-defined perhaps following a stereographic appraisal of all joint data.
3. Numerical analytical methods can be applied to simple wedges or blocks of rock but require the geometry to be pre-determined as for graphical methods.
4. Other methods include physical modelling using base friction or block models. Finite element and boundary element computer techniques are useful for calculating and illustrating stress distributions but have also been used to produce visual (and numerical) models of slopes failing.

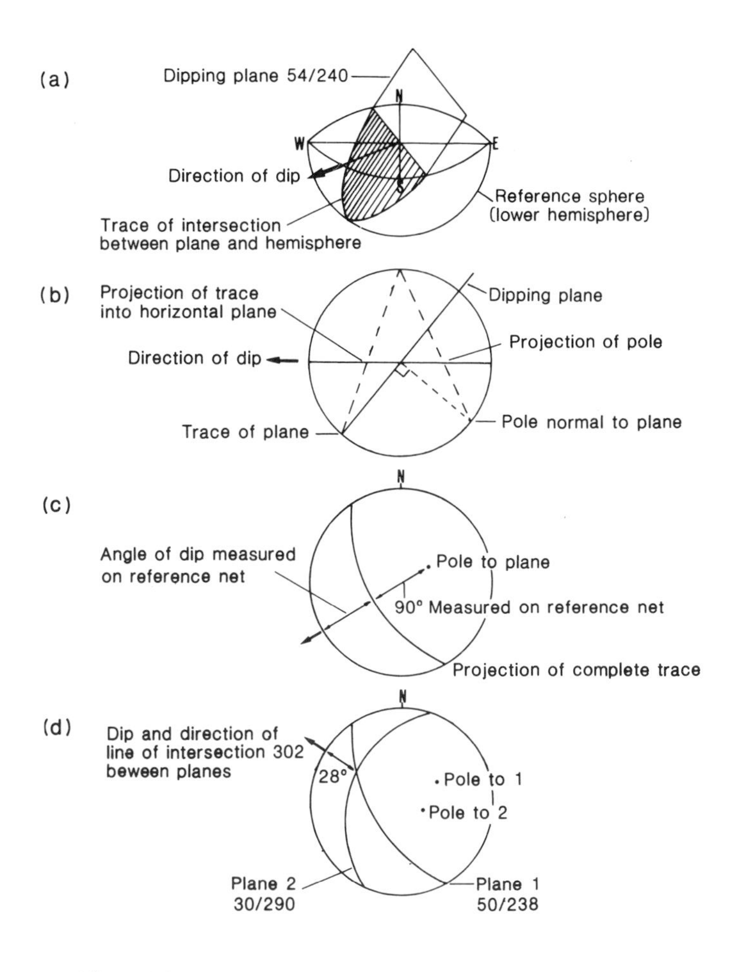

Figure 5.23 Representation of dipping planes as traces and poles

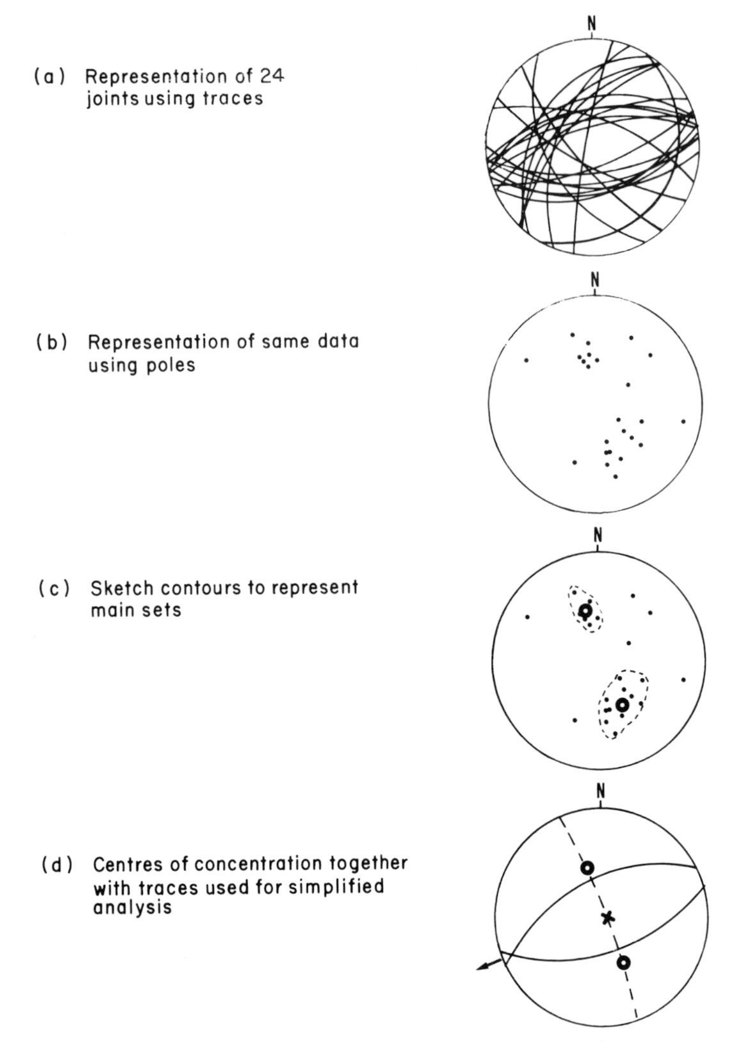

Figure 5.24 Principles for sorting large amounts of data

5.6.3 Stereographic Projection

Stereographic projection is a powerful tool for representing the geometrical relationships of inclined planes and is familiar to geologists through its use for evaluating problems in structural geology and crystallography. It has been adopted and developed by workers in rock mechanics to the stage where its use can provide direct analytical results for quite complex slope stability problems albeit with some difficulty in expressing input data with respect to cohesion, water pressures, and reinforcement (see, for example, Attewell and Farmer, 1976; Priest, 1985). Recent advances in the use of the technique for analysing the kinematics of moveable blocks are discussed by Goodman (1983).

The main use of stereographic projection in the stability analysis of jointed masses is for sorting otherwise unmanageable quantities of joint data (Pentz, 1971) and for the

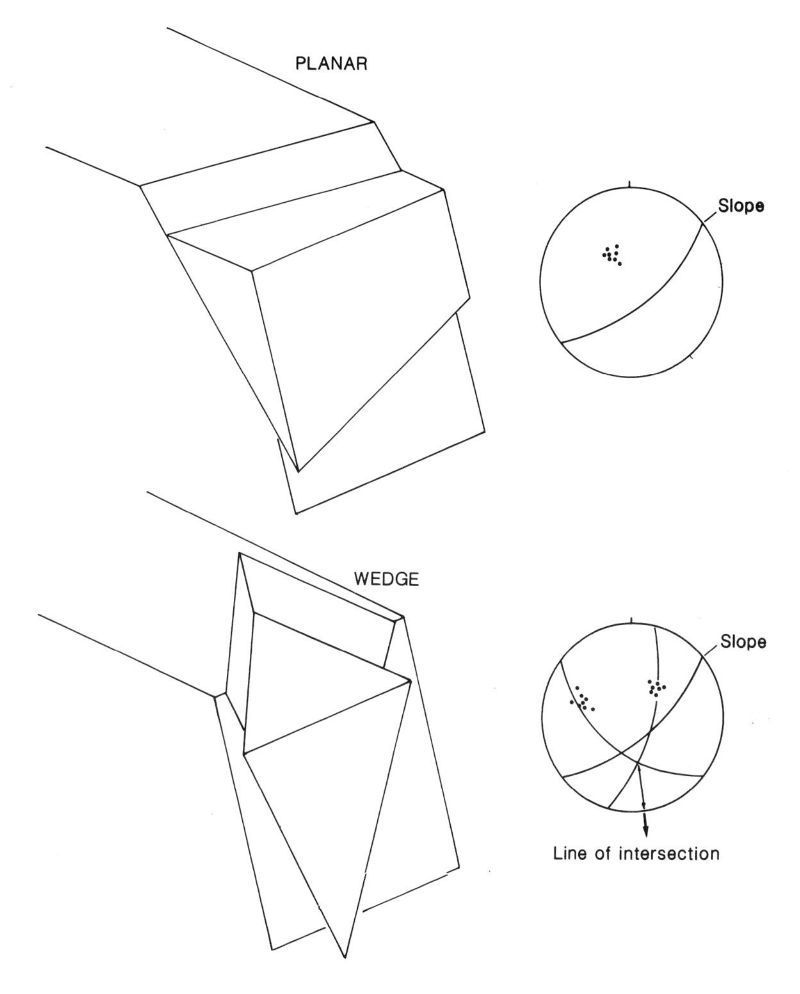

Figure 5.25 Typical modes of failure

preliminary analysis of those data with reference to other aspects of the problem that can be readily expressed geometrically (slope angle, friction angle). The stereographic projection allows identification of the main structural features and particular discontinuity combinations that require more detailed analysis.

The principles and technique for preparing a reference stereographic 'net', for plotting data and for analysis are covered by several excellent publications, notably Phillips (1971), Hoek and Bray (1981), Richards *et al.* (1978), and Priest (1985) the latter three with particular reference to slope stability, and the reader is referred to those references for a clear explanation of how to carry out this work. The principle, as illustrated in Figure 5.23(a), is to imagine all planes defined by their average dip angles and dip directions bisecting a reference sphere which is fixed in space. The planes intersect either the lower hemisphere (by convention) or more rarely the upper hemisphere. Either construction allows the same analysis. The trace marking the intersection of each plane with the hemisphere

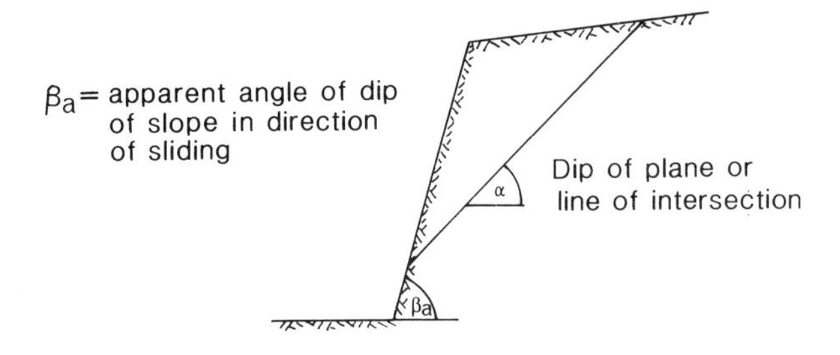

For sliding: $\beta_a > \alpha > \Phi$

where Φ is a representative friction angle

Note: Effects of water pressure, cohesion, impersistence and wedging not taken into account.

Figure 5.26 Conditions for sliding

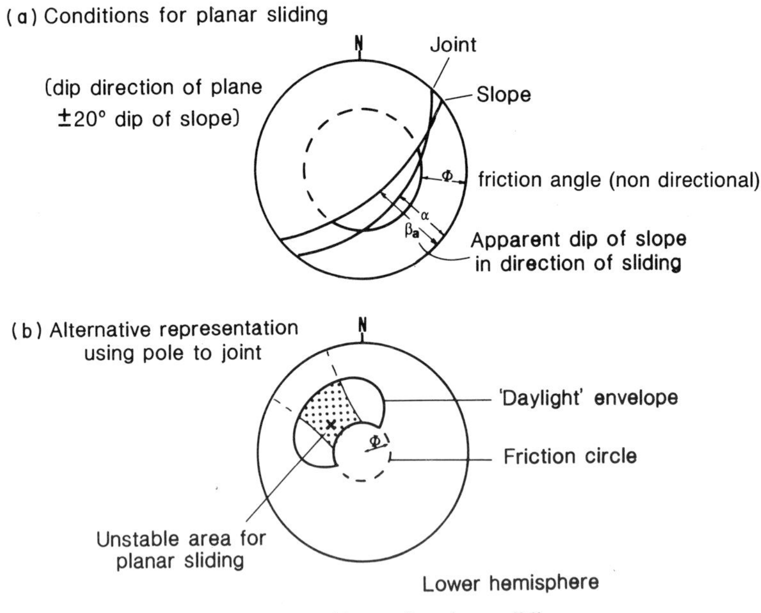

Figure 5.27 Stereographic test for planar sliding

is projected on to the horizontal section through the reference sphere as shown in Figures 5.23(b) and (c). As discussed later, it is often convenient to represent planes, not by their traces, but by a single pole normal to the plane (Figures 5.23(b) and (c)). Either construction may be used to represent the same unique information for a single plane as plotted in Figure 5.23(c). Figures 5.24(a) and (b) are plots of the same joint data presented as traces and as poles respectively. The advantage of using traces is that the points of intersection

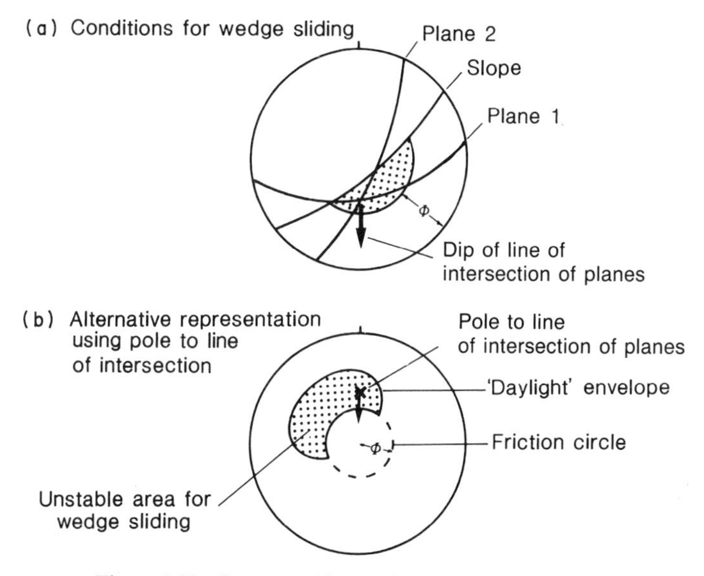

Figure 5.28 Stereographic test for wedge sliding

Table 5.4 Use of Stereographic Projections for Slope Analysis

USES FOR ROCK SLOPE STABILITY

- measuring angular relationships
- identifying sets of joints
- demonstrating relationships between discontinuities and slope face geometry thereby showing which individual discontinuities or wedges daylight and therefore might cause failure
- comparing inclinations of potential failure planes and wedge intersections with trial friction angles
- to demonstrate the degree of roughness of a single discontinuity
- can be used directly for analysis providing shear strength parameters c and Φ and water forces U and V are known (see Attewell and Farmer, 1976)

LIMITATIONS

- spatial relationships not shown (e.g. two joints apparently forming wedge on the stereoplot may be separated in the slope)
- nature of individual joints not represented
- discontinuities plotted as single poles
- cohesional component of strength not readily taken into account
- effects of water not readily taken into account (see discussion by Sekula, 1982)

COMMON MISUSES

- used directly for analysis in an oversimplistic manner rather than as tool to aid understanding
- contouring to show centres for joint sets can overshadow more important, adverse joints
- contouring of too few joints

HINTS

- only contour data where absolutely necessary or to get a better understanding of the problem; do not contour a limited amount of data
- always check the nature of any 'adverse' joints removed by contouring
- once adverse joints have been identified, then analyse potential failures individually, using limit equilibrium methods to take full account of strength, water pressures, and other characteristics of specific joints of interest
- critically assess the representativeness of your data

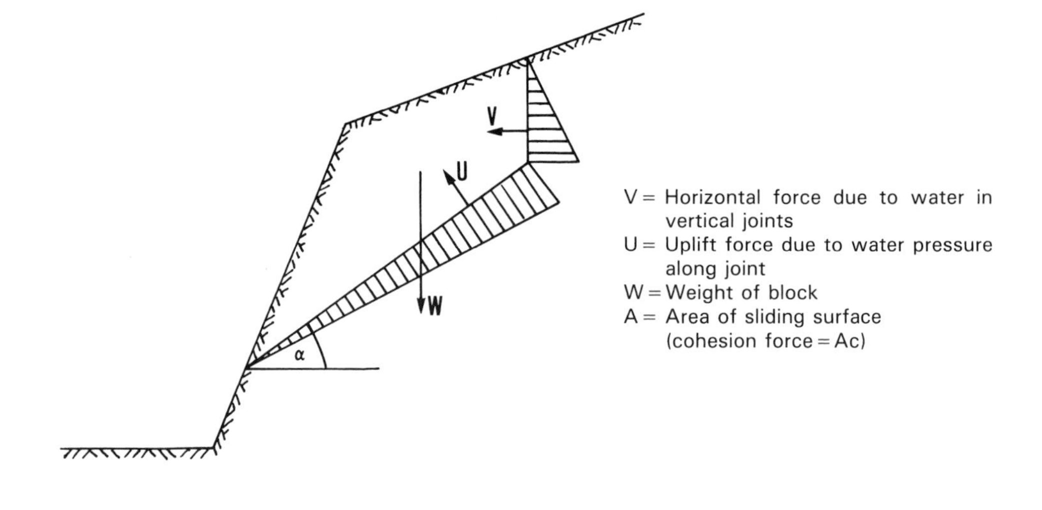

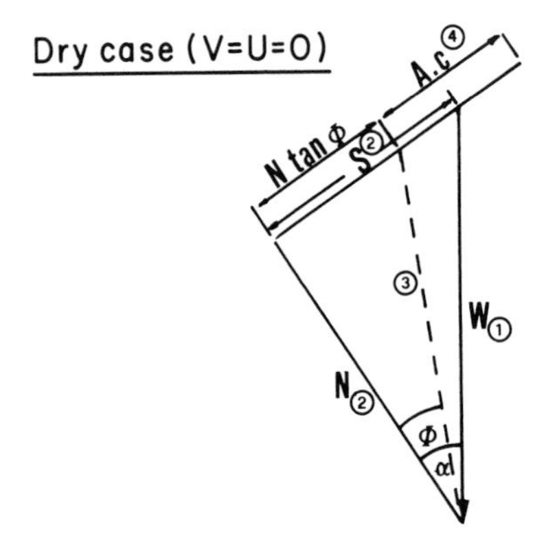

1. Construct weight vector
2. Construct shear and normal components of weight
3. Draw line offset Φ degrees from N to construct friction force N tan Φ
4. Draw on cohesive force

$$\text{F of S} = \frac{N \tan \Phi + Ac}{S}$$

(in case illustrated > 1)

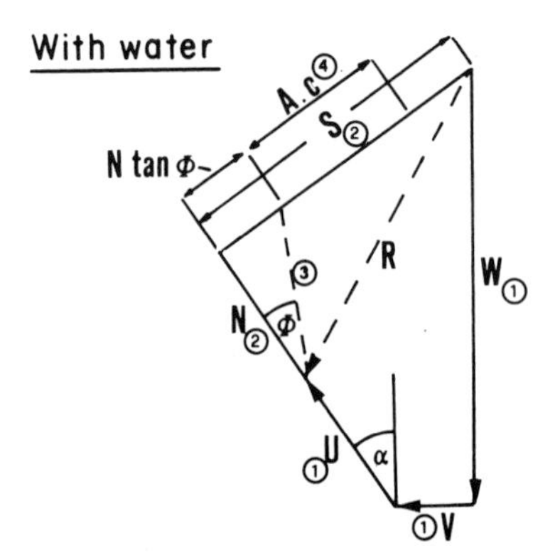

1. Draw vectors for weight and water forces as shown
2. Construct shear and normal components of resultant R
3. As above
4. As above

$$\text{F of S} = \frac{N \tan \Phi + Ac}{S}$$

(in case illustrated < 1)

Additional forces such as earthquake and anchor forces can be drawn at Stage 1 to produce a different resultant R

Figure 5.29 Force diagram for assessing stability

are all clear but the overall diagram is cluttered, the data are more difficult to plot and it is less easy to identify any 'sets' than on the plot of poles. Poles can be split into groups by contouring, the centres of concentrations being marked by single poles (Figure 5.24(c)). Techniques are discussed by Hoek and Bray (1981). Centres of concentration are taken as representative of 'sets' and can then be used for further analysis. It should be noted, however, that there is a danger that important individual or rare discontinuities may be overlooked as a result of contouring and this is discussed in detail by Hencher (1985).

Planes and wedges which may result in failure are typified by particular patterns on the stereogram (Figure 5.25) and are analysed by comparing their angles of dip with their angles of sliding and determining whether or not they will daylight (appear in the slope face). Figures 5.26 to 5.28 illustrate the procedures. It should be noted that although two planes appear to form an adverse wedge on the stereogram they may not do so in the field due to geographical separation. Wedges formed from sets of joints are much more likely to occur naturally. It is always necessary to check the results of the stereographic analysis in the field and a useful though time-consuming technique to aid in this check is to mark each joint with a reference number during the initial survey and to use those numbers during analysis.

Toppling failures are associated with joints dipping steeply back into the slope and such joints can be identified using stereographic techniques. Goodman and Bray (1976) discuss the criteria for such failures. Some of the most important uses and potential misuses of stereographic techniques for the analysis of slope stability in jointed rocks and soils are listed in Table 5.4, the misuses mostly stemming from the exclusive use of contoured data for interpretation.

5.6.4 Graphical Methods

Provided the geometry of the slope and failure surface are well defined then stability can be assessed graphically by constructing a force diagram as illustrated in Figure 5.29. Hoek *et al.* (1973) use the method to analyse the stability of a complex wedge.

5.6.5 Numerical Methods

The simplest slope failure type to analyse is where sliding occurs along a single plane. The factor of safety is calculated by resolving all forces along the failure plane and dividing the forces resisting sliding by the forces inducing sliding, as illustrated in Figure 5.30. Stimpson (1979) presents simplified equations for calculating factors of safety for various slope conditions. It has been noted earlier that cohesion is often of great significance in such calculations yet is very difficult to determine.

The analysis of wedges of rock such as that in Figure 5.31 sliding on two surfaces is complex despite the apparent simplicity of the situation. The reader faced with the need to analyse such a wedge is referred to Hoek and Bray (1981) who explain the analytical method very carefully and provide examples of calculation. Most such calculations are carried out using either a programmed calculator or a computer. Warburton (1983) has developed a program which can produce a three-dimensional block diagram of a slope and its critical discontinuities and can analyse the stability of individual wedges.

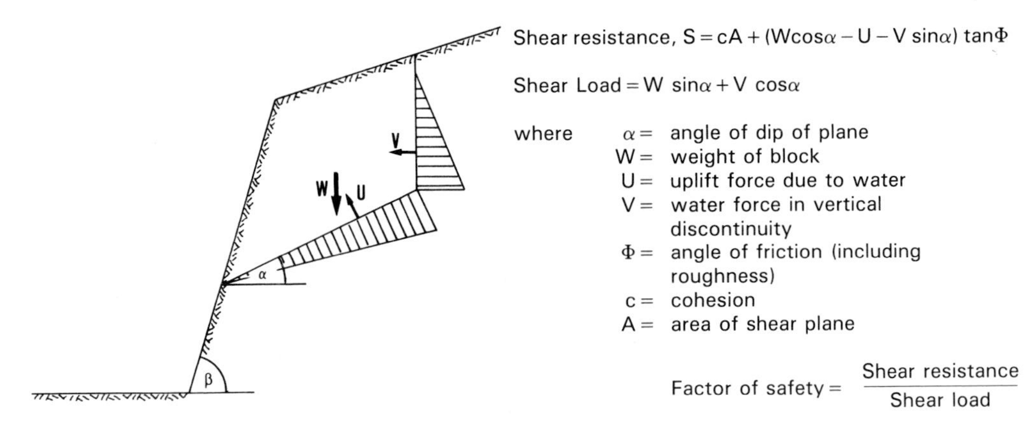

Figure 5.30 Analysis of sliding rock

Figure 5.31 Wedge due to intersection of two joint sets

Conditions for toppling failure (Figure 5.32) and its analysis are discussed in detail by Evans (1981) and Goodman and Bray (1976) and design charts have been presented by Zanbak (1983).

5.6.6 Other Methods

Physical models are useful both to illustrate geological structural relationships and actually to model loading conditions. One of the most illustrative methods is the base friction device

Figure 5.32 Potential toppling failure

which involves modelling slopes on their sides on a moving belt of rough sandpaper. The friction between the paper and the model provides an equivalent to gravitational loading and allows the importance of key blocks and joint orientations to be assessed (Bray and Goodman, 1981). Cundall (1971) developed a two-dimensional computer model which can similarly demonstrate the general mechanism of failure. Finite and boundary element techniques have not proved as useful for slope stability analysis as they have for underground excavations, particularly for assessing stress conditions. For rock slope analysis, stress conditions, though in reality quite complex, are generally treated in a very simplified manner.

5.6.7 Natural Slopes

Aerial photographic interpretation backed up by field checks is often the main method for investigating the stability of large hillslopes. The data obtained by such methods is unlikely to be of sufficient quality or quantity to allow detailed analysis. As a result several authors have developed broadly based classifications for assessing the risk of landslides based on such general parameters as degree of jointing, average rock strength, and slope angle (Selby, 1980 and Chapter 15 in this volume; Vecchia, 1977). Styles *et al.* (1984) have assessed the reliability of geotechnical risk maps based on such criteria by comparison with the results from 1,400 limiting equilibrium analyses along 32 km of cross-section and have shown reasonable agreement.

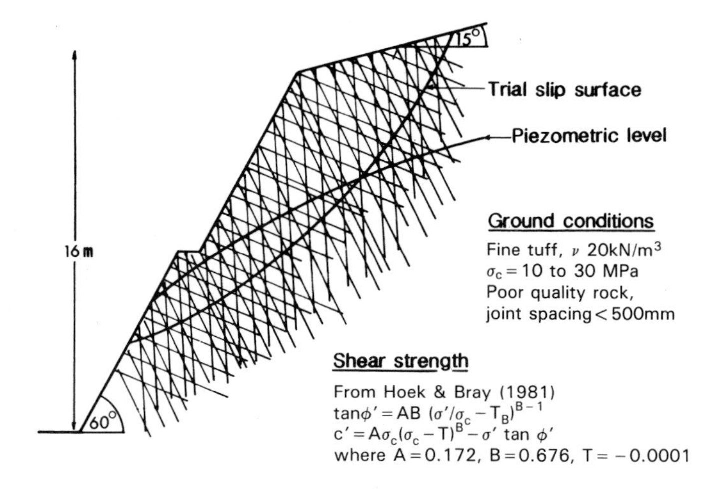

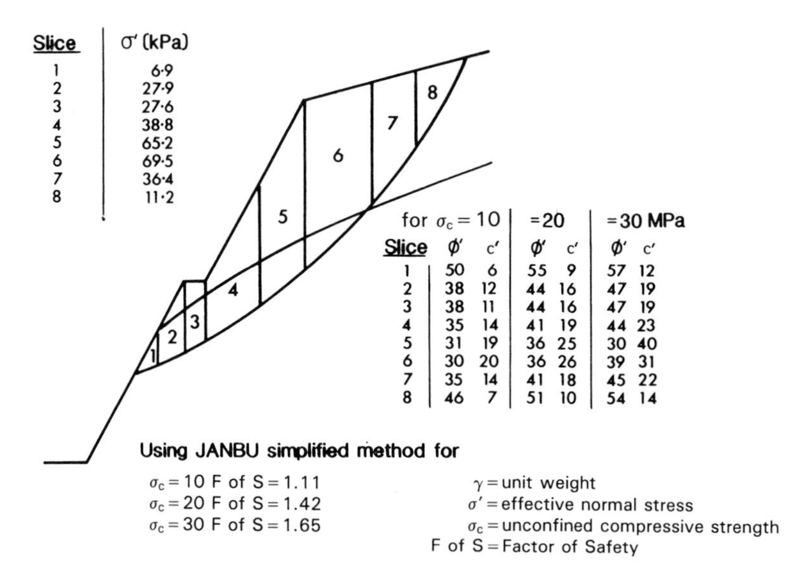

Figure 5.33 Example of analysis of closely jointed rock slope using Hoek–Brown empirical strength criteria

5.6.8 Closely Jointed Masses

Where the ground is so closely jointed that failure may occur along a generalized surface through the mass then soil mechanics methods of analysis such as those discussed in chapter 2 are the most applicable. Hoek and Bray (1981) give useful design charts against circular failure for different groundwater conditions and Hoek and Brown's (1980) empirical strength criteria can provide an estimate of the shear strength parameters to use in the analysis. Surfaces other than circular arcs can be dealt with most easily using the methods of Janbu (1973) and Sarma (1979). An example is given in Figure 5.33. Priest and

Brown (1983) give a well-documented example of the analysis of generalized failure surfaces using Janbus's method and Hoek and Brown's strength criterion. They vary the input data statistically within pre-defined limits and express the results in a probabilistic manner.

5.7 CONCLUSIONS

The presence of discontinuities in rock and soil masses considerably affects their engineering properties and generally imposes markedly anisotropic strength. Slope stability analysis in such materials is dominated by the need to identify, measure, and analyse potentially failing blocks delineated by discontinuities. The importance of geological knowledge to all site investigations lies in the geologist's ability to recognize and establish the geological structure on the basis of limited data. This allows him to predict and extrapolate the nature of materials away from the point of observation. A table of discontinuity types has been presented to aid in their recognition in the field and in assessing their probable nature.

The shear strength of persistent rock joints can be measured providing representative samples of joint surfaces are available. Impersistence and hence true cohesion of rock joints is, however, difficult to assess and further research is needed into how this property can be estimated. Groundwater pressures in jointed masses are often complex and require careful instrumentation to allow them to be measured with an acceptable degree of accuracy. This is probably one of the weakest aspects of most investigations and care should be taken to carry out preliminary analysis prior to instrumentation to ensure that pertinent information is collected. Methods of analysis for jointed rock and soil differ from those for 'homogeneous' and 'isotropic' soil in that the failure shape must be identified physically in the field rather than searched for mathematically. Stereographic analysis is particularly useful for identifying potential failure blocks from a mass of data on discontinuities. Numerical methods are now available to analyse quite complex wedge shapes allowing components of strength, weight, external loading, water pressure, and preventive or remedial works to be taken into account.

ACKNOWLEDGEMENTS

Thanks are due to Dr M. H. de Freitas of Imperial College and Mr A. C. Lumsden and Dr R. Knipe of Leeds University for reading the manuscript and suggesting useful improvements.

REFERENCES

Allum, J. A. E. (1966). *Photogeology and Regional Mapping*. Pergamon Press, 107pp.

Attewell, P. B., and Farmer, I. W. (1976). *Principles of Engineering Geology*. Chapman and Hall, London, 1045pp.

Bandis, S., Lumsden, A. C., and Barton, N. R. (1981). 'Experimental studies of scale effects on the shear behaviour of rock joints.' *Int. J. Rock Mech.* Mining Sci. Geomech. Abstr. **18**, 1–21.

Barton, N. R. (1971). 'Estimation of *in situ* shear strength from back analysis of failed rock slopes.' *Proc. Int. Symp. Rock Fracture*, Nancy, France, Paper II–27.

Barton, N. R. (1973). 'Review of a new shear strength criterion for rock joints.' *Eng. Geol.*, **7**, 287–332.

Barton, N. R., and Bandis, S. (1980). 'Some effects of scale on the shear strength of joints.' *Int. J. Rock Mech. Mining Sci. Geomech. Abstr.* **17**, 69–73.

Barton, N. R., and Choubey, V. (1977). 'The shear strength of rock joints in theory and practice.' *Rock Mech.* **10**, 1–54.

Ben Menahen, A. (1976). 'Dating of historical earthquakes by mud profiles of lake bottom sediments.' *Nature*, **262**, 200–2.

Brand, E. W., Borrie, G. W. and Shen, J. M. (1983). 'Field measurements in Hong Kong residual soils.' *Proc. Int. Symp. Field Measurements in Geomechanics*, Zurich, **1**, 639–48.

Brand, E. W., Premchitt, J., and Phillipson, H. B. (1984). 'Relationship between rainfall and landslides in Hong Kong.' *Proc. 4th Int. Symp. Landslides*, Toronto, 1, 377–84.

Bray, J. W., and Goodman, R. E. (1981). 'The theory of base friction models.' *Int. J. Rock Mech. Mining Sci. Geomech. Abstr.*, **18**, 453–68.

Brown, E. T., Richards, L. R. and Barr, M. V. (1977). Shear strength characteristics of the Delabole Slates.' *Proc. Conf. Rock Eng.*, Newcastle upon Tyne, 33–51.

Burford, A. E., and Dixon, J. M. (1978). 'Systematic fracturing in young clay of the Guyahoga River Valley, Ohio and its relation to bedrock jointing and drainage segments.' *Geotechnique*, **28**, 201–6.

Clark, M. M., Grantin, A., and Rubin, M. (1972). 'Holocene activity of the Coyote Creek fault as recorded in sediments of Lake Cahvilla.' *US Geol. Surv. Prof. Paper*, 787, 112–30.

Cowland, J. W. and Richards, L. R. (1985). 'Transient groundwater rises in sheeting joints in a Hong Kong granite slope.' *Hong Kong Engineer* **13**, 27–32.

Cundall, P. A. (1971). 'A computer model for simulating progressive, large scale movements in blocky rock systems.' *Proc. Int. Symp. Rock Fracture*, Nancy, France, Paper II–8.

Deere, D. U., and Patton, F. D. (1971). 'Slope stability in residual soils.' *Proc. 4th Pan American Conf. Soil Mech. Fndtn. Eng.*, Puerto Rico, **I**, 87–170.

de Sitter, L. U. (1964). *Structural Geology*. McGraw-Hill, 551pp.

Donovan, N. C. (1978). 'Uncertainties in seismic risk procedures.' *J. Geotech. Eng. Div., ASCE*, July, 869–87.

Douglas, P. M., and Voight, B. (1969). 'Anisotropy of granites: a reflection of microscopic fabric.' *Geotechnique*, **19**, 376–98.

Einstein, H. H., Veneziano, D., Baecher, G. B., and O'Reilly, K. J. (1983). 'The effect of discontinuity persistence on rock slope stability. *Int. J. Rock Mech. Mining Sci. Geomech. Abstr.* **20**, 227–36.

Evans, R. S. (1981). 'An analysis of secondary toppling rock failures—the stress redistribution method.' *Q. J. Eng. Geol.*, London, **14**, 77–86.

Fecker, E. (1978). 'Geotechnical description and classification of joint surfaces.' *Bull. Int. Assoc. Eng. Geol.*, **18**, 111–20.

Fecker, E., and Rengers, N. (1971). 'Measurement of large scale roughness of rock planes by means of profilograph and geological compass.' *Proc. Symp. Rock Fracture*, Nancy, France, Paper 1–18.

Fookes, P. G. (1965). 'Orientation of fissures in stiff overconsolidated clay of the Siwalik System.' *Geotechnique*, **15**, 195–206.

Fookes, P. G., and Denness, B. (1969). 'Observational studies on fissure patterns in Cretaceous sediments of South East England.' *Geotechnique*, **19**, 453–77.

Gamon, T. I., and Finn, R. P. (1984). 'The structure of the Hong Kong granite—a preliminary appraisal.' *Geol. Soc. Hong Kong Newsletter*, **2**(2), 5–10.

Goodman, R. E. (1983). 'Surface and near surface excavations, General Report Theme B.' *5th Int. Congr. Rock Mechanics*, Melbourne, Australia.

Goodman, R. E., and Bray, J. W. (1976). 'Toppling of rock slopes.' *ASCE Speciality Conf. Rock Eng. Fdtns and Slopes*, **2**, 201–34.

Harris, R. (1984). 'Rotten relics rule residual regoliths,' *Geol. Soc. Hong Kong Newsletter*, **2**(1), 1–7.

Hencher, S. R. (1985). 'Limitations of stereographic projections for rock slope stability analysis.' *Hong Kong Eng.*, **13**(7), 37–41.

Hencher, S. R. and Martin, R. P. (1982). 'The description and classification of weathered rocks in Hong Kong for engineering purposes.' *Proc. 7th SE Asian Geotech. Conf.*, Hong Kong, 125–42.

Hencher, S. R., and Martin, R. P. (1984). 'The failure of a cut slope on the Tuen Mun Road in Hong Kong.' *Proc. Int. Conf. Case Histories in Geotech. Eng.*, St Louis, Missouri, 2, 683–8.

Hencher, S. R., Massey, J. B., and Brand, E. W. (1984). 'Application of back analysis to Hong Kong landslides.' *Proceedings 4th Int. Symp. Landslides*, Toronto, 631–8.

Hencher, S. R., and Richards, L. R. (1982). The basic frictional resistance of sheeting joints in Hong Kong granite, *Hong Kong Engineer*, **11**(2), 21–5.

Herget, G. (1977). *Pit Slope Manual Chapter 2—Structural Geology*, CANMET (Canada Centre for Mineral and Energy Technology), CANMET Report 77-41, 123pp.

Hobbs, B. E., Means, W. D., and Williams, P. F. (1976). *An Outline of Structural Geology*. John Wiley & Sons, Inc., 571p.

Hoek, E. (1971). 'Influence of rock structure on the stability of rock slopes.' *Proc. 1st Int. Conf. Stability in Open Pit Mining*, 49–63.

Hoek, E. (1984). 'Strength of jointed rock masses.' *Geotechnique*, **33**, 187–223.

Hoek, E., and Bray, J. W. (1981). *Rock Slope Engineering* (3rd edn). Institution of Mining and Metallurgy, London, 358pp.

Hoek, E., Bray, J. W. and Boyd, J. M. (1973). 'The stability of a rock slope containing a wedge resting on two intersecting discontinuities.' *Q. J. Eng. Geol.*, **6**, 1–55.

Hoek, E., and Brown, E. T. (1980). *Underground Excavations in Rock*. Institution of Mining and Metallurgy, London, 527pp.

Hudson, J. A., and Priest, S. D. (1983). 'Discontinuity frequency in rock masses.' *Int. J. Rock Mech. Mining Sci. Geomech. Abstr.* **20**, 73–89.

ISRM (1978). 'Suggested methods for the quantitative description of discontinuities in rock masses.' *Int. J. Rock Mech. Mining Sci. Geomech. Abstr.*, **15**, 319–68.

Janbu, N. (1973). 'Slope stability computations.' In *Embankment Dam Engineering: Casagrande Volume*. Hirschfield, R. C. and Poulos, S. J. (eds.), Wiley, New York, pp. 47–107.

Jennings, J. E. (1970). 'A mathematical theory for the calculation of the stability of open cast mines.' *Proc. Symp. Theor. Background to Planning Open Pit Mines*, Johannesburg, pp. 87–102.

Koo, Y. C. (1982). 'The mass strength of jointed residual soils.' *Can. Geotech. J.*, **19**, 225–31.

Leach, B., and Herbert, R. (1982). 'The genesis of a numerical model for the study of hydrogeology of a steep hillside in Hong Kong.' *Q. J. Eng. Geol.* **15**, 243–59.

Leroueil, S., and Tavernas, F. (1981). 'Pitfalls of back-analysis.' *Proc. 10th Int. Conf. Soil Mech. Fndtn. Eng.*, Stockholm, pp. 185–90.

McGown, A., Salvidar-Sali, A., and Radwan, A. M. (1974). 'Fissure patterns and slope failures in fill at Hurlford.' *Q. J. Eng. Geol.*, **7**, 1–26.

McMahon, B. K. (1975). 'Probability of failure and expected volume of failure in high slopes.' *Proc. 2nd Australia–New Zealand Conf. Geomech.*, Brisbane, pp. 308–17.

Moore, J. F. A. (1974). 'Mapping major joints in the Lower Oxford Clay using terrestial photogrammetry.' *Q. J. Eng. Geol.*, **7**, 57–67.

Nichols, T. C. (1980). 'Rebound, its nature and effect on engineering works.' *Q. J. Eng. Geol.*, **13**, 133–52.

Norman, J. W., and Huntingdon, J. F. (1974). 'Possible applications of photography to the study of rock mechanics.' *Q. J. Eng. Geol.*, **7**, 107–19.

Patton, F. D. (1966). 'Multiple modes of shear failure in rock.' *Proc. 1st Int. Congr. Rock Mechn.*, Lisbon, vol. 1, pp. 509–13.

Patton, F. D. (1983). 'The role of instrumentation in the analysis of the stability of rock slopes.' *Proc. Int. Symp. Field Measurements Geotech. Eng.*, Zurich, pp. 3153–82.

Pentz, D. L. (1971). 'Methods of analysis of stability of rock slopes.' *Proc. 1st Int. Conf. Stability in Open Pit Mining*, pp. 119–41.

Phillips, R. C. (1971). *The Use of Stereographic Projection in Structural Geology* (3rd edn). Edward Arnold, 90pp.

Piteau, D. R. (1970). 'Geological factors significant to the stability of slopes cut in rock.' *Symp. Planning Open Pit Mines*, Johannesburg, 33–53.

Piteau, D. R. (1973). 'Characterising and extrapolating rock joint properties in engineering practice.' *Rock Mechanics*, Supplement 2, 5–31.

Piteau, D. R., and Martin, D. C. (1977). 'Slope stability analysis and design based on probability techniques at Cassiar Mine.' *Can. Mining Metallurg. J.*, March, 1b–12b.

Pope, R. G., Weeks, R. C., and Chipp, P. N. (1982). 'Automatic recording of standpipe piezometers.' *Proc. 7th SE Asian Geotech. Conf.*, vol. 1, 77–90.

Price, N. J. (1959). 'Mechanics of jointing in rock.' *Geol. Mag.*, **96**, 149–67.

Price, N. J. (1966). *Fault and Joint Development in Brittle and Semi-brittle Rock*. Pergamon Press, 176pp.

Priest, S. D. (1985). *Hemispherical Projection Methods in Rock Mechanics*. George Allen & Unwin, 124pp.

Priest, S. D., and Brown, E. T. (1983). Probabilistic stability analysis of variable rock slopes.' *Trans. Inst. Mining and Metallurgy (Section A: Mining Industry)*, vol. 92, A1–A12.

Priest, S. D., and Hudson, J. A. (1981). 'Estimation of discontinuity spacing and trace length using scanline surveys.' *Int. J. Rock Mech. Mining Sci. Geomech. Abstr.* **18**, 183–97.

Richards, L. R., and Cowland, J. W. (1982). 'The effect of surface roughness on the field shear strength of sheeting joints in Hong Kong granite.' *Hong Kong Engineer*, **10**(10), 39–43.

Richards, L. R., Leg, G. M. M., and Whittle, R. A. (1978). 'Appraisal of stability conditions in rock slopes.' Chapter 16 in *Foundation Engineering in Difficult Ground*. Bell, F. G. (ed.), Newnes-Butterworth, 598pp.

Ross-Brown, D. M., and Walton, G. (1975). 'A portable shear box for testing rock joints.' *Rock Mech.*, **7**, 129–53.

Sarma, S. K. (1979). 'Stability analysis of embankments and slopes.' *J. Geotech. Eng. Div., ASCE*, **105**, 1511–24.

Sekula, J. (1982). 'Stereographic projections for the engineering of rock.' *Hong Kong Eng.*, **10**, 59.

Selby, M. J. (1980). 'A rock mass strength classification for geomorphic purposes: with tests from Antarctica and New Zealand.' *Zeitschrift für Geomorphologie*, **24**, 31–51.

Sharp, J. C., Maini, Y. N. T., and Harper, T. R. (1972). Influence of groundwater on the stability of rock masses: 1—hydraulics within rock masses *Trans. Inst. Mining & Metallurgy*, London, vol. 81, A13–A20.

Sherard, J. L., Cluff, L. S., and Allen, C. R. (1975). 'Potentially active faults in dam foundations.' *Geotechnique*, **24**, 367–428.

Skempton, A. W., and La Rochelle, P. (1965). 'The Bradwell Slip: A short term failure in London Clay.' *Geotechnique*, **15**, 221–42.

Skempton, A. W., Schuster, R. L., and Petley, D. J. (1969). 'Joints and fissures in the London Clay at Wraysbury and Edgware.' *Geotechnique*, **19**, 205–17.

Starr, D. C., Stiles, A. P., and Nisbet, R. M. (1981). 'Rock slope stability and remedial measures for a residential development at Tsuen Wan, Hong Kong.' *Q. J. Eng. Geol.*, **14**, 175–93.

Stimpson, B. (1979). 'Simple equations for determining the factor of safety of a planar wedge under various groundwater conditions.' *Q. J. Eng. Geol.*, **12**, 3–7.

Styles, K. A., Hansen, A., Dale, M. J., and Burnett, A. D. (1984). 'Terrain classification methods for development planning and geotechnical appraisal: A Hong Kong case.' *Proc. 4th Int. Symp. Landslides*, Toronto, 2, 561–8.

Terzhagi, R. D. (1965). 'Sources of error in joint surveys.' *Geotechnique*, **15**, 287–304.

Twidale, C. R. (1973). 'On the origin of sheet jointing.' *Rock Mech.*, **5**, 163–187.

Vecchia, O. (1977). 'A simple terrain index for the stability of hillsides or scarps, from *Large Ground Movements and Structures*.' *Proc. Conf. University of Wales*, Cardiff, pp. 449–62.

Warburton, P. M. (1983). 'Applications of a new computer model for reconstructing blocky rock geometry, analysing single block stability and identifying keystones.' *Proc. 5th Int. Congr. Rock Mech.*, Melbourne, pp. F225–30.

Whitten, E. H. T. (1966). *Structural Geology of Folded Rocks*. Rand McNally & Co., 678pp.

Wittke, W., and Louis, C. (1966). 'Determination of the influence of ground water flow on the stability of slopes and structures in jointed rock.' *Proc. 1st Int. Congr. Rock Mech.*, Lisbon, vol. 2, 201–6.

Zanbak, C. (1983). 'Design charts for rock slopes susceptible to toppling.' *J. Geotech. Eng. Div., ASCE*, **109**, 1039–62.

Slope Stability
Edited by M. G. Anderson and K. S. Richards

Chapter 6

Vegetation and Slope Stability

D. R. GREENWAY
Geotechnical Control Office, Engineering Development Department, Hong Kong

6.1 INTRODUCTION

Incorporating the effects of vegetation in slope stability analyses is a relatively recent endeavour, having first been attempted during the 1960s. This endeavour has grown steadily in the last 20 years, and the influence that grass, shrubs, and trees may have on slope stability is now emerging. Vegetation–slope interactions are complex, however, and this has hampered the efforts to quantify them in stability analyses.

Vegetation that may be growing on a slope has traditionally been considered to have an indirect or minor effect on stability, and it is usually neglected in stability analyses. This assumption is not always correct, and for certain forested slopes with relatively thin soil mantles, it has been shown to be significantly in error. This discovery has largely resulted from observations of commercial timber harvesting activities in the mountainous regions of the United States, Canada, Japan, and New Zealand, and it has lent considerable impetus to the quantification of vegetative influences in stability assessments.

The interdisciplinary nature of this endeavour is particularly noteworthy. While detailed slope stability assessments are normally carried out by engineers and engineering geologists, the organic interactions between vegetation and soil that must now be evaluated are perhaps better understood by agriculturists, soil scientists, foresters, and hydrologists. It is hoped that this chapter will foster increased cooperation and exchange between disciplines, as the path to progress in this area is undoubtedly an interdisciplinary one.

6.2 VEGETATIVE INFLUENCES: SOME HISTORICAL PERSPECTIVES

Grass, shrubs, and trees have been used to control erosion on slopes and to stabilize landslide scars for many years. Lee (1985) reports the work of engineer Pan in 1591, who employed willow plantings to stabilize embankments during the Ming Dynasty of China. A long list of practitioners have through the years pioneered slope re-vegetation techniques to emulate the work of nature. The historic contributions from Europe are described by Schiechtl (1980) and Zaruba and Mencl (1982), including the 1834 writings of Duile on the repair of steep slopes, and the willow planting techniques of Demontzey from 1880 and Wang from 1903. Kraebel (1936) reported the use of vegetation to stabilize road slopes in California, and the pictures of the same slopes given by Schiechtl (1980) attest to the success of Kraebel's techniques. Vegetation has proved effective in sand dune stabilization (McLaughlin and Brown, 1942; Bache and MacAskill, 1984). The art and science of slope re-vegetation is continuing to evolve and expand (Gray and Leiser, 1982).

Quantitative analyses of slope stability are based on limit equilibrium concepts first developed in Sweden in the early 1900s. The effective stress principle, elucidated by Terzaghi in the 1920s (Terzaghi, 1925) profoundly influenced subsequent slope stability concepts, as it provided a direct link between pore-water pressures and soil shear strength. Modern analysis techniques are discussed by Chowdhury (1978) and in Chapter 2.

Also in the 1920s, the results of quantitative studies on soil erosion brought vegetation to prominence as a key defence against erosion (Hudson, 1971). It was observed in these studies that very little erosion occurred on land with an ample vegetation cover. The performance of low vegetation, such as grass, shrubs, and food crops, featured in this research. Thereafter, erosion research rapidly expanded because of its great agricultural significance. One outcome of this work has been the development of the Universal Soil Loss Equation to predict soil losses from cropland (Wischmeir and Smith, 1965).

The first attempts to quantify the influences of vegetation on stability were reported in the 1960s, but some earlier works exhibited a qualitative appreciation of vegetation influences. The Institution of Civil Engineers (London) Conference on Biology and Civil Engineering held in 1949 produced several contributions on this topic (Toms, 1949; Moran, 1949). In 1950, Terzaghi considered deforestation to be a possible cause of the landslide that occurred in 1915 at Hudson, New York (Terzaghi, 1950). Also, Croft and Adams (1950) concluded that the loss of mechanical support from roots led to increased landslide frequency in forested areas that had been harvested in Utah.

A number of countries reported quantitative research on vegetation influences in the 1960s. Major contributions arose from the USA (Bethlahmy, 1962; Bishop and Stevens, 1964; Kaul, 1965; Hallin, 1967), USSR (Turmanina, 1963, 1965; Ter Stepanian, 1963), Israel (Kassif and Kopelovitz, 1968), and Japan (Kitamura and Namba, 1966, 1968; Takahashi, 1968; Endo and Tusruta, 1969). Many of these contributions focused on the reinforcing role of tree roots. The works of Bethlahmy and Bishop and Stevens were the initial quantitative reports on timber harvesting effects, a field that has since contributed greatly to the state-of-the-art on vegetation influences.

Research on this topic continued to increase during the 1970s and 1980s, and it became more widespread world wide. Contributions were added from Brazil, Canada, New Zealand, Sweden, and Taiwan (e.g. Prandini *et al.*, 1977; O'Loughlin, 1974a,b; O'Loughlin and Pearce, 1976; Bjorkhem *et al.*, 1975; Chang, 1972; Yen, 1972). General reviews of the significance of vegetation influences have been given by Prandini *et al.* (1977), Gray (1970, 1974, 1978), Gray and Leiser (1982), Greenway *et al.* (1984), Wu (1984), and Lee (1985).

The research cited above clearly supports the conclusion that vegetation has a net stabilizing effect on most slopes. A few researchers have found a net destabilizing effect, however, including White (1949) who studied soil erosional processes in Hawaii, Ellison and

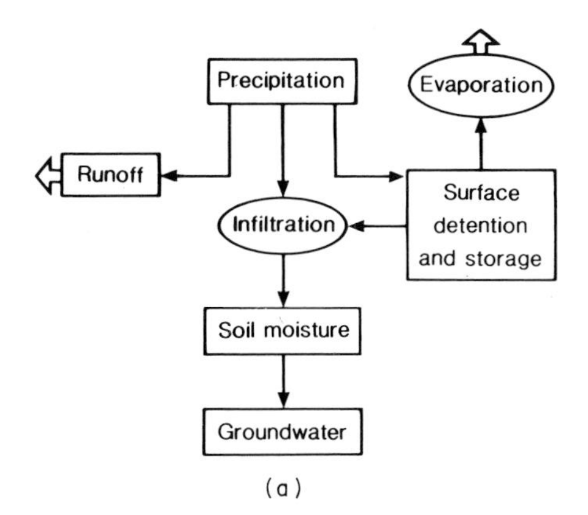

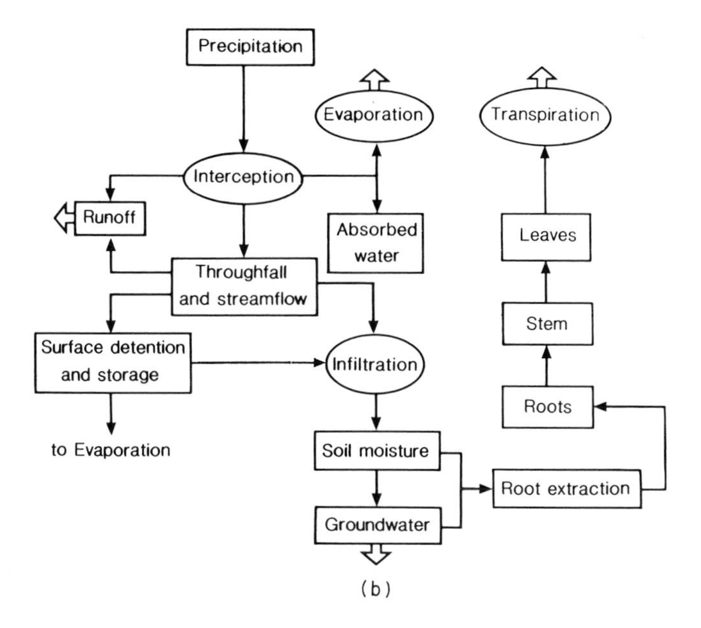

Figure 6.1 Schematic comparison of hillslope hydrological cycles; (a) on a bare slope, (b) on a vegetated slope

Coaldrake (1954) who studied timber harvesting effects in southeastern Australia, and Flaccus (1959) who made similar timber harvesting observations in eastern USA. So (1971) studied landslides during one severe rainstorm in Hong Kong and concluded that landslides occurred more frequently on heavily vegetated slopes. Shaw (1978) alleged that tree roots were a factor in the failure of the Kelley Barnes Dam in USA. These contributions are further discussed by Brown and Sheu (1975) and Gray (1978).

Another research topic, the influence of vegetation on ground movements affecting building foundations, has developed quite separately since the late 1940s. It nevertheless has contributed much to our knowledge of tree root systems and the significance of water consumed in transpiration. Desiccation and shrinkage of soil affected by trees was investigated by several early researchers (Ward, 1947, 1953; Felt, 1953; Skempton, 1954). Recent advances on this subject are covered by Cutler and Richardson (1981) and the Fourth Geotechnique Symposium, 'The influence of vegetation on the swelling and shrinking of clays' (ICE, 1983, 1984).

6.3 THE VEGETATION FACTORS

The mechanisms whereby vegetation influences slope stability may be broadly classified as either *hydrological* or *mechanical* in nature. The mechanical factors arise from the physical interactions of either the foliage or root system of the plant with the slope. The hydrological mechanisms are those intricacies of the hydrological cycle that exist when vegetation is present, as illustrated in Figure 6.1. Specific mechanisms are listed in Table 6.1 and illustrated on Figure 6.2.

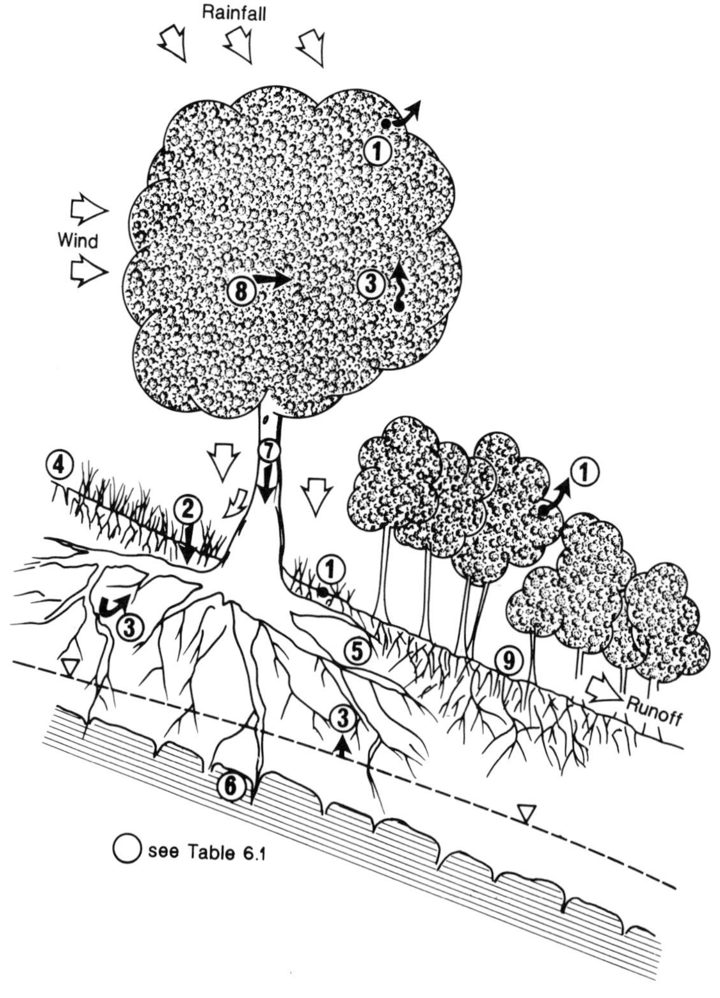

Figure 6.2 Slope–vegetation interactions influencing stability

Table 6.1 Effects of Vegetation on Slope Stability

	Hydrological Mechanisms	*Influence*
1	Foliage intercepts rainfall, causing absorptive and evaporative losses that reduce rainfall available for infiltration.	B
2	Roots and stems increase the roughness of the ground surface and the permeability of the soil, leading to increased infiltration capacity.	A
3	Roots extract moisture from the soil which is lost to the atmosphere via transpiration, leading to lower pore-water pressures.	B
4	Depletion of soil moisture may accentuate desiccation cracking in the soil, resulting in higher infiltration capacity.	A
	Mechanical Mechanisms	
5	Roots reinforce the soil, increasing soil shear strength.	B
6	Tree roots may anchor into firm strata, providing support to the upslope soil mantle through buttressing and arching.	B
7	Weight of trees surcharges the slope, increasing normal and downhill force components.	A/B
8	Vegetation exposed to the wind transmits dynamic forces into the slope.	A
9	Roots bind soil particles at the ground surface, reducing their susceptibility to erosion.	B

Legend: A – Adverse to stability
B – Beneficial to stability

6.3.1 Hydrological Factors

Rainfall on a vegetated slope is partly intercepted by the foliage, which leads to absorptive and evaporative losses of moisture that ultimately reduce the amount of rainfall available for infiltration. Vegetal debris lying on the ground surface may also have high absorptive capacities, contributing further to the water loss. In this context, water is considered 'lost' if it is evaporated back to the atmosphere or is otherwise held unavailable for infiltration.

The behaviour of moisture in snow may also be modified by the presence of vegetation. Shrubs and trees may affect snow accumulation patterns (drifts), and they may provide shade that delays the melting of the snow. The areal distribution of moisture on a slope and the rate at which the moisture is released for infiltration may thereby be affected.

The presence of roots and stems tends to increase the roughness of the slope surface, which may provide a greater capacity for infiltration. This surface roughness also reduces the velocity of surface runoff. The presence of roots and root channels left by decayed roots may increase the permeability and infiltration capacity of the surficial soil. Root channels may also contribute to the erosion of internal pipes in some soils.

Roots extract soil moisture that is ultimately lost to the atmosphere via transpiration from the foliage. In certain situations, prolonged extraction of moisture can lead to desiccation of the soil and to the formation of shrinkage cracks. Once formed, such cracks may permanently increase the permeability and infiltration capacity of the soil.

6.3.2 Mechanical Factors

Roots, due to their tensile strength and frictional or adhesional properties, reinforce the soil. Large roots, particularly of trees, may penetrate deeply and become anchored in firm strata, thereby forming a support (buttress) to the soil mantle upslope of the tree. By binding the soil particles at the ground surface, roots reduce the rate of soil erosion which may otherwise lead to undercutting and instability of slopes.

On rock slopes, tree roots may enter discontinuities and wedge blocks apart, possibly causing the detachment and fall of boulders. Conversely, trees may serve as a barrier to arrest the fall of boulders, or as avalanche barriers.

The weight of large trees surcharges the slope, increasing both normal and downhill force components on potential slip surfaces. Vegetation exposed to wind transmits dynamic forces into the slope, and if uprooting or overturning occurs, both increased erosion and infiltration may result.

6.3.3 Beneficial and Adverse Factors

The vegetation influences can be generally classified as either beneficial or adverse to stability, as shown in Table 6.1. The hydrological mechanisms that lead to lower pore-water pressures are beneficial, while those that yield increased pore pressures are adverse. Of the mechanical mechanisms, those that increase shear resistance in the slope are beneficial, while those that increase shear stress are adverse.

It is clear that not all the factors listed in Table 6.1 are universally applicable to all slopes. Additionally, the relative weighting of each factor would undoubtedly vary widely from slope to slope. It is not sufficient, therefore, simply to classify individual mechanisms; they must be quantified. Only then can the net influence of the vegetation be clarified and its significance to stability defined.

6.4 SIGNIFICANCE OF THE HYDROLOGICAL FACTORS

6.4.1 Interception

Interception losses that occur on vegetated slopes are controlled by many factors, including the type and species of vegetation present, the proportion of the slope area vegetated, rainfall intensity and duration, antecedent moisture conditions, and climatic or seasonal factors. Only general trends can be drawn from the available literature, but it appears that interception losses may account for a large proportion of gross rainfall in dense forests under certain circumstances.

Gregory and Walling (1973) report that approximately two-thirds of the total rainfall was lost in an evergreen rainforest in Brazil, based on data by Freise in 1936 (Table 6.2). Prandini *et al.* (1977) report the work of Sternberg, where a tree canopy intercepted from

Table 6.2 Approximate Average Rainfall Interception, Evergreen Rainforest of Brazil. (From Gregory and Walling, 1973. *Reproduced by permission of the authors.*)

Penetrating to rain gauge at 1.5 m 33%				
Evaporated directly from tree crowns 20%				
Running down trunks 46%→	evaporated from surface	9.2%		
	absorbed by bark	9.2%		
	reaching base of tress	27.6%→	absorbed by roots	20.7%
			reaching water table	6.9%

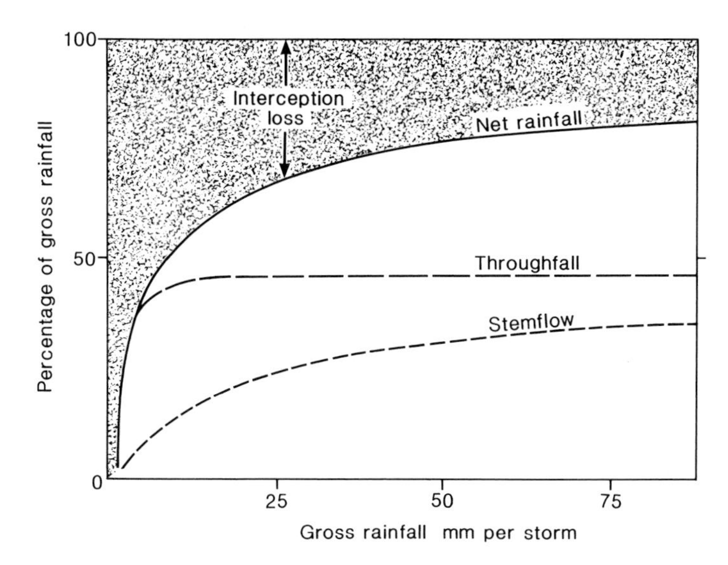

Figure 6.3 Interception losses from 5 m tall manuka stand in New Zealand. (*From Aldridge and Jackson, 1968*, NZ Journal Science, **II**. *Reproduced by permission of the Department of Scientific and Industrial Research, New Zealand*

10% to 25% of the precipitation, with up to 100% intercepted in periods of light rainfall. Similarly, Aldrich and Jackson (1968) found interception in a 5 metre high manuka (*Leptospermum scoparium*) stand in New Zealand to vary from 100% in very light rainfall to 20% in rainstorms of 75 mm gross rainfall (Figure 6.3).

A general trend of diminishing interception loss with increasing storm magnitude and intensity can be noted in these data. While interception losses would likely be greatest on densely forested slopes, further investigation of the factors controlling interception on all types of vegetated slopes would be most useful. As it now stands, interception is usually lumped together with runoff, infiltration or evapotranspiration when modelling hydrological processes relevant to stability, and the significance of interception cannot then be independently determined or predicted.

6.4.2 Infiltration

Increased capacities for infiltration have been experimentally observed on grassed slopes (Nassif and Wilson, 1975). As shown in Figure 6.4, a four-fold increase in infiltration rate was noted on a very gentle (9°) grassed slope, compared to a bare slope comprised of the same clayey sand. Similar data from agricultural land are given by Strahler (1969). The infiltration phenomenon is highly intricate, being influenced not only by vegetation, but also by rainfall characteristics, soil properties, and land management practices (Lull, 1964; Musgrave and Holtan, 1964; Dunne, 1978).

Increased permeability and infiltration capacity of the surface soil layers of vegetated slopes may be attributed to the presence of roots, vacant root channels, and increased macroscopic surface roughness (Gaiser, 1952; Aubertin, 1971). There is no satisfactory way to predict increases in permeability caused by vegetation, so recourse must be made to field permeability tests within the root zone to quantify this factor on particular sites. Once actual

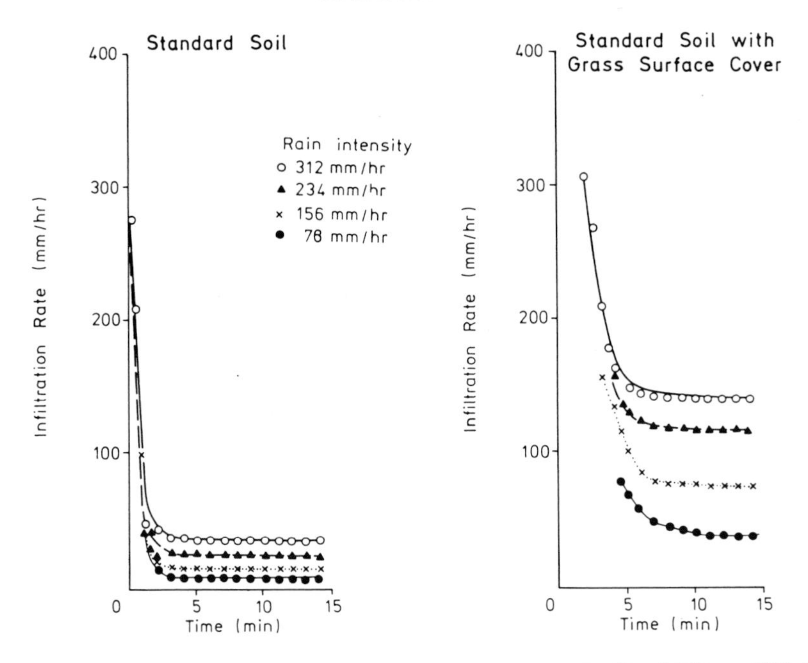

Figure 6.4 Comparative infiltration capacities from a 9° slope. (*From Nassif and Wilson, 1975,* Hydrological Sciences Bulletin, **20**. *Reproduced by permission of* Hydrological Sciences Bulletin.)

values for root zone permeability have been obtained, the pore-pressure response of a slope during infiltration can be predicted using a finite difference soil-water model (Anderson, 1984).

6.4.3 Transpiration

The rate at which a plant consumes soil moisture depends on a great number of factors, including the type, size and species of vegetation, weather, climatic and seasonal factors, and on features of the growing site (slope aspect, moisture availability, soil type). Table 6.3 illustrates the wide divergence between transpiration rates of grass and individual trees in South Africa, and between different tree species (Williams and Pidgeon, 1983). In broad terms, Scheichtl (1980) considers a typical annual transpiration loss in a deciduous forest in

Table 6.3 Transpiration Rates in South Africa (From Williams and Pidgeon, 1983, Geotechnique, **33**. *Reproduced by permission of the Institution of Civil Engineers.*)

Type of plant	Transpiration of moisture on a sunny day
Gum tree (*Eucalyptus macarthuri*)	500 l/day
Black wattle (*Acacia mollissima*)	250 l/day
Highveld grass (*Themeda*)	1 l/m^2/day

Table 6.4 Water Use and Rooting Habit of Trees on Clay Soils in England (From Binns, 1980, in: Trees and Water. *Reproduced by permission of HMSO.*)

Name	Species	Rooting Habit	Water Use
Broadleaves			
Alders	*Alnus spp*	Moderately deep	No information
Apples	*Malus spp*	Locally deep	Moderate
Ash	*Fraxinus excelsior*	Moderately deep	Moderate
Beech	*Fagus sylvatica*	Shallow	Low
Birches	*Betula spp*	Shallow	Low
Cherries	*Prunus spp*	Moderately deep	Low
Chestnut, horse	*Aesculus hippocastanum*	Moderately deep	Moderate
Hawthorn	*Crataegus spp*	Moderately deep	High
Limes	*Tilia spp*	Moderately deep	Moderate
Maples (large) and sycamore	*Acer spp*	Moderately deep	Moderate
Oaks	*Quercus spp*	Deep	High
Oak, turkey	*Quercus cerris*	Fairly shallow	No information
Plane, London	*Platanus × acerifolia*	Moderately deep	Moderate
Poplars	*Populus spp*	Deep, widespread, and intense	High
Willows	*Salix spp*	Moderately deep, widespread	High?
Conifers			
Cedars (true)	*Cedrus spp*	Moderately deep?	No information
Douglas fir	*Pseudotsuga menziesii*	Shallow	Moderate
Larches	*Larix spp*	Deep?	No information
Lawson cypress	*Chamaecyparis lawsoniana*	Moderately deep, intense	No information
Monterey cypress	*Cupressus macrocarpa*	Moderately deep, intense	High
Pine, Corsican and Austrian	*Pinus nigra*	Moderately deep	High
Pine, Scots	*Pinus sylvestris*	Shallow	High
Redwood, Sierra (Wellingtonia)	*Sequoiadendron giganteum*	Deep?	No information
Silver firs	*Abies spp*	Deep	No information
Spruces	*Picea spp*	Shallow	Moderate

Table 6.5 Water Use by Dense Growths of Phreatophytes (from Robinson, 1958. *Reproduced by permission of the U.S. Geological Survey.*)

Name	Species	Climate	Depth to Water (m)	Annual Water Consumption (l/m²)
Saltcedar	*Tamarix gallica*	Hot dry	1.23	2,700
		Hot dry	2.13	2,240
Greasewood	*Sacrobatus vermiculatus*	Cool dry	0.50	660
Willow	*Salix*	Hot dry	0.61	1,340
Cottonwood	*Populus*	Hot dry	2.20	2,380
Alfalfa	*Medicago savita*	Cool dry	0.91	800
		Hot dry	1.38	1,130
Alder	*Alnus*	Hot dry	—	1,620

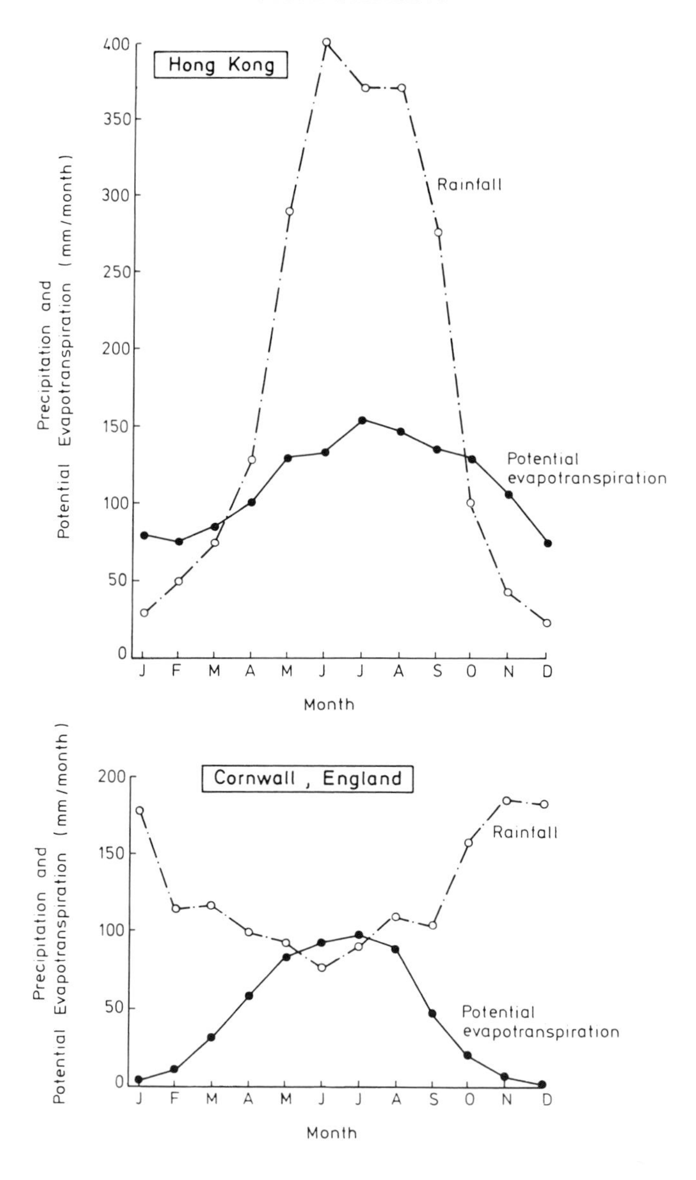

Figure 6.5 Potential evapotranspiration of grass in Hong Kong and Cornwall, England

Europe (1000 kg of water per square metre of forest) to be twice that of a spruce forest (500 kg/m^2). While the water demands of a large number of tree species are appreciated within localized areas (Table 6.4 for example) quantitative transpiration data on slope vegetation are generally lacking.

Plants that have a particular affinity for transpiring water are termed phreatophytes. Certain trees, shrubs, and grasses that have deeply penetrating roots habitually reaching the ground water table are in this category (Robinson, 1958). Water consumption rates for several phreatophytes are given in Table 6.5. Davis and DeWiest (1966) consider that

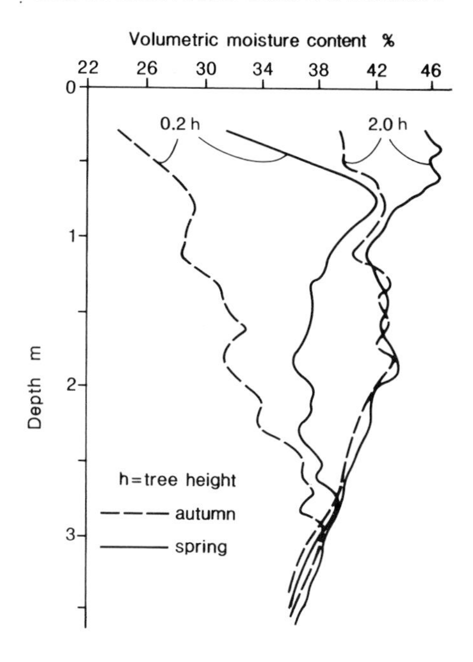

Figure 6.6 Soil moisture profiles in boulder clay near a poplar tree at different relative distances (*From Biddle, 1983*, Geotechnique, **33**. *Reproduced by permission of the Institution of Civil Engineeers.*)

trees of this type are capable of sending roots to a 30 m depth (if soil conditions are not otherwise limiting), and that shrubs and grasses are capable of sending roots down to 10 m and 3 m, respectively.

Figure 6.5 illustrates fluctuations in transpiration rates caused by seasonal and climatic factors. The potential evapotranspiration rate for grass in both Cornwall, England, and Hong Kong peaks in July, which is generally indicative of sites in the Northern Hemisphere. (For convenience, evaporation and transpiration moisture losses are often jointly reported as evapotranspiration; potential evapotranspiration denotes the full biological rate possible when soil moisture is freely available.) The overall evapotranspiration rate is markedly higher in sub-tropical Hong Kong than in the temperate climate of England. Rice and Krammes (1970) discuss the relationship between evapotranspiration and rainfall, and the implications this may have on slope stability.

On vegetated slopes the demands of the biological cycle for water are met by the extraction of soil moisture by roots; this directly lowers the moisture content of the soil within the root zone, and may generally alter the distribution of soil moisture (and pore-water pressures) well beyond the root zone. This is shown in detailed soil moisture studies near five types of trees growing on clayey soils in England (Biddle, 1983). As an example, Biddle found poplar trees caused significant changes in soil moisture content to a depth of 3.5 m near the tree, with the largest effect near the ground surface (Figures 6.6 and 6.7). A lesser effect was evident at increased distances from the tree. The average height of the poplar trees studied was 17 m.

Reductions in soil moisture content result in lower pore pressures within the slope, observable as increased matrix suctions in unsaturated soil and lower groundwater levels. Brenner (1973) reported model studies of vegetated and cutover slopes in which soil suction

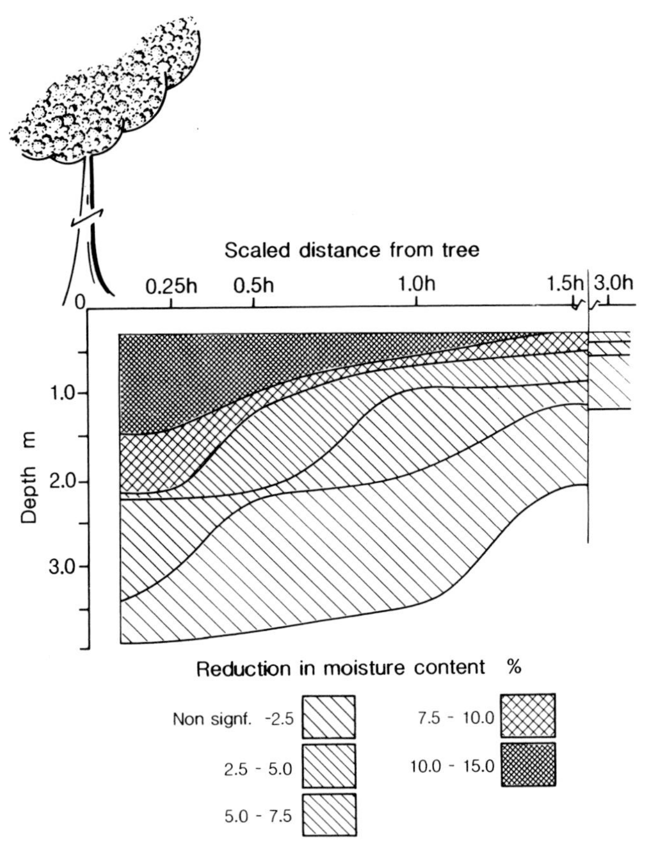

Figure 6.7 Reductions in soil moisture near a poplar tree on boulder/Oxford/Gault/London/clay. (*From Biddle, 1983*, Geotechnique, **33**. *Reproduced by permission of the Institution of Civil Engineers.*)

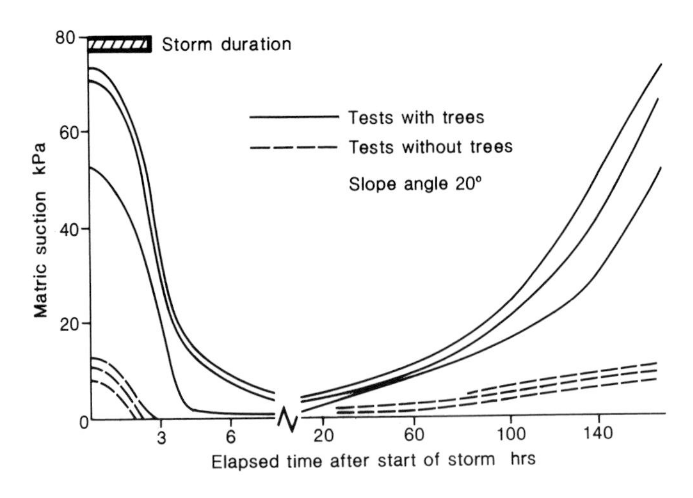

Figure 6.8 Soil suction response of a model slope. (*From Brenner, 1973*, Hydrological Sciences Bulletin, **18**. *Reproduced by permission of Hydrological Sciences Bulletin.)*

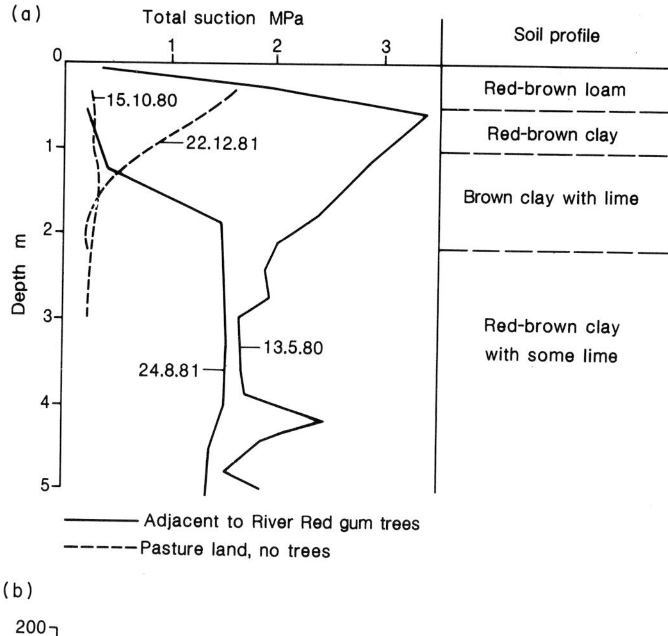

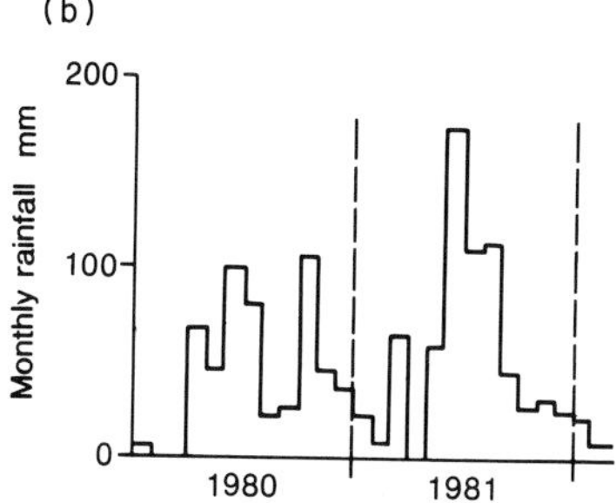

Figure 6.9 Influence of river red gum trees on soil suction profiles (a) Measured profiles of soil suction. (b) Monthly rainfall in Adelaide, Australia. (*From Richards* et al., *1983*, Geotechnique, **33**. *Reproduced by permission of the Institution of Civil Engineers.*)

was measured during and after a simulated storm. As shown in Figure 6.8, much higher suctions existed in the vegetated slope, although these suctions also dropped to relatively low values a short time after the storm.

Richards *et al.* (1983) report field measurements of soil suction for a site in Australia (Figure 6.9) that was partly affected by river red gum trees. Soil suctions were markedly higher near the trees than in adjacent pasture land without trees. Seasonal fluctuations in suction generally reached to a depth of 1.4 m in the adjacent grassed area, while near the trees the fluctuations persisted to a depth of 2.4 m.

Transpiration acts to lower groundwater levels; conversely, when vegetation is suddenly removed and transpiration ceases, a rise in the groundwater level can be observed. This behaviour has been widely reported in forest harvesting. For example, Hostener-Jorgensen (1967) gives groundwater data (Figure 6.10) for a beech forest in Denmark before and after tree removal (clearcutting). The water table was found to be several metres higher in the summer subsequent to clearcutting. Similar trends have been reported by Bethlahmy (1962) and Gray (1977).

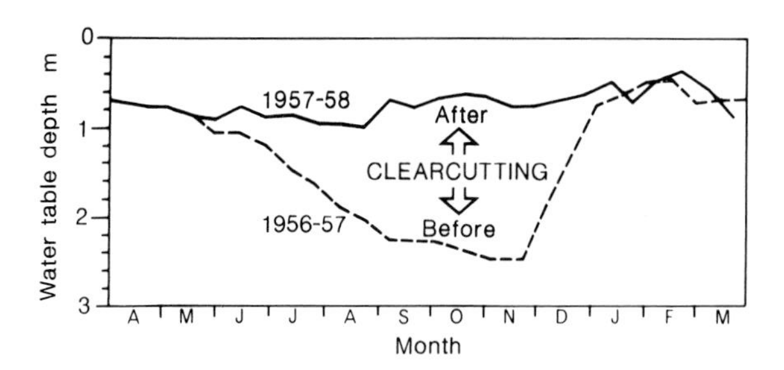

Figure 6.10 Groundwater fluctuations in a beech forest in Denmark. (*From Holstener-Jorgensen, 1967, in:* International Symposium on Forest Hydrology (*eds Sopper and Lull*). *Reproduced by permission of Pergamon Press.*)

Table 6.6 Clay Shrinkage Potential (From Driscoll, 1983, Geotechnique, **33**. *Reproduced by permission of the Institution of Civil Engineers.*)

Plasticity index (%)	Clay fraction (%)	Shrinkage potential
>35	>95	Very high
22–48	60–95	High
12–32	30–60	Medium
<18	<30	Low

Modelling of transpiration *per se* has been well reviewed by Penman (1963), Kramer (1969) and Muck and Hillel (1983). It may be possible in future to link transpiration models directly to the soil-water models used in stability analyses, which will then make it possible to determine fully the overall significance of transpiration to slope stability.

6.4.4 Desiccation Cracking

Transpiration by trees growing on clay soils may accelerate the formation of shrinkage cracks during the dry season. A particular combination of soil, vegetation, and climatic factors is required to produce such cracking. Anderson *et al.* (1982) and Anderson and Kneale (1984) describe the soil conditions leading to the formation of shrinkage cracks in a compacted clay embankment in England, where evaporative drying of the surface produced cracking to a depth of 200 mm. They reported that these cracks closed up during the following wet season, but the permeability of the clay had nevertheless been permanently increased by about two orders of magnitude.

Shrinkage cracks may, therefore, lead to increased infiltration capacity, higher pore pressures, and decreased stability of slopes. Natarajan and Gupta (1980) reported landsliding in decomposed shale in India, where desiccation cracking exacerbated by vegetation had been a causative factor. Soils most susceptible to cracking are the shrinkable clays, or soils with a high clay content. Shrinkable clays may be generally identified by their index properties (Holtz and Gibbs, 1956); see Table 6.6.

6.5 ROOT MORPHOLOGY AND STRENGTH

Large differences in the root systems of plants are intuitively obvious; a small garden weed that has hairlike roots extending a few centimetres into the ground readily contrasts with a large tree, whose structural roots may extend for many metres to provide firm anchorage. Between these extremes, an amazing variety of complex root systems have been observed. From a stability viewpoint, these variations are highly significant and will be briefly reviewed.

6.5.1 Root Types

Plants exhibit markedly different capacities for root growth (Kozlowski, 1971). Some plants produce *extensive* root systems, sending roots to considerable depth and spread, while others produce *intensive* systems of shorter and finer roots that are more localized. Extensive root systems of trees are comprised of *lateral roots* that radiate outward from the tree, and *sinker roots* that are oriented more vertically. A main vertical root, or *tap root*, is centrally located in many species.

The form of individual root systems is greatly dependent on soil and site conditions, as well as plant species. The relative hardness or compactness of ground stratum, the position and fluctuations of the groundwater table, the capacity of the soil to retain moisture, the availability of nutrients and air, and the presence of toxic elements in the soil all have a marked influence on the development of roots. Plants respond to these conditions by producing *adventitious roots* (Vanicek, 1973; Gray, 1978). For these reasons, a wide range of rooting habits has been reported in the literature; Table 6.7 lists a number of detailed studies of tree root morphology for reference, and some examples are cited below.

6.5.2 Depth of Root Systems

Roots, particularly of trees, have at times been observed at great depths. Coatsworth and Evans (1984) observed eucalypt roots at a depth of 22 m in South Africa, and Williams

Table 6.7 Selected Morphological Studies of Tree Roots

Reference	Trees Studied
Atkinson *et al.* (1976)	Apple
Bannan (1940)	Conifers
Curtis (1964)	Ponderosa pine
Henderson *et al.* (1983)	Sitka spruce
Heyward (1933)	Longleaf pine
Holch (1933)	Deciduous trees
Kalela (1949)	Pine, spruce
Kozlowski and Scholtes (1948)	Pine, hardwoods
Laitakari (1929)	Pine
McMinn (1963)	Douglas fir
McQuilkin (1935)	Pitch pine
Merritt (1968)	Red pine
Rogers and Booth (1960)	Fruit trees
Sutton (1969)	Conifers

Table 6.8 Damage Ranking of Trees in the Kew Root Survey (After Cutler and Richardson, 1981, in: Trees, Roots and Buildings. *Reproduced by permission of HMSO.*)

Damage Ranking	Tree Species	Max. tree height (m)	Root Damage Max. distance recorded (m)	Root Damage Max. distance for 90% of cases (m)
1	Oak	16–23	30	18
2	Poplar	25	30	20
3	Lime	16–24	20	11
4	Common ash	23	21	13
5	Plane	25–30	15	10
6	Willow	15	40	18
7	Elm	20–25	25	19
8	Hawthorn	10	11.5	9
9	Maple/sycamore	17–24	20	12
10	Cherry/plum	8	11	7.5
11	Beech	20	15	11
12	Birch	12–14	10	8
13	White beam/rowan	8–12	11	11
14	*Cupressus macrocarpa*	18–25	20	5

and Pidgeon (1983) reported gum tree rooting to 27.5 m. Roots were observed at a depth of nearly 30 m during excavations for the Panama Canal (Davis and DeWeist, 1966). Kozlowski (1971) notes that the desert shrub *Welwitschia mirabilis* may root to 18 m depth; conversely Kozlowski reports that 80% of the roots of some yellow birch trees in eastern Canada were located in the top 60 mm of the soil. The most numerous observations reported by Kozlowski range from 1 to 3 m, however, which may perhaps be considered typical of trees and shrubs in many vegetated slopes.

The depth of rooting is constrained by bedrock at relatively shallow depths in many slopes. The degree to which roots are able to penetrate rock stratum depends on the frequency and nature of discontinuities in the bedrock. A hand-held cone penetrometer has been developed in Japan to predict the extent of root penetration into soft rocks (Yamanaka, 1965).

In some localities, qualitative ratings of plant rooting habits may be available, such as shown in Table 6.4. Such ratings reflect local soil, water, and climatic conditions, and are not universally applicable.

6.5.3 Root Spread

Tree roots may spread laterally for considerable distances. Kozlowski (1971) reports poplar tree roots that extended more than 65 m in sandy soil in Wisconsin. Structural damage to buildings that was attributed to trees located up to 40 m away has been reported (Cutler and Richardson, 1981).

The extent of root spread is often expressed in relative multiples of the tree height or the radius of the crown (foliage). Kozlowski (1971) cites 10-year-old pine trees growing on sandy soil whose roots extended laterally about seven times the average height of the

Table 6.9 Tensile Strength of Shrub and Tree Roots (After Schiechtl, 1980, Bioengineering for Land. Reclamation and Conservation. *Reproduced by permission of University of Alberta Press.*)

Species	Common Name	Tensile Strength (MPa)
Acacia confusa	Acacia	11
Aleurites moluccana	Candlenut	6
Alnus firma v. multinervis	Alder	51
Alnus firma v. yasha	Alder	4–74
Alnus incana	Alder	32
Alnus japonica	Japanese alder	41
Betula pendula	European white birch	37
Cytisus scoparius	Scotch broom	32
Ficus microcarpa	Chinese banyan	16
Lespedeza bicolor	Shrub lespedeza	69
Meterosideros umbellata	Rata	53
Nothofagus fusca/truncata	Beech	36
Picea abies	European spruce	27
Picea sitchensis	Sitka spruce	23
Pinus densiflora	Japanese red pine	32
Pinus radiata	Radiata pine	18
Populus nigra	Black poplar	5–12
Populus deltoides (USSR)	Poplar	38
Populus deltoides (New Zealand)	Poplar	36
Populus euramericana ('1–78')	American poplar	46
Populus euramericana ('1–488')	American poplar	32
Populus yunnanensis	Yunnan poplar	38
Pseudotsuga menziesii (British Columbia)	Douglas fir	61
Pseudotsuga menziesii (Oregon)	Douglas fir	50
Psuedotsuga menziesii (Rocky Mountain)	Douglas fir	19
Quercus robur	Oak	32
Robinia psuedoacacia	—	68
Salix purpurea ('Booth')	Willow	36
Salix matsudana	Willow	36
Salix fragilis	Willow	18
Salix dasyclados	Willow	17
Salix elaeagnos	Willow	15
Salix helvetica	Willow	14
Salix hastata	Willow	13
Salix starkeana	Willow	12
Salix cinerea	Willow	11
Salix hegetschweileri	Willow	9
Thuja plicata	Western red cedar	56
Tilia cordata	Linden	26
Tsuga heterophylla	Western hemlock	27
Vaccinium	Huckleberry	16

trees, although this appears to be an extreme case. Fruit trees growing on clay were observed to have roots extending about 1.5 times the crown radius, while similar trees growing on loam extended to 2 times, and those on sand to 3 times the crown radius.

The Kew Root Survey conducted between 1971 and 1979 in England produced considerable data regarding the spread of tree roots and damage caused to buildings by transpiration (Cutler and Richardson, 1981). In this study, 2,285 claims involving subsidence of buildings were investigated that implicated trees as the cause, many of which arose in the 1975–76 drought.

Findings from the Kew Survey are summarized in Table 6.8. Many cases of damage were reported for elm, oak, poplar, and willow trees at separation distances of 18 to 40 m. The survey generally confirmed the '1H' rule arising from the work of Ward (1953), in which a separation distance equal to the height of the tree is prescribed between trees and buildings with shallow foundations resting on clay, in order to avoid possible structural damage from transpiration. Willow, however, appears to be an exception to the '1H' safety rule, as over 10% of all complaints against willow arose from beyond the '1H' distance.

Few of the foregoing examples refer directly to trees on slopes, but they provide an indication of possible rooting habits that can serve as a starting point for site-specific investigations.

6.5.4 Root Strength

The tensile strength of roots has been investigated by a number of researchers, beginning with Stiny in 1947 (Schiechtl, 1980). Several investigators have noted that it is difficult to test roots for tensile strength, as gripping the root during the test often damages it. Nevertheless, a sizeable amount of root strength data has now been amassed (Table 6.9). Wide variations in tensile strength (tensile stress at failure) can be noted between species. Some roots are very strong (up to 70 MPa), being one-quarter the tensile strength of mild steel.

Root tensile strength is known to vary with growing environment, season, root orientation, and root diameter, as well as species. Differences in growing environment are evident in the data from Burroughs and Thomas (1977), who tested Douglas-fir roots from both the Rocky Mountains and the Coast Range (western coast) of the United States. As shown in Table 6.10, roots from the Coast Range were about 2.5 times stronger than those from the Rocky Mountains.

Table 6.10 Tensile Strength Comparison of Douglas fir Roots from Different Locations. (From Burroughs and Thomas, 1977, Research Paper INT-190. *Reproduced by permission of US Forest Service.*)

Species/Location	Root Diameter (mm)	Tensile Strength (MPa)
Rocky Mountain Douglas fir	2	22.7
	4	22.3
	6	17.8
	8	16.2
	10	14.8
	average	18.7
Coastal Douglas fir	2	56.7
	4	58.7
	6	47.2
	8	44.7
	10	43.1
	average	50.0

Table 6.11 Tensile Strength of Roots With Different Stabilization Tissue. (From Schiechtl, 1980, in: Bioengineering for Land Reclamation and Conservation. *Reproduced by permission of University of Alberta Press.*)

Plant species	Tensile Strength (MPa) Minimum	Maximum	Mean	No. of samples
Alnus incana				
uphill	10.6	55.5	32.8	28
downhill	6.9	56.2	28.3	10
Alnus japonica				
uphill	12.5	90.5	42.0	24
downhill	17.2	73.8	40.1	25
Pinus densiflora				
uphill	30.9	71.2	47.6	6
downhill	12.7	33.8	24.8	9
horizontal	8.9	41.6	28.4	5

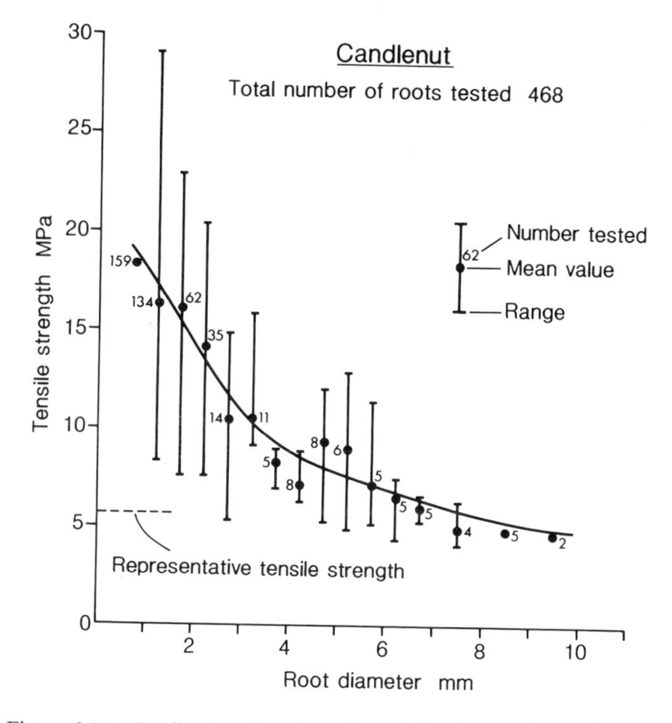

Figure 6.11 Tensile strength of candlenut (*Aleurites moluccana*) tree roots

Hathaway and Penny (1975) reported that variations in specific gravity and lignin/cellulose ratio within poplar and willow roots produced seasonal fluctuations in tensile strength. Schiechtl (1980) has shown for several tree species growing on slopes that roots extending uphill are stronger than those extending downhill (Table 6.11), apparently due to differences in structure of the root tissue. Several researchers have noted a decrease in tensile strength with increasing root diameter (Greenway *et al.*, 1984; Turmanina, 1965;

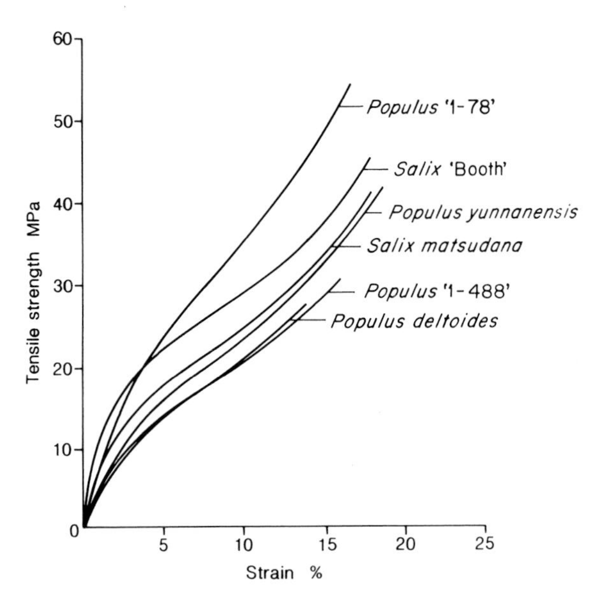

Figure 6.12 Typical stress–strain curves for poplar and willow roots. (*From Hathaway and Penny, 1975*, NZ Journal Botany, **13**. *Reproduced by permission of New Zealand Journal of Botany*.)

Table 6.12 Tensile Strength and Stress–Strain Behaviour of Some Poplar and Willow Roots. (From Hathaway and Penny, 1975, New Zealand J. Botany, **13**. *Reproduced by permission*.)

Species	Clone	Tensile Strength (MPa)	Young's Modulus (MPa)	Ultimate Strain (%)
Poplar	*Populus* 'I-78'	45.6	16.4	17.1
	P. 'I-488'	32.3	8.7	16.8
	P. yunnanensis	38.4	12.1	18.7
	P. deltoides	36.3	9.0	12.4
Willow	*Salix matsundana*	36.4	10.8	16.9
	S. 'Booth'	35.9	15.8	17.3

Wu, 1976; Burroughs and Thomas, 1977). Figure 6.11 shows the relationship between root tensile strength and diameter for candlenut trees (*Aleurites moluccana*) in Hong Kong.

From the foregoing discussion, it is clear that the tensile strength data summarized in Table 6.9 must be regarded as indicative for the species listed, rather than definitive. It should also be noted that a recognized standard test procedure for root testing is lacking, and that differences in test procedure may contribute to the scatter of the data.

Relatively little is known about the stress–strain behaviour of roots. Hathaway and Penny (1975) gave typical stress–strain curves obtained from several species of poplar and willow roots (Figure 6.12). They tested 50 mm long specimens, without bark, that had been air dried and then rewetted by soaking in water for one hour prior to testing. The ultimate breaking strains, Young's moduli, and tensile strengths measured in these tests are given in Table 6.12.

A limited amount of data is also available in the literature on the shear strength of individual roots (Ziemer, 1978, 1981). A number of researchers report data on the rate of decay in root strength after timber harvesting (O'Loughlin, 1974b; O'Loughlin and Watson, 1979, 1981; Burroughs and Thomas, 1977; Ziemer and Swanston, 1977).

6.6 QUANTIFYING THE MECHANICAL FACTORS

6.6.1 Root Reinforcement

There is direct evidence that root reinforced soil is stronger than the same soil without roots (Figure 6.13). These field data show that the shear strength of California coastal sand containing pine roots increased dramatically when increasing amounts of roots were present (Ziemer, 1981). (For Figure 6.13, biomass has been defined as the dry weight of live roots in a given soil volume.) Similar data have been reported by Endo and Tsuruta (1969), O'Loughlin (1974a), and Waldron and Dakessian (1981). It is extremely difficult to obtain reliable data from field tests on soil containing roots, but these results have been corroborated by laboratory studies (Kaul, 1965; Manbeian, 1973; Waldron, 1977; Brenner and Ng, 1977).

Roots provide a reinforcing effect to soil through their tensile resistance and frictional or adhesional properties. The mechanism is essentially that put forward by Vidal (1966) for reinforced earth, a combination of soil and steel reinforcing straps used to construct retaining structures. McGown *et al.* (1978) and Andrawes *et al.* (1980) review such earth reinforcement mechanisms.

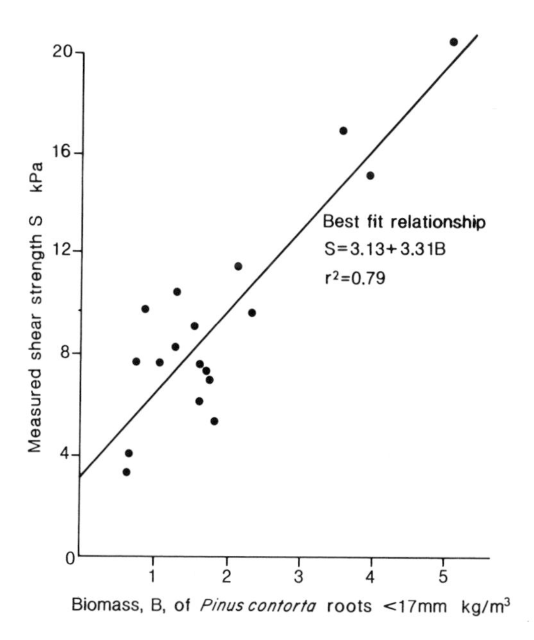

Figure 6.13 Shear strength of root permeated sand from field tests. (*From Ziemer, 1981, in*: Erosion and Sediment Transport in Pacific Rim Steeplands. *Reproduced by permission of International Association of Hydrological Sciences.*)

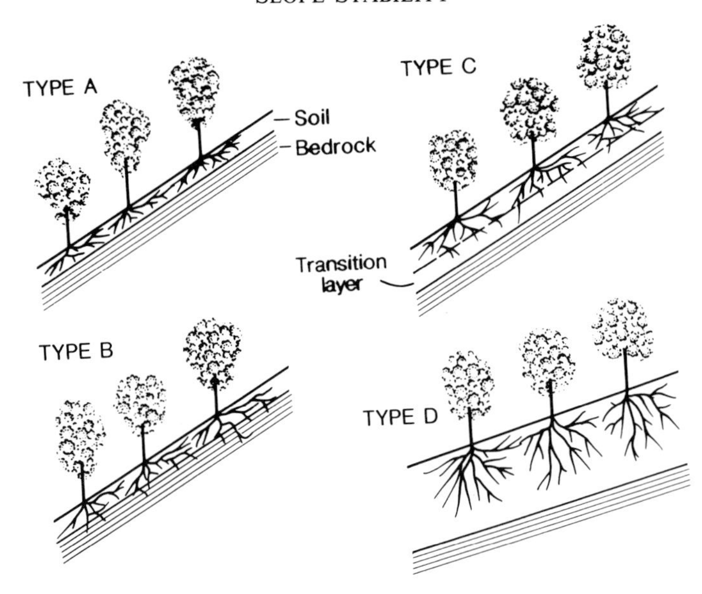

Figure 6.14 Slope classification scheme based on root reinforcement and anchoring (*From Tsukamoto and Kusakabe, 1984, Hawaii Symposium. Reproduced by permission of Forest Research Institute, NZ Forest Service.*)

Whether the reinforcing effect of roots is significant to slope stability depends primarily on the depth of potential slip surfaces within the slope. Tsukamoto and Kusakabe (1984) have developed a simple classification scheme for forested slopes in Japan, based on the depth of critical slip surfaces relative to the root zone (Figure 6.14). In this scheme, type A slopes have a relatively thin soil mantle that can be fully reinforced with tree roots, but massive bedrock which underlies the soil layer cannot be penetrated by roots. A plane of weakness may therefore exist at the bedrock interface that precludes the vegetation from having a major beneficial influence on stability. Type B slopes are similar to type A except that the bedrock layer now contains discontinuities that can be penetrated by roots. The presence of trees may have a major beneficial effect on these slopes. Type C slopes have thicker soil mantles that contain a transition layer in which soil density and shear strength increase with depth. Roots penetrating the transition layer provide a stabilizing force to the slope. Type D slopes have thick soil mantles where the potential for deep seated movement exists below the root zone. In this case, the trees are 'floating' and may have little mechanical influence on stability. Tsukamoto and Kusakabe's classification system may not be applicable in all geological conditions, but it does highlight the important relationship between root morphology and slope stability that should be considered in any stability assessment incorporating reinforcement by roots.

6.6.2 Root Reinforcement Models

Wu (1976) developed a simple theoretical model for predicting a shear strength increase due to the presence of roots. Similar models were developed independently by several other researchers (Gray and Leiser, 1982).

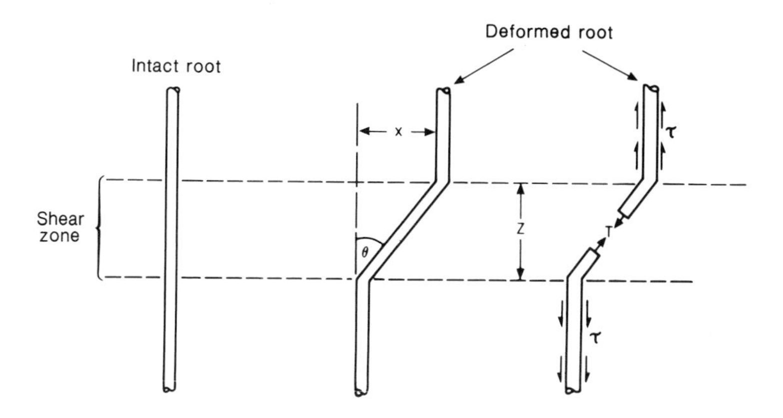

Figure 6.15 Perpendicular root reinforcement model. (*After Cary and Chasti, 1983. Reproduced by permission of the American Society of Civil Engineers.*)

This limit-state model assumes that a flexible elastic root extends across a shear zone as shown in Figure 6.15. Distortion of the root within the shear zone develops tension in the root; the component of this tension tangential to the shear zone directly resists shear, while the normal component increases the confining pressure on the shear plane. The root is assumed to be anchored in the soil on either side of the shear zone by friction or adhesion, such that it does not pull out.

If it is further assumed that all roots are oriented perpendicular to the slip plane, that the tensile strength of all roots is fully mobilized, and that the roots do not alter the angle of internal friction of the soil itself, then the resulting shear strength increase, Δs, predicted by the model is:

$$\Delta s = t_R(\sin\theta + \cos\theta\tan\phi) \tag{6.1}$$

where t_R is the average tensile strength of roots per unit area of soil, ϕ is the angle of internal friction of the soil, and θ is the angle of shear distortion in the shear zone. Angle θ (Figure 6.15) is given by:

$$\theta = \tan^{-1}(x/z) \tag{6.2}$$

where x is the shear displacement, and z is the shear zone thickness.

The average tensile strength of roots per unit area of soil (t_R) can be determined by multiplying the mean tensile strength of roots, T_R, by the fraction of soil cross-section occupied by roots (A_R/A):

$$t_R = T_R\,(A_R/\mathrm{A}) \tag{6.3}$$

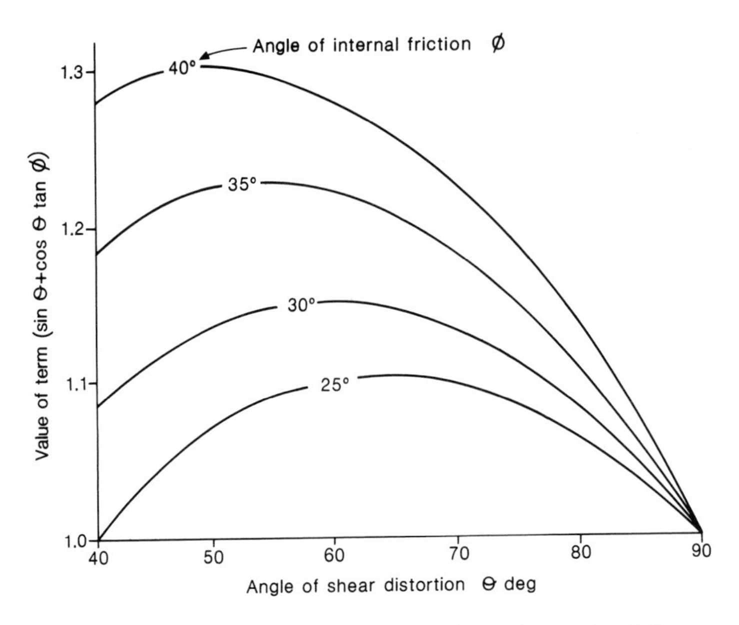

Figure 6.16 Range in value of bracketed term in equation 6.6.

where A_R is the total cross-sectional area of all roots in a given cross-section of soil, and A is the soil cross-section considered.

The fraction of soil cross-section occupied by roots, also termed the root area ratio, can be determined by counting roots by size classes within a given soil area:

$$\text{root area ratio} = \quad A_R/A = \frac{\Sigma n_i a_i}{A} \tag{6.4}$$

where n_i is the number of roots in size class i, and a_i is the mean cross-sectional area of roots in size class i.

Accounting for the variation in root tensile strength with root diameter (section 6.5.4), the mean tensile strength of roots (T_R) can be determined by:

$$T_R = \frac{\Sigma T_i \ n_i \ a_i}{\Sigma n_i a_i} \tag{6.5}$$

where T_i is the strength of roots in size class i.

Finally, substituting equation (6.3) into equation (6.1), the predicted shear strength increase may be found from:

$$\Delta s = T_R \ (A_R/A) \ [\sin\theta + \cos\theta \tan\phi] \tag{6.6}$$

Equation (6.6) shows that the shear strength increase due to roots can be predicted from the mean tensile strength of roots, the root area ratio, and a factor that depends on the shear distortion angle and the angle of internal friction of the soil. The range of the bracketed term in equation (6.6) is shown in Figure 6.16 for a range in ϕ between 25° and 40° and ϕ between 40° and 90°. Observations from the field and laboratory indicate that θ is most likely to fall within the range 40° to 70° in most practical situations (Gray and Leiser, 1982). The value of the bracketed term in equation (6.6) is relatively insensitive to normal variations in θ and ϕ, so Wu *et al.* (1979) chose an average value of 1.2 for the bracketed term. Equation (6.6) then simplified to:

$$\Delta s = 1.2\ T_R\ (A_R/A) \tag{6.7}$$

In this instance, the predicted shear strength increase depends entirely on the mean tensile strength of the roots and the root area ratio. The mean tensile strength is dependent on the vegetation species, as given in Table 6.9. Once the mean tensile strength is known, the prediction of shear strength increase essentially becomes a determination of the root area ratio, which will vary spatially in three dimensions within the slope (Figure 6.17).

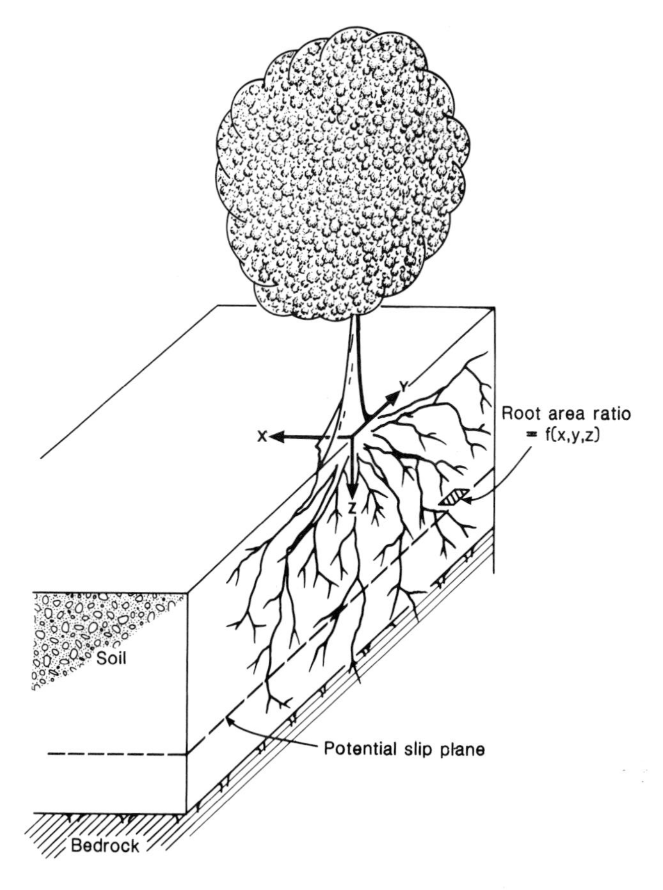

Figure 6.17 Schematic variation of root area ratio within a slope

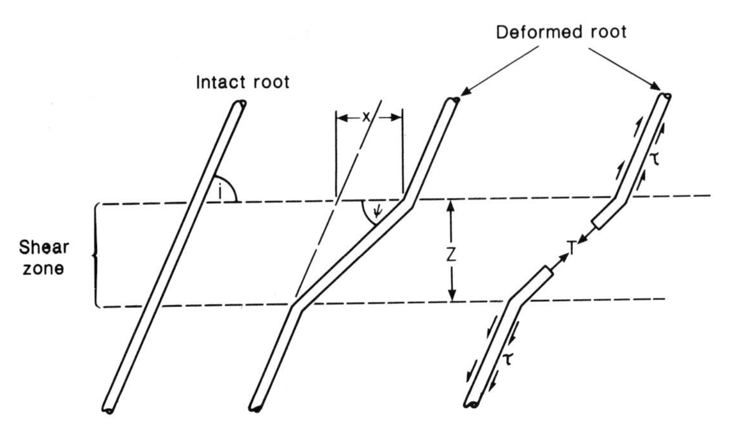

Figure 6.18 Inclined root reinforcement model. (*From Gray and Chastri, 1983. Reproduced by permission of the American Society of Civil Engineers.*)

Gray and Leiser (1982) have also considered the case of a root inclined with respect to the slip plane (Figure 6.18). A shear distortion ratio, m, can be computed:

$$m = x/z \tag{6.8}$$

the angle of shear distortion then becomes

$$\psi = \tan^{-1}\left[\frac{1}{m + (\tan i)^{-1}}\right] \tag{6.9}$$

where i is the initial angle of inclination of the root. The predicted shear strength increase becomes

$$\Delta s = t_R\,[\sin(90 - \psi) + \cos(90 - \psi)\tan\phi] \tag{6.10}$$

The range of values of the bracketed term in equation (6.10) are shown in Figure 6.19. It is apparent that roots having a 'backward' orientation ($i > 90°$) will tend to go into compression rather than tension, which negates their reinforcing effect. It is also apparent that a perpendicular orientation ($i = 90°$) is not the optimal orientation with respect to reinforcement; rather, an inclination between 40° and 70° is generally best.

The predicted shear strength increases can readily be incorporated in conventional stability analyses by modifying the soil strength:

$$s_R = s + \Delta s \tag{6.11}$$

where s_R is the shear strength of root reinforced soil, and s is the (unreinforced) soil shear strength as determined by conventional techniques. The simplifying assumptions made during the calculation of Δs should be critically examined, however.

The simplest root reinforcement model assumes the roots to be perpendicular to the slip plane. The extended model, which allows for inclined roots, demonstrates that a

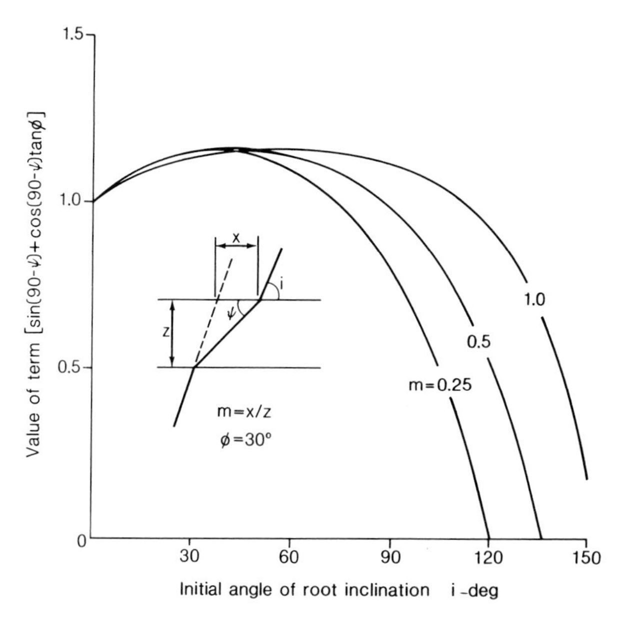

Figure 6.19 Range in value of bracketed term in equation 6.10. (*From Gray and Leiser, 1982.*)

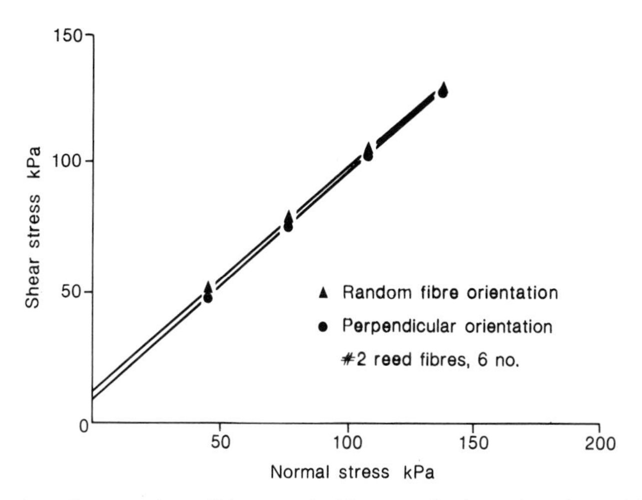

Figure 6.20 Shear strength comparison of dense sand with perpendicular and random reinforcing fibres. (*From Gray and Ohashi*, 1983. ASCE Geotechnical Journal, **109**. *Reproduced by permission of the American Society of Civil Engineers.*)

perpendicular orientation is not optimum, but that it may be a reasonably representative compromise between more and less optimum orientations. Furthermore, Gray and Ohashi (1983) have shown from laboratory tests that perpendicular orientations of reinforcing fibres provided comparable reinforcement to randomly oriented fibres (Figure 6.20). This lends support to the use of the simple perpendicular-root model in practical applications where it may be assumed that roots are oriented randomly within the slope.

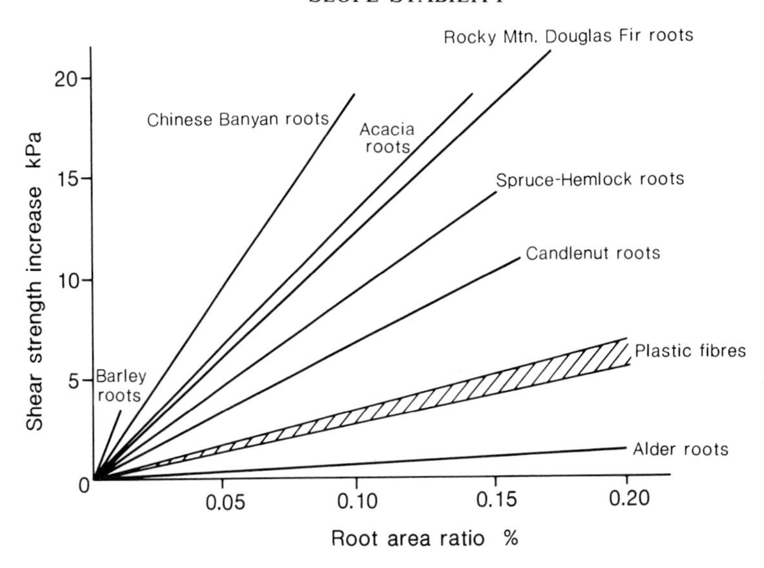

Figure 6.21 Comparison of shear strength increases due to fibre reinforcement. (*After Gray and Leiser, 1982.*)

A second assumption made in the root reinforcement models is that the full tensile strength of all roots is mobilized. This limit-state condition implies that full strain compatibility exists between individual roots, and between the roots and the soil, and this is clearly an idealization. Field observations would indicate that individual roots break at different applied strains, rather than simultaneously. While sufficient investigation has yet to be done to determine the full significance of this assumption, it would be prudent to adopt a conservative lower bound value for 'mean' root tensile strength (equation (6.6) or (6.7)) when using these models in stability analyses.

A third assumption made in the root reinforcement model is that the roots are well anchored and do not pull out of the soil when tensioned. If a simple uniform distribution of bond stress between soil and root is assumed, the minimum root length, L_{min}, required to prevent pull out is given by

$$L_{min} = \frac{T_R \, d}{4 \, \tau_b} \tag{6.12}$$

where T_R is the root tensile strength, d is the root diameter, and τ_b is the limiting bond stress between root and soil. Roots generally exceed this minimum length criteria, and this has been supported by field observations where a preponderance of broken roots (compared to roots that have pulled out) has been noted in landslide scars (Gray and Leiser, 1982).

Predictions of shear strength increase due to roots of different species are summarized in Figure 6.21. It is readily apparent that significant shear strength increases may be realized from root reinforcement, and that the magnitude of the strength increase depends on both the amount of roots present in the soil and the tensile strength of those roots.

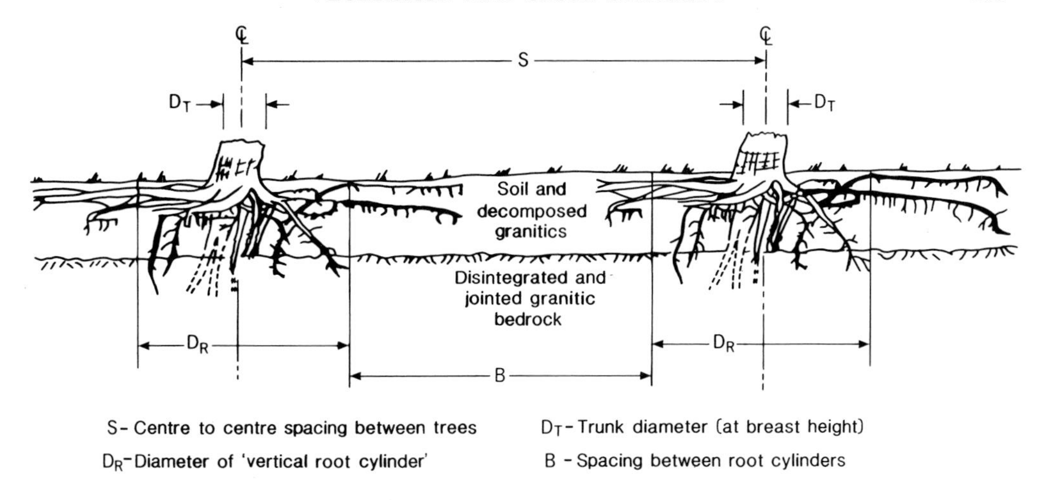

Figure 6.22 Vertical root cylinders of trees on granitic slopes. (*From Gray, 1978.*)

6.6.3 Anchoring, Buttressing, and Arching

The taproot and sinker roots of many tree species anchor them into the slope, causing them to act as slope stabilizing piles. Gray (1978) has illustrated this mechanism for forest slopes comprised of shallow residual soils of granitic origin (Figure 6.22). A vertical 'root cylinder' may be developed by sinker roots which penetrate into the jointed bedrock. This root cylinder supports, or buttresses, the soil mantle upslope of the tree (Figure 6.23). If the trees are not too widely spaced across the slope, arching may develop between the trees.

The mechanism of root anchoring is similar to that of root reinforcement, except that it occurs on a larger scale. This is illustrated by Figure 6.14, in which all four of Tsukomoto

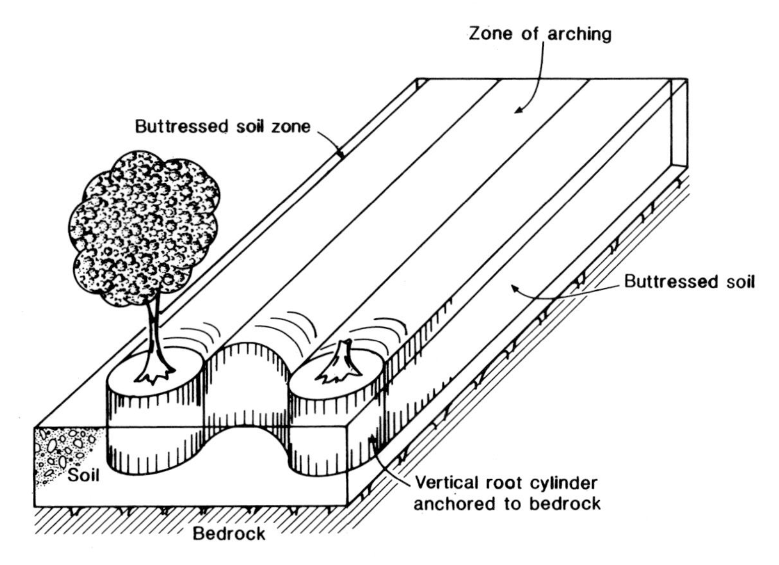

Figure 6.23 Schematic representation of anchoring, buttressing, and arching on slope with shallow soil mantle.

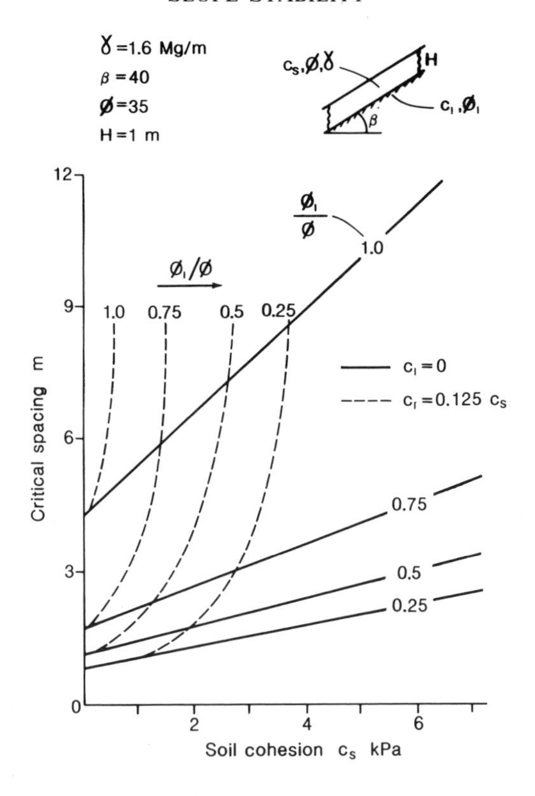

Figure 6.24 Theoretical critical spacing for arching between trees on steep sandy slopes. (*From Gray, 1978.*)

and Kusakabe's slope classes contain root reinforced soil, but only in slope types B and C is anchoring a significant mechanism. Anchoring of the upper portion of the soil mantle may occur to some extent in the type D slope, however.

Based on the theory by Wang and Yen (1974), Gray (1978) has shown that arching may occur between anchored trees at common tree spacings (Figure 6.24). The total force applied to each tree can also be predicted from Wang and Yen's theory of soil arching.

6.6.4 Surcharge

The weight of trees growing on a slope depends on their species, diameter, height, and spacing, but even a relatively dense forest represents a small surcharge when the total weight of the trees is considered to be uniformly distributed on the slope (Figure 6.25). Of course the weight of a tree is not distributed uniformly on the slope, but is transmitted to the area within the root spread. Gray (1978) estimated that the surcharge produced directly beneath large Douglas-fir trees (60 m typical height) might be as high as 70 kPa. For comparison, Wu *et al.* (1979) computed a value of 5 kPa for individual Sitka spruce trees in Alaska (6 m height).

Surcharge on a slope increases both the normal and downhill force components on potential slip surfaces (Figure 6.26). In basic terms, surcharge has a net stabilizing influence when the slope angle is less than angle of internal friction of the soil. Based on the

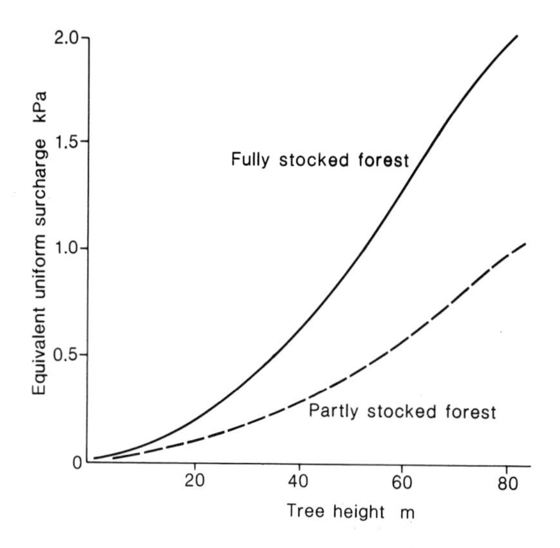

Figure 6.25 Equivalent uniform surcharge of conifer forest

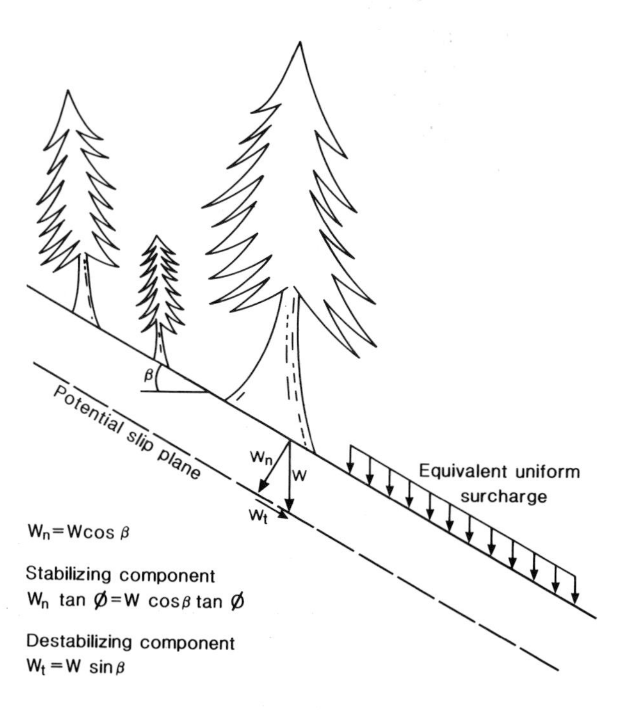

Figure 6.26 Surcharge effect of tree weight

work of Ward using an infinite slope model, Gray and Megahan (1981) state that surcharge is beneficial to stability when

$$c < \gamma_w H_w \tan \phi \cos^2 \beta \qquad (6.13)$$

where c is the soil cohesion, γ_w is the unit weight of water, β is the slope and slip plane angle and H_w is the groundwater rise in the slope above the slip plane. This equation shows that surcharge is beneficial on 'infinite' slopes for low cohesion values, high groundwater levels, high soil friction, and relatively gentle slopes.

6.6.5 Wind Effects

Hsi and Nath (1970) conducted wind tunnel tests on a simulated forest canopy and concluded that the shear stress, τ_w, generated by the wind was given by:

$$\tau_w = C(\tfrac{1}{2})\ \varrho V_a^2 \tag{6.14}$$

where C is the localized drag coefficient (Figure 6.27), ϱ is the mass density of air, and V_a is the ambient wind velocity. The drag coefficient is greatest near the edge of the forest, and a wind of 90 km/hour would produce a shear stress of approximately 1 kPa in this zone. Brown and Sheu (1975) developed equations for predicting both shear forces and overturning moments due to wind, assuming that the wind acted fully on individual trees. None of these analyses reflect the dynamic nature of the wind forces however.

If it were assumed that wind generated shear stresses acted directly downslope and that they affected a significant portion of the slope at a given instant, a significant destabilizing force could theoretically be imposed on a vegetated slope. These assumptions are rarely met, however. A more common phenomenon is the uprooting or overturning of individual trees during high winds. Trees with weak or shallow root systems are more susceptible to uprooting than deeply rooting species.

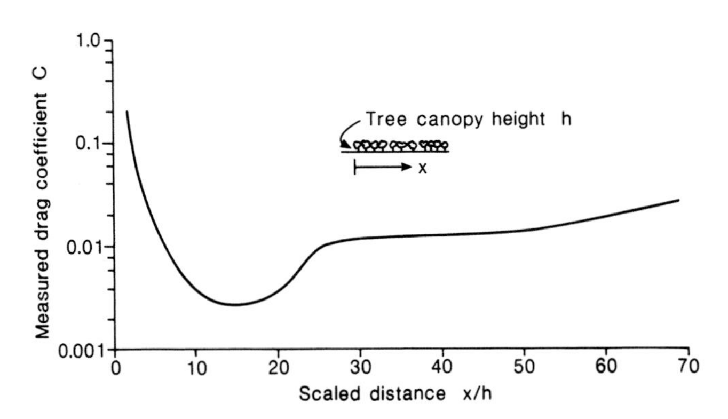

Figure 6.27 Local drag coefficient on a model tree-forest canopy. (*From Hsi and Nath, 1970.* Journal of Applied Meteorology, **9**. *Reproduced by permission of the American Meteorological Society.*)

6.7 CASE STUDIES

Several case studies have been reported in the literature in which the influence of vegetation on slope stability has been assessed (Table 6.13). These case records provide valuable insights into the net influence of vegetation in various locations, and they may also provide guidance regarding possible investigation methods. One particular case record is reviewed below (Greenway *et al.*, 1984).

Table 6.13 Selected Case Studies on Vegetated Slopes

Reference	Location	Tree Type/Soil Type
O'Loughlin and Pearce (1976)	North Westland, New Zealand	Hardwoods, pine/weathered siltstone, etc.
Wu (1976, 1984); Wu *et al.* (1979)	Alaska, USA	Sitka spruce, western hemlock/silty sand
Gray and Megahan (1981)	Idaho, USA	Pine, Douglas fir, spruce/ weathered granite
Riestenberg and Sovonick-Dunford (1983)	Ohio, USA	Sugar maple/clay
Greenway *et al.* (1984)	Hong Kong	Acacia, banyan, candlenut/ granitic fill

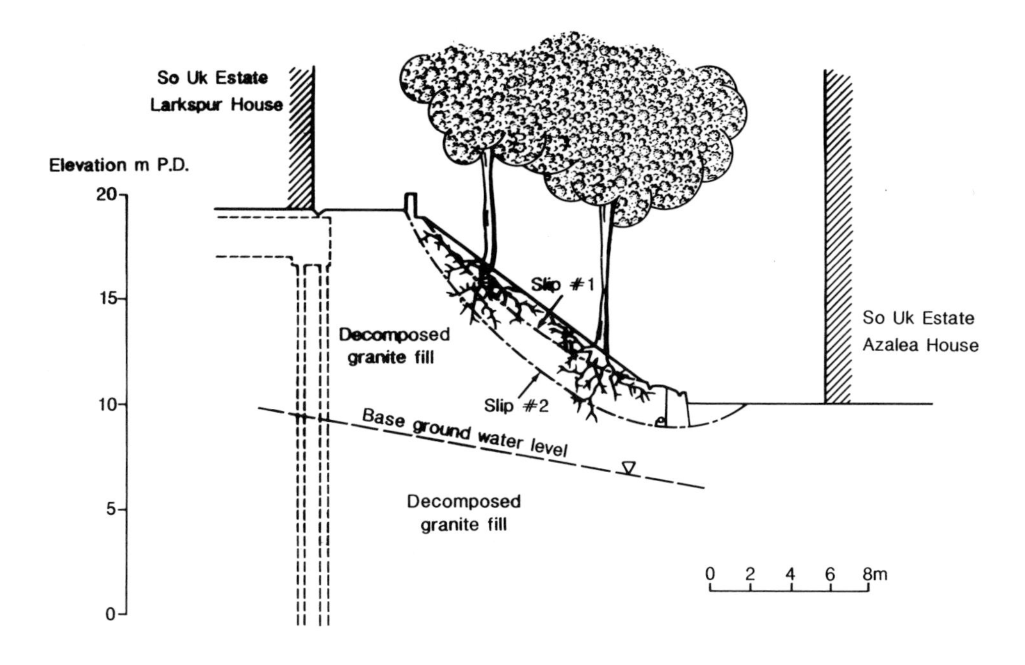

Figure 6.28 Cross-section of So Uk slope

6.7.1 So Uk Tree Study, Hong Kong

A relatively small fill slope in an urban Hong Kong setting was investigated in order to determine the stablizing effect of large trees growing on the slope. This 9 m high, 34° slope (Figure 6.28) has a crest length of 90 m. Three broad-leaf tree species grow on the slope, as listed in Table 6.14. The trees are about 23 years old, based on historical records. In addition, the slope surface is covered with a 50 mm thickness of a soil–cement plaster known locally as 'chunam' (see Chapter 7).

The slope was formed in 1958 from silty gravelly sand derived from weathered granite. The fill has an average dry density of 1.56 Mg/m^3, which is 86% of the maximum dry density obtainable in standard laboratory compaction tests. Consolidated undrained triaxial

Table 6.14 Range in Height, Trunk Diameter, and Spacing of Tree Species Studied at So Uk Estate, Hong Kong

Tree Species	Height (m)	Diameter (mm)	Spacing (m)
Acacia (*Acacia confusa*)	10–12	250–540	5–8
Chinese banyan (*Ficus microcarpa*)	10–15	350–470	5–7
Candlenut (*Aleurites moluccana*)	12–14	330–420	5–10

tests with pore-pressure measurement conducted on undisturbed samples of the fill gave effective stress strength parameters of $c' = 0$, $\phi' = 39°$.

Two standpipe piezometers installed beneath the crest of the slope were monitored for a period of 18 months. The groundwater level remained at a depth of 10 to 11 m below the slope crest throughout this period, and did not show marked daily or seasonal fluctuations. Rainfall at the annual rate of 1,710 mm/year occurred during the monitoring period. The nearly static groundwater behaviour of the slope is attributable to the relatively impermeable chunam slope surfacing and to the concrete paving on adjacent areas which limit infiltration into the slope.

The stability of this slope was analysed using Janbu's routine method on a programmable calculator. A number of slip surfaces were analysed and it was generally found that the factor of safety increased with the depth of sliding, shallow failures being the most critical therefore. Slips 1 and 2 (Figure 6.28) were found to have factors of safety of 1.23 and 1.56 respectively. The slope was then re-analysed to incorporate the effect of soil suction, tree root reinforcement, and other factors.

Soil suction (negative pore-water pressure) exists in the unsaturated soil above the groundwater table in this slope, and may be attributed to both the trees and the chunam slope surfacing. Two tensiometers were installed in the slope to depths of 1.6 and 2.1 m, and were monitored weekly through one wet season. The tensiometers were capable of measuring soil suctions up to 80 kPa by means of high air entry porous ceramic tips (see Chapter 3).

The minimum soil suction observed within the upper 2 m of the slope during the wet season was 15 kPa (Figure 6.29). The soil suction exceeded 80 kPa for a portion of the time, which caused the tensiometers to cavitate. The effectiveness of the chunam in contributing to the observed suctions was analysed following the procedures of Anderson (1984) as outlined in Chapter 7, and it was concluded that the chunam probably contributed about half (7 kPa) of the minimum observed suction. By inference, the remainder of the minimum suction arose from transpiration of the trees. Figure 6.5 shows that the potential evapotranspiration rate was relatively high at the time the minimum suction was observed (August).

As given by Anderson and Pope (1984) and discussed in Chapter 4, soil suction yields a shear strength increase of

$$\Delta s = (u_a - u_w) \tan \phi^b \tag{6.15}$$

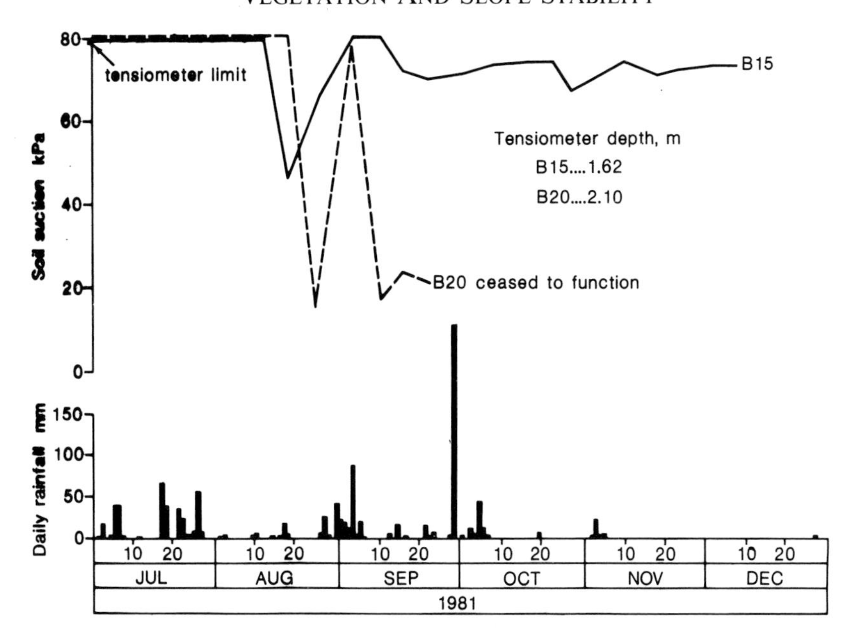

Figure 6.29 Observed soil suctions at So Uk

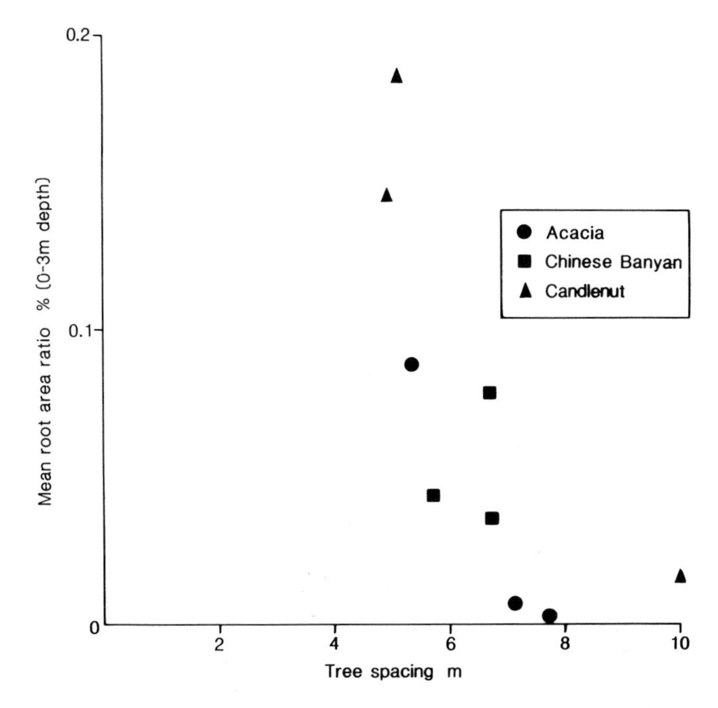

Figure 6.30 Relationship between root area ratio and tree spacing

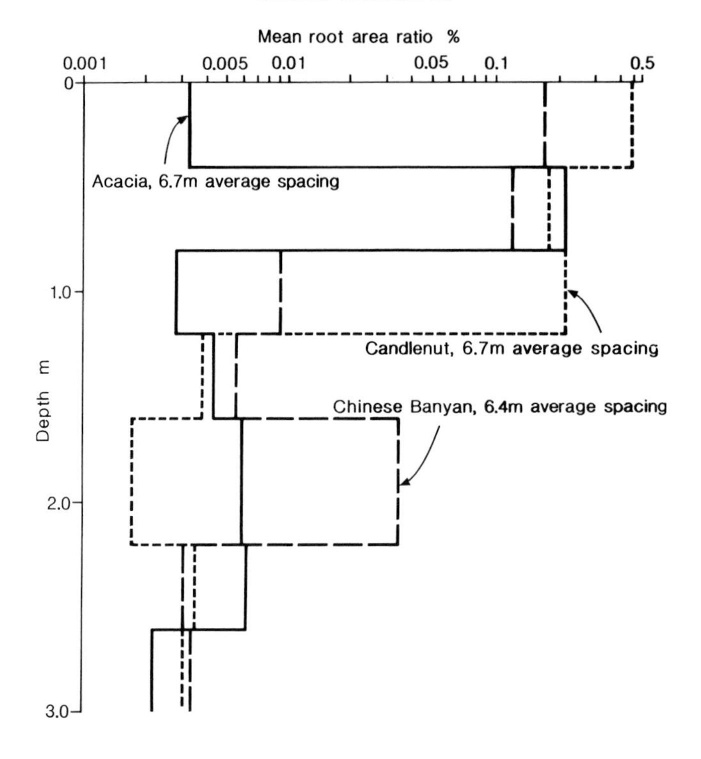

Figure 6.31 Average root area ratio distributions

where u_a is the pore-air pressure, u_w is the pore-water pressure, and ϕ^b is the angle of friction with respect to suction. Taking $u_a = 0$, $u_w = -15$ kPa (equivalent to the minimum observed suction), and $\phi^b = 15°$, then $\Delta s = 15 \tan 15 = 4$ kPa. As will be shown later, a shear strength increase of this magnitude has a major effect on the factor of safety of shallow slips.

The effect of root reinforcement was investigated by excavating nine hand dug trial pits positioned midway between trees. The trial pits were 1.4 m square in plan and 3 m deep. Each vertical face of the pit was examined over set depth intervals, and the number and diameter of roots was recorded. A root area ratio (equation (6.4)) was computed for each depth interval to quantify the relative amount of roots present in the soil. Each trial pit was located so that only roots of one type were encountered.

The mean root area ratio from each trial pit is shown in Figure 6.30 as a function of tree spacing. In general, closer tree spacings yielded markedly higher root concentrations. The distribution of roots also varied with depth and tree species, as shown in Figure 6.31. The root area ratio was greater near the surface, due to the many large roots (diameter > 10 mm) encountered there.

The tensile strength of over 1,000 roots encountered in the trial pits was tested in the field using a calibrated spring device similar to that described by Wu *et al.* (1979). Also, roots between 5 and 10 mm were tested in the laboratory using a Scott L6 tester. Tensile strength data obtained for one of the tree species are given in detail in Figure 6.11, and a summary of all three tree species is shown in Figure 6.32. Smaller roots were found to

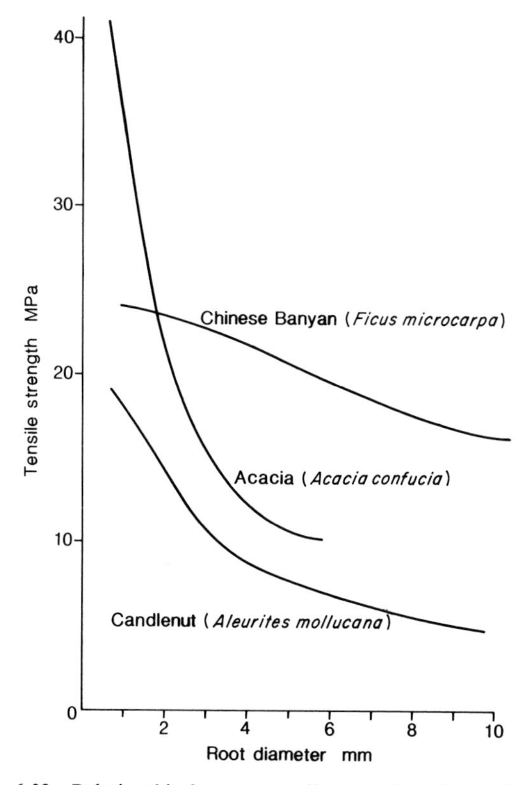

Figure 6.32 Relationship between tensile strength and root diameter

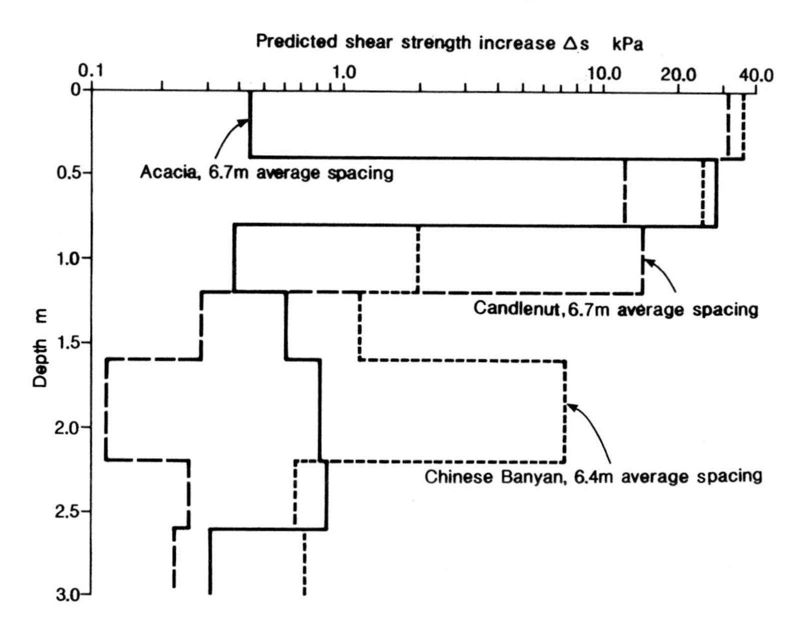

Figure 6.33 Predicted shear strength increases by tree species and depth

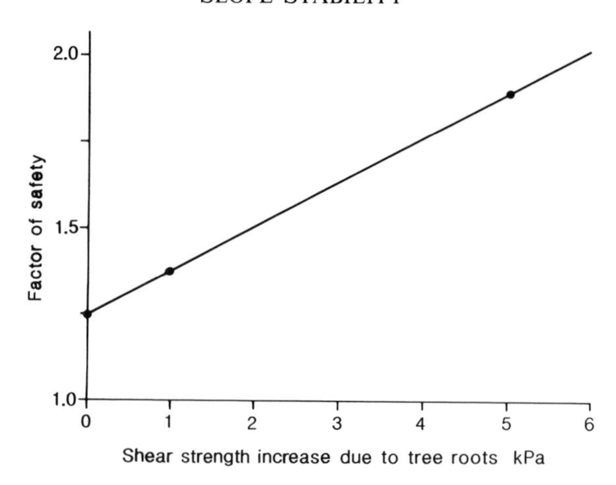

Figure 6.34 Effect of tree root reinforcement on factor of safety

be the strongest in all cases, although the diameter effect was less marked for Chinese banyan. Lower bound tensile strengths of 16.3, 10.6, and 5.5 MPa were adopted in subsequent calculations for Chinese banyan, acacia, and candlenut roots, respectively.

The simple perpendicular-root model given at equation (6.7) was used to predict a shear strength increase due to roots, except that a coefficient of 1.25 was adopted to reflect the high soil strength ($\phi = 39°$). The predicted shear strength increase due to roots is then given by:

$$\Delta s = 1.25\ T_R\ (A_R/A) \tag{6.16}$$

The predictions by tree species according to this equation have been included in Figure 6.21. When the observed root distributions shown in Figure 6.31 are incorporated into the model, the predicted shear strength increase profiles for each tree type are as shown in Figure 6.33. The predicted shear strength increases are significant, being up to 35 kPa at the ground surface and decreasing to 1 kPa or less below 2 m depth. These predicted increases are for trees spaced about 6.5 m apart, and would of course vary with other tree spacings.

Slip 1 (Figure 6.28) was then re-analysed incorporating the root reinforcement effect by selecting representative values for shear strength increase to be incorporated into the Janbu calculation. The average depth of Slip 1 is 1.3 m, but the shear strength increases shown on Figure 6.33 at 1.3 m depth are representative of points midway between trees, and therefore must be adjusted for spatial variations along the slip surface. In light of the observed spacing effect on root concentrations, much higher values would probably exist closer to and beneath the trees. Based on these considerations, a shear strength increase in the range 1 to 5 kPa was adopted in the re-analysis of Slip 1. The factor of safety then increased as shown in Figure 6.34, indicating that root reinforcement has a significant stabilizing influence on shallow slips in this slope.

Figure 6.33 shows that candlenut roots have a significant influence in the top 1.2 m of the slope, but a minor influence at greater depths. To check the significance of this distribution, a simple three-dimensional stability analysis was performed as outlined by

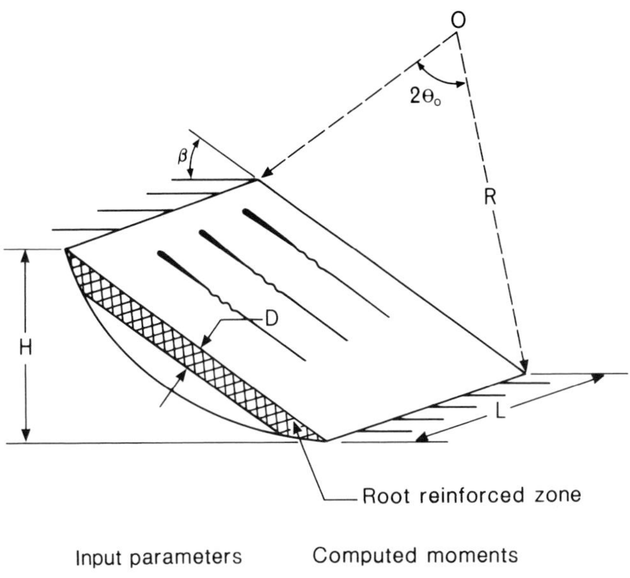

Input parameters	Computed moments
β = 34.5°	M_{R1} = 25.3 mN-m
H = 8.2m	M_{R2} = 25.4 mN-m
L = 8.0m	M_{R3} = 5.9 mN-m
R = 16.4m	M_{R4} = 8.9 mN-m
θ_o = 26.0°	M_D = 20.8 mN-m
D = 1.0m	
c = 0	
ϕ = 39.0°	
γ = 18kN/m^3	

Figure 6.35 Three-dimensional analysis model

Wu (1984). An 8 m segment of the slope (Figure 6.35) was analysed assuming a worst-case shear strength increase profile of 15 kPa in the uppermost 1.2 m of the slope and zero increase below 1.2 m depth. This analysis showed that the factor of safety increased by 29% when root reinforcement was included. The calculated three-dimensional factor of safety for the slope was 2.43 neglecting the tree roots, and 3.26 including them.

With respect to the shallow slips being investigated on this slope, it is likely that anchoring of the trees by vertical roots would restrict the lateral extent of any possible failures to the clear space between trees. As the average tree spacing on this slope is about 8 m (Table 6.14), the relatively high factors of safety computed in the three-dimensional model may be indicative of the possible significance of root anchoring on this slope.

The surcharge weight of the trees on this slope was found to be less than 0.5 kPa on the average, and had an insignificant effect on stability. While typhoon winds may possibly overturn individual trees that are not well anchored, it was considered unlikely that such an event would trigger landslips of significant volume on this slope.

Due to uncertainties over the adequacy of the tensiometer monitoring frequency, the influence of soil suction was not combined with the mechanical influences discussed above to determine the overall significance of the trees to slope stability. It is clear, however, that the trees provide a major stabilizing force on this fill slope.

ACKNOWLEDGEMENT

This chapter is published with the permission of the Director of Engineering Development of the Hong Kong Government.

REFERENCES

Aldrich, R., and Jackson, R. J. (1968). 'Interception of rainfall by Manuka (*Leptospermum scoparium*) at Taita, New Zealand.' *New Zealand J. Sci.*, **11**, 301–17.

Anderson, M. G. (1984). *Prediction of Soil Suction for Slopes in Hong Kong*. GCO Publication No. 1/84, Geotechnical Control Office, Hong Kong, 243pp.

Anderson, M. G., Hubbard, M. G., and Kneale, P. E. (1982). 'Influence of shrinkage cracks on pore-water pressures within a clay embankment.' *Q. J. Eng. Geol.*, **15**(1), 9–14.

Anderson, M. G., and Kneale, P. E. (1984). 'Discussion on the influence of vegetation on clays.' *Geotechnique*, **34(2)**, 150–1.

Anderson, M. G., and Pope, R. G. (1984). 'The incorporation of soil water physics models into geotechnical studies of landslide behaviour.' *Proc. 4th Int. Symp. Landslides*, Vol. 1, 349–53.

Andrawes, K. Z., McGown, A., Mashhour, M. M., and Wilson-Fahmy, R. F. (1980). 'Tension resistant inclusions in soils.' *J. Geotech. Eng. Div., ASCE*, **106**(GT12), 1313–26.

Atkinson, D., Naylor, D., and Coldrick, G. A. (1976). 'The effect of tree spacing on the apple root system.' *Horticulture Res.*, **16**, 89–105.

Aubertin, C. M. (1971). *Nature and Extent of Macropores in Forest Soils and their Influence on Subsurface Water Movement*. Research Paper NE-192, Northeast Forest and Range Experiment Station, US Forest Service, 33pp.

Bache, D. H., and MacAskill, I. A. (1984). *Vegetation in Civil and Landscape Engineering*. Granada Technical Books, 317pp.

Bannan, M. W. (1940). 'The root systems of northern Ontario conifers growing in sand.' *Am. J. Botany*, **27**, 108–114.

Bethlahmy, N. (1962). 'First year effects of timber removal on soil moisture.' *Int. Assoc. Sci. Hydrol. Bull.*, **7**(2), 34–8.

Biddle, P. G. (1983). 'Patterns of soil drying and moisture deficit in the vicinity of trees on clay soils.' *Geotechnique*, **33**(2), 107–26.

Binns, W. O. (1980). *Trees and Water*. Arboricultural Leaflet No. 6, Forestry Commission Research Station, Farnham, Surrey, England, 19pp.

Bishop, D. M., and Stevens, M. E. (1964). *Landslides on Logged Areas in Southeast Alaska*. Research Paper NOR-1, Northern Forest Experiment Station, US Forest Service, Juneau, USA, 18pp.

Bjorkhem, U., Lundeberg, G., and Scholander, J. (1975). *Root Distribution and Compression Strength in Forest Soils*. Research Note No. 22, Royal College of Forestry, Stockholm, Sweden.

Brenner, R. P. (1973). 'A hydrological model study of a forested and a cutover slope.' *Hydrol. Sci. Bull.* **18**(26), 125–44.

Brenner, R. P., and Ng, J. S. M. (1977). 'Effect of roots on the shear strength of a colluvial soil.' *Proc. 5th Danube Eur. Conf. Soil Mech. Fndtn. Eng.*, Bratislava, *CCSR*, 77–98.

Brown, C. B., and Sheu, M. S. (1975). 'Effects of deforestation on slopes.' *J. Geotech. Eng. Div., ASCE*, **101**(GT2), February, 147–65.

Burroughs, E. R., and Thomas, B. R. (1977). *Declining Root Strength in Douglas-fir after Felling as a Factor in Slope Stability*. Research Paper INT-190, Intermountain Forest and Range Experiment Station, US Forest Service, Ogden, Utah, USA, 27pp.

Chang, T. T. (1972). 'A study on the tensile stress of root systems for conservation grasses in Taiwan.' *J. Chinese Soil and Water Conserv. Taiwan*, **3**(1), 58–81.

Chowdhury, R. N. (1978). *Slope Analysis*. Elsevier, Amsterdam, 423pp.

Coatsworth, A., and Evans, J. (1984). 'Discussion on the influence of vegetation on clays.' *Geotechnique*, **34**(2), 154–5.

Croft, A. R., and Adams, J. A. (1950). *Landslides and Sedimentation in the North Fork of Ogden River, May 1949*. Research Paper INT-21, Intermountain Forest and Range Experiment Station, US Forest Service, Ogden, Utah, USA, 4pp.

Curtis, J. D. (1964). *Roots of a Ponderosa Pine*. Research Report INT-9, Intermountain Forest and Range Experiment Station, Ogden, Utah, USA, 10pp.
Cutler, D. F., and Richardson, I. B. K. (1981). *Tree Roots and Buildings*. Construction Press, London.
Davis, S. N., and DeWiest, R. J. M. (1966). *Hydrogeology*. John Wiley, New York, 463pp.
Driscoll, R. (1983). 'The influence of vegetation on the swelling and shrinking of clay soils in Britain.' *Geotechnique*, **33**(2), 93–105.
Dunne, T. (1978). 'Field studies of hillslope flow processes.' In *Hillslope Hydrology*. Kirkby, M. J. (ed.), John Wiley, London, pp. 227–94.
Ellison, L., and Coaldrake, J. E. (1954). 'Soil-mantle movement in relation to forest clearing in southeastern Queensland.' *Ecology*, **35**, 380–8.
Endo, T., and Tsuruta, T. (1969). 'The effect of tree roots upon the shearing strength of soil.' *Annual Report No. 18*, Hokkaido Branch, Tokyo Forest Experiment Station, Sapporo, Japan, pp. 167–82.
Felt, E. J. (1953). 'Influence of vegetation on soil moisture contents and resulting volume change.' *Proc., 3rd Int. Conf. Soil Mechan. Fndtn. Eng.,* Zurich, vol. 1, 24–7.
Flaccus, E. (1959). *Landslides and their Revegetation in the White Mountains of New Hampshire*. Ph.D. Thesis, Duke University, Durham, North Carolina, USA.
Gaiser, R. N. (1952). 'Root channels and roots in forested soils.' *Proc., Soil Sci. Soc. Am.*, vol. 16, 62–5.
Gray, D. H. (1970). 'Effects of forest clear-cutting on the stability of natural slopes.' *Bull. Assoc. Eng. Geol.*, **7**(1), 45–66.
Gray, D. H. (1974). 'Reinforcement and stabilization of soil by vegetation.' *J. Geotech. Eng. Div., ASCE*, **100**(GT6), 695–9.
Gray, D. H. (1977). *Creep Movement and Soil Moisture Stress in Forested vs. Cutover Slopes: Results of Field Studies*. Final Report, Grant No. ENG 74-02427, University of Michigan, Ann Arbor, Michigan, USA, 141pp.
Gray, D. H. (1978). 'Role of woody vegetation in reinforcing soils and stabilising slopes.' *Proc. Symp. Soil Reinforcing and Stabilising Techniques*, Sydney, Australia, 253–306.
Gray, D. H., and Leiser, A. J. (1982). *Biotechnical Slope Protection and Erosion Control*. Van Nostrand Reinhold, New York, 271pp.
Gray, D. H., and Megahan, W. F. (1981). *Forest Vegetation Removal and Slope Stability in the Idaho Batholith*. Research Paper INT-271, Intermountain Forest and Range Experiment Station, Ogden, Utah, USA, 23pp.
Gray, D. H., and Ohashi, H. (1983). 'Mechanics of fiber reinforcement in sand.' *J. Geotech. Eng. Div., ASCE*, **109**(3), 335–53.
Greenway, D. R., Anderson, M. G., and Brian-Boys, K. C. (1984). 'Influence of vegetation on slope stability in Hong Kong.' *Proc. 4th Int. Symp. Landslides*, Toronto, Canada, vol. 1, 399–404.
Gregory, K. J., and Walling, D. E. (1973). *Drainage Basin Form and Process: A Geomorphological Approach*. Edward Arnold Publishers, London, 456pp.
Hallin, W. E. (1967). *Soil-Moisture and Temperature Trends in Cutover and Adjacent Old Growth Douglas-Fir Timber*. Research Note PNW-56, Pacific Northwest Forest and Range Experiment Station, US Forest Service, Corvallis, Oregon, USA, 11pp.
Hathaway, R. L., and Penny, D. (1975). 'Root strength in some *Populus* and *Salix* clones.' *New Zealand J. Botany*, **13**, 333–44.
Henderson, R., Ford, E. D., Renshaw, E., and Decans, J. D. (1983). 'Morphology of the structural root system of Sitka spruce 1: analysis and quantitative description.' *Forestry*, **56**(2), 122–35.
Heyward, F. (1933). 'The root system of Longleaf Pine on the deep sands of western Florida.' *Ecology*, **14**, 136.
Holch, A. E. (1933). 'Development of roots and shoots of certain deciduous tree seedlings in different forest sites.' *Ecology*, **12**, 259–98.
Holtz, W. G., and Gibbs, J. J. (1956). 'Engineering properties of expansive clays.' *Trans. Am. Soc. Civil Eng.*, vol. 121, 641–63.
Hostener-Jorgensen, H. (1967). 'Influences of forest management and drainage on groundwater fluctuations.' In *Int. Symp. Forest Hydrol.*, Sopper, W. E. and Lull, H. W. (eds.), Pergamon, Oxford, pp. 325–34.
Hsi, G., and Nath, J. H. (1970). 'Wind drag within a simulated forest.' *J. Appl. Meteorol.*, **9**, 592–602.

Hudson, N. (1971). *Soil Conservation*. B. T. Batsford, London, 320pp.

ICE (1983). 'The influence of vegetation on the swelling and shrinking of clays.' *Geotechnique*, **33**(2), 85–164.

ICE (1984). 'The influence of vegetation on the swelling and shrinking of clays, Discussions.' *Geotechnique*, **34**(2), 139–72.

Kalela, E. K. (1949). 'On the horizontal roots in pine and spruce stand.' *Acta Forestalis Fenniae*, **57**(2), 1–79.

Kassif, G., and Kopelovitz, A. (1968). *Strength Properties of Soil Root Systems*. Technion Institute of Technology, Haifa, Israel, 44pp.

Kaul, R. (1965). *The influence of roots on certain mechanical properties of an uncompacted soil*. Ph.D. Thesis, Department of Agricultural Engineering, University of North Carolina, Raleigh, North Carolina, USA, 149pp.

Kitamura, Y., and Namba, S. (1966). 'A field experiment on the uprooting resistance of tree roots (I).' *Proc., 77th Meeting Jap. Forest Soc.*, 568–70.

Kitamura, Y., and Namba, S. (1968). 'A field experiment on the uprooting resistance of tree roots (II).' *Proc., 79th Meeting Jap. Forest Soc.*, 360–1.

Kozlowski, T. T. (1971). *Growth and Development of Trees*, vol. 2. Academic Press, New York, 520pp.

Kozlowski, T. T., and Scholtes, W. H. (1948). 'Growth of roots and root hairs of pine and hardwood seedlings in the Piedmont.' *J. Forestry*, **46**, 750–4.

Kraebel, C. J. (1936). *Erosion Control on Mountain Roads*. Circular No. 380, US Department of Agriculture, Washington, DC, 43pp.

Kramer, P. J. (1969). *Plant and Soil Water Relationships: A Modern Synthesis*. McGraw-Hill, New York.

Laitakari, E. (1929). 'The root system of pine (*Pinus silvestris*): a morphological investigation.' *Acta Forestalis Fenniae*, **33**(1), 1–30.

Lee, I. W. Y. (1985). 'A review of vegetative slope stabilisation.' *Hong Kong Engineer*, **13**(7), 9–21.

Lull, H. W. (1964). 'Ecological and silvicultural aspects.' In *Handbook of Applied Hydrology*. Chow, V. T. (ed.), McGraw-Hill, New York, Section 6.

Manbeian, T. (1973). *The Influence of Soil Moisture Suction, Cyclic Wetting and Drying, and Plant Roots on the Shear Strength of Cohesive Soil*. Ph.D. Thesis, Department of Soil Science, University of California, Berkeley, California, USA, 207pp.

McGown, A., Andrawes, K. Z., and Al-Hasani, M. M. (1978). 'Effect of inclusion properties on the behaviour of sand.' *Geotechnique*, **28**(3), 327–46.

McLaughlin, W. T., and Brown, R. L. (1942). *Controlling Coastal Sand Dunes in the Pacific Northwest*. Circular No. 660, US Department of Agriculture, Washington, DC, 46pp.

McMinn, R. G. (1963). 'Characteristics of Douglas-fir root systems.' *Can. J. Botany*, **41**, 105–22.

McQuilkin, W. E. (1935). 'Root development of Pitch Pine, with some comparative observations on Shortleaf Pine.' *J. Agric. Res.*, **51**, 983.

Merritt, C. (1968). 'Effect of environment and heredity on the root growth pattern of red pine.' *Ecology*, **49**, 34–40.

Moran, B. J. J. (1949). 'The use of vegetation in stabilising artificial slopes.' *Proc. Conf. Biol. Civil Eng.*, ICE, London, 113–23.

Muck, M. G., and Hillel, D. (1983). 'A model of root growth and water uptake accounting for photosynthesis, respiration, transpiration, and soil hydraulics.' In *Advances in Irrigation*, vol. 2, Hillel, D. (ed.), Academic Press, Orlando, Florida, USA, pp. 273–333.

Musgrave, G. W., and Holtan, H. N. (1964). 'Infiltration.' In *Handbook of Applied Hydrology*, Chow, V. T. (ed.), McGraw-Hill, New York, Section 12.

Nassif, S. H., and Wilson, E. M. (1975). 'The influence of slope and rain intensity on runoff and infiltration.' *Hydrol. Sci. Bull.*, **20**(4), 539–53.

Natarajan, T. K., and Gupta, S. C. (1980). 'Techniques of erosion control for surficial landslides.' *Proc. Int. Symp. Landslides*, New Delhi, India, vol. 1, 413–17.

O'Loughlin, C. L. (1974a). 'The effects of timber removal on the stability of forest soils.' *J. Hydrology (New Zealand)*, **13**, 121–34.

O'Loughlin, C. L. (1974b). 'A study of tree root strength deterioration following clear felling.' *Can. J. Forest Res.*, **4**(1), 107–13.

O'Loughlin, C., and Pearce, A. J. (1976). 'Influence of Cenozoic geology on mass movement and sediment yield response to forest removal, north Westland, New Zealand.' *Bull. Int. Assoc. Eng. Geol.*, **14**, 41–6.

O'Loughlin, C., and Watson, A. (1979). 'Root-wood strength deterioration in Radiata Pine after clearfelling.' *New Zealand J. Forestry Sci.*, **9**(3), 284–93.

O'Loughlin, C., and Watson, A. (1981). 'Note on rootwood strength deterioration in *Nothofagus fusca* and *N. truncata* after clearfelling.' *New Zealand J. Forestry Sci.*, **11**(2), 183–5.

Penman, H. L. (1963). *Vegetation and Hydrology*. Technical Communication 53, Commonwealth Bureau of Soils, Farnham Royal, England, 124pp.

Prandini, L., Guidicini, G., Bottura, J. A., Poncano, W. L., and Santos, A. R. (1977). 'Behaviour of the vegetation in slope stability: a critical review.' *Bull. Int. Assoc. Eng. Geol.*, no. 16, 51–5.

Rice, R. M., and Krammes, J. S. (1970). 'Mass wasting processes in watershed management.' *Proc. Symp. Interdisciplinary Aspects Watershed Management, ASCE*, 231–60.

Richards, B. G., Peter, P., and Emerson, W. W. (1983). 'The effects of vegetation on the swelling and shrinking of soils in Australia.' *Geotechnique*, **33**(2), 127–39.

Riestenberg, M. M., and Sovonick-Dunford, S. (1983). 'The role of woody vegetation on stabilising slopes in the Cincinnati area.' *Geol. Soc. Am. Bull.*, **94**, 506–18.

Robinson, T. W. (1958). *Phreatophytes*. Water-Supply Paper 1423, US Geological Survey, Washington, DC, 84pp.

Rogers, W. S., and Booth, G. A. (1960). 'The roots of fruit trees.' *Sci. Horticulture*, **14**, 27–34.

Schiechtl, H. M. (1980). *Bioengineering for Land Reclamation and Conservation*. University of Alberta Press, Edmonton, Alberta, Canada, 404pp.

Shaw, G. (1978). 'The search for dangerous dams.' *Smithsonian*, **9**(1), 36–45.

Skempton, A. W. (1954). 'A foundation failure due to clay shrinkage caused by Poplar trees.' *Proc. Inst. Civil Eng.*, part 1, vol. 3, January, 66–83.

So, C. L. (1971). 'Mass movements associated with the rainstorm of June 1966 in Hong Kong.' *Trans. Inst. Brit. Geogrs.*, no. 53, 55–65.

Strahler, A. N. (1969). *Physical Geography* (3rd edn). John Wiley, New York, p. 240.

Sutton, R. F. (1969). *Form and Development of Conifer Root Systems*. Technical Communication No. 7, Commonwealth Agricultural Bureau, England.

Takahashi, T. (1968). 'Studies of the forest facilities to prevent landslides.' *Bull. Faculty Agric.*, Shizuoka University, Japan, **18**, 85–101.

Ter Stepanian, G. (1963). *On the Long-Term Stability of Slopes*. Publication 52, Norwegian Geotechnical Institute, Oslo, Norway, pp. 1–15.

Terzaghi, K. (1925). *Erdbaumechanik auf bodenphysikalischer Grundlage*. Franz Deuticke, Vienna.

Terzaghi, K. (1950). 'Mechanisms of landslides.' *Engineering Geology (Berkey) Volume*, Geological Society of America, pp. 83–123. Reprinted in *From Theory to Practice in Soil Mechanics*. Bjerrum, L., Casagrande, A., Peck, R. B., and Skempton, A. W. (1960), John Wiley, New York, pp. 202–45.

Toms, A. H. (1949). 'The effect of vegetation on the stabilization of artificial slopes.' *Proc. Conf. Biol. Civil Eng.*, ICE, London, 99–112.

Tsukamoto, Y., and Kusakabe, O. (1984). 'Vegetative influences on debris slide occurrences on steep slopes in Japan.' *Proc. Symp. on Effects Forest Land Use on Erosion and Slope Stability*, Environment and Policy Institute, Honolulu, Hawaii.

Turmanina, V. I. (1963). 'The magnitude of the reinforcing role of tree roots.' *Moscow University Herald, Sci. J.*, series V, no. 4, 78–80.

Turmanina, V. I. (1965). 'On the strength of tree roots.' *Bull. Moscow Soc. Naturalists, Biol. Sect.*, **70**(5), 36–45.

Vanicek, V. (1973). 'The soil protective role of specially shaped plant roots.' *Biol. Conserv.*, **5**(3), 195–180.

Vidal, H. (1966). 'La terre armée.' *Anns Inst. tech Batim*, Paris, no. 223–229, July–August, 888–939.

Waldron, L. J. (1977). 'Shear resistance of root permeated homogeneous and stratified soil.' *J. Soil Sci. Soc. Am.*, **41**, 843–9.

Waldron, L. J., and Dakessian, S. (1981). 'Soil reinforcement by roots: calculation of increased soil shear resistance from root properties.' *Soil Sci.*, **132**(6), 427–35.

Wang, W. L., and Yen, B. C. (1974). 'Soil arching in slopes.' *J. Geotech. Eng. Div., ASCE*, **100**(GT1), 61–78.

Ward, W. H. (1947). 'The effect of fast growing trees and shrubs on shallow foundations.' *J. Inst. Landscape Architects*, **11**, 7–16.

Ward, W. H. (1953). 'Soil movement and weather.' *Proc. 3rd Int. Conf. of Soil Mech. Fndtn. Eng.*, Zurich, **1**, 477–82.

White, S. E. (1949). 'Processes of erosion on steep slopes of Oahu, Hawaii.' *Am. J. Sci.*, **247**, 168–86.

Williams, A. A. B., and Pidgeon, J. T. (1983). 'Evapo-transpiration and heaving clays in South Africa.' *Geotechnique*, **33**(2), 141–50.

Wischmeir, W. H., and Smith, D. D. (1965). *Predicting Rainfall–Erosion Losses From Cropland East of the Rocky Mountains.* Handbook 282, US Department of Agriculture, Washington, DC, 47pp.

Wu, T. H. (1976). *Investigation of Landslides on Prince of Wales Island.* Geotechnical Engineering Report 5, Civil Engineering Department, Ohio State University, Columbus, Ohio, USA, 94pp.

Wu, T. H. (1984). 'Effect of vegetation on slope stability.' *Transportation Research Record 965*, Transportation Research Board, Washington, DC, pp. 37–46.

Wu, T. H., McKinnell, W. P., and Swanston, D. N. (1979). 'Strength of tree roots and landslides on Prince of Wales Island, Alaska.' *Can. Geotech. J.*, **16**, 19–33.

Yamanaka, K. (1965). 'Measurement method of soil consistency, determination of bond, crushing degree and hardness.' *Soil Physical Conditions and Plant Growth, Japan*, nos. 11 and 12, 1–8.

Yen, C. P. (1972). 'Study on the root system form and distribution habit of the ligneous plants for soil conservation in Taiwan (preliminary report).' *J. Chinese Soil and Water Conserv.*, Taiwan, **3**(2), 179–204.

Zaruba, Q., and Mencl, V. (1982). *Landslides and Their Control* (2nd edn). Elsevier, Amsterdam, 320pp.

Ziemer, R. R. (1978). 'An apparatus to measure the cross-cut shearing strength of roots.' *Can. J. Forestry Res.*, **8**(1), 142–4.

Ziemer, R. R. (1981). 'Roots and the stability of forested slopes.' *Erosion and Sediment Transport in Pacific Rim Steeplands*, Publication 132, International Association of Hydrological Sciences, London, pp. 343–61.

Ziemer, R. R., and Swanston, D. N. (1977). *Root Strength Changes after Logging in Southeast Alaska.* Research Note PNW-306, Pacific Northwest Forest and Range Experiment Station, US Forest Service, Portland, Oregon, USA, 9pp.

Slope Stability
Edited by M. G. Anderson and K. S. Richards

Chapter 7

Modelling the Effectiveness of a Soil–Cement Protective Cover for Slopes

M. G. ANDERSON
Department of Geography,
University of Bristol, Bristol BS8 1SS

and

J. M. SHEN
Geotechnical Control Office,
68 Mody Road, Kowloon, Hong Kong

7.1 INTRODUCTION

In an ever-increasing area of the world, man is able to effect externally induced changes to hillslopes. While much of this activity is generally aimed at increasing slope stability, the long-term performance of slopes in certain cases illustrates that we still have much to learn with regard to such modifications. It is those regions with the more extreme slope stability problems that provide the most rigorous validation and verification of models related to slope stability. In this context, Brand (1985) has provided a summary of knowledge of aspects of slope stability for Hong Kong (Table 7.1). It is to be noted that the area of greatest current deficiency is that of hydrology (both soil water and groundwater conditions). Attention to this relative deficiency has been given by recent workers (Brand, 1982; Anderson, 1983; Anderson and Howes, 1985). Nevertheless, the refinement of hydrological aspects of slope stability problems while being a relatively complex issue both in terms of model coupling and field validation, is clearly deserving of attention; the more so since external artificially induced changes to slopes inevitably change the hydrological regime of the slope. To define the nature of external slope changes, we need to establish a model formulation that will enable us to make an assessment of the effects of such changes.

The purpose of this chapter is to formulate and discuss a model that the authors have developed to predict the effectiveness of covering a slope with a material designed to reduce

Table 7.1 State of Knowledge of Aspects of Slope Stability for Hong Kong Conditions (modified from Brand 1985)

Aspect	Current State of Knowledge for Hong Kong Conditions	Overall Rating of Knowledge
Methods of stability analysis	Janbu (1954, 1973) method of analysis for non-linear surfaces thought satisfactory. Recommended factors of safety of 1.2 to 1.4 are satisfactory (GCO, 1984). Experience has shown that data are often poorly handled (Lumsdaine and Tang, 1982).	very good
Geometry of failure	Pre-failure geometry is easily defined. Sometimes difficult to decide critical potential failure surface for design.	good to very good
Geology	Site investigation procedures are adequate, but descriptions often poor. Complex weathering profiles are difficult to describe (Hencher and Martin, 1982). Understanding of influence of geological details on hydrogeology is poor.	fair
Shear strength	Mass strength as distinct from SAMPLE strength is poorly understood. Laboratory tests are commonly used to determine saturated strengths of SAMPLES, in terms of effective stress, but doubt exists about applicability of test results (Brand, 1982) Limited amount of *in situ* strength testing carried out. Weakening effect of relict joints recognised (Koo, 1982a, 1982b). Effects of boulder and corestone content unknown (Hencher and Martin, 1982).	fair to poor
Groundwater and pore pressures	Useful correlations available between landslides and rainfall (Lumb, 1975; Brand *et al.*, 1984). Rapid changes in pore pressure with rainfall are very difficult to predict for design (Anderson *et al.*, 1983). Only limited attempts made to model groundwater (Leach and Herbert, 1982). Extrapolation of *in situ* measurements seems best design approach (Koo and Lumb, 1981; Endicott, 1982). Some progress made with field instrumentation (Pope *et al.*, 1982) Erosion pipes are important in transmitting water.	poor

Figure 7.1 (a) A chunam covered slope on Clearwater Bay Road, Hong Kong. (b) A possible site for chunaming in East Nepal (photograph by J. W. F. Dowling)

Figure 7.2 Chunam in a poor state of repair

or eliminate surface infiltration and hence improve slope stability. While the model was developed for these specific artificial surfaces, the general framework can also be applied to slopes having naturally developed surface crusts.

Slope failures in Hong Kong (Lumb, 1975; Brand, 1984) and more recent failures on the Dharan–Dhankuta Road, East Nepal (Hearn and Jones, 1985) require long-term solutions in often difficult slope terrain. In Hong Kong, a soil–cement plaster known locally as 'chunam' (Wood, 1981) is often used as a slope protection against infiltration and erosion. It is a traditional material based on local skills and historical practice, consisting of one part of cement, three parts of hydrated lime, twenty parts of soil and water mixed together to the required consistency and applied to surfaces of slopes. Although other forms of surface protection such as hydroseeding and sprayed concrete are becoming popular, chunam is still used quite often on steep slopes. The Geotechnical Control Office (1984) *Manual for Slopes* states that 'chunam should be applied to the surface in two layers not less than 20 mm thick'. Figure 7.1(a) shows a chunam covered slope in Hong Kong and Figure 7.1(b) a slope in Nepal that may be suitable for chunaming. For many developing areas suffering severe slope stability problems, a protective cover similar to chunam may provide a relatively inexpensive solution to such problems. Before making recommendations of this type, it is instructive to examine the Hong Kong experience in this regard. Cracks can occur in chunam and in severe cases sections can become detached (Figure 7.2). It is desirable, therefore, to investigate not only the possibility of a model which will predict water flow through a homogeneous cover (intact chunam), but, rather more importantly, a model that can establish flow rates through cracked chunam (which may be induced through incorrect mix proportions or an incorrect depth of chunam being applied, as we will discuss below). It is this latter model which is formulated first and then coupled with a finite difference vertical infiltration model for layered soil systems. A programme of data acquisition is then described to enable the models to be paramaterized and subsequently applied. Due to the fact that certain of the field and laboratory tests had not been previously undertaken on chunam, the procedures we adopted for such tests are fully described in the appropriate sections.

7.2 SETTING UP THE FLOW MODELS

7.2.1 Establishment of a Model to Predict Flow Through Chunam Cracks

A seepage model, outlined by Harr (1962) may be adopted and used for this purpose. Here a crack of width B and vertical extent (chunam thickness) H is assumed. The flux passing through the crack (which is assumed water filled) can then be determined by:

$$B = \frac{8q}{K\pi^2} \sum_{n=1}^{\infty} \frac{e^{-(2n-1)\alpha}}{(2n-1)^2} \tag{7.1}$$

where $\alpha = \pi \, KH/q$

and where
$K =$ soil permeability m s^{-1}
$B =$ crack width m m^{-1}
$q =$ flux m^2 s^{-1}
$H =$ chunam thickness (head) m

Since α is a function of q, q can be found only by iteration.

Of major interest in terms of effective chunam protection is the effective crack width (B_1—see Figure 7.3) and the effective width at depth when the seepage becomes parallel (B_∞). The former spread immediately under the chunam can be given as:

$$B_1 = \frac{4q}{K\pi^2} \sum_{n=1}^{\infty} \frac{1 + e^{-2(2n-1)\alpha}}{(2n-1)^2} \tag{7.2}$$

while the spread at depth becomes:

$$B_\infty = q/K \tag{7.3}$$

Table 7.2 illustrates solutions to equations (7.1) to (7.3) with $H = 0.05$ m, $K = 1 \times 10^{-6}$ m s^{-1} and selected values of crack width B.

However, before such a model can be considered suitable for application, it is necessary to determine the crack permeability with respect to the soil permeability below, since in the above solution it is assumed that there is no head loss within the crack. The following parallel plate flow model (Massey, 1970) can be used:

$$Q = \frac{b\,B^3}{12}\,\mu \left(\frac{dp^*}{dx} \right) \tag{7.4}$$

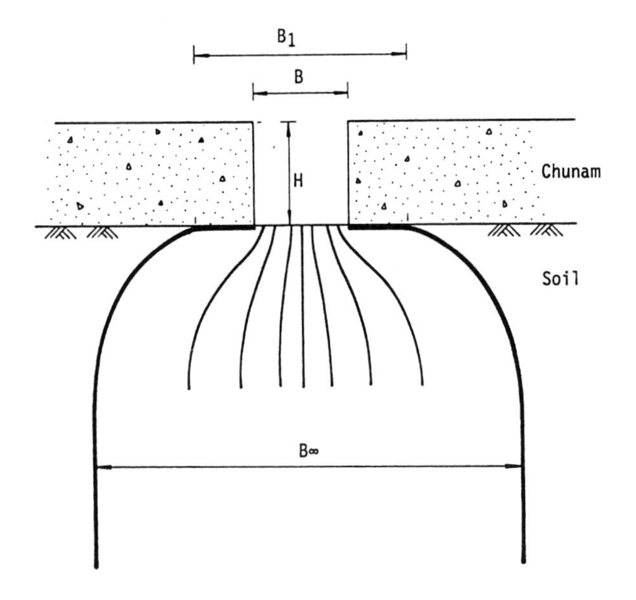

Figure 7.3 Definition diagram for crack model (see equations 7.1–7.3)

Table 7.2 Illustrative relationships of crack geometry

Crack width, B (m per metre)	Under surface effective crack width, B_1 (m per metre)	Flux, q ($m^2 s_{-1}$)	Effective crack width at depth, B_∞ (m per metre)
0.0005	0.019	3.75×10^{-8}	0.038
0.001	0.022	4.40×10^{-8}	0.044
0.002	0.026	5.14×10^{-8}	0.051
0.003	0.029	5.73×10^{-8}	0.057
0.004	0.031	6.20×10^{-8}	0.062
0.005	0.033	6.60×10^{-8}	0.066

Notes (1) Solutions based on $K=1\times10^{-6}$ m s^{-1} and $H=0.05$ see equations (7.1)–(7.3)
(2) See Figure 7.3 for definition diagram.

where $b=$ length of crack
$B=$ crack width
$p^*=$ piezometric head
$\frac{dp^*}{dx}=$ piezometric head gradient along direction of flow

$= \varrho gi$
$\varrho=$ density of fluid
$g=$ gravitational acceleration
$i=$ hydraulic gradient
$\mu=$ viscosity
(see Figure 7.3)

By Darcy's law equation (7.4) reduces to

$$K = \frac{b\,B^3}{12\mu}\,\varrho\,g \tag{7.5}$$

for unit area calculation.

If the crack spacing is regular and of distance S, then the total length of transverse and longitudinal cracking per unit area is

$$b = \frac{2}{S}$$

and thus

$$K = \frac{B^3 \varrho g}{S 6\mu} \tag{7.6}$$

Taking $\mu = 1.0 \times 10^{-3}\,\mathrm{kg\,m^{-1}\,s^{-1}}$ at 20 °C then (7.6) can be evaluated for different crack widths (B) and crack spacing (S), thereby providing estimates of crack permeability. Selected estimations are provided in Table 7.3. Calculations show the unrestricted permeability of cracks in chunam is likely to approximate $10^{-3}\,\mathrm{m\,s^{-1}}$, which is far greater than the underlying soils. Thus, it can be assumed that the flow in cracks is governed by the soil permeability, and by equations (7.1), (7.2) and (7.3).

On this basis, then, values of flux for different crack widths and soil permeabilities, assuming a crack depth of 50 mm, have been calculated (utilizing equations (7.1)–(7.3)). Figure 7.4 shows the change in flux for changing crack widths. These results show that the influence of crack width is small, since the effective crack width B_∞ is controlled principally by the head (H).

The solutions presented in Figure 7.4 confirm the minimal head loss that occurs in the crack prior to entry. Taking a 1 mm crack at 1 m spacing with a soil permeability of $K = 1 \times 10^{-5}\,\mathrm{m\,s^{-1}}$, then from Figure 7.4 $q = 4.38 \times 10^{-6}\,\mathrm{m^2\,s^{-1}}$ and from Table 7.3 $K = 1.64 \times 10^{-3}\,\mathrm{m\,s^{-1}}$. Since head $H = q/K$ by Darcy's law, then the actual head loss $= 1.3 \times 10^{-4}$ m, or 0.27% of the total head of 0.05 m.

During rainfall the total flux Q_T passing through unit area of the surface is given by

$$Q_T = Q_i + Q_c \tag{7.7}$$

Table 7.3 Predicted permeability of cracks from equation (7.5)

Crack Spacing (m)	Crack width, B (m)					
	5.00×10^{-4}	1.00×10^{-3}	2.00×10^{-3}	3.00×10^{-3}	4.00×10^{-3}	5.00×10^{-3}
0.250	8.18×10^{-4}	6.54×10^{-3}	5.23×10^{-2}	1.77×10^{-1}	4.19×10^{-1}	8.18×10^{-1}
0.500	4.09×10^{-4}	3.27×10^{-3}	2.62×10^{-2}	8.83×10^{-2}	2.09×10^{-1}	4.09×10^{-1}
1.000	2.04×10^{-4}	1.64×10^{-3}	1.31×10^{-2}	4.41×10^{-2}	1.05×10^{-1}	2.04×10^{-1}
2.000	1.02×10^{-4}	8.18×10^{-4}	6.54×10^{-3}	2.21×10^{-2}	5.23×10^{-2}	1.02×10^{-1}
5.000	4.09×10^{-5}	3.27×10^{-4}	2.62×10^{-3}	8.83×10^{-3}	2.09×10^{-2}	4.09×10^{-2}

Note Permeability values in $\mathrm{m\,s^{-1}}$.

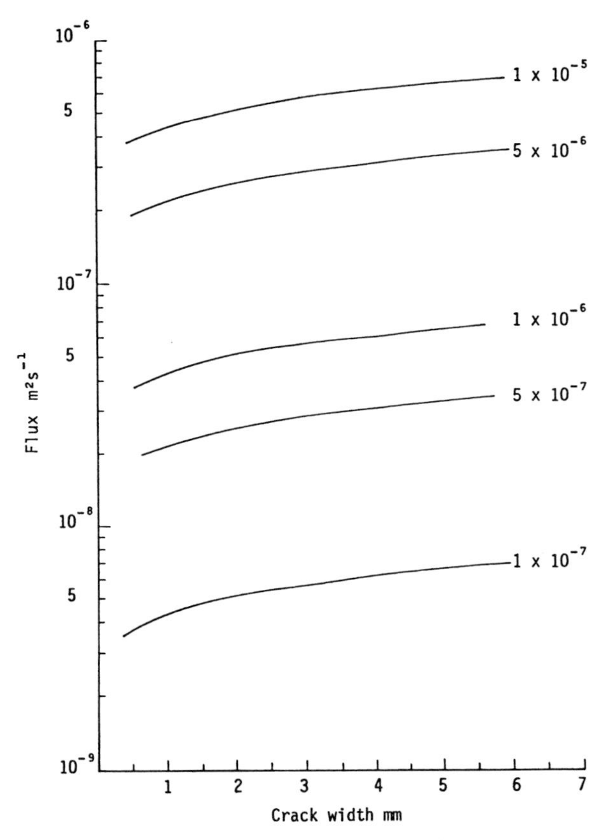

Figure 7.4 Seepage predicted by the crack model (with water head (H) = 50 mm) for different soil permeabilities (m s^{-1}).

where $Q_c =$ flux through the crack
and $Q_i =$ flux through the intact chunam

For a unit area the effective chunam permeability (K_e) is

$$K_e = K_i + q_o \left(\frac{2}{S} \right) \tag{7.8}$$

where K_i is the permeability of intact chunam, with crack spacing S as defined in equation (7.6). Figure 7.5 illustrates the effective permeability of chunam (K_e), based on solutions to equations (7.1)–(7.3) and 7.8 with a crack of 12 mm and uniform spacing. This illustrates the dependency of the effective chunam permeability on the permeability of the underlying soil.

Thus, the effective permeability of chunam can be estimated by use of the crack model (equations (7.1)–(7.3)), knowledge of the intact permeability of chunam (K_i—the methods of determination for which we discuss in section 7.3.1) and the crack spacing (see equation (7.8)).

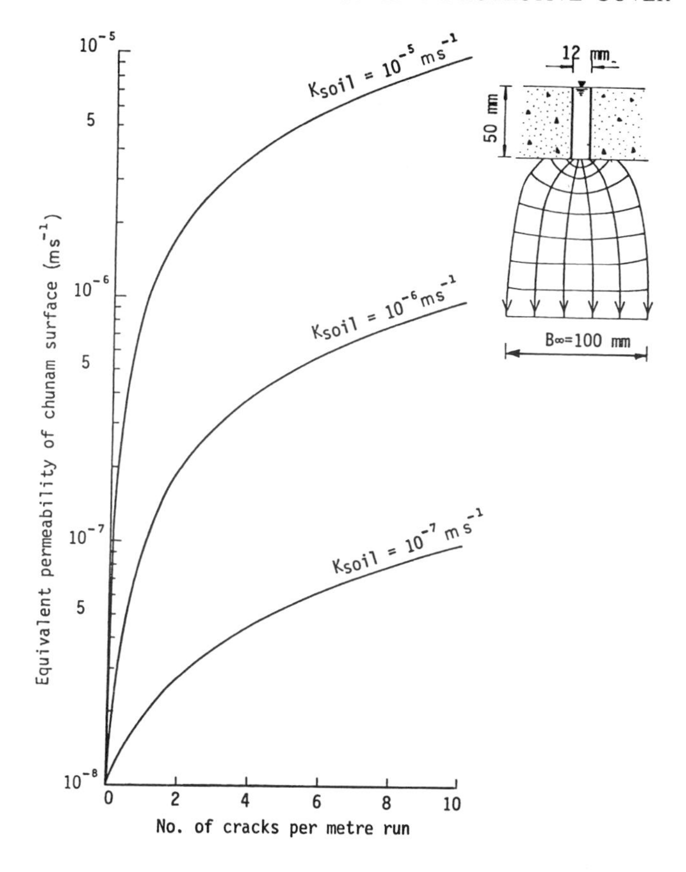

Figure 7.5 Effective permeability of chunam predicted by the crack model

7.2.2 Coupling of model for crack flow with vertical infiltration model

To provide a means of estimating chunam effectiveness, the crack flow model outlined in section 6.2.1 needs to be coupled to a vertical infiltration scheme. An explicit finite difference scheme may be selected for this purpose. The basic model structure employed for the soil water finite difference scheme is shown in Figure 7.6. The profile used is divided into NL components of potentially different thicknesses, and the model allows for four layers to be incorporated within which the material properties are taken to be uniform. Thus the top layer comprises the effective permeability model of chunam as developed in section 6.2.1 above.

In the field, the description of infiltration is highly complicated since the initial and boundary conditions are usually not constant while the soil characteristics may vary with time and space. Haverkamp *et al.* (1977) have illustrated the variety of finite difference

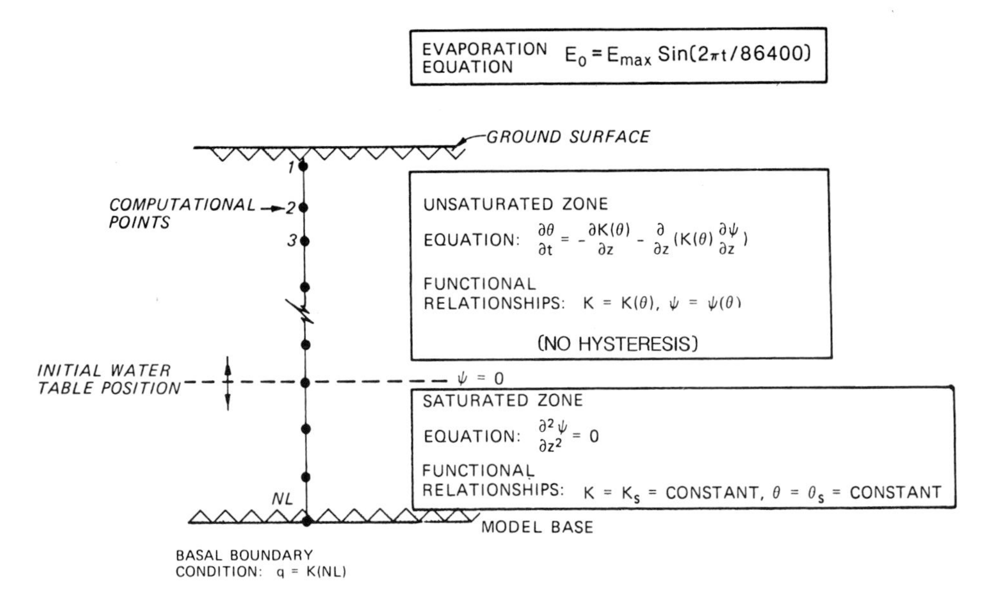

Figure 7.6 Governing equations used in the finite difference model

solutions employing different forms of the non-linear Fokker–Planck equation and different ways of discretization. They were able to show that the six numerical models examined yielded comparable results, which were not significantly different from the measured water contents. The scheme followed here is an explicit scheme, essentially following that of Hillel (1977). While Haverkamp *et al.* showed the CSMP method to require a larger execution time than the implicit models they examined, it has a definite advantage in the relative ease of programming. The execution time may be reduced by increasing the layer thickness and adjusting the time step as a function of the soil water diffusivity.

The soil moisture-hydraulic conductivity and soil moisture–matric suction relationships have to be defined to allow modelling of the water flux through both saturated and unsaturated conditions. Either empirical curves can be used for this purpose, or a suction–moisture curve can be assumed and the hydraulic conductivity-soil moisture (K–θ) relationship derived. This latter procedure ensures physical consistency and is now well established. It is considered sensible to avoid any experimental errors in this area, and thus the K–θ curve is derived using the Millington–Quirk (MQ) method (Millington and Quirk, 1959). Hysteresis in the suction (ψ)-moisture curve is not incorporated into the current model.

A simple isothermal evaporation routine is used as a forcing function in the top cell. This function allows potential evapotranspiration E_o (the maximum and limiting rate of evaporation possible from the soil as long as the surface is kept sufficiently wet) to vary as a size function during the day:

$$E_o = E_{max} \sin(2\pi t/86{,}400) \tag{7.9}$$

where t = time in seconds from sunrise
and E_{max} = the maximum midday evaporativity.

When the soil moisture content in the top cell exceeds the air dry value θ_d, evaporation takes place at rate E_o. After the top cell dries to its minimum matric potential, the

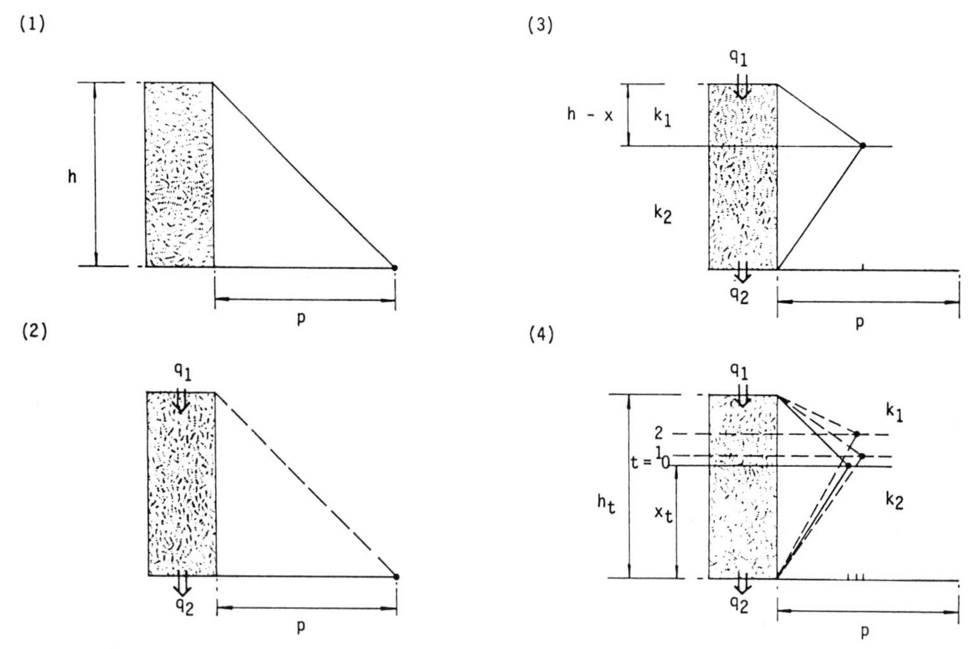

Figure 7.7 Four methods for considering pore pressures induced by perched water tables

evaporation rate becomes equal to the rate of upward transmission of moisture by the profile or to E_o, whichever is the smaller. The obvious attraction of this procedure is its simplicity, with only the requirement of specifying E_{max}. The night time evaporation rate is set equal to 1/100 of the maximum daytime value.

In terms of the application requirements it is a necessary prerequisite that perched water tables should be included in the modelling framework, and that pore pressures be evaluated at selected points throughout the perched zone. The mechanisms that can induce positive pore pressures are either a reduction in permeability or, theoretically, entrapment of air during infiltration. This latter process is ignored in virtually all soil water physics approaches to the estimation of soil water conditions.

Figure 7.7 illustrates four possible approaches to the generation of positive pore pressures in the context of perched water tables. First is the condition that as soon as part of the column is saturated, hydrostatic conditions are developed. This is an extreme situation and one which would represent a worst condition, reflecting, for example, a transient condition of perching above boulders in colluvium. Second is a similar situation, but one in which hydrostatic conditions are only generated if $q_1 > q_2$. Both of these conditions are considered within a homogeneous soil, and may prove significantly less common than other conditions and so would be inappropriate to base further inference on. The third and fourth conditions (Figure 7.7) reflect the occurrence of perching at a K_1/K_2 interface occasioned by a permeability reduction. In the third condition we can set the criterion that $q_1 = q_2$ and that $K_2 < K_1$. The maximum pore pressure (P_{max}) at the interface can then be expressed as:

$$\frac{P_{max}}{h-x} = 1-(K_2/K_1)^{0.5} \tag{7.10}$$

from the following analysis.

At time t a water column has descended a distance x into a medium with permeability K_2. Let p be the pore-water pressure at the K_1/K_2 interface. Then the hydraulic gradients are:

$$i_1 = \frac{h-x-p}{h-x} = 1 - \frac{p}{h-x} \tag{7.11}$$

$$i_2 = \frac{p+x}{x} = 1 + \frac{p}{x} \tag{7.12}$$

For continuity, $\text{flux} = K_1 i_1 = K_2 i_2 = \dfrac{\mathrm{d}x}{\mathrm{d}t}\, n$

where n = effective porosity.

Let $\alpha = K_2/K_1$ and $\beta = x/h$ then from the continuity of flux and after simplification we have

$$\frac{p}{h} = \frac{(1-\alpha)\ \beta\ (1-\beta)}{\alpha\ (1-\beta)+\beta} \tag{7.13}$$

$$\text{and} \quad \frac{p}{h_{\max}} \quad \text{occurs at } \beta = (\alpha^{0.5})/(1+\alpha^{0.5}) \tag{7.14}$$

$$\text{and} \quad \frac{p}{h_{\max}} = (1-\alpha)/(1+\alpha^{0.5})^2 \tag{7.15}$$

$$\text{thus} \quad \frac{P_{\max}}{h-x} = (1-\alpha^{0.5}) \tag{7.16}$$

Condition 3 in Figure 7.7, therefore, considers the solution to equation (7.16), in which the maximum pressure at the interface during the passage of the column of water is generated. Of course this again represents a 'worst' solution, since p is a function of time. This dynamic condition is reflected in equation (7.13), which can be rewritten:

$$\frac{P_t}{h_t} = \frac{(1-\alpha)\beta_t(1-\beta_t)}{\alpha(1-\beta_t)+\beta_t} \tag{7.17}$$

and a solution given in terms of time t (condition 4, Figure 7.7). Figure 7.8 illustrates the solution in terms of the dimensional parameters P/h and $T = \dfrac{tK_1}{nh}$. Under the conditions of K_1, h and n specified in Figure 7.8, $t = 10^4 T$ seconds, i.e. $T = 1$ represents 2.78 hours, $T = 10 = 27.8$ hours. Thus as α reduces, i.e. K_2 becomes much less than K_1, the relatively high pore pressures at the K_2/K_1 interface are sustained for increasingly longer periods.

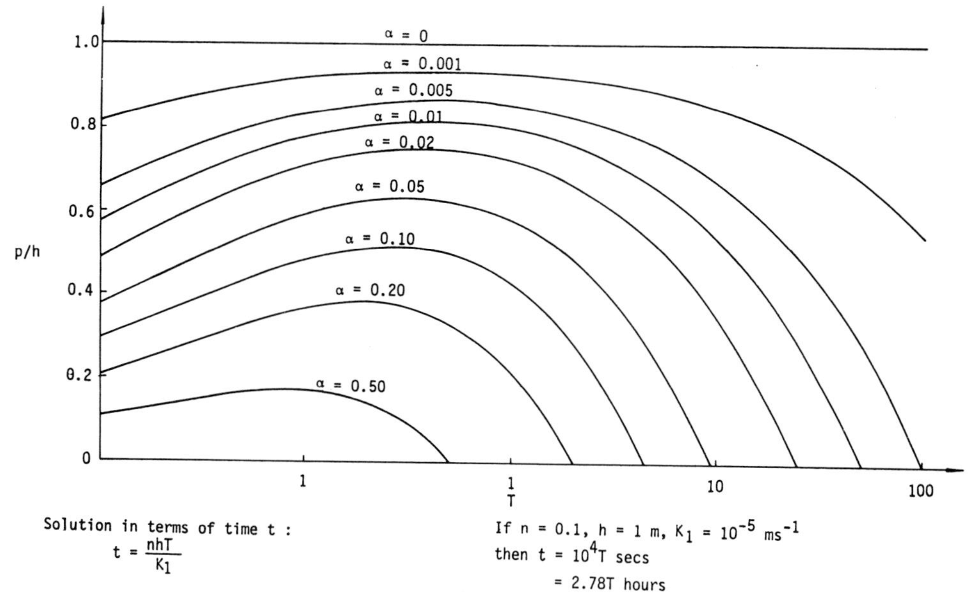

Figure 7.8 Solution to equation 7.17

Table 7.4 Summary of model equations

Condition	Equation	Application
Upper boundary condition	Evaporation (7.9)	Sine based forcing function for evaporation at top computation point
Chunam cover	Seepage (7.1)–(7.3) and parallel plate flow (7.4)	Determination of effective chunam permeability to be used in top model layer
Unsaturated flow	Millington–Quirk	Estimation of unsaturated permeability for specified soil moisture content
Saturated flow	Darcy's Law	Basis of soil water movement extended to unsaturated flow
	Pore water pressures in perched water table (7.17)	Estimation of transient pore pressures throughout perched water table
Lower boundary condition	Model base	Set drainage to 0 (impermeable base) or to any selected condition up to permeability of base computation point

This dynamic condition (equation (7.17)) reflected in Figure 7.8 is considered the most realistic one to be incorporated into the finite difference model, and, accordingly, this equation is adopted to provide solutions to p, for all times t. Pore-water pressure conditions above and below the interface to the saturated/unsaturated boundary are thus calculated by multiplying P_t/h_t by the previously determined distances from the end points of saturation to the interface, thus pressures decline *pro rata* from the interface to zero at the top and base of the saturated column.

Table 7.4 summarizes the main model equations with associated ancillary definitions being given in Figure 7.6.

7.3 DATA ACQUISITION FOR CHUNAM FLOW MODEL

7.3.1 Permeability Determinations of Chunam

While very limited empirical data testify to the higher maintained suctions under properly maintained chunam (McFarlane, 1980), the permeability of chunam has not been evaluated. This is an important prerequisite for modelling (see Table 7.4) as well as for comparisons with rainfall intensity. It is necessary here to outline the procedures we adopted. Two studies were undertaken with regard to chunam permeability. Firstly, samples were taken from field sites, and the results of those analyses are presented in this section. Secondly, the effect of the mix proportions on the permeability of newly formed chunam was examined. As specimens of 100 mm diameter were required, field samples had to be larger. Their sizes were more or less 300 mm × 300 mm. Cracks were deliberately avoided. Various methods

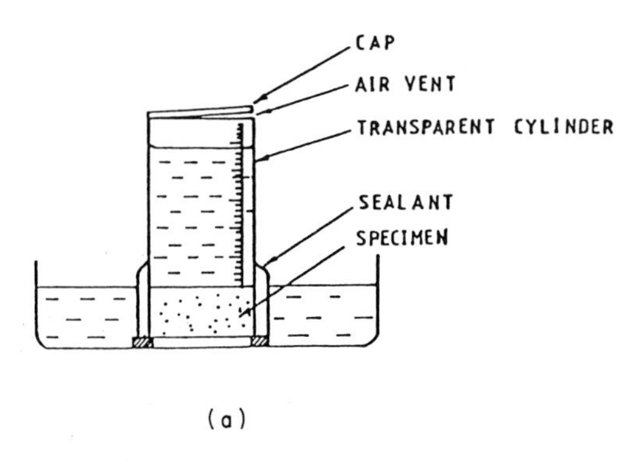

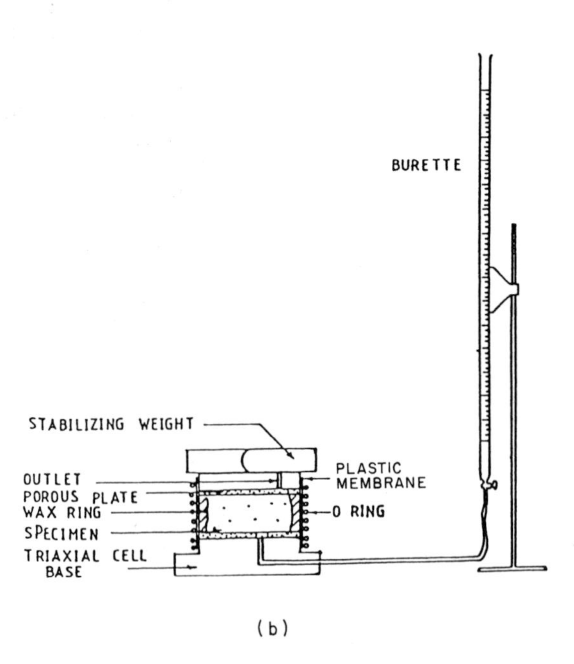

Figure 7.9 Ponding and burette methods used to determine the permeability of chunam

were used for final trimming. Due to the brittle nature of chunam, coring was not successful and most of the samples were prepared by sawing.

Two tests were undertaken to determine chunam permeability. Firstly, a ponding test was undertaken. In this test, water was allowed to pond over the test specimen. The variation of water head with time was monitored. In effect, it was a falling head permeability test. The apparatus set up (Figure 7.9) was extremely simple. A transparent plastic cylinder of 100 mm internal diameter was mounted over the specimen which was trimmed to a diameter of 100 mm. The contact was kept watertight by a sealant which also provided some structural strength during handling.

The specimen with the cylinder was then soaked in water for saturation. Afterwards, the cylinder was filled with water which was then allowed to seep through the soil to an outside trough which held water up to the top of the specimen, thereby avoiding withholding of water by suction. The outflow head could be maintained by overflowing or manual removal of excess water if the flow rate was slow. The water head used was limited by the length of the cylinder which was 0.2 m long. Under a head of that order of magnitude, a week was usually required to complete the test. Over such a period, evaporation might account for a large quantity of water loss. A cap with a small air vent was provided to minimize evaporation.

The second permeability test employed was a falling head test. The apparatus for this test was a triaxial cell and a burette. Water was allowed to run from the burette through the base of the triaxial cell into the test specimen. Flow was confined to the specimen by

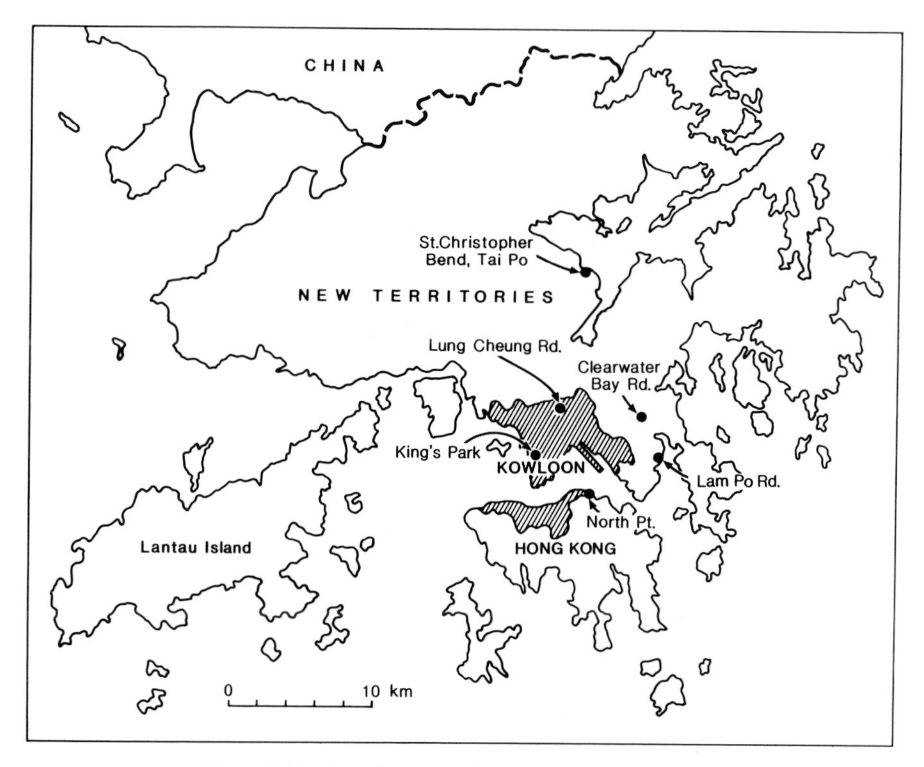

Figure 7.10 Location map of study sites in Hong Kong

Table 7.5 Permeability of chunam from field samples

	Sample Location	Permeability ($m\,s^{-1}$)	
		Using Burette	Using Ponding
Decomposed granite	King's Park	1.5×10^{-7}	
	Upper Clearwater Bay Road	1.4×10^{-7}	5.2×10^{-8}
	Lower Clearwater Bay Road		6.9×10^{-8}
	Lung Cheung Road		2.3×10^{-7}
Decomposed volcanics	Po Lam Road (P1)	6.0×10^{-6}	1.6×10^{-8}
	Po Lam Road (P2)	7.5×10^{-6}	2.2×10^{-8}
	Tai Po (MB)—New	8.9×10^{-7}	2.6×10^{-9}

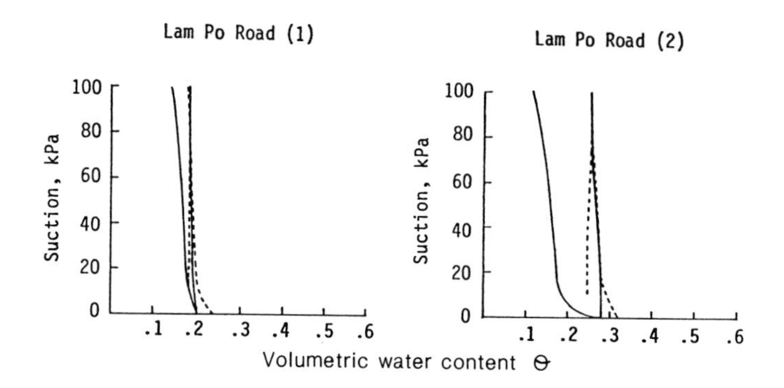

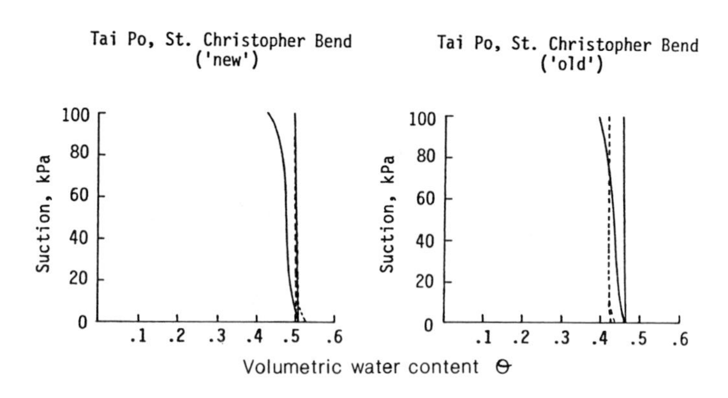

Legend :
———— 100 kPa → 0 → 100 kPa (Wetting - Drying)
- - - - - 0 kPa → 100 → 0 kPa (Drying - Wetting)

Figure 7.11 Suction moisture curves for volcanic chunam (see Figure 7.10 for site locations)

a plastic membrane together with the use of O-rings (Figure 7.9). A smooth-sided paraffin wax ring was cast around the specimen to shield off the irregularities. Figure 7.10 shows the locations in Hong Kong at which field samples for chunam were taken.

Generally, the burette method gave a higher value for the coefficient of permeability (see Table 7.5). This can be attributed to two facts. Firstly, in this method, the water head was higher ranging from 0.8 m to 0.2 m, whereas in the alternative ponding method the water head ranged from 0.2 m to zero. The permeability of chunam may not be constant over these two ranges of water head. Secondly, there seemed to be a substantial amount of side leakage between specimen and wax ring in the burette method. This is reflected in the results: the burette method gives a permeability 300 times that given by the ponding method. The ponding method is a close simulation of possible site conditions. The actual ponding depth will seldom exceed 0.2 m especially on slope surfaces.

Except for one specimen, the permeability values were quite constant and repetition of results was high. The only exception is the sample from Lower Clearwater Bay Road.

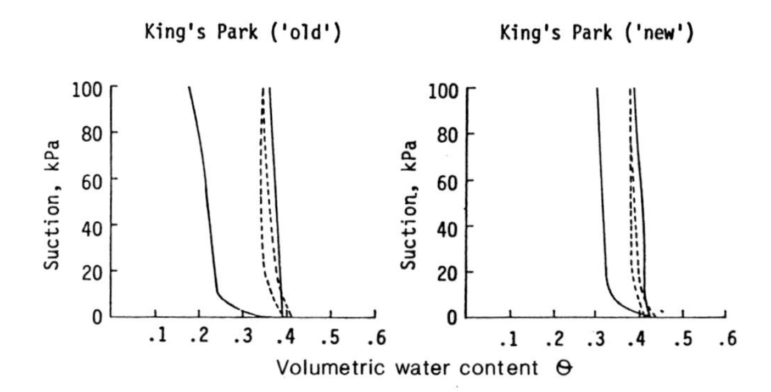

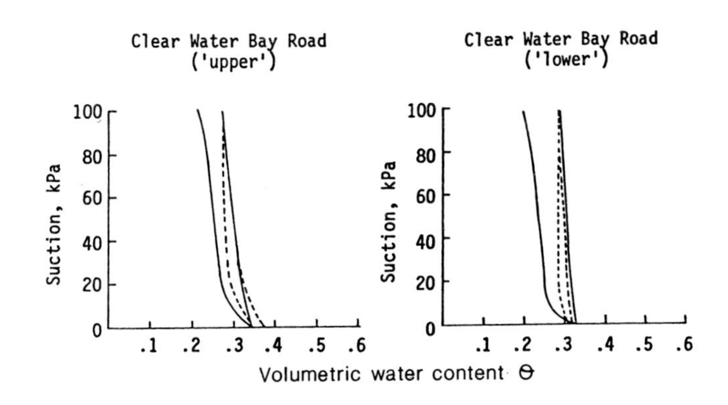

Legend :

________ 100 kPa → 0 → 100 kPa (Wetting - Drying)
- - - - - - - 0 → 100 kPa → 0 (Drying - Wetting)

Figure 7.12 Suction moisture curves for granitic chunam (see Figure 7.10 for site locations)

Over a period of 10 days, the permeability varied from 5×10^{-7} m s^{-1} to 7×10^{-8} m s^{-1} and finally settled down to the latter value. This is probably due to silting up of pores by fines.

For chunam made from decomposed granite, the coefficient of permeability was recorded to be in the range of 10^{-7} to 10^{-8} m s^{-1}; an average of 1.2×10^{-7} m s^{-1}. For chunam made from decomposed volcanics, the coefficient of permeability ranged from 10^{-8} to 10^{-9} m s^{-1}, with an average of 1.4×10^{-8} m s^{-1}. The lower value reflects a higher proportion of fines in decomposed volcanics.

The modelling approach outlined in section 7.2 above allows chunam cover to be included, providing that it can be assigned a suction–moisture curve and a saturated permeability. Since no suction–moisture curve for chunam had previously been determined, it was necessary to acquire such data. Samples of granite and volcanic chunam were taken in the field from the locations shown in Figure 7.10.

Figures 7.11 and 7.12 show the derived suction–moisture curves. The single most noteworthy feature of these curves is of course the steepness of both the drainage and wetting curves. This represents a point of convenience in modelling since, in effect, the only discriminating element is the volumetric moisture content at saturation.

7.3.2 Effect of Chunam Mix on Permeability

If the specification for chunam mix is always followed, then the hydrological properties would be expected to be relatively constant. Construction practice may be expected to vary, however, and it is important, therefore, to determine the control on permeability. Given, too, that no such tests have previously been undertaken, it is considered necessary that the experimental procedure we adopted for this purpose be fully described.

A total of 30 trial mixes were made in this study. Laboratory reconstituted chunam from King's Park Decomposed Granite made up 12 of the mixes and another 12 mixes came from Tai Po Decomposed Volcanics. King's Park Decomposed Granite and Tai Po Decomposed Volcanics were preferred because significant amounts of suction related engineering data were already available from previous investigations. The 12 mixes for each soil type were made according to the mix proportions given in Table 7.6. Water content was varied accordingly in each case to produce the necessary workability. The remaining six mixes were made with Clearwater Bay Road Decomposed Granite. Again water content was varied accordingly in each mix to produce the necessary workability. Figure 7.13 shows the particle size distributions of the soils used.

Cylindrical moulds of nominal internal diameter 73 mm and nominal thickness 42 mm were used in the preparation of samples. Moulds for the chunam samples were removed one day after mixing. The samples were left indoors and air cured.

The formation of cracks in chunam is directly related to its shrinkage property. The area shrinkage is relevant in respect of crack formation in view of the fact that chunam plaster on slopes is very thin when compared to its spatial extent. It is noted that other factors as well may contribute to crack formations. Field observations show that chunam made of coarse sand generally has cracks wider than those of fine soils. High water–cement ratio will also lead to severe cracking in the chunam.

During curing of the samples, measurements of the diameter of the samples were made daily for seven days. Three diameters, spaced evenly, were marked on the

Table 7.6 Trial Mixes for Laboratory Constituted Chunam Samples of Tai Po Decomposed Volcanics and King's Park Decomposed Granite ○ and Clearwater Bay Road Decomposed Granite □

			Parts by weight of *Hydrated Lime* per 20 parts of soil			
			0	1	3	5
Parts by weight of *cement* per 20 parts of soil	0	Test Base-line	1	2	3	4 □
	1		5	6	7 Standard PWD Specification	8
	2		9 □	10	11 □	12 □

Note: Numbers refer to mix number.

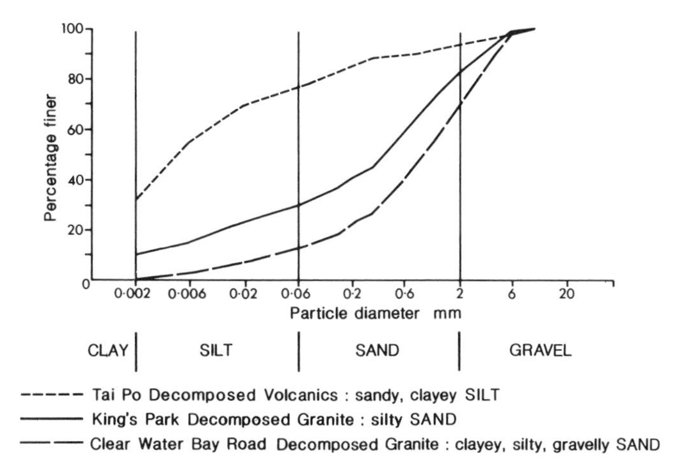

Figure 7.13 Particle size distributions for soils used in chunam mixes

samples. The diameters were measured by a screw gauge. The percentage shrinkage was calculated from the following formula:

$$\text{Percentage shrinkage} = \left(1 - \frac{\text{average measured area}}{\text{average original area}}\right) \times 100\% \qquad (7.18)$$

Since the cement and lime contents of the trial mixes vary and there may be cases in which cement and lime contents are nil, the water to soil ratio was adopted for comparison and analysis instead of the usual approach of water to cement ratio. The amount of water in each trial was adjusted so that all the trial mixes were consistent with their workabilities. Since there is not a quantitative measurement of workability for chunam, the required workability is a matter of personal judgement. Due to the silty nature of the Tai Po

Table 7.7 Variation of area shrinkage with age for chunam of Tai Po Decomposed Volcanics

Trial Mix No.	% shrinkage at ages (in days)						
	1	2	3	4	5	6	7
1	—	11.0	15.0	17.3	—	—	17.9
2	—	10.1	11.9	12.5	—	—	13.1
3	—	8.1	—	10.5	—	11.1	11.7
4	3.3	8.1	—	10.5	—	11.0	11.0
5	2.3	—	4.6	5.2	5.8	5.9	6.1
6	—	4.0	4.6	4.5	4.8	—	4.9
7	2.2	4.8	5.4	—	6.0	6.1	6.1
8	3.2	5.9	6.3	—	6.7	6.8	6.8
9	1.2	2.9	3.2	—	4.2	4.4	4.5
10	1.4	1.9	—	2.3	2.4	2.5	2.6
11	2.0	3.5	—	4.4	4.4	4.4	4.4
12	2.2	3.7	—	4.2	4.4	—	4.6

Note (1) Trial Mix No. 7 is EDD specification.

Decomposed Volcanics, the mixes were very cohesive. Compaction of the mixes by the method of tamping was not practicable. King's Park Decomposed Granite is sandy and much less cohesive. Clearwater Bay Road Decomposed Granite is gravelly sand and chunam made from it has properties approaching that of concrete. The determination of the required workabilities depends largely on experience.

Measurement of the area shrinkage of the chunam samples began one day after the preparation of the mixes when the moulds were removed. All the samples were left indoors and air cured. The temperature and humidity changes were very small and their effects on the curing of the samples were minimal. Table 7.7 shows the variation in area shrinkage with age for the chunam for all the trials undertaken. Further comparisons revealed that chunam mixes of Tai Po Decomposed Volcanics were generally more susceptible to shrinkage than that of King's Park Decomposed Granite with chunam mixes of Clearwater Bay Road Decomposed Granite the least susceptible. A comparison of the particle size distribution curve for the three soil types indicates a gradual transition from high silt content (Tai Po Volcanics) to high sand content with the presence of gravel (Clearwater Bay Road granites). A significant portion of the shrinkages occur in the first seven days of curing, although Ho (1978) reported that more than 70% of the shrinkage was obtained at the end of 60 days. The comparatively large proportion of shrinkage obtained in the first seven days for the trial mixes here may be attributed to the small size of the samples and the curing condition. Less humid conditions and smaller samples both lead to drying of the sample at a faster rate. Cracks were generally not observed in the samples except for a few cases which have low cement and lime content.

It is desirable to isolate and identify the principal controls on shrinkage. The obtained shrinkage data in association with the cement and lime content and water to soil ratio (as given by the trial numbers—Table 7.6—were subjected to multiple regression analysis, as detailed in Table 7.8. For both materials, the negative values of the particle regression coefficients for cement and lime contents (X_1 and X_2 respectively) imply that the increase

Table 7.8 Shrinkage area relationships

(a) Chunam of Tai Po Decomposed Volcanics (7.19)

$\hat{Y} = 5.3 - 71.8X_1 - 24.6X_2 + 20.2X_3$

adjusted $R^2 = 0.81$

standard error = 2.0% area shrinkage

standardized regression coefficients:

$\beta_1 = -0.67$

$\beta_2 = -0.54$

$\beta_3 = 0.55$

(b) Chunam of King's Park Decomposed Granite (7.20)

$\hat{Y} = -0.49 - 31.2X_1 - 10.8X_2 + 16.1X_3$

adjusted $R^2 = 0.57$

standard error = 0.83% area shrinkage

standardized regression coefficients:

$\beta_1 = -1.06$

$\beta_2 = -0.86$

$\beta_3 = 0.87$

where Y = % area shrinkage at age of 7 days
X_1 = cement to soil ratio
X_2 = hydrated lime to soil ratio
X_3 = water to soil ratio

Table 7.9 Permeability of trial mix chunam samples

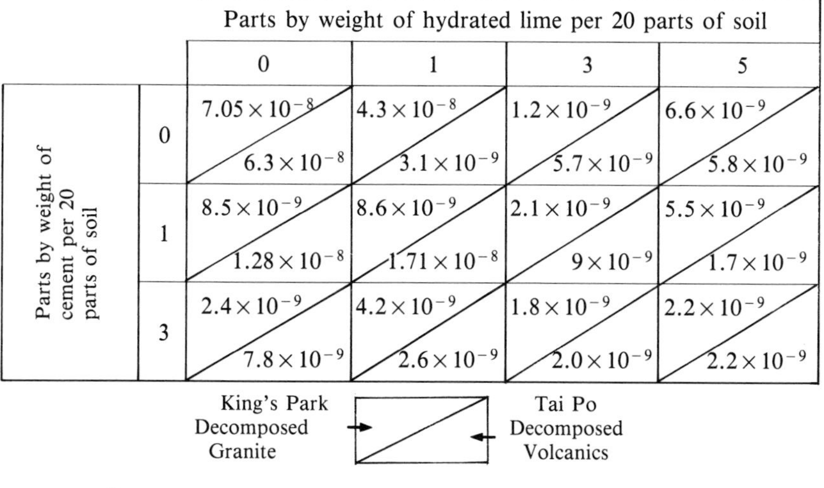

Parts by weight of cement per 20 parts of soil	Parts by weight of hydrated lime per 20 parts of soil: 0	1	3	5
0	7.05×10^{-8} / 6.3×10^{-8}	4.3×10^{-8} / 3.1×10^{-9}	1.2×10^{-9} / 5.7×10^{-9}	6.6×10^{-9} / 5.8×10^{-9}
1	8.5×10^{-9} / 1.28×10^{-8}	8.6×10^{-9} / 1.71×10^{-8}	2.1×10^{-9} / 9×10^{-9}	5.5×10^{-9} / 1.7×10^{-9}
3	2.4×10^{-9} / 7.8×10^{-9}	4.2×10^{-9} / 2.6×10^{-9}	1.8×10^{-9} / 2.0×10^{-9}	2.2×10^{-9} / 2.2×10^{-9}

Key: upper-left value — King's Park Decomposed Granite; lower-right value — Tai Po Decomposed Volcanics

in cement or lime content will reduce the area shrinkage. Other things being equal, the standardized partial regression coefficients β indicate that one standard deviation unit change of the cement to soil ratio (β_1) would introduce the greatest change in shrinkage and one unit change in lime to soil ratio (β_2) the least. It is, however, important to recall that the water to soil ratio and the cement and lime content are not strictly independent. Regression equations show that increases in lime and cement content reduce shrinkage. At the same time, increase in lime and cement content lead to an increase in water content which acts to nullify the effect.

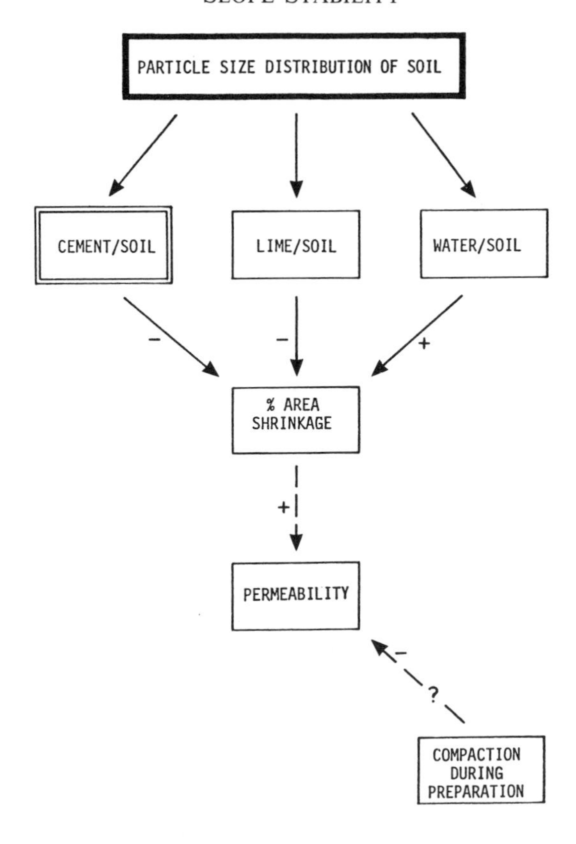

Figure 7.14 Controls on the shrinkage and permeability of chunam

The second stage of the laboratory experiment sought to determine the control that mix proportions have on permeability. The falling head method was used to measure the permeability of the chunam samples. The results obtained (Table 7.9) show that laboratory reconstituted samples have permeability values up to an order of magnitude lower than the site samples for some soil types.

However, no statistically significant relationships were found between permeability and per cent of area shrinkage, probably because compaction of the chunam mix during preparation is a stronger influence on permeability than mix proportions. Figure 7.14 summarizes the controls on chunam performance in the context of mix variation which can be stated as follows:

1. The standard specification 1:3:20 mix, i.e. mix No. 7, compared with other mixes has low shrinkage and medium to low permeability.
2. The permeability of intact chunam is unrelated to the mix properties.
3. The principal effect of mix variation is thus related to chunam shrinkage.
4. Particle size distribution of the soil part of chunam mixes is the most important determining factor of the properties of the mixes. Chunam mix of fine soil is associated with large drying shrinkage and lower permeability.

5. Both cement and lime stabilize the soil. Increase in cement or lime content alone reduces the shrinkage. Increase in water content results in larger shrinkage. Statistical correlation indicates that the effect of change in cement content is most prominent while lime content is the least.
6. Since variation of cement and lime content necessitate a corresponding change in water content, there exists an optimum cement content. The existence of such an optimum content for lime is not so obvious regarding the influence on shrinkage. For fine material, high cement content is desirable (lime content proportions are irrelevant). For coarser material, cement and lime content assume equal importance. The actual proportions are unimportant *providing* the content is 1:20 or greater.

7.4 MODEL VERIFICATION AND APPLICATION

7.4.1 Laboratory Verification of Crack Model

It is necessary to obtain some verification of the effective chunam permeability predicted by the model outlined in section 7.2.1. Accordingly, a laboratory column was set up, being 300 mm diameter and 225 mm high. The lower 50 mm contained a filter above which was placed 125 mm of soil. Chunam was then formed on top of the soil according to EDD specification. A crack forming strip of desired width (within the range 1 mm to 6 mm) was placed on the soil before the chunam was poured. The complete apparatus could be used to measure: (a) the permeability of the soil prior to the chunam pouring; (b) the discharge through the 'cracked' chunam; and (c) the final permeability of the soil (with the chunam removed).

In the initial trials undertaken crack widths of approximately 1, 3, 4 and 6 mm were formed and the crack and chunam discharge measured through a soil of $K_s = 6.9 \times 10^{-5}$ m s^{-1}. A further trial on a soil of $K_s = 1 \times 10^{-6}$ m s^{-1} with a 6 mm wide crack was also completed. Table 7.10 summarizes the measured discharge through the crack widths stipulated. In addition, the flows predicted by equations (7.1)–(7.3) are also shown. It can be seen that the measured flow exceeds the predicted flow in the case of the more permeable soil. This difference is of the order of half an order of magnitude. It is to be noted that for the less permeable soil the disparity between predicted and measured flow is somewhat larger, and in this instance the predicted flow exceeds the measured flow.

By reference to Figure 7.3 and the analysis in section 7.2, it is clear that the initial spread below the chunam (B_1) is a critical determinant in the resulting discharge. The nature of the bonding between the soil and chunam is obviously important in this regard. In the context of the laboratory experiment it was observed that in the less permeable soil the chunam, rather than remaining as a distinct layer above the soil, had bonded very strongly with the underlying soil during the curing process. For the case of the less permeable soil, the depth of bonding within the soil was observed to be much less.

Permeability tests confirmed this, as Table 7.10 indicates. The question upon which interpretation of the results hinges is, therefore, the nature of the chunam/soil bond under *field* conditions. If the field application of chunam results in a somewhat superficial degree of bonding (by which it is implied that the chunam layer and soil layer are readily distinct in permeability terms, and without the permeability gradation shown in less permeable soil—Table 7.10), then the Harr model (equations (7.1)–(7.3)) slightly under-estimates the

Table 7.10 Results of Chunam Crack Discharge Tests

Soil permeability ($m\,s^{-1}$)			Crack width (mm)			
before test	after test	Flow ($m^3\,s^{-1}\,m^{-1}$)	1.4	2.9	4.4	6.0
6.9×10^{-5}	5.9×10^{-5}	measured	1.6×10^{-5}	1.9×10^{-5}	2.2×10^{-5}	2.4×10^{-5}
		predicted	4.4×10^{-6}	6.6×10^{-6}	7.2×10^{-6}	7.9×10^{-6}
1.2×10^{-6}	2.1×10^{-7}	measured				5.2×10^{-9}
		predicted				3.1×10^{-8}

discharge and effective chunam permeability. The nature of this degree of bonding is judged to be that commonly exposed in field conditions where substantial areas of cracked chunam are exposed. If this association is accepted, then the predictions based upon the Harr model are not over-estimating the effective chunam permeability required to prevent infiltration according to the selected criteria.

7.4.2 Applications of the Model of Effective Chunam Permeability to Field Conditions

A field survey of chunam cracks was undertaken in order to parameterize the crack model (equations (7.1)–(7.3)) and thereby to facilitate an estimation of effective chunam permeability. Eight sites were selected for this analysis. The quality of the chunam was classified on the basis of visual inspection only (Brand and Hudson, 1982).

The field survey, carried out for these eight sites, followed the procedures below:

1. An area with a crack density representative of the chunam surface was selected.
2. A 1 m^2 area was marked on the surface.
3. The widths of the cracks and their corresponding cumulative length within the area were measured.
4. For large extents of chunam, two areas were selected for measurement in order to improve the representative aspect of the survey. These were designated areas A and B.

Table 7.11 summarizes the data obtained, and Figure 7.15 illustrates a typical site.

The crack length and width data obtained can then be used to estimate the flow through the cracks, according to the relationships shown in Figure 7.4. Ignoring hairline cracks, this summation of total seepage was evaluated for each site and is shown in Table 7.12. The seepage, together with the intact chunam permeability, taken to be $1\times10^{-8}\,m\,s^{-1}$ (see Table 7.5) allows the effective chunam permeability, K_e, to be calculated by equation (7.15), where regular crack spacing is now replaced by total length estimation (Table 7.11), as shown in Figure 7.16

Figure 7.17 summarizes the effective permeability predictions made on the basis of the field survey and the crack model as reported in Table 7.12. These results suggest that for soils of high permeability ($1\times10^{-4}\,m\,s^{-1}$) the condition of the chunam needs to be good for the chunam to be considered effective. In Figure 7.17, good protection is considered

Table 7.11 Summary of Chunam Crack Survey

Chunam condition	Bad	Bad	Bad	Poor	Poor A	Poor B	Average A	Average B	Average	Average
Crack widths (mm)										
Hair crack	3,350		1,530	8,840	720	1,170	1,080	1,780	1,890	1,750
0.5	2,790			350			320	330	670	630
1.0	650	3,620	1,890		610	1,270	510			
1.5	410					380				
2.0			2,210		490	3,050				
3.0					750	510				
4.0		50	250			1,330				
5.0										
6.0			85							

Note Numbers in table are the total length of cracks (mm) of given crack width in a 1 m^2 area.

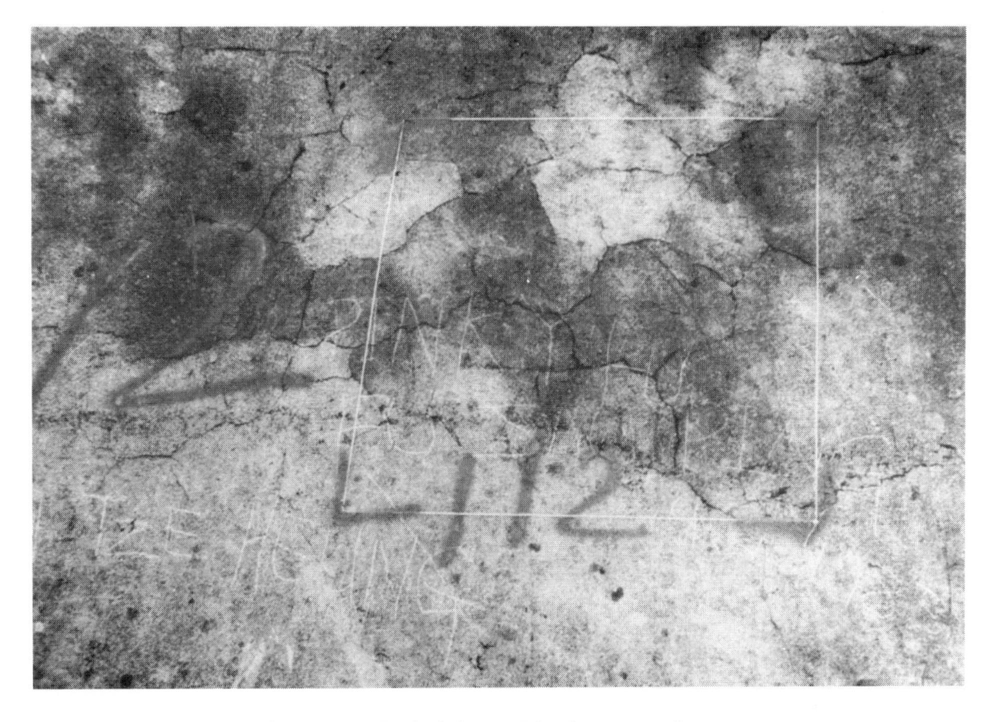

Figure 7.15 Typical site used in chunam crack survey

to be an effective chunam permeability of less than $1.0 \times 10^{-6}\,m\,s^{-1}$. In this latter case, the *duration* of rainfall is the only important factor.

In the interpretation of Figure 7.17 and Table 7.12 several factors must be noted:

1. The effective permeability, in association with crack condition (poor–average) is predicted by equations (7.1)–(7.3).

Table 7.12 Summary of Crack Seepage and Effective Chunam Permeability

Chunam Condition			Bad	Bad	Bad	Poor	Poor A	Poor B	Average A	Average B	Average	Average
Chunam Permeability	Soil Permeability											
1.0×10^{-8}	1.00×10^{-6}	q	1.53×10^{-7}	1.43×10^{-7}	2.10×10^{-7}	1.33×10^{-8}	9.45×10^{-8}	3.40×10^{-7}	4.00×10^{-8}	1.40×10^{-8}	2.50×10^{-8}	2.30×10^{-8}
		K_e	1.63×10^{-7}	1.53×10^{-7}	2.20×10^{-7}	2.33×10^{-8}	1.45×10^{-7}	3.50×10^{-7}	5.00×10^{-8}	2.40×10^{-8}	3.50×10^{-8}	3.30×10^{-8}
	1.00×10^{-5}	q	1.53×10^{-6}	1.43×10^{-6}	2.10×10^{-6}	1.33×10^{-7}	9.45×10^{-7}	3.40×10^{-6}	4.00×10^{-7}	1.40×10^{-7}	2.50×10^{-7}	2.30×10^{-7}
		K_e	1.54×10^{-6}	1.44×10^{-6}	2.11×10^{-6}	1.43×10^{-7}	9.55×10^{-7}	3.41×10^{-6}	4.10×10^{-7}	1.50×10^{-7}	2.60×10^{-7}	2.40×10^{-7}
	1.00×10^{-4}	q	1.53×10^{-5}	1.43×10^{-5}	2.10×10^{-5}	1.33×10^{-6}	9.45×10^{-6}	3.40×10^{-5}	4.00×10^{-6}	1.40×10^{-6}	2.50×10^{-6}	2.30×10^{-6}
		K_e	1.53×10^{-5}	1.43×10^{-5}	2.10×10^{-5}	1.33×10^{-6}	9.45×10^{-6}	3.40×10^{-5}	4.00×10^{-6}	1.40×10^{-6}	2.50×10^{-6}	2.30×10^{-6}

Notes (1) q is seepage through cracks based on crack model (equations (7.1)–(7.3)) ($m^3 s^{-1}$).
(2) K_e is effective chunam permeability based on equation (7.8) ($m s^{-1}$).
(3) All permeability values are $m s^{-1}$.

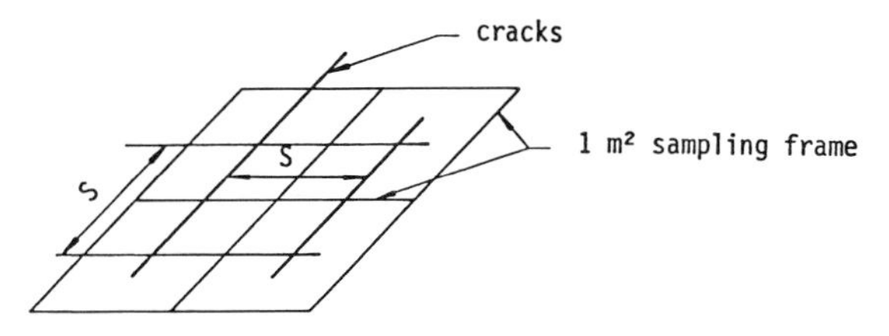

Total crack length per m² = $\frac{2}{S}$

where : S = crack spacing, m

$$K_e = K_i + \sum_{i=1}^{j} q_i \frac{2}{S_i}$$

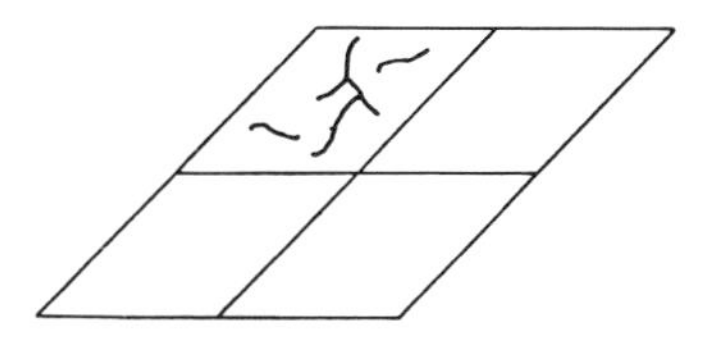

$$K_e = K_i + \sum_{i=1}^{j} q_i L_i$$

Where : L_i = field measured crack length per m² for crack width, i.

Figure 7.16 Alternative approaches to the estimation of K_e in terms of regular and irregular crack spacing

2. The model used for the prediction assumes no void space below the chunam. Field conditions frequently suggest the presence of such a void (hence a likely increase in effective crack width B_∞—Figure 7.3). We might, therefore, expect the protection level of chunam to be lower than Figure 7.17 would suggest.
3. The crack model assumes the crack to be formed by two parallel plates. Departures from this condition in the field may lower head losses within the crack and this factor may increase the protection shown in Figure 7.17.

7.4.3 Predictions of Chunam 'Effectiveness' using the Vertical Infiltration Model

The analysis of chunam permeability and suction–moisture curves in section 7.4.1, together with the crack model summarized in Figure 7.17, facilitates a more detailed assessment of chunam performance by making use of the vertical infiltration model.

Here, layer 1 of the model was set to comprise the chunam surface, with three computational points or cells (see Figure 7.6) and two subsequent soil layers to 9 m depth. Of course there are innumerable combinations of permeability, rainfall, and start conditions

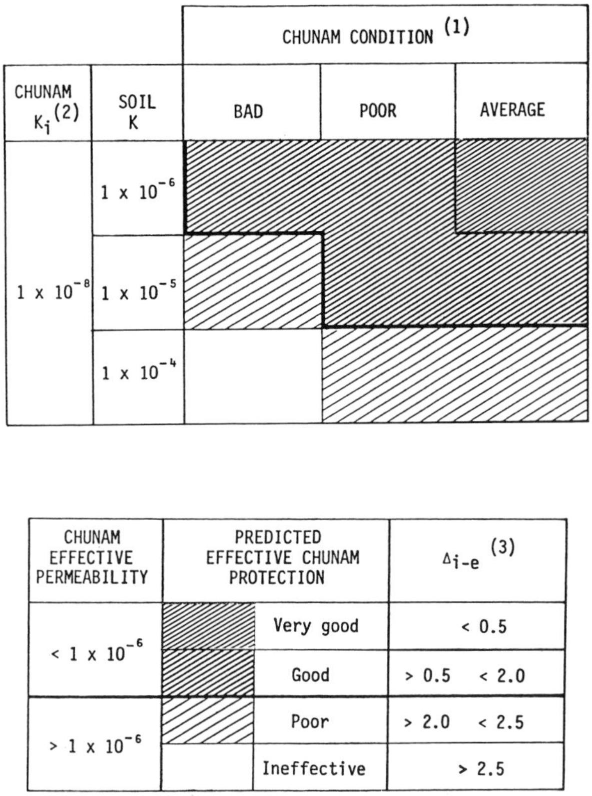

CHUNAM K_i (2)	SOIL K	CHUNAM CONDITION (1)		
		BAD	POOR	AVERAGE
1×10^{-8}	1×10^{-6}			
	1×10^{-5}			
	1×10^{-4}			

CHUNAM EFFECTIVE PERMEABILITY	PREDICTED EFFECTIVE CHUNAM PROTECTION	Δ_{i-e} (3)
$< 1 \times 10^{-6}$	Very good	< 0.5
	Good	> 0.5 < 2.0
$> 1 \times 10^{-6}$	Poor	> 2.0 < 2.5
	Ineffective	> 2.5

Notes : (1) As assessed by 'CHASE' report
(2) Permeability of intact chunam K_i ms^{-1}
(3) Order of magnitude difference between intact and effective chunam permeability

Figure 7.17 Predicted effective chunam protection

that can be used in such a modelling scheme. In addition, it must be noted that the analysis of chunam performance summarized in Figure 7.17 assumes a constant permeability of soil with depth. It might, therefore, be expected that chunam effective performance could be revised by the results of modelling these conditions of soil heterogeneity and various start conditions, which so far have not been considered.

A standard suction–moisture curve for chunam was adopted, that of Figure 7.12(c), and the suction–moisture curve adopted for the soil was that corresponding to completely decomposed granite (CDG). Start conditions in the profile were a mean start suction of approximately 20 kPa. A 1 in 10 year three-day rainfall event was generated from Figure 2.4 in Geotechnical Control Office (1984). Figure 7.18 shows the results of simulations using the conditions outlined, and portraying the lowest recorded suction at each depth in a 200 hour simulation period (storm length 72 hours).

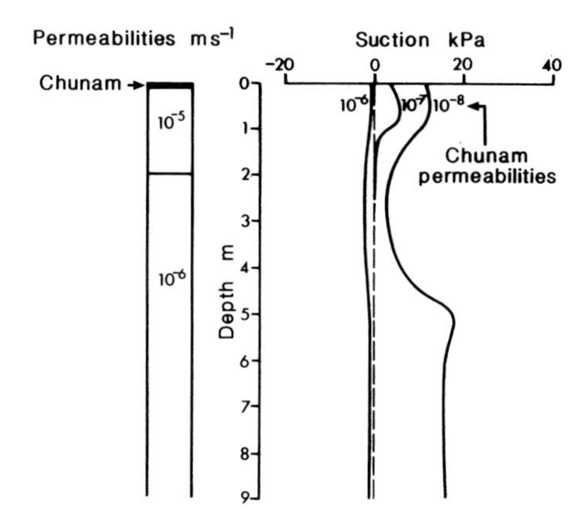

Figure 7.18 Simulation of soil water conditions for a three-day 1 in 10 year event (lowest suctions shown)

Figure 7.18 shows the improvement rendered by a decrease in chunam effective permeability of one order of magnitude to 10^{-7} m s^{-1}. Under the 1 in 10 year event the perched water table, present when $K_e = 10^{-6}$ m s^{-1}, is absent when $K_e = 10^{-7}$ m s^{-1}. This result is consistent with the summary plot of Figure 7.17 and Table 7.12. In that table the conditions in Figure 7.18 plot as poor protection ($K_e = 10^{-6}$ m s^{-1}) and good protection ($K_e = 10^{-7}$ m s^{-1}) respectively. Although in Figure 7.17 this latter condition is classified as good protection (since K_e is two orders of magnitude less than the underlying soil), it is observed from Figure 7.18 that it is just insufficient to prevent saturation occurring. An increase of K_e to 10^{-8} m s^{-1} under the same storm conditions ('very good' protection as classified by Table 7.13 and Figure 7.17) renders the soil profile safe from saturation under the 1 in 10 year event used. The rainfall conditions used in the simulation represent the worst possible from the standpoint of the destruction of soil suction, since an even rain intensity was used in the model for each of the three-day rainfall totals. It is well known that there is considerable intensity variation which would serve better to maintain soil suctions than Figure 7.18 would suggest.

From these simulations, as well as others undertaken, it is appropriate to consider $K_e = 10^{-7}$ m s^{-1} as a level of protection that under normal field start conditions and typical 1 in 10 year events should maintain at least low suctions. The large number of permutations of conditions that are possible in relation to this problem cannot all be simulated, but the indications from this analaysis are that $K_e = 10^{-7}$ m s^{-1} is an acceptable and realistic boundary value for practical purposes.

It is appropriate in the light of this analysis to refine the associations shown in Figure 7.17 to establish good protection (i.e. $K_e < 10^{-6}$ m s^{-1}), and protection likely to ensure the maintenance of some degree of suction (i.e. $K_e < 10^{-7}$ m s^{-1}). Figure 7.19 shows the appropriate refinement. The associations here were established by the following procedure which obviously has the potential to be more thoroughly researched:

1. Determine soil permeability (or estimate from available sources).
2. Determine permeability of intact chunam (an estimate from this study should be adequate—see Table 7.5).
3. Undertake field crack survey (as outlined in section 7.4.2).

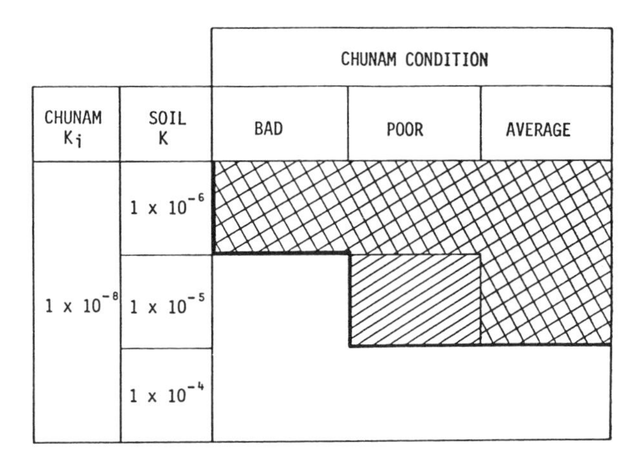

Legend :

Conditions likely to ensure maintenance of some degree of soil suction ($K_e < 10^{-7}$ ms^{-1})

Good protection ($K_e < 10^{-6}$ ms^{-1})

Figure 7.19 Predicted effectiveness of chunam using coupled crack flow and finite difference models (see section 7.2)

4. Use the crack model to determine the flow through the total recorded cracks (as outlined in section 7.2)
5. Make a preliminary assessment of chunam effectiveness (as shown in Table 7.12 and Figure 7.17).
6. If desired, input the effective permeability determined in (4) (equation (7.8)) into the soil water finite difference model in association with the soil permeability (1) to examine the protection under storm types of particular interest.

7.5 DISCUSSION

Many of the chapters in this book concern the natural processes of mass movement on slopes (e.g. Chapter 9) and currently available schemes for modelling such phenomena (e.g. Chapters 2 and 4). The increase in interest relating to slopes that present extreme problems of stability necessitates a reappraisal of modelling assumptions and site investigation procedures (Anderson, 1983; Hearn and Jones, 1985). It is the investigation of difficult terrain (where classical engineering solutions may be inadequate) that the authors argue necessitates the inclusion of process-based model structures to yield more appropriate analytical frameworks.

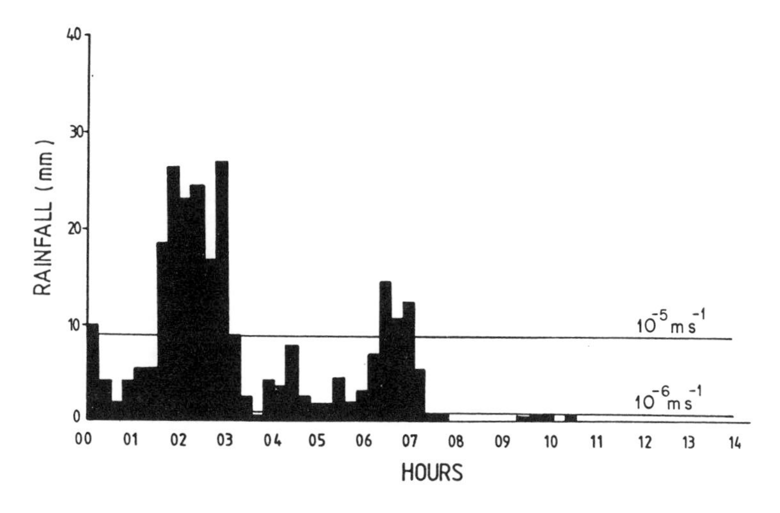

Note: Rainfall intensity
10^{-5} m s^{-1} = 9 mm in 15 minutes
10^{-6} m s^{-1} = 0.9 mm in 15 minutes

Figure 7.20 Illustration of rainfall intensities in relation to permeability values. Shown is a storm at North Point, Hong Kong (see Figure 7.10) on 29 May 1982

This chapter has sought to achieve this goal in terms of one aspect—that of surface covering of slopes using chunam to reduce infiltration-induced landslide risk. We can summarize the principal findings as follows:

1. A model has been outlined which predicts flow through cracks under specified conditions (equations (7.1)–(7.3)).
2. This model can be used to estimate the effective permeability of chunam with cracks (Figures 7.5 and 7.16, equation (7.8)).
3. From a field crack survey, chunam effective permeability was predicted for poor to average chunam condition cover (Figure 7.17).
4. From the standpoint of rainfall intensity, an effective chunam permeability of 1×10^{-6} m s^{-1} or less needs to be achieved (Figure 7.20). This is the distinction made in the summary graph (Figure 7.19).
5. Selected factors have been outlined which may serve to reduce or increase the effective permeability of chunam for a specific crack distribution. These are factors such as vertical crack irregularity and void space at the chunam–soil interface that are not modelled.
6. The effective permeability of 1×10^{-6} m s^{-1} renders the slope soil water conditions independent of rainfall intensity (Figure 7.20) and, in consequence, only dependent on rainfall duration.

7. Whether $1 \times 10^{-6}\,m\,s^{-1}$ is a sufficiently low effective permeability to render the underlying soil free from saturation under specific rainfall duration recurrence intervals, can only be answered by reference to simulation studies undertaken by the soil water model outlined in section 7.2.2.

There are three, progressively less restrictive criteria by which to judge the effectiveness of chunam:

1. *Maintenance of soil suction beneath the cover* Both field and modelling evidence (Figure 7.19) suggests the necessity of having the effective permeability (K_e) of the chunam less than $10^{-7}\,m\,s^{-1}$ to achieve maintenance of suctions for 1 in 10 year rainfall events. Effective permeability is defined by equation (7.8).
2. *Providing significant protection in terms of percentage surface runoff generated (suctions not now necessarily maintained)* An effective permeability of $10^{-6}\,m\,s^{-1}$ in relation to rainfall provides significant protection (see Figure 7.20). This condition is in fact achieved by a wide range of chunam quality and underlying soil permeabilities (see Figure 7.17).
3. *Prevention of infiltration under conditions of high soil permeability ($10^{-4}\,m\,s^{-1}$) and chunam in 'bad' condition* Even under these extreme conditions chunam affords some protection against the volume of water allowed to infiltrate into the slope. Table 7.12 shows typically that in those circumstances $K_e = 1 \times 10^{-5}\,m\,s^{-1}$, which still prevents of the order of 40% of rainfall infiltrating in storms typified by that shown in Figure 7.20.

While this model structure has been developed and applied to an artificial slope covering (chunam), there is the possibility of extending such a scheme to circumstances where similar conditions occur naturally on slopes. In strongly weathered andesite in the Caribbean, for example, a baked less permeable skin has been observed to be acting as a natural chunam, maintaining soil suctions even in highly intense storms (Anderson and Kemp, 1985). In such conditions, more research needs to be undertaken on both soil crusting models of the form in which there are time dependent material property changes with depth (e.g. Whisler *et al.*, 1979), and the type of 'fixed property' model outlined in this chapter. Both approaches will allow a better understanding of slope hydrology, and hence stability, in tropical regions.

ACKNOWLEDGEMENTS

Much of the work reported in this chapter was undertaken by the authors at the Geotechnical Control Office, Hong Kong, and is published with the permission of the Hong Kong Government. The authors acknowledge the support and encouragement given to the study by Dr E. W. Brand. Certain of the laboratory work was undertaken by Mr C. Chan and Mr W. Kwan and their assistance is gratefully acknowledged. Mr R. R. Hudson provided helpful comments both during the project and on initial drafts of selected sections of the material reported here.

REFERENCES

Anderson, M. G. (1983). 'The prediction of soil suction for slopes in Hong Kong.' Final Technical Report to Geotechnical Control Office, Hong Kong, on CE 3/81, 244pp.

Anderson, M. G., and Howes, S. (1985). 'Development and application of a combined soil water slope stability model.' *Q. J. Eng. Geol.*, **18**, 225–36.

Anderson, M. G., and Kemp, M. J. (1985). 'The prediction of pore water pressure conditions in road cut slopes, St Lucia, West Indies.' Final Technical Report to Overseas Development Administration, London, on R3426C, 208pp.

Anderson, M. G., McNicholl, D. P., and Shen, J. M. (1983). 'On the effect of topography in controlling soil water conditions, with specific regard to cut slope piezometric levels.' *Hong Kong Engineer*, **11**, 35–41.

Brand, E. W. (1982). 'Analysis and design in residual soils.' *Proc. Am. Soc. Civil Eng.* Speciality Conference on Engineering and Construction in Tropical and Residual Soils, Honolulu, 89–143.

Brand, E. W. (1984). 'Landslides in Southeast Asia: a state of the art report.' *Proc. 4th Int. Symp. Landslides*, Toronto, **1**, 17–59.

Brand, E. W. (1985). 'Geotechnical engineering in tropical residual soils.' *Proc. Int. Conf. Geomech. in Tropical Lateritic and Saprolitic Soils*, Brasilia, Brazil, 3.

Brand, E. W., and Hudson, R. R. (1982). 'CHASE—an empirical approach to the design of cut slopes in Hong Kong soils.' *Proc. 7th SE Asia Geotech. Conf.*, Hong Kong, **1**, 1–16 and **2**, 61–72, 77–9.

Brand, E. W., Premchitt, J., and Phillipson, H. B. (1984). 'Relationship between rainfall and landslides in Hong Kong.' *Proc. 4th Int. Symp. Landslides*, Toronto, vol. **1**, 377–84.

Endicott, L. J. (1982). 'Analysis of piezometer data and rainfall records to determine groundwater conditions.' *Hong Kong Engineer*, **10**, 53–6.

Geotechnical Control Office (1984). *Geotechnical Manual for Slopes* (2nd edn). Geotechnical Control Office, Hong Kong, 295pp.

Janbu, N. (1954). 'Application of composite slip surfaces for stability analysis.' *Proc. Eur. Conf. Stability Earth Slopes*, Stockholm, 43–9.

Janbu, N. (1973). 'Slope stability computations.' In *Embankment Dam Engineering* (Casagrande Volume). Hirschfeld, R. C., and Poulos, S. J. (eds.), Wiley, New York, pp. 47–107.

Harr, M. E. (1962). *Groundwater Flow and Seepage*, McGraw Hill, New York, 315pp.

Haverkamp, R., Vauclin, M., Touma, J., Wierenga, P. J., and Vachand, E. (1977). 'A comparison of numerical simulation models for one-dimensional infiltration.' *Proc. Soil Sci. Soc. Am.*, **41**, 285–94.

Hearn, G., and Jones, D. K. C. (1985). 'An appraisal of slope and drainage instability along the Dharan—Dhankuta Road in East Nepal caused by the storm of 15–16 September 1984, and proposals for road remedial and associated stabilization work.' Report to Transport and Road Research Laboratory, England, Contract 842/489, 86pp.

Hencher, S. R., and Martin, R. P. (1982). 'The description and classification of weathered rocks in Hong Kong for engineering purposes.' *Proc. 7th SE Asian Geotech. Conf.*, Hong Kong, **1**, 125–42.

Hencher, S. R., Massey, J. B., and Brand, E. W. (1984). 'Application of back analysis to some Hong Kong landslides.' *Proc. 4th Int. Symp. Landslides*, Toronto, **1**, 631–8.

Hillel, D. (1977). *Computer Simulation of Soil Water Dynamics*. IDRC, Ottawa, 214pp.

Ho, K. H. (1978). 'The properties and application of chunam surfacing.' Department of Civil Engineering, Hong Kong University. Unpublished dissertation.

Koo, Y. C. (1982a). 'Relict joints in completely decomposed volcanics in Hong Kong.' *Can. Geotech. J.*, **19**, 117–23.

Koo, Y. C. (1982b). 'The mass strength of jointed residual soils.' *Can. Geotech. J.*, **19**, 225–31.

Koo, Y. C., and Lumb, P. (1981). 'Correlation between rainfall intensities and groundwater levels in the natural slopes of Hong Kong.' *Hong Kong Engineer*, **9**, 49–52.

Leach, B., and Herbert, R. (1982). The genesis of a model for the study of a steep hillside in Hong Kong. *Quarterly Journal Engineering Geology*, **15**, 243–259.

Lumb, P. (1975). 'Slope failures in Hong Kong.' *Q. J. Eng. Geol.*, **8**, 31–65.

Lumsdaine, R. W., and Tang, K. Y. (1982). 'A comparison of slope stability calculations.' *Proc. 7th SE Asian Geotech. Conf. Hong Kong*, **1**, 31–8.

Massey, B. S. (1970). *Mechanics of Fluids*, Van Nostrand Reinhold, New York, 543pp.

McFarlane, J. (1980). 'Soil suction measurements in Hong Kong soil—instrumentation and preliminary results.' Public Works Department, Hong Kong. Unpublished paper.

Millington, R. J., and Quirk, J. P. (1959). 'Permeability of porous media.' *Nature*, **183**, 387–8.

Pope, R. G., Weeks, R. C., and Chipp, P. N. (1982). 'Automatic recording of standpipe piezometer.' *Proc. 7th SE Asian Geotech. Conf., Hong Kong*, **1**, 77–89.

Whisler, F. D., Curtis, A. A., Niknam, A., and Ronkens, M. J. M. (1979). 'Modelling infiltration as affected by surface crusting.' In *Surface and Subsurface Hydrology*. Morel-Seytoux, H. J., Salas, J. D., Sanders, T. G., and Smith, R. E. (eds.), Water Resource Publications, Fort Collins, pp. 400–13.

Wood, D. M. (1981). 'To chunam or not to chunam—interaction of politics and slope stability.' *Ground Engineering*, **14**, 28–30.

Slope Stability
Edited by M. G. Anderson and K. S. Richards

Chapter 8

Groundwater Models for Mountain Slopes

K. OKUNISHI
Disaster Prevention Research Institute, Kyoto University, Japan

and

T. OKIMURA
Faculty of Engineering, Kobe University, Japan

8.1 GROUNDWATER AND MASS MOVEMENT

Most mountain slopes are mantled by soils rendered more permeable than the substrate by weathering and root penetration. Subsurface lateral water flow occurs within this mantle at shallow depths, and exerts a significant influence on superficial mass movement. This saturated and unsaturated throughflow (Kirkby, 1978) follows infiltration of rainfall, and may result in a perched water table. Deeper groundwater occurs in the underlying substrate which may be a weathered saprolite, a sedimentary deposit, or bedrock. Hydraulic properties of these materials depend on their pore properties and on fracture characteristics, and as a result groundwater occurrence in mountain regions is highly variable and difficult to generalize.

The groundwater exerts gravitational, buoyancy, and drag forces on the soil mass, and a vector sum of these forces is

$$F = \varrho_s \, g \text{ grad } z - \text{grad } u \tag{8.1}$$

where F is the volumetric force, ϱ_s the soil bulk density, g the acceleration of gravity, and u the pore pressure (de Wiest, 1965, pp. 192–6). Conditions are hydrostatic where there is no flow. Equation (8.1) can be written

$$F = -(\varrho_s - \varrho_w) \, g \text{ grad } z - (1/K) v_f \tag{8.2}$$

where ϱ_w is the density of water, K the soil permeability, and v_f the seepage velocity. A quantitative evaluation of the effect of seepage force requires observation of the spatial distribution of pore pressure, using techniques outlined in Chapter 3. The seepage force is directed downwards relative to the soil mass in upslope locations, but upwards in slope-foot areas. Confined groundwater exerts an upward pressure on overlying beds (Záruba and Mencl, 1982, pp. 31–2).

The mechanical stability of a slope is affected by the reduction of effective normal stress ($\sigma_n - u$) caused by positive pore pressures (Terzaghi, 1925). The Mohr–Coulomb criterion of failure on a slip surface is written in effective stress terms as

$$\tau = c' + (\sigma_n - u) \tan \phi' \tag{8.3}$$

where τ is shear stress, σ_n is normal stress, c' is cohesion and ϕ' the angle of internal friction. When measured with respect to effective stress, c' and ϕ' are the critical state soil mechanical properties (see Chapter 2). In total stress analyses, c and ϕ vary with the water content. Marui *et al.* (1978) showed that cohesion of clay soil reaches a maximum at a critical water content above which it decreases, and Hulla *et al.* (1984) demonstrated that the cohesion of a clay loam with a plasticity index of 35% decreased by a half for a 5% increase in moisture content. Similarly, Brand *et al.* (1983) showed that cohesion may decrease to one-third of the original value when moisture content increases from 25% to 28%, in certain soils. In a sandy soil with a void ratio of 1.0, simple shear tests indicate that an increase of percentage saturation from 43% to 90% results in decreases of c and ϕ, from 1.0 to 0.5 t m^{-2} and from 35° to 30° respectively (Kazama *et al.*, 1980).

An additional mechanical influence occurs when groundwater flow is concentrated and seepage velocities are high. The shear strength of soil is reduced by lateral translocation of soil matrix, and by leaching of cementing material (Záruba and Mencl, 1982, pp. 31–2). Subsurface eluviation and erosion may cause subsidence of the ground surface and induces secondary mass movement (Stocking, 1981), especially when sudden subsidence causes instantaneous increase in pore-water pressure (undrained loading) which effectively 'liquefies' the soil mass (Sassa *et al.*, 1981). Chemical weathering by groundwater flow also stimulates mass movement, especially where active reaction occurs between rock-forming minerals and water containing HCO_3, $SO_4{}^{2-}$, and NO_3^- ions in solution. Weathering is dependent on groundwater circulation and supply of these ions, and slow mass movement is often evident in areas where the partial pressure of CO_2 in the soil is particularly high (Yoshioka and Kanai, 1975). The rate of weathering, in terms of alteration of rock-forming minerals to clay minerals, can be estimated by analysing the chemical composition of groundwater at springs or seepages (Yoshioka and Okuda, 1972).

A range of mass movement types exists, but all are strongly affected by groundwater conditions. A convenient classification is based on speed of motion (Oyagi, 1974), and involves the following main types:

1. *Creep displacement* After an initial deep-seated failure, landslide blocks may creep progressively on the shear plane at rates controlled by the stress–strain rate behaviour of the soil, and the pore-water pressure at the shear surface. In clay failures, creep occurs when viscous behaviour is apparent in the shear zone (van Asch, 1984). In some cases, rock mass creep may also precede shear failure.
2. *Landslides* Most failures which involve marked downslope displacement of the slipped mass reflect extreme groundwater conditions in heavy rain or rapid snowmelt. It is,

therefore, more difficult than in the case of creep displacement to monitor the pore pressure conditions responsible for failure, or to extrapolate from measurement in normal periods. Furthermore, the water in the slipped mass has no connection with the soil mass at rest, and springs are often observed in landslide scars. After a failure it is, therefore, difficult to recreate the initial groundwater characteristics. However, abnormal turbidity of spring water prior to a landslide is sometimes a warning of imminent instability (Okunishi, 1983).

3. *Debris avalanches, debris flows, earth flows and mudflows* These are multi-phase flows of solid, liquid, and gas (when air entrapment occurs). When a rigid sliding mass becomes a flow, fluidization involves a large energy conversion from the released potential energy, and Ashida *et al.* (1983) have evaluated this energy conversion to discriminate conditions in which a slide may become a debris avalanche. Debris flows can occur when surface runoff is applied to loose debris on a steep channel bed (Takahashi, 1977), or when sudden overburden collapse of loose saturated debris occurs (Sassa *et al.*, 1981). In all cases, it is clearly impossible to identify the groundwater conditions immediately prior to, and responsible for, failure. The mechanics of debris flows are discussed further in Chapter 16.

Since the more rapid modes of failure are characteristic of mountain regions, novel approaches are required for the analysis and prediction of groundwater-induced slope failure. The following sections therefore outline the basic general principles involved in the analysis of groundwater flow, and some methods developed in Japan for the prediction of slope failure due to extreme groundwater conditions in mountain areas.

8.2 GENERAL PRINCIPLES GOVERNING GROUNDWATER

Groundwater in mountain regions obeys the same general principles as all subsurface water, and it is appropriate to review those principles as a background for the interpretation of models developed for the simulation of groundwater influences on slope stability.

An equation of continuity can be applied to groundwater, although in hydrology it is usually referred to as the water-balance equation. This may be written

$$\frac{\mathrm{d}M}{\mathrm{d}t} = \int_{\mathrm{s}} V_{\mathrm{fn}}\,\mathrm{d}S + R - L \qquad (8.4)$$

where M is the groundwater mass defined by the closed surface S, t is time, V_{fn} is the component of the seepage velocity V_{f} normal to S (positive when outward), and R and L are rates of gain and loss. For an infinitesimal soil volume with no gains and losses, this is

$$\mathrm{div}\ V_{\mathrm{f}} = 0 \qquad (8.5)$$

for an unconfined aquifer, and

$$S'\,\frac{\partial p_{\mathrm{s}}}{\partial t} = -\mathrm{div}\,(\varrho_{\mathrm{w}}\,V_{\mathrm{f}}) \qquad (8.6)$$

for a confined aquifer, S' being the coefficient of compressibility of the aquifer system including the skeleton and pore water (de Wiest, 1965, pp. 179–83).

Evaluation of the water balance of an area is most conveniently achieved using equation (8.4), and is simplest when the surface S coincides with the groundwater divide (which may differ from the surface divide), since there V_{fn} is theoretically zero and needs no evaluation. The water balance method provides a useful check on the status of groundwater variables.

The flow of groundwater in porous media obeys Darcy's law

$$V_f = -K \text{ grad } h \tag{8.7}$$

where h is the piezometric head defined as

$$h = u/(\varrho_w g) + z \tag{8.8}$$

provided flow is laminar. Field evaluation of flow velocity and direction usually requires tracer experimentation, for example by measuring the dilution of a tracer injected into a well. The time change of tracer concentration, C, is

$$\frac{dC}{dt} = \frac{Cq}{A} \tag{8.9}$$

when perfect mixing occurs (A is the well cross-section area, and q the groundwater discharge per unit height). For an initial concentration C_o, this results in the expression

$$\ln (C/C_o) = -qt/A \tag{8.10}$$

and the discharge q which is thereby defined is proportional to the seepage velocity V_f by a relationship dependent on aquifer properties. The method, therefore, identifies the vertical distribution of groundwater flow, rather than the absolute flow velocity.

The relationships between pore pressure, piezometric head, and groundwater level are important in the analysis of the effects of groundwater on mass movement. At the water table, pore pressure is atmospheric. In the case of confined groundwater the water level in an observation well is sometimes treated as the equivalent of a water table. Usually the pore pressure is required at a given depth beneath the groundwater level H, and for horizontal flow and hydrostatic conditions, the pore pressure at height z is

$$u = \varrho_w g(H - z) \tag{8.11}$$

A theoretical expression for the spatial distribution of pore pressure is derived by combining equation (8.5) or (8.6) to give the three-dimensional expression

$$S' \frac{\partial h}{\partial t} = \text{div } (\varrho_w K \text{ grad } h) \tag{8.12}$$

for a confined aquifer, and

$$\text{div grad } h = 0 \tag{8.13}$$

for unconfined acquifers. This is difficult to solve because the boundary conditions are not fully known, so a two-dimensional differential equation, similar to the basic equation of horizontal groundwater flow and applicable to unconfined, inclined, and uniform flow, is often used (de Wiest, 1965, pp. 318–48; Okunishi and Esumi, 1975). This takes the form

$$n_e \frac{\partial h}{\partial t} = K \frac{\partial}{\partial x} \left\{ (h - x \tan \theta) \frac{\partial h}{\partial x} + \tan \theta \right\} + r - l \tag{8.14}$$

where x is distance downslope, θ is slope angle, and r and l are gains (R) and losses (L) defined for a unit area of slope. Effective porosity n_e is defined as

$$n_e = n - n_w \tag{8.15}$$

where n is the porosity and n_w the water content just above the water table.

Other factors may be relevant to slope stability in special circumstances. For example, in clays of low permeability, gravity and pore pressure may be less important than osmotic phenomena, with differences in electrical potential, temperature, and groundwater solute concentrations driving the water flow. Veder (1981) has analysed the effect of electro-osmosis on creep displacement.

Normally the necessary boundary and initial conditions for solution of equations (8.12) and (8.13) are incompletely known, and simplifying assumptions are made. For steady state groundwater regimes, and constant density ϱ_w, equation (8.6) reduces to equation (8.5) and in the x–z section is written

$$\frac{\partial^2 h}{\partial x^2} + \frac{\partial^2 h}{\partial z^2} = 0 \tag{8.16}$$

if there is no flow in the direction y. This equation is solved graphically if the water table is parallel to the ground surface (Hubbert, 1940), or analytically under the assumptions that the rectangular boundary is impermeable, the water table coincides with the ground surface, and the surface is represented by a combination of straight lines and sine waves (Tóth, 1962). Freeze and Witherspoon (1967) used numerical methods to model groundwater flow for arbitrary water table forms and heterogeneous permeability (see Figure 8.1). The equipotential lines (broken lines in Figure 8.1) are perpendicular to the streamlines (solid lines), forming a 'flow net'. A stream function ψ is defined as

$$V_{fz} = \frac{\partial \psi}{\partial x} \text{ and } V_{fx} = \frac{\partial \psi}{\partial z} \tag{8.17}$$

and ψ is constant along any streamline, while the difference in ψ between two streamlines represents the flow between them.

Figure 8.1 illustrates both general flow along the slope and local circulation reflecting ground surface undulation. The occurrence of groundwater springs in hollows is relevant to the spatial location of mass movement, it being evident that landslides take place preferentially in hollows (Tsukamoto *et al.*, 1982). However, any quantitative consideration

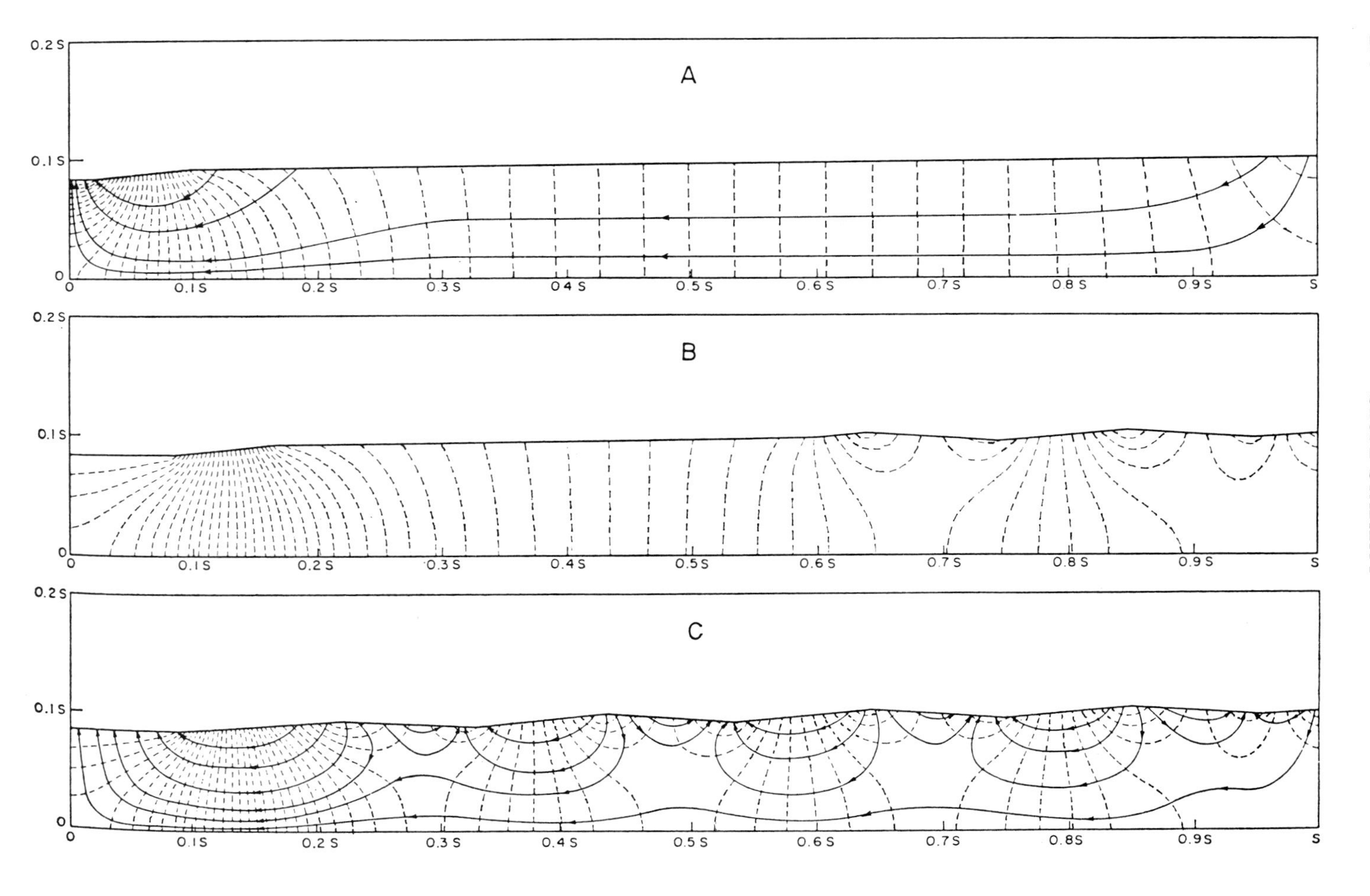

Figure 8.1 Flow net in a vertical cross-section across an undulating terrain. (*After Freeze and Witherspoon, 1967, in:* Water Resources Research, *vol. 3. Reproduced by permission of the American Geophysical Union.*)

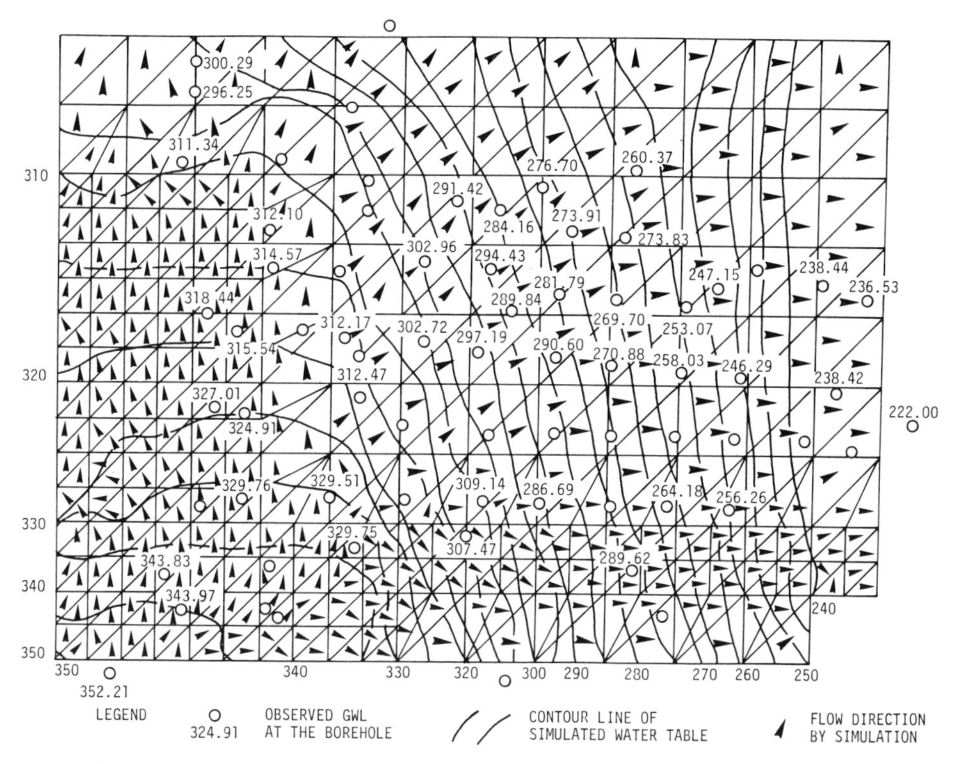

Figure 8.2 Equipotential lines (contour lines on the water table) and the vector of groundwater flow in a region of creep displacement. (*After Masuko and Makino, 1982*. J. Japan Landslide Society, **69**. *Reproduced by permission of the authors.*)

of groundwater influences on *superficial* mass movement demands detailed information on soil horizon properties and structures (Iida and Okunishi, 1981), which are not incorporated in the simplified theoretical models. General flow patterns may be more relevant to *deep-seated* creep displacement, as modelled by Masuko and Makino (1982) using a finite element method to define the flow net for a case in which geological structure was necessarily taken into consideration (Figure 8.2).

8.3 SIMULATION MODELS OF GROUNDWATER INFLUENCES ON MASS MOVEMENT

Analysis and prediction of landslides in mountainous terrain in Japan has necessitated the development of simulation models such as the conceptual 'tank' model originally proposed by Sugawara (1969). These provide useful insights into slope instability associated with extreme and often unmeasurable groundwater conditions, in a range of circumstances considered in this section.

8.3.1 The Prediction of Superficial Slides and Debris Flows

Superficial failures with a slip surface 1–2 m below the surface are often induced by abnormal increases of throughflow, and the sites and timing of failure can be predicted by

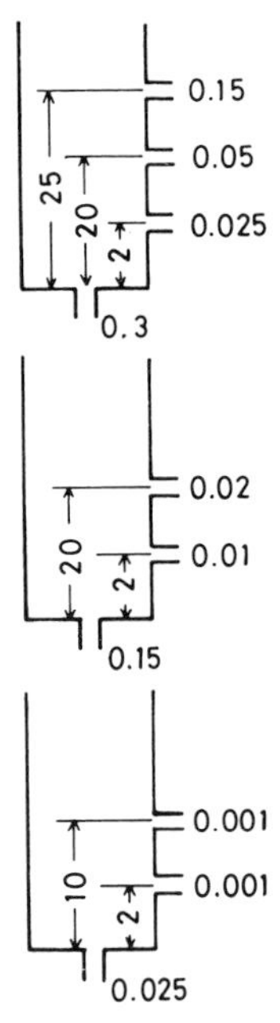

Figure 8.3 The tank model of throughflow developed by Suzuki *et al.*, 1979. **110**. (*Reproduced by permission of the authors.*)

simulating throughflow mechanics. Shallow slides and flows reflect high rainfall intensity which causes a sudden increase in throughflow discharge (Seo and Funazaki, 1974; Okuda *et al.*, 1980). Although prediction of the time of occurrence of a failure is impossible because rainfall intensity cannot be forecast, the danger of landslide activity can be assessed by relating throughflow discharge to rainfall.

Suzuki *et al.* (1979) simulated runoff on granite slopes in the Rokko Mountains, Japan, using Sugawara's (1969) tank model. This consists of three tanks in cascade which represent storages of surface and subsurface water and groundwater (Figure 8.3). Each tank has a side orifice simulating runoff from the relevant storage, and a bottom orifice representing percolation. The model simulated hydrographs following heavy rainfalls responsible for disasters in 1938, 1961, and 1967, and the timing of landslides and debris flows in these years could be evaluated relative to the hydrographs (Figure 8.4). Slope failures occurred when the water depths were 35 mm in the upper tank and 50 mm in the middle tank, and the specific runoff exceeded 10 mm h^{-1}. This simulates the critical storage of water in topsoil horizons required to affect the mechanical stability of the slopes, and Suzuki *et al.* (1979) suggest that landslide hazard in the Rokko Mountains can be predicted by monitoring the response of the model tanks to rainfall. The model is applicable to other locations after modification of the tank parameters. It is, however, a lumped model of slope response to rainfall input, and significant improvements can be made by modelling spatial and temporal variability.

For example, Matsuo and Ueno (1981) have assumed that the soil mechanical properties c and ϕ vary spatially on a mountain slope and follow the normal distribution. Values of c decrease linearly with moisture content, but are distributed statistically for a given moisture content such that c^* $(=c/c_{\max})$ is distributed as

$$f(c^*) = \frac{1}{\sqrt{2\pi}\sigma_c} \exp\left\{-\frac{1}{2}\left(\frac{c^* - \mu_c}{\sigma_c}\right)^2\right\} \tag{8.18}$$

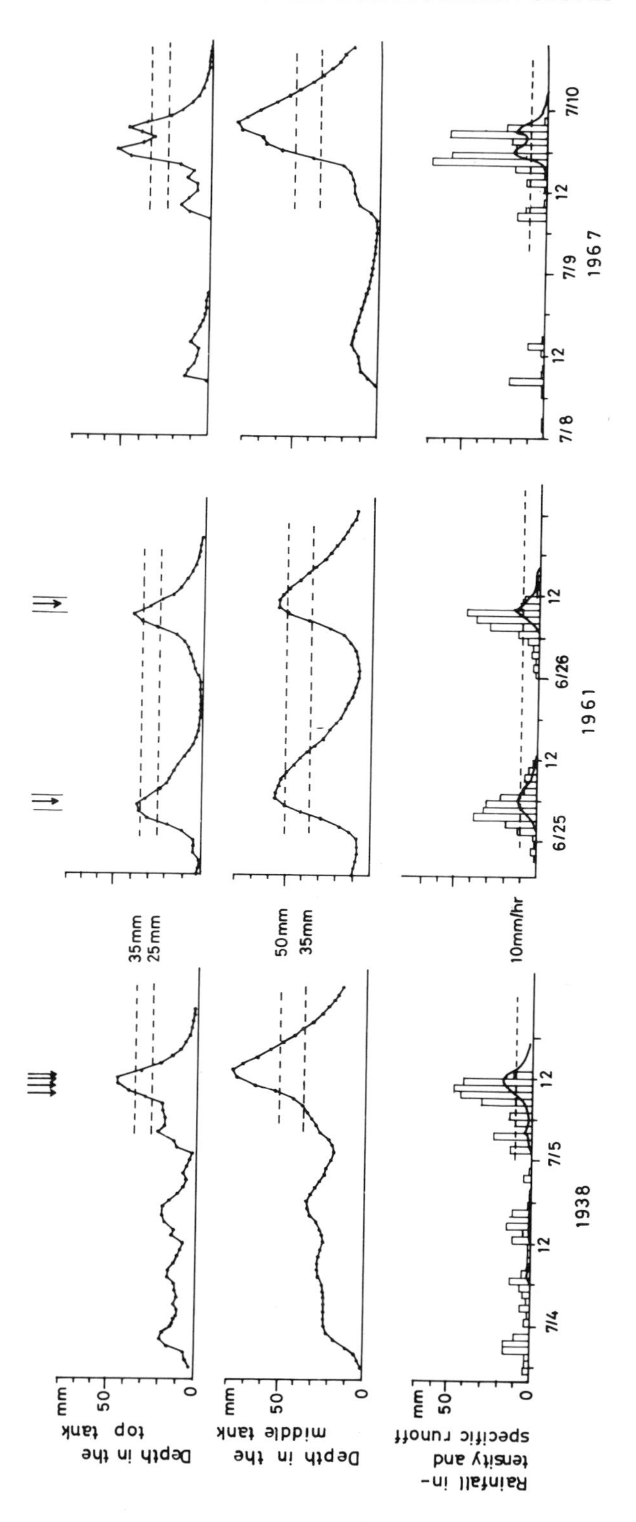

Figure 8.4 Simulated water level in the tanks of Figure 8.3, and the occurrence of debris flows (*After Suzuki* et al., *1979. Reproduced by permission of the authors.*)

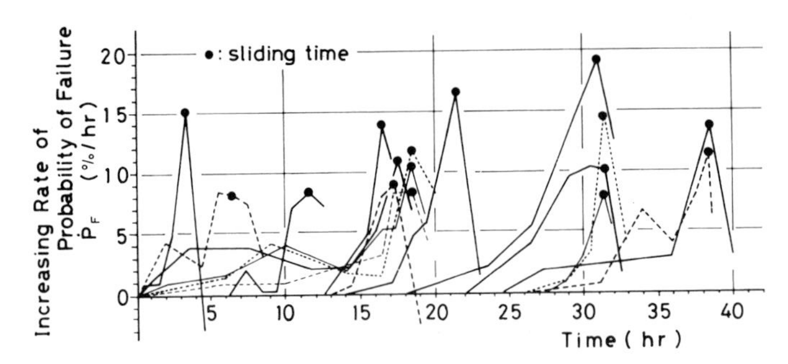

Figure 8.5 The relationships between dP_F/dt and the occurrence of landslides. P_f is the probability of failure. (*After Matsuo and Ueno, 1981*, JSSMFE, **21**. *Reproduced by permission of the authors.*)

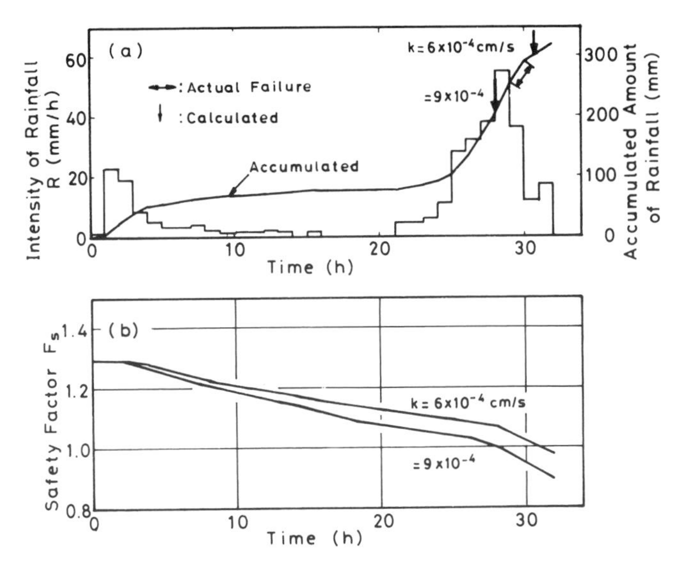

Figure 8.6 Rainfall and the time change in the safety factor. (*After Akutagwa* et al., *1983*. Ann Proc JSCE. *Reproduced by permission of the authors.*)

where μ_c and σ_c are respectively the mean and standard deviation of c^* and $f(c^*)$ is the probability density function of c^*. The friction angle ϕ is independent of moisture content, and Matsuo and Ueno (1981) therefore assumed a normal distribution about a constant mean value. They then calculated the probability of failure (P_F) on a cylindrical slip surface, and assessed its change with rainfall as moisture content increased, pore pressure increased, soil unit weight increased, and cohesion decreased. Landslide occurrence was found to be concentrated at the time when dP_F/dt is a maximum notwithstanding the value of P_F itself (Figure 8.5). A similar study by Akutagawa *et al.* (1983) considered sandy slopes on which both c and ϕ decreased with increasing soil moisture content. They also found that the danger of landslide occurrence was a function of accumulated rainfall and the ratio of mean rainfall intensity to permeability (Figure 8.6).

Concentration of groundwater flow into a limited area can induce slope instability and is influenced by the hydraulic conductivity and bedrock geometry (Reneau *et al.*, 1984). The effect of concentration of throughflow by three-dimensional bedrock geometry has been modelled using a digital terrain model with a 10 m grid spacing by Okimura and Ichikawa (1985). Finite difference methods were used to solve the basic equation

$$n_e \frac{\partial H}{\partial t} + \frac{\partial (HV_{fx})}{\partial x} + \frac{\partial (HV_{fy})}{\partial y} = r \tag{8.19}$$

where H is throughflow depth or water table height above an impermeable substrate, V_{fx} and V_{fy} are the components of seepage velocity V_f in x and y directions, and r is the effective rainfall intensity. The infinite slope stability analysis was applied to each cell of the grid to calculate the factor of safety at time t, $F(t)$, defined according to Simons *et al.* (1978) as

$$F(t) = \frac{c_s + c_r + \{(\varrho_{sat} - \varrho_w)\, H(t) + \varrho_s(H_s - H(t))\} \cos^2\theta \tan\phi}{\{\varrho_{sat}\, H(t) + \varrho_s\, (H_s - H(t))\} \sin\theta \cos\theta} \tag{8.20}$$

where c_s is soil cohesion, c_r is effective root cohesion, ϱ_{sat} is the bulk density of saturated soil and H_s is the depth of the soil layer.

This method was applied to granite mountain slopes in Gifu Prefecture, Japan. The soil depth was assumed to be 1.2 m except where bedrock exposure occurs. The cell slope was evaluated by linear trend surface analysis (Davis, 1973). The safety factor was calculated

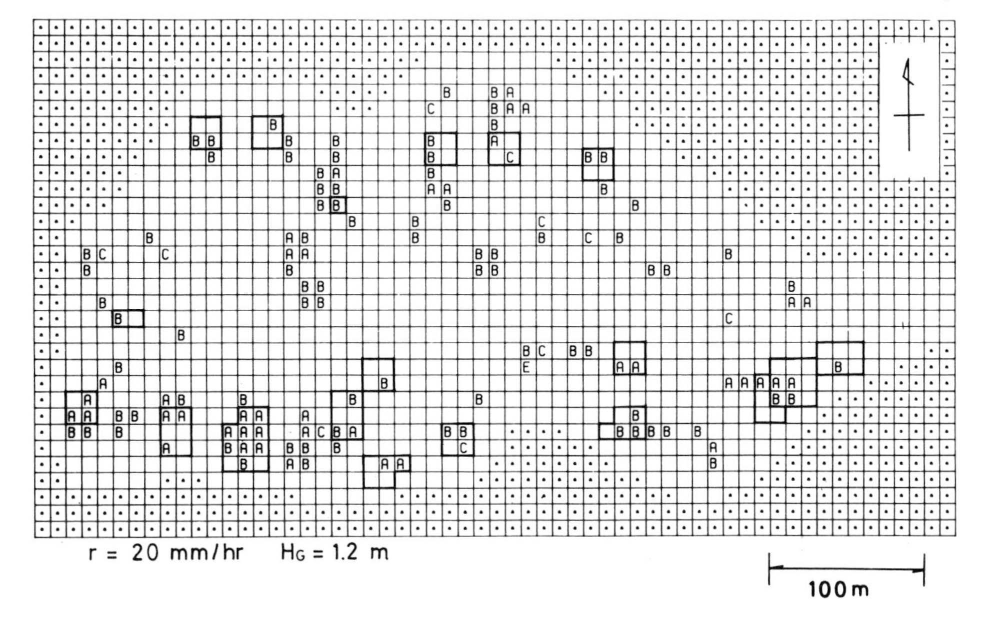

Figure 8.7 Distribution of t_{cr} (time to failure) in a test region. The range of t_{cr} in hours is; A<10, B<20, C<30, D<40 and E<50. (*After Okimura and Ichikawa, 1985.* Natural Disaster Science. *Reproduced by permission of the authors.*)

every hour under the condition of continuous rainfall for 50 hours at an effective intensity of 20 mm h^{-1}. The degree of danger for a cell was represented by the time (t_{cr}) taken for the safety factor to become less than unity, after the beginning of rainfall, and Figure 8.7 shows the distribution of t_{cr} values in the study area. Cells enclosed by the heavy line are sites of landslide occurrence during heavy rainfall in 1972, and the most dangerous cells (marked A or B) overlap well with the failed sites.

8.3.2 Simulation of Confined Fissure Water Stimulating Landslides

In the mountains, fracture systems are well developed because of intense crustal movement, and confined fissure water causes instability of slopes. Shallow slides of intensely weathered topsoil over fractured bedrock, and deep-seated slides in bedrock deeply weathered because of the fracturing, are both experienced (Tanaka *et al.*, 1973). Okimura (1981) has examined the effect of fissure water on the timing of superficial slides, using the Sugawara (1969) tank model to simulate the hydraulics of fissures crossing an observation borehole on a granite mountain slope in the Rokko Mountains (Figure 8.8). Model parameters were estimated from two years of water level observations from the borehole (1975–77), then rainfall records for 1938, 1961, and 1967 were used as model inputs. Borehole water level behaviour was compared with the occurrence of swarms of landslides which took place in those years (Figure 8.9). Arrows in Figure 8.9(a) indicate the timing of landslides, which relate to peaks in the water level acceleration (d^2h/dt^2) shown in Figure 8.9(e) and the exceedance by the water level itself of a critical height (Figure 8.9(a)). Figure 8.10 shows the quantitative dependence of the occurrence of landslides on the water level and its acceleration. The dotted line represents the critical condition for landslides, since it discriminates cases when landsliding occurred (closed circles) from data for the period 1968–79 when no landslides occurred (open circles).

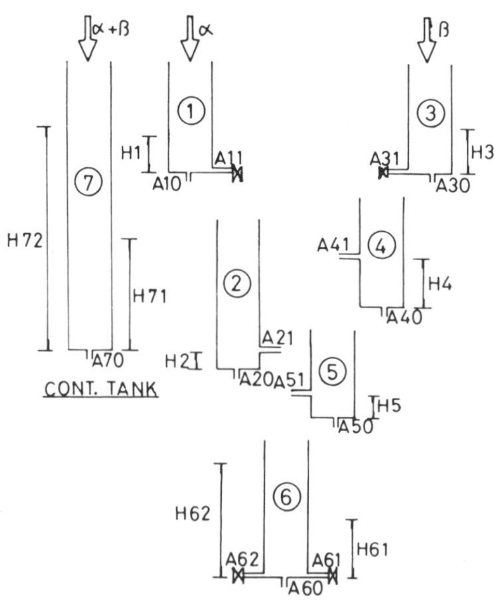

Figure 8.8 The tank used by Okimura (1981) for the simulation of the hydraulics of a borehole which is affected by fissure water. (After *Okimura, 1981.* Trans. Jap. Geomorph. Union, **2**. *Reproduced by permission of the author.*)

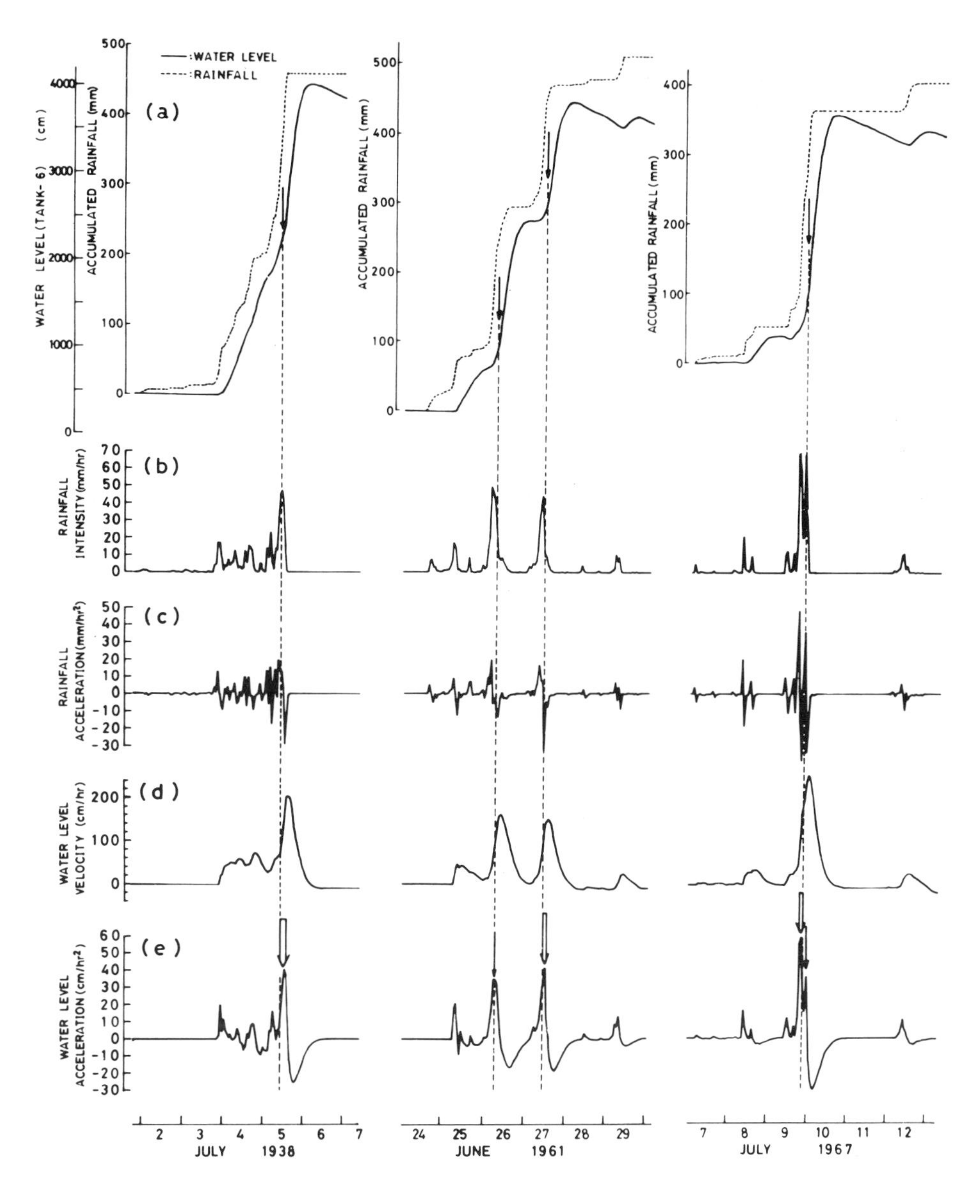

Figure 8.9 The results of calculation by the model of Okimura (1981) shown in Figure 8.8. (*After Okimura, 1981.* Trans. Jap. Geomorph. Union, **2**. *Reproduced by permission of the author.*)

Hydrological interpretation of these findings is based on the assumption of quasi-steady flow of fissure water under the slope (Okimura, 1983). The water level acceleration reflects an increasing water pressure in the fissure which itself relates to the change in rainfall intensity. Water level in the borehole, however, represents a component of water pressure in the fissure which responds more slowly to rainfall and reflects antecedent conditions.

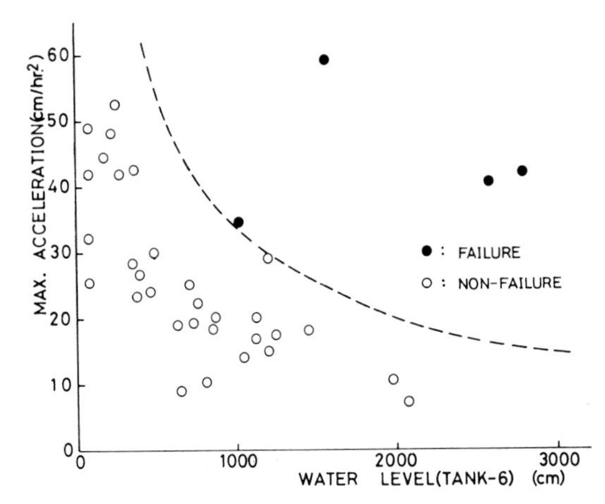

Figure 8.10 Discrimination of the occurrence conditions for landslides, based on the groundwater simulation by Okimura (1981). (Trans. Jap. Geomorph. Union **2–2**, *1981. Reproduced by permission of the author.*)

8.3.3 Simulation of Geological Structure as a Control of Groundwater Regime and Rockslide Activity

Rockslides (Vernes, 1978) occur under particular structural conditions and groundwater regimes (Voight and Pariseau, 1978). Faults, joints, and bedding planes all encourage rockslides, although the latter are associated largely with phenomena known as rockglides (Hirano *et al.*, 1984). Concentrated groundwater flow is encouraged by the combination of open cracks (faults, joints) in which high discharges can occur, and aquicludes formed by fault clay or other thin impermeable layers.

Figure 8.11 illustrates the geological structure of slopes in the coastal region of Shizuoka Prefecture, Japan. Bedrock is permeable because of well-developed fracture systems, but near-horizontal thin layers of black shale are impermeable, intercept percolating rainwater, and cause lateral flow. This structure was simulated by Kimiya (1980) using a tank model in which fractured rock masses were represented by three water tanks. Lateral flow over impermeable layers is simulated by side orifices, and downward leakage recharging groundwater in the next lower rock mass is simulated by the bottom orifice. Calibration of the model involved additional simulation of discharge from a horizontal borehole. Figure 8.12 compares observed and calculated groundwater discharge, and the tank model (with appropriate empirical parameters) appears to simulate the effect of rainfall on groundwater flow successfully except on the recession, when the lumped parameters do not adequately represent spatially-distributed groundwater discharge.

Another example of the application of such a groundwater model to rockslides is the 1972 Shigeto landslide in Kochi Prefecture, Japan (Nakagawa and Okunishi, 1977). About 10^5 m^3 of rock and debris failed, killing 60 people. Figure 8.13 shows a representative longitudinal section based on field investigation and borehole logging. Chert at the top of the slope is heavily fractured and permeable, but infiltrated rainwater cannot readily percolate downward through the alternating sandstone, shale, and mudstone in the middle and lower slope. After the landslide it was evident that a fault crossing the slope had a significant influence on the failure. The fault fracture zone is clayey where it passes through

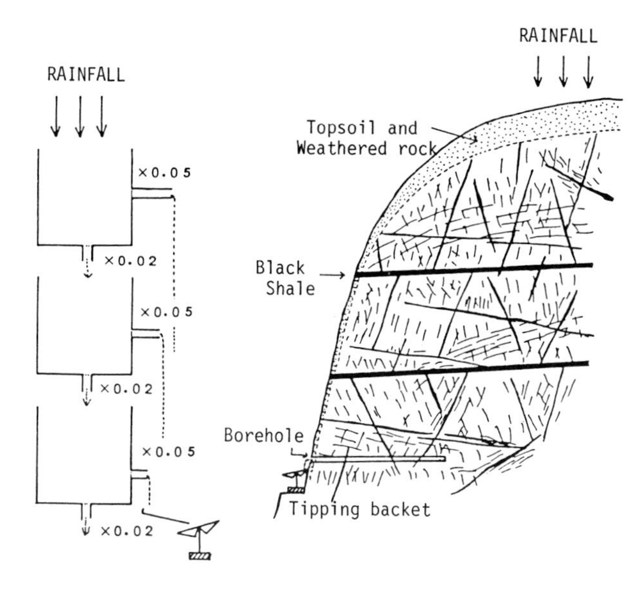

Figure 8.11 Schematic illustration of the geological structure of coastal slopes in Shizuoka Prefecture, and the corresponding tank model. (*After Kimiya, 1980.* Natural Hazards and Water. *Reproduced by permission of the author.*)

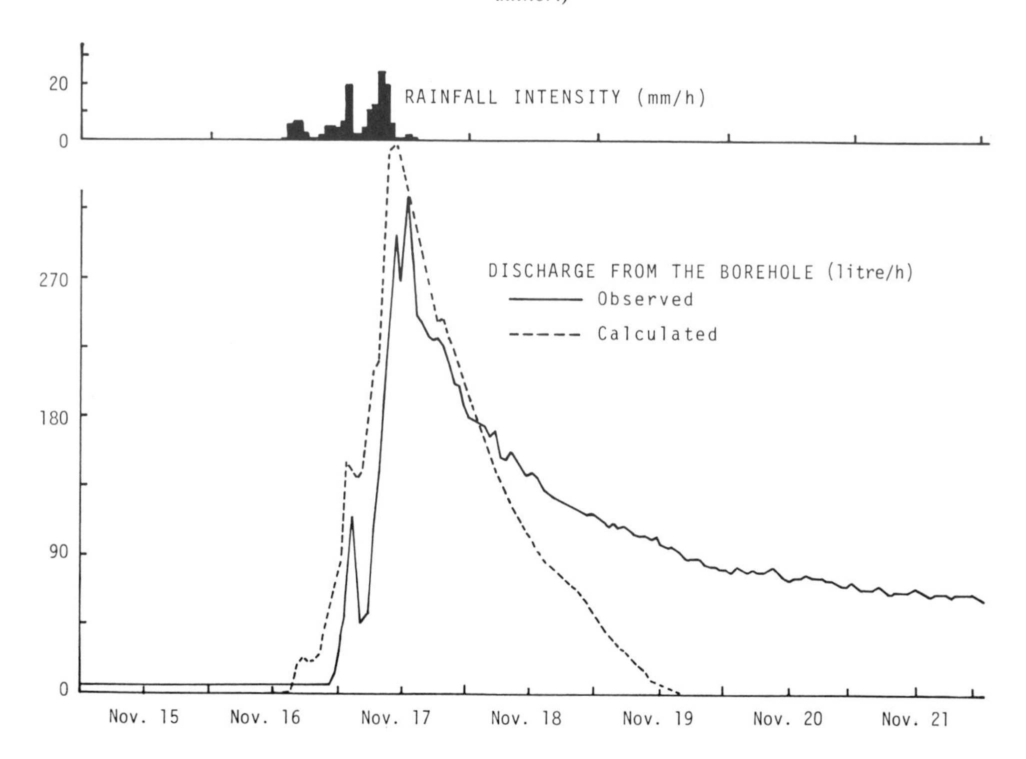

Figure 8.12 Observed groundwater discharges and those predicted by the tank model of Figure 8.11. (*After Kimiya, 1980.* Natural Hazards and Water. *Reproduced by permission of the author.*)

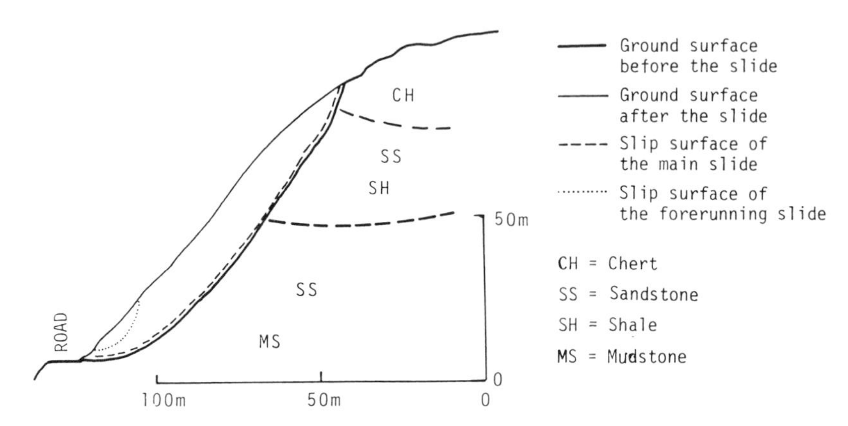

Figure 8.13 Geological structure of the Shigeto rockslide. (*After Nakagawa and Okunishi, 1977.* Ann. Disaster Prevention Res. Inst. Kyoto Univ., **20B-1**. *Reproduced by permission of the authors.*)

shale or mudstone, but pebbly where it crosses sandstone. Thus water conduits occur between the groundwater storage in the chert and the lower slope surface, and one of these feeds the most powerful spring in the area. Springs in the hollows at each side of the slope also drain the chert, and downward leakage is indicated by flow from horizontal boreholes.

These hydrogeological conditions were summarized by the tank model shown in Figure 8.14. The two kinds of spring are located at the same elevation, so their combined discharge was lumped into a single orifice discharge from the side of the main tank. Downward leakage is again simulated by the bottom orifice, and a secondary tank represents the time delay of groundwater recharge. The model parameters were determined by trial so that the discharges of side and bottom orifices were proportional to those of the main spring and the horizontal borehole respectively, and the relative water level in the main tank matched

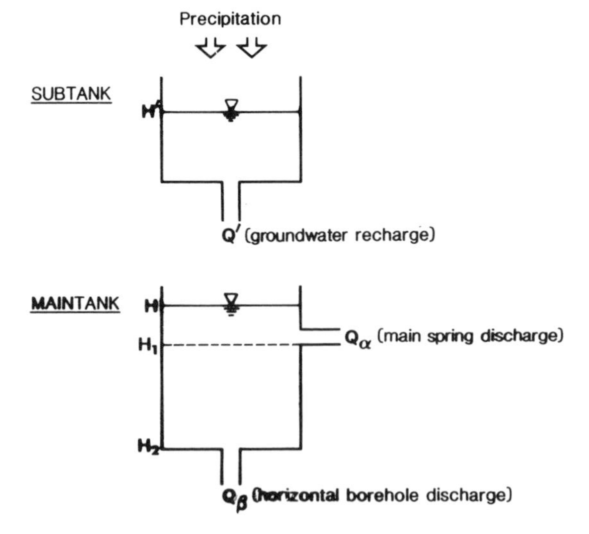

Figure 8.14 The tank model of *Okunishi and Nakagawa, 1977.* Ann. Disaster Prevention Res. Inst. Kyoto Univ., **20B-1**. (*Reproduced by permission of the authors.*)

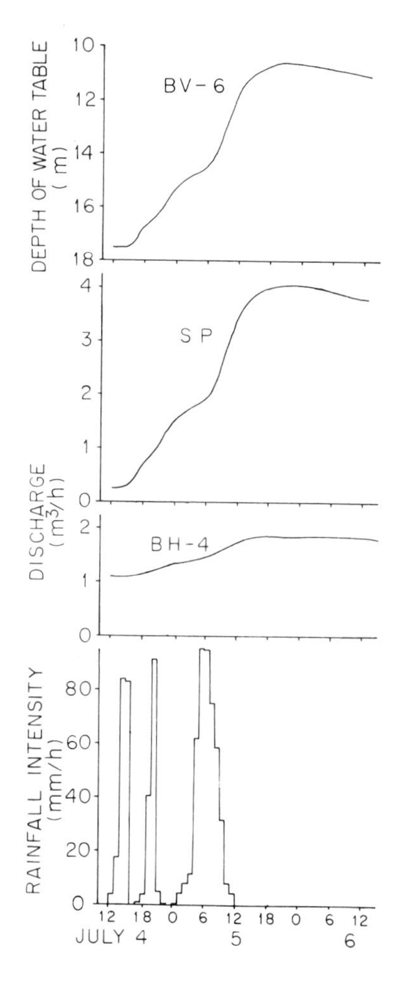

Figure 8.15 Response of the groundwater system simulated by the tank model of Figure 8.14, for the rainfall which induced a rockslide at Shigeto. (*After Nakagawa and Okunishi, 1977.* Ann. Disaster Prevention Res. Inst. Kyoto Univ., **20B-1**. *Reproduced by permission of the authors.*)

that in a vertical borehole in the bedrock behind the slide (Okunishi and Nakagawa, 1977). The response of the groundwater system to the rainfall which induced the landslide was simulated with this model (Figure 8.15). A heavy rainfall of about one day in duration is evidently effective in increasing groundwater level and discharge in this case. Quantitative examination of the mechanical effect of the increase in groundwater on the rockslide is difficult because representative c' and ϕ' parameters for a slip surface cutting different rock types cannot be evaluated. However, the simulation suggests that the conduit formed in the fracture zone supplied water to the failure and increased the pore pressure along it. Furthermore, the gradient of piezometric head in the slipped mass increased when the pore pressure at the slip surface was rapidly increased.

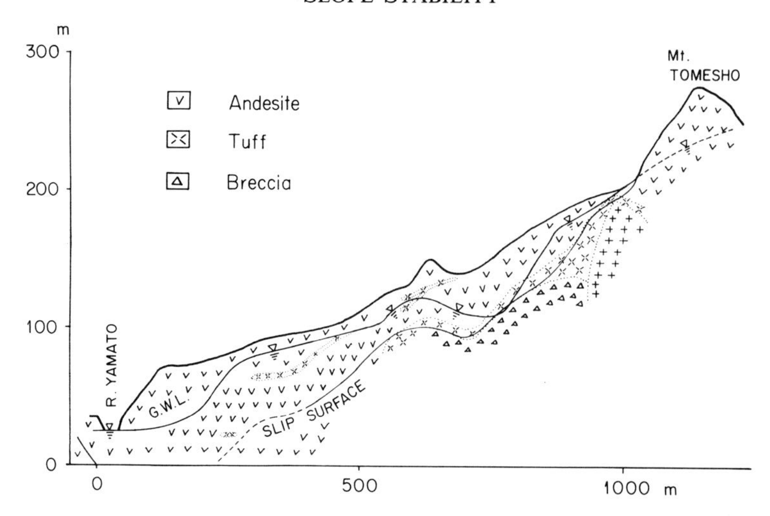

Figure 8.16 A hydrogeological cross-section of a zone of landslide creep displacement at Kamenose, Japan. (*After Okunishi, 1970,* Spec. Contrib. Geophys. Inst. Kyoto Univ. *Reproduced by permission of the author.*)

8.3.4 Groundwater and Creep Displacement

Large-scale progressive creep displacement often involves complex groundwater regimes; the Kamenose creep failure in Japan has been intensively investigated, and a representative longitudinal section is shown in Figure 8.16. The groundwater level is defined by a curve connecting the water surfaces in boreholes. Between 300 and 500 m (horizontal axis) the water level appears to represent a perched water table on a thin layer of impermeable tuff, since it is discordant with other water table segments. At 700 m a pronounced concavity reflects the existence of a buried valley. Generally, however, groundwater flow is dependent on geological structure and topography in this section. Hydrological investigations suggest that rain infiltration is sufficient to recharge groundwater, but only about 40% of the groundwater is drained by the stream and springs. Creep displacement on this slope has been stabilized by drainage works involving two-dimensional tunnel arrays to drain the groundwater (Okunishi, 1970; Murata, 1980).

The complexity of geological and topographical conditions here results in sensitive response of groundwater to rainfall which then induces creep movement on the slip surface; this complexity is difficult to examine in simple quantitative models. However, it is possible to combine several simple tank models to simulate the structure of groundwater processes. Yoshimatsu (1980) developed the tank model by combining several such models as shown in Figure 8.14 along the slope to simulate complex groundwater response to rainfall. The most promising approach to analysis of complex landslide sites is that of Terakawa *et al.* (1982), in which a composite of tanks and connecting tubes represents the groundwater behaviour in the catchment of the Yachi creep failure zone, Akita Prefecture, Japan. Here, rock masses move as a block slide which opens cracks in the bedrock and colluvium and encourages rapid groundwater flow. The mass movements which occur may then close the conduits again. The model illustrated in Figure 8.17 accounts for the complex groundwater relations in this catchment. As this model is developed, observation of groundwater levels in boreholes will enable determination of model parameters so that prediction of creep

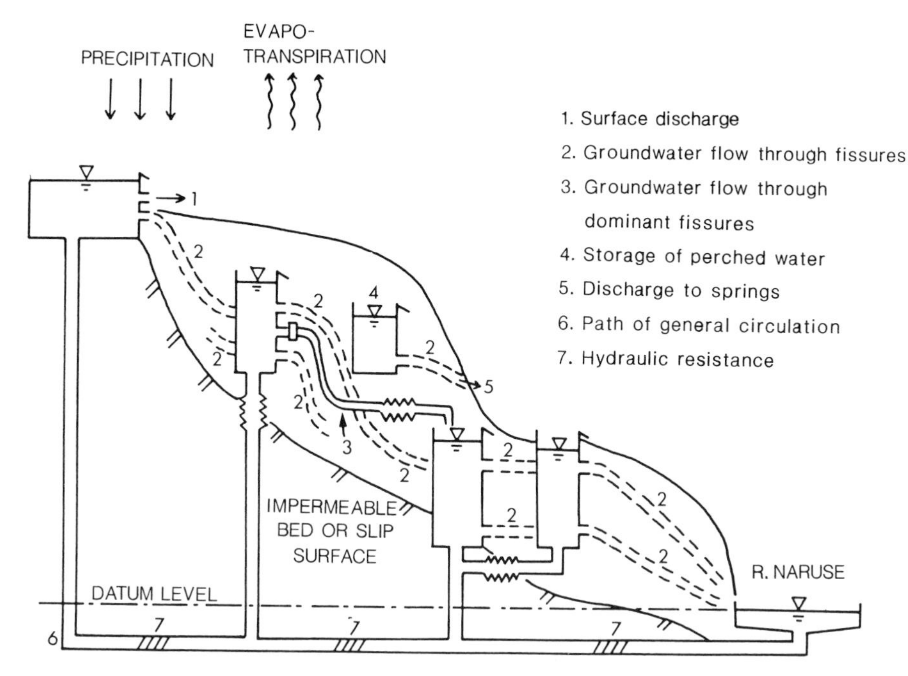

Figure 8.17 The tank model of Terakawa *et al.* (1982) developed to represent the groundwater regime of a zone of creep displacement at Yachi, Akita Prefecture, Japan. (*After Terakawa* et al., *1982..* J. Jap Landslide Soc., **69**. *Reproduced by permission of the author.*)

displacement from the piezometric head, calculated for anticipated or observed rainfall and snowmelt, is possible.

This final example illustrates the potential of the simple tank model to be extended to simulate complex groundwater regimes in the mountain areas of Japan. These regions are subject to a range of landslide hazards which must be regularly monitored and predicted, and the tank models provide a simple aid to disaster prevention on slopes where the complexity of structure and groundwater regime and the nature of the failure processes often preclude more sophisticated modelling.

REFERENCES

Akutagawa, S., Kazama, H., and Nakajima, K. (1983). 'Relationship between failures on sandy slopes and rainfall characteristics.' *Ann. Proc. JSCE*, **III-150**, 299–300 (in Japanese).

Ashida, K., Egashira, S., and Ohtsuki, H. (1983). 'Dynamic behaviour of a soil mass produced by slope failure.' *Ann. Disaster Prevention Res. Inst., Kyoto Univ.*, **26B-2**, 315–27 (in Japanese).

Brand, E. W., Phillipson, H. B., Berrie, G. W., and Clover, A. W. (1983). '*In situ* tests on Hong Kong residual soils.' *Soil and Rock Investigations by In Situ Testing, II*, Paris, 13–17.

Davis, J. C. (1973). *Statistics and Data Analysis in Geology*. John Wiley, Chichester, p. 550.

de Wiest, J. M. (1965). *Geohydrology*. John Wiley, Chichester, p. 366.

Freeze, A., and Witherspoon, P. A. (1967). 'Theoretical analysis of regional groundwater flow: 2. Effect of water-table configuration and subsurface permeability variation.' *Water Resources Res.*, **3**, 623–34.

Hirano, M., Suwa, H., Ishii, T., Fujita, T., and Gocho, Y. (1984). 'Reexamination of the Totsukawa Hazard in August 1889 with special reference to geologic control of large-scale landslides.' *Ann. Disaster Prevention Res. Inst., Kyoto Univ.*, **27B-1**, 369–86 (in Japanese).

Hubbert, M. K. (1940). 'The theory of ground-water motion.' *J. Geol.*, **48**, 785–944.

Hulla, J., Turćek, P., and Ravinger, R. (1984). 'Water movement in landslide slopes.' *4th Int. Symp. Landslides*, Toronto, vol. I, 405–10.

Iida, T., and Okunishi, K. (1981). 'Evolution of hillslopes including landslides.' *Trans. Jap. Geomorph. Union*, **2**, 291–300.

Kazama, H., Kŏnidokoro, K., and Mashita, S. (1980). 'Slope stability by shear strength decrease caused by submergence of sandy soil.' *15th Ann. Proc. JSSMFE*, **275**, 1097–100 (in Japanese).

Kimiya, K. (1980). 'Landslides and the water.' *Natural Hazards and Water*, Nagoya, 63–74 (in Japanese).

Kirkby, M. J. (ed.) (1978). *Hillslope Hydrology*. John Wiley, Chichester, p. 389.

Marui, H., Kobashi, S., and Nakano, M. (1978). 'Studies of mud-flows and their occurrence conditions (3)—On the density properties and shearing strength properties of debris.' *Shin-Sabo* (Erosion-Control Eng. Soc. Japan), **109**, 1–8 (in Japanese).

Masuko, K., and Makino, H. (1982). 'On a method to decide on the equipment as a counterplan against landslides.' *J. Jap. Landslide Soc.*, **69**, 17–26 (in Japanese).

Matsuo, M., and Ueno, M. (1981). 'Mechanical significance of probability of failure as an index for prediction of slope failure.' *Soils and Foundations, JSSMFE*, **21**, 19–34.

Murata, Y. (1980). 'Kamenose landslide.' *J. Jap. Landslide Soc.*, **60**, 25–32 (in Japanese).

Nakagawa, A., and Okunishi, K. (1977). 'A large-scale landslide at Shigeto, Kochi Prefecture—Part 1. Characteristics of ground structure on the landslide.' *Ann. Disaster Prevention Res. Inst. Kyoto Univ.*, **20B-1**, 209–22 (in Japanese).

Okimura, T. (1981). 'Analysis of mountainslope failures based on the simulated groundwater level in bore hole.' *Trans. Jap. Geomorph. Union*, **2**, 253–62.

Okimura, T. (1983). 'Rapid mass movement and groundwater level movement.' *Z. Geomorph. N.F., Suppl.-Bd.*, **46**, 35–54.

Okimura, T., and Ichikawa, R. (1985). 'A prediction method for surface failures by movements of infiltrated water in a surface soil layer.' *J. Natural Disaster Sci.*, **7**, 41–51.

Okuda, S., Suwa, H., Okunishi, K., Yokoyama, K., and Nakano, M. (1980). 'Observation on motion of a debris flow and its geomorphological effects.' *Z. Geomorph. N.F., Suppl.-Bd.*, **35**, 142–63.

Okunishi, K. (1970). 'Ground water regime of the Kamenose Landslide Area, Osaka Prefecture.' *Special contrib., Geophys, Inst., Kyoto Univ.*, **10**, 105–26.

Okunishi, K. (1983). 'Refuge from large-scale landslides—A case study: Nishiyoshino Landslide, Nara Prefecture.' *J. Natural Disaster Sci.*, **4**, 79–85.

Okunishi, K., and Esumi, S. (1975). 'Hydrological studies on small mountainous drainage basin (III) Characteristics of local ground water formed at the head of the first-order valleys,' *Ann. Disaster Prevention Res. Inst., Kyoto Univ.*, **16B**, 411–23 (in Japanese).

Okunishi, K., and Nakagawa, A. (1977). 'A large-scale landslide at Shigeto, Kochi Prefecture—Part 2. Effect of ground water on the landslide.' *Ann. Disaster Prevention Res. Inst., Kyoto Univ.*, **20B-1**, 223–36 (in Japanese).

Oyagi, N. (1974). 'Disaster of landslide and their related phenomena.' *Technology for Disaster Prevention*, **13**, 159–77.

Reneau, S. L., Dietrich, W. E., Wilson, C. J., and Rogers, J. D. (1984). 'Colluvial deposits and associated landslides in the northern San Francisco Bay area, California, USA.' *4th Int. Symp. Landslides*, Toronto, **I**, 425–30.

Sassa, K., Takei, A., and Kobashi, S. (1981). The mechanism of liquefied landslides and valley-off type debris flows. *Mitteilungen der Forstlichen Bundes-Versuchsanstalt*, Wien, **138**, 151–62.

Seo, K., and Funazaki, M. (1974). 'The relation between total rainfall amounts and sand disasters.' *Shin-Sabo* (Erosion-Control Eng. Soc. Japan), **89**, 22–8 (in Japanese).

Simons, D. B., Li, M. M., and Ward, T. J. (1978). 'Mapping of potential landslides area in terms of slope stability.' *USDA Forest Service, Rocky Mountain Forest and Range Experiment Station*, **75**.

Stocking, M. A. (1981). 'A model of piping in soils.' *Trans Jap. Geomorph. Union*, **2**, 263–78.

Sugawara, M. (1969). 'Report of the researches about the circulation system of water resources.' *Rep. of the Meeting Resources Res., Sci. and Technol. Agency*, **47**, 562 (in Japanese).
Suzuki, M., Fukushima, Y., Takei, A., and Kobashi, S. (1979). 'The critical rainfall for the disasters caused by debris movements.' *Shin-Sabo* (Erosion-Control Eng. Soc. Japan), **110**, 1–7 (in Japanese).
Takahashi, T. (1977). 'A mechanism of occurrence of mud-debris flow and their characteristics in motion.' *Ann. Disaster Prevention Res. Inst. Kyoto Univ.*, **20B–2**, 405–35 (in Japanese).
Tanaka, S., Kawatani, T., and Okimura, T. (1973). 'Characteristics of water in fractured rocks at Kobe in Japan.' *Atti Del 2° Convegno Internazionale Sulle Acque Sotterranee, Palermo*, 1–7.
Terakawa, T., Mizutani, N., and Nishida, S. (1982). 'The behaviour of groundwater at Yachi Landslide, a rock block slide in Japan.' *J. Jap. Landslide Soc.*, **69**, 34–43 (in Japanese).
Terzaghi, K. (1925). *Erdbaumechanik*, F. Deutiche, 140–52.
Tóth, J. (1963). 'A theoretical analysis of groundwater flow in small drainage basins.' *J. Geophys. Res.*, **68**, 4795–812.
Tsukamoto, Y., Ota, T., and Noguchi, H. (1982). 'Hydrological and morphological studies of debris slides on forested hillslopes in Japan.' *IAHS Publ. No. 137*, 89–98.
van Asch, Th. W. J. (1984). 'Creep processes in landslides.' *Earth Surf. Proc. Landforms*, **9**, 573–83.
Veder, C. (ed.) (1981). *Landslides and Their Stabilization*. Springer-Verlag, New York, p. 247.
Vernes, D. J. (1978). 'Slope movement types and processes.' *Landslide Analysis and Control, T.R.B., Spec. Rep.*, **176**, 11–33.
Voigt, B., and Pariseau, W. G. (1978). 'Rockslides and avalanches: An introduction.' In *Rockslides and Avalanches, 1. Natural Phenomena*. Voigt, B. (ed.), Elsevier, Amsterdam, 1–67.
Yoshimatsu, H. (1980). 'Fluctuation behaviour of ground water analysis in landslide area.' *J. Jap. Landslide Soc.*, **62**, 20–5 (in Japanese).
Yoshioka, R., and Kanai, T. (1975). 'Partial pressure of dissolved carbon dioxide gas of ground water in landslide areas.' *Ann. Disaster Prevention Res. Inst. Kyoto Univ.*, **18B**, 271–82 (in Japanese).
Yoshioka, R., and Okuda, S. (1972). 'Estimate of the amounts of clay minerals formed by weathering process through the chemical compositions of water in the Kamenose Landslide Area.' *Ann. Disaster Prevention Res. Inst., Kyoto Univ.*, **15B**, 171–82 (in Japanese).
Zaruba, Q., and Mencl, V. (1982). *Landslides and Their Control* (2nd edn), Elsevier, Amsterdam, p. 324.

Slope Stability
Edited by M. G. Anderson and K. S. Richards

Chapter 9

Earthquake-prone Environments

V. COTECCHIA
Instituto di Geologia Applicata e Geotecnica,
Universita di Bari, Italy

9.1 INTRODUCTION

Stories of large landslides caused by earthquakes have been passed down from remote times. One of the first recorded mass movements occurred in BC 372–73, when Helice, then a large, flourishing Greek city on the northern shores of the Peloponnese, slid into the sea after having been razed to the ground (Marinatos, 1960; Seed, 1968)

A typical early report of landslides triggered by an earthquake in Italy is by Vivenzio, who described some of the geomorphological disturbances produced by a catastrophic 3-minute earthquake that struck Calabria (southern Italy) in 1783:

> 'Nello stato di Oppido fu così grande lo sconvolgimento, cagionato dall'orrendo terremoto de' cinque Febbraio 1783, che all'istante in molti luoghi sprofondò il terreno, furono le colline intere trasportate con moto orizzontale . . .'
> ('The shaking that occurred in the State of Oppido as a result of the terrible earthquake of 5 February 1783 was so great that in many places the earth sank immediately and entire hills were shifted horizontally . . .')
> '. . . Repente, per cosi dire, s'intenerì tutto il materiale degli oliveti, della conca, e de' monti terminali, e come pasta liquida e molle, rendutasi fluente e scorrevole, cadde in tale universale rivolgimento che in pochi minuti se condi . . . tutti que' fondi perdettero interamente l'antica loro consistenza.'
> ('The ground forming the olive groves, the hollow and the surrounding hills suddenly softened, so to speak. The slopes were liquefied and started to flow; during the tremors lasting only a few seconds . . . all the lands were completely bereft of their former consistency.')
> 'Ne' monti poi osservammo una perpetua alterazione. La loro faccia . . . è quasi tutta da cima a fondo rabbiosamente scorticata.'
> ('Then we saw that the mountains had undergone a permanent change. Their face . . . had been angrily scarified from virtually top to bottom.')
> (Vivenzio, 1788).

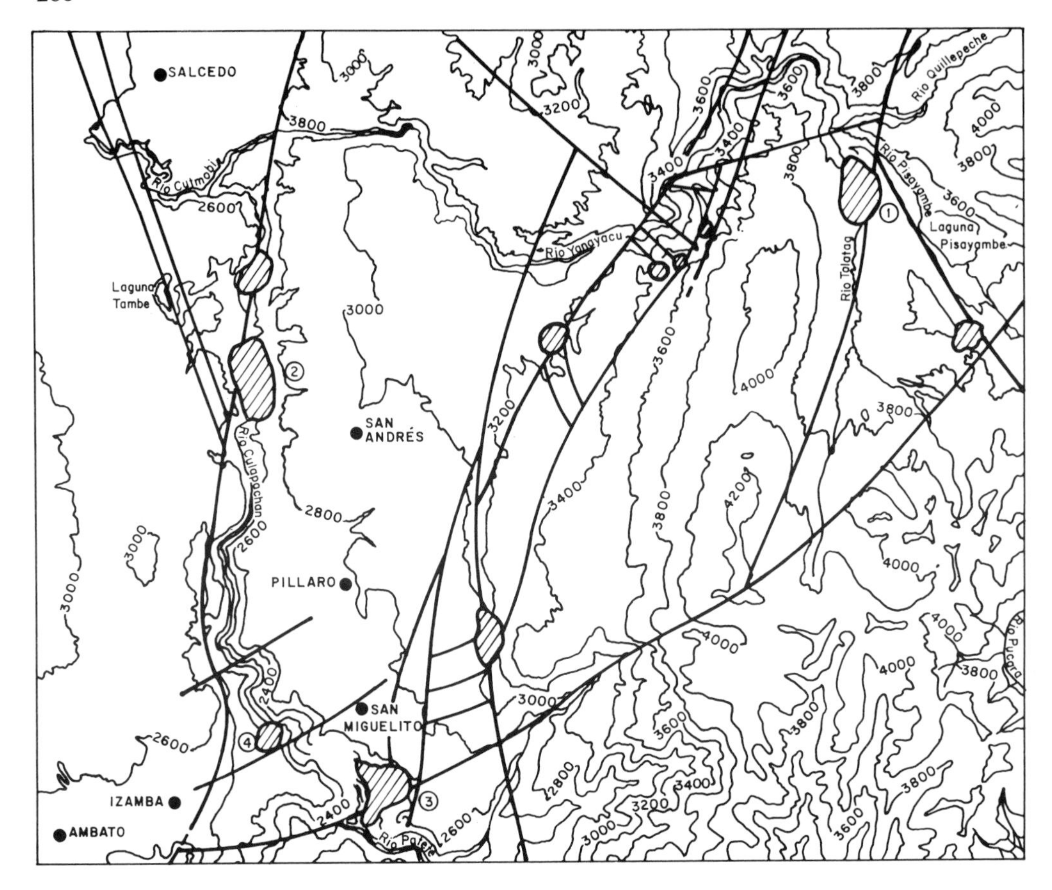

Figure 9.1 Andean Cordillera of Ecuador: relation between landslides and active faults. Sedimentary clastic and loose pyroclastic deposits cover the region; the slide material is often volcanic ash, the absolute age of which dates back 5,300 ± 113 years. (1) Samanga dry flow; (2) and (3) Quillan and Cunchibamba slumps; (4) Rumicucho slip

However, information on the majority of such historical events is generally scanty and poorly substantiated. Yet the role of earthquakes in the geomorphological history of some regions can be readily distinguished from that of other possible agents of morphogenetic slope evolution. Where such phenomena are frequent and extensive they are certainly a prime factor in the development of landscape. The regions where they are encountered most frequently include all those lying on the edge of the crustal plates, e.g. the Andean Cordillera, where evidence of the role played by earthquakes in slope evolution is by no means lacking today. For example, the destruction of the town of Pelileo (Ecuador) occurred by landslides during the earthquake of 5 August 1949, while in the Rio Patate Valley there are innumerable signs of the river having been dammed and diverted as a result of huge mass movements produced by earthquakes in recent centuries. The main landslide had a front of at least 2 km and involved a mass of earth consisting of lacustrine and volcanic ash deposits up to 300 m thick. The earthquake occurred in an active fault area where similar events have taken place many times in the past (Cotecchia and Zezza, 1974) (Figures 9.1 and 9.2).

Figure 9.2 The Samanga (Ecuador) rock slide photographed on the 4 December 1968 during a fourth degree earthquake shock according to the Mercalli scale. During the earthquake a great quantity of 'cangagua' (similar to loess) and pumice was eroded in a storm of thick whirling dust clouds

Other examples have left even more vivid impressions, such as the movements produced by the earthquake at New Madrid on the banks of the Mississippi (Missouri) in 1811, at Kansu (China) in 1920, at Chait in the Surchob and Yasman valleys (USSR) in 1949, San Francisco in 1957, Alaska in 1964. Fragile lithological constitution in Italy, and the geomorphological and structural configuration of the country, are such that earthquakes are often accompanied by mass movements and by changes in the natural drainage system. For example, the earthquake at Arzino in the pre-Alps on 27 March 1928 triggered off some 400 landslides.

In the study of relationships linking neotectonics, earthquakes and mass movements reference is frequently made not only to phenomena induced by recent tremors but also to those that occurred centuries ago. The aim is to ascertain the original geomorphological, geohydrological, and geotechnical environments and the changes produced by seismic events. The information derived in this way can then be used to extend and check on the catalogue of historical earthquakes, which is inadequate at present for exact seismic zoning. This approach is currently adopted in Italy, for instance, when assessing sites for nuclear installations and major civil engineering works. One of the first such studies concerned the phenomena produced as a result of the Calabrian earthquake of 5–6 February 1783. The intensity and catastrophic effects of this event were such that today it is still known as the 'Calabrian Earthquake' despite the numerous shocks of considerable intensity that have shaken the region since then. For example, in 1905 devastation was caused on Magisano Hill, in the Monachello district on the Amato River and at several places around Caraffa,

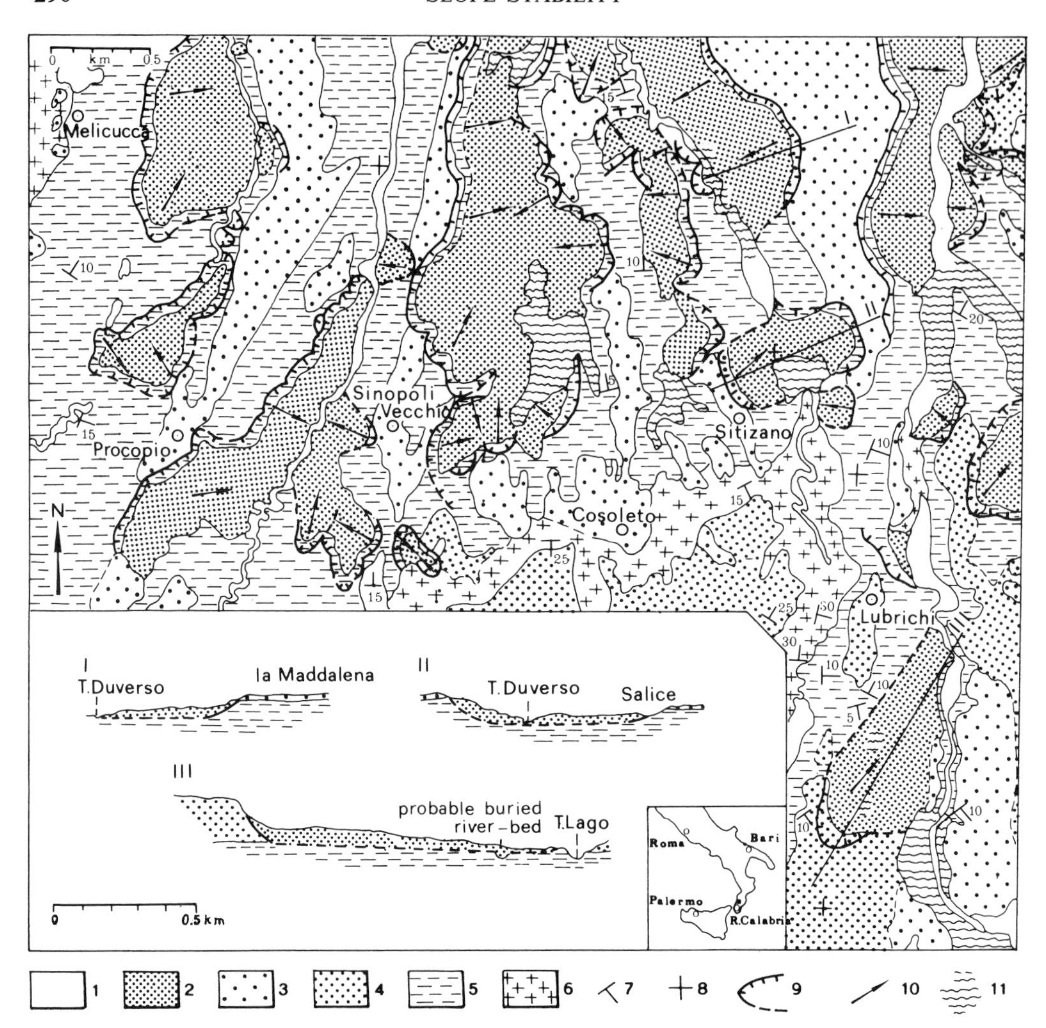

Figure 9.3 1783 Calabria (southern Italy) earthquake: detail of geological map and landslides, most of them following liquefaction phenomena. (1) Fluvial deposits (Holocene); (2) Landslide detritus; (3) Sands and conglomerates (Pleistocene); (4) Sands and sandstones (Pleistocene–Pliocene); (5) Clays and marls (Lower Pleistocene–Pliocene); (6) Metamorphites: gneisses and schists (Palaeozoic); (7) Attitude of strata; (8) Horizontal strata; (9) Main scarp and limit of landslide; (10) Main direction of mass movement; (11) Lake produced by landslides

while the November 1849 earthquake caused two large landslides in the Miocene sandstones at Cinidà and other smaller ones at Scilla and Santa Cristina d'Aspromonte.

However, the effect of the 1783 earthquake on the stability of slopes in 'Further Calabria' was far greater (Figure 9.3). Very strong tremors started at 7 p.m. on 5 February and lasted about three minutes, being followed by other equally strong tremors some twenty-four hours later. These shocks had been preceded by two of 'moderate' and 'slight' intensity in January, which heralded a period of seismic activity in Calabria that lasted about four years from 1783 to 1786. During the Calabrian Earthquake numerous large landslides swept away entire towns carrying them into the valley below. Two hundred and fifteen lakes were formed when rivers and streams were dammed, as illustrated by a number of classical prints of that time (Figure 9.4). A complete study of the area affected by the earthquake-initiated landslides, whose 'scars' are still quite evident, has enabled the original geology of part of the epicentre zone on the northern side of the Aspromonte to be mapped (Cotecchia *et al.*, 1969). Strips of land up to 4 or 5 km long broke away from the hillsides, as in the Duverso Valley. Though the slip surfaces here are mainly in clayey deposits, there are reports of mass movements that occurred only in sandy sediments outside the area studied in detail. The mass movement mechanism in the majority of cases is a combination of slumping and slipping. However, there are instances of pure sliding phenomena due to detachment of sedimentary formations from the crystalline bedrock.

Other interesting characteristics observed in many of the slides surveyed are the huge size of the slip surface area, and the extent of horizontal movement in the direction of travel, which far exceeds that commonly observed in landslides produced by gravity action alone.

The earthquake caused so much devastation partly because of the amount of energy it released, and partly because of the particular geological and geotechnical situation in the areas affected.

9.2 INTRODUCTION TO EARTHQUAKES

9.2.1 General

Underlying the most common definition of the word *earthquake*—a shaking of the ground caused by rupture processes and the consequent release of stresses due to the action of endogenous forces on the more rigid strata of the earth—is the theory of plate tectonics. Plates form the parts into which the outermost portion of the earth's mantle and crust (lithosphere) are divided. They float on the underlying viscous part (asthenosphere) and move relative to one another at a velocity of a few centimetres per year. There is as yet no universal agreement on the 'driving force' for this movement, but it is probably bound up with thermal convection currents in the upper part of the molten mantle. Together with the dynamic effects of the earth's rotation, these currents move the overlying lithosphere.

There may be three types of relative movement among plates: divergence, sideways slip, and convergence. In the latter case the plates meet and subduction occurs, with one plate disappearing beneath another. In the last two cases there is an enormous build-up of potential energy, and when the induced stresses exceed the strength of the material the strata involved rupture and a shear or slip surface (fault) is produced or reactivated. The two blocks or plates move along this fault in an attempt to return to their original configuration. The point where the initial fracture occurs is the hypocentre (or focus) of the earthquake, while the point on the earth's surface directly above it is called the epicentre. The release of energy does not occur at one single point but affects a more or less extensive area around the focus.

9.2.2 Seismic Waves

When fracturing takes place the stored energy is immediately released, partly as heat and partly as elastic rebound of the rock masses, where a series of elastic waves is generated.

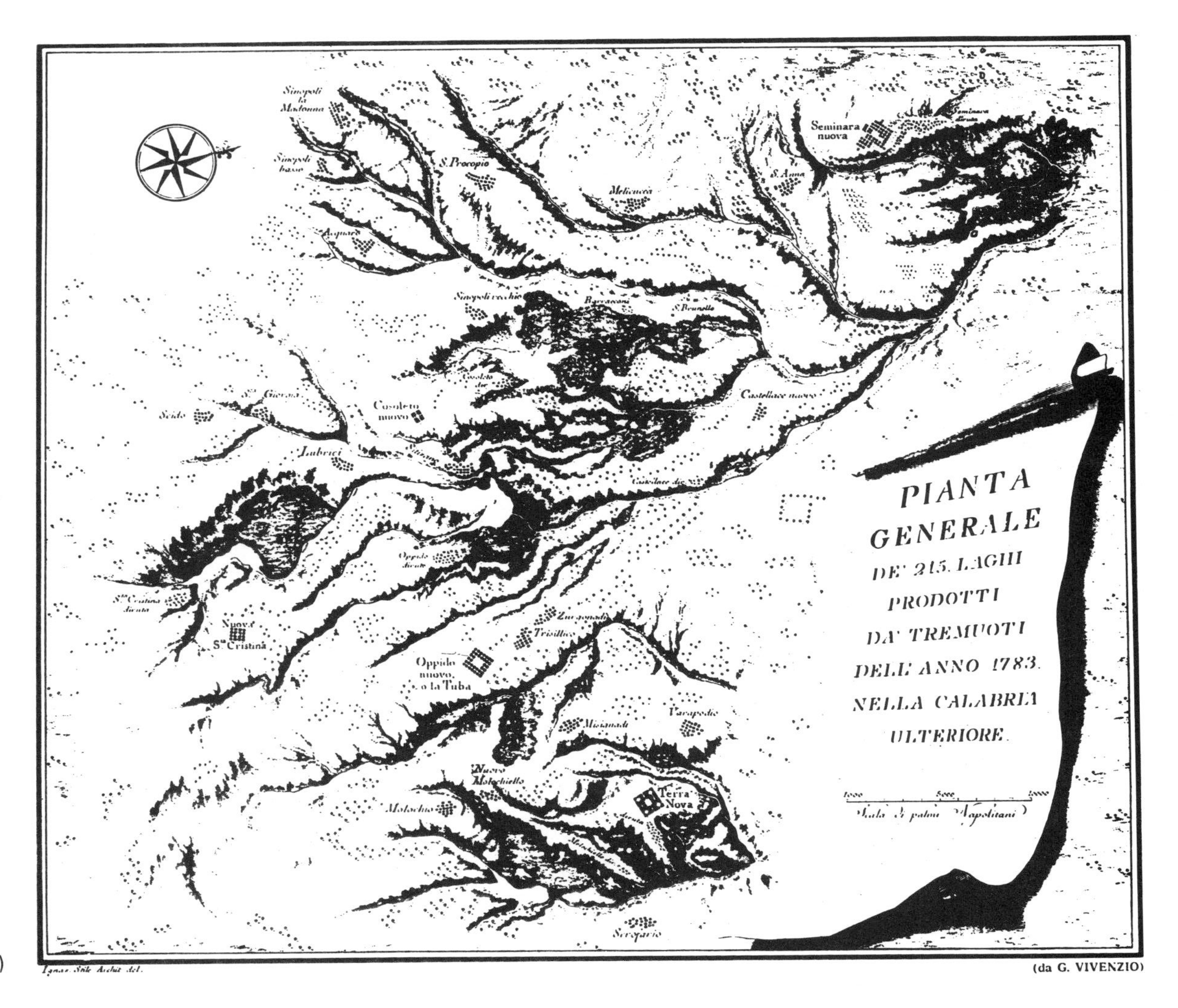
PIANTA
GENERALE
DE' 215. LAGHI
PRODOTTI
DA' TREMUOTI
DELL'ANNO 1783.
NELLA CALABRIA
ULTERIORE.
Seminara nuova
S. Procopio
Oppido nuovo, o la Tuba
Cosoleto nuovo
Terra Nova
(da G. VIVENZIO)

(a)

(b)

Figure 9.4 1783 Calabria (southern Italy) earthquake. (a) General plan of the 215 lakes, each 5 km long at most, which formed due to the landslides which followed the earthquake. (*From Vivenzio, 1788.*) (b) An old drawing showing one of the several landslides that dammed a river to form a lake. (*From Sarconi, 1784.*)

Because of their origin, these are known as seismic waves, and they propagate in all directions.

By reference to the velocity and mode of propagation, various types of seismic wave can be distinguished. These can be grouped into two categories: body waves and surface waves. The former propagate within the rock mass and can be either:

- P waves (primary or longitudinal or compressional and dilational waves) in which each particle vibrates in the direction of propagation. These are characterized by alternating compression and dilation in the medium through which they pass. P-waves are the fastest of all seismic waves and propagate in every type of medium: solid, liquid, and gaseous.
- S waves (secondary or transverse or shear waves) in which each particle vibrates at right angles to the direction of propagation, inducing shear stresses in the materials through which they pass. These waves do not propagate in fluid media.

Surface waves, instead, are formed and propagate only along the earth's surface, starting from the epicentre. They can be classed as:

- L waves (Love waves) in which each particle vibrates transverse to the direction of propagation, shaking the ground in horizontal planes parallel to the earth's surface. They do not propagate in fluids and are the fastest, most destructive of the surface waves.
- R waves (Rayleigh waves) shake the ground both horizontally and vertically in a plane parallel to the direction of propagation.
- G waves (Gutenberg waves) are similar to R waves but of lesser importance.

The velocity of propagation of seismic waves can be calculated by applying the equations for velocity of propagation of elastic waves in general, on the basis of the elastic characteristics of the medium traversed (density, elasticity, compressibility, etc.).

A whole series of effects originate from combinations of the various wave components, and their impact on the ground. These effects may take the form of damage to structures or initiation of various forms of instability. When seismic waves reach the surface of the ground, most of their energy is reflected, so the surface itself is hit simultaneously by waves directed upwards and downwards, causing marked amplification of the vibrations.

The earthquakes described so far are commonly referred to as tectonic earthquakes. They account for some 90% of all earthquakes affecting our planet and are certainly the most important. Volcanic earthquakes also occur: but even these really originate from global geodynamic activity. Though they, too, may be destructive they are generally restricted to a small area around the volcano. There are also earthquakes attributable to local causes. These can be natural, due to physical and chemical phenomena or collapse in the subsurface or subaerial environments, or they may be man-made, resulting from underground explosions, seepage of water into the subsurface from reservoirs, and the extraction of oil and gas from oil-fields (induced seismicity).

9.2.3 Earthquake time sequence

An earthquake is never an isolated event, it generally involves a temporal distribution of shocks that can be grouped schematically under three main headings depending on the physical state of the medium and the range of stresses generated:

1. Mainshock of high energy and a series of aftershocks, the number of which decreases gradually according to a hyperbolic law, in materials of homogeneous structure.

2. Foreshocks–mainshock–aftershocks sequence, with the number and intensity of shocks increasing exponentially up to the mainshock, in materials of somewhat heterogeneous structure.
3. Swarm earthquakes with shocks of virtually equal energy increasing in frequency up to a maximum and then decreasing, in materials of extremely inhomogeneous structure.

9.2.4 Earthquake Measurement

The best known parameter for indicating the size of an earthquake is intensity (I). This is a subjective measurement based on the effects caused by the earthquake on the environment. The local destructive potential of the event is indicated by an adjective which is linked to a given degree of intensity on an empirical scale. The first known scale was that of De Rossi-Forel in 1883. This was followed by many others in an attempt to characterize the effects in ever-greater detail: Mercalli, Modified Mercalli (MM), Mercalli–Cancani–Sieberg (MCS), Medvedev–Sponheuer–Karnik (MSK), and Japan Meteorological Agency (JMA).

Lines drawn through all places with the same intensity are isoseismal lines. Maps with such lines provide a graphic picture of areas with the same earthquake intensity. However, when the effects of an earthquake on a given site are analysed, account must be taken of a large number of variables to ascertain the 'local seismic response' of the site. The variables to be considered depend both on the characteristics of the vibrations on arrival and on local geological and geomorphological conditions. It thus follows that a given earthquake can produce damage of very different magnitude even in places close together or which lie the same distance from the epicentre. Consequently isoseismal lines are often very irregular in shape, and there may be nuclei of high intensity within areas of generally lower intensity.

A parameter that is, instead, physically defined and unequivocally measurable is magnitude (M), introduced by Richter in 1935. This is the index of total energy released by an earthquake and is the logarithm of the maximum amplitude (A) of the seismic wave (P, S or L) recorded at any station, compared with the amplitude (A_o) of a sample shock which is attributed a magnitude of nil:

$$M = \log A - \log A_o$$

The standard earthquake $M = 0$ is defined as the earthquake which produces a trace with a maximum amplitude of 1 micron on a Wood–Anderson seismograph at a distance of 100 km from the epicentre. Magnitude is measured in whole units and fractions ranging from 0 to a peak fixed by the condition of maximum deformation that can be supported by the rocks of the earth's crust. During this century the peak value measured has been $M = 8.9$, recorded on 31 January 1906 off the coast of Ecuador and on 3 March 1933 in the sea off Japan.

Depending on the various phases and the components of the motion, magnitude is described as local (M_L), S-wave type (M_S), body-wave type (M_B) or duration type (M_D) determined on the basis of the recorded duration of the oscillations. The two parameters—intensity and magnitude—are correlated only by relations of a purely statistical nature that cannot be generalized.

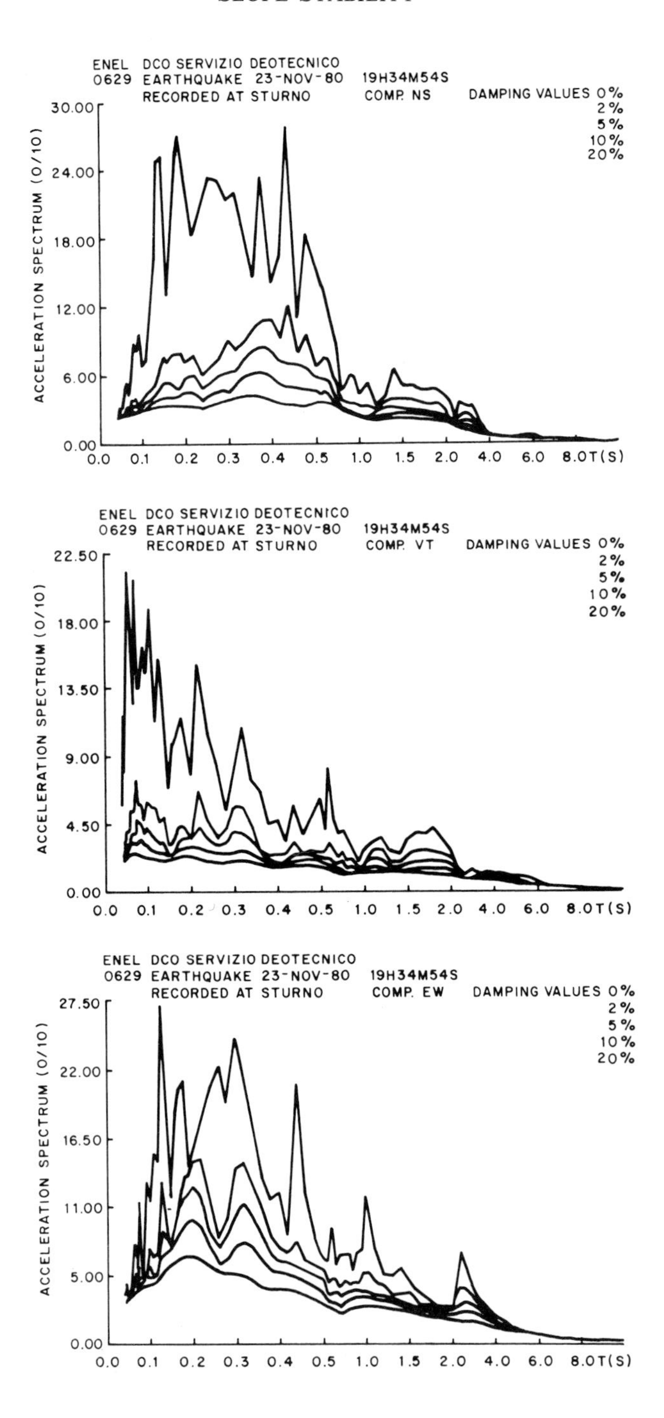

Figure 9.5 1980 Southern Italy earthquake: response spectra of the three orthogonal components of acceleration recorded at Sturno (Avellino). (*Reproduced from Berardi* et al., *1981, by permission of Commissione Enea-Enel.*)

Ground motion during an earthquake is defined by a whole range of parameters relating to the characteristics of the arriving waves. The most important of these is maximum ground acceleration (a_m), mainly because of the effects produced on structures by the horizontal forces. This parameter is measured by means of an accelerograph which plots ground acceleration versus time on an accelerogram, generally as a fraction of acceleration due to gravity ($1\ g = 9.81\ m\ s^{-2}$).

In addition to maximum or peak acceleration, the characteristics of an earthquake are also determined by frequency and duration. The latter feature is of particular importance as regards damage to man-made and natural structures, since motion of a relatively small amplitude which continues for a fairly long time at uniform frequency can result in greater damage than that caused by a much more intense shock lasting only a short time.

When undertaking earthquake-resistance analysis on any structure, starting from the excitation function formed by the accelerogram and the Fourier spectrum which indicates the dominant frequencies, a response spectrum is plotted for the earthquake. On this diagram, the response of an elementary dynamic structure (a perfectly elastic simple oscillator) subjected to a given accelerogram is given on the ordinate in terms of maximum acceleration, velocity or displacement, while the natural period of vibration of the structure, as a function of the relevant damping factor, is given on the abscissa. As an example, the response spectra relating to the three components of motion recorded at Sturno (Avellino) during the $M = 6.8$ earthquake in southern Italy on 23 November 1980 are shown in Figure 9.5.

9.3 EARTHQUAKES AND LAND INSTABILITY

Assessment of ground stability during earthquakes is a vast, complex matter involving such fields as geology, geophysics, seismology and geotechnics. There is not always the necessary interchange of information among these fields, but improved knowledge of structural geology, geophysics, and geotechnics reveals ever more clearly the decisive role of earthquakes in the making of landscape in some earthquake-prone regions.

Earthquakes can cause various forms of rupture and instability in the ground. These can include ground displacements both horizontal and vertical with tectonic uplift or subsidence of large areas, the formation of enormous devastating seismic seawaves (tsumani) resulting from sudden vertical movements of the seabed, the radical alteration of hydrogeological regimes, liquefaction of unconsolidated, water-saturated deposits, and mass movements.

With the extension that has taken place recently in the network of seismographs capable of recording even instrumental tremors there is now a much broader knowledge of the relationship between earthquakes and landslides. It has thus even been possible to detect various cases of earthquakes—even quite strong ones—that may have been *caused* by landslides: the most obvious case being that which occurred on Hawaii on 29 November 1975 when a mass of several thousand square kilometres slid down the side of Kilauea Volcano towards the sea and was the probable cause of an earthquake with magnitude of 7.2.

Though there are many who still doubt whether landslides can really cause big earthquakes (Radbruch-Hall and Varnes, 1976), there can be no question that earthquakes are the cause of huge landslides, often with disastrous consequences for the people and property concerned. For instance, more than 90% of the victims of the 1974 and 1978 earthquakes on Izu Peninsula (Japan) were killed by landslides.

Some mass movements can occur due to disturbance of equilibrium already close to instability prior to the tremors. Frequently, an earthquake accelerates ongoing movement or reactivates slipped masses that had previously stabilized: ample evidence of such events was provided by the earthquake that hit a vast area along the border between Campania and Basilicata (southern Italy) in 1980. Mass movements on slopes close to failure can be triggered off even by very weak shocks.

In other situations an earthquake can initiate landslides on hitherto intact slopes, rather than simply reactivate old movements. This was the case with most of the 40 historical earthquakes (1811 to 1980) considered in a recent study (Keefer, 1984) designed to determine the effect of geological environment on landslide type. Fourteen different types were recognized, initiation of which depends on the magnitude of the earthquake and the distance from the point of rupture. The most susceptible materials include poorly cemented rocks, residual and colluvial sands, granular alluvial sediment and man-made granular fills.

Falls and slips, and mass movements on steep slopes in general are triggered even by weak shocks of short duration but high frequency. Landslides in cohesive materials occur only as a result of strong, long-duration shocks, while even stronger tremors are needed to mobilize flows or complex movements such as debris avalanches produced by rock falls.

A statistical correlation between earthquake magnitude and the size of the area affected by mass movements shows that the latter increases with the former. For instance, following the 1964 Alaska earthquake, an area of more than 200,000 km^2 was affected by mass movements.

In seismically-active areas, the preparation of landslide hazard maps is an essential prerequisite for proper seismic zoning of a region. Zoning of an area according to the landslide hazard calls for consideration of a wide variety of factors affecting slope stability, including lithological conditions, geological structure, seismicity, climatic conditions, vegetation cover, groundwater conditions, soil strength parameters, presence of old slides, and the actions of man.

9.3.1 Mass Movements

The behaviour of a slope affected by an earthquake depends on the nature of the ground motion, the slope geometry, and its composition. Indications as to the likelihood of mass movements under non-seismic conditions do not always hold good for landslides triggered off by earthquakes. Ground shaking can produce accelerations within the soil mass, accompanied by a system of stresses that can completely alter the strength of the materials. Thus when considering the response of a slope to accelerations that vary in amplitude and direction the dynamic properties of the ground cannot be ignored.

Slope failure following an earthquake can occur because of an increase in shear stresses in the soil mass or because of a decrease in strength under dynamic loading conditions. The probability of failure because of a given deviator stress increases with the number of load cycles and the deformation produced during a given number of cycles is proportional to the deviator stress applied.

The aim of analysing slope stability under seismic conditions is to check compatibility of the strength characteristics of the materials in the total stress state, with the superimposition of the earthquake-induced system of forces, directed alternately in favour of instability and stability, on the pre-existing stress due to static loads (dead load and any live loads).

Soil shear strength (τ) can be expressed, of course, by the Mohr–Coulomb equation:

$$\tau = c' + (\sigma_n - u) \tan \phi'$$

where c' and ϕ' are effective cohesion and effective angle of internal friction, σ_n is the total stress normal to the plane of failure, and u is the pore pressure (see Chapter 2). Shear strength generally attains a maximum (peak strength) during the initial rupture phase and then decreases, more or less rapidly, to attain a constant minimum value (residual strength).

In reality, it is difficult to predict the earthquake stability of slopes and, indeed, slopes which have seemed relatively stable have failed, while slopes in limit equilibrium conditions have not suffered significant deformation as a result of tremors. This is particularly the case where, in addition to an unfavourable seismotectonic state, there is predisposition to mass movement due to the presence of highly weathered, deformed and fissured rock formations and rapid, continual slope activity. In such instances, most of the slopes have already reached the state of minimum resistance (residual strength), so most mass movements are actually renewals of old movements that have not always been recognized previously. A delay can occur between the occurrence of the mainshock and the start of mass movement. It often happens, in fact, that mass movement, frequently in the form of flows, begins some time after the original rupture when rain falls or other factors occur that reduce the factor of safety of the slope.

As regards dynamic stability analysis and the strength parameters to be considered, it is necessary to know the behaviour of the soil under loading conditions that simulate seismic action as closely as possible (cyclic-type laboratory tests). It should be observed, however, that knowledge of the dynamic behaviour of clays is still rather limited, despite the great progress made in recent years and the large amount of published data available.

The residual behaviour is generally most important because the main feature underlying the predisposition of a mass to move is the pre-existence of active or quiescent landslide conditions. Under such conditions, there exists only residual or near-residual resistance to movement on the pre-existing slip surfaces.

It is known that the concept of residual strength for soils containing essentially platy clay materials, is especially significant in plastic over-consolidated clays in which the decrease in strength from peak to residual is often considerable. For example, Chattopadhyay (1972) quotes a decrease from a 28° peak angle of internal friction to a residual value of 2.6° to 3.6° in the Bearpaw Shales of the area of the South Saskatchewan Dam (Canada). Remarkable decreases have also been measured on smectitic varicoloured clays of southern Italy. Yet reference to average residual angles is not very useful because the ϕ'_r value is not necessarily independent of normal stress. In fact, a phenomenon usually observed is the decrease in residual friction angle or residual strength as the effective normal stress increases. Such dependency is more marked with low values of effective normal stress (Figure 9.6). On the basis of the data of Skempton and Petley (1967) and Chattopadhyay (1972), independence of ϕ'_r from normal effective stress occurs over 200 kN/m^2; however,

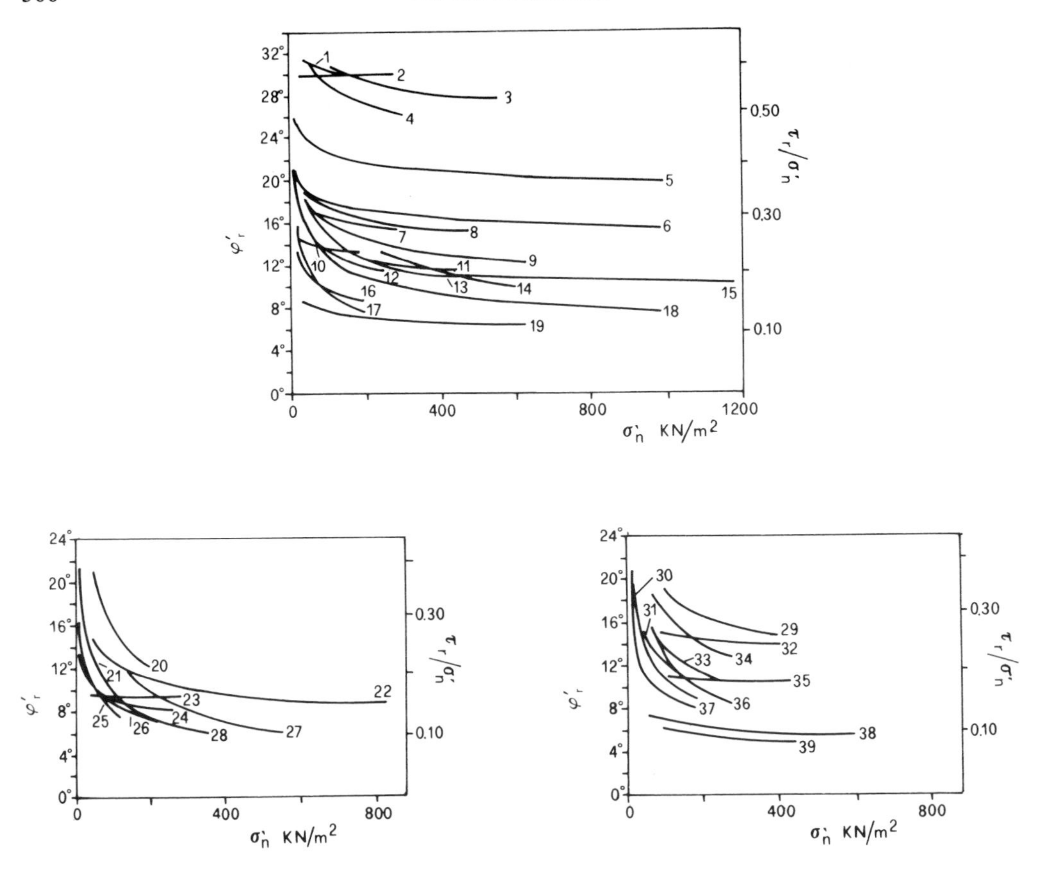

Figure 9.6 Residual friction angle (φ'_r) and residual friction coefficient (τ_r/σ'_n) versus effective normal stress (σ'_n) for various natural soils. (1) and (22) Bishop *et al.* (1971); (2) Skempton (1964); (3) Chandler (1966); (4) and (20) Chowdhury and Bertoldi (1977); (7) and (17) Hutchinson *et al.* (1973); (8), (9), and (12) Skempton and Petley (1967); (10) and (16) Petley (1980); (11) and (35) Petley (1969), in Bishop *et al.* (1971); (13) and (19) Hutchinson *et al.* (1980); (14) and (15) Esu and Calabresi (1969); (21), (25), (26) and (31) Chandler (1976); (23) and (24) Garga (1970), in Bishop *et al.* (1971); (27) Insley *et al.* (1977); (28) Blondeau and Josseaume (1976); (29) Calabresi *et al.* (1980); (30) Maugeri (1976); (32) Cancelli (1979); (33), (34) and (36) Al-Layla (1970); (37) Hutchinson and Gostelow (1976); (5), (6), (18), (38) and (39) Cotecchia and Federico (1980)

some materials show dependence up to and over 500 kN/m² (Cotecchia and Federico, 1980). Observations by scanning electron microscopy have allowed clarification of the prevailing mechanism of the phenomenon as the formation of different levels of microtextural organization on the shear plane. A reasonable formal explanation of the phenomenon has been suggested in the increase of the product $d*\sigma'$ or $w*\sigma'$ (where: d = interparticle distance, w = water content, σ' = effective normal stress) as σ' increases (Cotecchia and Federico, 1980). Because of this phenomenon, in many natural soils, it is arbitrary—and sometimes unsafe—to refer to linearization of Mohr–Coulomb failure envelopes for residual states (Chandler, 1973). This is particularly important when studying the stabilization of landslides for which resistance conditions are already governed by residual values. In fact, among the possible corrective measures are loading of the toe zone and/or lowering of groundwater or of neutral stresses along the shear surface. The result of such operations is to increase the effective normal stress, so the correct evaluation of their effectiveness must consider the possibility of a curved failure envelope. Consequently, in determining the residual strength to be adopted in the analysis of particular problems, it is necessary to refer to stress intervals as near those *in situ* as possible.

A further issue has been revealed by recent experimental findings (Skempton, 1985) which could be extremely important for earthquake-related stability. The rates of displacement on pre-existing natural shear surfaces can vary considerably, ranging from very slow movements in some reactivated slides to quite rapid ones induced by ground shaking. Though residual strength is relatively unaffected by variations in low rates of displacement encountered in reactivated slides and in the usual laboratory tests, experiments by Skempton (1985), carried out in connection with seismic analysis of the Kalabagh Dam project, show that changes in the pattern of behaviour can occur at displacement rates of about 100 mm/min. A marked increase in strength as a result of shear strain is followed by a drop to a minimum. The strength increase is probably associated with disturbance of the originally ordered texture of the soil, and perhaps with viscous effects. Generation of negative pore pressures is also likely and it is possible that their dissipation, as deformation proceeds, leads to loss of strength. In clays and in low-clay silts this minimum strength is not lower than the slow or static residual value, but in clayey silts (around 15–25% clay) the minimum can be as low as one-half the static value. Though more research is needed to define and explain these phenomena better—and to check on their occurrence in even faster tests—it may prove advisable to consider their existence in seismic geotechnical engineering.

The links between dynamic loading, progressive effects, initial stress states, and delayed failure are still not entirely clear. Problems also remain regarding the quality and pertinence of input data in stability analysis. All these factors tend to lead to a low level of accuracy where predictions are concerned.

9.3.2 Settlement and Liquefaction

Of the factors that contribute to progressive instability in earthquake-prone environments, settlement and packing of surface soils induced by seismic shaking play an important role. A distinction must be drawn between cohesionless and plastic unconsolidated soils.

In the case of granular soils in various states of saturation the amount of settlement produced by cyclic stresses can be assessed theoretically by reference to variation in the void ratio. Alternatively (Ivanov *et al.*, 1981) the maximum variation in relative density as a result of seismic shock can be evaluated on the basis of the stratification of the deposit and the initial relative density. The presence of a structure superimposed on a deposit of loosely-packed cohesionless soils causes an increase in total settlement, so that the actual settlement in free sands for instance marks the lower limit expected for the settlement of a structure resting thereon, when an earthquake occurs. Such settlement can have a disastrous effect on the stability of the structure.

When cohesionless soils are saturated the major hazard arises with finer particles for which loading is undrained. Seismic shaking can result on the one hand in loss of strength due to an increase in pore pressure where drainage is impeded, and on the other in the creation of a system of shear stresses within the soil mass forming the slope, that together cause failure. In the former case, there can be a complete loss of soil strength when total pore-water pressure equals the external load. Under such extreme conditions the soil behaves like a fluid, which explains why the phenomenon is known as 'liquefaction'.

As reported by Newmark and Rosenblueth (1971), some workers distinguish between initial, partial, and total liquefaction. Others prefer to speak of 'cyclic mobility' in respect of loss of strength due to undrained cyclic loading, while using the term liquefaction for the development of excess pore-water pressure under static conditions.

When soils are subject to strong shaking, the loss of strength due to increase in pore pressure may also affect the stability of relatively more consolidated materials or those with a particle-size distribution that extends to the finest of grains. Not even clay bodies with high pre-consolidation values are excluded from the phenomenon, which can occur along fracture planes or fissures, sometimes deep, subject to local variations of the structure of the cohesive soil (Newmark and Rosenblueth, 1971).

The three most important variables that define the behaviour of saturated cohesionless soils under seismic excitation are relative density, confining pressure, and drainage conditions. When confining pressures are low, if such a soil is surrounded by materials that cannot liquefy, it is sufficient for the relative density to be greater than 0.7 to avoid liquefaction even in strong earthquake tremors, and whatever the grading and drainage conditions. However, if drainage is inadequate, this value must increase to 0.8 when there is a strong, long-lasting earthquake, e.g. *MM* intensity of 10 or more and duration in excess of 3–4 minutes.

With lower relative densities or high confining pressures, the soil may liquefy if the grain format prevents dissipation of excess pore pressures, if the boundary conditions are such as to make drainage difficult, or if a neighbouring formation liquefies, transmitting excess pore pressures.

Precise assessment of potential liquefaction of a soil calls for thorough examination of its dynamic behaviour. Yet the complexity of the investigation and its cost in the case of detailed zoning on a regional scale makes it preferable to predict liquefaction by reference to synthetic criteria rather than to utilize numerical models.

For detailed seismic zoning studies, three methods are available for assessing potential liquefaction: empirical, semi-empirical, and analytical. The choice of method depends on the amount and quality of data available and on the precision required. Although analytical methods may seem to be the most suitable, a preliminary evaluation based on empirical methods is always useful. Empirical assessment of the liquefaction hazard is performed by reference to particle-size distribution, relative density, or penetrometer test results. In addition, there are parameters characterizing ground motion (magnitude, maximum acceleration, duration, frequency, etc.), plus the depositional characteristics and stress history of the deposit, its size and drainage conditions, boundary conditions, and confining pressure. However, the crucial factor as regards the liquefaction susceptibility of a soil is its relative density (Rd):

$$\mathrm{Rd} = \frac{e - e_{min}}{e_{max} - e_{min}}$$

where e = effective void ratio and e_{max} and e_{min} = void ratio at maximum and minimum compaction, respectively. Other things being equal, the possibility of soil liquefaction decreases with increase in relative density, which can be measured *in situ* by the standard penetration test (SPT). It has also been shown that the liquefaction resistance of saturated sands increases with the coefficient of earth pressure at rest (K_o). This could explain why old flows are so frequently reactivated in silty–clayey–sandy soils.

Diagrams have been proposed based on empirical data (Sherif and Ishibashi, 1978; Botea *et al.*, 1980) for a rough step-by-step evaluation of the liquefaction susceptibility of a given deposit. Though empirical methods are quick and easy to use, they suffer from a number of limitations. For instance, they are calibrated by reference to a limited number of cases, and do not take account of the characteristics of the seismic excitation involved. They cannot be employed, therefore, for definitive evaluation of the problem where major structures are concerned.

Semi-empirical methods, instead, are based on comparison of earthquake-induced stresses and those needed to produce liquefaction. This comparison can be performed at various levels of precision. It follows that at one end of the scale, the approach may merely involve evaluation of the shear stresses developed in a soil element at a given depth, solely by reference to knowledge of earthquake magnitude and distance from the epicentre, compared with shear stresses causing liquefaction in the same element derived by reference to grading characteristics and penetration resistance. At the other extreme, the approach may involve comparison of stresses induced by an earthquake, derived analytically by dynamic analysis performed by the Finite Element Method (FEM), with those needed to cause cyclic liquefaction in the laboratory. The fundamental advantage of this approach is that it can evaluate both the seismic event and the real dynamic behaviour of the soils, and provides a kind of factor of safety against liquefaction. The method thus seems to be the best one at the present moment for dealing with the complexities of the phenomenon.

The problems posed by plastic (or cohesive) soils, especially those with little consistency, are no less complex than those concerning granular soils. Two phenomena, having opposing influences on strength properties, are usually involved since application of a rapid load increases strength, while alternating load reduces it. The final result seems to depend on a great number of factors, with sensitivity playing a leading role. In soils of high sensitivity the decrease in strength is so marked that it can be expressed in terms of liquefaction phenomena alone. This loss of strength after shaking depends also on reorientation of the particles after destruction of the secondary bonds and on the increase in neutral stress caused by cyclic loading. It also seems to be a decreasing function of the number of cycles, cyclic deformation, and frequency.

Liquefaction effects seen on the surface include fissures, sandblows, and sand 'volcanoes', while extensive areas of quicksand may be formed. Any structure built there 'sinks', even though its resisting members may remain intact. For instance, during the 16 June 1964 Niigata earthquake in Japan ($M = 7.5$) textbook examples of liquefaction of unconsolidated sand deposits were observed. Sandblows occurred and vast areas of quicksand developed, swallowing up cars, buildings, and other structures. Many buildings suffered settlement of more than 1 m, often accompanied by marked tilting sometimes exceeding 45° from the vertical. Figure 9.7 illustrates the results of standard penetration tests performed immediately prior to and just after the earthquake. It can be seen that some lithological levels at diverse depths have become compacted (increasing N_{SPT} values) while others have lost cohesion (with the opposite effect on the N_{SPT} values).

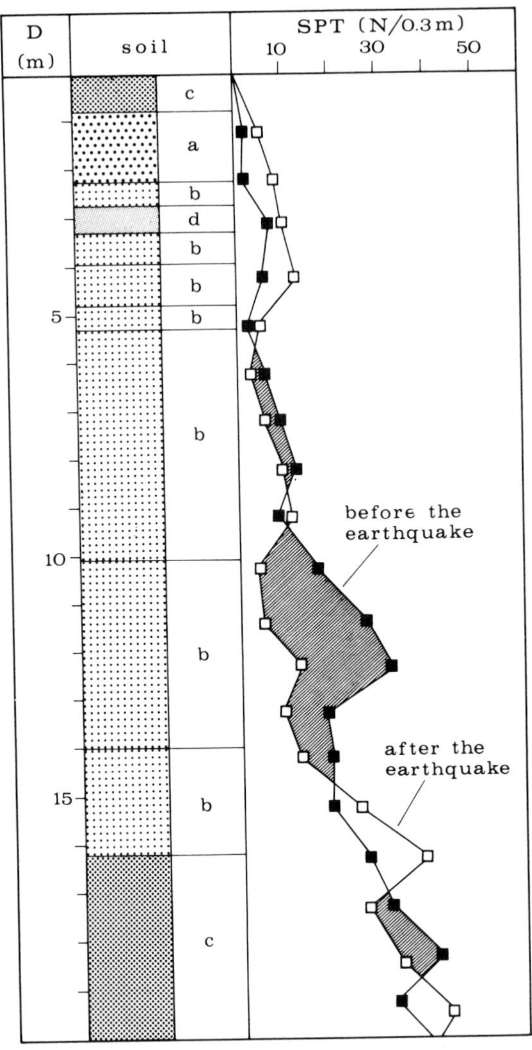

Figure 9.7 1964 Niigata (Japan) earthquake: comparison of standard penetration tests in sands before and after the earthquake. (*From Japan National Committee on Earthquake Engineering, 1965.*)

Similar effects have been observed on numerous other occasions, for example, during the 10 October 1980 El Asnam earthquake in Algeria ($M=7.3$), the disastrous 1964 Alaska earthquake, where mainly clayey deposits were involved, and the San Fernando earthquake on the 9 February 1971 ($M=6.6$) where an earth fill dam on alluvium failed. The effects of liquefaction are felt over vast areas extending beyond the individual structures hit by the earthquake. They may also be felt at a considerable distance from the epicentre. For instance, following the $M=7.2$ earthquake in Rumania on 4 March 1977 with its epicentre in the Vrancei Mountains, liquefaction phenomena were found in the alluvial area of the Danube as far as 250 km from the epicentre.

Figure 9.8 1980 Southern Italy earthquake: saturated landslide materials (talus mixed with patches of clayey, limey, sandy flysch and varicoloured clays, whose original position is uncertain, and pyroclastic blocks) involved in the liquefaction phenomenon at Senerchia (Avellino)

Liquefaction is often responsible for instability in the form of flows. When liquefaction occurs in slopes with an initial angle of even 20° the whole mass tends to flow until a very low angle of repose is attained ($\leqslant 4°$). One of the largest flows ever observed happened in the Kansu province of China during the $M = 8.5$ earthquake of 16 December 1920; the phenomenon affected a good 260 km^2 area of löess.

Liquefaction mainly affects certain classes of silty soil. However, mass movements in clayey soils are often attributable to liquefaction of underlying lenses and beds of sand. This was found frequently in the case of the 1980 earthquake that hit Campania and Basilicata in Italy. One example is the complex slide in the area immediately downstream of Senerchia (Avellino) which is built partly or entirely on scree, talus cones and alluvium with groundwater circulation at shallow depth. These materials generally overlie clayey flysch formations, frequently saturated (Figure 9.8). Here, as at Buoninventre (Caposele, Avellino), liquefaction led to the reactivation of enormous earthflows (28 Mm^3 at Senerchia and 25 Mm^3 at Buoninventre) that had lain quiescent for dozens of years. At Lioni (Avellino) liquefaction was again the main cause of the collapse of many buildings and of the cracks that appeared in others.

Liquefaction was observed even at a distance of more than 70 km from the epicentre (with peak accelerations of only a few hundredth *g*) accompanied by the outburst of jets of water and sandblows. As local conditions are not particularly favourable for liquefaction, the origin of such a phenomenon must certainly be ascribed to the length of the earthquake (about 80 seconds) which caused a great number of cyclic loads (not less than twenty) albeit of low acceleration (Da Roit *et al.*, 1981). Also frequent even in areas far from the epicentre was the appearance of deep fissures in the ground when the earthquake occurred, followed several weeks later by multiple slumps or other more or less extensive types of mass movement, as rain found an easy passage into the subsurface where the water produced unusually high pore pressures.

As has been seen, soil liquefaction is one of the most insidious phenomena an earthquake may cause and it is extremely important to adopt countermeasures to reduce the hazard and thus protect lives and property. Some of the methods available today are:

1. Vibroflotation, which is a process for compacting sand by means of a very heavy vibratory device that is introduced into the ground together with a sufficient quantity of water to cause liquefaction of the sand which is then compacted. The last step is to add sand to fill the voids.
2. Subsurface drainage to remove water, since drained soils are not subject to liquefaction.
3. Excavation and removal of layers potentially susceptible to liquefaction.

There are no *technical* difficulties in applying any of these methods: the main obstacle remains their high cost. The most obvious suggestion, therefore, is to avoid building on soils that have a potential liquefaction hazard.

To conclude, while a large amount of literature has accumulated on liquefaction processes, pertinent regulations and relevant research on the detection and diffusion of the phenomenon still lag behind.

9.3.3 Influence of Earthquakes on Submarine Landslides

A large amount of evidence is now available on instability of submarine slopes in a great variety of offshore environments ranging from shallow-water near-shore areas to continental slopes and the ocean deeps. These forms of instability were defined first in 1908 by Heim (Prior and Coleman, 1984) as 'subsolifluction or subaqueous solifluction'.

The commonest type of submarine instability, however, is the slide involving large volumes of material, often far greater than in terrestrial slides. For instance, the Agulhas slide off the South African coast involved a volume of rock exceeding 20,000 km^3; it was 750 km long and 106 km wide. Submarine slides, moreover, take place on very flat slopes, even less than one degree. For example, Mississippi delta front slides occur on submarine slopes with an angle of only 0.5° (Prior and Coleman, 1984).

Submarine slides are often associated with crustal deformations and seismic shocks. The earthquake which hit south-central Alaska on 27 March 1964 (the Good Friday earthquake) is one of the most closely studied and completely documented events. With a magnitude of 8.4 it was one of the strongest earthquakes ever recorded, the energy liberated at the time of the mainshock being at least twice that of the 1906 San Francisco earthquake. Owing to the sparsely populated nature of the area involved the number of victims was small (114 dead and missing) considering the size of the event. However, its effects on the landscape were so diverse, so numerous, and so extensive that they attracted attention from students of all earth sciences.

Figure 9.9 1964 Alaska earthquake: view of the landslide at Turnagain Heights

In some parts the ground was raised by up to 10 m, while in others there was subsidence of as much as 2 m. Spectacular liquefaction of soils caused the collapse of a clay bank more than 20 m high on Turnagain Heights (Figure 9.9). Enormous mass movements affected a total area of more than 200,000 km^2, while the largest number of submarine slides ever recorded in any previous earthquake made their presence felt in many places, causing great damage, as in the harbours of Seward and Valdez for example. At Seward a strip of land including the harbourworks and other facilities slid into the sea in a huge submarine slide which, in turn, raised enormous seawaves that, together with the tsunami generated directly by the earthquake, hit the coast causing further serious damage. Before the earthquake the harbour-front at Seward had been extended by loose sand and gravel fill. This surcharge was probably one of the factors that led to the slide, together with the long duration of ground shaking (3 to 4 minutes) and perhaps liquefaction of the finer materials.

Though static conditions in the Seward area now appear stable, the presence of fractures along the coastline creates a notable risk of landslides in the event of further earth tremors, so the authorities have been advised not to rebuild there but to keep the area as a nature park.

At Valdez (Figure 9.10) in Alaska ground shaking was amplified by the soft sediments of the delta on which the town stood. Liquefaction phenomena were observed. A sudden submarine slide at the seaward end of the delta, during and immediately after the earthquake, carried away the harbour region and raised seawaves which swept into the town. Valdez has now been relocated a few kilometres away on a safer site (Mineral Creek) protected on the seaward side by a buttress of bedrock.

Figure 9.10 1964 Alaska earthquake: sketch of the submarine landslide that carried away the harbour region of the city of Valdez.

In conclusion, it should be noted that the real triggering mechanism of these mass movements and the magnitude–frequency relationships thereof are still not completely understood. One reason for this is that direct observation of the phenomenon has only recently become possible, thanks to the sophisticated equipment developed lately for scanning the seabed. This is a realm, therefore, that has still to be investigated by parallel development of offshore studies and geological interpretation of the results obtained.

9.4 STABILITY ANALYSIS UNDER SEISMIC LOADING CONDITIONS

9.4.1 Introduction

There are two basic lines of approach to the study of slope stability under the influence of seismic shock, namely the pseudostatic method and dynamic analysis. A brief exposition of the advantages and limitations of the two methods is given below.

9.4.2 Pseudostatic Method

'Limit equilibrium' systems of analysis are involved. The methodology simulates seismic action by adopting an equivalent static horizontal force applied in the centre of the potentially unstable earth mass (Figure 9.11(a)). The force is directed so as to be destabilizing. It is usually expressed as a percentage of the force of gravity, according to a seismic coefficient (k) the size of which is related to the seismicity of the zone, which is generally codified by national norms. It thus follows that:

$$S = kgW$$

where: S = pseudostatic earthquake force
k = seismic coefficient
W = weight of the potentially unstable mass
g = acceleration due to gravity

Using this approach, slope stability can be calculated by the usual limit equilibrium static methods (see Chapter 2), it merely being necessary to introduce the equivalent static force S among the other forces in play.

The usual operative approach consists of selecting a hypothetical failure surface within the slope, assumed to be in the 'plane strain' state. In this manner the earth mass whose stability is to be examined is established, after which it is divided into a given number of vertical slices. The equilibrium of the whole mass is thus calculated by summation of the equilibrium conditions of each slice (Figure 9.11(b)). The most unfavourable position of the potential failure surface (critical slip surface), with the lowest factor of safety, is found by trial and error. The factor of safety (F) is usually defined as:

$$F = \frac{\tau_f}{\tau_m} = \frac{\text{available shearing resistance}}{\text{mobilized shearing stress}}$$

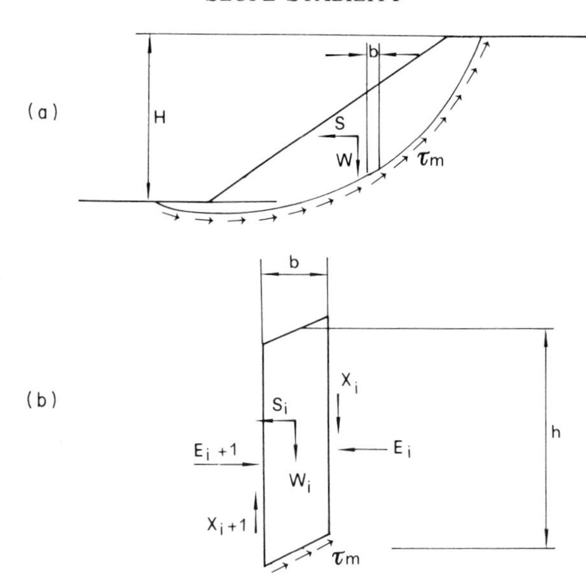

Figure 9.11 Slice method in pseudostatic analysis: schematization of slope (a) and forces acting on a given slice (b). S = pseudostatic earthquake force, w = weight of the potentially unstable mass, b = slice width and X and E are interslice forces (see Figure 2.20)

The method of slices has seen many improvements since it was first proposed by Fellenius (1927, 1936) many years ago. Other methods are those proposed by Bishop (1955), Morgenstern and Price (1965), Janbu (1969), and Sarma (1973); these are reviewed and compared in Chapter 2.

The differences among the various methods are not, however, great. They all have in common the equations of equilibrium and use more or less realistic and simplifying assumptions to render the problem determinate (Cotecchia *et al.*, 1979). Despite the doubtless practical advantages of the pseudostatic approach it must be considered simplistic for a variety of reasons:

1. It cannot take account of the complex nature of seismic phenomena that involve cyclic actions alternating in time and space, changing in intensity and sign from moment to moment and from point to point.
2. It does not adequately characterize the behaviour of soils, ascribing static resistance parameters to these, while in fact, the geotechnical characteristics of a site subjected to alternating seismic stresses vary in time (depending on the degree of strain) and may be reduced drastically.
3. All limit equilibrium methods provide only one statement of the degree of *overall* stability along a given slip surface, while offering no indications of the stress and strain states at various points within the medium concerned.

Despite these shortcomings, however, the pseudostatic method is by far the most widely used. The reasons for its popularity are basically practical, namely:

1. Its adoption does not involve significant computing costs.

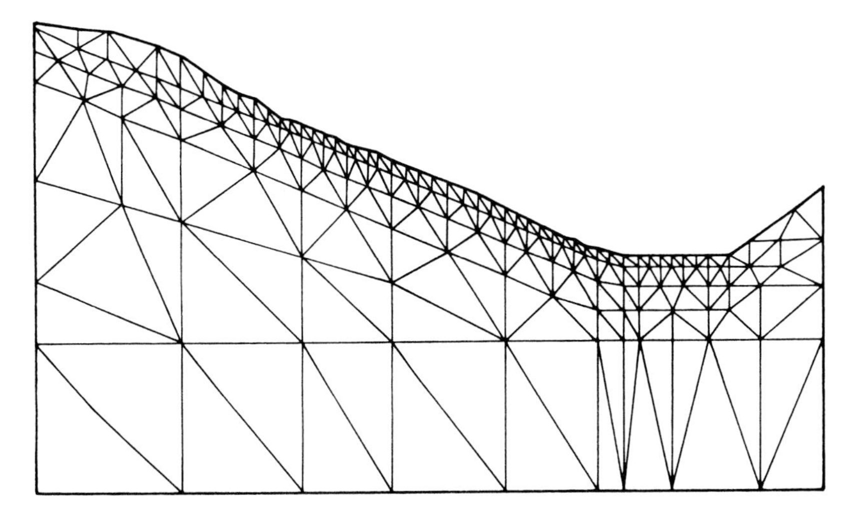

Figure 9.12 Discretization of a natural slope with a mesh of triangular finite elements

2. Much has been written about it, and more or less sophisticated refinements have been suggested, so there are many possibilities for comparison.
3. For soil characterization it uses parameters readily obtainable from static laboratory tests that are widely known and easy to perform.

9.4.3 Dynamic Analysis

The limitations involved in the simplifying assumptions required in the pseudostatic method can be overcome by dynamic analysis generally performed by means of a mathematical model composed of finite elements (FEM analysis).

The main steps in this approach can be summarized as follows:

1. Discretization of the slope by means of a finite element mesh (generally consisting of triangles or quadrilaterals) (Figure 9.12).
2. Dynamic characterization of the materials, in which the parameters generally used to describe the behaviour of a soil subject to seismic action are the shear modulus (G) and the damping factor (D). The shear modulus at small strains (G_{max}) can be derived by simple *in situ* geophysical tests. However, to obtain G for large strains (such as those that may be caused by an earthquake) cyclic laboratory tests are needed, correlating the shear modulus with strain. It is clear from Figure 9.13 that G decreases as the amplitude of the shear strain cycle increases. Lacking direct investigations of this kind, reference can be made both for cohesionless soils and clay, to the interpolated design curves obtained by various workers from laboratory tests on samples representative of a wide range of soils (Seed and Idriss, 1970). In the dynamic approach the damping is hysteretic, reflecting energy dissipation resulting from each loading cycle to which the ground is subject and proportional to the area enclosed by the cycle concerned (Figure 9.13). Here, too, in the absence of direct data, reference may be made to standard laboratory curves, by correlating the damping factor D with strain.

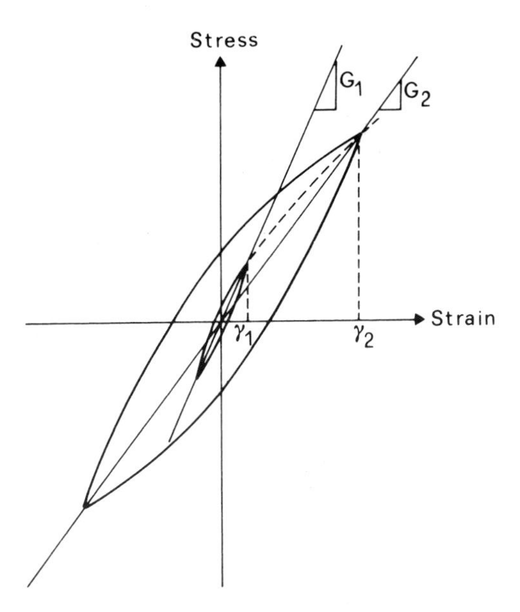

Figure 9.13 Stress–strain relations at different amplitudes of cyclic strain

3. Choice of design accelerogram: in dynamic analysis, seismic action is introduced in accelerogram form, supplied pointwise as program input and applied at the base of the slope. The accelerogram can be a recording of a real event or a 'design earthquake' constructed artificially taking account of geological and seismological factors. Hence all available earthquake recordings for the area concerned or for areas with similar geological features are analysed. The accelerogram is selected whose 'form' (namely the frequency content, which can be revealed by plotting the relevant 'response spectrum'), is such as to heighten local amplification and whose intensity is in keeping with the seismic hazard.
4. FEM analytical procedure, in which the fundamental equation of motion is:

$$M\ddot{u} + C\dot{u} + Ku = R(t)$$

where: M = system mass matrix
C = damping matrix
K = rigidity matrix
$u, \dot{u}, \ddot{u}$ = displacement, velocity, and acceleration vectors
$R(t)$ = loading vector simulating the earthquake (design accelerogram).

The equation is usually solved by direct integration using the step-by-step method. The values of u, $\dot{u}$, $\ddot{u}$, as well as the state of stress and strain, are derived for each of the time intervals (Dt) into which the input accelerogram is divided, at the points of each element. Then through the constitutive relations of the various materials (point 2), the values of shear modulus G and damping factor D are established and linear dynamic analysis is again performed using these new values. The process is iterative to the point of desired convergence.

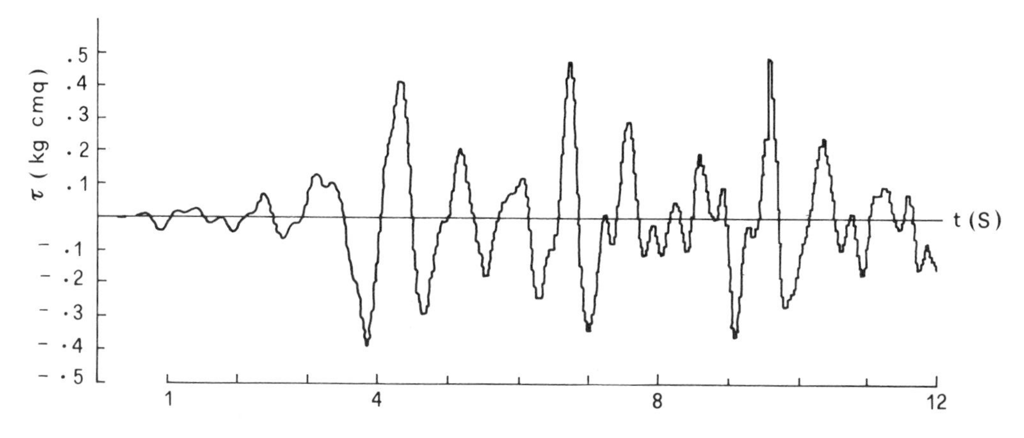

Figure 9.14 τ_{xy} stress history at a point in the FEM mesh

5. Analysis of results generally involves laboratory checks on various representative samples of loose soil. The samples are subjected to cyclic tests where the amount of stress and the number of cycles depends on the stress history derived from the dynamic analysis (Lee and Chan, 1972; Seed *et al.*, 1975). In a similar manner information can be obtained on local and overall liquefaction hazard, which is particularly to be feared in saturated, granular materials.

The advantages of dynamic analysis are evident from the foregoing. The introduction of dynamic stress through a design accelerogram permits all aspects of complex earthquake phenomena to be taken into account. FEM analysis—generally involving iterative procedures—permits use to be made of time-variable strength parameters so as to keep in line with the successive stress and strain states that occur due to the various shocks.

Another particular advantage of the FEM method, is that it can embrace all the local aspects of the phenomenon, thanks to discretization that enables any portion of the slope to be dealt with as desired from the point of view of geometric and geotechnical parameters. A wide range of information can be gathered including the temporal trend of displacements, velocities, and accelerations; moment by moment variations in stress and strain histories (Figure 9.14); variation of material characteristics; local response as regards damping or amplification of design accelerogram (Figure 9.15).

In this manner it is possible to ascertain zones of plasticization and the build-up of high stress gradients or—in the case of saturated granular materials—zones with a liquefaction hazard. By way of example, Figure 9.16 illustrates the lines of equal peak acceleration and equal maximum strain obtained during a dynamic analysis on Conza della Campania Dam (Avellino, Italy). The potentially hazard-prone zones revealed by dynamic analysis are also indicated.

There is certainly a number of practical and theoretical difficulties involved in this type of analysis, the most obvious being the large amount of computer time needed. Despite the availability of ever more powerful computers, this aspect often discourages adoption of the method. Another difficulty concerns characterization of the materials by dynamic

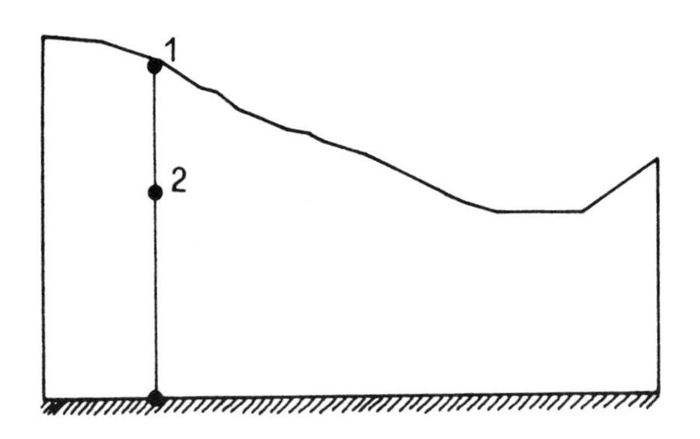

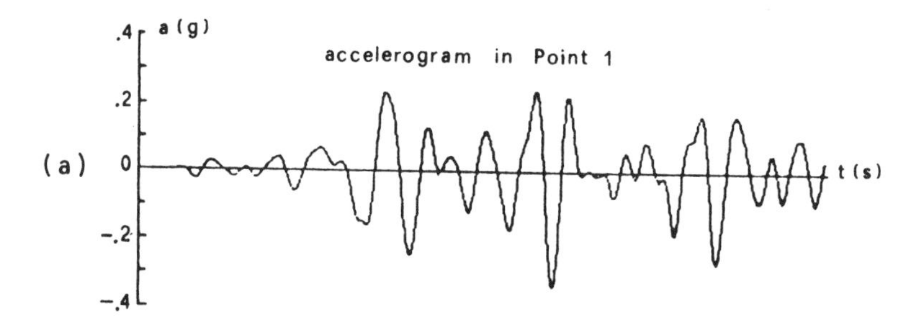

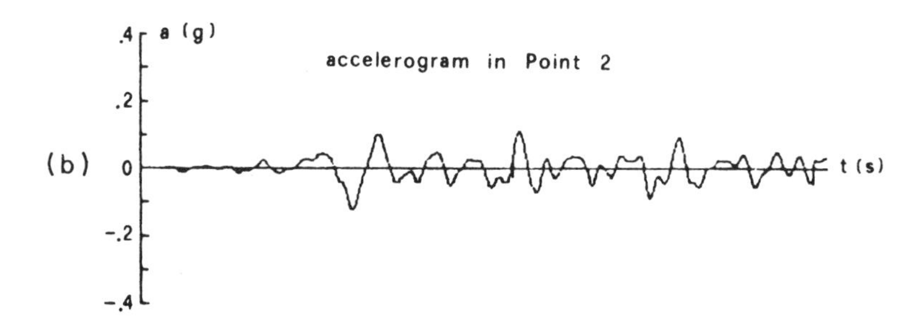

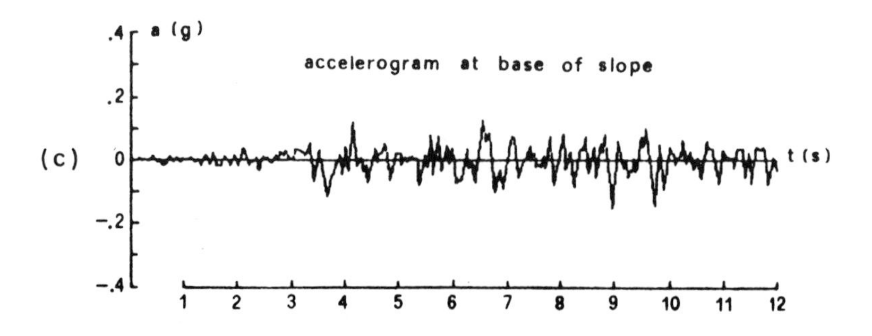

Figure 9.15 Dynamic analysis of a natural slope; accelerograms at various points on the same vertical line. (a) point 1; (b) point 2; (c) rigid base. t = time, a = acceleration in units of g

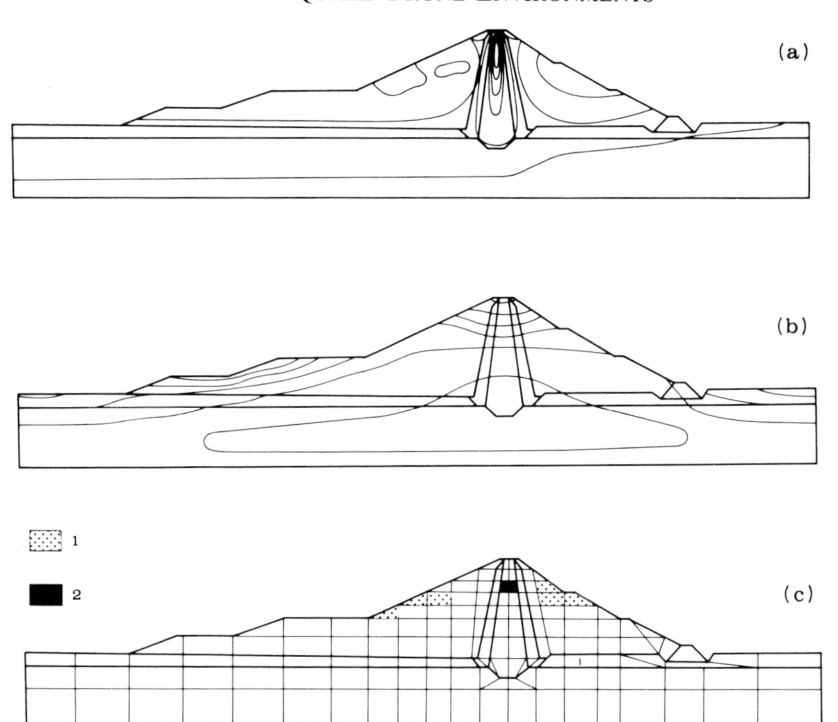

Figure 9.16 Results of dynamic analysis of an earth dam. (a) Lines of equal peak acceleration. (b) Lines of equal maximum strain. (c) Zones with potential liquefaction (1) and excessive strain hazard (2)

parameters. This requires the use of fairly sophisticated laboratory techniques and, despite the considerable developments that have occurred in recent years (Baldi and Silver, 1983), accuracy of the results is still open to some doubt. Reasonable care must also be paid to selecting the most suitable 'design earthquake' and this necessitates geo-seismological investigation of the formations concerned.

It will be readily appreciated, therefore, that what with the laborious nature of the procedure and the uncertainty attaching to some relevant elements, dynamic analysis is still not widely adopted for ordinary purposes, especially where the network of seisomographs is still somewhat thin on the ground.

At the present time dynamic analysis can be considered as a research tool essential for understanding the complex phenomena associated with earthquakes. It also provides pointers to the reliability of the simplifying assumptions adopted in common practice. This being the case, research should aim at clarifying the influence of the main variables involved (geometry, dynamic characteristics of the soil, and design earthquake).

A parametric study is required to ascertain the simplest techniques for realistic representation of seismic phenomena. In this regard the 'variable coefficient pseudostatic method' (Seed and Martin, 1966) provides an example of a possible combination of the two lines of approach. Though this solution still involves interpretation of the seismic coefficient as the only parameter capable of representing the maximum force of inertia induced in the ground by the earthquake, a more refined line is taken, the value being

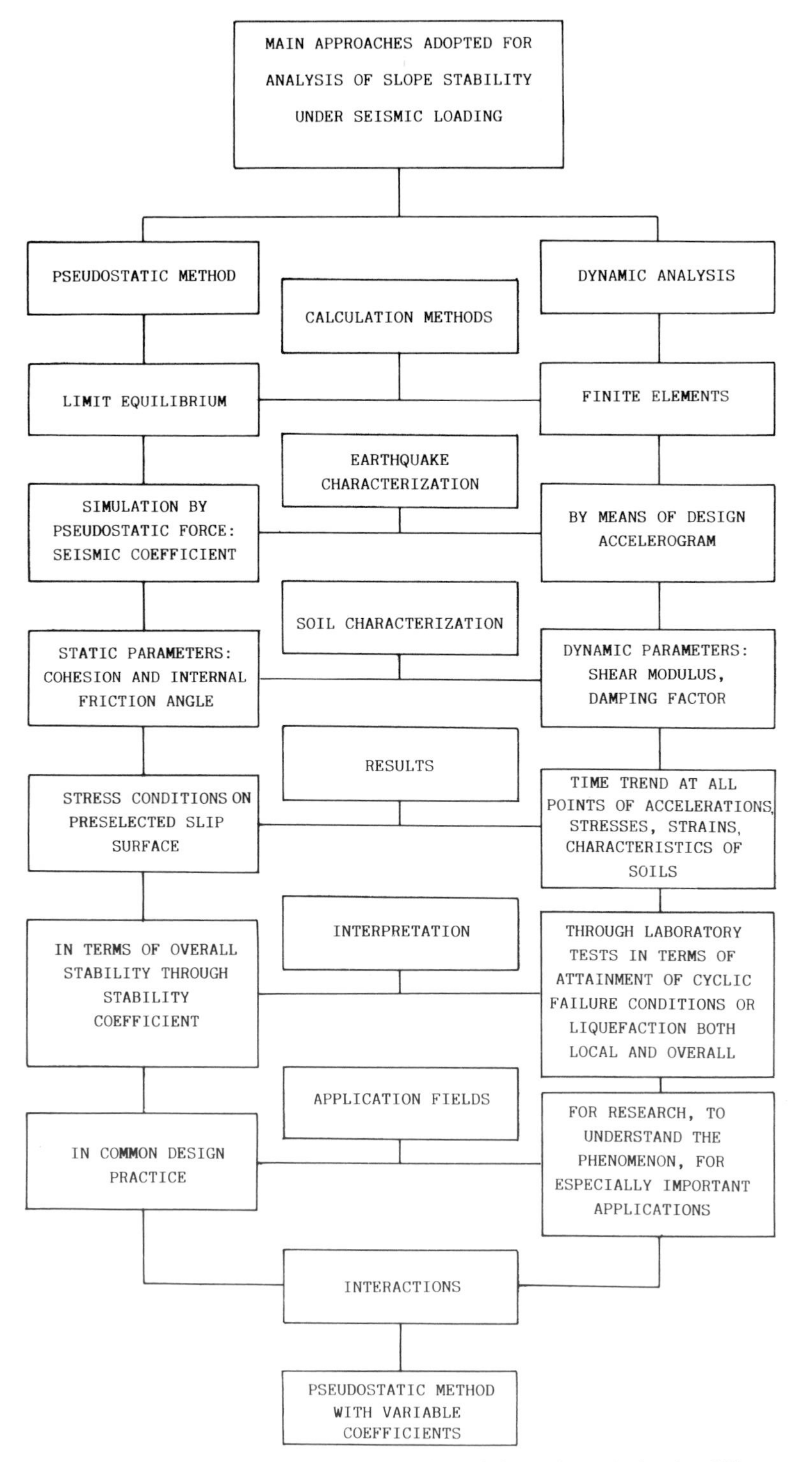

Figure 9.17 Block diagram illustrating pseudostatic and dynamic methods of stability analysis

varied in relation to the distribution of the maximum horizontal accelerations suggested by dynamic analysis. While still suffering from the conceptual limitations of the pseudostatic method, this solution represents a considerable improvement on it. The approach also shows how research can assist the correct utilization of rapid methods of analysis generally adopted in current practice.

The block diagram in Figure 9.17 summarizes the main characteristics of the two lines of approach.

9.5 PAST AND PRESENT PROCESSES

Numerous examples of landslides produced by earthquakes are to be found in the literature. Mention has already been made of some representative cases. Following the 1971 San Fernando earthquake about 1,000 mass movements with fronts ranging from 15 to 300 m were found. More than 250 landslides were initiated by the $M = 6.4$ Friuli (Italy) earthquake of 6 May 1976, affecting areas with active faults already hit in the past by strong earthquakes. The tremors reactivated movements on slopes formed by detritus and old landslides; they also caused numerous mass movements even on slopes formed of sound rock (Figure 9.18). Numbers of failures triggered by single earthquakes may therefore be large, but single failures can also be catastrophic. One of the world's most disastrous

Figure 9.18 Landslide which hit Braulins after the 1976 Friuli (Italy) earthquake

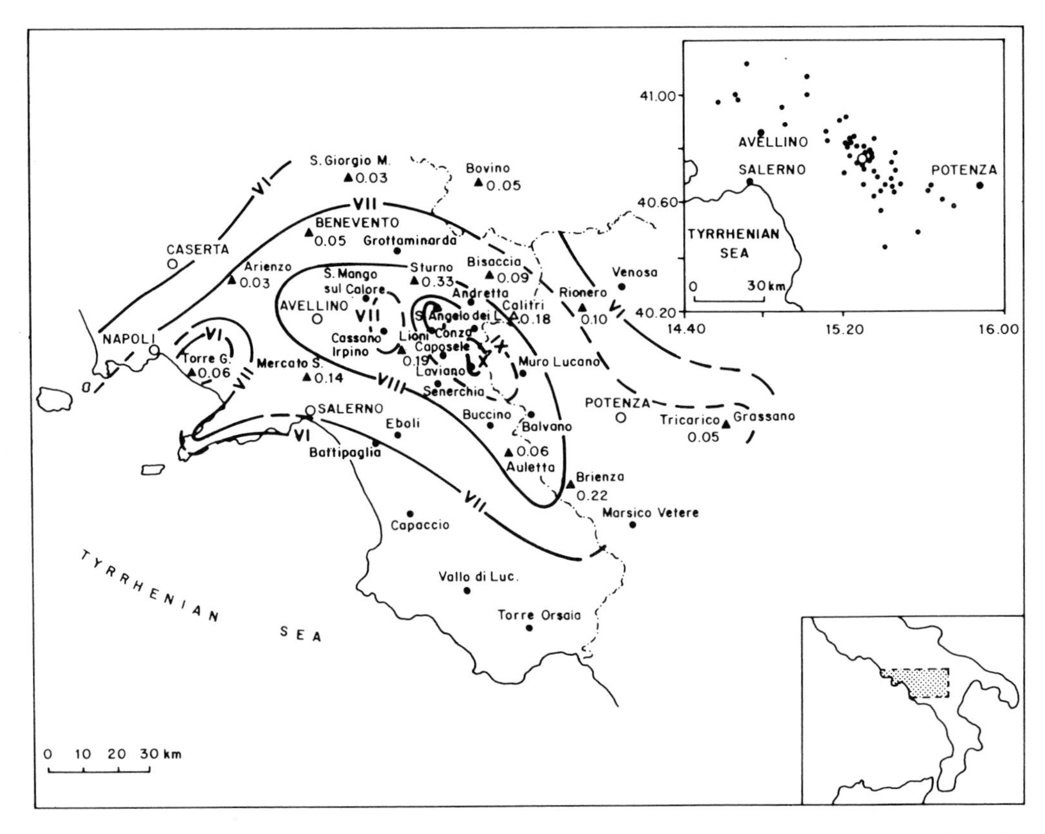

Figure 9.19 23 November 1980 earthquake in Campania and Basilicata (southern Italy). Isoseismal lines (MSK scale) with indications of accelerations recorded at various monitoring stations. The epicentral map of the principal earthquake (large circle) and the main aftershocks (small circles) up to 30 April 1981 is shown in the top right-hand corner

earthquakes ($M = 7.9$) happened in Peru on 31 May 1970. This caused a giant avalanche of rock and ice some 130 km from the epicentre which was on the bed of the Pacific Ocean some 25 km offshore. The avalanche tore down the slopes of Mount Huascarán at over 200 km/h becoming bigger and bigger, destroying the towns of Yungay and Ranrahirka and claiming over 25,000 victims.

The earthquake that hit southern Italy on 23 November 1980 (Figure 9.19) provided a great quantity of information thanks to which it has been possible to evolve models of soil behaviour that are of general validity, and to ascertain equally meaningful relationships between earthquakes and geomorphological conditions (V. Cotecchia, 1986). The geomorphological, structural, drainage, hydrogeological, and climatic characteristics of most of the areas affected by the ground tremors were such that there existed the potential danger of widespread, deep-seated mass movements being initiated. Many towns and villages were thus threatened by destruction, with the ensuing collapse of the economy in the districts concerned.

Most of the geological formations in this region are still affected by intensive neotectonic disturbance. The overall outcome was the remobilization of huge mass movements that took the form of falls, translational failures, and flows. The latter certainly produced the

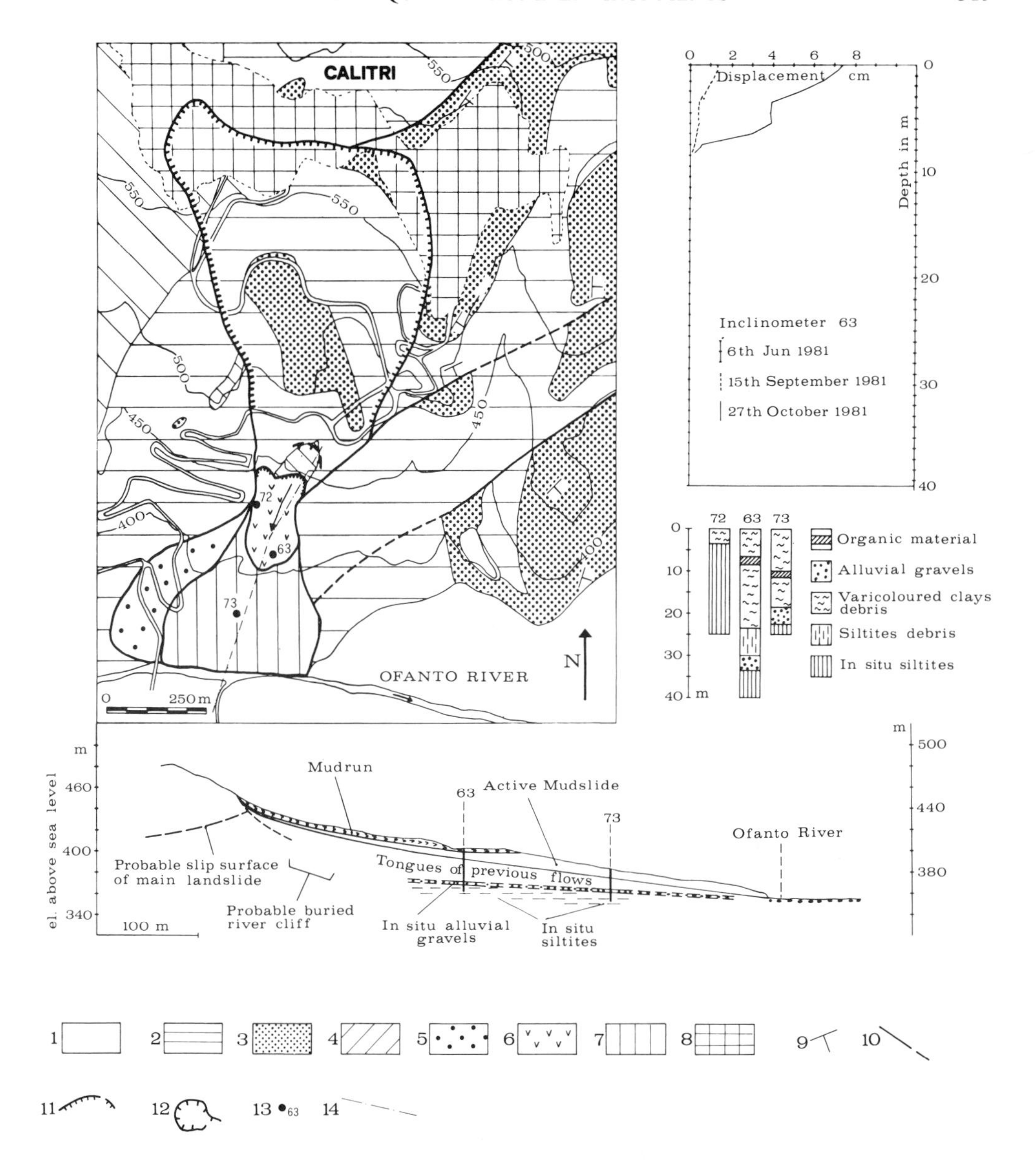

Figure 9.20 1980 Southern Italy earthquake: geological map of the Calitri (Avellino) mudslide. (1) Present, recent, and terraced, fluvial deposits (Quaternary); (2) Grey-blue siltites (Middle–lower Pliocene); (3) Sands in lenses interbedded in siltites; (4) Olistostromes of varicoloured clays in siltites; (5) Old mudslide body not removed by earthquake; (6) Mudrun; (7) Mudslide body removed by earthquake; (8) Urbanized area; (9) Attitude of beds; (10) Faults; (11) Main Calitri landslide boundaries; (12) Mudrun scar; (13) Boreholes; (14) Section line

most spectacular geomorphological effects. In the case of landslides following an earthquake it is often difficult to establish how much of the damage is attributable to the direct action of ground shaking, how much to landslides, and how much to a combination of the two. Where the 1980 earthquake is concerned, however, most of the damage was certainly caused

a)

b)

Figure 9.21 1980 Southern Italy earthquake. (a) General view of landslide which hit Calitri (Avellino). (b) Mudslide materials that reached the left bank of the Ofanto River

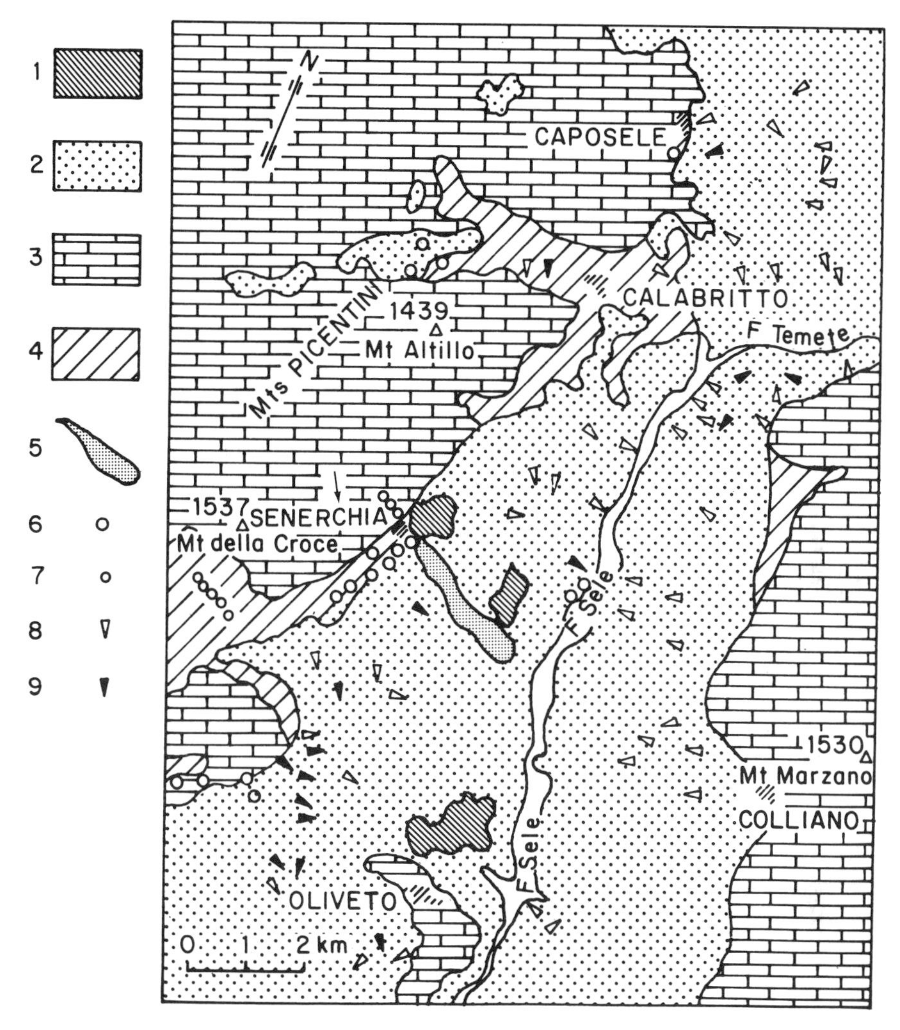

Figure 9.22 1980 Southern Italy earthquake: lithological map and slope instability phenomena in the upper valley of Sele River. (1) Terraced alluvial deposits; (2) Flysch and varicoloured clays; (3) Limestones; (4) Dolomites and dolomitic limestones; (5) Senerchia landslide; (6) Main springs; (7) Minor springs; (8) Main landslide masses existing before the 23 November 1980 earthquake; (9) Main landslide masses reactivated after the earthquake

by the mass movements, since these occurred in areas that had already been subject to landslides of one kind or another during previous earthquakes, always along the same slip surfaces. Damage also occurred in towns and villages several weeks after the earthquake, which had apparently caused no immediate harm. This was particularly the case with Calitri (Avellino) (Figure 9.20), which was found to be more or less intact the day after the earthquake, but was subsequently hit by rototranslational failure and by a mudslide which started from the foot of the main landslide and reached the left bank of the Ofanto River (Figure 9.21). The destruction affected the southern part of the town, built on Pliocene blue clays with patches of varicoloured clays which had been affected by mass movements

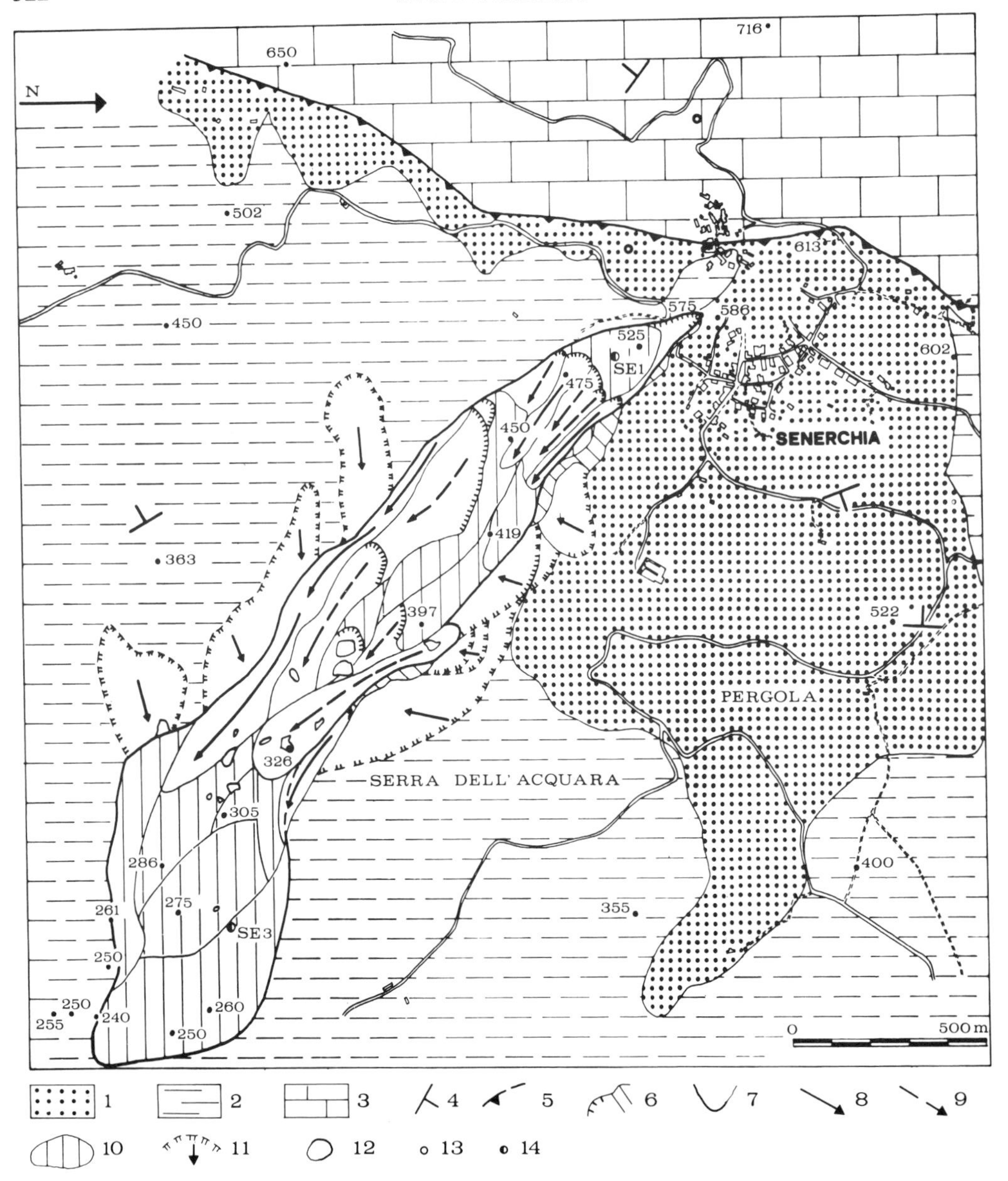

Figure 9.23 1980 Southern Italy earthquake: geological map of Senerchia (Avellino) earthflow. (1) Detrital mantles (Quaternary); (2) Varicoloured clays alternating with thin beds of calcilutites and arenites—Sicilide Complex (Eocene); (3) Limestones and dolomitic limestones of Campano-Lucana Platform (Jurassic); (4) Attitude of beds; (5) Normal fault; (6) Scarps; (7) Main shear boundaries of the landslide; (8) Mudruns; (9) Secondary slides evolving to mudruns; (10) Highly disturbed slipped bodies; (11) Lateral slips; (12) Ponds; (13) Springs; (14) Boreholes

in the past. From what can be seen in the field and from aerial photos, there are signs of recent changes in the valley that are certainly attributable to Quaternary mass movements mainly of the rototranslational type. The slip surfaces occur at some depth essentially in grey–blue siltites with olistostrome inclusions. Bearing witness to this are the numerous

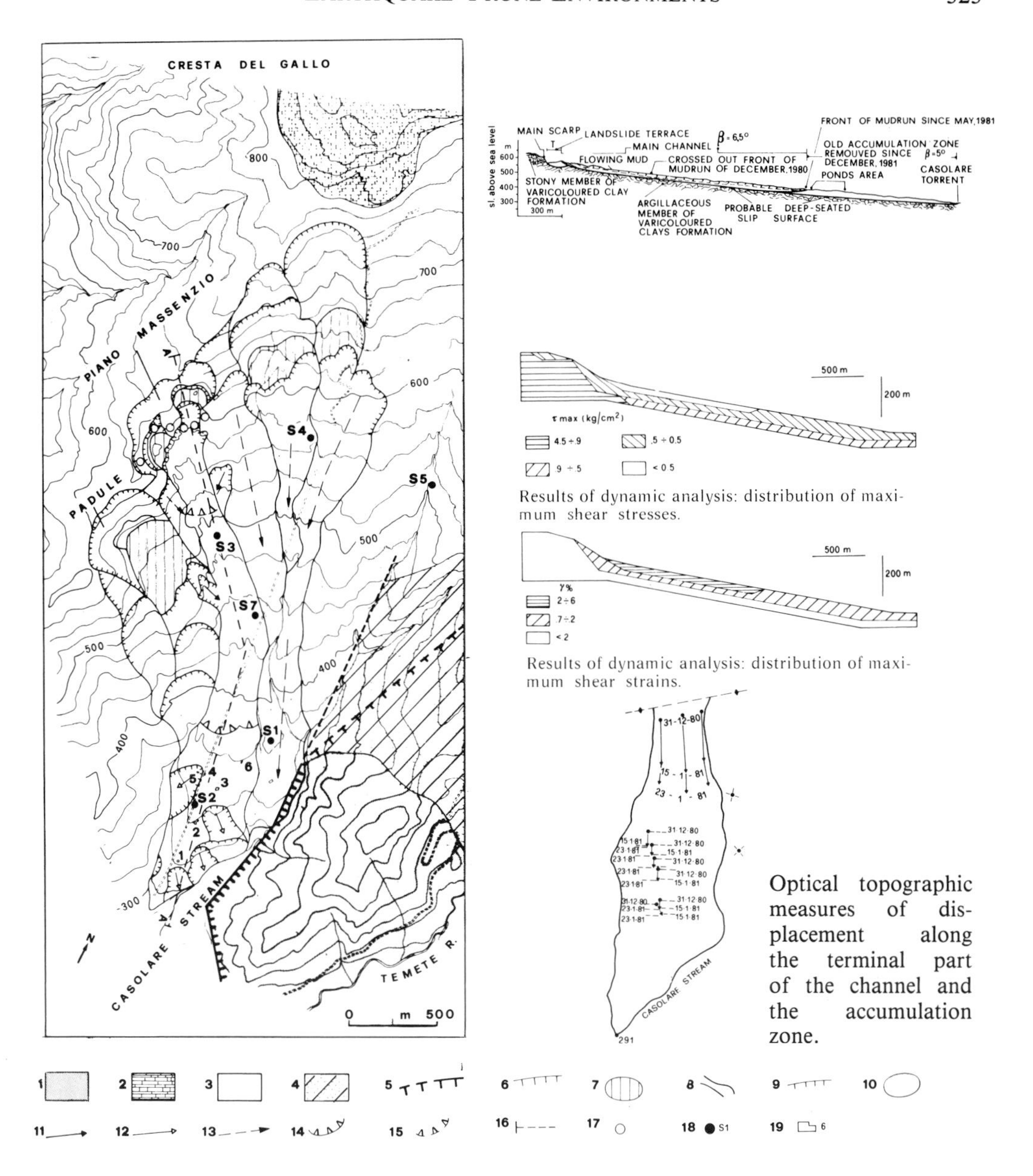

Figure 9.24 1980 Southern Italy earthquake: geological map of Caposele (Avellino) mudslide. (1) Conglomerates and sandstones (Irpinian Units); (2) Calcareous breccias and calcarenites (incertae sedis); (3) Marly limestones, marls, subordinate arenaceous and calcarenitic beds (Variegated Clays Unit — Upper Member); (4) Scaly variegated clays (Variegated Clays Unit — Lower Member); (5) Faults; (6) Main scarps of the landslides; (7) Landslides terraces; (8) Main shear boundaries; (9) Present crown of secondary scarps; (10) Ponds; (11) Lateral slip; (12) Secondary landslides activated along the accumulation zones; (13) Old landslides; (14) Mudrun front on May 1981; (15) Crossed out front of mudrun on December 1980; (16) Section; (17) Springs; (18) H.M.D. Boreholes; (19) Damaged houses

Figure 9.25 1980 Southern Italy earthquake: Buoninventre (Caposele, Avellino) mudslide. (a) Trend of slickensides in end part of mudslide channel. (b) Ponds formed in the accumulation zone.

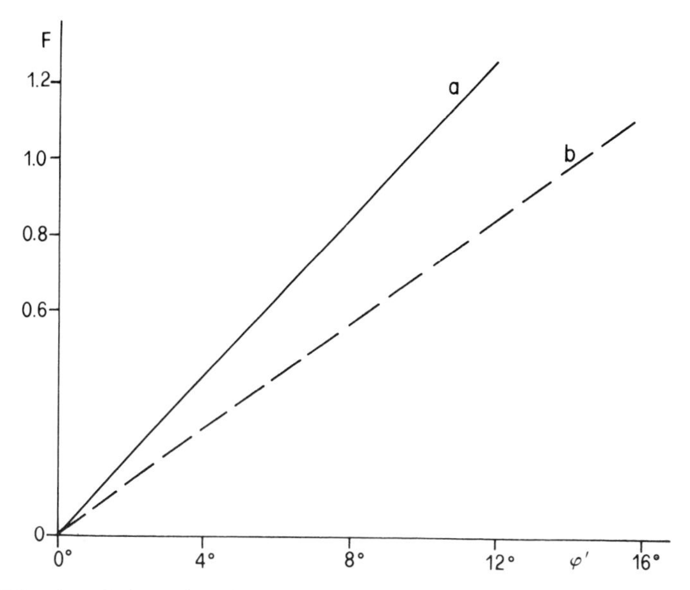

Figure 9.26 1980 Southern Italy earthquake: safety factors F versus φ'-values for the accumulation zone of Caposele (Avellino) mudslide (a) before earthquake ($r_u = 0.47$), and (b) after remobilization of the channel ($r_u = 0.55$)

terraced blocks that occur at various levels, denoting repeated mass movements some of which are quite recent. The mass that was remobilized by the 1980 earthquake measured about 23 Mm^3 and the assumed depth is 100 m, the width varying between 400 and 600 m and the length around 850 m (Del Prete and Trisorio Liuzzi, 1981).

The collapse that affected Senerchia (Avellino) on 23 November 1980 was also the result of the reactivation of an old slide. The geological setting of Senerchia area is given in Figure 9.22 (Agnesi *et al.*, 1982). Most of the landslides in the area originate at the tectonic contact between the carbonate massif and the varicoloured clays where the latter plug the large aquifer in the limestones and dolomites, and are subject to high pore pressures. The earthflow that starts on the southern edge of the village is the largest mass movement in the area (Figure 9.23). Its boundaries are clear from aerial photographs taken prior to the earthquake. The numerous farm buildings present on the landslide area before the earthquake testify to the long quiescent period which lasted at least 30 years.

The earthflow started uphill about four hours after the main shock with a slow movement of separate slabs on which there were buildings and abundant tree cover. Next day the disconnected slabs were surrounded by surficial secondary mudruns, triggered on the slip planes laid bare along the mass by the slippage of the individual slabs and along the main lateral shear boundaries. Movement of the accumulation zone started 3 days later, after a strong uplift at the foot of the channel (325 m a.s.l.). The deep-seated slip plane of the earthflow, plotted on the basis of recent borings and now being completed by inclinometer measurements, is roughly parallel to the hillslope at a maximum depth of around 40 m (Cotecchia and Del Prete, 1984).

In a different geological setting another huge mass movement occurred at Buoninventre near Caposele (Avellino) (Cotecchia *et al.*, 1986) (Figure 9.24). Observations, made in the period

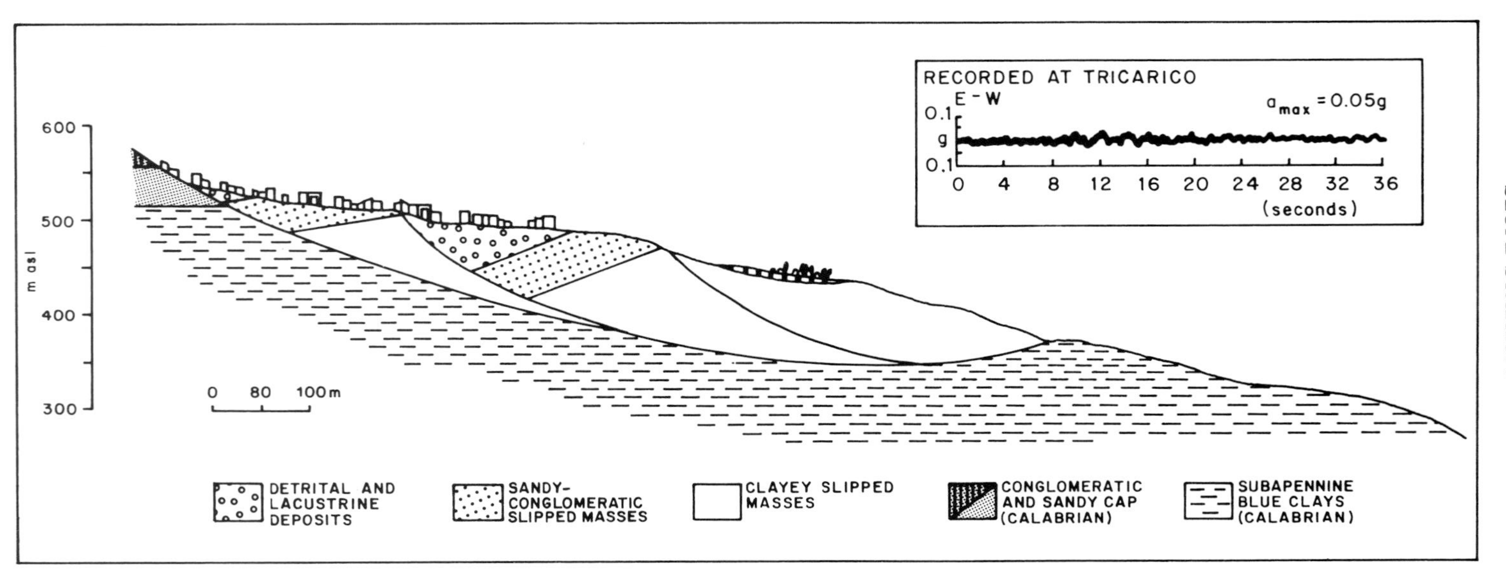

Figure 9.27 1980 Southern Italy earthquake: Geological section of Grassano (Matera) landslide. The slide which affected the whole series of Calabrian (probably pre-Wurmian) clays, sands and conglomerates is a typical example of failure and collapse that has occurred in the geomorphological evolution of the Basento River Valley. During the recent Quaternary, variations in river base levels connected with eustacy, neotectonic processes and earthquakes have all played a role in this evolution. A stability analysis showed that the slide could be reactivated by acceleration as low as 0.02 g

November 1980–May 1981 by repeated field investigation and two air-photo surveys, indicate that the main phases of the reactivation mechanism can be summarized as follows:

1. Rotational slip in the source area initiated a few hours after the main shock.
2. Formation of shear boundaries and slow translational sliding of highly disturbed masses along the channel (from 24 to 26 November).
3. Transformation of the material upstream of the channel into mud over 600 m, and complete remobilization of the channel body (from 26 November to 7 December).
4. Steady loading of the accumulation zone with the formation of small ponds, owing to the rapid increase in pore pressure and remobilization of secondary slides therein: overriding at the neck of the accumulation zone by channel materials, and an attenuated wave-like yield from upstream of the whole of the accumulation zone (from 7 December to 8 January 1981).
5. Transformation of all channel surface materials into mud and complete demolition of the buildings in the accumulation zone (January to February).
6. Gradual extinction of movement proceeding upstream (February to May).

A geomorphological feature of great interest is the presence on the hillside of other similar old mudslides with more or less the same slopes. These mudslides apparently remained quiescent probably because there was no reactivation of movement in source areas.

The above phenomena provide some indication of the slide reactivation mechanism. This can be assumed to have involved the sliding of the channel mass and progressive loading of the resistant toe, with maximum concentration of stresses behind the stable accumulation zone. Transformation of the channel material into mud occurred only in the most surficial levels as a result of remoulding caused by the high rate of movement. Two distinct movements thus occurred in the channel after formation of the mud: surficial mudflows, and a basal translational slide. Where the slope decreases at the neck of the accumulation zone, the channel mass overrode the accumulation mass, as can be seen from the trend of the slickensides dipping at about 10° upstream (Figure 9.25).

In assessing the stability of the accumulation zone, knowledge of the forces acting on the channel is of particular importance, because not only are these responsible for the movement of the upper part of the slope, they are also the main cause of the successive movement of the downstream part. Figure 9.26 gives the safety factor of the accumulation zone in two different situations:

a) before the earthquake, in a simplified slope model disregarding upstream forces, with a water level 6 m below ground level as deduced from the wells present in the area; and
b) after the earthquake, with the water level at the ground surface, and considering the passive earth-pressure after remobilization of the channel (Cotecchia and Del Prete, 1984).

During the earthquake clear signs of renewed mass movements were recorded in areas that were decidedly on the margins of the epicentre or even some distance away. However, it has already been stressed that instability on the surface and in subterranean cavities is possible in some geological environments even as the result of microseismic activity resulting from earthquakes thousands of kilometres away (Cotecchia, 1978).

For instance, at Grassano (Figure 9.27), 80 km in a straight line from the epicentral area, acceleration was 0.05 g. Stability analysis run on the slide affecting the town shows that this could be reactivated by accelerations as low as 0.02 g, a value which was attained and exceeded even at great distances from the epicentre of the 1980 earthquake. Thus, another slide was triggered by the earthquake at Paola (Cosenza) at a distance of about 200 km from the epicentral area.

It can be concluded from the few examples outlined that the earthquake of 23 November 1980, combined with a particularly active neotectonic situation and a fragile lithological composition, greatly increased the geological hazard in areas where seismic effects were superimposed on a geomorphology in rapid evolution. There are also still many areas apparently unharmed by the shocks that could well be damaged in the future by mass movements in slow evolution.

The study of relationships linking neotectonics and seismic hazard finds its main application in the field of physical planning, especially as regards sites for major structures or those that could cause public danger in case of rupture or collapse. One of the central problems facing the physical planner is evaluation of the degree of stability or potential instability of vast seismically-active areas, often with an already high landslide hazard.

REFERENCES

Agnesi, V., Carrara, A., Macaluso, T., Monteleone, S., Pipitone, G., and Sorriso-Valvo, M. (1982). 'Osservazioni preliminari sui fenomeni di instabilità dei versanti indotti dal sisma del 1980 nell'alta valle del Sele.' *Geol. Appl. e Idrogeol.*, **17**(1), 79–93.

Al-Layla, M. T. H. (1970). *Study of certain geotechnical properties of Beaumont clay.* Ph.D. Thesis, Texas A. & M. University, College Stn, Texas, USA.

Baldi, G., and Silver, L. M. (1983). 'La caratterizzazione dinamica dei terreni in laboratorio: prove standard e nuovi sviluppi,' *Atti del XV Conv. Naz. Geot.*, 1, Spoleto.

Berardi, R., Berenzi, A., and Capozza, F. (1981). 'Campania-Lucania earthquake on 23 November 1980; accelerometric recordings of the main quake and related processes.' *Annual Meeting of the CNR Geodynamic Project*, Udine, 1–103.

Bishop, A. W. (1955). 'The use of the slip circle in the stability analysis of slopes.' *Géotechnique*, **5**, 7–17.

Bishop, A. W., Green, G. F., Garga, V. K., Anderson, A., and Brown, J. D. (1971). 'A new ring shear apparatus and its application to the measurement of residual strength.' *Géotechnique*, **4**, 273–328.

Blondeau, F., and Josseaume, H. (1976). 'Mesure de la résistence au cisaillement résiduelle en laboratoire.' *Bulletin de Liaison des Laboratoires des Ponts et Chaussées*, **2**, 90–106.

Botea, E., Perlea, V., and Perlea, M. (1980). 'Liquefaction susceptibility of sand deposits in the Danube flood plain.' *Proc. 6th Danube-European Conf. Soil Mechanics and Foundation Engineering*, Varna.

Calabresi, G., Caruana, R., Pagliano, C., and Tonnetti, G. (1980). 'Una frana in terreni argillosi eterogenei di origine lacustre,' *Rivista Italiana di Geotecnica*, **14**(2), 86–100.

Cancelli, A. (1979). 'Sulla misura della resistenza residua dei terreni mediante prove di taglio su contatto suolo/roccia'. *Atti Conv. 'Movimenti Franosi e Dinamica dei Versanti*, Salice Terme (PV).

Chandler, R. J. (1966). 'The measurement of residual strength in triaxial compression.' *Géotechnique*, **16**(3), 181–186.

Chandler, R. J. (1973). 'Discussion: Residual strength and landslides in clay and clay shales,' *J. Geotech. Eng. Div., ASCE*, GT5, 560–1.

Chandler, R. J. (1976). 'The history and stability of two clay slopes in the Upper Gwash valley, Rutland.' *Phil. Trans Roy. Soc., London*, **A283**, 463–91.

Chattopadhyay, P. K. (1972). *Residual strength of some pure clay minerals*. Ph.D. Thesis, University of Alberta, Edmonton, Canada.
Chowdhury, R. N., and Bertoldi, C. (1977). 'Residual shear tests on soil from two natural slopes.' *Austral. Geomech. J.*, **G7**, 1–9.
Cotecchia, V. (1978). 'Systematic reconnaissance mapping and registration of slope movements.' *Bull. Int. Ass. Eng. Geol.*, **17**, 5–37.
Cotecchia, V. (1986). 'Ground displacement and slope instability triggered off by the earthquake of November 23, 1980 in Campania and Basilicata.' *Proc. Int. Symp. Engineering Geology Problems in Seismic Areas*, IAEG, Bari, *4* (in press).
Cotecchia, V., and Del Prete, M. (1984). 'The reactivation of large flows in the parts of Southern Italy affected by the earthquake of November 1980, with reference to the evolutive mechanism.' *Proc. Int. Symp. Landslides*, Toronto.
Cotecchia, V., Del Prete, M., Federico, A., and Trisorio Liuzzi, G. (1979). 'Sugli apporti odierni della geologia e della geotecnica nella problematica dei movimenti franosi.' *Annali della Facoltà di Ingegneria*, 4, Bari.
Cotecchia, V., and Federico, A. (1980). 'Sulla dipendenza della resistenza al taglio residua drenata di terreni coesivi dal livello di sforzo normale efficace.' *Atti XVI Conv. Naz. di Geotecnica*, Firenze, **2**, 317–324.
Cotecchia, V., Lenti, V., Salvemini, A., and Spilotro, G. (1986). 'Reactivation of the large Buoninventre slide by the Irpinia earthquake of 23 November 1980.' *Proc. Int. Symp. Engineering Geology Problems in Seismic Areas*, IAEG, Bari, *4* (in press).
Cotecchia, V., Travaglini, G., and Melidoro, G. (1969). 'I movimenti franosi e gli sconvolgimenti della rete idrografica prodotti in Calabria dal terremoto del 1783.' *Geol. Appl. e Idrogeol.*, **4**, 1–24.
Cotecchia, V., and Zezza, F. (1974). 'Geology and geotechnics of the sedimentary and volcanic formation of the Tungurahua and Cotopaxi Provinces (Ecuadorian Andes) referring to engineering works for water and power supply of the region.' *Proc. Int. Seminar 'Land Evaluation in Arid and Semi-arid Zones of Latin America'*, IILA, Rome, 531–623.
Da Roit, R., Fontanive, A., Lojelo, L., Muzzi, F., and Spat, G. (1981). 'Terremoto campano-lucano del 23 novembre 1980: evidenze di liquefazione di terreni non coesivi saturi,' *Att Conv. Ann. CNR-PF Geodinamica*, Udine, 1–15.
Del Prete, M., and Trisorio Liuzzi, G. (1981). 'Risultati dello studio preliminare della frana di Calitri (AV) mobilitata dal terremoto del 23 novembre 1980.' *Geol. Appl. e Idrogeol.*, **16**, 153–156.
Esu, F., and Calabresi, G. (1969). 'Slope stability in an overconsolidated clay,' *Proc. 7th Int. Conf. SMFE*, Mexico City, **2**, 555–63.
Fellenius, W. (1927). *Erdstatische Berechnungen mit Reibung und Kohesion*. Ernst, Berlin.
Fellenius, W. (1936). 'Calculation of stability of earth dams.' *Proc. 2nd Congr. on Large Dams*, Washington, **4**, 445–63.
Garga, V. K. (1970). *Residual shear strength under large strains and the effect of sample size on the consolidation of fissured clay*. Ph.D. Thesis, University of London.
Hutchinson, J. N., Bromhead, E. N., and Lupini, J. F. (1980). 'Additional observations on the Folkestone Warren landslides.' *Q. J. Eng. Geol.*, **13**, 1–31.
Hutchinson, J. N., and Gostelow, T. P. (1976). 'The development of an abandoned cliff in London Clay at Hadleigh, Essex.' *Phil. Trans. Roy. Soc., London*, **A283**, 557–604.
Hutchinson, J. N., Somerville, S. H., and Petley, D. J. (1973). 'A landslide in periglacially disturbed Etruria marl at Bury Hill, Staffordshire.' *Q. J. Eng. Geol.*, **6**, 377–404.
Insley, A. E., Chatterji, P. K., and Smith, L. B. (1977). 'Use of residual strength for stability analyses of embankment foundations containing pre-existing failure surfaces.' *Can. Geotech. J.*, **14**(3), 408–28.
Ivanov, P. L., Sinitsin, A. P., and Musaelyan, A. A. (1981). 'Characteristics of soils at cyclic and shock loads.' *Proc. X ICSMFE*, Stockholm, **3**, 239–42.
Janbu, N. (1969). 'An advanced method of slope stability analysis.' *Lecture Suppl. on Soil Stability*, Univ. of California, Berkeley.
Japan National Committee on Earthquake Engineering (1965). 'Niigata earthquake of 1964.' *Proc. Third World Conf. Earthq. Eng.*, Auckland and Wellington, N. Zealand, S78–S109.

Keefer, D. K. (1984). 'Landslides caused by earthquakes,' *Geol. Soc. Am. Bull.*, **95**, 406–21.
Lee, K. L., and Chan, K. (1972). 'Number of equivalent significant cycles in strong motion earthquakes.' *Proc. Int. Conf. Microzonation*, Washington, 2.
Marinatos, S. N. (1960). 'Helice submerged town of classical Greece'. *Archaeology*, **13**(3).
Maugeri, M. (1967). 'L'apparecchio di taglio anulare nella determinazione della resistenza residua di terreni sottoposti a ridotte tensioni normali.' *Rivista Italiana di Geotecnica*, **10**, 114–24.
Morgenstern, N. R., and Price, V. E. (1965). 'The analysis of the stability of general slip surfaces.' *Géotechnique*, **15**, 79–93.
Newmark, N. W., and Rosenblueth, E. (1971). *Fundamentals of earthquake engineering*. Prentice Hall, Englewood Cliffs, NJ.
Petley, D. J. (1969). *Unpublished data from Imperial College of Science and Technology, London.*
Petley, D. J. (1980). *Unpublished data.*
Prior, D. B., and Coleman, J. M. (1984). 'Submarine slope instability.' In *Slope Instability*. Brundsen, D., and Prior, D. B. (eds.), Wiley. pp. 419–45.
Radbruch-Hall, D. H., and Varnes, D. J. (1976). 'Landslides. Cause and effect,' *Bull. Int. Ass. Eng. Geol.*, **14**, 205–16.
Sarconi, M. (1784). *Osservazioni fatte nelle Calabrie e nella frontiera di Valdemone sui fenomeni del terremoto del 1783 e sulla geografia fisica di quelle regioni.* R. Acc. Sc. e Belle Lett., Napoli.
Sarma, S. K. (1973). 'Stability analysis of embankments and slopes.' *Géotechnique*, **23**, 423–33.
Seed, H. B. (1968). 'Landslides during earthquake due to soil liquefaction.' *Proc. ASCE*, **94**, SM5, 1053–1122.
Seed, H. B., and Idriss, J. M. (1970). 'Soil moduli and damping factors for dynamic response analyses.' *Earthquake Engineer Res. Ctr. Rep. EERC 70–10*, Univ. of California, Berkeley.
Seed, H. B., Idriss, J. M., Makdisi, F., and Banerjee, N. (1975). 'Representation of irregular stress time histories by equivalent uniform stress series in liquefaction analysis.' *Rep. EERC 75–29*, Univ. of California, Berkeley.
Seed, H. B., and Martin, G. R. (1966). 'The seismic coefficient in earth dam design.' *J. Soil Mech. Fnd. Div. ASCE*, **92**, SM3, 59–83.
Sherif, M. A., and Ishibashi, I. (1978). 'Soil dynamics considerations for microzonation.' *Proc. 2nd Int. Conf. Microzonation*, San Francisco, **1**, 81–110.
Skempton, A. W. (1964). 'Long-term stability of clay slopes (Fourth Rankine Lecture)'. *Géotechnique*, **14**(2), 77–102.
Skempton, A. W. (1985). 'Residual strength of clays in landslides, foliated strata and the laboratory.' *Géotechnique*, **35**(1), 3–18.
Skempton, A. W., and Petley, D. J. (1967). 'The strength along structural discontinuities in stiff clays.' *Proc. Geotech. Conf.*, Oslo, **2**, 29–46.
Vivenzio, G. (1788). *Istoria de' terremoti avvenuti nella Provincia della Calabria Ulteriore, e nella città di Messina nell'anno 1783. E di quanto nella Calabria fu fatto per lo suo risorgimento fino al 1787. Preceduta da una teoria e istoria generale dei terremoti.* Stamperia Reale, Napoli.

Slope Stability
Edited by M. G. Anderson and K. S. Richards

Chapter 10

Basal Erosion and Mass Movement

K. S. RICHARDS
Department of Geography
University of Cambridge, Cambridge CB2 3EN

and

N. R. LORRIMAN
Department of Geography,
The University, Hull

10.1 INTRODUCTION

Terzaghi (1950) classified the major controls of slope instability and mass movement as passive or inherent, and active or initiating. The former included lithology, stratigraphy and structure, and topography, while the latter included loss of shear resistance because of weathering, and increased pore-water pressure. Such a distinction assumes that the passive factors change only over time periods several orders of magnitude longer than the triggering active factors. Slopes being actively undercut by river or marine erosion can, however, experience steepening slope angles or increased vertical extent at rates in excess of the rates of weathering or even pore pressure change, so that inherently unstable topographies can be actively generated. In this case clearly the critical *active* factor causing mass movement depends on *relative* rates of basal erosion and weathering.

Basal erosion is particularly significant as an active, triggering mechanism of mass movement on slopes in soft rocks and sediments. Cliffs are often created by marine, river, or glacial erosion in hard cohesive rocks, in which the *ultimate* limiting cliff heights in intact rock may be defined by

$$H_c = q_u/\gamma \tag{10.1}$$

where q_u is unconfined compressive strength and γ is unit weight (Terzaghi, 1962). However, observed maximum cliff heights are considerably less than this limiting value

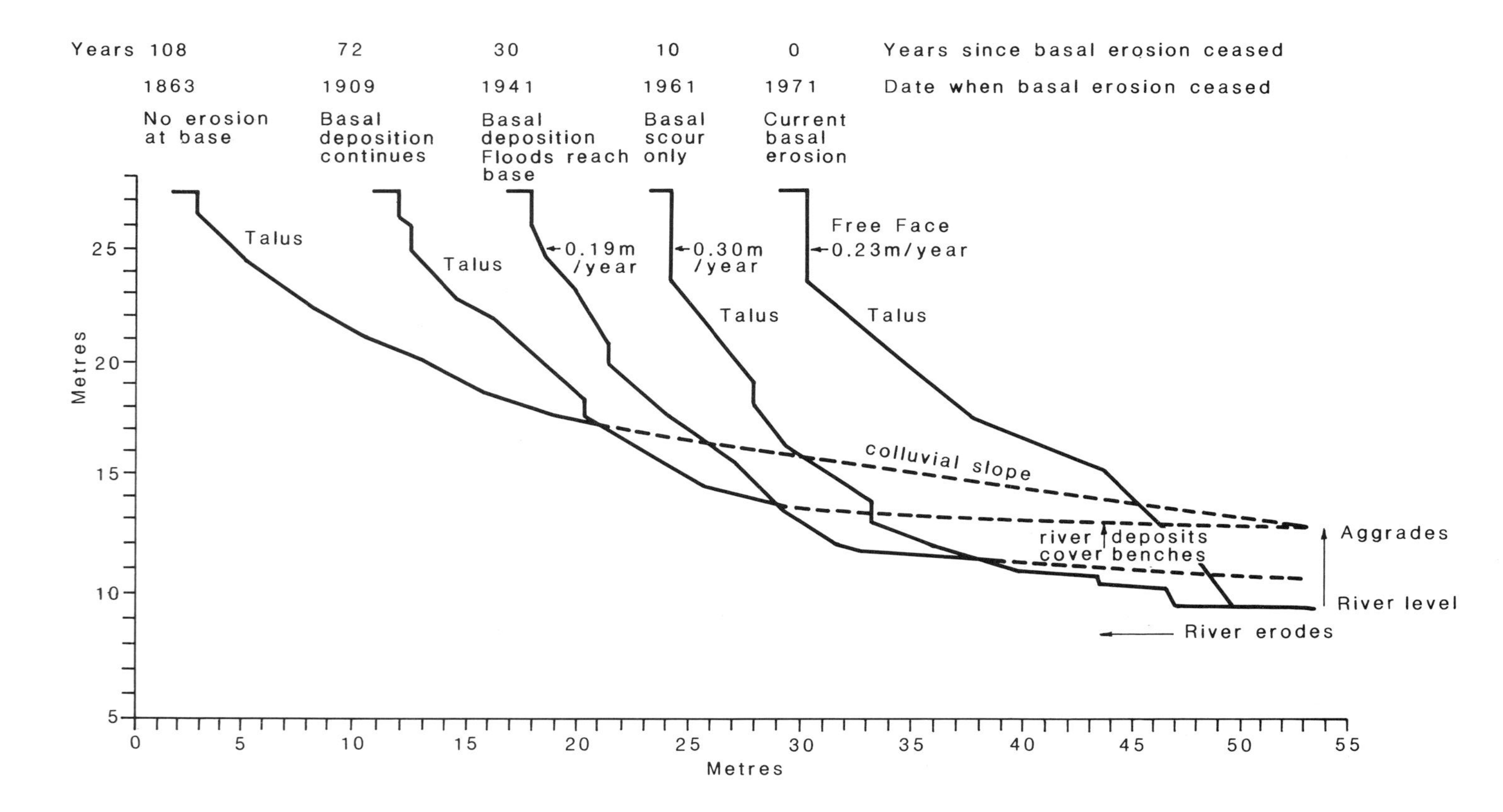

Figure 10.1 The evolution of a Mississippi River bluff with changing river activity at the base. (*After Brunsden and Kesel, 1973*, J. Geol. **81**. *Reproduced by permission of the University of Chicago Press.*)

because cliff stability reflects joint strength rather than intact rock strength. As cliffs are formed, stress release jointing develops and mass strength is reduced (Gerber and Scheidegger, 1973; Eisbacher, 1979). A lag occurs between cliff formation and joint development, and undercutting or incision rates exceed the rate of this dilation 'weathering' process. Thus the cliffs persist, although a gradual evolution occurs as the rock slopes develop an equilibrium with the rock mass strength (Selby, 1982; and Chapter 15 this volume). The critical active factor causing rock slope failure is thus generally the weathering process rather than the basal erosion process.

The role of slope-foot conditions in relation to slope development has been demonstrated particularly in the context of *declining* slope angles associated with the *withdrawal* of basal removal processes. Savigear (1952) identified a series of cliff profiles in south Wales exhibiting progressive reduction of the cliffed segments and lower slope angles, and interpreted this as the result of successively longer periods of protection from direct basal marine erosion, by salt marsh development. Similar investigations which place observed slope profiles in a temporal sequence include those by Brunsden and Kesel (1973) and Nash (1980). The former demonstrates the expected decline in the free face segment and overall angle of a Mississippi River bluff protected from basal scouring (Figure 10.1), while the latter shows a similar pattern of upper slope retreat and basal accumulation resulting in lower angles of progressively older abandoned lake-shore cliffs in Michigan. Given an *increase* in basal erosion, it might be expected that the sequence of profiles in Figure 10.1 could be reversed. However, the mechanisms of slope steepening and failure during active undercutting are not necessarily those of slope decline in reverse. 'Free degradation' of London Clay cliffs when basal erosion ceases involves a sequence of mass movement processes which lower the slope angle (Hutchinson, 1967). Initially, shallow rotational slides reduce angles from 20° to about 15°, then retrogressive slips work upslope to reduce angles to about 8°. These processes would not necessarily occur in reverse during a phase of renewed basal erosion, since they represent the response of the slope to weathering and pore-water pressure adjustment.

These mechanisms of slope angle decline after withdrawal of basal erosion are not entirely irrelevant to the case of active basal erosion, since this is rarely a continuous process. Between major slope failure events the slope foot is protected by failed debris in which strength and pore-water adjustments occur, and the next major failure, therefore, awaits removal of this material. Slope or cliff recession under conditions of active but intermittent basal erosion is continuous in the long term but pseudo-cyclic in the short term. The relationship between long- and short-term behaviour, and the cyclic process of failure and temporary stabilization, both vary with the efficiency of basal erosion and with the stratigraphy and lithology of the eroded slope. Four distinctive sets of conditions are discussed below in order of progressively increasing basal erosion rates.

10.2 SLOW BASAL EROSION IN RIVER VALLEYS

On transport-limited slopes where a weathered mantle and soil covers the bedrock, a frequency distribution of the angles of straight segments of slope profiles in a drainage basin commonly displays modal angular classes, and intervening angles which are relatively rare. The modes have been defined as threshold slope angles (Carson and Petley, 1970; Francis, Chapter 19 this volume), and matched to limiting angles predicted using the infinite slope

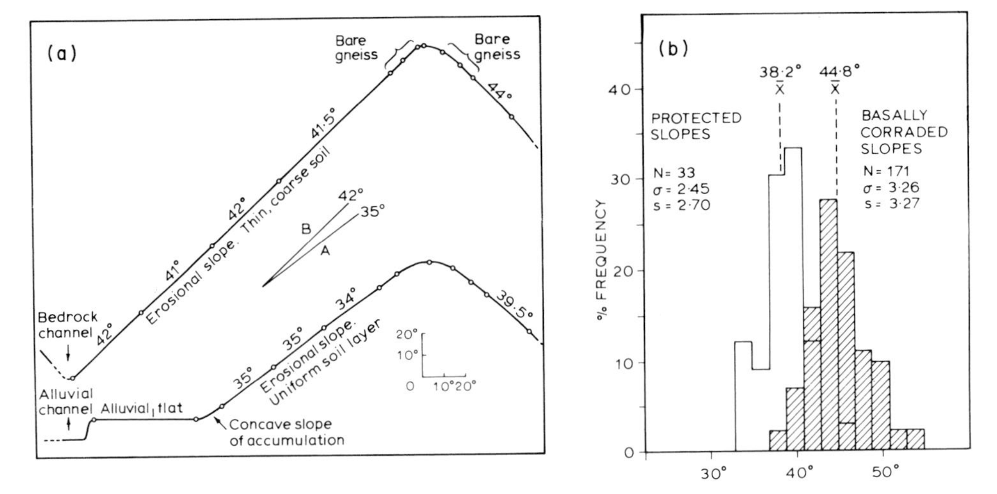

Figure 10.2 Examples of basally-protected and basally-eroded slopes in the Verdugo Hills, California, and frequency distributions of the angles of these two slope categories. (*After Strahler, 1950,* Am. J. Sci., **248**. *Reproduced by permission of the American Journal of Science.*)

stability analysis (Skempton and de Lory, 1957) for varying pore-water pressure states. This interpretation is applicable to local slope angle variations affected by river undercutting, and more generally to the response of slopes to river incision and valley development.

Local asymmetry of river valleys occurs when one valley wall is actively undercut and the opposite wall is basally protected by floodplain deposition and colluvium. This differential in basal corrasion is associated with contrasts of slope angle, as Strahler (1950) noted in a study of the Verdugo Hills in California (Figure 10.2). Carson (1971) suggests that this difference of angle reflects the mechanical properties of material in the densely packed and loose states, quoting the evidence of Silvestri (1961) for density-related variation in the angle of internal friction for similar sediments. Thus, in Figure 10.2, the average angle for basally-corraded slopes reflects the friction angle (on these dry slopes) of densely packed and relatively unweathered regolith, while the average angle of protected slopes reflects slow weathering and a loose state of packing.

More generally, the existence of threshold slopes in fluvial landscapes is dependent on the rate of downcutting and valley formation. Arnett (1971) demonstrated for a Queensland basin that mean and maximum slope profile angles increase to a peak value downstream then decline, and the frequency of occurrence of evidence of mass movement features matches this pattern (Figure 10.3). Similar morphological variations have been observed in drainage basins at various scales (e.g. Richards and Anderson, 1978). This suggests that threshold slopes controlled by shallow sliding in the regolith are confined to those parts of drainage basins in which a critical depth and rate of valley deepening have occurred, but floodplain development has not begun to protect extensive stretches of the valley sides from the effects of basal erosion (Richards, 1977). Thus, inevitably, the correspondence between modal slope angles and the limiting angles (β_c) predicted from

$$\tan \beta_c = \left(\frac{\gamma_s - m\gamma_w}{\gamma_s} \right) \tan \phi' \qquad (10.2)$$

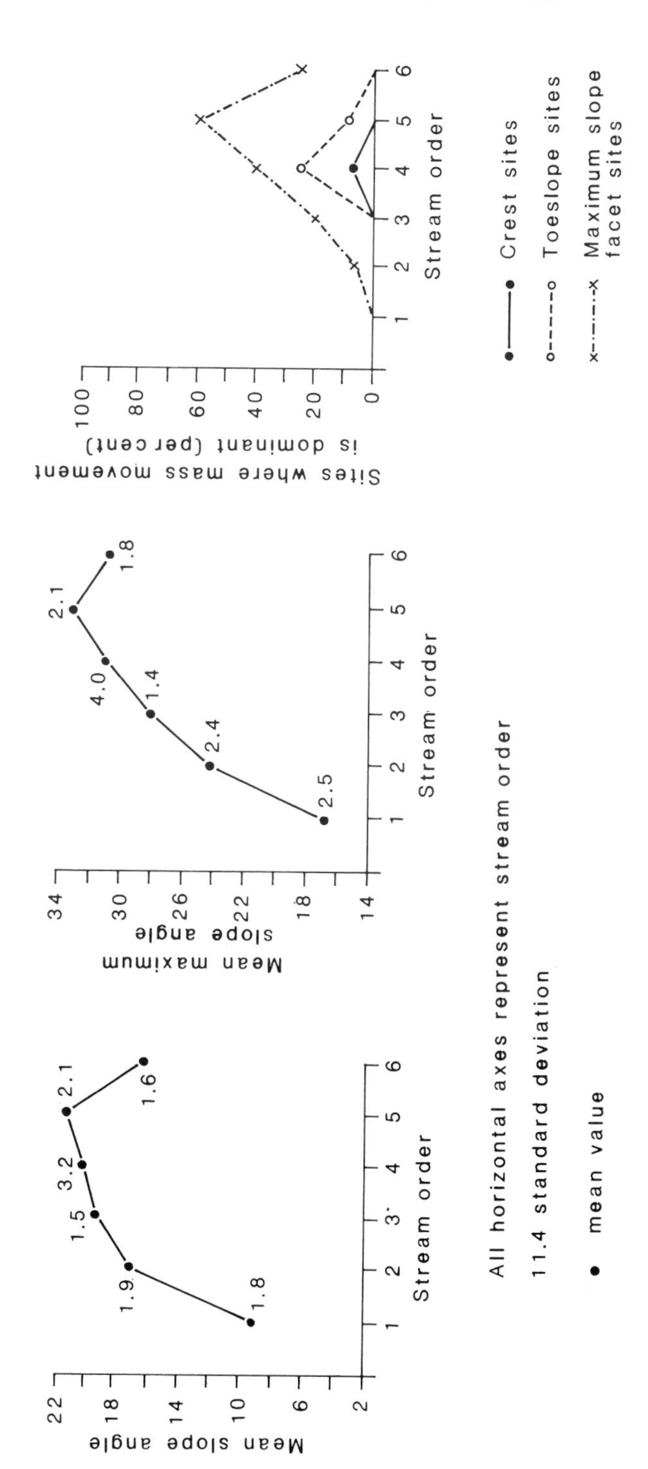

Figure 10.3 Mean slope angles, mean maximum slope angles, and the percentage of slope sites exhibiting evidence of mass movement in the Rocksberg basin, Queensland. Stream order increases downvalley as tributary streams join. (*After Arnett, 1971*, Inst. Br. Geog. Spec. Publ. No. 3. *Reproduced by permission of the Institute of British Geographers.*)

for appropriate values of the relative water table height parameter m, is considerably improved by screening the slope angle data initially to incorporate only those profiles showing evidence of regolith instability (Rouse, 1976).

Valley development and river erosion can be associated with a combination of shallow regolith slides and deep-seated failures. For example, Skempton (1953) considered slope development in sandy clay Shotton till in Durham. Surface weathering of the till resulted in maintenance of slope angles at 30–35° by shallow translational slides as valley deepening progressed through the late Quaternary, until at a valley depth of about 45 m the slopes became unsafe against deep-seated rotational failure. As in the case of the Selset slide analysed by Skempton and Brown (1961), in which the River Lune appeared to erode the toe of the slope during flood periods, these failures involve mobilization of peak or near-peak strengths, and are first-time slides. A somewhat different sequence of events is evident in Glencullen, County Wicklow, Eire, where *rapid* river erosion into glacial till is occurring (Statham, 1975). Here, first time deep-seated failures occur when steep bluffs are created by basal river erosion. These may be planar (Culmann) or rotational failures depending on slope height and angle at failure. Following those failures, the till is remoulded, loses shear resistance to the residual state (i.e. $c' = 0$), and pore-water pressures rise in the collapsed debris. Superficial translational slides then occur to reduce the slope angle to 16–18° (that is, to approximately $1/2 \tan^{-1} \phi'_r$). Such a pattern of slope development is similar to that observed in sea cliffs on the south coast of England formed in over-consolidated Mesozoic and Cenozoic mudrocks with permeable caprocks, and actively eroded at their base (see below; Brunsden, 1974). Hillslope development in river valley locations inland also reflects changes in the intensity of mass movement and basal erosion during the Quaternary, as the absence of cambering features on some Northamptonshire slopes suggests. Chandler (1976) interprets this as the result of Ipswichian river erosion which steepened slopes, removed the evidence of Wolstonian cambering, and prepared them for a phase of intense periglacial landsliding in the Devensian.

10.3 MODERATE TO RAPID BASAL EROSION OF OVERCONSOLIDATED CLAY CLIFFS

The southeast and south coasts of England and the Isle of Wight provide extensive stretches of cliff where active basal marine erosion occurs in heavily over-consolidated Mesozoic and Cenozoic clays and clay shales. These erosion-susceptible rocks are locally overlain by more competent and better drained beds which form caprocks. On the north coast of Kent, London Clay is overlain by Tertiary sands and gravels (the reading and Bagshot Beds), on the south coast the Gault Clay is succeeded by Greensand and Chalk, and further west in Dorset these Cretaceous beds cap Lias Clay. Several mass movement types occur on these cliffs, but they may be broadly classified into shallow translational mudslides (MS) and deep-seated rotational slides (DR) following the investigations of Hutchinson (1969) and Bromhead (1979). The translational mudslides are often associated with bilinear cliff-profiles, in which the steeper backslope provides a feeder zone from which the gentler debris-covered bench beneath receives mudslide inputs. These cause undrained loading in the debris, and the inability of pore pressures to dissipate rapidly enough to prevent artesian pressures results in mobilization of the debris even on gentle foot-slope gradients (Hutchinson and Bhandari, 1971). The deep-seated rotational slides may be multiple, retrogressive failures in the presence of a thick caprock. Several interacting factors determine the type of failure, and the rates of mass movement and cliff recession.

The intensity of basal marine erosion is clearly one important influence. Hutchinson (1967, 1968, 1969) has distinguished cliffs experiencing medium rates of toe erosion and mudslide activity from those where strong marine undercutting results in deep-seated failure. However, the circumstances which control the rate of basal erosion are also important, and these include the structural and stratigraphic properties of the cliffs. Where the geological strike is perpendicular to the coastline and the strata dip along the coast, the spatial pattern of mass movement types reflects the nature of the outcrop at sea level (Hutchinson, 1983). If competent strata occur at sea level and the base of the clay is above wave influence, cliff retreat is slow and underdrainage occurs in the clay, so that only limited mudsliding occurs. Where the dip brings the base of the clay to about 1–3 m above mean sea level, or in the upper quarter of the range of spring tides, moderate cliff-foot erosion is associated with active mudsliding. Finally, when the base of the clay stratum is below sea level, cliff-foot recession is rapid and deep rotational slides occur. The rate of basal erosion is thus itself structurally controlled. More subtle basal influences are suggested in the case of the Folkestone Warren landslide complex (Hutchinson *et al.*, 1980). Here, it appears that successive phases in the development of Folkestone Harbour pier from 1810 to 1905 were associated with an interruption of longshore drift which depleted beach accretion to the east beneath the Warren, accelerating landslide activity through this period and culminating in a major event in 1915 in which 30 m of displacement occurred on a deep-seated curved failed surface.

Although basal erosion conditions clearly distinguish between cliffs which are actively receding and those which are declining in angle during free degradation, the intensity and type of mass movement involved in the recession process reflect additional influences. Bromhead (1979) emphasizes the importance of cliff-top materials and groundwater conditions. A resistant or well-drained caprock or soil tends to promote deep-seated rotational failures. For example at Herne Bay, on the north Kent coast, the London Clay cliffs experience mudsliding, except where a 3 m thick sandy head (solifluction deposit) helps to drain the backslope and thereby reduces the supply of material to the mudslide debris and, therefore, the occurrence of undrained loading (Bromhead, 1978). By contrast Conway (1979) describes upper slope head deposits in west Dorset that inhibit drainage and contribute to mudsliding such that cliff-top retreat is largely independent of cliff-toe erosion, at least in the short term. In some cases, a sea cliff is topped by a bench, and an upper rear scarp above this bench retreats through active mass movement which is only indirectly controlled by toe erosion. Hutchinson *et al.* (1981) describe such a case on the Isle of Wight where mass movement is triggered by seepage erosion by groundwater outflow from a sandy layer at the foot of the rear scarp. The entire cliff system is in balance in the long term in that retreat of the rear scarp and sea cliff occur at similar rates—in this case about 0.2 m yr^{-1} from 1861 to 1907, and 0.6 m yr^{-1} from 1907 to 1980. This balance arises because superficial mass movements transport the failed material from the rear scarp over the sea cliff, a process also described by Brunsden (1974). The rear scarp experiences multiple rotational failures, while translational mudsliding occurs in the brecciated failed blocks on the bench beneath.

In the long term, therefore, the critical controls of cliff retreat are the rate of basal erosion and the recent geological history of the cliffs. As Pitts (1983) notes, older Quaternary

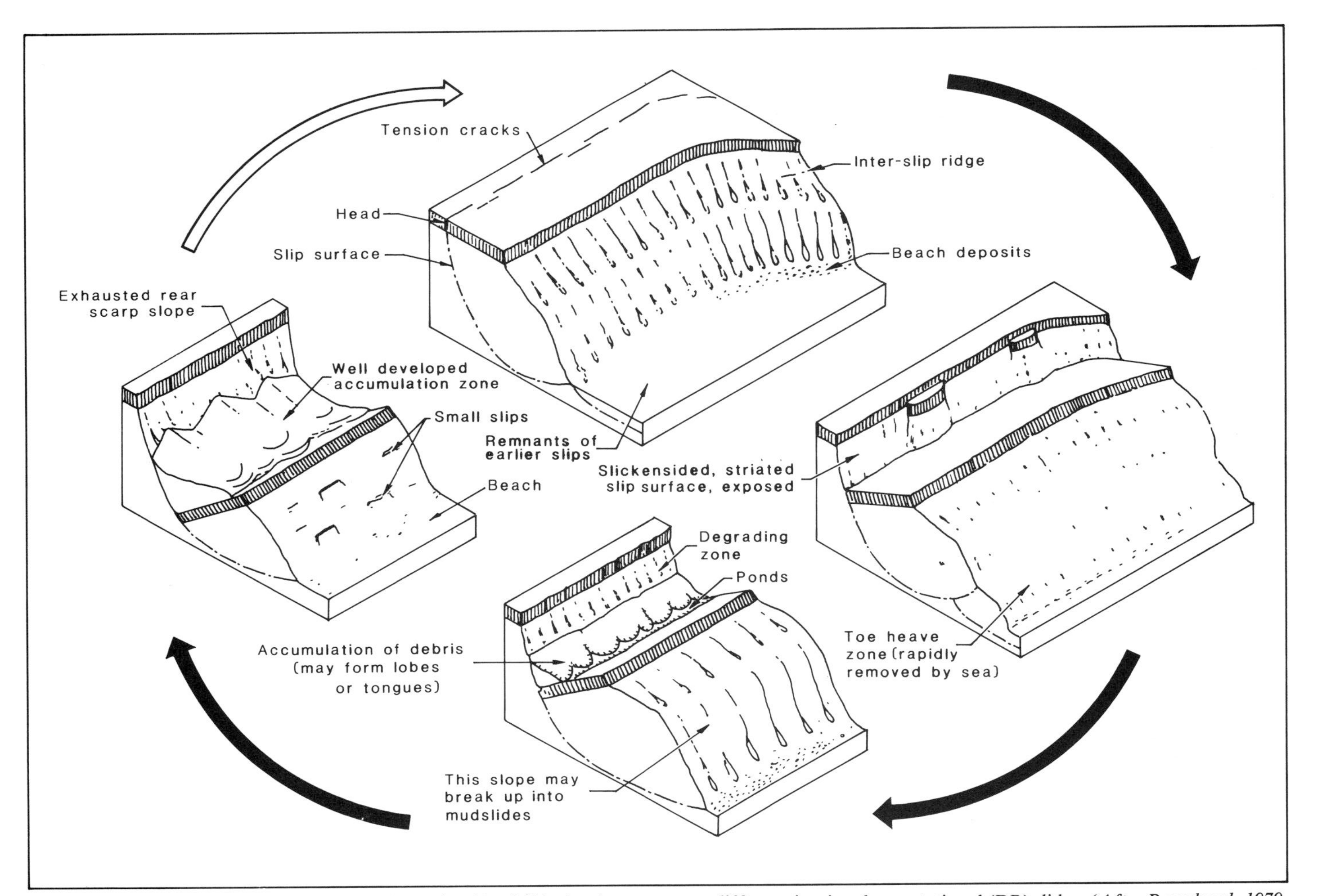

Figure 10.4 Block diagrams summarizing the cycle of landslide development on a cliff experiencing deep rotational (DR) slides. (*After Bromhead, 1979,* Q. J. Eng. Geol., **12**. *Reproduced by permission of the Geological Society of London.*)

multiple rotational slides on the Devon coast appear to have been planed and reactivated by post-glacial rising sea level, with particular periods having experienced accelerated slope instability generally (e.g. 11,000–9,000, 7,000–5,500, and 1,500–550 years BP). Such reactivation may partly explain the general tendency of the large south coast landslides to involve mobilization only of residual strengths (Bromhead, 1978). In the shorter term other factors are of greater significance, particularly in controlling the cyclic behaviour of cliff erosion. Where translational mudslides predominate because of the generally moderate rates of basal retreat, seasonal variations in mudslide advance reflect upper slope factors. As Bromhead (1979) notes, a typical mudslide requires extreme toe erosion amounts of 37 m in summer and 20 m in winter (i.e. for different pore-water pressure regimes), to produce a safety factor of unity. The prime factor in restarting motion of the slide each winter is the input of debris from the feeder slope at the head, producing undrained loading. This supply is critically dependent on upper slope soil and groundwater conditions, and if inadequate, will cease to 'drive' the mudslide which accordingly is eroded at the toe. In the case of deep-seated rotational slides (DR) a cycle of cliff retreat also occurs, but on a longer time-scale (Figure 10.4). After a rotational slide, the sea erodes the toe, removing support, and rear scarp falls (such as the chalk falls at Folkestone Warren; Hutchinson *et al.*, 1980) provide undrained loading. These are sufficient to overcome the residual strength mobilized on the shear surface, and block motion continues as it is degraded until another rotational failure occurs. Brunsden (1974) estimated that superficial mass movement in the remoulded, brecciated rotated block at Stonebarrow Hill, Dorset would take about 40 years to prepare the rear scarp for the next major failure. It is clear from those studies that controlling factors vary with the time-scale involved, and that cliff-top and cliff-foot recession are not necessarily in phase in the shorter term, and not necessarily controlled by the same phenomena.

10.4 RAPID BASAL EROSION IN QUATERNARY SEDIMENTS

Normally-consolidated or lightly over-consolidated cohesive Quaternary sediments include loess and glacial till. Cliffs and steep slopes formed in such sediments by efficient slope-foot erosion are subject to mass movement by wedge, slab or steeply inclined circular arc failures. For example, river erosion in thick loess deposits in Iowa results in successive stages of slope instability analysed by Lohnes and Handy (1968) using the Culmann wedge stability analysis. In this model a planar shear surface is assumed to pass through the toe of the slope at an angle θ, where

$$\theta = (i + \phi)/2 \tag{10.3}$$

i being the slope angle. This stability analysis defines the limiting slope height (H_c) as

$$H_c = \frac{4c}{\gamma} \frac{\sin i \cos \phi}{[1 - \cos(i - \phi)]} \tag{10.4}$$

which indicates that a steeper slope has a smaller limiting height. Thus river downcutting which increases the slope height, or slope-foot erosion which steepens the angle, may both cause slope failure. If a vertical cliff is maintained by basal erosion (i.e. $i = 90°$),

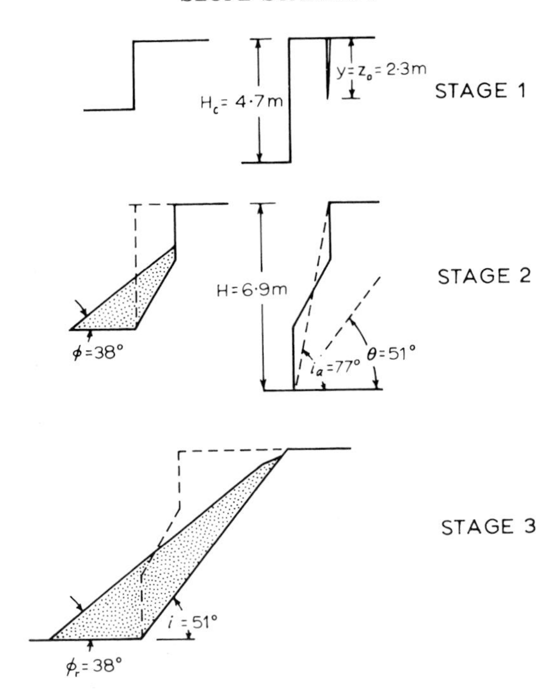

Figure 10.5 Hypothetical sequence of loess slope failure caused by downcutting, with depths and angles based on mechanical properties of Iowan loess. (*After Lohnes and Handy, 1968,* J. Geol., **81**. *Reproduced by permission of the University of Chicago Press.*)

$$H_c = \frac{4c}{\gamma} \frac{\cos \phi}{1 - \sin \phi} \tag{10.5}$$

Before this limiting cliff height is reached, stability may be reduced by the formation of a tension crack behind the crest which penetrates to the failure surface, converting the wedge to a slab failure (Stage 1 in Figure 10.5). Given that the maximum depth of tension crack formation is about half the limiting cliff height, this can significantly reduce maximum cliff heights unless the tension crack only forms just prior to the shear failure.

In Lohnes and Handy's (1968) analysis the mechanical properties of the Iowan loess were defined as

$$c = 8.97 \text{ kN m}^{-2}, \ \phi = 25° \text{ and } \gamma = 11.85 \text{ kN m}^{-3}$$

Figure 10.5 illustrates three stages of cliff evolution during active basal erosion. In Stage 1 a vertical cliff is predicted to fail when a limiting height of 4.73 m is reached. Observed cliff heights are about 4.5 m, suggesting that tension crack formation is delayed until just before failure, but does convert the wedge failure to a slab failure. The shear surface resulting from this failure is inclined at 57.5°, but evidently basal removal of collapsed loess and continued vertical erosion occur to produce the Stage 2 failure. As the first failure reduces the slope angle, the Stage 2 limiting height is increased, vertical incision steepens the average angle and increases the height until the height is critical for the

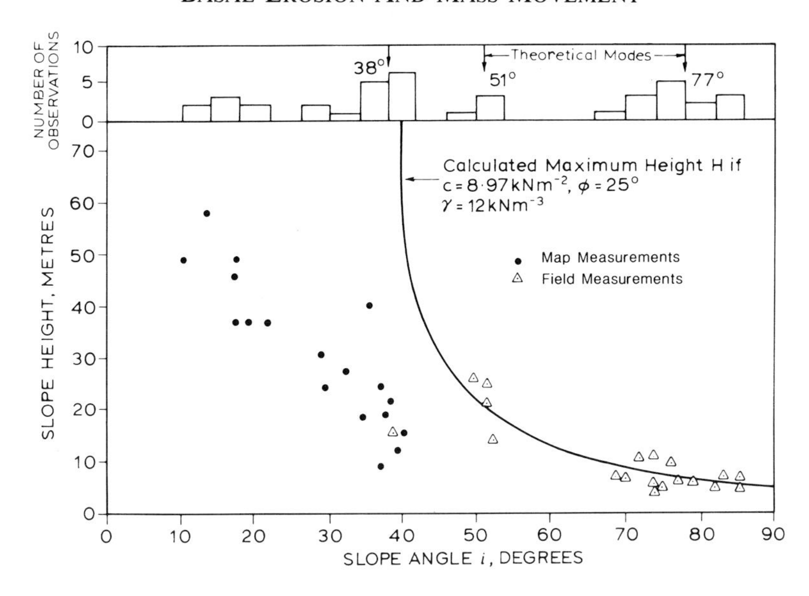

Figure 10.6 Heights and slopes of natural and artificial loess cuts in western Iowa. The curve predicts limiting heights for slopes of varying angle, and is based on equation 10.4; the theoretical modal angles are explained in the text. (*After Lohnes and Handy, 1968,* J. Geol. **81**. *Reproduced by permission of the University of Chicago Press.*)

increased angle. The Stage 2 failure thus occurs when the average slope angle is 77° and the height is about 6.9 m, and produces a failure surface inclined at 51°, which may be maintained at this angle by wash processes, or may be mantled by dry cohesionless loess debris after the Stage 2 failure, thereby producing an angle of 38°. Figure 10.6 shows that the observed slope angle distribution has modes at 77°, 51°, and 38°, and that until the 38° slope is created the observed heights and angles are related according to the Culmann model.

This analysis ignores pore-water pressures, because the Iowan loess is highly permeable. It can, however, easily be generalized to incorporate groundwater conditions and effective stresses (Stimpson, 1979). In glacial tills of low permeability, this modification is essential, and analysis is additionally complicated by the development of a weathering profile generated by pedogenesis and stress release in lightly over-consolidated lodgement tills. The following case study illustrates some of this complexity.

10.4.1 Cliff Recession on Holderness, East Yorkshire

The eroding coastal cliffs of Holderness extend for about 50 km between Bridlington and Spurn Head in East Yorkshire, England, and range up to 40 m in height. They are formed in unlithified soft rocks of Late Devensian age (i.e. post-18,000 years BP) and display considerable variation in stratigraphic and sedimentological detail. However, they are characteristically composed of glacial sediments which are matrix-dominant tills (with less than 40% clast content and a clay matrix), being ice-marginal deposits containing interstratified glacio-fluvial sands and gravels. Sladen and Wrigley (1983) estimate that

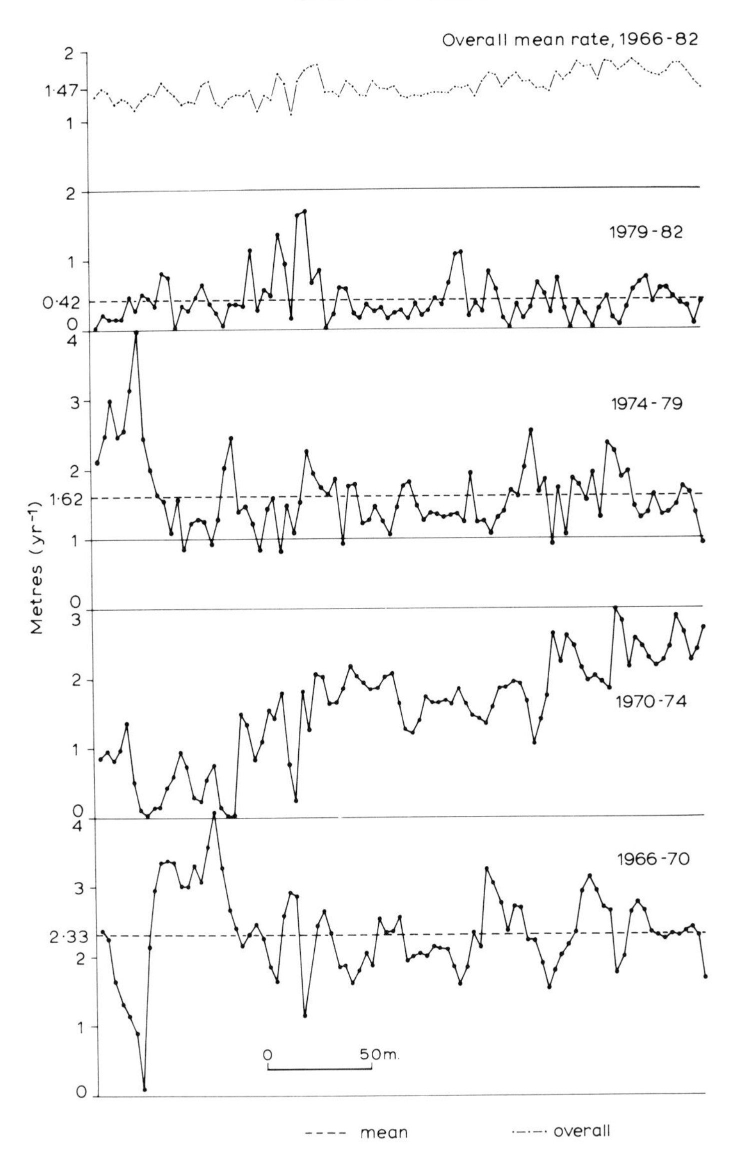

Figure 10.7 The pattern of cliff retreat along a 300 m stretch of cliff-top at Easington, Holderness, over the period 1966–82 (top) and for four consecutive 3–4 year periods

they have been slightly over-consolidated by excess overburden pressures of approximately 2,000 kN m^{-2}, typical of relatively thin ice cover at ice-sheet margins. Evidence of consolidation pressures is provided by glaci-tectonic structures and streaked-out chalk fragments, but there are few macro-discontinuities beneath 4–5 m below the ground surface (Marsland and Powell, 1985).

The cliffs experience very rapid rates of coastal recession, which averaged 1.2 m yr^{-1} over the period 1852–1952 (Valentin, 1954). This process of recession involves: (a) toe erosion by marine action which both undercuts and steepens intact clay and removes failure deposits; and (b) a range of mass movement mechanisms including relatively deep-seated wedge and rotational failures, slumps, spalling, and superficial mudflows. The latter are often accelerated by drainage outflow from old tile drains 1–2 m below ground surface which have been intersected by the retreating cliff line. Such processes typically result in considerable spatial and temporal variations in the rate of cliff retreat (Cambers, 1976). For example, Valentin's (1954) data suggest a systematic north–south increase in retreat rate from 0.29 m yr^{-1} in the northern 10–15 km section to 1.75 m yr^{-1} in the most southerly section. Within shorter sections, however, this directional trend may be completely reversed. Temporal variability of coastal retreat is particularly evident in short sections of coast. Figure 10.7 shows that 300 m of cliffline at Easington, where the cliffs are about 8 m high, have experienced a sharp decline of their recession rate from 2.33 m yr^{-1} to 0.42 m yr^{-1} when this is averaged for successive 3–4 year periods between 1966 and 1982. The overall average of 1.47 m yr^{-1} is associated with a reduced variance because specific sites receding rapidly in one sub-period then recede more slowly when protected by adjacent projecting headland sites, until excessive exposure of the latter leads to their rapid retreat. The short-term recession rates plotted in Figure 10.7 also illustrate significant variations in the small-scale spatial pattern of cliff retreat. In the 1966–70 period, the data suggest a periodicity of about 60 m, approximately the scale of the persistent mass-movement embayments controlled by tile-drain emissions. In 1970–74, however, there was a pronounced south–north increase which reverses Valentin's larger scale, longer term pattern.

Several spatial and temporal controls of these varying recession rates exist, and modelling their interaction is impracticable. Spatial variations occur in cliff materials and morphology, and temporal influences include seasonal changes in pore-water pressure, seasonal and non-seasonal alteration of the basal conditions (steepening the angle and increasing cliff height), and secular sea level variation.

Although stratigraphic and sedimentological complexity causes *spatial* variation in the mechanical properties of the tills, *vertical* zonation of the till units is particularly important geotechnically. The traditional Late Devensian till sequence for Holderness included the Drab, Purple and Hessle Tills, but Madgett and Catt (1978) concluded that the Hessle represents a weathered upper horizon in the Purple Till, and defined the sequence as: Skipsea (= Drab) and Withernsea (= Purple plus Hessle) Tills. Geotechnical aspects of weathering profiles in over-consolidated lodgement tills such as these have been discussed by Eyles and Sladen (1981) and Russell and Eyles (1985); see also Chapter 13. The surface Zone IV in such a profile is a reddened, oxidized, and partly pedogenic profile incorporating rotted stones and a high clay content, and with well-developed stress-release fissures. Zones III and II are successively less weathered with increasing depth and overlie the unweathered Zone I till. There are significant depth- and weathering-related trends in moisture content, plasticity, brittleness, and undrained strength, which may be abrupt if the weathered zones are underdrained by a sand or gravel layer. Shear strength parameters for Late Devensian till units vary with sampling and testing methods (Marsland and Powell, 1985), but the consolidated undrained triaxial tests performed on 38 mm undisturbed samples by Joyce (1969) are clearly classified with respect to the main stratigraphic units, and yield the following peak strength parameters in terms of effective stresses:

Weathered Withernsea Till; $c' = 90\ \text{kN m}^{-2}$, $\phi'_p = 14°$, $\gamma = 21.5\ \text{kN m}^{-3}$
Unweathered Withernsea Till; $c' = 41\ \text{kN m}^{-2}$, $\phi'_p = 25°$, $\gamma = 23.1\ \text{kN m}^{-3}$
Unweathered Skipsea Till; $c' = 34\ \text{kN m}^{-2}$, $\phi'_p = 29°$, $\gamma = 23.4\ \text{kN m}^{-3}$

Since the lateral retreat rate of the cliffs is rapid relative to rates of weathering and pore pressure adjustment in the unweathered till, it is likely that peak strengths are mobilized in these units, while shear surfaces passing through the weathered zone exploit fissures along which the cohesion is near residual.

Figure 10.10 demonstrates the variability of cliff height between Easington and Withernsea. This reflects the variation of till unit thickness, and controls the critical angles at which failure occurs. Furthermore, this morphological control influences the magnitude–frequency characteristics of cliff mass movement, and the relationship between cliff-top and cliff-foot retreat. Larger failures in high cliffs may cause sudden, rapid cliff-top retreat, but a longer interval then passes before the next such event because basal removal takes longer. Cliff height is, however, not simply a static control of instability. At retreat rates of more than $1\ \text{m yr}^{-1}$, recession itself can encounter different topographic conditions within engineering time-scales. Nevertheless, of more significance are the seasonal and other

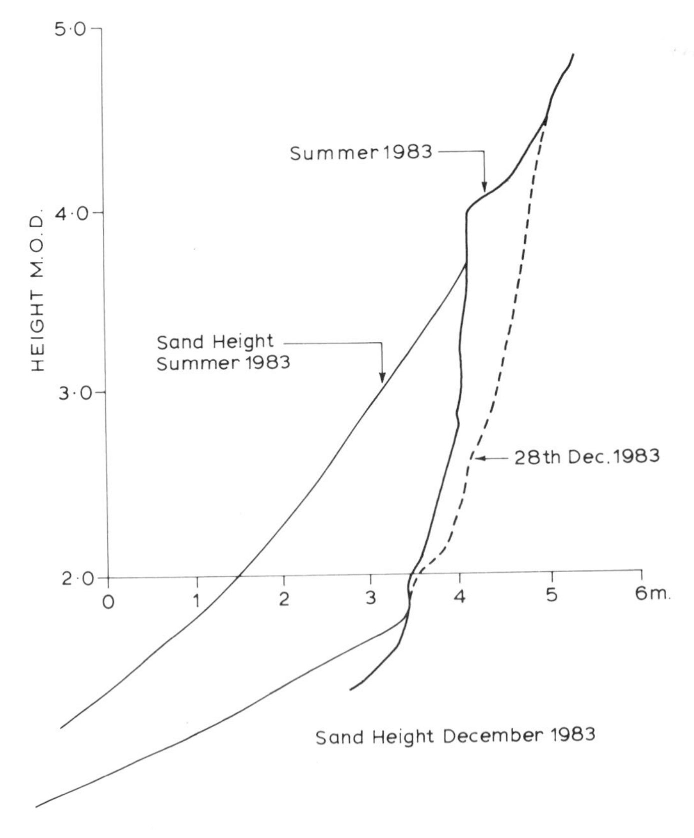

Figure 10.8 Profiles of the cliff foot and beach at Easington, Holderness, illustrating 2 m of exposure of the cliff foot between summer and winter 1983

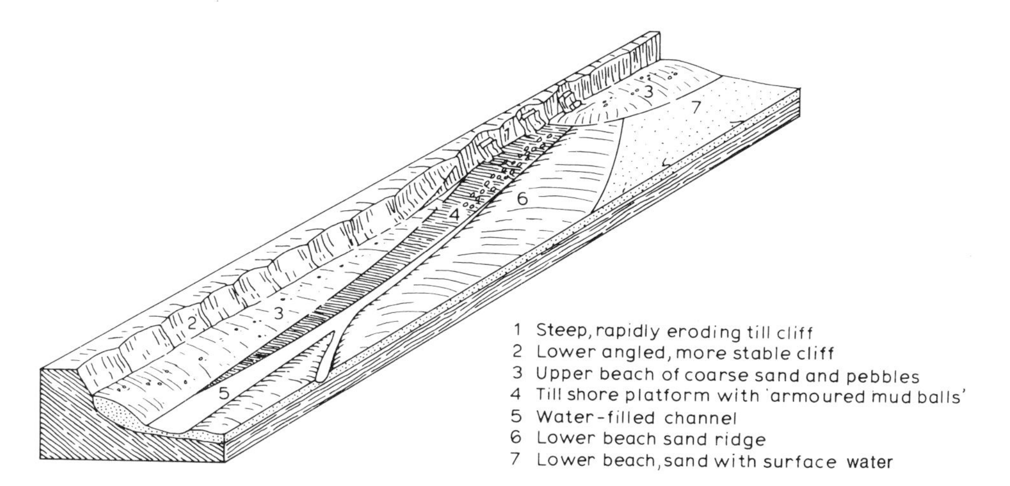

Figure 10.9 The characteristic features of a Holderness 'ord' illustrating the relationship between ord channel and eroding cliff. (*After Pringle, 1985*, Earth Surf. Proc. Landforms, **10**.)

short-term factors which influence cliff height and steepness. Wave conditions cause variation in the beach profile at the cliff foot, with the plunging breakers of steep storm waves scouring the beach sediment seawards, and the spilling breakers of longer constructive waves building an upper beach berm. This process tends on average to produce a seasonal cycle of high summer and low winter beach levels, and on the Holderness coast this can cause exposure of an additional 2–2.5 m of lower cliff during winter (Figure 10.8). This coastline is also characterized by features locally known as 'ords' (Pringle, 1985), which are rather like rip-current channels which run seawards from the cliff foot, exposing the till wave-cut platform (Figure 10.9). Where they meet the cliff the beach level is effectively lowered by 2–4 m and basal erosion of the cliff is more efficient, steepening the cliff profile and removing failed debris. As Cambers (1976) notes for sites on the similar eroding till cliffs of Norfolk, a strong relationship exists between cliff recession rates and the frequency with which high tides reach the cliff base. The most interesting aspect of the Holderness ord, however, is that it *appears* to migrate southwards along the coast with the longshore drift direction, at a rate of about 500 m yr^{-1}. This is associated with locally-accelerated cliff retreat in its wake (Figure 10.10), although cause and effect are here difficult to distinguish. This increased cliff recession is associated with the combined effects of increased cliff height and steepening of cliff angle which are both caused by lowered beach level and intensified wave attack at the cliff foot. Superimposed on these short-term variations in the basal conditions is the secular trend of relative sea-level change. Valentin (1954) estimated a 1.5 mm yr^{-1} rise in sea level relative to land on the Holderness coast from 1852 to 1952, and Suthons (1963) has summarized data for east coast tide guages as far north as Kings Lynn which suggest a 3 mm yr^{-1} rise in annual maximum high water level between about 1900 and 1955. The combined effect of glacial retreat caused by atmospheric warming in this period, and subsidence of coastal areas around the North Sea basin, is responsible for this sea-level trend, which encourages continuing coastal erosion by maintaining efficient basal removal of the products of mass movement.

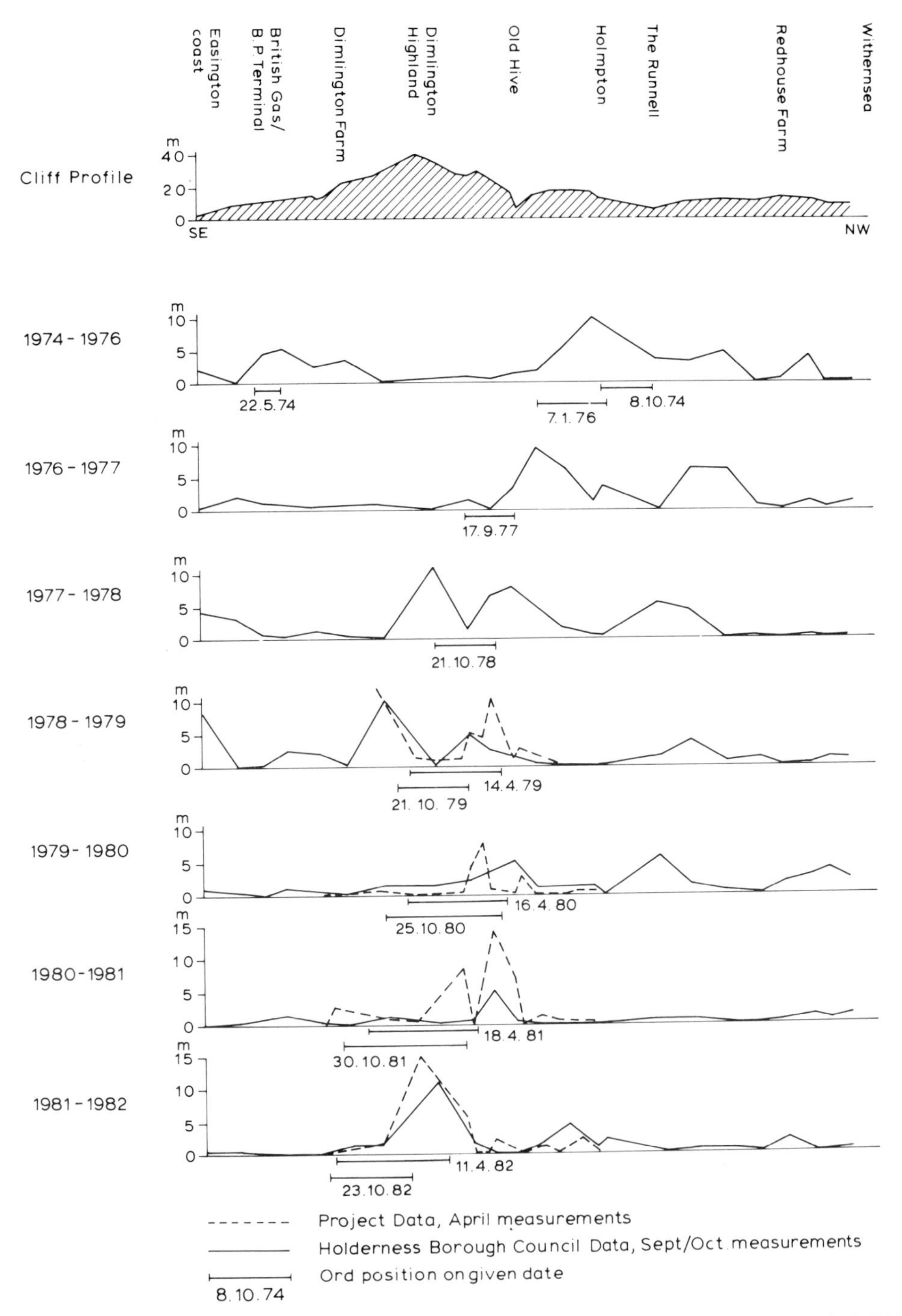

Figure 10.10 Cliff-top retreat between Withernsea and Easington, Holderness, over the period 1974–82, illustrating the association between ord position and areas of rapid retreat. (*After Pringle, 1985.*) Earth Surf. Proc. Landforms, **10**

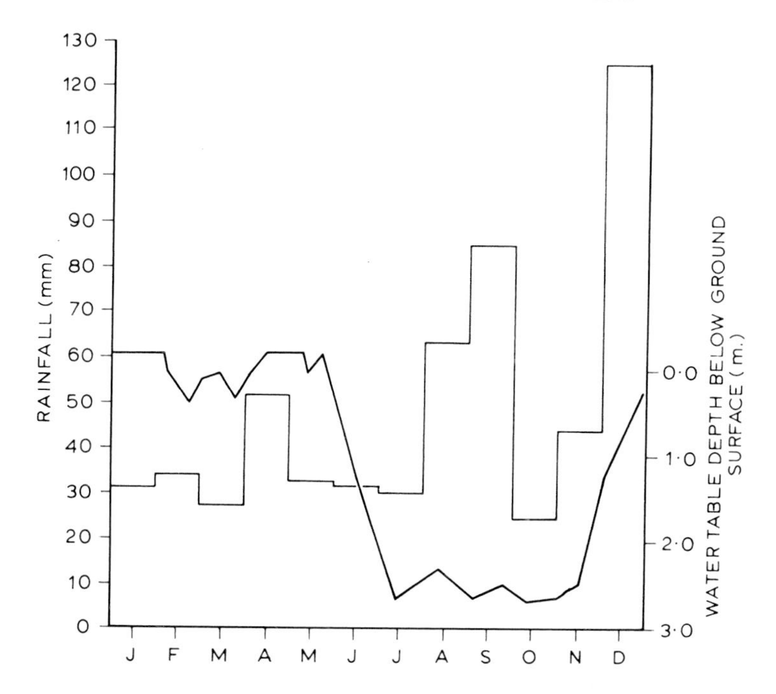

Figure 10.11 Response of water level in an unlined borehole at Easington, Holderness, and monthly rainfall during 1983

Pore-water pressure changes provide an additional temporal influence over cliff stability. Figure 10.11 shows that the water table at an unlined borehole site at Easington was within 0.5 m of the flat surface at the cliff top for six months during 1983, dropping to 2.5 m below the surface between July and October. Groundwater fluctuations are, therefore, confined to the fissured, weathered Zone IV horizon, and result in complete saturation of the cliff materials at the time of year when the beach level is lowest, the cliff height is greatest, and wave attack is steepening the cliff foot.

This summary indicates that several interacting factors control mass movement on these cliffs, and it is difficult to isolate a single *critical* influence. Because of the spatial and temporal variation of the controls of failure, individual mass movement events may be triggered by increased cliff height, steepened cliff angle, increased pore-water pressure, and reduced shear strength in Zone IV of the weathering profile. Most failures are wedge (Culmann) or rotational types with shear surfaces steeply inclined and passing through the toe of the cliff, sometimes with a tension crack extending to a depth of 1.5–2.5 m and situated 1.5–2.5 m behind the crest of the cliff (Figure 10.12). Occasionally, however, the failure is confined to the weathered Zone IV unit, and a bench forms at the top of the unweathered till. Given the evidence of these failure types, it is possible to apply appropriate stability analyses in order to assess the sensitivity of the cliffs to change in the various controls of failure.

Stimpson (1979) has developed simplified stability analyses based on the Culmann wedge model but generalized for various groundwater conditions. Figure 10.13 illustrates the two cases of bi-linear and horizontal water tables, each of which can be examined with or without

Figure 10.12 A typical example of cliff failure by slumping, near Easington on the Holderness coast

a water-filled tension crack, and Figure 10.14 shows how the factor of safety varies with cliff angle for nine separate cases. This analysis is based on a cliff 8 m high at Easington, assumed to be composed of uniform material for which consolidated drained direct shear tests indicated $c' = 31$ kN m^{-2} and $\phi'_p = 24°$. As such a cliff is steepened by basal erosion, the factor of safety follows one of the curves on Figure 10.14. However, this process is more gradual than the effect of a rising water table from the Ai and Ci to Di conditions, which suggests that pore-water pressure changes are potentially more critical. As Figure 10.11 demonstrates, an unlined borehole response is sufficiently rapid to display a 1 m rise of the water table in the one month of December 1983. A similar percentage change in the factor of safety produced solely by steepening would necessitate a cliff angle increase of about 10° and more than 2 m of basal retreat.

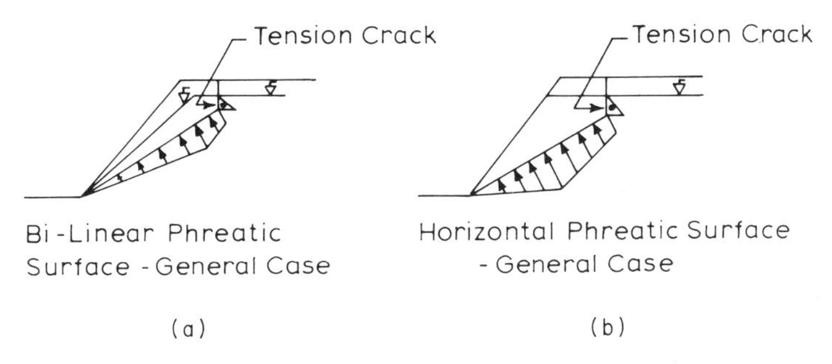

Figure 10.13 Definition diagrams for the two groundwater models (bi-linear and horizontal water tables) used to generate simplified stability analyses for planar wedge failures. (*After Stimpson, 1979*, Q. J. Eng. Geol., **12**. *Reproduced by permission of the Geological Society of London.*)

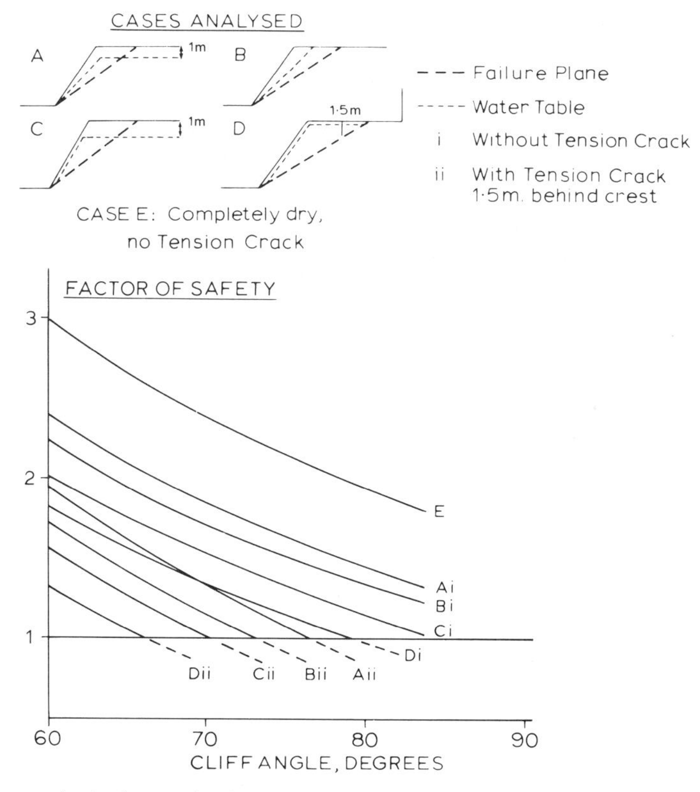

Figure 10.14 Changes in the factor of safety of planar wedges for an 8 m high cliff at different angles, for several different groundwater conditions

This analysis is simplified in that it assumes a single material. It is also likely that more critical conditions are encountered on curved and steeply inclined slip surfaces. Accordingly a series of tests using Bishop's (1955) simplified method of slices on circular arc failure surfaces was performed for an 8.2 m high cliff profile, again from Easington, with the upper 3.2 m weathered, and the strength parameters for Withernsea Till units taken from

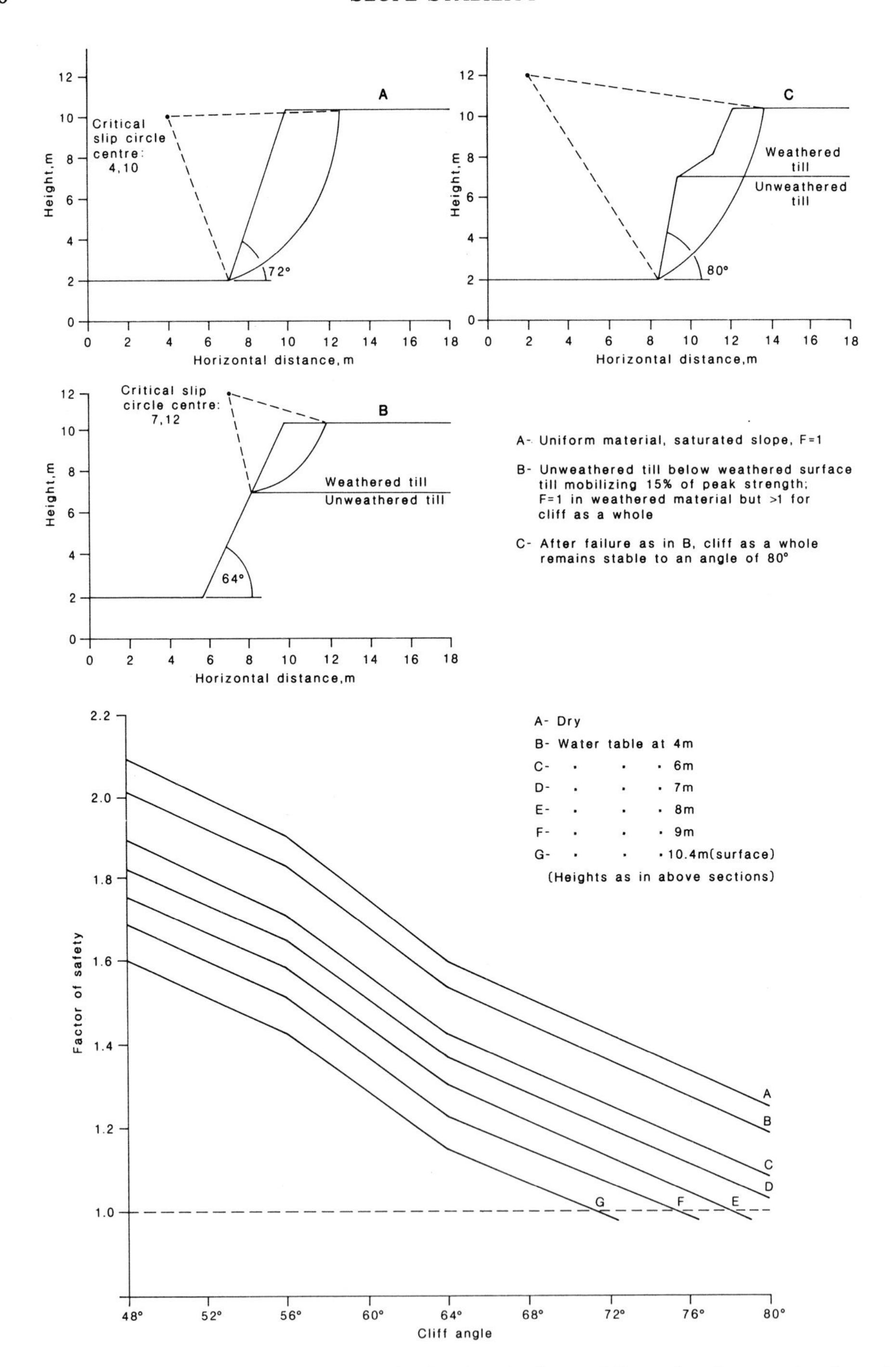

Figure 10.15 Changes in the minimum factor of safety for circular arc failures (i.e. those passing through the toe) on an 8.2 m high cliff at different angles, assuming uniform material properties. The inset diagrams show the critical slip circles for three different conditions

Joyce (1969), as quoted above. The results indicate that the critical circles pass through the toe of the cliff and 2 m behind the crest, and are centred from 6–8 m seaward of the crest and 0–2 m above the crest elevation. Under conditions of saturation to the surface, the factor of safety is unity for a 72° slope, and at this slope a 20% reduction of the factor of safety from 1.2 to 1.0 requires a rise in the water table from 3.2 m below the surface (that is, through the full extent of the weathered unit). A similar reduction of the factor of safety to failure under saturated conditions by steepening alone would require a change from 62° to 72°, while an extra 2 m of height at the cliff base would cause failure at 62°. These calculations assume that the weathered zone mobilizes peak strength. On a 64° slope which is fully saturated, the factor of safety is marginally stable if the upper weathered unit only mobilizes *c.* 15% of its peak cohesion. However, under these conditions the upper unit itself would already have failed on a slip circle passing through the outcrop on the cliff face of the base of the unit. This removal of the crest of the cliff in a small bank failure would then increase the stability of the whole cliff against a toe failure, such that when saturated it would need basal steepening to 80°. These critical conditions are illustrated in Figure 10.15. It is evident from these stability analyses that no simple model of the retreat process can be formulated. Cliff-top and cliff-foot recession are in the long term balanced, but the 'layered' nature of these till cliffs may cause short-term discrepancies which reflect the differing mobilization of peak strength in the weathered and unweathered units. Nevertheless, the overall rapidity of retreat of these cliffs reflects their sensitivity to a range of different factors all of which may trigger failure, either through changing basal conditions which modify the cliff morphology, or through seasonal alteration of pore-water pressures.

10.5 UNDERMINING AND COLLAPSE OF STEEP COMPOSITE BANKS

Stratigraphic arrangements in which cohesive silts and clays overlie non-cohesive sands and gravels are common in alluvial and coastal sediments, and the difference of shear and tensile strengths between upper and lower units is often increased by the surface vegetation root-mat. In river valleys, Holocene alluviation (Bell, 1982) has often involved fine silt–clay overbank deposits accumulating to depths of 1–3 m, depending on catchment size and lithology, over older channel lag deposits of sand or gravel. The initiation of this aggradation in lowland Britain has been dated to the late Bronze Age and explained by deforestation, ploughing, and soil erosion (Shotton, 1978), but also may reflect climatic or base level changes. In the lower Mississippi valley, for example, gravels are succeeded (at *c.* 10,000 years BP) by deposits of silty fill incorporating two palaeosols (Grissinger 1982). Thus river banks are frequently sedimentological composites, in which bank retreat involves entrainment by direct fluid shear of the basal sands or gravels until the resulting cantilever overhang, formed by such undercutting, collapses. The *erosional* retreat of the two units may differ by an order of magnitude (about 30 mm yr^{-1} in the upper silty unit, and about 300 mm yr^{-1} in the lower sands). This process has been described by Klimek (1974) and Thorne and Tovey (1981), and is illustrated in Figure 10.16. Similar processes of bank or 'cliff' retreat occur at the edge of coastal salt marshes where cohesive surface units of sandy loam to clayey silt overlie sand, and small cliffs of 0.25–1.00 m in height are undermined by wave action then collapse when a critical degree of undercutting is attained (Van Eerdt, 1985).

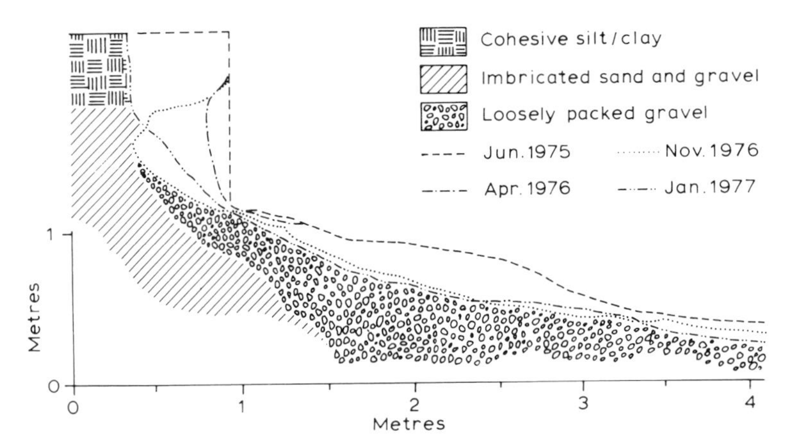

Figure 10.16 River bank retreat on the River Severn, resulting in formation of a cantilever overhang; undercutting of a composite river bank. (*After Thorne and Tovey, 1981*, Earth Surf. Proc. Landforms, **6**.)

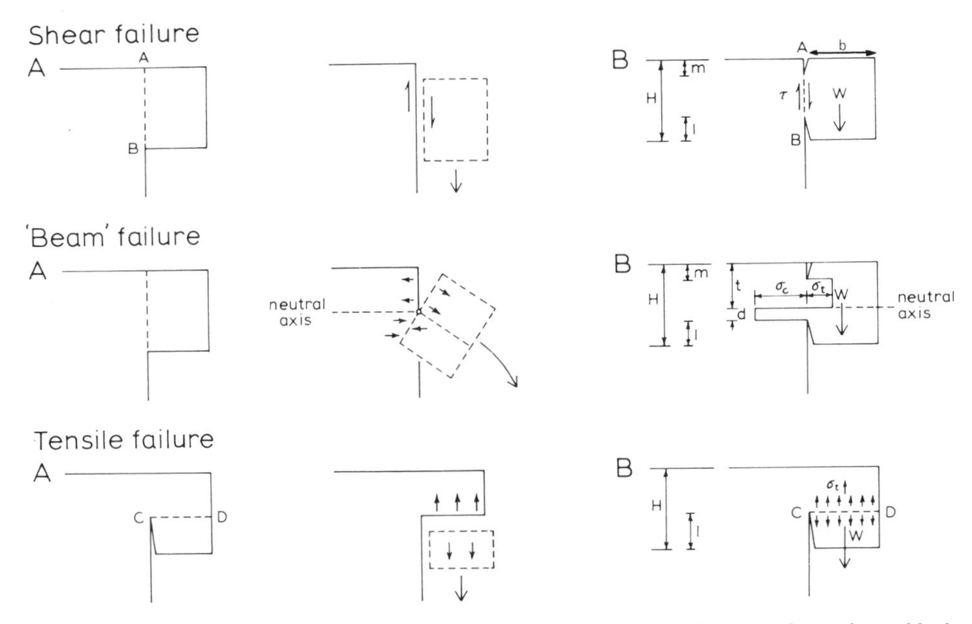

Figure 10.17 Definition diagrams for shear, beam and tensile failures of cantilever overhangs formed by basal undercutting. (After *Thorne and Tovey*, 1981, Earth Surf. Proc. Landforms, **6**.)

The collapse of cantilever overhangs in composite banks is by one of three principal failure modes whose stability conditions have been analysed by Thorne and Tovey (1981) using static equilibrium of forces and the elementary theory of the bending of beams. These are shear, beam, and tensile failures (Figure 10.17). Shear failure occurs when the overhanging block is displaced along the vertical plane AB, beam failure involves rotation of the block about a neutral axis with tension occurring above and compression beneath the neutral point, and tensile failure arises when the tensile stress due to the weight of the

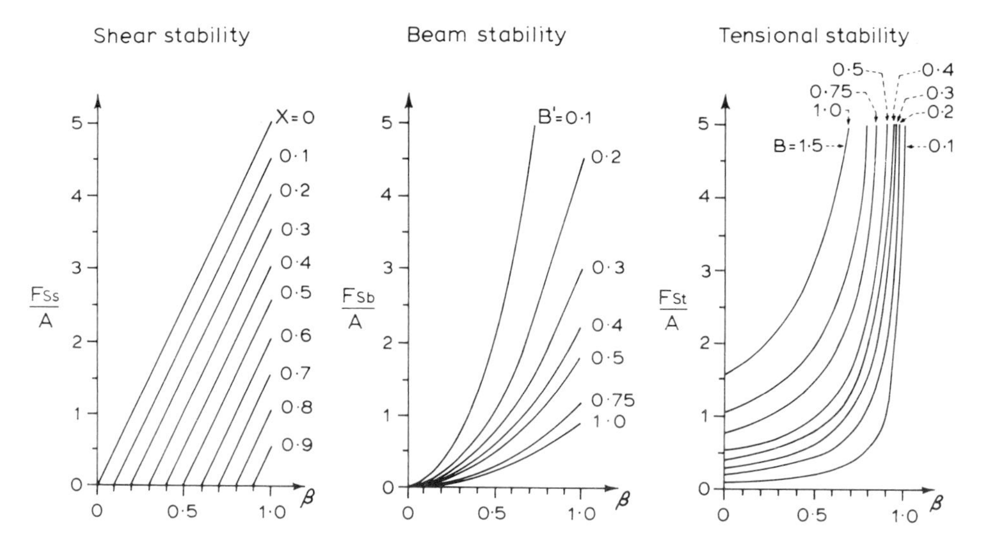

Figure 10.18 Dimensionless stability charts for cantilever stability, with F_{ss}, F_{sb}, and F_{st} being the factors of safety for shear, beam, and tensile failure modes. The parameters defining curves are defined in the text. (*After Thorne and Tovey, 1981*, Earth Surf. Proc. Landforms, **6**.)

lower part of the block exceeds the tensile strength. Failure surfaces may exploit planes of weakness in the soil, such as vertical inter-ped surfaces in the case of shear failure and the zone beneath the root-mat in tensile failure.

These stability conditions have been summarized in the form of dimensionless stability charts (Figure 10.18) defining the factors of safety against failure by each mode. These are simplified by assuming the ratio (r) of tensile to compressive strengths of soil to equal 0.1, a reasonable average value but one to which the factor of safety is in any case insensitive. Thorne and Tovey (1981) suggest that the factor of safety changes by only *c.* 10% for a doubling of the ratio, r. Each factor of safety (F_{ss} for shear, F_{sb} for beam, and F_{st} for tensile failure) is expressed in relation to the general stability parameter A, where

$$A = \sigma_t / \gamma b \tag{10.6}$$

Here, σ_t is the tensile strength (kPa), γ is the unit weight of soil, and b is the overhang width. The unit weight may be taken as 19–20 kN m^{-3} for saturated soil after heavy rain or the passage of a flood wave associated with bank storage, as *c.* 10 kN m^{-3} when submerged, or as *c.* 16 kN m^{-3} when dry. Since the weight of the block is the main force contributing to instability, it is apparent that the worst condition is post-flood but pre-drainage. The relative factor of safety, F_{sx}/A, is for each failure mode a function of the parameter β which measures the degree of tension crack development and is defined as

$$\beta = \frac{H - l}{H} \tag{10.7}$$

(see Figure 10.17). Stability against shear failure is influenced by crack development through the parameter X where

$$X = m/H \tag{10.8}$$

and stability against tensile failure reflects the cantilever geometry

$$B = b/H \tag{10.9}$$

Finally, beam failure is controlled by the parameter B' which also measures cantilever geometry and takes the form

$$B' = B\left(\frac{\beta}{\beta - X}\right)^2 \tag{10.10}$$

For any particular cantilever formed by removal of non-cohesive basal sediment from a composite bank, the three factors of safety can be calculated and the likely mode of failure predicted. For example, Thorne and Tovey (1981) analysed one overhang on the River Severn, of height 0.4 m and depth of overhang 0.3 m, and found that $F_{sb} = 0.92$ when $F_{ss} = 7.22$ and $F_{st} = 6.1$. Thus beam failure was predicted, and was indeed the mode of failure observed just after formation of a tension crack at the surface. Pizzuto (1984) noted that many composite river banks in southern Minnesota collapsed through a beam failure process, evidenced by the forward rotation of blocks into the channel. However, when the thickness of the upper silty unit exceeds 0.5 m the tensile failure mode is more common, because the tensile strength of the bank material decreases systematically with increasing bank height as the vegetation root-mat contributes less to the total strength. For higher banks with thicker silty units, therefore, a complex failure sequence is envisaged in which undercutting first promotes tensile failure of the lower part of the overhang, with beam failure of the upper part of the silty unit then following. Continued retreat of a river bank depends, of course, on the persistence of the collapsed block. Klimek (1974) observed that bank failure in a Polish valley occurs after bankfull flood events saturate the bank sediments, and that the collapsed block then dries and crumbles to be rapidly removed in smaller flood events. However, successive cycles of undercutting and failure may be significantly delayed if the rotation of the block during beam failure presents a resistant vegetated surface to the flow.

Van Eerdt (1985) has also emphasized the predominance of beam failure in the undermined composite banks at the seaward edge of Dutch salt marshes. Her analysis differs from that of Thorne and Tovey (1981), however, in that it emphasizes that the compressive strength of the sediment is sufficiently greater (3–12 times) than the tensile strength so that at failure the compressive stress is less than the compressive strength. Thorne and Tovey (1981) assume that the compressive and tensile strengths are mobilized simultaneously, and accordingly over-estimate the factor of safety by up to 50%. Van Eerdt (1985) derives an expression for the critical undermining overhang width of a rectangular block:

$$b^2 = \frac{\tfrac{1}{3}\sigma_{da}d^2 - \tfrac{1}{3}\sigma_{tmax}t^2}{\tfrac{1}{2}H\gamma} \tag{10.11}$$

Here, d and t are defined in Figure 10.17, σ_{da} is the compressive stress at tensile failure, and σ_{max} is a normalized measure of the tensile strength integrated over the distance

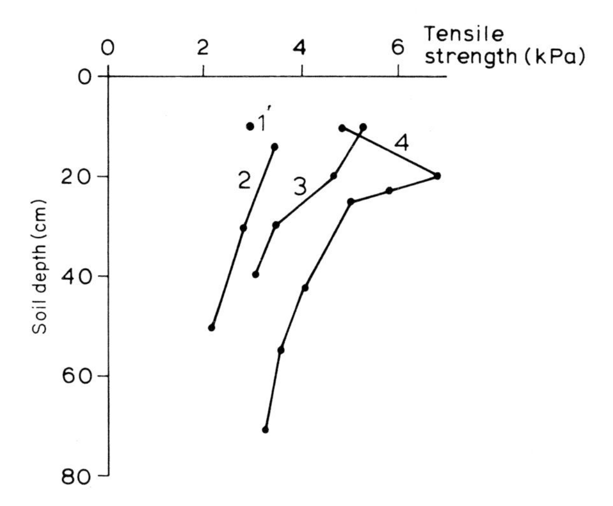

Figure 10.19 The variation with depth of the tensile strength in four salt marsh cliff profiles. (*After Van Eerdt, 1985*, Earth Surf. Proc. Landforms, **6**.)

t, and accounting for depth variation of tensile strength associated with root systems (Figure 10.19).

The stability criteria developed for the cases of composite banks outlined here may also prove relevant to overhangs at the top of higher (over 5 m) coastal and other cliffs subject to the basal erosion and deeper-seated failure processes discussed in previous sections. For example, Barton *et al.* (1983) describe 'spalling' of such overhangs as a secondary mechanism of failure in cliffs predominantly affected by 'slumping'.

ACKNOWLEDGEMENTS

The authors would like to thank Dr J. S. Pethick and BP Ltd for permission to use some material gathered originally for an unpublished report on coastal processes on the Holderness coast.

REFERENCES

Arnett, R. R. (1971). 'Slope form and geomorphological process; an Australian example.' In *Slopes: Form and Process*. Brunsden, D. (ed.), Inst. Brit. Geogrs. Spec. Publ. no. 3, pp. 81–92.

Barton, M. E., Coles, B. J., and Tiller, G. R. (1983). 'A statistical study of the cliff top slumps in part of the Christchurch Bay coastal cliffs.' *Earth Surf. Proc. Landforms*, **8**, 409–22.

Bell, M. (1982). 'The effects of land-use and climate on valley sedimentation.' In *Climatic Change in Later Prehistory*. Harding, A. F. (ed.), Edinburgh Univ. Press, pp. 127–42.

Bishop, A. W. (1955). 'The use of the slip circle in the stability analysis of slopes.' *Geotechnique*, **5**, 7–17.

Bromhead, E. N. (1978). 'Large landslides in London Clay at Herne Bay, Kent.' *Q. J. Eng. Geol.*, **11**, 291–304.

Bromhead, E. N. (1979). 'Factors affecting the transition between the various types of mass movement in coastal cliffs consisting largely of overconsolidated clay, with special reference to Southern England.' *Q. J. Eng. Geol.*, **12**, 291–300.

Brunsden, D. (1974). 'The degradation of a coastal slope, Dorset, England.' In *Progress in Geomorphology*. Brown, E. H. and Waters, R. S. (eds.), Inst. Brit. Geogr. Spec. Publ. no. 7, pp. 79–98.

Brunsden, D., and Kesel, R. H. (1973). 'Slope development on a Mississippi River bluff in historic time.' *J. Geol.*, **81**, 576–97.

Cambers, G. (1976). 'Temporal scales in coastal erosion systems.' *Inst. Brit. Geogr., New ser.*, **1**, 246–56.

Carson, M. A. (1971). *The Mechanics of Erosion*. Pion, London, p. 174.

Carson, M. A., and Petley, D. J. (1970). 'The existence of threshold hillslopes in the denudation of the landscape.' *Trans. Inst. Brit. Geogrs.* **49**, 71–95.

Chandler, R. J. (1976). 'The history and stability of two Lias clay slopes in the upper Gwash valley, Rutland.' *Phil. Trans. Roy. Soc., London,* **A283**, 463–90.

Conway, B. W. (1979). 'The contribution made to cliff instability by Head deposits in the west Dorset coastal area.' *Q. J. Eng. Geol.*, **12**, 267–75.

Eisbacher, G. H. (1979). 'Cliff collapse and rock avalanches (Sturzstroms) in the Mackenzie Mountains, north western Canada.' *Can. Geotech. J.*, **16**, 309–34.

Eyles, N., and Sladen, J. A. (1981). 'Stratigraphy and geotechnical properties of weathered lodgment till in Northumberland, England.' *Q. J. Eng. Geol.*, **14**, 129–41.

Gerber, E. K., and Scheidegger, A. E. (1973). 'Erosional and stress-induced features on steep slopes.' *Zeitschrift fur Geomorphologie* Supp. 18, 38–49.

Grissinger, E. H. (1982). 'Bank erosion of cohesive materials.' In *Gravel Bed Rivers*. Hey, R. D., Bathurst, J. C. and Thorne, C. R. (eds.), Wiley, Chichester, 273–87.

Hutchinson, J. N. (1967). 'The free degradation of London Clay cliffs.' *Proc. Geotech. Conf.*, Oslo, **1**, 113–18.

Hutchinson, J. N. (1968). 'Field meeting on the coastal landslides of Kent, 1st–3rd July 1966.' *Proc. Geol. Assoc., London,* **79**, 227–37.

Hutchinson, J. N. (1969). 'A reconsideration of the coastal landslides at Folkestone Warren, Kent.' *Geotechnique*, **19**, 6–38.

Hutchinson, J. N. (1983). 'A pattern in the incidence of major coastal mudslides.' *Earth Surf. Proc. Landforms*, **8**, 409–22.

Hutchinson, J. N., and Bhandari, R. K. (1971). 'Undrained loading, a fundamental mechanism of mudflows and other mass movements.' *Geotechnique*, **21**, 353–8.

Hutchinson, J. N., Bromhead, E. N., and Lupini, J. F. (1980). 'Additional observations on the Folkestone Warren landslides.' *Q. J. Eng. Geol.*, **13**, 1–31.

Hutchinson, J. N., Chandler, M. P., and Bromhead, E. N. (1981). 'Cliff recession on the Isle of Wight, SW coast.' *Proc. 10th Inst. Conf. Soil Mech. Fndtn. Engng.*, 429–34.

Joyce, M. D. (1969). *A geotechnical study of the boulder clay cliffs of Holderness, East Yorkshire.* Unpubl. Msc Thesis, University of Leeds.

Klimek, K. (1974). 'The retreat of alluvial river banks in the Wisloka Valley (South Poland).' *Geogr. Polonica*, **28**, 59–75.

Lohnes, R. A., and Handy, R. L. (1968). 'Slope angles in friable loess.' *J. Geol.*, **76**, 247–58.

Madgett, P. A., and Catt, J. A. (1978). 'Petrography, stratigraphy and weathering of Late Pleistocene tills in east Yorkshire, Lincolnshire and north Norfolk.' *Proc. Yorks. Geol. Soc.*, **42**, 55–108.

Marsland, A., and Powell, J. J. M. (1985). 'Field and laboratory investigations of the clay tills at the Building Research Establishment test site at Cowden, Holderness.' *Proc. Int. Conf.*, 12–14 March 1985. *Construction in Glacial Tills and Boulder Clays*. Forde, M. C. (ed.), Edinburgh Technics Press, pp. 147–68.

Nash, D. (1980). 'Forms of bluffs degraded for different lengths of time in Emmett County, Michigan, USA.' *Earth Surf. Proc. Landforms*, **5**, 331–45.

Pitts, J. (1983). 'The temporal and spatial development of landslides in the Axmouth–Lyme Regis undercliffs National Nature Reserve, Devon.' *Earth Surf. Proc. Landforms*, **8**, 589–603.

Pizzuto, J. E. (1984). 'Bank erodibility of shallow sandbed streams.' *Earth Surf. Proc. Landforms*, **9**, 113–24.

Pringle, A. W. (1985). 'Holderness coast erosion and the significance of ords.' *Earth Surf. Proc. Landforms*, **10**, 107–24.

Richards, K. S. (1977). 'Slope form and basal stream relationships: some further comments.' *Earth Surf. Proc.*, **2**, 87–92.
Richards, K. S., and Anderson, M. G. (1978). 'Slope stability and valley formation in glacial outwash deposits, North Norfolk.' *Earth Surf. Proc.*, **3**, 301–18.
Rouse, W. C. (1976). 'An interpretation of valley slopes in South Wales.' *Cambria*, **3**, 11–32.
Russell, D. J., and Eyles, N. (1985). 'Geotechnical characteristics of weathering profiles in British overconsolidated clays (Carboniferous to Pleistocene).' In *Geomorphology and Soils*. Richards, K. S., Arnett, R. R., and Ellis, S. (eds.), George Allen & Unwin, London, pp. 417–36.
Savigear, R. A. G. (1952). 'Some observations on slope development in South Wales.' *Trans. Inst. Brit. Geogrs.*, **18**, 31–51.
Selby, M. J. (1982). 'Controls on the stability and inclinations of hillslopes formed on hard rock.' *Earth Surf. Proc. Landforms*, **7**, 449–67.
Shotton, F. W. (1978). 'Archaeological inferences from the study of alluvium in the lower Severn and Avon valleys.' In *The Effect of Man on the Landscape: The Lowland Zone*. Limbrey, S. and Evans, J. G., (eds.), Council for British Archaeology Res. Report 21, 27–32.
Silvestri, T. (1961). 'Determinazione sperimentale de resistenza meccanica del materiale constituente il corpo di una diga del tipo "rockfill".' *Geotechnica*, **8**, 186–91.
Skempton, A. W. (1953). 'Soil mechanics in relation to geology.' *Proc. Yorks. Geol. Soc.*, **29**, 33–62.
Skempton, A. W., and Brown, J. D. (1961). 'A landslide in boulder clay at Selset, Yorkshire.' *Geotechnique*, **11**, 280–93.
Skempton, A. W., and de Lory, F. A. (1957). 'Stability of natural slopes in London Clay.' *Proc. 4th Int. Conf. Soil Mech. Fndtn. Eng.*, **2**, 378–81.
Sladen, J. A., and Wrigley, W. (1983). 'Geotechnical properties of lodgement till.' In *Glacial Geology: An Introduction for Engineers and Earth Scientists*. Eyles, N. (ed.), Pergamon, Oxford.
Statham, I. (1975). 'Slope instabilities and recent slope development in Glencullen, Co. Wicklow.' *Irish Geog.*, **8**, 42–54.
Stimpson, B. (1979). 'Simple equations for determining the factor of safety of a planar wedge under various groundwater conditions.' *Q. J. Eng. Geol.*, **12**, 3–7.
Strahler, A. N. (1950). 'Equilibrium theory of erosional slopes approached by frequency distribution analysis.' *Am. J. Sci.*, **248**, 673–96, 800–14.
Suthons, C. T. (1963). 'Frequency of occurrence of abnormally high sea levels on the east and south coasts of England.' *Proc. Inst. Civ. Eng.*, **25**, 433–49.
Terzaghi, K. (1950). 'Mechanism of landslides.' *Geol. Soc. Am., Berkey Volume*. pp. 83–123.
Terzaghi, K. (1962). 'Stability of steep slopes on hard unweathered rock.' *Geotechnique*, **12**, 251–70.
Thorne, C. R., and Tovey, N. K. (1981). 'Stability of composite river banks.' *Earth Surf. Proc. Landforms*, **6**, 469–84.
Valentin, H. (1954). 'Der Landverlust in Holderness, Ostengland von 1852–1952.' *Erde*, **6**, 296–315.
Van Eerdt, M. M. (1985). 'Salt marsh cliff stability in the Oosterschelde.' *Earth Surf. Proc. Landforms*, **10**, 95–106.

Slope Stability
Edited by M. G. Anderson and K. S. Richards

Chapter 11

General Models of Long-Term Slope Evolution Through Mass Movement

M. J. KIRKBY
School of Geography, University of Leeds, Leeds LS2 9JT

11.1. THE CONTEXT OF LONG-TERM SLOPE DEVELOPMENT

On the geological time span over which slope profiles evolve, landslides and other rapid mass movements occur almost instantaneously. Although it may, in principle, be possible to aggregate the effects of individual movements, there are severe difficulties with this approach. Perhaps the most important is that work on rapid mass movements has concentrated on stability analysis, so that forecasting of destinations for slide debris is very inexact, even for an individual slide. The second major problem lies in aggregating from the individual slide, for which soil moisture may in principle be known, to the assemblage of slides over a long period, generally without detailed meterological records. Even with long-term records the modelling task is disproportionate to its probable payoffs. In other words, the change of scale, particularly of time-scale requires models which are built from somewhat different premises than those of short-term stability analysis, although plainly they must be compatible with it.

Alternative approaches to the long-term evolution of slopes have generally concentrated on the slope profile as the unit of study, although some steps have been taken towards the simulation of hillslopes with two horizontal dimensions. Generally slope models have been based on budgets of sediment passing down the length of the profile, with a variety of 'process laws' to describe the rates at which material is transferred through or removed from each of a number of sections down the length of the profile. Simple models of this kind generally describe either 'transport limited' or else 'weathering limited' removal. The former applies to slopes with a soil cover thick enough to allow sediment transport to proceed without constraint by material availability, so that the transport process(es) operate(s) at full capacity, and control(s) the rate of slope evolution. Seasonal soil creep within the top half metre of the soil is commonly considered as a good example of a transport limited removal process. Detachment or weathering limited removal occurs where sediment transport is constrained by the supply of material, which thus controls the actual transport rate. Rock-fall from a cliff face is a good example, since the transporting capacity for sediment transport is evidently large, whereas the actual rate is small, and limited by the weathering of joint blocks, etc., until they are loose enough to fall.

It may be seen that both of the examples quoted are for mass movement processes, which can thus fall into either category in suitable cases. However, most forms of rapid mass movement are thought to lie in or close to the weathering limited category, since movement is dictated by stability criteria which are concerned with their detachment. Some mass movements may be intermediate between these categories, behaving as weathering limited in their source area, but as transport limited in their destination area. This possibility is also explored below.

When measured rates of rapid mass movement processes are compared with those for flow erosional processes like solution, water-produced soil erosion, rainsplash and soil creep, it may be seen that mass movements are generally characterized by a lower threshold of stability, below which their rate is negligible and above which it rises rapidly. The slower processes are generally thought to have no lower threshold gradient, but to increase steadily with gradient over a wide range of values. The slower processes are thus dominant on gradients up to, and very slightly above the stability threshold, and mass movements strongly dominate at steeper gradients. Since most slope profiles comprise elements at gradients below any relevant thresholds, as well perhaps as steeper elements, it is relevant to construct and use models which incorporate a range of erosional processes. It may be particularly interesting to consider areas at near-threshold gradients, where slow and rapid processes are of similar magnitude, so that both may play an effective role in shaping the profile.

A particular problem in simulating slope development by mass movements is their large individual size. A single movement can have a large enough impact on the slope profile to influence strongly the course of subsequent development, by changing topography and thence loadings and pore-water distributions. In an aggregate model, the influence of individual events will inevitably be largely suppressed, so that their particular influence on the subsequent course of development will be lost. Long-term models are, therefore, likely to be most effective for mass movements in which individual events are small. In the context of landslides on London Clay cliffs, Hutchinson (1967, 1973) has demonstrated a sequence from large deep-seated rotational slides towards smaller and shallower slides while the profile declines in average gradient towards a rather uniform gradient at an ultimate threshold value. In this kind of sequence the deep-seated slides occur rapidly, over a relatively brief period, whereas the shallow slides occur over a much longer period. It is, therefore, argued that imprecise modelling of the deep-seated phase will tend to have rather little influence on the long-term evolution of the landscape, so that the imprecision may, in practice, be acceptable.

Although models have not been developed to date for many kinds of mass movements, and flow-slides in particular are poorly represented, there have been a number of attempts at modelling slope evolution through mass movements, either on their own or in combination with associated slower processes. The greatest amount of work has been on rock fall and the associated talus accumulation, both for humid and arid slopes, and this is described in the next section. Section 11.3 describes alternative approaches to soil landslides and a model which combines them with slow processes. The final section (section 11.4) considers the relationships between landslides and weathering rates, both as a guide to long-term landslide rates and through the development of changing geotechnical properties over time.

11.2 ROCK FALL AND TALUS ACCUMULATION

The earliest model for cliff retreat was proposed by Fisher (1866), and most subsequent developments have generalized his formulation. In this family of models, the cliff is assumed to be vertical or at a constant angle, and the part exposed above the talus is assumed to retreat uniformly, parallel to itself. No assumption is made about rate of retreat, which is not required to produce the predicted forms, but only to date them. Material falling from the cliff is assumed to accumulate in a talus of uniform gradient which remains constant over time. As the talus builds up it progressively protects a core of *in situ* rock which develops a parabolic shape, initially steep and eventually at the same angle as the talus burying it. Much of the theory has concentrated on forecasting the form of this rock core. Rock fall is perhaps the type of rapid mass movement which comes closest to meeting the modelling requirement for individually small events.

The work of Lehmann (1933) and Bakker and Le Heux (1946, 1947, 1950, 1952) considered a wide range of possible variants, perhaps most significantly in introducing the notion of a bulking factor, initially to allow for the increase in cross-section of material in the scree resulting from its more open packing. They also used the same constant to allow for removal of material from the talus either by a basal river or the sea, or by weathering of talus material to fines which were then assumed to be washed out. These processes do not in general lead to the same sequence of forms as are forecast by using a bulking factor much less than 1.0, even if the end-products are similar. The model presented below allows both for bulking and weathering, although with some approximation unless one or other factor is ignored.

Figure 11.1 shows a diagrammatic cross-section of a retreating cliff of gradient angle β and constant height h, initially at OT. At some subsequent time, its position is at PQ, projected down to N. The curve OP shows the form of the rock core, protected beneath the talus surface PM, inclined at constant angle α. The development of the rock core is traced using inclined axes as indicated, with origin at O, to represent the point P (x, y). In retreating a horizontal distance Dx the rock core advances from P to P′. The cross-section PQQ′P′ is removed from the cliff cross-section, and after bulking by a factor k is added to the talus cross-section. At the same time a thickness of $m.Dx$ (measured horizontally) is removed from the talus either by basal undercutting or by weathering of the talus material.

Budgeting for the cross-section PNM, the input from the cliff, allowing for bulking is $k(h-y)Dx \sin\beta$, and the output from weathering is $mDx \sin\beta$. The resulting change in cross-section PMM′P′ is (PM $\cos\alpha$).PL, where PL is the vertical height change (see Figure 11.1(b)) $[Dy \sin\beta - (Dx + Dy \cos\beta) \tan\alpha]$. The overall storage equation for PNM is then;

$$[k(h-y) - my]Dx \sin\beta = y \sin\beta \cot\alpha [Dy \sin\beta - (Dx + Dy \cos\beta)\tan\alpha] \qquad (11.1)$$

Cancelling $\sin\beta$ terms and grouping terms in x and y;

$$\frac{y\, Dy}{kh - y(k + m - 1)} = \frac{Dx}{\sin\beta \cot\alpha \cos\beta} \qquad (11.2)$$

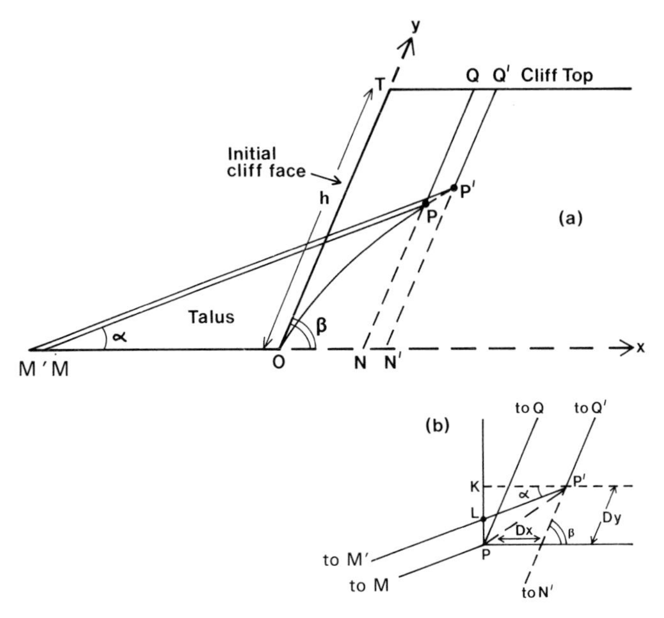

Figure 11.1 (a) Schematic diagram for analysis of cliff retreat and talus accumulation. (b) Detail of pp′ area

In general this may be solved as a differential equation when Dx and Dy are small enough, leading to the solution;

$$(k+m-1)^2x/[kh(\sin\beta\ \cot\alpha-\cos\beta)] = \ln[1-(k+m-1)y/(kh)]-(k+m-1)y/(kh) \tag{11.3}$$

There are two important special cases, which are more readily derived directly from equation (11.2).

For $k=0$: $$y=(m-1)x/(\sin\beta\ \cot\alpha-\cos\beta) \tag{11.4}$$

and for $k+m=1$: $$y^2=2khx/(\sin\beta\ \cot\alpha-\cos\beta) \tag{11.5}$$

The previous models quoted above all ignore the weathering constant, m. Fisher's original model is obtained from equation (11.5) for no change in bulk ($k=1$) and for a vertical cliff ($\beta=90°$). It describes a parabolic rock core, initially rising vertically but reclining gradually until, as the cliff top is reached, the rock core is tangential to it, as is shown in Figure 11.1(a). The various papers of Bakker and Le Heux (ibid.) generalized to the cases of inclined cliffs and various bulking factors (k not 1). They derived equation (11.3) with $m=0$, which describes a curve which initially rises at the angle of the cliff and finally at the angle of the talus (Figure 11.2(a)). They also drew attention to the special case of $k=0$ (equation (11.4) with $m=0$), relating it to desert 'boulder controlled slopes' like those described by Bryan (1922). This is incorrect, both because it does not correctly allocate the material eligible for weathering (i.e. the talus material and not the cliff material), and because it does not permit the possibility of considerable lateral retreat of the cliff and boulder

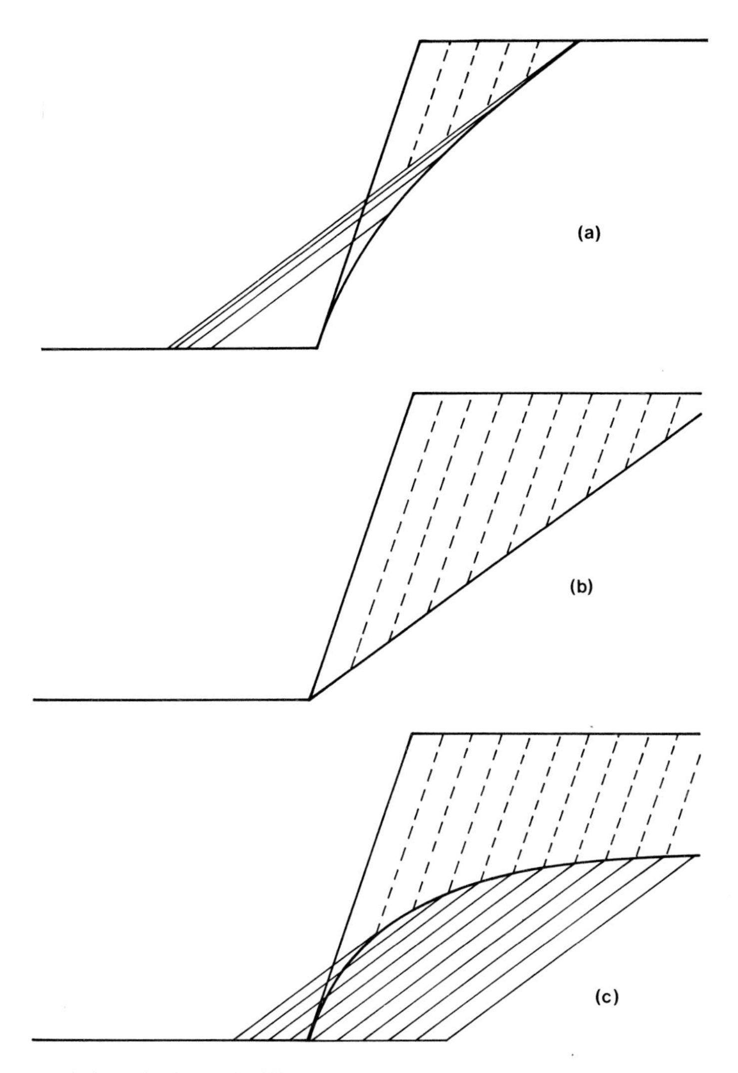

Figure 11.2 Accumulation of talus and cliff retreat models, for various rates of bulking (k) and weathering (m) for a 70° cliff. (a) $k=1$: $m=0$: Simple talus accumulation. (b) $k=0$: $m=0$: 'Richter' slope. (c) $k=1$: $m=1.5$: Mesa retreat with back-wearing of boulder-controlled slope

controlled slope unit without consuming the cliff, as is evident from the presence of many buttes, like those in the American south-west and elsewhere. Their $k=0$ case appears to be appropriate only to the case of a cliff retreating from a static river bank or coast at which all new rock fall material is removed (Figure 11.2(b)).

Weathering is required to export material from the talus, and it may be seen that for all non-zero rates of weathering ($m>0$), the cliff form necessarily continues to retreat, even after the talus has buried the original cliff. In many cases, it may be seen from equation (11.5) that the cliff is never consumed, but retreats indefinitely towards a rock core height of $y=hk/(k+m-1)$, which is less than the cliff height h if m, the ratio of scree weathering rate to cliff retreat rate, is greater than 1. This case is illustrated in Figure 11.2(c). It shows the initial development of a normal talus and rock core, and a subsequent period during which the talus gradually eats into the rock core to produce a retreating boulder controlled slope with a constant ratio of cliff height to boulder controlled slope height. The model thus allows for the formation and long survival of buttes with both slope elements present, and compares well with forms in the Colorado plateau area, such as sandstone cliffs with boulder controlled slopes of up to about half the total butte height, corresponding to ready weathering of the sandstone blocks ($m>2$). On granite outliers with highly resistant boulders ($m<1$), there is generally no cliff element and the entire slope is boulder veneered at a talus gradient. It should be noted that the model requires further refinement where bulking factors differ notably from 1.0, since no allowance is made for possible differences of weathering release from true talus and from exposed rock core. This model for talus weathering applies most directly to arid and semi-arid conditions, since it is assumed that the weathered material is readily washed from the talus slopes. For a humid area, weathering would instead largely contribute to an *in situ* soil, equivalent to the case of $m=0$ if transport limited removal of the soil is ignored. In this case the same model is relevant to describing the geometry of slope replacement where there is more than one threshold gradient.

Simple cliff/talus models assume constancy of both cliff and talus gradients, and both may be questioned, even though differences do not affect the qualitative results of the simple model. Cliff gradients depend very strongly on the spacing and frictional properties of joints, bedding planes and other discontinuities in the rock face. The controlling factors are more fully discussed in Chapter 15 below, but it is relevant to note here the variation of cliff gradient with height, and its possible decline over time. Figure 11.3(a) shows a simple wedge failure for a straight cliff of height h at angle β, with failure along a plane at angle α. The maximum stable height is then given by:

$$\gamma h/(2c)=\sin\alpha \cos\phi/[\sin(\beta-\phi) \sin(\alpha-\beta)] \tag{11.6}$$

This is the relevant expression where the failure direction is controlled by discontinuities. Where they are not dominant, the angle β is determined by minimizing the safety factor, when it takes the value $(\alpha+\phi)/2$ and

$$\gamma h/(2c)=\sin\alpha \cos\phi/\sin^2[(\alpha-\phi)/2] \tag{11.7}$$

The forms of expressions (11.6) and (11.7) are shown in Figure 11.3(b) for $\phi=30°$. The decline in mean gradient with cliff height is paralleled by real cliffs (e.g. Grant-Taylor, 1964, quoted in Selby, 1982, Figure 7.7; see also Chapter 10 of this book). This variation of mean gradient with height leads in principle to the formation of concave cliff profiles, although other factors generally obscure any clear empirical relationship. The second possible complication for cliff angles in the simple cliff talus model is that they may recline with time towards a stable value. Rates of cliff retreat may be assumed to increase in some manner with gradient, perhaps in proportion to their gradient in excess of a threshold value. As cliff retreat proceeds, there may be an explicit dependence on elevation, as proposed in some of Scheidegger's (1961) models, which produces more rapid retreat of the cliff top and a direct decline in angle. For sufficiently high cliffs in suitable environments, differences in for example freeze–thaw intensity may have such an effect. Cliff decline may

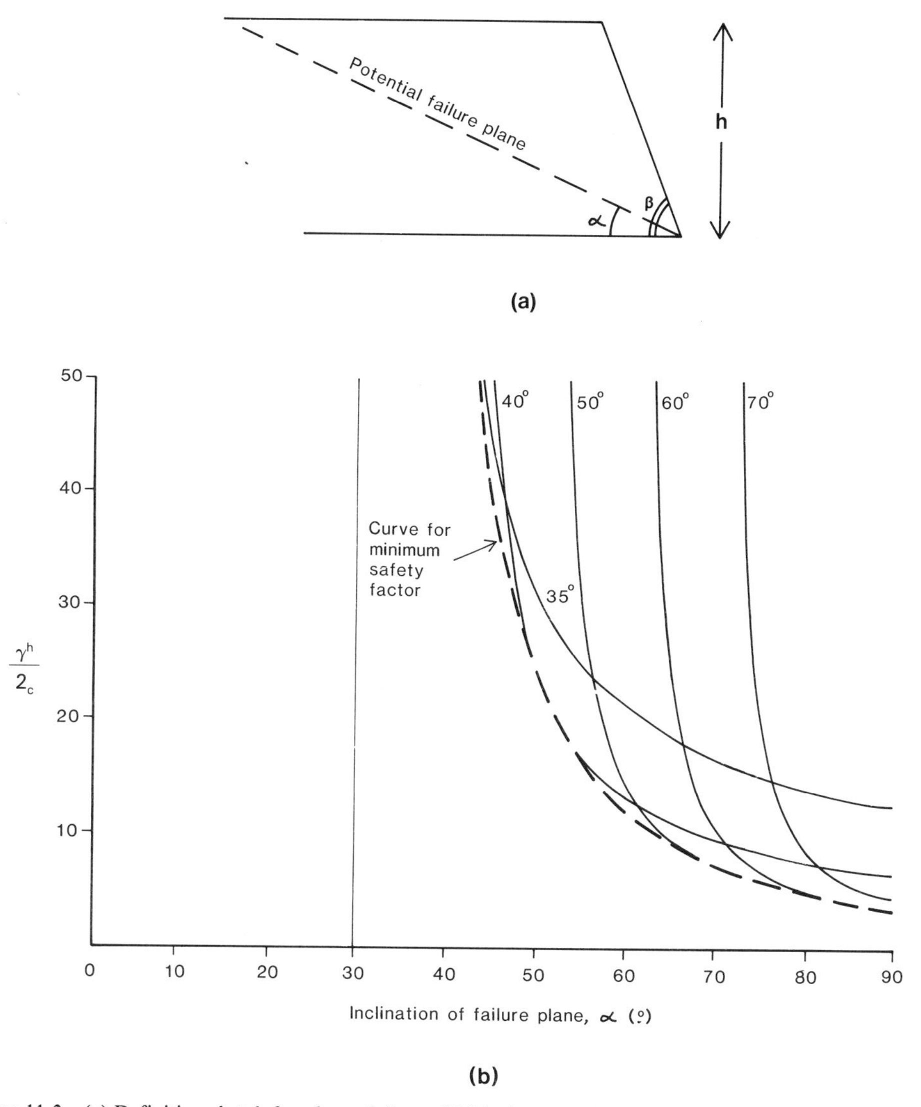

Figure 11.3 (a) Definition sketch for planar failure. (b) Maximum safe height for given failure direction (solid curve labelled with β), and curve where β is unconstrained by discontinuities (dotted curve)

also occur firstly through interaction with slower processes above the cliff, and secondly through some limitations on transport rate, the lower parts of the cliff being partially protected from retreat by debris from above, temporarily resting on it. These possibilities, and appropriate formulations for rates of retreat are discussed in the context of a landslide model below.

There have been a number of models for the retreat of cliffs or other slopes by weathering limited processes, but which have not been concerned with the problem of basal deposition. Scheidegger (1961) modelled slope development for rates of lowering constant, or

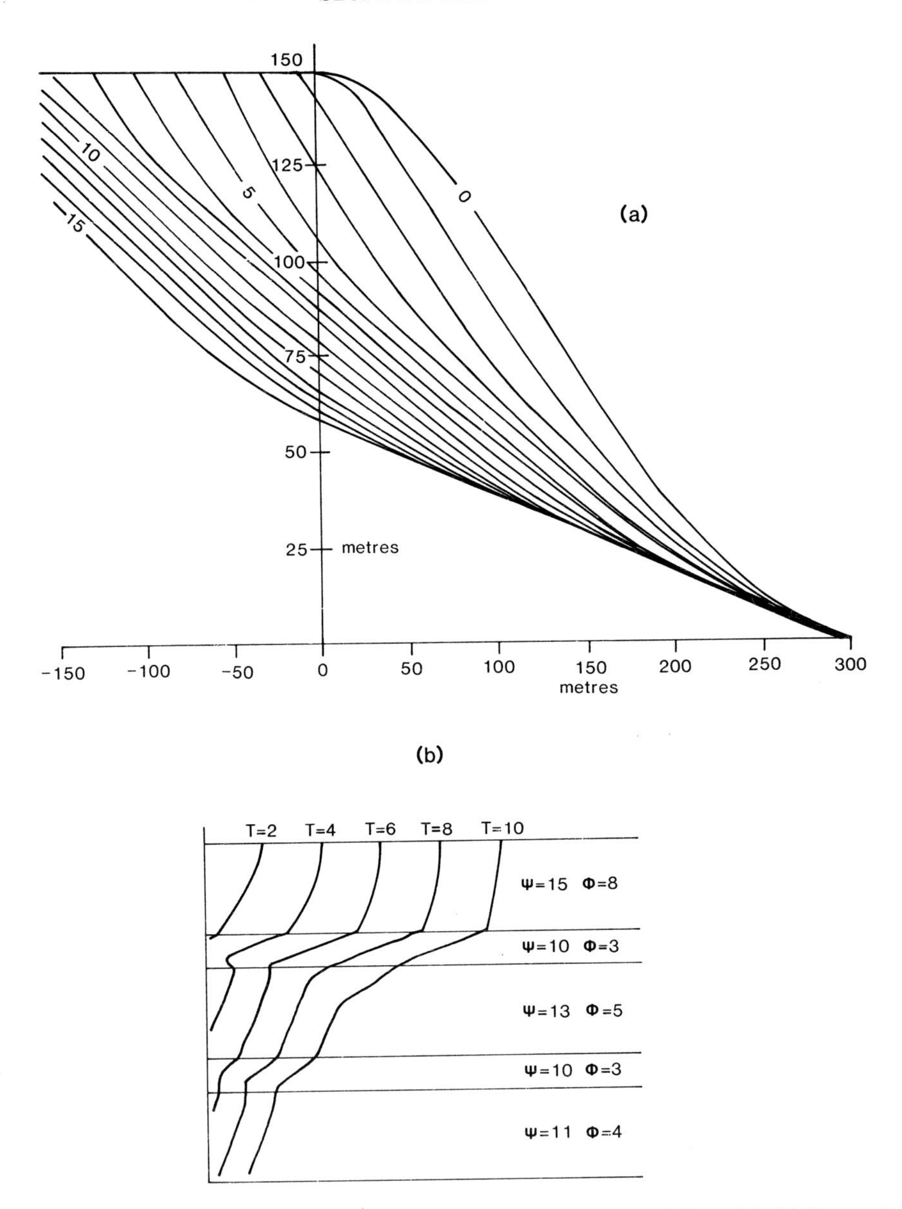

Figure 11.4 (a) Weathering limited retreat of a uniform material, following Luke's model. (b) Retreat for a stratified material. (*From Aronsson and Linde, 1982*, Earth Surf. Proc. Landforms, **7**.)

proportional to gradient or elevation; with the lowering rate taken to be either vertical or perpendicular to the slope. For the vertical lowering case, these lead respectively to simple vertical or horizontal retreat of existing forms, and to an exponential (Davisian) decline of elevation at all points. For perpendicular lowering, these forms are slightly modified. Initial forms consisting of a cliff with a plateau above it lead to slope sequences which preserve the sharp break in slope at the cliff top, with cliff gradient declining only in the

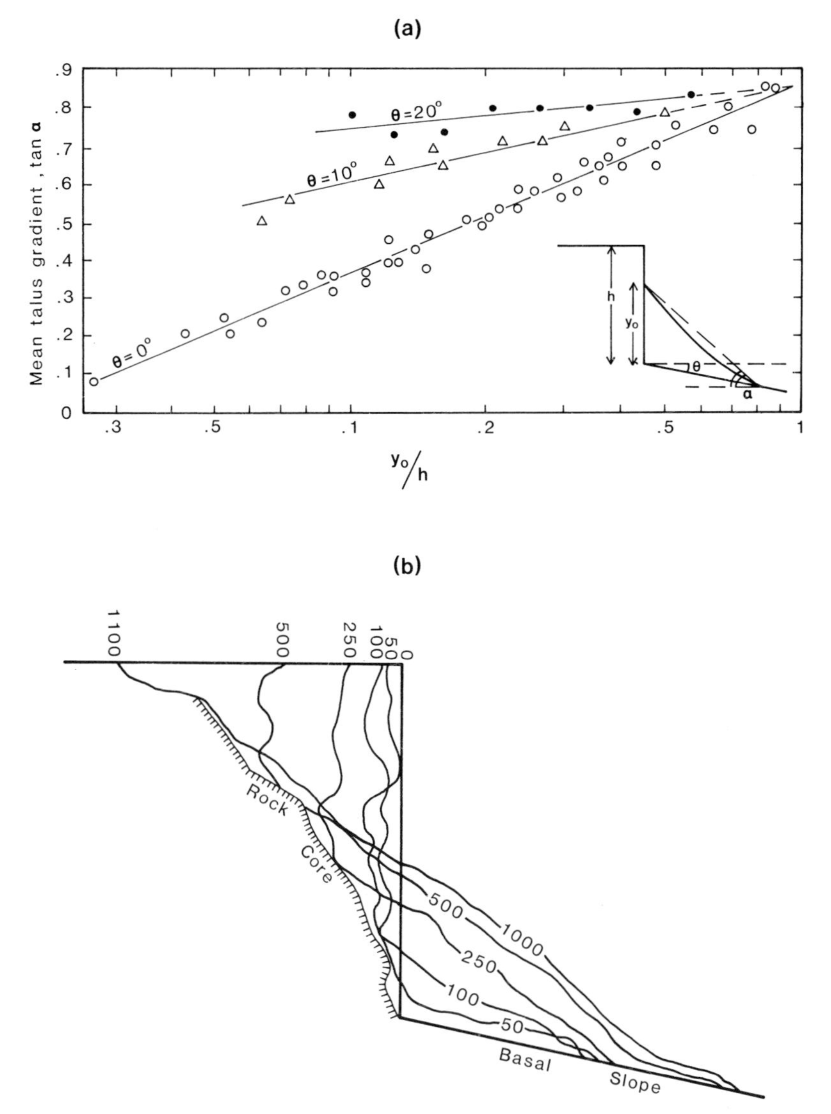

Figure 11.5 (a)Mean talus gradient as a function of relative height of cliff and talus. (b) Rock core formation with cliff retreat, including grain size and drop height effects. (*Both from Kirkby and Statham, 1975*, J. Geol., **83**.)

Davisian case. A more flexible approach to the simulation of weathering limited removal has been presented by Luke (1972, 1974, 1976) and illustrated in the context of the Grand Canyon by Aronsson and Linde (1982). This makes use of a kinematic wave solution to generate 'characteristics' along which gradient remains constant during slope evolution. Where characteristics intersect, 'shocks' may develop which generate sharp breaks in slope in previously smooth surfaces. The characteristics may be plotted from the equations:

$$\partial X/\partial t = \partial f/\partial g \tag{11.8}$$

$$\partial Z/\partial t = g\ \partial f/\partial g - f \tag{11.9}$$

where g is the slope gradient, X, Z are, respectively, the horizontal and vertical coordinates along the characteristic, measured from the slope base and $f(X,Z,g)$ is the rate of lowering at any point.

Figure 11.4(a) illustrates the construction of characteristics and simulated slope evolution for a simple case where the rate of lowering increases with gradient in relation to two thresholds g_0 and g_1:

$$\begin{aligned} f &= 0 \text{ for } g < g_0 \\ &= k_0(g - g_0) \text{ for } g_0 < g < g_1 \\ &= k_1 \end{aligned}$$

Figure 11.4(b) shows the application of this method to a stack of strata of differing resistance, broadly representative of conditions in the Grand Canyon.

Constancy of talus gradient may also be questioned for the simple cliff/talus models. Talus profiles are commonly slightly concave, and evidence for experimental talus slopes suggests that their average gradient is least and their concavity greatest when the scree is low relative to the cliff height. Figure 11.5(a) exemplifies the results obtained. The effect on overall simulated cliff development is shown for a stochastic model in Figure 11.5(b). Neither this example nor parallel analyses suggest that there are very significant differences from the simpler models described above. The direction of the slight differences is to reduce the gradient of the rock core near its base.

11.3 SLOPE EVOLUTION MODELS INCORPORATING LANDSLIDES

If landslides or other mass movements act alone, the previous section has described ways to model them as weathering limited processes. This approach breaks down either if the mass movements are not wholly weathering limited, or if they are combined with other, transport limited processes. Both of these cases may be dealt with by using the concept of surplus capacity, which postulates that lowering is proportional to the excess of transporting capacity over actual transport rate; and vice versa for deposition provided that the actual transport remains positive. This is a first order reaction model, proposed in various geomorphological contexts by Kirkby (1971), Foster and Meyer (1972), and Bennett (1974). In combination with a mass balance equation, it provides a framework for slope modelling within which transport- and weathering-limited removal are special cases.

The mass balance or continuity equation may be written, for a simple slope profile (with one horizontal dimension), as:

$$\partial S/\partial x + \partial z/\partial t = 0 \tag{11.10}$$

where S is the actual sediment transport per unit contour length
x is distance measured downslope from the divide
z is elevation
and t is elapsed time

This equation states that net erosion (or deposition) is associated with an increase (or decrease) in actual sediment transport. One simple approach used by Ahnert (1973) and Armstrong (1976) consists of using equation (11.10) to describe transport limited removal ($S = C$) by slow processes, and to combine this with instantaneous removal of all material on gradients above a threshold value of, say 45°. This method reproduces the threshold concept of mass movement, although, in an over-simple way.

Pursuing the concept of surplus capacity, a second equation expresses the rate of erosion in terms of the difference $(C - S)$:

$$-\partial z/\partial t = (C - S)/h = D - S/h \tag{11.11}$$

where C is the capacity transport rate,
D is the potential rate of detachment $(= C/h)$
and h is mean distance travelled by the moving material.

In this equation, h is the rate constant which converts surplus capacity into erosion, but may also be identified as the average (horizontal) distance travelled by detached material. In general both D and h may depend on the process and climate operating, and on slope topographic and soil factors. In some contexts it is convenient to exchange horizontal and vertical axes, in which case D and h should be re-interpreted as rate of lateral slope retreat and vertical distance of travel. The ratio of horizontal and vertical values is equal to the slope gradient.

Two special cases may usefully be distinguished, which correspond to transport- and weathering-limited removal. First, if the travel distance h is small, it follows that $S \simeq C$, so that the continuity equation converges on its transport-limited form, with C replacing S in equation (11.10). Second, if h is large, $S << C$ and the equations converge on $-\partial z/\partial t = D$, which is the weathering-limited case described above. In physical terms, the transport-limited processes, such as soil creep or wash, are associated with travel distances that are short relative to the length of the slope (though they may be important within an erosion plot); whereas landslides and removal in solution are associated with much longer travel distances, so that it is consistent to consider them as primarily weathering limited. Below, the general case is explored, on the assumption that landslides are significantly transitional between the simple extreme cases.

The simplest functional forms which have been found to explain some of the variation in rates of lowering and travel distance assume that the rate of lateral retreat increases linearly or as a power law above a threshold of ultimate stability; and that the travel distance is that due to frictional sliding to rest from a fixed initial downslope velocity. These assumptions are clearly great simplifications of the complex processes which determine slope stability and slide travel after failure, but they provide an initial estimate of the gradient dependence of these two factors which is rationally related to two important gradient thresholds in the landscape; of ultimate stability and of talus stability. The expression for detachment capacity (or rate of unconstrained lowering) is:

$$D = \alpha(i - i_*)^m i \tag{11.12}$$

where i is tangent slope gradient
i_* is its ultimate lower threshold of stability
m is an exponent, usually 1–2
and α is a rate constant.

The horizontal travel distance is:

$$h = h_0/(i_0 - i) \tag{11.13}$$

where $i_0 = \tan\phi$ is the tangent of the effective angle of friction for surface material. It is the angle below which moving material comes to rest rather than continuing downslope indefinitely; that is the talus angle for clastic materials. It is normally a steeper gradient than the ultimate lower threshold of sliding, i_*.

Figure 11.6(a) shows some values for rates of lowering, compared with curves following equation (11.12). The data for London Clay is derived from Hutchinson (1973) for degradation of coastal cliffs. A curve may be fitted through his data (shown in Figure 11.6(b) for $\alpha = 10$ m yr^{-1}, $i_* = 0.14 = \tan(8°)$, $m = 2$, $h_0 = 1$ m and $i_0 = 0.4 = \tan(22°)$. The threshold gradient is that found for long-term stability of inland slopes in London Clay. The values for Old Red Sandstone on Exmoor (N. Devon) are approximate ones, obtained from a consideration of solute denudation rates and the state of soil weathering (Kirkby, 1973). The lowest curve, for ORS in south Wales, is obtained by optimizing the parameters of equation (11.12) to fit observed coastal slope profiles (Kirkby, 1984). The parameters are $\alpha = 0.001$ m yr^{-1}, $i_* = 0.4 = \tan(22°)$, $m = 1$, $h_0 = 20$ m and $i_0 = 0.7 = \tan(35°)$. Inland slopes in the study area rarely have maximum gradients of less than 24° and it is argued that this 2° excess over the ultimate threshold is appropriate given the response time for this relatively resistant bedrock. In the south Wales study, the form of equation (11.12) was found preferable to more complex forms, either with non-linear dependence or multiple thresholds. This result may not be general, and the fit to the London Clay data in Figure 11.6 suggests a stronger rate of increase with gradient above about 25°.

The expression for travel distance (equation (11.13)) shows a correct response at gradients above the talus gradient (or appropriate comparable threshold for coming to rest in non-clastic materials), in that material keeps going indefinitely. At low gradients material, once started, will always travel some distance, although decreasing with gradient, and equation (11.13) again provides this qualitative response. A more doubtful element of the expression is the assumption of a constant initial velocity in the downslope direction. It may be more rational to consider dependence on local gradient and elevation, but no suitable expression has been proposed to date. For the south Wales study, the parameters in equation (11.13) were optimized to $i_0 = 0.7 = \tan(35°)$, corresponding to a normal talus gradient for a lithology with a high clast content; and $h_0 = 20$ metres, corresponding to free fall from this height, and giving the rather higher initial velocity of 20 m s^{-1}.

Figure 11.7(a) shows the expressions used for lowering and travel distance from the south Wales data. For gradients of between i_* and i_0, both D and h are defined, and consequently their product, the transporting capacity C also has a defined valued, which is shown in Figure 11.7(b) for comparison with accepted rates of solifluction transport. It may be seen that there is a narrow band of gradients from about 0.45 (24°) to 0.58 (30°) where landslides are transporting material at rates comparable to those for solifluction.

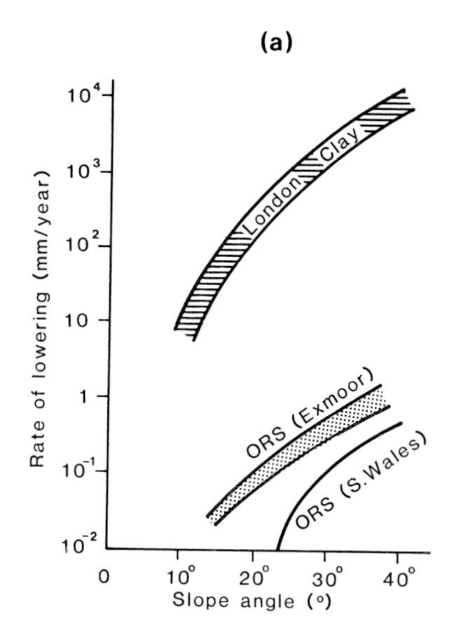

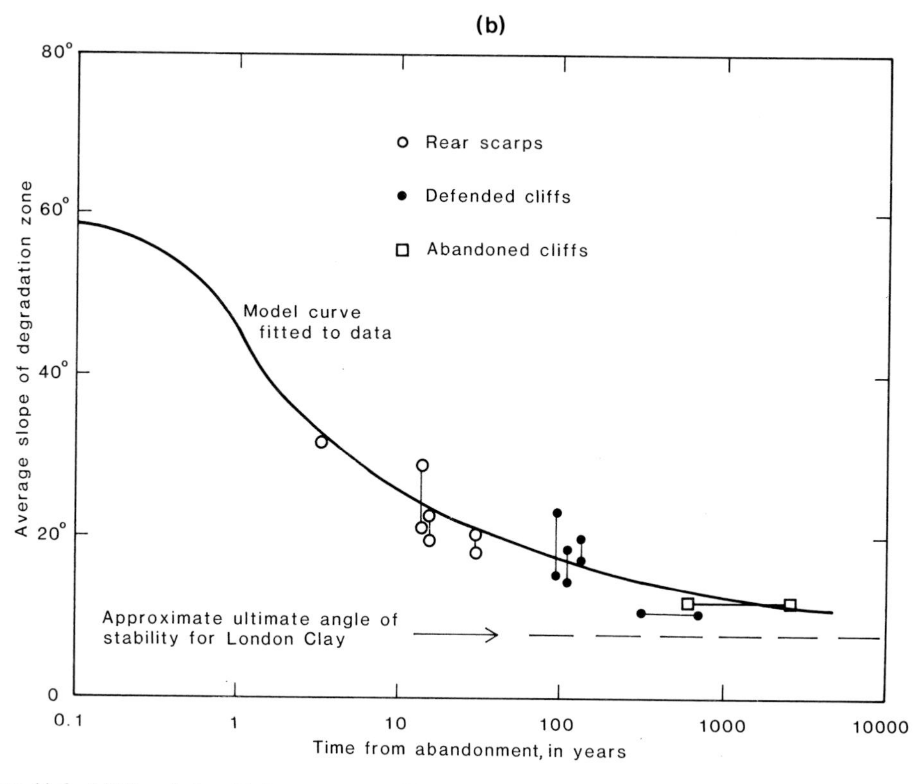

Figure 11.6 (a) The relationship between rates of lowering by mass movements and slope angle. (*Modified from Kirkby, 1973*, Geologia Applicata e Idrogeologia Bari, **8**.) (b) Measured and simulated values for gradient decline over time for London Clay cliffs. (*Based on Hutchinson, 1973*, Geologia Applicata e Idrogeologia, Bari. **8**.)

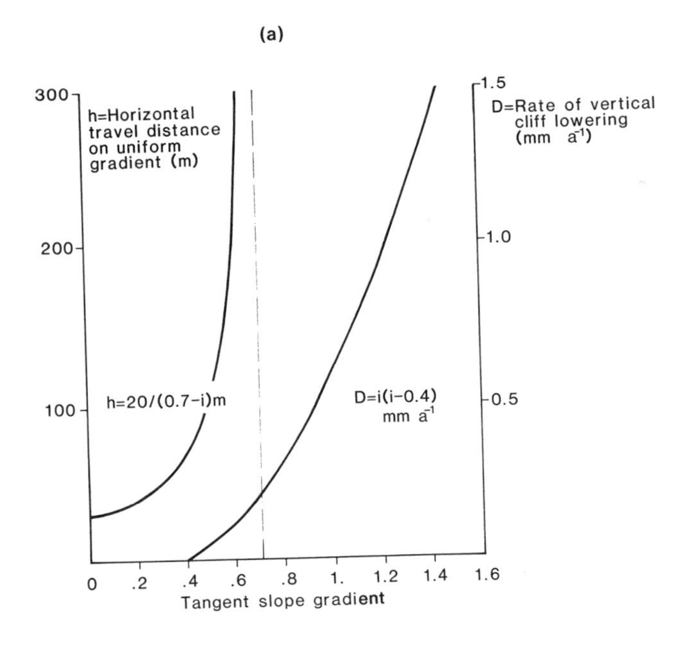

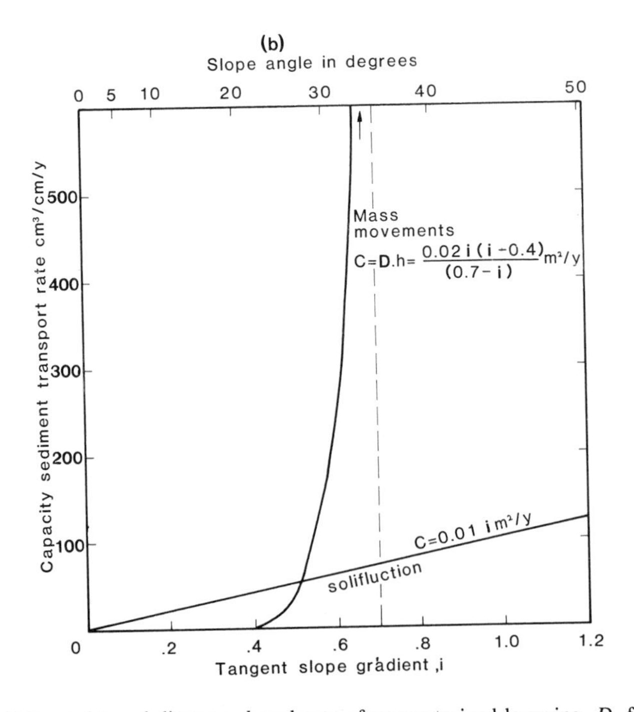

Figure 11.7 (a) Estimated travel distance, h and rate of unconstrained lowering, D, for Old Red Sandstone cliffs in south Wales. (b) Comparison of estimated capacity transport by mass movement (where defined) and solifluction for south Wales cliffs. (*From Kirkby, 1984*, Zeitschrift fur Geomorphologie **28**.)

(a)

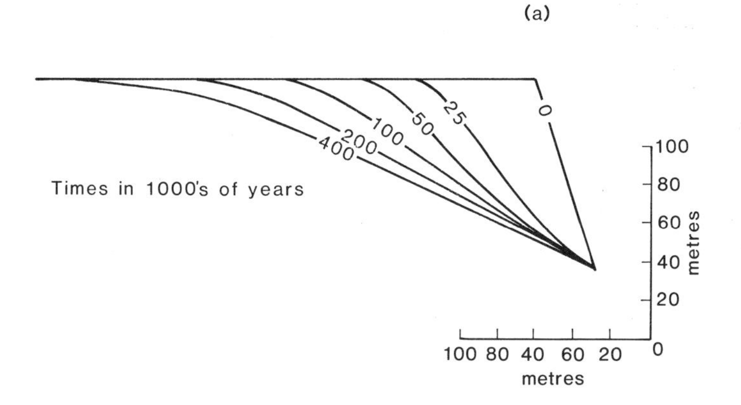

(b)

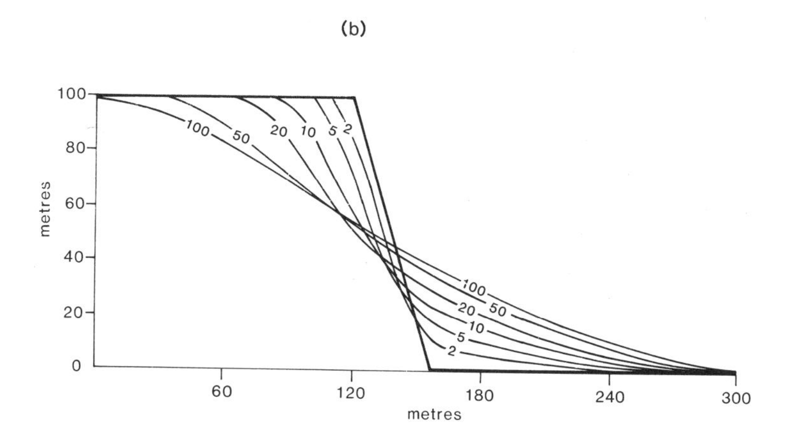

(c)

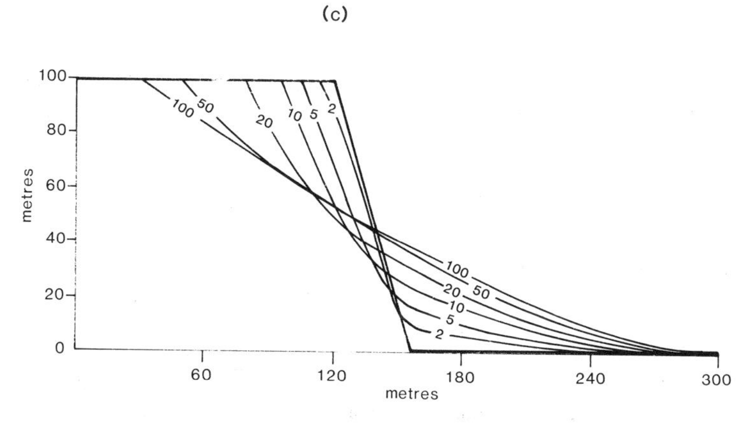

Figure 11.8 Simulated slope evolution for south Wales cliffs (times in thousands of years): (a) with basal removal. (*From Kirkby, 1984*, Zeitschrift fur Geomorphologie, **28**.); (b) with basal accumulation and solifluction; (c) as (b) but with creep replacing solifluction

Outside this range, one or other set of processes is very strongly dominant. If the rates of lowering for London Clay are substituted for the ORS values, the transitional zone between processes spans about 0.1°. Figure 11.8 shows example simulations for the ORS values. In 11.8(a) a 100 m cliff, initially at 70°, and backed by a broad horizontal plateau, evolves under conditions of basal removal, showing progressive decline of maximum slope gradients towards the ultimate threshold, together with some apparent tendency towards basal slope replacement. This replacement may also be modelled, in a simpler framework, using the kind of model illustrated in Figure 11.2 above, though with lower gradients. In Figure 11.8(b) the same initial form evolves, but with basal accumulation of material on a horizontal surface at the cliff foot. It may be seen that the broad features of simpler cliff and talus models are reproduced, with the formation of a convex rock core, but with concave talus profiles and a reclining cliff angle. In both cases, the forms broadly reflect relevant forms and processes, which are assumed to include solifluction given the time spans involved. In Figure 11.8(c), conditions are the same as in (b), but with creep replacing solifluction as the dominant 'slow' process, at 10% of the solifluction rate. It may be seen that the effect of the more rapid solifluction is to increase the rate of rounding of the cliff top convexity, and consequently the rate of decline of the overall cliff gradient.

Figure 11.9 illustrates the forecast convexities developed at the head of a constant gradient slope section produced by landslides, in combination with solifluction; again using the south Wales ORS process parameter values. The profiles shown are for equilibrium with a range of rates of horizontal slope retreat. Up to three sections of the curves are apparent. At elevations close to the plateau level, there is a linear increase of gradient with drop in height, giving an equilibrium profile of inverse exponential form. For low rates of retreat (e.g. for 0.1 and 0.2 mm yr^{-1} curves in Figure 11.9), the profile then straightens out to a

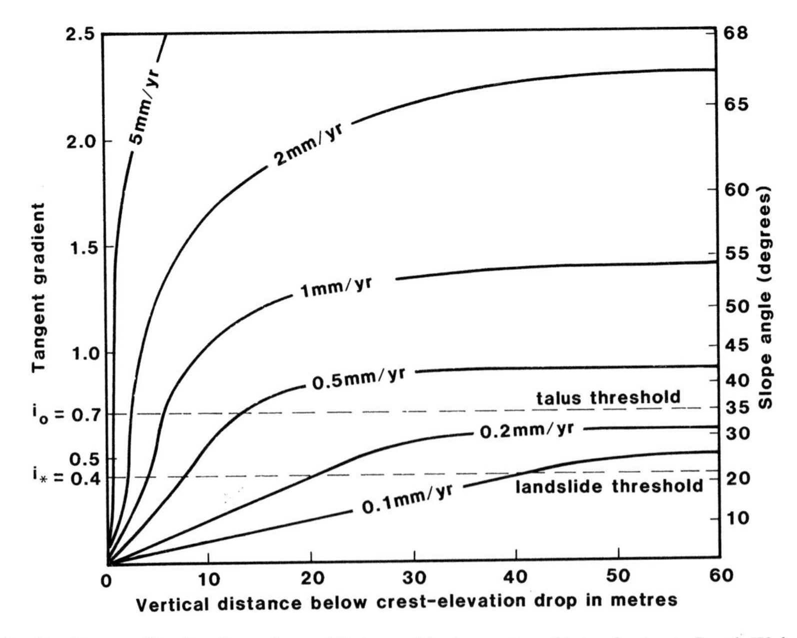

Figure 11.9 Gradient profiles for slopes in equilibrium with given rates of lateral retreat. South Wales parameter values. (*From Kirkby, 1984*, Zeitschrift fur Geomorphologie, **28**.)

constant gradient intermediate between the ultimate and talus threshold gradients. For higher rates of retreat (curves for 0.5 mm yr^{-1} or more), there is an increase in convexity between the ultimate and talus threshold gradients, giving a definite shoulder in the profile. The profiles then straighten out towards a constant gradient in excess of the talus threshold. Although many cliff tops exhibit a convexity, its development must depend on the history of recent mass movements, and the equilibration times are long ($>10^4$ yr) except at the highest rates of retreat.

It is plain from the models described in this section that forecasts of long-term evolution by mass movements are very imperfect. This is partly because existing models do not attempt to cover the full range of possible processes, concentrating mainly on shallow soil landslides; partly because existing models aim at average evolution rather than identifying individually large movements either stochastically or analytically; and partly because existing mass movement theory is largely concerned with stability analysis rather than subsequent movement of failed material. There is scope for a more developed theory for each process, analysing detailed slope stability to determine when events occur, and following the subsequent stability history of the moving mass. At the present state of the art, however, a satisfactory explanation at that level appears to be beyond our grasp.

11.4 LANDSLIDES AND SOIL WEATHERING

In association with landform evolution by mechanical processes including mass movements, there is evolution of the regolith which interacts strongly with the mechanical processes. As the regolith weathers through solutional removal, its geotechnical properties change, particularly its angle of internal friction and drainage characteristics. If weathering proceeds far enough, the direction of change is usually from properties of a highly interlocking clastic material towards those of a residual clay. At the same time, the degree of weathering achieved in the soil is a balance between the rates of solutional loss and of mechanical stripping of the surface. Where the latter is dominant the resulting hillside will be only slightly weathered, as for example on a cliff face. Where gradients and mechanical removal are low, weathering can proceed with minimal competition so that soils eventually become very highly weathered and clay rich. This section briefly explores the influences of weathering on landslide rates and vice versa. Some of this material is expanded below in Chapter 13.

Figure 11.10 shows a generalized relationship between grain size composition and angle of internal friction. The relationship is thought to depend mainly on two indirect cause and effect chains. First, it reflects the normal association of grain size with mineralogy, from quartz and feldspars, through micas to clay minerals in a descending sequence of both angles of internal friction and typical grain sizes. Second, it reflects the increase in interlocking and dilation angles associated with mixtures of disparate sized grains (Kenney, 1967; Kirkby, 1973). Thus maximum angles of internal friction are not associated with simple talus accumulations, but with weathered talus, or 'taluvium' in which some grain breakdown and weathering has increased the range of grain sizes present and their interlocking. At some stage in this weathering process, enough fines will be present to retard drainage of the material sufficiently for saturation to occur during wet periods. Near this point the stable angle for the material (ignoring cohesion for long-term slope development) drops to approximately half its 'dry' value. *In situ* weathering of a gravel talus, therefore, tends to produce an initial rise and a subsequent sharp fall in the maximum stable slope

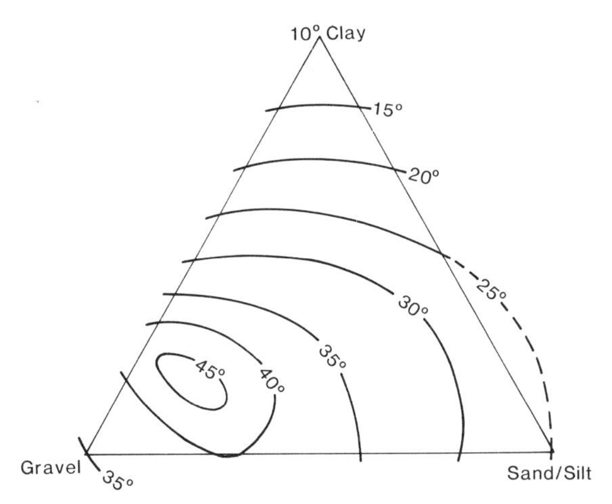

Figure 11.10 Generalized relationship between soil grain size and angle of internal friction. (*From Kirkby, 1973,* Geologia Applicata e Idrogeologia, Bari, **8**.)

gradient for the material. The exact course of this change is influenced by the manner in which the talus material breaks down; whether discontinuously like a sandstone, from blocks to sand grains; or continuously like a shale, to a wide range of sizes. For a given parent material, however, it is plausible to expect a more or less single valued relationship between degree of weathering, measured by chemical criteria, and angle of internal friction or maximum stable slope angle.

Turning to the converse relationship, the degree of soil weathering is a result of mechanical slope processes interacting with solutional removal. Regoliths generally develop a profile, with little weathering close to the parent material at depth, and increasing weathering upwards towards a maximum near the base of the organic horizon. The degree of weathering at this maximum is usually a good indicator for the soil as a whole, and may be measured as the proportion 'by volume' of bedrock material remaining (or the reciprocal of the enrichment ratio for a stable mineral). Comparisons based on this proportion are usually reliable within a single parent material. A second useful measure of overall weathering and its depth is the total 'soil deficit', defined as the total mass of material required to convert the existing soil back to its parent material, and expressed in volumetric terms as an equivalent depth of bedrock.

The soil deficit may be augmented directly by solution, and depleted somewhat less directly through mechanical erosion of the surface. The mass balance for soil deficit may be written in the form:

$$\partial w/\partial t = \partial/\partial x \ [V - (1 - p_s)/p_s S] \tag{11.14}$$

where w is the soil deficit (defined above)
p_s is the near-surface proportion of bedrock remaining
V is the solute transport at distance x downslope
and S is the mechanical sediment transport at x.

The coefficient of S in the final term is obtained by considering the equivalent depth of soil stripped by removing S (equal to S/p_S), and the loss of deficit per unit soil depth (equal to $1-p_S$). For low rates of mechanical removal, this expression describes indefinite thickening of the regolith through weathering. For the high rates of removal often associated with rapid mass movements, it can also describe the degree of weathering associated with equilibrium between mechanical and chemical removal. For this equilibrium;

$$p_S = S/(S+V) \tag{11.15}$$

Although not a fruitful approach for clay parent materials, the concept of equilibrium soil provides a useful way of estimating long-term rates of erosion by mass movements for residual soils on bedrock. Rates of solute loss are thought to be relatively independent of slope gradient within a uniform parent material, so that equation (11.15) provides a direct relationship between degree of weathering and erosion rate, provided the latter is high enough for equilibrium to be attained.

Since it has been argued above that maximum stable angle may also be directly related to degree of weathering (that is to p_S), it follows that there is a simple relationship between gradient and rate of lowering of natural slopes by rapid mass movements of whatever kind, of the type illustrated in Figure 11.6(a). The relationship should be unique for a given parent material and solute denudation rate; and its dependence on climate consists primarily of a scaling of mass movement rate in proportion to the rate of solution. Thus in comparing, say, similar granites between two areas, in the wetter of which solution rates are double that of the drier area, it is anticipated that similar gradients will be associated with similar degrees of weathering, but that the wetter area will be undergoing mass movement denudation on those gradients at twice the rate for the drier area. This analysis makes no assumptions about driving forces or process mechanisms, but essentially concerns the rate of release of material by weathering of its parent material. In order to meet the assumption of approximate equilibrium, the rates of mechanical lowering should be at least 5 × that of solutional loss; and it is plain that observations of degree of weathering are in practice confused by the particular history of mass movements for any site. With these reservations, some estimates may be made of lowering rates from observations of degree of soil comminution and weathering as it varies with gradient in an area: the curve in Figure 11.6(a) for Exmoor Old Red Sandstone has been derived in this way as an example.

This discussion of the relationship between landslide rates and weathering goes some way to justifying the use of a relationship between gradient and potential rate of retreat (for example equation (11.12)) for modelling long-term slope evolution. It also suggests the way in which climatic dependence should be built into this type of forecasting model. If more sophisticated models for solution are developed, and soil equilibrium is no longer assumed, then it is further suggested that a better route to long-term forecasting is through simulating the development of weathering extent (p_S) or soil deficit, and developing an explicit relationship between p_S and stable slope gradient. Some progress has been made on this problem (Kirkby, 1985), though much remains to be done. Figure 11.11 shows an example simulation for a slope intersected by a band of lower solubility and lower rates of retreat (α in equation (11.12)), although with the same ultimate threshold (22°). The upper curves (a) shows the development and later decline of a marked escarpment in the

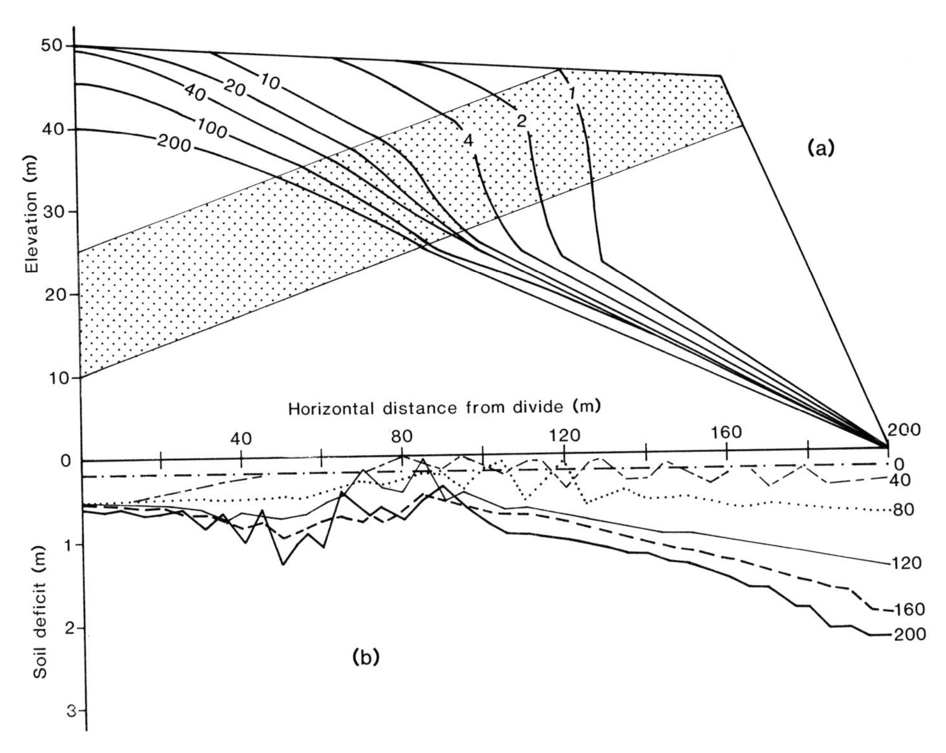

Figure 11.11 Simulated slope evolution for a stratified material, by solifluction, solution and mass movements. Time in thousands of years. Dotted band is more resistant. (*From Kirkby, 1985,* Models in Geomorphology, *M. Woldenberg, ed.*)

resistant stratum. The lower curves (b) show a general trend to thickening soils downslope after sufficiently long time spans, with thinning across the resistant band.

The state of the art in long-term modelling of slope profile evolution through mass movements is, as can be seen from this brief survey, far from adequate. It is severely restricted by a lack of studies on the factors controlling travel distances for mass movements, which are crucial to the development of an overall mass budget in a forecasting context. It is also hampered by the uneven development of relevant concepts across the full range of mass movement types, so that while some principles may remain valid across the spectrum of types, others differ considerably and are to varying degrees unknown. Perhaps the most generally relevant concepts are those of surplus capacity and travel distance; and the process non-specific argument about the existence of relationships between stable angles and rates of release of fresh material through solution and weathering.

REFERENCES

Ahnert, F. (1973). 'COSLOP2—a comprehensive model program for simulating slope profile development.' *Geocom Programs*, **8**, 24 pp. London.

Armstrong, A. C. (1976). 'A three-dimensional simulation of slope forms.' *Zeitschrift fur Geomorphologie, Suppl. Band* **25**, 20–28.

Aronsson, G., and Linde, K. (1982). 'Grand Canyon—A quantitative approach to the erosion and weathering of a stratified bedrock.' *Earth Surf. Proc. Landforms*, **7**(6), 589-600.

Bakker, J. P., and Le Heux, J. W. N. (1946). 'Projective-geometric treatment of O. Lehmann's theory of the transformation of steep mountain slopes.' *Koninklijke Nederlandsche Akademie van Wetenschappen, Series B*, **49**, 533-47.

Bakker, J. P., and Le Heux, J. W. N. (1947, 1950). 'Theory on central rectilinear recession of slopes.' *Koninklijke Nederlandsche Akademie van Wetenschappen, Series B*, **50**, 959-66, 1154-62; **53**, 1073-84, 1364-74.

Bakker, J. P. and Le Heux, J. W. N. (1952). 'A remarkable new geomorphological law.' *Koninklijke Nederlandsche Akademie van Wetenschappen, Series B,* **55**, 399-410, 554-71.

Bennett, J. P. (1974). 'Concepts of mathematical modelling of sediment yield.' *Water Resources Research*, **10**(3), 485-92.

Bryan, K. (1922). 'Erosion and sedimentation in the Papago country, Arizona.' *US Geol. Survey Bull.*, 730.

Fisher, O. (1866). 'On the disintegration of a chalk cliff.' *Geol. Mag.*, **3**, 354-56.

Foster, G. R., and Meyer, L. D. (1972). A closed form soil erosion equation for upland areas. In *Sedimentation: Symposium to honour Professor H. A. Einstein*. Shen, H. W., (ed.), Fort Collins, Colorado, pp. 12.1-12.19.

Grant-Taylor, T. L. (1964). 'Stable angles in Wellington greywacke.' *New Zealand Eng.*, **19**, 129-30.

Hutchinson, J. N. (1967). 'The free degradation of London Clay cliffs.' *Proc. Geotech. Conf., Oslo*, **1**, 113-18.

Hutchinson, J. N. (1973). 'The response of London Clay cliffs to different rates of toe erosion.' *Geologia Applicata e Idrogeologia, Bari,* **8**(1), 221-39.

Kenney, T. C. (1967). 'The influence of mineral composition on the residual strength of natural soils.' *Proc. Geotech. Conf.*, Oslo, 123-9.

Kirkby, M. J. (1971). 'Hillslope process-response models based on the continuity equation.' *Trans. Inst. Brit. Geogrs., Special Publication 3*, 15-30.

Kirkby, M. J. (1973). 'Landslides and weathering rates.' *Geologia Applicata e Idrogeologia, Bari*, **8**(1), 171-83.

Kirkby, M. J. (1984). 'Modelling cliff development in South Wales: Savigear re-viewed.' *Zeitschrift fur Geomorphologie*, **28**(4), 405-26.

Kirkby, M. J. (1985). 'A model for the evolution of regolith-mantled slopes.' In *Models in Geomorphology*. Woldenberg M. (ed.), George Allen & Unwin, pp. 213-37.

Kirkby, M. J., and Statham, I. (1975). 'Surface stone movement and scree formation.' *J. Geol.*, **83**, 349-62.

Lehmann, O. (1933). 'Morphologische theorie der verwitterung von steinschlag wanden.' *Vierteljahrschrift der Naturforschende Gesellschaft in Zurich*, **87**, 83-126.

Luke, J. C. (1972). 'Mathematical models for landform evolution.' *J. Geophys. Res.*, **77**(14), 2460-4.

Luke, J. C. (1974). 'Special solutions for non-linear erosion problems.' *J. Geophys. Res.*, **79**(26), 4035-40.

Luke, J. C. (1976): 'A note on the use of characteristics in slope evolution models.' *Zeitschrift fur Geomorphologie, Supp. Band* 25, 114-19.

Scheidegger, A. E. (1961). 'Mathematical models of slope development.' *Bull. Geol. Soc. Am.*, **72**, 37-50.

Selby, M. J. (1982). *Hillslope Materials and Processes*. Oxford University Press, 264 pp.

Slope Stability
Edited by M. G. Anderson and K. S. Richards

Chapter 12

Modelling Interrelationships Between Climate, Hydrology, and Hydrogeology and the Development of Slopes

R. ALLAN FREEZE
Department of Geological Sciences
University of British Columbia, Vancouver, Canada V6T 1W5

12.1 INTRODUCTION

Most of the land surface of the earth is formed by valley slopes. Young (1972) noted that even in the Great Plains region of North America, which is surely one of the flattest places on earth, 93% of the land area can be classified as dissected relief and only 7% can be considered as level plain. In this light, it is clear that a primary concern in the study of landform evolution must lie with hillslopes.

It has long been recognized that the process of slope formation involves a complex set of interactions between the soils and rocks of the earth's surface and the hydrological regime that exists there in response to climatic events. This interrelationship between geomorphology, hydrology, and climate is discussed in most modern geomorphology texts (Leopold *et al.*, 1964; Carson and Kirkby, 1972; Derbyshire, 1976; Ritter, 1978; Embleton and Thornes, 1979). In this volume emphasis is placed on one of the most important slope-forming processes: mass movement; and in this chapter emphasis is placed on one of the most important controls on mass movement: hillslope hydrogeology. The chapter presents an idealized quantitative model of slope stability that traces the controlling pore pressures to their source as climatic inputs, through the mechanisms of hillslope runoff generation, and the development of hillslope groundwater flow systems.

The output from the model is in the form of slope-angle predictions for various climatic and geologic environments. These predictions are not quantitative absolutes, but are rather presented as an indication of the relative sensitivity of slope angle to various controls, both individually and in concert. The results emphasize the importance of saturated hydraulic conductivity as a parameter influencing slope development.

The model is based on analytical representations of hillslope processes. It is carried out in a deterministic framework, for a very simple topographic configuration, and a homogeneous geological environment. The purpose is not to provide a methodology for the investigation of specific slopes, but rather to carry out a generic analysis by which it can be shown that climatic and hydrogeologic parameters exert a degree of control on slope-angle development comparable to the recognized influence of soil-strength parameters.

In that the emphasis here is on mass movements, it would first seem wise to place the role of mass movement into context within the rather extensive suite of processes that lead to slope development. Slope formation results from the dual processes of weathering and surface transport; the material made available at the land surface by means of weathering becomes available for downslope transport. It is usual (see Chapter 11) to differentiate between slope formation under transport-limited conditions wherein the mechanisms of downslope transport cannot remove all the material produced by the weathering process and weathering-limited conditions wherein the production of material by weathering cannot keep pace with the rate of removal. On transport-limited slopes the mechanisms of downslope sediment transport can be grouped into five categories:

1. Slow mass movements such as creep, solifluction, and heave.
2. Rapid mass movements such as slumps, landslides, and earthflows.
3. Erosion by surface-water runoff.
4. Piping due to subsurface-water runoff.
5. Chemical solution in surface and subsurface runoff.

The model presented in this chapter is limited to the analysis of rapid mass movements on transport-limited slopes.

Carson (1969) has suggested that slope development may take place in two distinct phases. In the first phase, slope development occurs primarily by rapid mass movement. In this phase, steep slopes are reduced to gentle slopes, and these gentle slopes presumably have long-term stability with respect to rapid mass movement. In the second phase, surface-water erosion becomes the dominant hillslope-forming process. The results of the model are germane to the first phase rather than the second.

Pore pressures on hillslopes are controlled by the height of the water table and its fluctuations (see Chapter 8). The water-table position in a slope with given hydrogeological properties is in turn controlled by the infiltration rates that occur under the prevailing climatic regime. An understanding of the pore-pressure field, therefore, requires an understanding of the mechanisms of infiltration at the land surface. These mechanisms are innately bound up with the mechanisms of overland-flow generation. In the following sections, the model is described after a review of the mechanisms by which infiltration and overland flow are generated on hillslopes. Finally, the implications of this work for mass movements in more complex hydrogeological environments are discussed.

12.2 INFILTRATION AND OVERLAND FLOW ON HILLSLOPES

The question of how rainfall is partitioned into infiltration and overland flow has been rather carefully addressed during the past decade from the perspective of understanding the various mechanisms of runoff generation from hillslopes. An excellent summary of

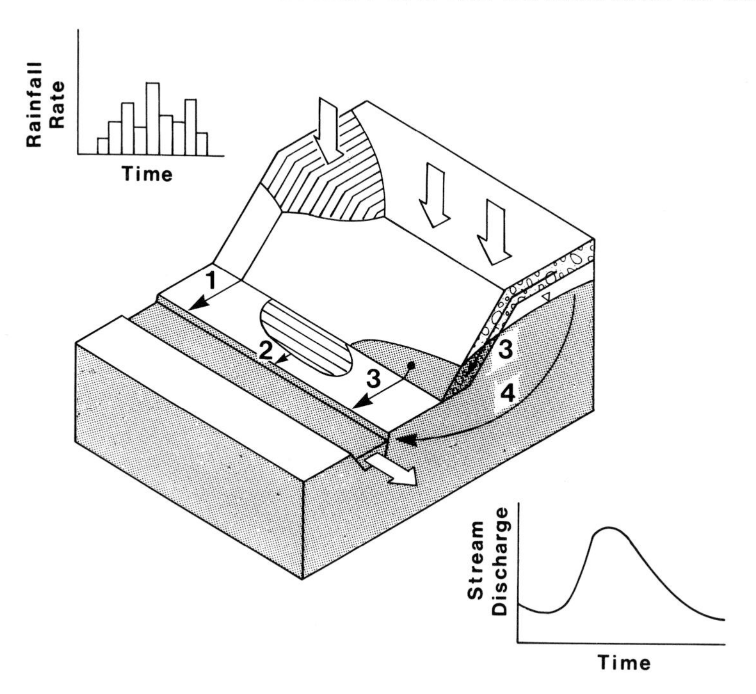

Figure 12.1 Mechanisms of delivery of rainfall to a stream channel from a hillslope: (1) Horton overland flow, (2) Dunne overland flow, (3) subsurface stormflow, and (4) groundwater flow

current knowledge can be found in the collection of papers edited by Kirkby (1978a), and particularly that by Dunne (1978).

Consider the schematic diagram shown in Figure 12.1 for a topographic configuration consisting of a flood plain, a hillslope, and an upland. The stream that drains the basin is fed by lateral inflows from the valley slopes, and these lateral inflows may arrive at the stream as groundwater flow (4) or as overland flow (1, 2, 3). It has long been recognized that groundwater flow provides the base-flow component of streams that sustains their flow between storm periods. The flashy response of streamflow to individual precipitation events must be ascribed to overland flow that enters the stream from the flood plain. The overland flow may have its source on the uplands (1), on the flood plain (2), or as subsurface stormflow from the hillslope (3). Overland flow due to (1) or (2) is generated at a point only after surface ponding takes place. Ponding cannot occur until the surface soil layers are saturated. It is now recognized that surface saturation can be introduced by two quite distinct mechanisms.

The classic mechanism, which was first espoused by Horton (1945) and placed in a more scientific framework by Rubin and Steinhardt (1963), is for a precipitation rate, r, that exceeds the saturated hydraulic conductivity, K, of the surface soil. As illustrated in Figure 12.2(a), a moisture content versus depth profile during such a rainfall event will show moisture contents that increase at the surface as a function of time. At some point in time (t^3 in Figure 12.2(a)), the surface becomes saturated and an inverted zone of saturation begins to propagate downward into the soil. The time, t^3, is called the ponding time. The necessary conditions for the generation of overland flow by the Horton mechanism

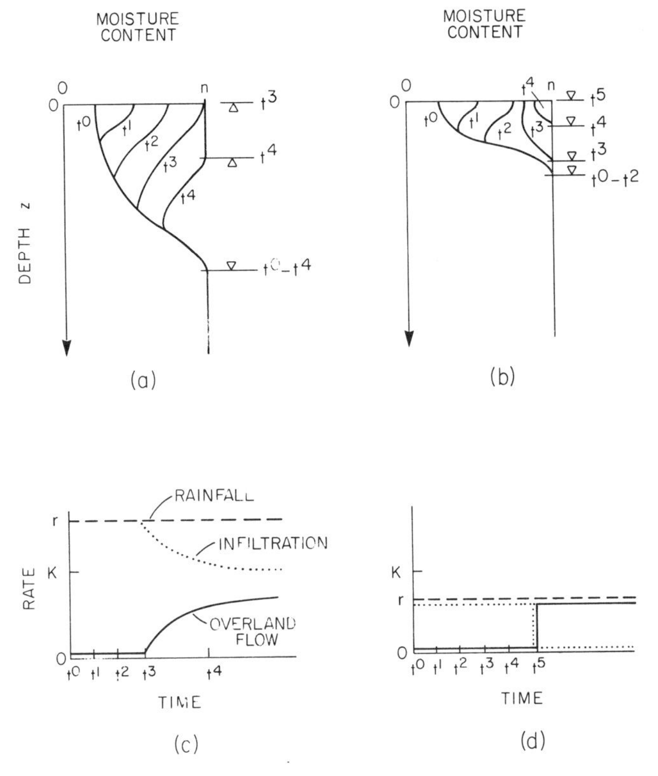

Figure 12.2 Moisture content v. depth profiles for the (a) Horton mechanism and (b) Dunne mechanism. Overland flow generation for (c) Horton overland flow and (d) Dunne overland flow

are: (a) a rainfall rate greater than the saturated hydraulic conductivity of the soil; (b) a rainfall duration longer than the required ponding time for a given initial moisture profile.

The second mechanism, as described by Dunne (1978), is illustrated in Figures 12.2(b) and 12.2(d). In this case, $r<K$, and the initial water table is shallow. Surface saturation is due to a rising water table; ponding and overland flow occur at time, t^5, when no further soil-moisture storage is available.

Dunne (1978) has summarized the environmental controls on the various mechanisms of streamflow generation. Within a given watershed, the Horton mechanism is more common on upslope areas. The Dunne mechanism is more common on near-channel wetlands. Horton overland flow is generated from partial areas of the hillslope where surface hydraulic conductivities are lowest. Dunne overland flow is generated from partial areas of the hillslope where water tables are shallowest. Both mechanisms lead to variable source areas that expand and contract throughout wet and dry periods. In Figure 12.1, the overland flow generated on the upland (1) is due to the Horton mechanism; the overland flow generated on the flood plain (2) is due to the Dunne mechanism.

In the model presented in the following section, where our interest lies in the stability of hillslopes above the flood plain, only the Horton mechanism will be invoked in the consideration of hillslope hydrological response. As noted earlier, interest actually centres on that portion of the rainfall that infiltrates rather than on the portion that becomes overland flow. In the model, the infiltration rates drive the water-table fluctuations and these, in turn, control the pore pressures that influence slope stability.

12.3 AN IDEALIZED MODEL OF SLOPE DEVELOPMENT BY MASS MOVEMENT

We are now ready to consider a simple quantitative geomorphic model of slope development. It is an idealized model that assumes the simple slope geometry of Figure 12.1: a straight slope connecting a level flood plain and a level upland. The goal is to calculate the maximum stable slope angle, α, under a variety of climatic, hydrological, and hydrogeological conditions. There are four main elements to the model: climate, hydrology, hydrogeology, and slope stability. Each is now discussed in turn. The assumptions and limitations invoked in each of the elements are summarized at the end of this section.

12.3.1 Climate

Climate is represented by the simplest possible step function of rainfall and evapotranspiration (Figure 12.3(a)). There are N rainfall events per year, each of duration, t_r, and intensity, r. If t_r is measured in s and r in ms^{-1} and the average annual rainfall intensity in ms^{-1} is represented by R, then R, N, r, and t_r are related by:

$$r = \frac{3.15 \times 10^7 \, R}{N t_r} \tag{12.1}$$

The coefficient in the numerator is the number of seconds in a year.

If R is allowed to take on the values 10^{-9}, 10^{-8}, and 10^{-7} m/s (approximately 1, 10, and 100 inches per year), N to take on the values, 1, 10, and 100 events per year, and t_r to take on the values 3.15×10^a s, where $a = 2$, 3, 4, and 5 (approximately 5 minutes, 1 hour, 10 hours, and 4 days), the various combinations can produce a wide range of climates. Not all members of the set are equally likely (or perhaps even possible), but the most realistic possibilities are all included.

If R, N, and t_r are set, then the rainfall intensity, r, of each event is specified by equation (12.1). The stippled sloping plane of Figure 12.3(b) provides a diagrammatic representation of the set of possible combinations of event properties, r and t_r, that can arise for various values of N, for the specific average annual rainfall intensity, $R = 10^{-8}$ m s^{-1}. Similar constructions can be made for $R = 10^{-7}$ and $R = 10^{-9}$ m s^{-1}. In reality, observed event intensities rarely exceed 10^{-4} m s^{-1}, so the climate diagram (Figure 12.3(b)) has been truncated at this value.

There are also N periods between rainfall events each year (Figure 12.3(a)), each of duration t_b, where $t_b = (3.15 \times 10^7/N) - t_r$ s. During these periods the potential evapotranspiration rate, e_p, is given by:

$$e_p = \frac{3.15 \times 10^7 \, E}{N t_b} \tag{12.2}$$

where E is the average annual potential evapotranspiration rate in m s^{-1}.

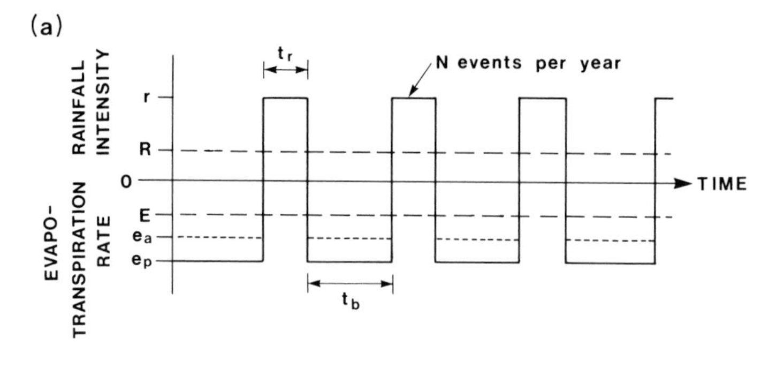

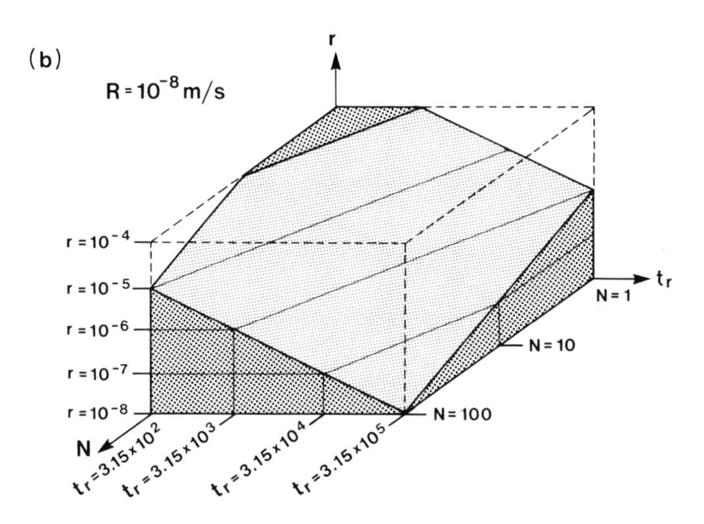

Figure 12.3 Climate representation. (a) Step function of rainfall intensity and evapotranspiration rate. (b) Possible combinations of event properties, r and t_r, for various N for $R = 10^{-8}\ \mathrm{m^{-1}\ s}$

The actual evapotranspiration rate, e_a, must be less than or equal to e_p. For the purposes of this study, it will be assumed that all evapotranspiration takes place from flood plains (Figure 12.1), where water tables and soil moisture contents are high, and none takes place from uplands and hillslopes where water tables and moisture contents are lower. In effect, we are setting the ratio e_a/e_p equal to unity on the flood plains and zero on the uplands. This assumption is relatively good on high-permeability soils and relatively poor on low-permeability soils. For the idealized purposes of this chapter, it has the effect of removing evapotranspiration from our consideration.

12.3.2 Hydrology

In the absence of evapotranspiration, the groundwater recharge rates that control water-table elevations and hence pore pressures will be equal to the infiltration rates at the surface. We must, therefore, identify those climates for which all rainfall becomes infiltration and those for which some of the rainfall is removed from access to the subsurface system by overland flow.

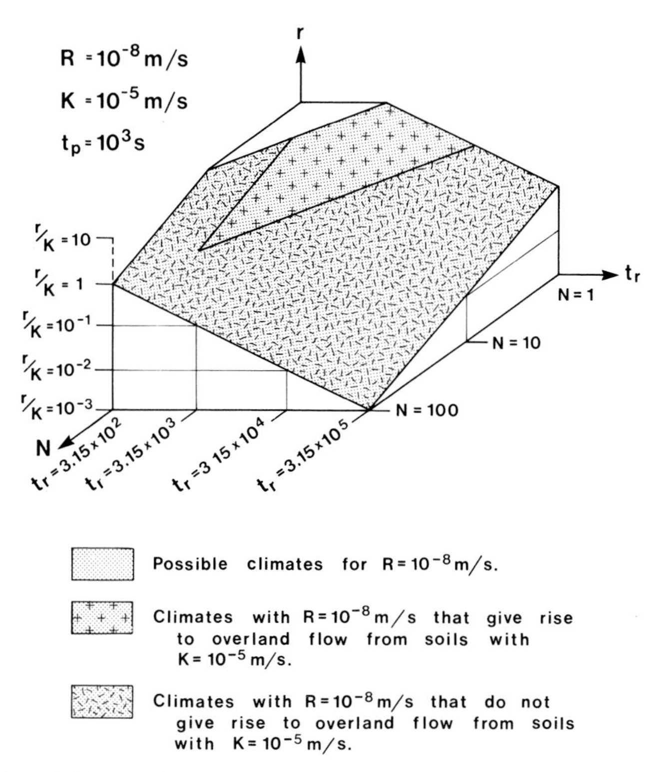

Figure 12.4 Partition of climates with $R = 10^{-8}$ m s^{-1} into those that produce overland flow from soils with $K = 10^{-5}$ m yr^{-1} and those that do not

As previously noted, overland flow will be generated from uplands by the Horton mechanism whenever the rainfall intensity, r, exceeds the saturated hydraulic conductivity, K, of the hillslope soils, and the duration of the rainfall event, t_r, exceeds the ponding time, t_p. For our simple deterministic representation of climate and for hillslopes that are homogeneous with respect to hydraulic conductivity, these criteria imply that overland flow will be produced from an entire hillslope for each and every rainfall event or it will not be produced anywhere on the hillslope for any of the events. In other words, each climate–soil combination is binary; it either produces overland flow or it does not. In Figure 12.4 this binary partition is graphically illustrated for the specific set of climates with $R = 10^{-8}$ m s^{-1} and for soils with $K = 10^{-5}$ m s^{-1}. In the cross-hatched trapezoidal region, $r > K$ and $t_r > t_p$, and overland flow will be generated. In the remainder of the field all rainfall will become infiltration. It is possible to prepare similar diagrams for $R = 10^{-7}$ and $R = 10^{-9}$ ms^{-1} and for various values of K.

It is clear that the time until ponding, t_p, is an important parameter in the control of overland-flow occurrences. In general, a saturated–unsaturated analysis of the infiltration process would be necessary to produce accurate t_p values for each climatic representation for each soil. However, as the ratio of the rainfall rate, r, to the saturated hydraulic conductivity, K, increases, the value of t_p decreases (Figure 12.5(a)). In fact, for $r/K \geqq 10$, one would expect $t_p \rightarrow 0$. In this idealized study, where values of r and K are specified on an order-of-magnitude basis, r–K ratios will take on values . . , 10^{-2}, 10^{-1}, 1, 10,

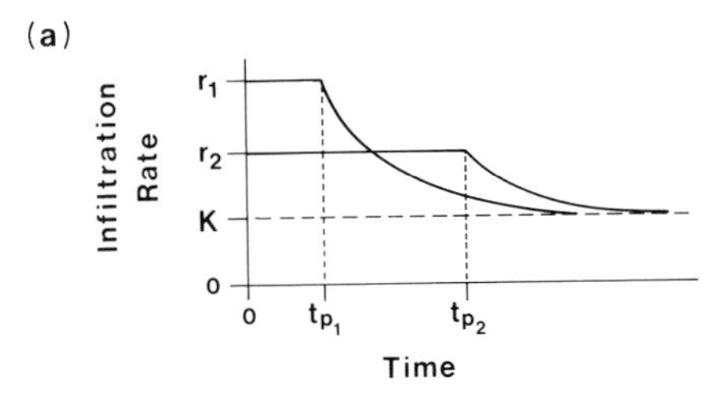

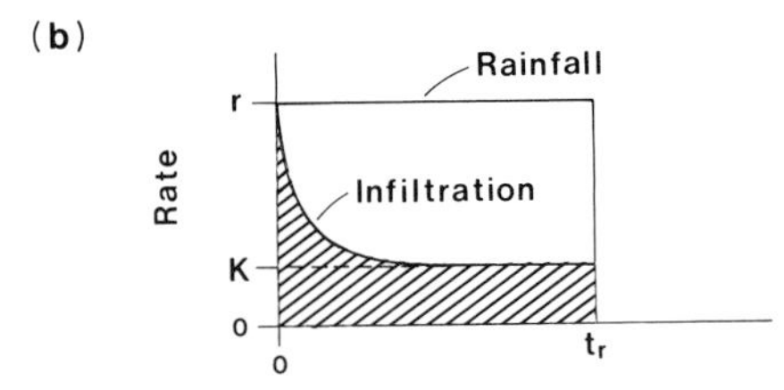

Figure 12.5 Infiltration. (a) Ponding time, t_p, as an inverse function of r/K. (b) Schematic explanation of the partitioning coefficient, a

10^2, . . . In other words, either $r/K \leqq 1$, in which case no overland flow occurs, or $r/K \geqq 10$, in which case we will assume $t_p = 0$.

The average annual groundwater recharge rate, R', which is equal to the average annual infiltration rate, is related to the average annual rainfall rate by the simple relationship:

$$R' = aR \tag{12.3}$$

where a is the partitioning coefficient that denotes the percentage of annual rainfall that becomes infiltration rather than overland flow. For climates that do not produce overland flow, $a = 1$; for climates that do produce overland flow, $a < 1$. For these latter cases it would once again be necessary to turn to a full saturated–unsaturated infiltration analysis to determine a accurately. In this study, where overland flow-producing rainfall events have $r/K \geqq 10$ and ponding occurs very quickly, it will be assumed that:

$$a = \frac{Kt_r}{rt_r} = \frac{K}{r} \tag{12.4}$$

Figure 12.5(b) identifies this expression as an approximation of the actual infiltration regime.

The event-based rainfall parameters, r and t_r, are used in this hydrologic analysis only for the purpose of separating those climatic representations that produce overland flow from those that do not and in the calculation of the partitioning coefficient, a. The only output from this section that will be used in the next is the average annual groundwater recharge rate, R', from equation (12.3).

12.3.3 Hydrogeology

Consider the groundwater-flow situation represented by the boundary-value problem shown in Figure 12.6(a). The rectangular region is a two-dimensional cross-section of width,

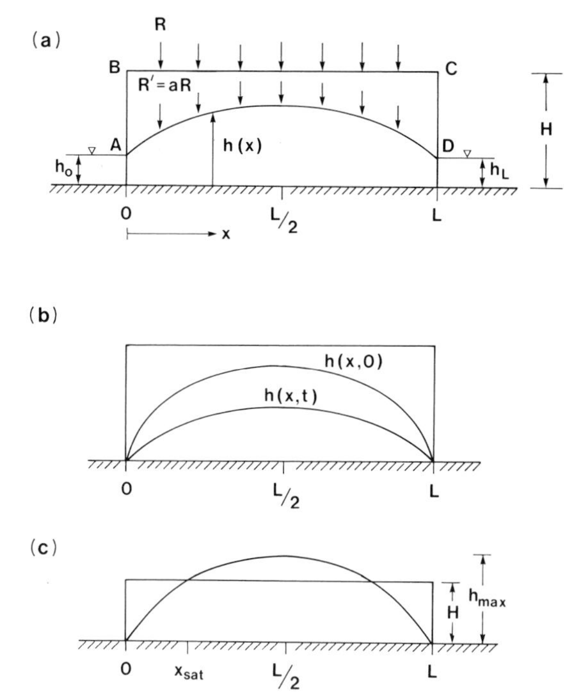

Figure 12.6 Groundwater flow. (a) Steady-state water-table configuration. (b) Transient water-table configurations. (c) Definition of x_{sat}

L, and height, H, through a hill separating two parallel valleys. The topographic configuration ABCD is not invoked in the mathematics that follow; it can be considered to take on any shape including the more realistic hillslope configurations of Figure 12.1. The region of groundwater flow is bounded by the water table AD on the top and an impermeable boundary on the bottom. Within this region the Dupuit–Forchheimer (DF) assumptions will be invoked. They state that: (a) flow lines are horizontal and equipotential lines are vertical; and (b) the hydraulic gradient at a point of distance x from the origin is equal to the slope of the water table above that point and is invariant with depth. The DF assumptions have the effect of reducing a two-dimensional problem to one dimension. The hydraulic head in the region of flow is now given by $h(x)$, where $h(x)$ is the height of the water table across the region of flow, $x=0$ to $x=L$. The DF assumptions are valid when the water-table slope is small and the depth of the flow field is shallow. Within a flow field that satisfies these conditions, results are least accurate at the water divide where vertical gradients can be important.

For steady-state flow, the equation of flow in a homogeneous medium under DF assumptions (Bear, 1972) is:

$$\frac{K}{2}\left[\frac{\mathrm{d}^2(h^2)}{\mathrm{d}x^2}\right]+R'=0 \tag{12.5}$$

For the boundary conditions:

$$h(0) = h_o$$

$$h(L) = h_L$$

the solution is:

$$h(x) = \left[h_o^2 - \frac{h_o^2 - h_L^2}{L} x + \frac{R'}{K} (L-x)x \right]^{1/2} \tag{12.6}$$

For the special case of interest where $h_o = h_L = 0$, we have:

$$h(x) = \left[\frac{R'}{K} (L-x)x \right]^{1/2} \tag{12.7}$$

The maximum water-table height occurs at the midpoint, $x = L/2$, where:

$$h\left(\frac{L}{2}\right) = \left[\frac{R'L^2}{4K} \right]^{1/2} \tag{12.8}$$

Under the DF assumptions the $h(x)$ values from equation (12.7) hold at all elevations, including the specific elevation where a possible slip surface may exist.

Equation (12.7) provides an estimate of the steady-state position, or average-annual position, of the water table for a given climate (R'), soil (K), and hillslope width (L). However, for the purpose of determining the maximum stable slope that can exist under a given climatic regime, it is not sufficient to use the steady-state pore pressures. Failure is much more likely to occur when water tables are at the highest elevation of their annual range of fluctuation. To calculate the annual range we must address the question of transient water-table behaviour. This is most easily done in the context of the water-table declines that occur between rainfall events. For the case of a declining water table (Figure 12.6(b)) the equation of flow under Dupuit–Forchheimer assumptions is:

$$\frac{K}{2} \left[\frac{\partial^2 (h^2)}{\partial x^2} \right] + R' = n \frac{\partial h}{\partial t} \tag{12.9}$$

where n is the porosity of the soil. For $R' = 0$ and boundary conditions as before, the solution (Bear, 1972) is:

$$h(x,t) = h(x,0) \left[1 + \frac{4.48Kt\ h(L/2,0)}{nL^2} \right]^{-1} \tag{12.10}$$

where $h(x,0)$ is the initial water-table configuration at time $t = 0$ and $h(L/2,0)$ is the specific initial value of h at $x = L/2$.

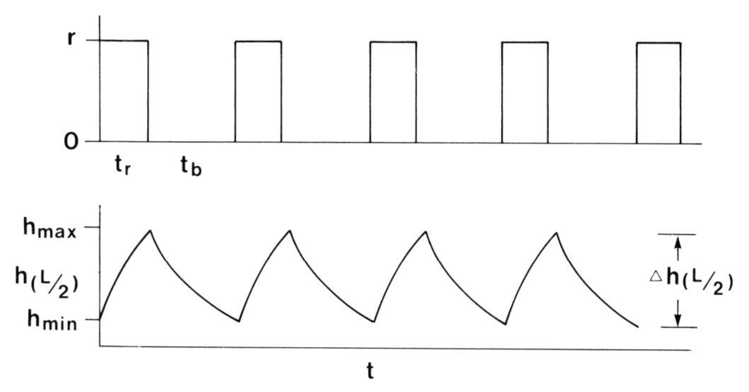

Figure 12.7 Transient water-table fluctuations at $x = L/2$

Figure 12.7 illustrates the type of transient water-table fluctuation that might be expected at $x = L/2$. The range at $L/2$ is denoted by Δh, and h_{max} is defined as:

$$h_{max} = h(L/2) + \tfrac{1}{2}\,\Delta h \qquad (12.11)$$

where $h(L/2)$ is the steady-state value from equation (12.8). It is clear that h_{max} represents the maximum hydraulic head on the hillslope both in time and space. The value of Δh is calculated as $h(L/2) - h(L/2, t_b)$ where the first term is given by equation (12.8) and the second by equation (12.10). In equation (12.10), $h(x,0) = h(L/2,0) = h(L/2)$, and $t = t_b$, where t_b is the time between rainfall events. This method of calculating Δh involves a slight approximation but is easily calculated.

In situations where $h_{max} > H$, the maximum water-table configuration will intersect the upland plateau surface at a distance $x = x_{sat}$ from the origin as shown in Figure 12.6(c). In such cases, the value of x_{sat} can be calculated from manipulation of equation (12.10). The ratio, B:

$$B = \frac{x_{sat}}{H} \qquad (12.12)$$

which is here termed 'the intercept ratio', is required in the slope stability calculation that follows.

With the equations presented in this section, it is possible to calculate the maximum water-table height, h_{max}, or the intercept ratio, B, that will occur under a given climate–soil combination. As input, we require the average annual groundwater recharge, R', the hillslope parameters, L and H, and the hillslope soil properties, K and n.

12.3.4 Slope Stability

Consider the slope shown in Figure 12.8(a). The maximum angle, α, at which such a slope is stable will depend on the geometry of the slope, the soil properties, and the pore-pressure distribution. Methods for calculating the factor of safety against failure along the slip surface ACB are presented more fully in Chapter 2. A brief summary is sufficient here.

(a)

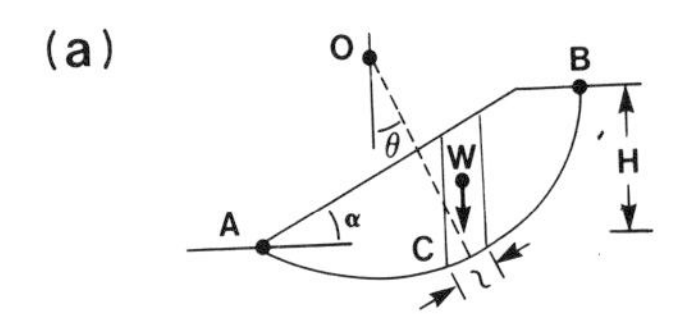

(b)

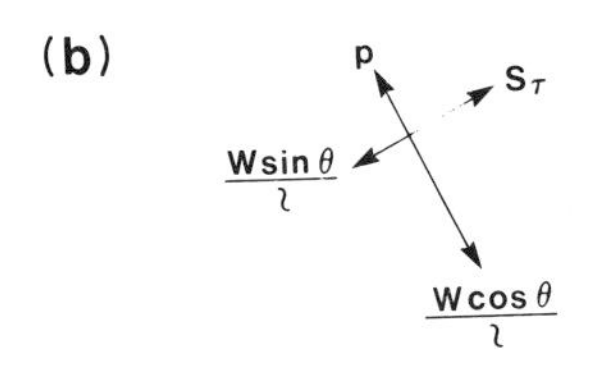

(c)

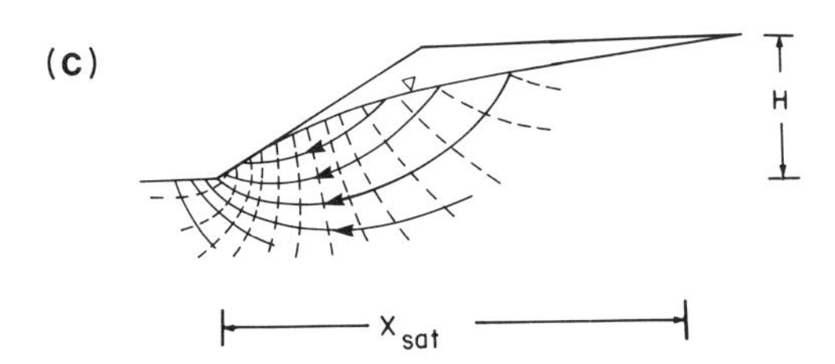

Figure 12.8 Slope-stability analysis by the conventional method of slices. (a) Geometry. (b) Stress equilibrium at point C. (c) Flow net for determination of pore pressures along slip surface

The factor of safety, F, is defined as the ratio of the shearing strength, S_τ, mobilized along the failure plane, to the shear stress, τ, along the plane. Movement will occur only if the shear stress exceeds the shear strength and $F<1$. The shear strength is usually expressed in terms of the empirical Mohr–Coulomb failure law:

$$S_\tau = c' + (\sigma - u) \tan \phi' \tag{12.13}$$

where σ is the normal stress across the failure plane and c' and ϕ' are two mechanical properties of the material, c' being the cohesion and ϕ' being the angle of internal friction. The u in equation (12.13) is the pore pressure on the failure plane. At any point with elevation z, the pore pressure is given by:

$$u = \gamma_w (h - z) \tag{12.14}$$

where γ_w is the specific weight of water and h is the hydraulic head. Equation (12.13) makes it clear that increases in pore pressure tend to decrease shear strength on failure planes.

The simplest practical method of slope stability analysis is the conventional method of slices (Lambe and Whitman, 1969; Morgenstern and Sangrey, 1978). In the conventional method the slope is divided into a series of vertical slices. Figure 12.8(a) shows the geometry of an individual slice, and Figure 12.8(b) indicates the conditions of stress equilibrium that exist at point C on the slip surface at the base of the slice. At C the shearing stress, τ, is given by $W\sin\theta/\ell$ and the normal stress, σ, across the failure plane is $W\cos\theta/\ell$. Invoking equation (12.13) for S_τ, the factor of safety is given by

$$F = \sum_A^B \frac{S_\tau}{\tau} = \frac{\sum_A^B [c'\ell + (W\cos\theta - u\ell)\tan\phi']}{\sum_A^B W\sin\theta} \qquad (12.15)$$

The weight of the slice, W, is determined from the specific weight of the soil, γ, and the slice geometry. The slice geometry is dependent on the slope height, H, and slope angle, α.

In practice, neither the centre of the slip circle, O, nor the position of the slip circle ACB in Figure 12.8(a) is known *a priori*. A large number of slip circles must be analysed on a trial-and-error basis in order to locate the minimum factor of safety. The critical slip surface may or may not go through the toe, A.

To calculate F for a simple slope with homogeneous soil, one must know the geometrical properties, H, and α, the soil properties, c', ϕ', and γ, and the pore pressure, u, along the various slip circles under analysis. In this study, where the interest lies in calculating maximum stable slope angles, the method will be applied inversely; we will set $F=1$ and solve for α.

To simplify slope-stability calculations for engineering purposes, several sets of charts and nomographs have been produced to allow relatively quick calculations of factors of safety. They are based on various methods of slope-stability analysis, similar in principle to the conventional method of slices but more sophisticated in many cases. The charts of Bishop and Morgenstern (1960) are perhaps the most widely used. For the purposes of the present study, a set of charts presented by Hoek and Bray (1977) have proved the most suitable. These charts are based on the analysis of a set of slip circles that pass through the toe and whose location is fixed at the upper end just above the crest of the slope. For a given chart, if c', ϕ', γ, and H are known and F is set equal to unity, α can be calculated. Each chart applies to a particular water-table configuration and hence to a particular pore-pressure distribution on the slip circles, as indicated by the steady-state flow net construction of Figure 12.8(c). The water-table configuration for which a given chart applies is identified by its intercept ratio, $B=x_{sat}/H$, as given by equation (12.12) and shown on Figures 12.6(c) and 12.8(c). For a water-table configuration that intersects the surface at a distance $x=4H$ behind the toe of the slope, $B=4$. For complete saturation, $B=0$; for a fully drained slope, $B=\infty$. It is possible to interpolate between charts for intermediate water-table configurations.

12.3.5 Summary of Assumptions and Limitations

1. The model is limited to slope development by rapid mass movements on transport-limited slopes.
2. The climate is treated as a simple deterministic sequence of identical storms.
3. Partition of precipitation into overland flow and infiltration is based on the Horton mechanism.

4. Evapotranspiration from upland soils is considered negligible. In the absence of evapotranspiration, groundwater recharge rates equal infiltration rates.
5. The hydrogeological analysis assumes hillslope soils that are homogeneous with respect to hydraulic conductivity, K, and porosity, n.
6. The hydrogeological analysis utilizes analytical solutions that invoke Dupuit–Forchheimer assumptions.
7. The time until ponding, t_p, and the partitioning coefficient, a, are calculated by means of simple parametric expressions rather than a full saturated–unsaturated analysis.
8. The slope stability analysis assumes simple slope geometry, a circular failure surface, and hillside soils that are homogeneous with respect to cohesion, c', friction angle, ϕ', and specific weight, γ.

Assumptions of this type would be totally unsuitable for modelling specific slopes encountered in the field in complex hydrogeological environments. In such cases it would be necessary to use more sophisticated computer codes based on numerical simulation of saturated–unsaturated conditions in heterogeneous soils (see Freeze, 1971; Stephenson and Freeze, 1974). For the present purpose, however, which is limited to a generic analysis of the sensitivity of slope angles to climatic and hydrogeological factors, a simple model of the type outlined has considerable power.

12.3.6 Summary of Quantitative Methodology

1. Given R, N, and t_r, calculate r from equation (12.1).
2. Given K and r, calculate a from equation (12.4).
3. Given R and a, calculate R' from equation (12.3).
4. Given R', K, n, L, and H, calculate $h(L/2)$, Δh, h_{max}, and x_{sat} from equations (12.8), (12.10), and (12.11).
5. Given H and x_{sat}, calculate B from equation (12.12).
6. Given B, c', ϕ', γ, and settling $F=1$, calculate α from the Hoek and Bray (1977) charts.

The necessary input is:

Climate: R, N, t_r
Geometry: L, H
Soil properties: K, n, c', ϕ', γ

Table 12.1 lists the ranges of values that have been utilized in this study. The soil-property values have been selected with the aid of Lambe and Whitman (1969), Carson and Kirkby (1972), Attewell and Farmer (1976), and Freeze and Cherry (1979). The soil properties c', ϕ', n, and γ have been grouped in five representative combinations denoted as Soils A, B, C, D, and E. For each soil, four K values have been utilized. In all, there are 36 climatic combinations, 4 geometric combinations, and 20 soil combinations, for a total of 2,880 calculations of α. The calculations have been carried out with a simple computer program. Selected results are presented in the following section.

Table 12.1 Range of Input Data Values

Climate: $R = 10^{-7}$, 10^{-8}, 10^{-9} m s^{-1}
$N = 1$, 10, 100 events per year
$t_r = 3.15 \times 10^2$, 3.15×10^3, 3.15×10^4, 3.15×10^5

Geometry: $L = 100$, 1,000 m
$H = 10$, 20 m

Soil Properties:

		Soil A (sand)	Soil B (sand)	Soil C (clay)	Soil D (clay)	Soil E (rock)
c'	kN m^{-2}	10	0	40	20	10
ϕ'	degrees	30	40	10	20	30
n	dec. fract.	0.30	0.30	0.50	0.50	0.01
γ	kN m^{-3}	15	15	15	15	25
K	m s^{-1}	10^{-3}	10^{-3}	10^{-6}	10^{-6}	10^{-6}
		10^{-4}	10^{-4}	10^{-7}	10^{-7}	10^{-7}
		10^{-5}	10^{-5}	10^{-8}	10^{-8}	10^{-8}
		10^{-6}	10^{-6}	10^{-9}	10^{-9}	10^{-9}

12.4 RESULTS

Table 12.2 and Figures 12.9 and 12.10 provide a sample of the results that have been generated in this study.

Table 12.2 displays a set of step-by-step calculations for 12 climatic combinations of N and t_r, each based on an average annual rainfall rate of $R = 10^{-7}$ m s^{-1}. The calculations shown on the table are for a geometric configuration with $H = 10$ m and $L = 1{,}000$ m and for a soil with hydraulic conductivity values of 10^{-3}, 10^{-4}, 10^{-5}, and 10^{-6} m s^{-1} and the strength properties of Soil A from Table 12.1. Table 12.2 provides information on the occurrence of overland flow (check marks), the steady-state water-table height at the midpoint, $h(L/2)$, the maximum transient water-table height, h_{max}, and the intercept ratio, B. Note that for all cases where $h_{max} < H$, $B = \infty$.

On Figure 12.9, a plot has been prepared relating the maximum stable slope angle, α, and the intercept ratio, B, for each of the five soils listed in Table 12.1, for two values of H. These curves are independent of K, n, and L. They have been determined directly from the Hoek and Bray charts. A comparison of the curves in Figure 12.9 with the strength parameters in Table 12.1 confirms the well-known fact that soils with high cohesion can support steeper slope angles than soils with low cohesion. Comparison of Figures 12.9(a) and 12.9(b) clarifies the fact that slopes of great height do not develop slope angles as steep as those of slopes of lesser height for a given soil.

Given the B values from Table 12.2 and curves of Figure 12.9, it is possible to relate each climate–soil combination to its maximum stable slope angle. Figure 12.10 presents some selected results in diagrammatic form. Figures 12.10(a) and 12.10(b) are based on the data from Table 12.2 for Soil A, $R = 10^{-7}$ m s^{-1}, $H = 10$ m, and $L = 1{,}000$ m. Figure 12.10(a) shows the slope angle, α, as a function of rainfall event duration, t_r, for four hydraulic conductivities. Slope angles range from 40° to 60° and are greater for higher

Table 12.2 Overland flow occurrences; and values of $h(L/2)$, h_{max}, and B for various combinations of climate and hydraulic conductivity. In all cases, the soil is Soil A, $H=10$ m, $L=1{,}000$ m, and $R=10^{-7}$ m s^{-1}

Climate		Overland Flow Soil K				Steady $h(L/2)$ Soil K			
N	t_r	10^{-3}	10^{-4}	10^{-5}	10^{-6}	10^{-3}	10^{-4}	10^{-5}	10^{-6}
1	3.15×10^2	√	√	√	√	1.6	1.6	1.6	1.6
	3.15×10^3	X	√	√	√	5.0	5.0	5.0	5.0
	3.15×10^4	X	X	√	√	5.0	16.0	16.0	16.0
	3.15×10^5	X	X	X	√	5.0	16.0	50.0	50.0
10	3.15×10^2	X	√	√	√	5.0	5.0	5.0	5.0
	3.15×10^3	X	X	√	√	5.0	16.0	16.0	16.0
	3.15×10^4	X	X	X	√	5.0	16.0	50.0	50.0
	3.15×10^5	X	X	X	X	5.0	16.0	50.0	160.0
100	3.15×10^2	X	X	√	√	5.0	16.0	16.0	16.0
	3.15×10^3	X	X	X	√	5.0	16.0	50.0	50.0
	3.15×10^4	X	X	X	X	5.0	16.0	50.0	160.0
	3.15×10^5	X	X	X	X	5.0	16.0	50.0	160.0

Climate		Transient h_{max} Soil K				B Soil K			
N	t_r	10^{-3}	10^{-4}	10^{-5}	10^{-6}	10^{-3}	10^{-4}	10^{-5}	10^{-6}
1	3.15×10^2	1.9	1.6	1.6	1.6	∞	∞	∞	∞
	3.15×10^3	6.8	5.5	5.1	5.0	∞	∞	∞	∞
	3.15×10^4	6.8	19.	16.	16.	∞	7.3	10.	11.
	3.15×10^5	6.7	19.	55.	51.	∞	7.3	0.84	1.0
10	3.15×10^2	5.5	5.1	5.0	5.0	∞	∞	∞	∞
	3.15×10^3	5.5	16.	16.	16.	∞	10.	11.	11.
	3.15×10^4	5.5	16.	51.	50.	∞	10.	1.0	1.0
	3.15×10^5	5.4	16.	51.	160.	∞	11.	1.0	0.10
100	3.15×10^2	5.1	16.	16.	16.	∞	11.	11.	11.
	3.15×10^3	5.1	16.	50.	50.	∞	11.	1.0	1.0
	3.15×10^4	5.1	16.	50.	160.	∞	11.	1.0	0.10
	3.15×10^5	5.0	16.	50.	160.	∞	11.	1.0	0.10

conductivity soils. Figure 12.10(b) shows α as a function of K for the three values of N used in the study. For soils with conductivity in the range 10^{-5}–10^{-6} m s^{-1}, greater slope angles can be supported in climates where total annual rainfall is concentrated in a few events than in climates where the same annual rainfall is spread more evenly through the year. This result, which may appear counterintuitive, reflects the fact that in climates that release their rainfall in a few large events, the events exhibit high intensities, and much of the water that in more even climates infiltrates and maintains pore pressures, is lost to overland flow.

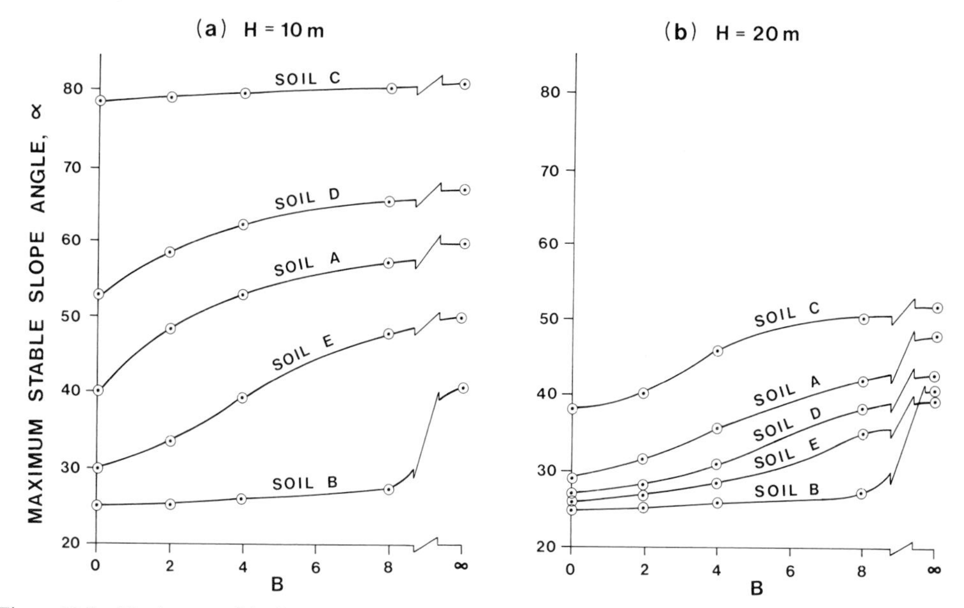

Figure 12.9 Maximum stable slope angle, α, as a function of the intercept ratio, B, for five soils and two slope heights, H. Soil properties are listed in Table 12.1

Tables similar to Table 12.2 can be prepared for different values of R, H, and L and for Soils B, C, D, and E. On the basis of such tables and Figure 12.9, figures similar to Figures 12.10(a) and 12.10(b) can be constructed to display the functional dependence between various combinations of geometric properties, soil properties, climatic variables, and slope angles. Figures 12.10(c) to 12.10(f) are based on data that are not included on Table 12.2. Figure 12.10(c) compares the stable slope angles for arid ($R=10^{-9}\,\mathrm{m\,s^{-1}}$), semi-arid ($R=10^{-8}\,\mathrm{m\,s^{-1}}$), and humid ($R=10^{-7}\,\mathrm{m\,s^{-1}}$) climates for four different hydraulic conductivities for a soil with the strength parameters of Soil A. Figures 12.10(d) and 12.10(e) do the same for soils E and C. Figure 12.10(f) shows the influence of soil types on slope angle for the only hydraulic conductivity value common to all five soils. Figure 12.10(g) displays the influence of slope geometry on slope angle for a particular climate–soil combination.

The results plotted on Figure 12.10 provide a quantitative indication of the wide variation in slope angles that can be expected in similar soils under different climatic regimes and in soils of different hydraulic conductivity under similar climatic regimes. They can also be viewed as a sensitivity analysis. Each curve tends to show regions of sensitivity and regions of insensitivity. For example, consider the following:

1. In Figure 12.10(a) for soils with $K=10^{-5}\,\mathrm{m\,s^{-1}}$, the slope angles are sensitive to the climate only within the range $t_r=3.15\times10^2$ to $t_r=3.15\times10^3$ s. For all storm durations less than 3.15×10^2 s, $\alpha=60°$, and all storm durations greater than 3.15×10^3 s, $\alpha=45°$.
2. A comparison of Figures 12.10(c), 12.10(d), and 12.10(e) shows that slope angles for Soil A are more sensitive to hydraulic conductivity in humid climates, whereas those for Soils E and C are more sensitive in arid climates. Slope angles for Soils A and E show a range of almost 20°, whereas the slope angle for Soil C varies over only 5°.

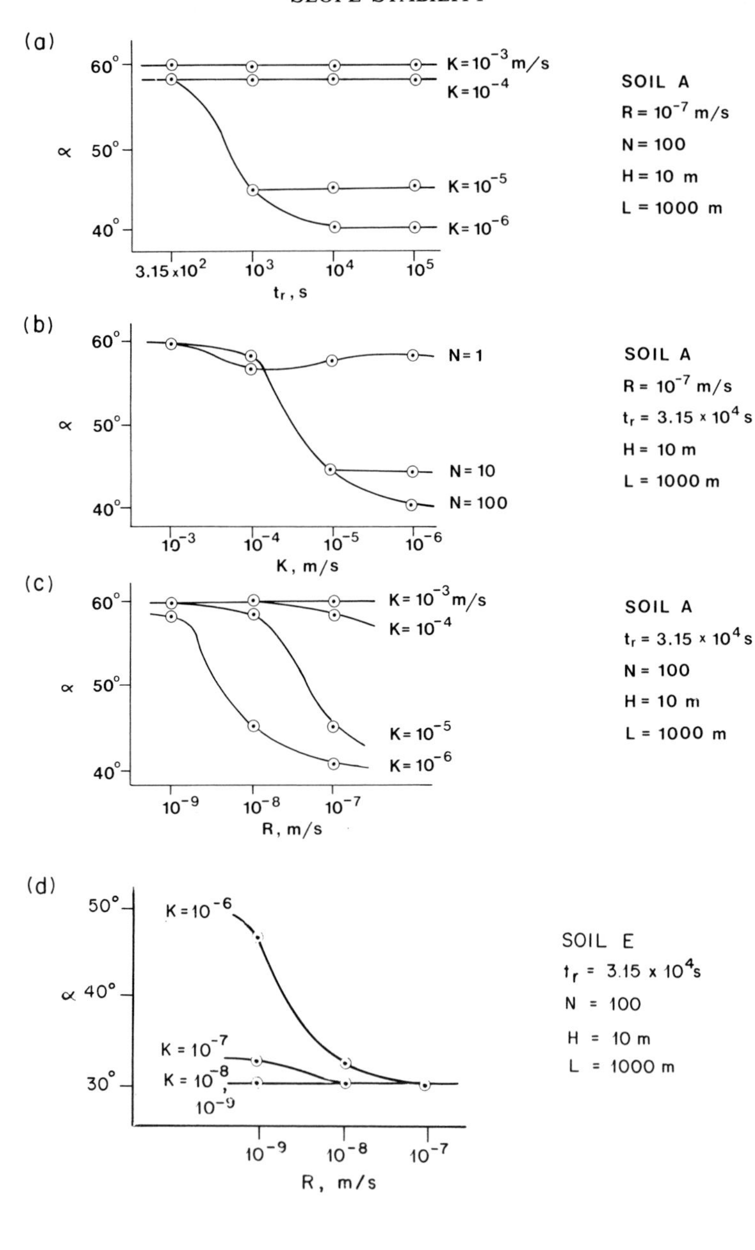

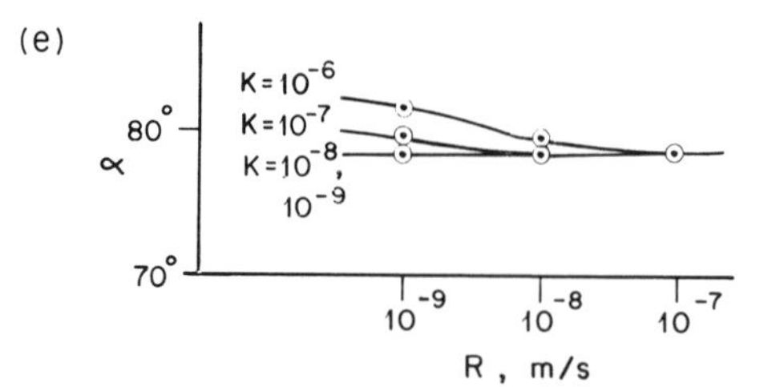

SOIL C

t_r = 3.15 x 10^4s

N = 100

H = 10 m

L = 1000 m

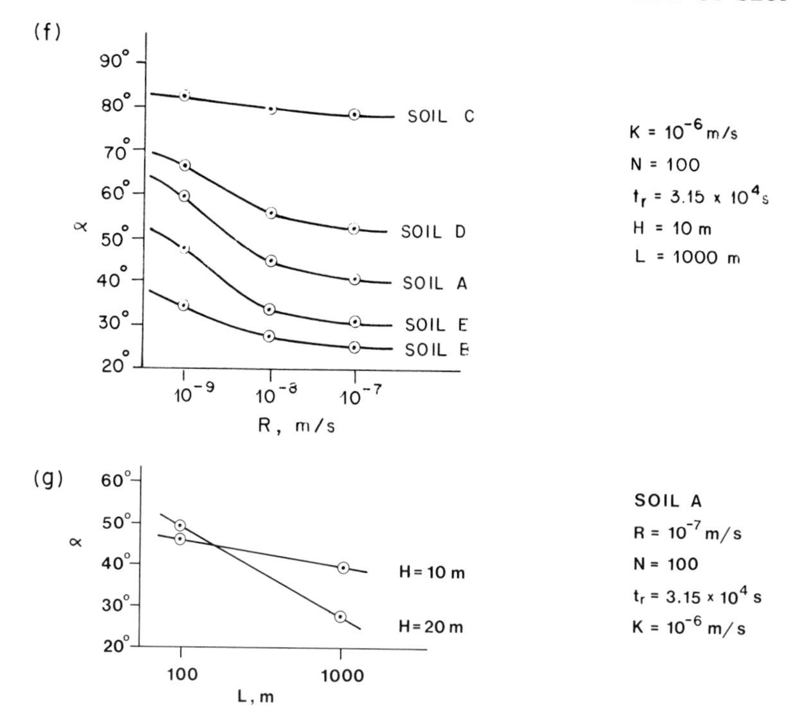

Figure 12.10 Selected results. Maximum stable slope angles for various combinations of soil types; climatic parameters, R, N, and t_r; geometric parameters, H and L; and soil hydraulic conductivity, K

Slopes represent an integrated reaction between past climates and existing soil properties. Considering the zones of sensitivity and insensitivity displayed in Figure 12.10, it may be possible to use slope measurements as an indication of past climate in regions of insensitivity to soil type or as an indirect measure of soil properties in regions of insensitivity to climate.

12.5 IMPLICATIONS FOR MORE COMPLEX HYDROLOGICAL ENVIRONMENTS

The analysis included in the previous sections is for the simplest of all hydrogeological environments: a deep homogeneous soil (or geological formation) whose saturated hydraulic conductivity is everywhere the same. In reality, large spatial variations in K, and in the mechanical properties, c' and ϕ', can be expected to occur over relatively small distances in the field. Considering the sensitivity shown in this study to these parameters, it is not surprising that slope angles for slopes on which failures occur commonly cover a wide range of values within relatively small areas (Moser and Hohensinn, 1983; Rouse, 1975; Rouse and Farhan, 1976).

It is also true, because the hydraulic conductivity of surface soils varies so greatly, that individual hillslopes often exhibit different runoff-generating mechanisms at different places during the same storm or at the same place during different storms (Freeze, 1980). Such complexities could have significant impact on infiltration rates and the growth and decay of pore pressures through time.

With respect to the more structured spatial variability in saturated hydraulic conductivity caused by geologic structure and stratigraphy, there is a large literature that shows the influence of such variability on hydraulic-head patterns at a regional scale. Hodge and Freeze (1977) presented a suite of numerically simulated, steady-state, saturated flow nets for a variety of hydrogeological environments and discussed the implications of the hydraulic-head patterns to considerations of regional slope stability.

On the scale of an individual hillslope, there are particular complications that are introduced by layered stratigraphy. Rulon and Freeze (1985) have carried out a transient saturated–unsaturated, finite-element analysis of such systems. They showed that layered slopes feature multiple seepage faces, perched water tables, and wedge-shaped unsaturated zones. The pore-pressure distributions and the locations of the seepage faces are strongly dependent on the positions of the impeding layers and their hydraulic properties. They noted that predictions of pore-pressure fields based on homogeneous saturated analyses may be significantly in error when applied to problems on layered slopes.

One of the most common hydrogeological environments is a thin soil regolith of high conductivity overlying bedrock of much lower conductivity. The mechanisms of infiltration, overland flow, and runoff generation on such slopes are treated by Kirkby and Chorley (1967), Freeze (1972), and Dunne (1978). Mass movements in such environments will take the form of shallow soil slips rather than deep rotational slides. Recent field studies of this type of mass movement have been reported by Rice *et al.* (1969), Rice and Foggin (1971), and Moser and Hohensinn (1983). All these authors noted that failures are always associated with heavy rainfall events. Moser and Hohensinn (1983) actually classified the mass movements they observed with respect to the type of climatic events that caused them. The climatic events are classified as: (a) cloudbursts ($r = 50\text{–}500\ \mathrm{mm\,h^{-1}}$, $t_r = 0.5\text{–}5\ \mathrm{h}$); (b) prolonged rainstorms ($r = 10\text{–}100\ \mathrm{mm\,h^{-1}}$, $t_r = 5\text{–}50\ \mathrm{h}$); and (c) long heavy rainfalls ($r = 0.1\text{–}10\ \mathrm{mm\,h^{-1}}$, $t_r = 50\text{–}500\ \mathrm{h}$). Measured soil conductivities are in the range of $10^{-4}\text{–}10^{-6}\ \mathrm{m\,s^{-1}}$. A plot of scar frequency versus slope inclination shows a peak around 40° and a range of 20°–50°. Although the hydrogeological environment is different from the one invoked in the model presented here, the approach of Moser and Hohensinn (1983) shows the potential for an encouraging convergence between theory and practice.

12.6 IMPORTANCE OF SATURATED HYDRAULIC CONDUCTIVITY IN THE DEVELOPMENT OF SLOPES

Perhaps the primary conclusion that can be drawn from this study of the interrelationships between climate, hydrology, hydrogeology, and geomorphology lies in recognition of the importance of the saturated hydraulic conductivity of hillslope soils and geological formations in the development of slopes and landforms. In the most general context, for a given climatic environment with its specific suite of representative rainfall intensities, it is the saturated hydraulic conductivity values that control whether the primary agent of erosion will be surface runoff or subsurface pore pressures.

With reference to mass movements, it is widely recognized that the soil-strength parameters, c' and ϕ', exert an important controlling influence on slope development. It seems to be less widely recognized that saturated hydraulic conductivity exerts an equally important control, yet this fact emerges clearly from the results shown in Figures 12.10(a), 12.10(b), and 12.10(c). The role of hydraulic conductivity is a dual one. On the one hand, it controls the runoff mechanism and hence the percentage of rainfall that becomes infiltration, and on the other, it controls the pore-pressure values that will arise from a given infiltration rate. As an example, consider the dip in the $N = 1$ line in Figure 12.10(b). It is a result of this dual role of hydraulic conductivity. More pronounced effects of this type may be expected elsewhere in the full suite of curves that can be produced from this type of study.

A particular concept of equilibrium, which might be termed 'hydrogeological equilibrium', rests on the relative values of rainfall intensity and soil hydraulic conductivity.

In areas where saturated hydraulic conductivities are less than rainfall intensities, overland flow will take place and such soils will be removed by erosion due to surface-water runoff. Once a higher conductivity formation from lower in the stratigraphic sequence is uncovered (or a higher permeability soil regolith is developed by weathering processes), the runoff mechanism will shift from overland flow to subsurface delivery. One can picture as an end product, at least in principle, an equilibrium slope profile in which overland flow is not generated and valley slopes are stable with respect to mass movements.

12.7 SUMMARY

This chapter outlines an approach to the quantitative investigation of the influence of climate on geomorphology. Relationships are derived that show the effect of rainfall rates and durations, hillslope geometry, and soil properties on the maximum stable slope angle of geomorphic landforms. The analysis includes token consideration of the mechanisms of surface runoff and detailed consideration of the mechanisms of subsurface pore-pressure development. This idealized analysis invokes deterministic climates, homogeneous soils, simple slope geometries, and analytical hydrogeological solutions. The results show that a wide variation of response by slope angle can be expected in similar soils to differences of climatic regime, and to differences of soil properties in similar climatic regimes. Saturated hydraulic conductivity is identified as a parameter of some importance in that it plays the dual role of controlling the runoff mechanism on the surface and the pore-pressure development at depth. In many of the relationships between soil properties, climatic variables, and slope angles, there are regions of great sensitivity and others of relative insensitivity. In that slopes represent an integrated reaction between past climates and existing soil properties, it may be possible to use slope-angle measurements as an indication of past climate in regions of insensitivity to soil type or as an indirect measure of soil properties in regions of insensitivity to climate.

ACKNOWLEDGEMENTS

The material presented in this chapter was first presented at the Symposium on Hydrology, Geomorphology and Climate, held in Caracas, Venezuela in 1980 during a decennial celebration at Simón Bolívar University. I am indebted to Ignacio Rodriquez-Iturbe, Peter Eagleson, Juan Valdes, and John Schaake for discussions held at that time.

NOTATION

a = partitioning coefficient denoting percentage of annual rainfall that becomes infiltration rather than overland flow

c = cohesion (F/L^2)
e_a = between-storm actual evapotranspiration rate (L/T)
e_p = between-storm potential evapotranspiration rate (L/T)
r = rainfall event intensity (L/T)
h = hydraulic head (L)
h_{max} = maximum hydraulic head (L)
Δh = range in hydraulic head at $x=L/2$ (L)
ℓ = width of slice in slope stability analysis (L)
n = porosity (decimal fraction)
t = time (T)
t_b = time between storms (T)
t_p = ponding time (T)
t_r = rainfall event duration (T)
u = pore pressure
x = distance (L)
x_{sat} = distance at which water table intersects upland peneplain (L)
z = elevation (L)

B = intercept ratio (dimensionless)
F = factor of safety (dimensionless)
E = average annual potential evapotranspiration rate (L/T)
H = length of hillslope (L)
K = saturated hydraulic conductivity (L/T)
L = width of hillslope (L)
N = number of rainfall events per year (dimensionless)
R = average annual rainfall intensity (L/T)
R' = average annual groundwater recharge rate (L/T)
S_τ = shearing strength (F/L^2)
W = weight of slice (F)

α = maximum stable slope angle (degrees)
γ = specific weight of soil (F/L^3)
γ_w = specific weight of water (F/L^3)
ϕ = angle of internal friction (degrees)
σ = normal stress (F/L^2)
τ = shearing stress (F/L^2)
θ = angle to the perpendicular of base of slice (degrees)

REFERENCES

Attewell, P. B., and Farmer, I. W. (1976). *Principles of Engineering Geology*. John Wiley, New York.
Bear, J. (1972). *Dynamics of Fluids in Porous Media*. Elsevier, New York.
Bishop, A. W., and Morgenstern, N. R. (1970). 'Stability coefficients for earth slopes.' *Geotechnique*, **10**, 29–150.
Carson, M. A. (1969). 'Models of hillslope development under mass failure.' *Geog. Anal.*, **1**, 76–100.

Carson, M. A., and Kirkby, M. J. (1972). *Hillslope Form and Process*. Cambridge University Press.
Derbyshire, E. (ed.) (1976). *Geomorphology and Climate*. John Wiley, New York.
Dunne, T. (1978). 'Field studies of hillslope flow processes.' In *Hillslope Hydrology*. Kirkby, M. J. (ed.), John Wiley, New York, pp. 227–93.
Embleton, C., and Thornes, J. (eds.) (1979). *Process in Geomorphology*. Edward Arnold, London.
Freeze, R. A. (1971). 'Three-dimensional transient, saturated–unsaturated flow in a groundwater basin.' *Water Resources Res.*, **7**, 347–66.
Freeze, R. A. (1972). 'Role of subsurface flow in generating surface runoff. 2. Upstream source areas.' *Water Resources Res.*, **8**, 1271–83.
Freeze, R. A. (1980). 'A stochastic-conceptual analysis of rainfall–runoff processes on a hillslope.' *Water Resources Res.*, **16**, 391–408.
Freeze, R. A., and Cherry, J. A. (1979). *Groundwater*. Prentice-Hall, Englewood Cliffs, New Jersey.
Hodge, R. A., and Freeze, R. A. (1977). 'Groundwater flow systems and slope stability.' *Can. Geotech. J.*, **14**, 466–76.
Hoek, E., and Bray, J. (1977). *Rock Slope Engineering*. Institute of Mining and Metallurgy, London.
Horton, R. E. (1945). 'Erosional development of streams and their drainage basins: Hydro-physical approach to quantitative morphology.' *Geol. Soc. Am. Bull.*, **56**, 275–370.
Kirkby, M. J. (ed.) (1978a). *Hillslope Hydrology*, John Wiley, New York.
Kirkby, M. J., and Chorley, R. J. (1967). 'Throughflow, overland flow and erosion.' *Int. Assoc. Sci. Hydrol. Bull.*, **12**, 5–21.
Lambe, T. W., and Whitman, R. V. (1969). *Soil Mechanics*. John Wiley, New York.
Leopold, L. B., Wolman, M. G., and Miller, J. P. (1964). *Fluvial Processes in Geomorphology*. W. H. Freeman, San Francisco, New York.
Morgenstern, N. R., and Sangrey, D. A. (1978). 'Methods of stability analysis.' In *Landslides: Analysis and Control*. National Academy of Sciences, Transport Research Board, Washington, DC, Spec. Rept. 176, pp. 155–71.
Moser, M., and Hohensinn, F. (1983). 'Geotechnical aspects of soil slips in alpine regions.' *Eng. Geol.*, **19**, 185–211.
Rice, R. M., Corbett, E. S., and Bailey, R. G. (1969). 'Soil slips related to vegetation, topography and soil in California.' *Water Resources Res.*, **5**, 647–59.
Rice, R. M., and Foggin, G. T. (1971). 'Effect of high-intensity storms on soil slippage on mountainous watershed in southern California.' *Water Resources Res.*, **7**, 1485–96.
Ritter, D. F. (1978). *Process Geomorphology*. Brown, New York.
Rouse, W. C. (1975). 'Engineering properties and slope form in granular soils.' *Eng. Geol.*, **9**, 221–35.
Rouse, W. C., and Farhan, Y. I. (1976). 'Threshold slopes in South Wales.' *Q. J. Eng. Geol.*, **9**, 327–38.
Rubin, J., and Steinhardt, R. (1963). 'Soil water relations during rain infiltration. I. Theory.' *Soil Sci. Soc. Am. Proc.*, **27**, 246–51.
Rulon, J. J. and Freeze, R. A. (1985). 'Multiple seepage faces on layered slopes and their implications for slope stability analysis.' *Can. Geotech. J.*, **22**, 347–356.
Stephenson, G. R., and Freeze, R. A. (1974). 'Mathematical simulation of subsurface flow contributions to snowmelt runoff, Reynolds Creek Watershed, Idaho.' *Water Resources Res.*, **10**, 284–94.
Young, A. (1972). *Slopes*. Oliver and Boyd, London.

Slope Stability
Edited by M. G. Anderson and K. S. Richards

Chapter 13

Weathering Effects: Slopes in Mudrocks and Over-consolidated Clays

R. K. TAYLOR
School of Engineering & Applied Science
University of Durham, Durham DH1 3LE

and

J. C. CRIPPS
Department of Geology
University of Sheffield, Sheffield S1 3JD

13.1 INTRODUCTION

Rock weathering is customarily regarded as the result of two sets of processes: physical disintegration and chemical decomposition (for example, Whitten and Brooks, 1972). Both occur under the direct influence of the hydrosphere and atmosphere. The effects of physical and chemical weathering are complementary, but in so far as physical processes are more rapid and increase surface area, thus accelerating chemical weathering, physical disintegration can be viewed as acting as a control on chemical decomposition (Taylor and Spears, 1970).

Five sets of variables were recognized by Jenny (1951) as being instrumental in weathering and soil formation. They are: climate (*cl*), biological activity, or organisms (*b*), topography, which influences drainage (*r*), parental rock type (*p*), and the time during which soil formation takes place (*t*). For a particular soil property (*S*) a functional relationship will obtain with these variables. As explained by Carroll (1970), Jenny's incorporation of moisture content (*m*) and temperature (*T*) as climatic variables (i.e. $S=f(m)_{T,b,r,p,t}$ and $S=f(T)_{m,b,r,p,t}$) can be used to explain why a particular soil type should form under given climatic conditions.

The above approach is sensibly related to longer-term chemical weathering in which pore-water composition and mineral stability are concerned. Put simply, minerals will either

Table 13.1 Major Soil Groups with Reference to USA (Largely from Carroll, 1970)

Soil Group		Mudrock/Clay cited as a Parental Material—among others (Yes√ No x)	Principal Clay Minerals present (on world-wide basis)
(a) ZONAL SOILS			
Tundra	1. Cold zone	—	illite
DESERT	2. Light coloured soils of arid regions	√ shales	mixed-layer clay, smectite
RED DESERT		x	illite
SIEROZEM		√ shales	—
BROWN		√ marl (Ogallata Fm.) shales	—
REDDISH BROWN		√ calcareous clay, sandy clay	—
CHESTNUT	3. Dark-coloured soils. Semi-arid, humid and sub-humid grassland	√ Tertiary shales	smectite
REDDISH CHESTNUT		√ clays, shales	—
CHERNOZEM (black earth)		√ Cretaceous shales (Pierre Shale)	smectite, mixed-layer clay
PRAIRIE		√ calcareous shale	smectite, mixed-layer clay
REDDISH PRAIRIE		√ clay	—
Degraded Chernozem	4. Forest-grassland transition soils	— —	—
NON CALCIC BROWN		x	illite, mixed-layer clay
PODZOL	5. Light-coloured podzolized soils of timbered regions	x	illite
Grey wooded or grey podzolic		— —	—
BROWN PODZOLIC		√ shale	
GREY-BROWN PODZOLIC		x	illite, kaolinite
RED-YELLOW PODZOLIC (lateritic)		√ slate, shale, sandy clay, clays	kaolinite
Reddish brown lateritic	6. Lateritic soils of forested warm temperate and tropical regions	— —	kaolinite
Yellowish brown lateritic		— —	kaolinite
Laterite soils		— —	kaolinite

continued

Table 13.1 *(continued)*

Soil Group		Mudrock/Clay cited as a Parental Material—among others (Yes√ No x)	Principal Clay Minerals present (on world-wide basis)
(b) INTRAZONAL SOILS			
SOLONCHAK OR SALINE	1. Halomorphic (saline and alkali)	√ clayey materials	similar to surrounding soils
SOLONETZ		√ clayey materials	similar to surrounding soils
Soloth		– –	variable
HUMIC GREY SOILS		√ clays, sandy clays, calc. clays, marls.	smectite
Alpine Meadow	2. Hydromorphic—marshes, swamps, etc.	— —	—
BOG SOILS		x	—
HALF-BOG SOILS		√ clays, sandy clays, calc. clays, marls.	—
Low-humic gley		— —	—
PLANOSOLS		√ calcareous clays, marls.	illite, mixed-layer clay, kaolinite
GROUNDWATER PODZOL		√ clays, sandy clays, calc. clays, marls.	smectite
Groundwater laterite		— —	—
Brown forest soils	3. Calcimorphic soils	— —	—
RENDZINA		√ calcareous shale, calc. clays, marls.	smectite
(c) AZONAL SOILS			
LITHOSOLS (active erosion)		x	—
Regosols (incl. dry sands) etc.		—	—

Note Those soils found in USA shown in capital letters.

be dissolving or precipitating under a given climatic regime. At one extreme evapotranspiration will lead to an increase in ionic strength, so that -cretes or duricrusts will form, while at the other extreme, continuous leaching of ions will result in the accumulation of hydrolysates and resistates such as kaolinites, laterites, and bauxites. With respect to clay minerals, prolonged leaching leads to simplified forms, such that in traversing southwards to the equator, illite and mixed-layer mica–montmorillonites are replaced by kaolinites and ultimately by gibbsite.

One of the problems pertaining to weathering studies is the difficulty in distinguishing between contemporary and ancient weathering. In this context it is not just the 8,000–10,000 years of post-glacial weathering in the United Kingdom which is of concern, but rather more protracted (geological) time intervals. For example, lateritic bauxites in India, Africa, South and North America were formed (world-wide) in Cretaceous and Tertiary coastal plains. Many were subsequently uplifted and dissected at altitudes ranging from a few hundred to up to 2,000 metres (Valeton, 1983). However, these residual deposits outcrop in areas which are still largely conducive to the same weathering styles.

This chapter is specifically aimed at the weathering of over-consolidated clays and mudrocks, and the manner in which weathering processes are believed to affect the stability and morphology of slopes. Over-consolidated clays and their indurated equivalents, mudrocks, are deposits which have been subject to overburden pressures in excess of those which prevail at present.

Man-made slopes are included in the following discussion so that short-term effects can be demonstrated. In this chapter the mudrocks and clays remain recognizable as argillaceous sediments; slopes formed under various other (often extreme) climatic conditions, and in other rock types, are dealt with in other chapters. Nevertheless, data extracted largely from Carroll (1970) demonstrate that in the USA alone nearly all pedological types can develop from clays and mudrocks (Table 13.1). Also included in Table 13.1 are the principal clay minerals associated with these major soil groups.

The effect of climatic variation on the *chemical* weathering of clay minerals has already been referred to, but changes which occur during burial diagenesis are outlined in section 13.2. Subsequent rebound (exhumation) and weathering processes detailed in sections 13.3 and 13.4 can be viewed as being the converse of burial diagenesis.

The impact of weathering processes on the shearing resistance of argillaceous rocks and sediments, and thus on the stability of natural and cut slopes, is discussed in the penultimate section of this chapter.

13.2 COMPOSITION

Over-consolidated clays and mudrocks are dominantly composed of (siliciclastic) detritus derived from a pre-existing landmass. Modern geological classifications (e.g. Potter *et al.*, 1980) stipulate that they should contain more than 50% siliciclastic grains of less than 63 μm in size (that is, silt and clay sizes). It will be observed from Table 13.2 that, *on average* siliciclastics make up about 94% of the total, of which clay minerals account for over 60% and quartz about 26%. Carbonates and other non-detrital (cementing) minerals customarily make up less than 10% of the total.

Geological classifications do not readily differentiate between lithified mudrocks and over-consolidated clays, but a three-fold sub-division is favoured by many investigators,

Table 13.2 Average mineral composition of mudrocks and over-consolidated clays

	Yaalon (1962)	Shaw and Weaver (1965)	Pettijohn (1975)	Smith (1978) UK	Smith (1978) N. America
Clay minerals	59	66.9	58*	70*	61
Quartz	20	36.8	28	29	36
Feldspar	8	4.5	6	1	3
Caronates	7	3.6	5	4	3
Iron oxide	3	0.5	2	3	1
Organic carbon	—	1.0	—	1	1
Miscellaneous	—	0.2	—	—	—

*includes Fuller's earths.

viz. >2/3 silt-size (2–63 μm) = silt/siltstone; 2/3 to 1/3 silt-size = mud/mudstone/mudshale; >2/3 clay-size (<2 μm) = clay/claystone/clayshale. Shale is used to signify a laminated and/or fissile structure. Parting frequency (Anon, 1970; Potter *et al.*, 1980) is important in terms of physical weathering (section 13.4.2).

Attempts to distinguish mudrocks from clays have been made in geotechnical practice (for example, Deo, 1972; Morgenstern and Eigenbrod, 1974). However, they are not altogether successful. As pointed out by Taylor and Spears (1981) only certain calcareous mudstones of Lower Lias age, and indurated mudrocks of Carboniferous age and older, would be designated mudrocks on Morgenstern and Eigenbrod's criteria. Well-known argillaceous beds of Jurassic, Cretaceous and Eocene age fall into the over-consolidated clay category. Contrary to this, however, undrained triaxial tests conducted on many of these 'clays' indicate shear strengths ($s_u = c_u$, $\phi_u = 0$) which conform with uniaxial compressive strengths (UCS = $2c_u$) which are commonly assigned to *weak* and *very weak* rocks (0.6–1.25 MN/m^2, *very weak*; 1.25–5.0 MN/m^2, *weak rocks*)—see Table 6 in Cripps and Taylor (1981).

The majority of over-consolidated clays were deposited under marine conditions. Post-depositional chemical changes and the genesis of non-detrital (cementing) minerals are detailed by Curtis (1980). It should be appreciated that during diagenesis structural modifications to clay minerals may take place. These modifications are governed by temperature or heat flow rates as a function of depth of burial, by pore-water chemistry, and time. Figure 13.1 illustrates the increase in illitic mica and chlorite at the expense of kaolinite and expandable clays with geological time in mudrocks from both the UK (Shaw, 1981) and the USA (Weaver, 1967). It is reasonable to conclude from both of these stratigraphic or diagenetic sections that mudrock disintegration is unlikely to be exacerbated by *intra*particle swelling of *expandable* clay minerals (*per se*) in mudrocks older than the Silurian in the UK, or Lower Mississippian in the USA.

The progressive illitization (transformation to mica) of smectite (montmorillonitic minerals) below a depth of about 500 m leads to increased mechanical stability because of the loss of expandable layers from the clay mineral structure. On the other hand, the formation of chlorite at high diagenetic levels (Figure 13.1) might result in a reduction in stability on weathering if the chlorite formed is rich in *ferrous* iron (Evans and Adams, 1975). The physical changes such as the reduction in porosity and permeability which occur during compaction (see Figure 13.3) help to retard chemical reactions during weathering

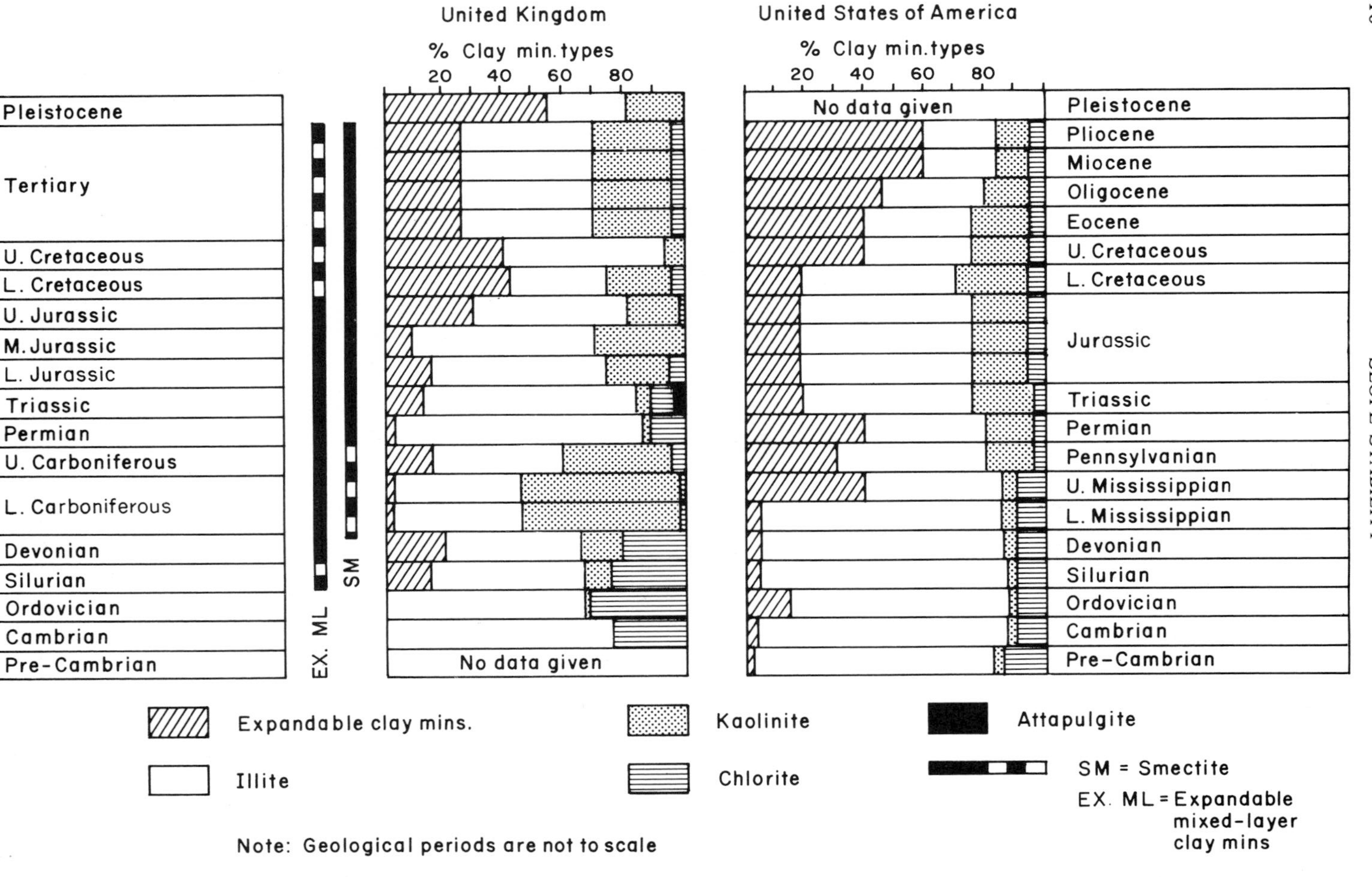

Figure 13.1 Variation in clay minerals with geological age. (*After Shaw, 1981, UK, 1050 samples. After Weaver, 1967, USA, c. 70 000 samples. Shaw reproduced by permission of the Geological Society and Blackwell Scientific Press. Weaver reproduced by permission of Pergamon Press.*)

by restricting water access, as well as reducing physical breakdown mechanisms. Interparticle forces (bonds) also increase with increased diagenetic level.

13.3 COMPACTION, EXHUMATION, AND REBOUND

High horizontal stresses in over-consolidated clays and mudrocks are believed to be an important factor in exacerbating weathering, with the consequent reduction in the mass shear strength of near-surface clays and mudrocks (see for example, Bjerrum, 1967).

Horizontal stresses are commonly equal to or greater than the vertical stresses in the upper part of the earth's crust, irrespective of rock type. Hoek and Brown (1980) have plotted the ratio, K, of average horizontal to vertical stress against depth below the surface for 116 measurements taken from the literature. Values of K were found to lie within the following limit lines:

$$(100/z) + 0.3 < K < (1500/z) + 0.5$$

where z is the depth below surface in metres.

For depths of less than 500 m the results show the horizontal stresses to be significantly greater than vertical stresses, whereas at 1 km depth or more, vertical and horizontal stresses are more equal. A value of $K = 5.56$ is the highest figure quoted, although Hoek and Brown (1980) recognize that values of up to about 10 occur locally in tectonically-active mountain belts. Analysis of plate motions has indicated that in general present continental areas are in compression (Dewey, 1984). It is therefore probably realistic to consider the cumulative effects of both current and residual stress fields that have been locked into a rock when parts of the earth's crust are uplifted and eroded, or deep excavations are engineered by man.

In mudrocks and over-consolidated clays the highest measured and calculated K_o values occur very close to the ground surface. The symbol K_o is used in clays to indicate that the horizontal (σ_h') to vertical (σ_v') stress ratio is in terms of effective stresses. The K_o distributions shown in Figure 13.2, which are for London Clay, Gault Clay, and Keuper Marl, indicate that values in excess of 3 generally lie within the depth range 2–12 m below ground level. It will be observed that in Gault Clay a K_o value of 3 at 24.7 m depth has been calculated. At 85.3 m K_o was still equal to 1.4 at this site (Samuels, 1975).

Laboratory experiments by Brooker (1967) have shown that past geological loading increases the stored strain energy and thus the consequent K_o value in an over-consolidated clay.

An explanation of the way in which these high horizontal stresses may arise in clays and mudrocks as a consequence of the release of strain energy stored during compaction can be obtained from a laboratory consolidation test analogy. This is illustrated in Figure 13.3(A) where the effective (overburden) pressure is shown on a logarithmic scale.

Given an increasing sedimentary overload there is a rapid decrease in water content and porosity. This also obtains in natural systems as shown by Hedberg (1936) and Skempton (1970a). Clays lying on the line a–b–c in Figure 13.3(A) are 'normally consolidated'. When the clay is unloaded to present overburden pressure (say, point e) it becomes 'over-consolidated'. Although under the same effective overburden pressure as its normally-consolidated equivalent the water content (and porosity) will be less. Since the mineral

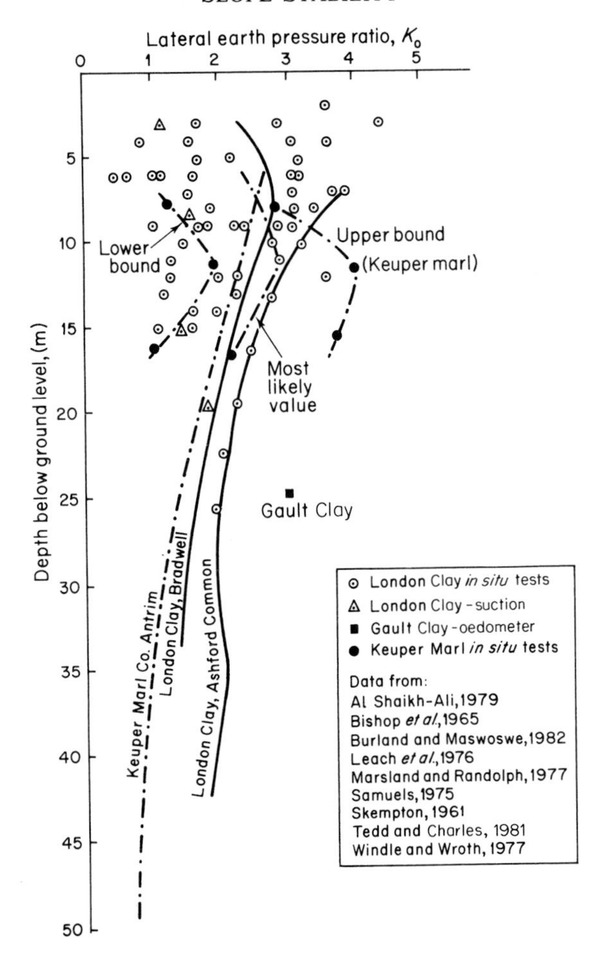

Figure 13.2 Distribution with depth of high horizontal stresses (expressed as K_0) in UK over-consolidated clays

particles are in a denser packing configuration the shear strength of the over-consolidated clay will be higher than in the normally-consolidated state (Figure 13.3(B)).

Bjerrum (1967) postulated that if at (say) point c the sediment remained under the same sedimentary overload for a considerable period of time there would be some reduction in moisture content (c–d), which represents the secondary or creep compression of the sediment. During a protracted time interval, diagenetic bonds would develop at this stage. These may comprise authigenic cements, cationic adhesion at particle to particle contacts, or 'cold welds' between grains (see also Rosenqvist, 1968; Andersland and Douglas, 1973). Consequently the sediment will become even more brittle and stronger.

During unloading, the deposit will have a tendency to expand and increase in water content (d–e–f or d–g–h, Figure 13.3(A)). This is a function of the recovery of the strain energy stored in compaction by the bending and compression of clay minerals and so on. The expansion on unloading of clays and mudrocks with strong diagenetic bonds will be restricted. In nature, for example, the estimated rebound of up to 10% in valley downcutting (unloading is universally reported to consist of both an elastic and a time-dependent

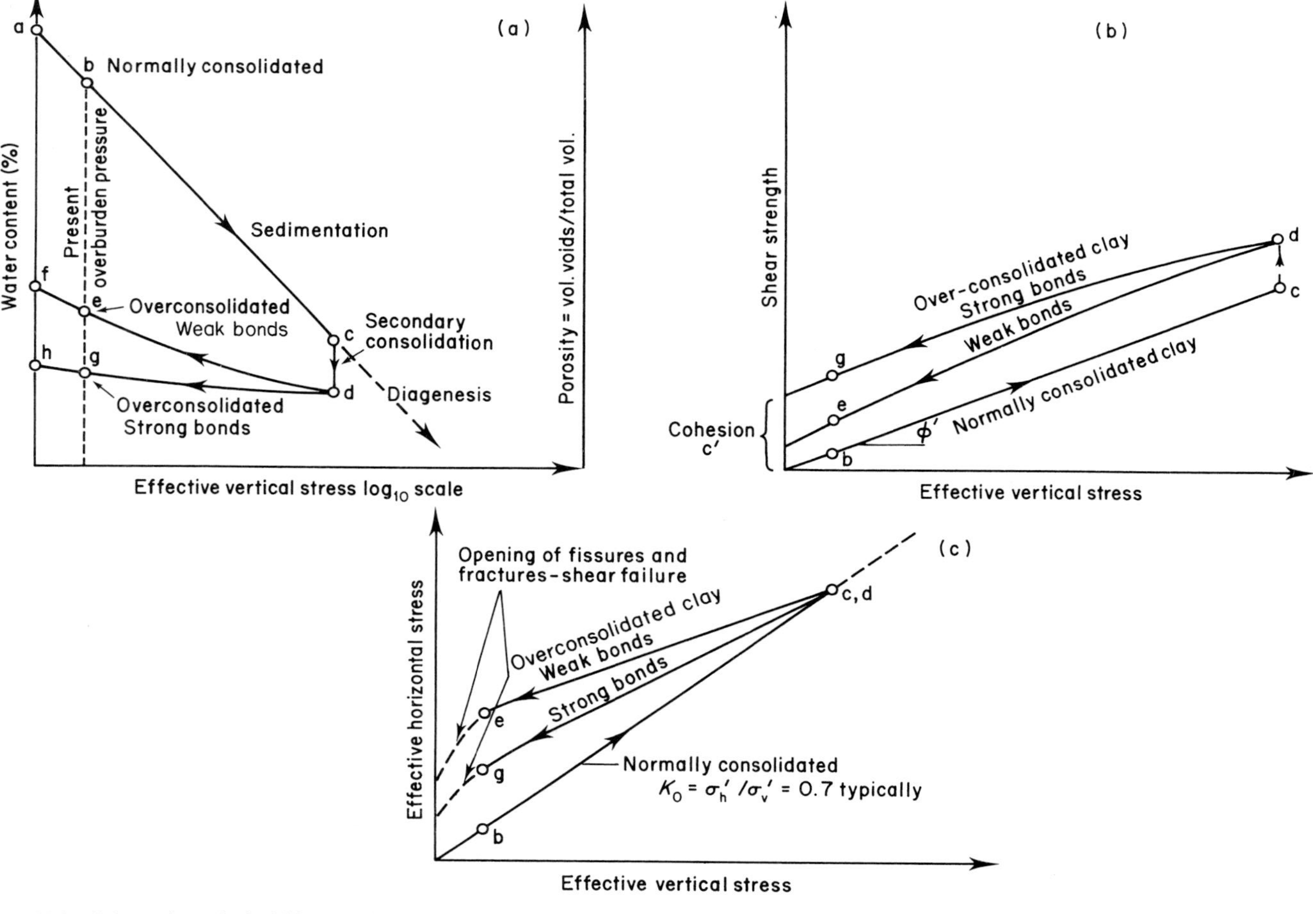

Figure 13.3 Schematic geological history of an over-consolidated clay or mudrock. (A) Compaction and unloading. (B) Shear strength as a function of compaction and unloading. (C) Vertical and horizontal stresses in response to compaction and unloading. (*Data modified after Bjerrum, 1967 and Fleming* et al., *1970. Bjerrum reproduced by permission of the American Society of Civil Engineers.*)

component (Fleming *et al.*, 1970; Matheson and Thomson, 1973; Nichols, 1980). Jointing may well be an expression of the elastic response to unloading during uplift and erosion on a geological scale (see, for example, Chapter 15).

The breakdown of bonds in a particle structure will also be time dependent so there will be a secondary time effect in the expansion process for *strongly bonded* mudrocks modelled in Figure 13.3(A). Nevertheless, bonds will be further stressed and will fail in increasing numbers closer to the ground surface.

Modelling the horizontal to vertical stresses in Figure 13.3(C) indicates that in clays with weak bonds the horizontal stress would be larger than for strong clays. The stress differences are developed on unloading because the clay is free to expand in the vertical direction but not in the horizontal direction. Hence the drop in effective vertical stress is greater than the change in effective horizontal stress. In the case of a strongly bonded clay or mudrock a greater amount of latent strain energy is considered by Bjerrum (1967) to be retained within the clay structure. However, this will be released progressively, to the detriment of the long-term stability of slopes in mudrocks and heavily over-consolidated clays.

The mechanical effects of high horizontal stresses in engineering are manifest by 'rock squeeze' phenomena in excavations (e.g. Harper and Szymanski, 1983), and more generally, by the development of additional extensional fractures and fissures which are commonly sub-parallel to the local topography. Fractures of this type, which are usually without filling and are contained within specific lithological units, have been reported by numerous authors (for example, Hofmann, 1966; Ferguson, 1967; Nichelson and Hough, 1967; Nichols, 1980).

It is of interest that Terzaghi (1961) observed that within the same depth range for London Clay, K_0 was of the same order as the passive earth pressure coefficient, K_p (which can be derived from Rankine's earth pressure theory). This means that, in essence, the maximum shear stress is equal to the shearing resistance (strength) of the clay, so that failure of the clay would be occurring at shallow depths. This observation is in accordance with the views relating to high horizontal stresses which have found wide acceptance in soils engineering. Moreover, a number of investigators have suggested that slickensided surfaces in mudrocks occur in this way. Banks *et al.* (1975) comment on this question in respect of the slickensided mudrocks of the Panama Canal.

Pertinent to the expansion of mudrocks and the development of a weathering profile, is Peterson's (1954, 1958) detailed study of rebound in the Bearpaw Shale of North America. Figure 13.4 illustrates the average rebound curve for the shale based on water content determinations and calculated effective stress measurements taken from borings. The expansion (swelling) at two levels of shale disintegration is matched by increasing water content. The rebound curve for the unweathered shale is indicated in Figure 13.4 by the curve A–B. Superimposed on the field data are the laboratory consolidation test compression and rebound curves for a disaggregated sample of the shale remoulded to the same water content. The maximum loading in the laboratory is the same as the field value. These latter curves can be construed as being representative of the shale without its diagenetic bonds.

Although substantial expansion has occurred in the natural material it will be observed in Figure 13.4 that the water content of the completely weathered *in situ* material is still less than that of the rebounded remoulded clay (compare points *C* and *D*). The *in situ* material has still retained some latent strain energy, presumably in the form of bonded aggregates of particles. It should be mentioned that the Bearpaw Shale is a formation in which rock squeeze effects in excavations and tunnels can be dramatic.

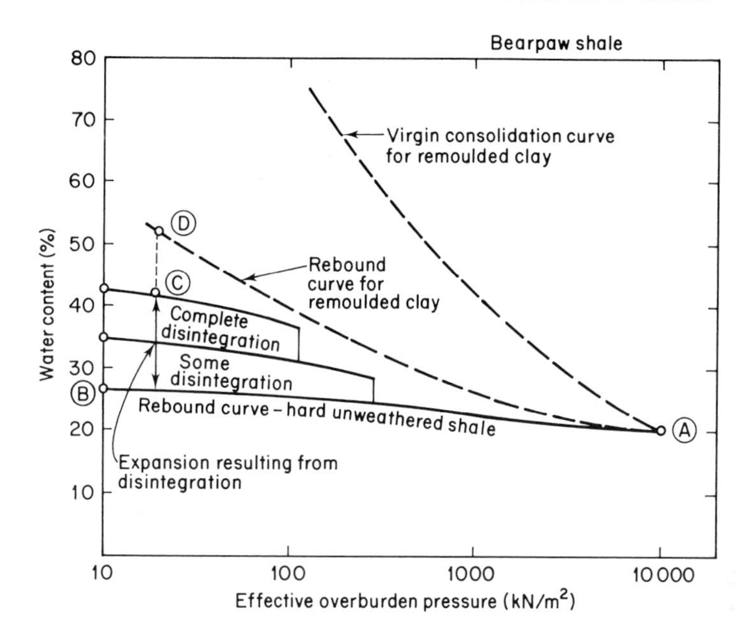

Figure 13.4 Rebound curves for non-weathered and weathered Bearpaw shale, constructed from field data. (*After Bjerrum, 1967. Reproduced by permission of the American Society of Civil Engineers.*)

13.4 MUDROCK DEGRADATION

13.4.1 Swelling Processes

Swelling and disintegration of mudrocks, accompanied by increasing water content (Figure 13.4), are a hallmark of weathering profiles (Figure 13.6). Swelling is a result of: (a) stress relief, particularly when high horizontal stresses are involved; (b) *intra*particle swelling of expandable clay minerals; and (c) *inter*particle (osmotic) swelling between clay minerals.

Stress relief induces a negative pressure, or suction pressure, u_s, in the pore water. At the simplest level, the numerical value of negative pore-water pressure will be approximately equal to the reduction in mean stress that is acting on an element of ground. That is, $u_s \simeq 1/3(\sigma'_1+\sigma'_2+\sigma'_3)$, where σ'_1 is the maximum vertical stress (σ_v'), and $\sigma'_2=\sigma'_3=\sigma'_h$ are the principal horizontal stresses acting on an element of ground prior to it being stress relieved (see for example, Chenevert, 1970 for mudrocks; Driscoll, 1984 for over-consolidated clay). Reference to Driscoll's (1984) calculation indicates that if $K_o>1$ the negative pressure, u_s, will be greater than the effective overburden pressure. Consequently, swelling will occur as these depressed pore pressures equilibriate.

In addition to mechanical concepts of rebound and volume increase, physico-chemical swelling effects at the scale of clay minerals are relevant (see Bolt, 1956). *Intra*particle rehydration (Figure 13.5(A)) of expandable clay minerals—smectites (montmorillonites), vermiculites, mixed-layer clays—will occur in response to uplift and erosion, or following desiccation at ground level during weathering.

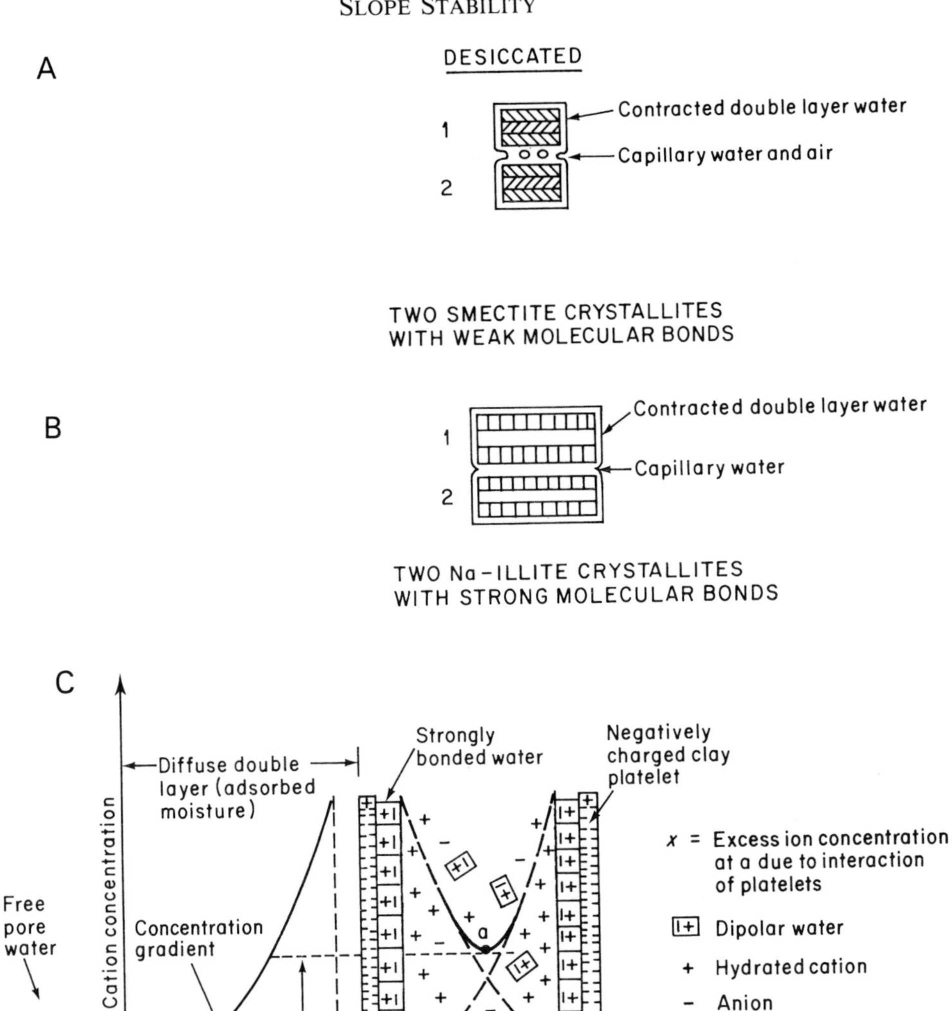

Figure 13.5 Clay mineral swelling. (A) *Intra*particle swelling.

The swelling of some of the other clay minerals, like illite, is governed to a greater or lesser degree by *inter*particle osmotic processes (Figures 13.5(B), and (D)). Here, the concentration of cations between two clay mineral surfaces can exceed that of the pore water outside the system (Figures 13.5(C) and (D)). There will be a net repulsion between the particles (swelling) and water will be drawn in until the concentrations are in equilibrium. Unloading from depth will undoubtedly help to promote osmotic swelling on a geological scale, since the spacing between the mineral surfaces must be about 15 Å or more for it

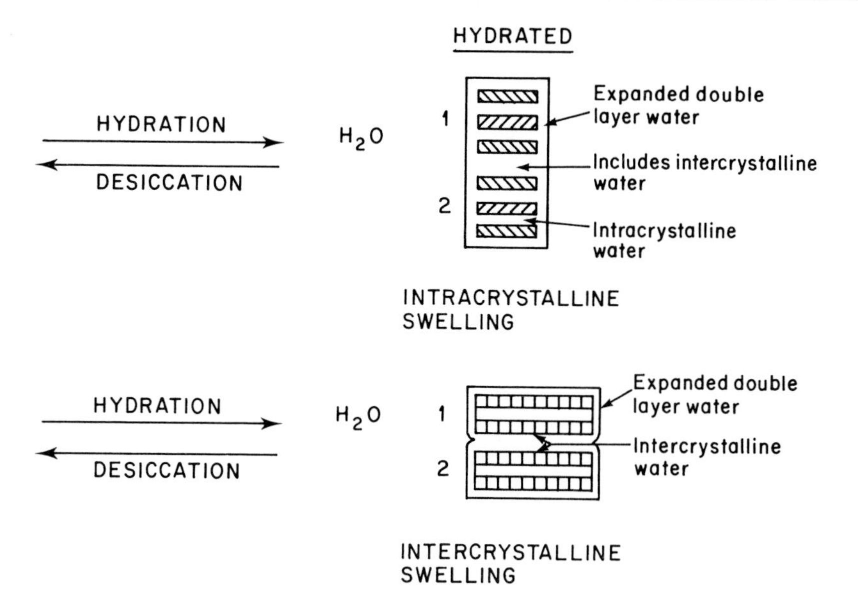

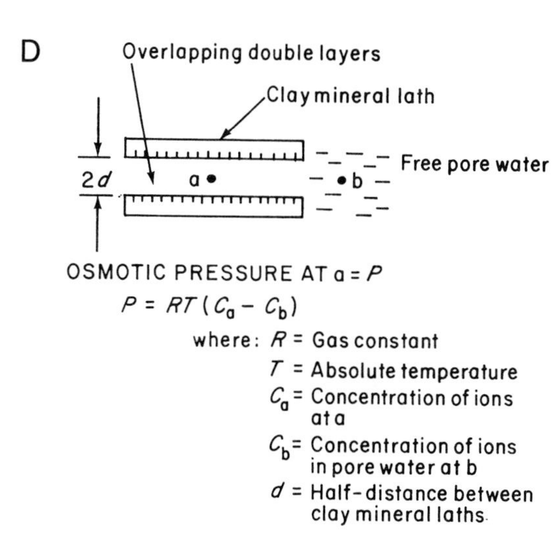

(B) *Inter*particle swelling. (C) and (D) Nature of double layer swelling.

to occur. At smaller spacings (< 15 Å) prevailing under high overburden pressures, there is a net attraction between particles due to the short-range van der Waals' force which will then be operative.

Importantly, *low* concentrations of *monovalent* cations (e.g. sodium, potassium) will generate the greatest diffuse double layers (Figure 13.5(C)), and the overall swelling will be greater. In contrast, *divalent* cations (e.g. calcium and magnesium) and *high* concentrations will reduce swelling. Since cations can replace one another in the double

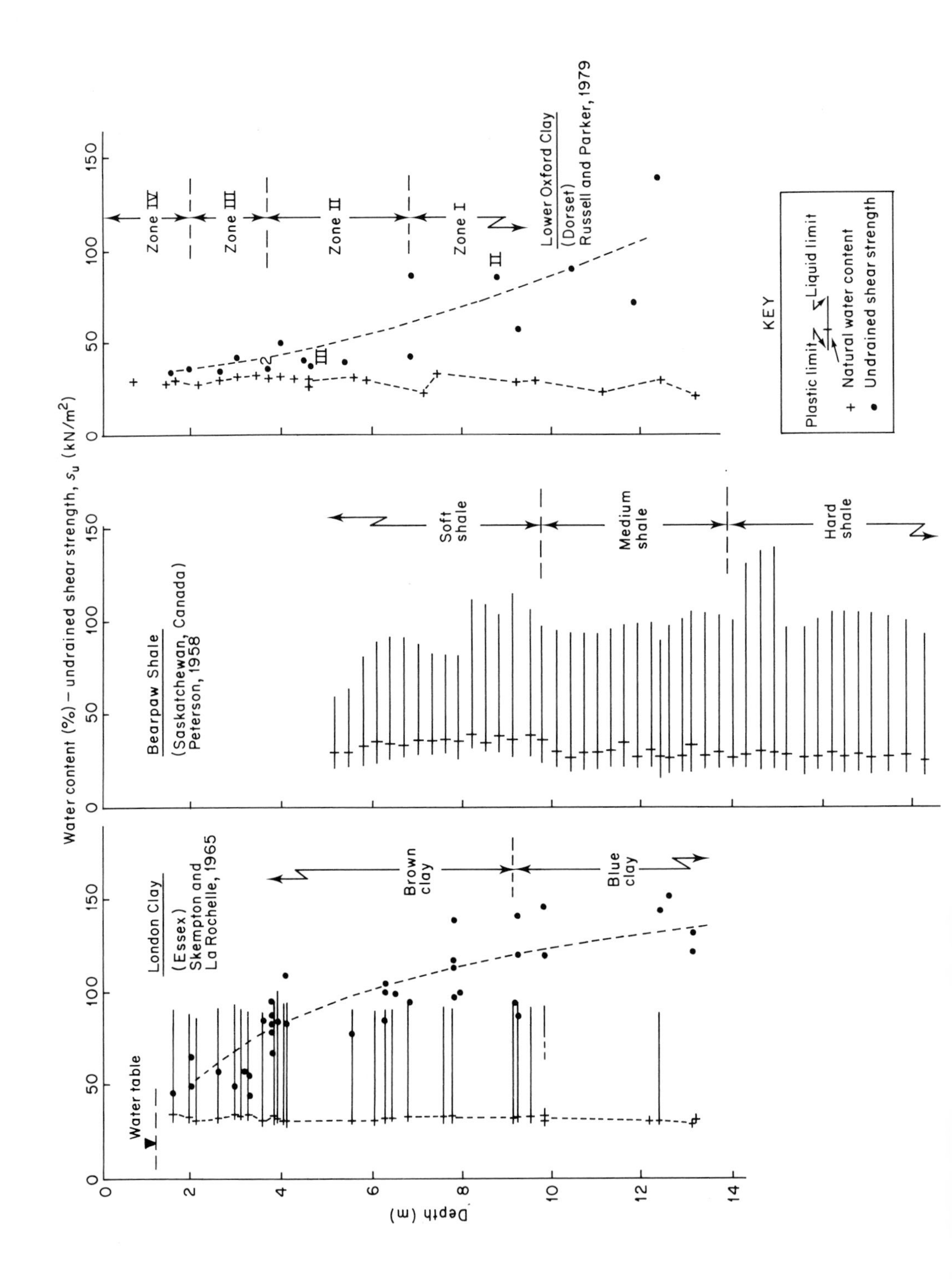
Water content (%) – undrained shear strength, s_u (kN/m^2)
Depth (m)
London Clay
(Essex)
Skempton and
La Rochelle, 1965
Water table
Brown clay
Blue clay
Bearpaw Shale
(Saskatchewan, Canada)
Peterson, 1958
Soft shale
Medium shale
Hard shale
Lower Oxford Clay
(Dorset)
Russell and Parker, 1979
Zone IV
Zone III
Zone II
Zone I
KEY
Plastic limit
Liquid limit
Natural water content
Undrained shear strength

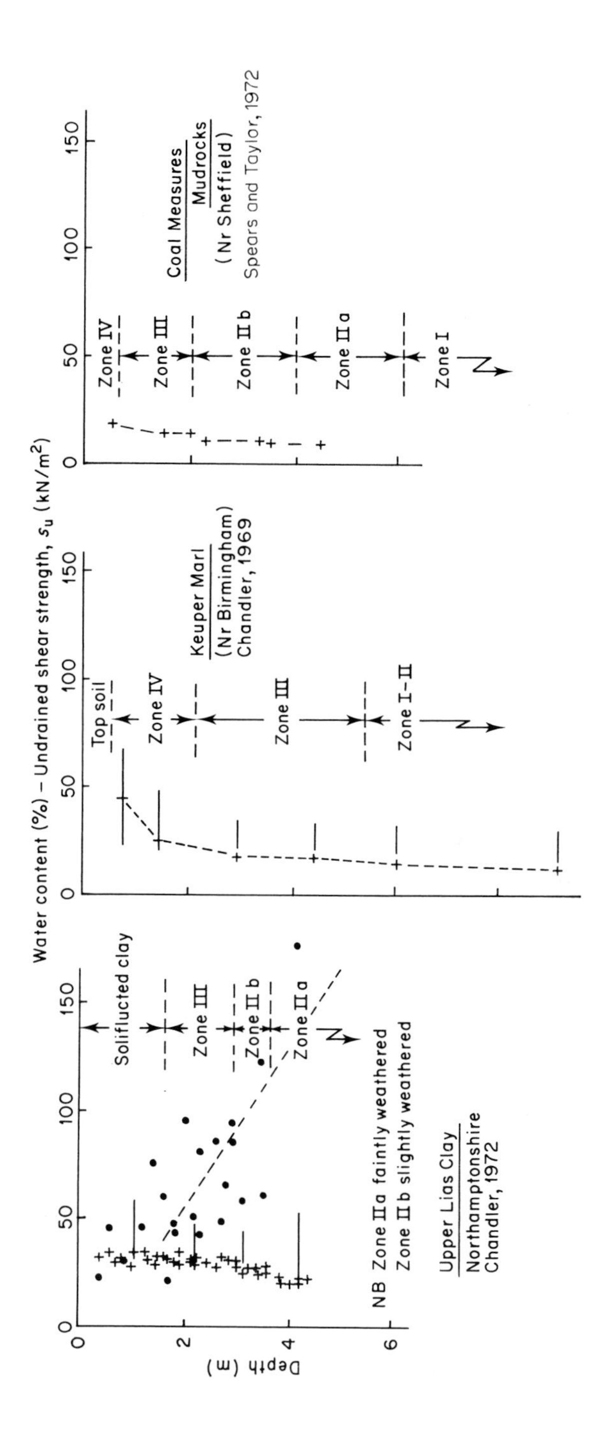

Figure 13.6 Geotechnical weathering profiles for certain clays and mudrocks

layer (Figure 13.5(C)), or within the structure of expandable clay minerals (Figure 13.5(A)), the system chemistry in a weathering profile will have a major influence on clay mineral swelling.

Weathering is a process which is at a maximum at or near the ground surface. In the context of physico-chemical swelling the water percolating downwards through a soil profile will have fallen as precipitation and is thus likely to be different in composition from the pore water within the strata. If the ionic concentration of the soil water is less than that of the double-layer (strata) water, swelling will continue until equilibrium is attained. Bolt (1956) also makes the important point that immediately after a load has been removed, osmotic pressures will be out of balance within the system; water will consequently be drawn in to redress the equilibrium.

Osmotic pressures are not necessarily of importance for all clays and clay minerals. For example, they are considered to be small in randomly orientated clay fabrics (Yong and Warkentin, 1975). Similarly, a number of workers (Olson and Mesri, 1970; Sridharan and Venkatappa Rao, 1973) have shown that the behaviour of kaolinite (like that of granular materials) is dominated by mechanical rebound phenomena, with the exchangeable cations in the pore water having little or no effect on swelling. With expandable clay minerals, the swelling of sodium montmorillonite (rare in nature) accords well with double-layer theory and can be dramatic. However, with Ca-montmorillonite, 'dead volume' corrections have to be made because the crystallites compact into domains when loaded (see Blackmore and Miller, 1961; Aylmore and Quirk, 1959).

Reasonable agreement between measured swelling pressures on clays and those predicted by theory was reported many years ago by Verwey and Overbeek (1948). More recently, Madsen (1979) has demonstrated good compatibility for Alpine mudrocks of Miocene and Jurassic age.

13.4.2 Physical and Chemical Weathering of Mudrocks

In the context of weathering profiles (Figure 13.6) developed within the last 8,000–10,000 years the location of the water table is important. A low water table is more likely to result in a greater degree of disintegration and chemical weathering (oxidation) in both mudrocks and over-consolidated clays.

Relative humidity has been shown to be a major factor in the development of unconfined swelling strains in mudrocks. Harper *et al.* (1979) found that Ordovician mudrock cores achieved an absolute extension at high relative humidity following a short drying cycle with subsequent slaking. The effect of fluctuating humidity is to cause expansion–contraction effects which lengthen internal cracks. Van Eeckhout (1976) concludes that the increased water content which is then permitted within the discontinuities will reduce the fracture energy required to fail the rock. In other words, breakdown is accelerated by cycles of wetting and drying.

The effect of drying is seen commonly as a polygonal crack pattern which develops on bedding planes. Kennard *et al.* (1967) concluded that these discontinuities are a desiccation feature rather than a stress relief phenomenon, since they do not develop if the freshly excavated mudrock is immersed and stored under water.

The disintegration of mudrocks of Carboniferous age has been investigated for over 30 years by the National Coal Board Shale Panel and later research workers, e.g. Badger *et al.*, 1956;

Beckett *et al.*, 1958; Berkovitch *et al.*, 1959; Horton *et al.*, 1964; Taylor and Spears 1970; Taylor 1984. The main controls on the physical disintegration of these indurated rock types can be identified as follows:

1. Incidence of sedimentary structures and discontinuities such as bedding planes, laminations, joints, slickensides, and fissures.
2. Slaking (or 'compressed air breakage'), which is a function of capillarity and suction characteristics.
3. Expansion of those constituent clay minerals with expanding layers: this would seem to be more important when the percentage of exchangeable sodium (Na) cations is high.

In terms of slope stability, Skempton and Petley (1967) recognize two groups of discontinuity in over-consolidated stiff clays. These are bedding surfaces of depositional or diagenetic origin, and structural discontinuities consisting of systematic joints, non-systematic 'fissures', minor shears and principal displacement shears. The latter shears comprise the principal slip surfaces in landslides, faults, and bedding-plane slips.

Dark grey laminations, which are part of the alternating sequence of dark–light grey units in both marine and non-marine Carboniferous strata, have been observed to swell in water (Spears, 1969, Taylor and Spears, 1970). Splitting occurs along the interface between the dark laminations with floc-type fabric and the lighter grey laminae. This imparts a fissility to the rock. Importantly in the context of weathering, Spears (1980) has concluded that the fissility which is found in stress-relieved shales at outcrop is the surface expression of the aforementioned varve-like laminations recorded in borehole cores at depth.

Slaking effects have already been referred to in relation to relative humidity variations. Laboratory slaking test data shown in Table 13.3 indicate that *de-airing* has restricted breakdown in five out of the eight specimens cited. Samples 6–8, in which breakdown is less restricted, contain increasing quantities of expandable mixed-layer clay, with the Stafford tonstein comprising up to 78% mixed-layer illite–montmorillonite. All three specimens have high exchangeable Na-ion percentages as well (Taylor, 1984). Earlier work by NCB's Shale Panel scientists showed that mudrock breakdown versus slaking time did not conform to a first order decay law. Expandable clay mineral species are now considered to be the other major control, sometimes to the exclusion of capillary effects, such as 'compressed air breakage' which occurs in the cores of desiccated mudrock fragments when they are immersed in water.

Physical breakdown of rock particles and clay aggregates in the partly saturated zone above the water table will expose an ever increasing surface area to *chemical* weathering processes. Detailed studies have highlighted the oxidation of ferrous sulphide minerals (pyrite, marcasite, and pyrrhotite) as the most important chemical decomposition process in UK weathering profiles, e.g. Spears and Taylor, 1972: Coal Measures; Chandler, 1972: Lias Clay; Russell and Parker, 1979: Oxford Clay. The highly exothermic, biogenic oxidation of pyrite is probably along the following lines (Penner *et al.*, 1973):

Stage 1: $2FeS_2 + 7O_2 + 2H_2O \rightarrow 2FeSO_4$ (ferrous sulphate) $+ 2H_2SO_4$
Stage 2: (Bacteria as catalyst):
$4FeSO_4 + O_2 + 2H_2SO_4 \rightarrow 2Fe_2(SO_4)_3$ (ferric sulphate) $+ 2H_2O$
Stage 3: $7Fe_2(SO_4)_3 + FeS_2 + 8H_2O \rightarrow 15FeSO_4 + 8H_2SO_4$

Table 13.3 Static Slaking Tests
(Air dried aggregate passing through 1/4 in. and retained on 3/16 in. BS sieve immersed in water on No. 14 BS sieve for 1/2 hour—see Taylor and Spears, 1970).

Carboniferous rock/type	Diagenetic coal* rank code number of associated coal	Slaking in water, % retained on No. 14 BS sieve	Slaking in water† *in vacuo*, % retained on No. 14 BS sieve
1 Dark grey shaly MUDSTONE, Blaina, S. Wales	101	95.8	99.1
2 Dark grey shaly MUDSTONE Cwmaman, S. Wales	102	97.6	99.0
3 Grey laminated silty MUDSTONE Gt Lumley, Co Durham	502	98.3	99.9
4 Dark grey fissile SHALE Warsop Colliery E. Midlands	802	91.6	97.7
5 Dark grey carbonaceous MUDSTONE Yorkshire Main Colliery	600	97.5	100.0
6 Grey carbonaceous MUDSTONE (Park) Littleton Colliery Staffs.	802	78.5	94.5
7 Dark grey MUDSTONE (Brooch) Littleton Colliery Staffs	802/902	0	27.1
8 Grey MUDSTONE band (tonstein) Stafford Colliery	602/701	0	2.0

*Low number—deeply buried (anthracite-100's)
High number – shallow burial (highly bituminous coals – 802/902).
†Slaking tests conducted entirely within a closed, evacuated system.

The reaction of the sulphuric acid produced by oxidation with carbonates and cations in clay minerals leads to the formation of secondary sulphates; the other resultant products are mainly hydrated iron oxides, usually in an amorphous state.

Both pyrite decomposition itself (Winmill, 1915–1916) and the sulphates formed by subsequent chemical reactions can involve large volume increases with consequential pressures of up to 500 kN/m^2 (see Taylor and Cripps, 1984). However, the precipitation of sulphates rarely occurs at the site of mineral reactions. For example, Chandler (1972) concludes that gypsum (selenite) produced ahead of the oxidizing front in Lias weathering profiles is a consequence of the local reaction of downwards percolating sulphuric acid with calcium carbonate near the Zone II a–b (Table 13.4) boundary. Comparison of rock disintegration versus pyrite content with depth led Spears and Taylor (1972) to deduce that although pyrite decomposition may contribute to mudrock breakdown it is not the prime cause.

Gypsum ($CaSO_4.2H_2O$) was not present in the Coal Measures section referred to above. This is because calcium carbonate was quantitatively unimportant in these cores. The more resistant mineral, siderite ($FeCO_3$), was the main carbonate present, both as a cement and in the form of nodules. In contrast, calcium carbonate (calcite) is the principal carbonate

Table 13.4 *Weathering classification for mudrocks* (After Engineering Group Working Party; Anon, 1977)

Description	Zone	Definition
Fresh	IA	No visible evidence of weathering
Faintly weathered	IB	Discolouration of major discontinuity surfaces
Slightly weathered	II	Discolouration of discontinuity weathered surfaces and some degradation material on discontinuity surfaces
Moderately weathered	III	Less than half rock material in a degraded condition
Highly weathered	IV	More than half rock material in a degraded condition
Completely weathered	V	All rock material in a degraded condition but original mass structure still discernible
Residual soil	VI	All rock material in a degraded condition and original rock structure destroyed

Note: For Upper Lias Clay, Chandler (1972) has modified the numbering so that Zone IB above is essentially his Zone IIa and Zone II above is his Zone IIb. The latter is slightly extended to include 'brown staining usually in areas parallel to fissures'.

present in mudrocks containing secondary gypsum (e.g. Chandler, 1972; Parry, 1972; Penner *et al.*, 1973; Russell and Parker, 1979).

The dissolution of calcite cement was cited by Tourtelot (1962) and Kennard *et al.* (1967) as the principal cause of disintegration in the shales studied by them. Calcite dissolution is also identified by Russell and Parker (1979) as a significant weakening process in weathering of Oxford Clay. Because calcite is commonly present in variable quantities due to changing depositional and diagenetic conditions, it usually exhibits a non-systematic distribution with depth. Consequently, the mineral is of limited value as a sensitive weathering indicator in mudrocks.

Chandler (1972) shows that in the Lias the *mean* oxidation ratio (expressed as Fe_2O_3/FeO) increases upwards from 0.6 in Zone I to 6.0 in landslipped/soliflucted materials at the surface. The most notable change in the ratio occurs between Zone IIa and Zone IIb (from 0.8 to 3.0). Zonal classifications are described in Table 13.4.

The effect of weathering on clay mineralogy is generally small in the United Kingdom. This is not entirely surprising since clay minerals were derived from a palaeo-land source where the climate was commonly more extreme than it is today. Consequently, they will have achieved some stability towards modern weathering.

The potash/alumina ratio (K_2O/Al_2O_3) can be used as an index of illite content. Alumina is largely confined to the clay minerals in mudrocks and potassium to micaceous types. The ratio also has the advantage of being an index of weathering of micas, in that it measures the preferential upwards loss of potassium (by leaching) relative to alumina. It was used in this way by both Spears and Taylor (1972) and Russell and Parker (1979). In the Oxford Clay, however, the sulphate jarosite ($K\,Fe_3(OH)_6(SO_4)_2$) was considered to act as a 'sink' for some of the potassium leached from the illite. Jarosite is a common (secondary) sulphate formed in over-consolidated clays and mudrocks, including Carboniferous and older types.

Acid weathering of micaceous minerals can increase the amount of clay with expanding lattices (e.g. vermiculite formed from illite)—see, for example, Quigley (1975). An increase in expandable clay minerals is commonly recorded in mudrocks and clays (even indurated types) which have been thoroughly reworked by surface processes. Small increases in vermiculite and montmorillonite were found to be statistically significant in weathered Oxford Clay by Russell and Parker (1979), whereas Spears and Taylor (1972) found that the fine clay fraction apparently decreased in the surface layers. In the Coal Measures weathering profile there was an increase in quartz content at the surface and a relative increase in kaolinite at the expense of chlorite and the fine-grained mixed-layer clay. An increase in the 1M (degraded) mica polymorph and decrease in the 2M variety accompanied a fall in the K_2O/Al_2O_3 ratio in the upper metre of strata.

The Coal Measures site was chosen purposely because it was free from drift or landslide colluvium. It could be significant that other sites mentioned in this section were blanketed to some extent.

The conclusion drawn is that physical disintegration of mudrocks and clays is more rapid and acts as a control of chemical weathering, particularly in indurated mudrocks. Pyrite decomposition, and to a lesser extent, calcite dissolution are significant processes in the scheme of chemical weathering.

13.4.3 Classification of *in situ* Weathering

Although it is convenient to consider swelling, physical weathering, and chemical weathering as separate agents of degradation, there are difficulties in separating them in natural weathering systems. At first sight a consideration of the individual agents of degradation would appear to be useful for prediction purposes. Unfortunately, although chemical weathering, for example, can be simulated in laboratory experiments, the results can be ambiguous because only the end-product of the process remains for analysis. Moreover, most weathering systems include a high degree of chemical openness. The problem is further complicated in that changes recorded for pure clay minerals will not necessarily be comparable with those in naturally-occurring multicomponent mixtures. However, the extent to which the physical character of mudrocks can be modified during weathering may be ascertained by reference to typical weathering profiles of the type presented in Figure 13.6. Thus, in moving upwards from fresh, unweathered material at depth, successively more weathered material is usually encountered. In most profiles, the increase in weathering corresponds with changes in geotechnical properties, including reductions in shear strength and increases in water content.

It is useful to describe the condition of weathered geological materials (including mudrocks) in a uniform style, such as that recommended by the Engineering Group Working Party (Anon, 1977)—Table 13.4. Weathering classifications have been devised with reference to particular geological formations, so several variants of the scheme appear in the literature. It is important to appreciate that the definition of weathering zones refers to the amount of degraded material rather than the extent of alteration experienced by the material. For example, in the case of a typical partly-weathered mudrock, the clay or silty-clay product surrounding the lithorelicts may itself have been subject to further degrees of weathering.

Weathering changes can be demonstrated by reference to Table 13.5. With the exception of the Bearpaw shale from North America, all the remaining mudrocks and clays are UK deposits. Although for a number of reasons, the precise weathering environment of these examples may be different they are all typical of temperate humid conditions and, in these terms, climate is unlikely to be a significant cause of variation in weathering effects. Many of the data presented in Table 13.5 have been extracted from Cripps and Taylor (1981) who discuss their derivation and the variations that can be attributed to sampling and testing methods. Mineralogical data have been obtained from Perrin (1971) and Shaw (1981). Information relating to the Bearpaw Shale has been extracted from Peterson (1954, 1958), Ringheim (1964), Bjerrum (1967), Scott and Brooker (1968) and Fleming *et al.* (1970). Figure 13.6 illustrates in finer detail geotechnical changes through some of the weathering profiles for which broad details are given in Table 13.5.

The examples range from lightly over-consolidated London Clay of Tertiary age to very heavily over-consolidated mudrock belonging to the Triassic System and indurated Carboniferous mudrock. In terms of composition, the clay mineralogy of Bearpaw Shale is dominated by smectite compared with predominantly illite and kaolinite in the other formations.

Weathering changes are, of course, accompanied by gradational decreases in strength and increases in water content as indicated in Figure 13.6. However, the Zone II/III boundary has been chosen as the transition from weathered to unweathered strata in Table 13.5. It is not an arbitrary choice but is based on the observation that degradation (*per se*) is more significant in the zones above this boundary.

Following many separate studies of London Clay from a number of localities, Skempton (1977) concludes that where the formation extends to the surface, or is overlain by only thin drift deposits, the junction between weathered brown London Clay (Zones III and IV) and the unweathered blue variety (Zones I and II) occurs at a depth of between 5 m and 15 m (Table 13.5). In the case of Bearpaw Shale, Bjerrum (1967) identifies zones in which different degrees of shale fragmentation have occurred and weathering effects are observed to extend to depths of about 22.9 m. From Peterson's (1954, 1958) descriptions, however, it seems reasonable to suggest for the purposes of Table 13.5 that Zone III weathering extends to a depth of about 14.9 m. Weathering zones for Lower Oxford Clay have been identified by Russell and Parker (1979) using Chandler's (1972) classification which is essentially similar to that of Table 13.4. Their data indicate that Zone III weathering of Lower Oxford Clay extends to depths ranging from about 6 m to 10 m. The transition from Zone II to Zone III material occurs at depths varying between 4 m and 8 m, with an overlying thickness of up to 2 m of Zone IV clay.

Difficulties which arise in defining the depths of various weathering zones can be demonstrated by considering Lias Clay and Keuper Marl examples. Permafrost activity has been suggested by Chandler (1972) to be one of the causes of variation in the depth of weathering observed in the Upper Lias Clay of the East Midlands of England. At a site near Stamford, Lincolnshire, although oxidation was confined to depths of only 3 m to 4 m, the ground consisted of small lithorelicts in a softer clay matrix to depths of about 11 m. At this level the number of horizontally orientated lithorelicts exceeded rotated ones. In terms of the weathering classification, Zone II clay extended to a depth of approximately 5.5 m at this location.

In addition to anomalies caused by the presence of overlying recent deposits which were mentioned in section 13.4.2, differences in weathering depth also arise because of variation in the character of the original mudrock. This point can be demonstrated by reference to Keuper Marl which is a rather variable, generally water-lain, terrestrial deposit, formed under hot arid climatic conditions during Triassic times. In most locations this formation consists of red–brown mudstone which is the subject of present attention, although

Table 13.5 Engineering properties of selected fresh and weathered mudrocks and over-consolidated clays

Formation		London Clay	Bearpaw Shale	Lower Oxford Clay	Upper Lias Clay	Keuper Marl	Coal Measures Mudrock
Geological age		Tertiary	Upper Cretaceous	Upper Jurassic	Lower Jurassic	Triassic	Carboniferous
Laboratory pre-consolidation pressure (kN/m^2)		1,436–4,137	>16,100	9,583–14,504	9,540–10,510	11,850–21,450	c.48,500
Depth to base of Zone III weathering (m)		5.0–15.0	≃14.9	6.0–10.0	2.9–5.5	5.6–8.5	2.0–4.0
Clay mineralogy (%)	Kaolinite	10–40 (27)	—	20–35 (15)	15–85 (20)	6–10 (5)	(35)
	Illite/mica	20–60 (43)	0–39	65–80 (52)	5–90 (58)	28–93 (70)	(43)
	Smectite/ mixed layer	0–70 (27)	44–100	(29)	0–55 (15)	(13)	(15)
	Chlorite	0–20 (3)	0–17	0–5 (1)	0–10 (5)	0–23 (7.5)	(7)
Natural water content (%)	Fresh	19–28	19–27	15–25	11–23	5–15	3–8
	Weathered	23–49	29–36	20–33	20–38	12–40	3–32
Liquid limit (%)	Fresh	50–105 (70)	80–150	45–75	20–38	12–40	30–51 (42)
	Weathered	66–100 (82)	80–150	—	56–72 (54)	25–35	27–72 (57)
Plasticity index (%)	Fresh	41–65 (47)	62–123	12–50 (32)	19–40 (33)	10–15*	18–33
	Weathered	36–55 (44)	62–123	—	24–42 (32)	10–35	18–42 (30)

Clay size fraction ($<2\ \mu m$, %)	Fresh		48–61	($<5\ \mu m$) 30–65	30–70 (52)	18–65 (38)	10–35	12–29
	Weathered			($<5\ \mu m$) 30–65	—	28–68 (48)	10–50	25–45 (31)
Bulk density Mg/m^2	Fresh		1.92–2.04	1.93–2.06	1.84–2.05 (2.03)	1.87–2.09	2.20–2.50	2.15–2.76 (2.23)
	Weathered		1.70–2.00 (1.90)	1.85–1.96	—	1.79–1.96 (1.90)	1.80–2.30	1.86–2.18 (1.92)
Undrained shear strength kN/m^2	Fresh		80–800 (100–400)	276–874	96–12,000 (360–1,100)	40–1,200 (110–240)	130–2800 (300–1,200)	9,000–103,000 (38,000–50,000)
	Weathered		40–190 (100–175)	24–276	52–93	20–180 (30–150)	70–200 (100–150)	15–335 (80–150)
Effective shear strength	Fresh	c'	31–252	10–152	10–216	27	>30	$c' \simeq 131$ $c_a \simeq 2{,}000–3{,}200$
		ϕ'	20–29	25–30	23–40	24	>40	$\phi' \simeq 46$ $\phi'_a \simeq 28–39$
c', c_a kN/m^2 ϕ', ϕ^a Degrees	Weathered	c'	1–18	0–41	0–20	1–17	2–80	0–25
		ϕ'	17–23	20–28	21.5–28	18–25	25–42	26–39

*May be non-plastic. Average values bracketed.

sandstones and silty beds also occur in the sequence. Lithological variations (and mantling effects) can produce significant differences in zone depths. A comparison of Chandler's (1969) weathering profile in the English Midlands with that of Marsland (1977) for a site in Cheshire exemplifies this point. Indeed, lithological variations can result in complex profiles in which highly weathered marl underlies less weathered material.

The non-blanketed Carboniferous section from near Sheffield, described by Spears and Taylor (1972) comprises mudstone, shale, coal, seatearth and siltstone. At this location, moderate weathering (Zone III) extends to depths ranging from 2.4 m to 4.0 m in mudstone (Table 13.5) but to only about 2.0 m (1.98 m) in shale. Slight weathering (Zone II) extends to depths of about 4.2 m but there is no evidence of weathering below about 6.1 m. It should be recognized that weathering profiles can be affected by the presence of geological structures including, in particular, faults and fracture zones. Observations of unfavourably situated Carboniferous Coal Measures rocks indicate that in association with a high frequency of rock mass discontinuities moderate weathering can extend to depths of 20 m or 25 m. Variation in the degree of weathering of Oxford Clay due to the presence of a fault has been described by Parry (1972).

The increases in water content towards the ground surface illustrated in Fig. 13.6 are more pronounced in the indurated mudrocks. Hence, although very high water contents have been measured by Hutchinson (1970) for mudflow material derived from London Clay, the natural water content of undisturbed brown London Clay (Zones III and IV) is only a few per cent higher than the value of the blue (Zones I and II) material. On the other hand, the increases are much more dramatic for Keuper Marl (Figure 13.6) and Coal Measures examples, since the water content of the fresh material is much lower than in the younger geological horizons of Table 13.5. Compared with British examples with similar pre-consolidation pressures (e.g. Keuper Marl in Table 13.5), fresh Bearpaw Shale has a high average water content of 22% at a depth of 14.9 m, increasing upwards to about 33% at a depth of 6.1 m. It is significant that in this formation smectite is the predominant clay mineral.

Although in some profiles shown in Figure 13.6, little consistent variation appears to occur in the values of the Atterberg limits, average values do increase generally with the degree of weathering degradation. In addition, the results exhibit scatter attributed to minor variations in the composition of the mudrock in the profile. Chandler (1969) observed from Figure 13.7 that little change attributable to weathering occurs in Keuper Marl until degradation has advanced as far as Zone IV. This change coincides with a pronounced reduction of particle size in Zone IV marl compared with Zones I, II, and III material. Behaviour of this type contrasts with that observed for the less indurated over-consolidated clays in which any changes in plasticity are less severe and occur progressively as weathering intensity increases.

Quoted values of undrained shear strength ($s_u = c_u$ where $\phi_u = 0$) indicate that large reductions may be attributed to weathering effects. For example, Table 13.5 indicates that degradation of blue London Clay to the brown equivalent can cause reductions in s_u of about 50%. Shear strength reduction as a consequence of increasing fissure (discontinuity) frequency, has long been recognized in over-consolidated clays. Significantly, the average fissure spacing increases with depth (Ward *et al.*, 1965; Chandler, 1972; Burland *et al.*, 1978). In this respect Chandler (1972) records the following zone relationship with fissure spacing in Upper Lias Clay: Zone III, spacing 10–20 mm; Zone II, spacing 10–100 mm;

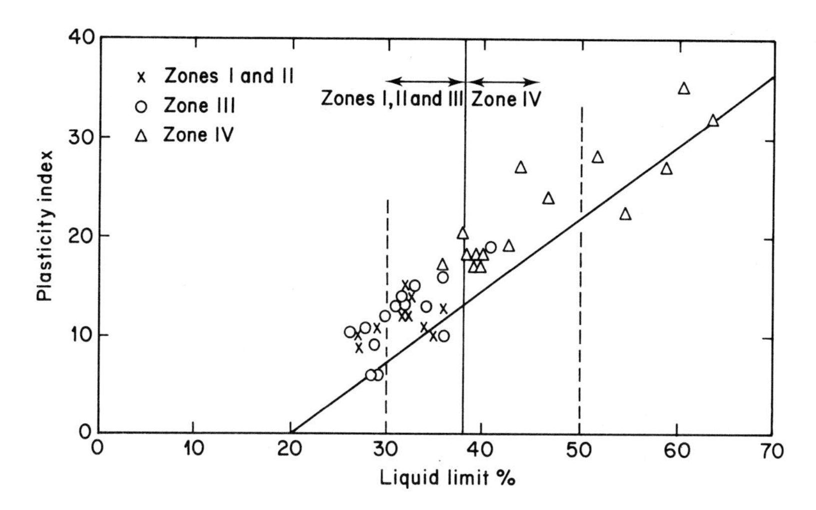

Figure 13.7 Plasticity chart for Keuper Marl (*After Chandler, 1969.*)

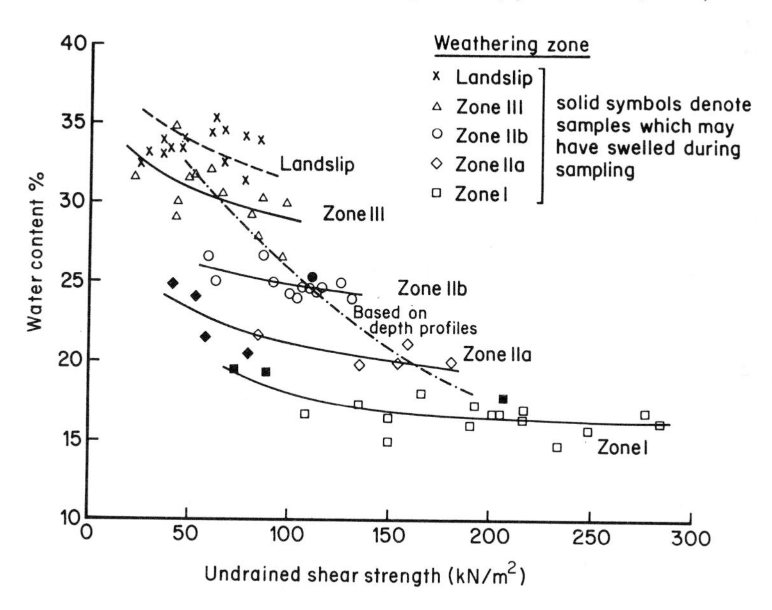

Figure 13.8 Water content–strength relationships for Upper Lias Clay from Northamptonshire (*After Chandler, 1972.*)

Zone I, spacing >100 mm. His water content versus undrained shear strength values for a site near Corby are shown in Figure 13.8. Different water content–strength curves have been applied to each weathering zone, and for landslip colluvium at the highest water content. Superimposed on the above data is the water content–strength curve constructed from the depth profiles for Upper Lias Clay illustrated in Figure 13.6. Chandler concludes that the large variation in strength of the clay specimens is a consequence of both the random variation in water content and fissure spacing.

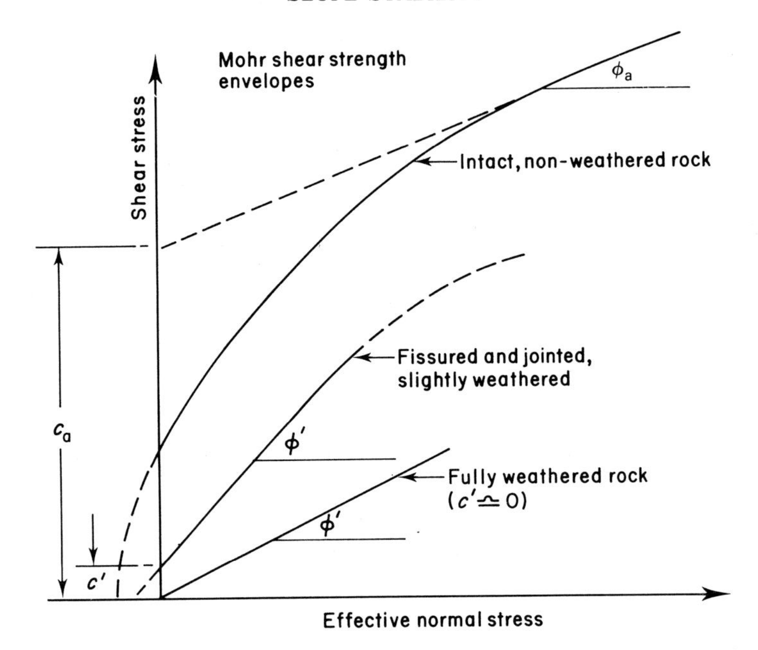

Figure 13.9 Interpretation of effective shear strength parameters (c', ϕ') and apparent effective shear strength parameters (c_a, ϕ_a) for intact, fissured and fully weathered Carboniferous mudrocks

Peak effective shear strength parameters of the over-consolidated clays listed in Table 13.5 also show a drop from non-weathered to weathered. The change customarily involves a large drop in cohesion intercept with a smaller change in angle of shearing resistance (ϕ').

Interpretations of shear strength parameters in the geologically older, more indurated mudrocks are strongly influenced by discontinuities and by curvature of the Mohr envelope. Furthermore, many triaxial tests on intact cores are carried out without pore pressure measurement or back-saturation. A linear shear strength envelope is also commonly fitted as shown in Figure 13.9. This can lead to further difficulties in interpretation. To avoid confusion with other geological horizons Coal Measures shear strength parameters obtained from tests which may not be on a truly effective stress basis are given the subscripts ϕ_a, c_a in Table 13.5. There are a few results for conventional effective stress triaxial tests on jointed and fissured Zone IB/II rock (Figure 13.9). These will be considered, together with specimens in more advanced stages of weathering in section 13.5.2. Nevertheless, the Coal Measures shear strength parameters in Table 13.5 show a similar trend to those of other geological horizons, but in this case the drop in cohesion (c_a) is much more dramatic.

In summary, zonal weathering schemes are useful for comparative purposes in mudrocks and clays. The upwards reduction in undrained shear strength is seen to be gradual in over-consolidated clays (Figure 13.6). Water content, together with fissuring, would appear to exert a strong control over the undrained strength–depth relationships. However, there is a sharper increase in water content near the surface in the more indurated mudrocks than in over-consolidated clays. There is marked drop in cohesion with weathering on an effective stress basis; it is dramatic in indurated mudrocks. Other things being equal, expandable clays, in particular smectites, probably have a greater effect on the depth of weathering than is commonly recognized.

13.5 SLOPE DEVELOPMENT AND SHEAR STRENGTH

13.5.1 Slope Angle

A simple form of slope instability classification described by Skempton and Hutchinson (1969) provides a useful reference for studies of slope angle. It includes both morphological and genetic criteria and has, in fact, been criticized on this score (see Prior, 1978). The actual roles of the various styles of mass-wasting activity in slope profile development are considered by Carson (1976) who also discusses the control exerted by liquidity index (or water content) in determining whether instability occurs by flow or by sliding action. Generally speaking, the styles of mass movement can be placed in an order compatible with the increasing water content of the failed material, viz. falls, deep rotational slips, shallow rotational slips, translational or slab slides, earth flows, and mudflows. The parallel trend with decreasing slope angle is discussed later in this section.

Consideration of instability in the latter terms is advantageous since the type of failure relates directly to the method of predicting the stability condition. Reference should be made to Chapter 2 regarding the appropriate stability analysis for slopes subject to different styles of failure (e.g. rotational or translational, deep-seated or shallow).

Factors controlling stability are discussed by Bromhead (1984) who explains that the stability condition is customarily expressed in terms of limiting equilibrium in which a factor of safety (F) is defined as the ratio of forces inhibiting movement to those promoting it. Hence, values of F greater than unity signify a stable slope, whereas values less than this signify instability. Hence, if the properties (unit weight and shear strength) of the soil are known, together with the dimensions of the unstable mass and the pore-water pressure distribution, then the stability equations can be solved to determine the slope angle for limiting stability ($F=1$). Manipulation of known parameters and values in this way, in order to find unknowns, is referred to as 'back-analysis'.

This principle can be demonstrated by reference to the stability formula for a translational (slab) slide of infinite extent, viz.

$$F=\frac{c' + (\gamma z \cos^2\beta - u)\tan\phi'}{\gamma z \sin\beta \cos\beta} \tag{13.1}$$

where c' and ϕ' are the *effective* shear strength parameters of the soil, γ is the unit weight of the soil, z is the vertical depth to the failure surface from ground level, β is the angle of the slope, and u is the pore-water pressure operating on the failure surface.

For reasons explained in section 13.5.2, the effective cohesion intercept of degraded over-consolidated clays and mudrocks is at most a very small value indeed. If zero cohesion ($c'=0$) is assumed, then the above stability equation can be simplified. Carson (1969) suggests that further simplification is possible, since on a geomorphological time-scale, stability can be considered in terms of two pore-water pressure conditions:

1. Pore-water pressures equivalent to the groundwater table at the surface.
2. Groundwater table below the level of the slip surface.

In the first case, if the pore-water pressure is in equilibrium with the groundwater regime and flow is parallel to the slope, then $u=\gamma_w z \cos^2\beta$, where γ_w is the unit weight of water. Hence, for limiting equilibrium:

$$\tan\beta = \left(\frac{\gamma-\gamma_w}{\gamma}\right)\tan\phi' \tag{13.2}$$

For values of $\gamma=20\,\text{kN/m}^3$ and $\gamma_w=9.81\ (\simeq 10)\,\text{kN/m}^3$, this equation becomes:

$$\tan\beta \simeq \tfrac{1}{2}\tan\phi'$$

If the groundwater table is below the slip surface (case 2), then $u=0$; the condition for limiting equilibrium is then given by:

$$\tan\beta \simeq \tan\phi'$$

The applicability of these relationships depends on the validity of the assumptions made regarding cohesion, pore-water pressure, and soil unit weight. The value of pore-water pressure is particularly important since it is possible for artesian conditions to occur (see Weeks, 1969). Similarly, it is feasible for depressed pore-water pressures to persist or even become negative (see Chapter 4). Furthermore, the tacit assumption made by Carson (1976) and others, that the pore-water pressure in a failure zone is equivalent to the height of the water table may not be valid, especially in areas of varied topography with mixed strata of variable permeability.

These simple limit equilibrium deductions have led to the prediction of 'threshold' angles for slopes. Carson (1975, p. 19) gives the following definition of threshold slope angle . . . 'For a natural slope in earth material there is a single angle above which rapid mass movement will occur from time to time and below which the slope material is stable with respect to mass wasting processes, although subject to slower processes of creep.' It is assumed in this definition that $c'=0$ and also that the pore-water pressure conditions are constant within the area of interest.

Carson and Petley (1970) suggest that observations by Young (1961), that slopes occur at particular 'characteristic' angles can be interpreted using the threshold slope concept. At the initial stages of instability these authors envisage the formation of a coarse regolith which will attain stability at a particular threshold angle. The regolith (colluvium) will then become subject to induced shear strength reductions because of weathering and/or changes in pore-water pressures. It thus becomes unstable and fails ultimately to attain a new threshold angle. In this way it is envisaged that slope development occurs as a sequence of phases of instability with a particular threshold slope being established at each interval. Young (1975) observes that although slope failure can result in a steeper slope being formed (i.e. the back-scarp of the failure), it will not be as high as the original one. However, unless slopes are being steepened by basal erosion as considered in Chapter 9, a shallow regolith angle will replace a steep one until an ultimate limiting slope angle for mass movement is attained. It is important to appreciate that with mudrocks the shear strength of fresh, undisturbed material is much greater than that of the degraded material (for example, Figure 13.9). Moreover, in section 13.5.2 it will be explained that the *lowest* shear strength is only mobilized when a principal slip surface is formed. Suffice it to say, the shearing resistance will then be at the 'residual' value (ϕ_r', where $c'_r \rightarrow 0$). Hence, the ultimate threshold slope angle will be:

$$\beta = \tan^{-1} (\tfrac{1}{2} \tan \phi_r') \text{ if } u = \gamma_w z \cos^2 \beta$$

A number of authors quote slope angle data which correspond closely with predicted threshold slopes. Carson and Petley (1970) consider regolith-mantled slopes in Exmoor and the southern Pennine areas of England where they found that measured slope angles correspond with threshold slopes predicted for the residual shear strength (ϕ_r') of the regolith and the likely pore-water pressures obtaining in the area. Further examples analysed by Carson (1975) include mudrock slopes subjected to contrasting climatic conditions in which fairly good agreement was found to exist between predicted threshold angles based on ϕ_r' and field observations. In the case of some Namurian shales from the southern Pennine area of England some difficulty occurred in explaining slope angles of less than 8°. However, this probably reflects an insufficiency of knowledge regarding pore-water pressures and geotechnical properties of the materials concerned in this generalized study. Some of these problems and limitations of the threshold slope concept are considered in Chapter 19.

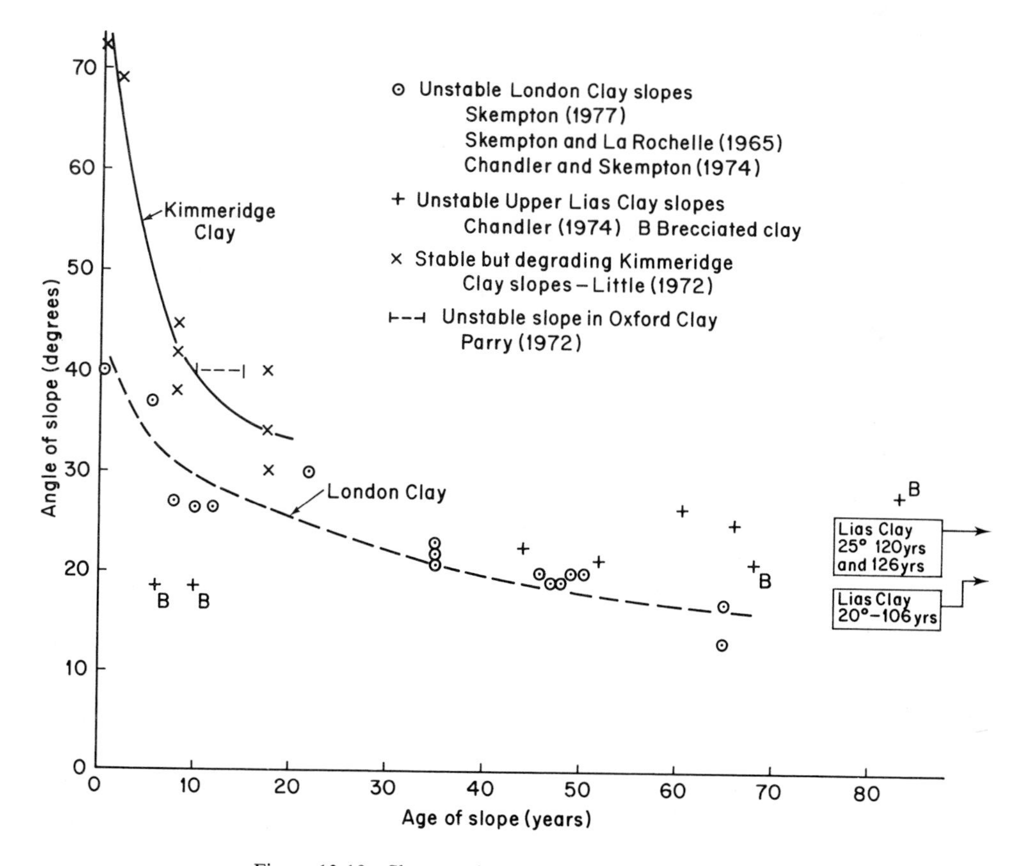

Figure 13.10 Slope angle data for slopes of known age

A measure of both time-scale and the role of the style of instability in slope development can be gained from a study of unstable slopes of known age. Early work by Skempton (1948) relates to 6 m high vertical, 25° and 18° slopes in London Clay which failed respectively after several weeks, 10–20 years, and 50 years. These and additional data for London Clay are presented in Figure 13.10, together with information for slopes in other formations. From this diagram a trend of decreasing slope angle with age can be discerned in both unstable London Clay slopes and currently stable, but degrading slopes in Kimmeridge Clay. The reason why the Lias Clay data do not appear to follow this pattern is not clear. Some of these slopes occur in periglacially disturbed brecciated clay, and in any case all angles fall within a narrow range.

Clearly, mass instability can follow very soon after the formation of a steep slope. In the case of actively-eroding cliffs considered by Skempton and Hutchinson (1969) (see also Bromhead, 1978), slopes of about 30° can be produced by strong marine erosion. These rapidly become unstable so that successive stages of slope angle reduction take place in increasing orders of time. In the case of the steeper slopes instability occurs as deep seated rotational sliding. Cliffs which have not been subjected to erosion for periods of between 30 and 150 years are found to rest at angles of between 13° and 20°. Hutchinson (1967) records that these cliffs are affected by shallow rotational slips. Freedom from basal erosion for periods of some hundreds of years, produces cliffs with inclinations of between 8° and 13° of which unstable ones are subject to successive shallow rotational slips or minor surface processes. Unstable inland slopes which lie at angles of 8° to 12° are probably several thousand years old. For these slopes, instability takes the form of markedly non-circular rotational slips or shallow translational slab slides.

13.5.2 Effective Shear Strength Parameters and Instability

Shear strength parameters in relation to weathering grades are listed in Table 13.5, section 13.4.3. The composite plot shown in Figure 13.11 is a further demonstration of the significant drop in shear strength in highly indurated mudrocks as a consequence of *in situ* weathering. These data (shown diagrammatically in Figure 13.9) have been replotted from the weathering profile of Spears and Taylor (1972). It is the *effective stress* shear strength parameters which are important in any investigation or design of a slope on a long-term basis.

The shear strength envelope fitted to the jointed and fissile mudrocks of Zone IB/II is markedly curved with a cohesion intercept which must be relatively small. In the original paper several rather more *intact* seatearth samples were also included. These additional specimens resulted in a cohesion intercept ($c' = 131\ \mathrm{kN/m^2}$) with possibly a very small tensile strength component (Figure 13.11). At outcrop, rocks of this type have little or no tensile strength 'in the mass' because of the influence of joints, fissile bedding planes, and other extensional discontinuities. However, if all samples in Figure 13.11 had been taken from similar, but completely unweathered strata at a greater depth, a larger cohesion intercept may have resulted (see Figure 13.9).

For the specific mudrocks illustrated in Figure 13.11 there is a large drop in shear strength, such that at the highest level of weathering the lower bound effective stress shear strength parameters are $\phi = 26°$, $c' \simeq 0$. Specimens at intermediate stages of weathering (including some that were slaked) are depicted in Figure 13.11 by their $\bar{p}q$ coordinates, rather than

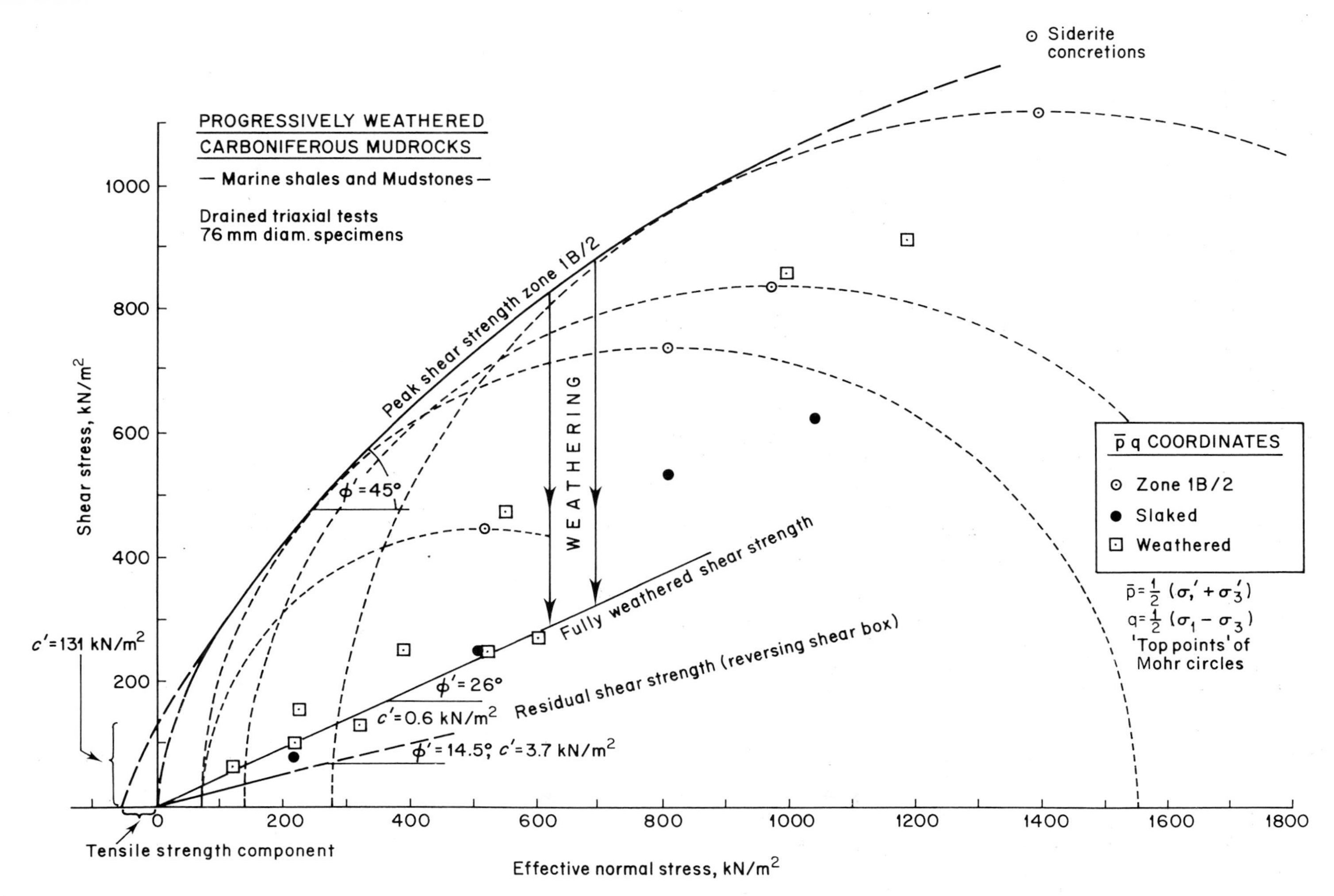

Figure 13.11 Shear strength parameters for progressively weathered indurated mudrock of Carboniferous age

by complete Mohr circles. The drop in peak effective shearing resistance to a lower bound value because of weathering processes marks a return towards a normally consolidated state. In the context of landslides and shearing processes, however, over-consolidated clays and mudrocks are materials which dilate to a *critical voids ratio* or porosity, in which state deformation can continue. When sheared beyond their maximum shearing resistance they experience a further loss of strength. The ultimate constant resistance attained is the residual strength. The residual shear strength envelope shown on Figure 13.11 is for the fissile marine shale, with parameters $\phi_r' = 14.5°$, $c_r' = 3.7\ \text{kN/m}^2$. In this case the residual strength was determined in a reversing shear box on samples split along the laminations. The envelope is, in fact, probably curved at low normal stresses, such that c_r' is essentially not greatly different from zero. A curved residual shear strength envelope is commonly obtained in over-consolidated deposits—see, for example, the ring shear data on London Clay in Figure 13.13.

The large percentage drop in strength displayed by many over-consolidated clays and mudrocks can be signified by a Brittleness Index, I_B, which is explained in Figure 13.12. In addition to dilation, brittleness in these mudrocks and clays is mainly a result of the forced alignment of clay minerals on the shear plane in the direction of shear.

Investigations of slope failures in natural and man-made (cut) slopes in over-consolidated clays have indicated that the *average* shear stress along the failure surface is commonly much less than the peak shear strength (point A, Figure 13.12) of the material. Also, long-term stability of some natural slopes (Figure 13.10) demands flatter slopes than might have been anticipated from design slopes for cuttings.

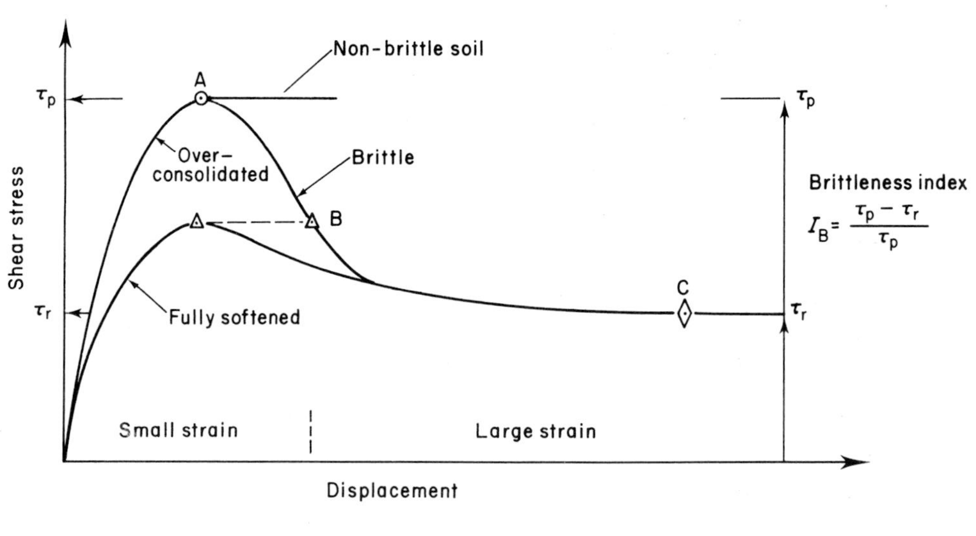

Figure 13.12 Shear stress/displacement curves (peak to residual) for over-consolidated and fully softened clays

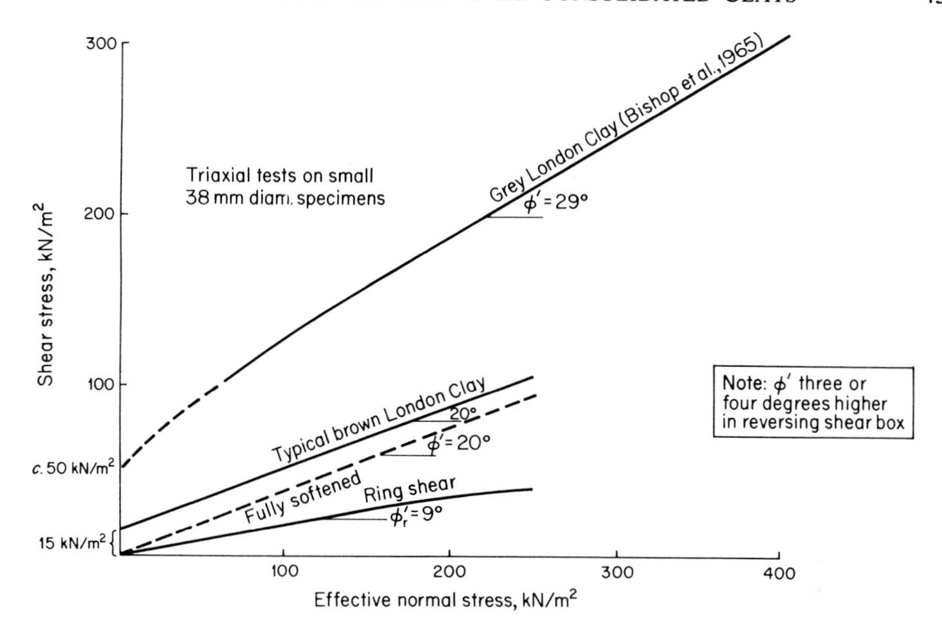

Figure 13.13 Effective stress–shear strength envelopes for unweathered grey and brown London Clay. Also shown are fully softened and residual envelopes

When a slope is overstressed, because it is too steep or too high, elements of clay or mudrock must shed part of the load they are supporting to neighbouring elements. In other words, their carrying capacity is diminished as they deform under increasing shear strain or displacement. Load shedding is one of the implications of brittle materials with strain softening characteristics (Figure 13.12). It has long been concluded (e.g. Terzaghi and Peck, 1948), that given a non-uniform stress and strain situation, shear stress redistribution will take place in this way. Joints, fissures, and other discontinuities are considered to help facilitate the process by their weakening effect on the clay and by acting as stress concentrators. The term *progressive failure* is given to the process; the *average* shear strength along a developing shear surface will lie somewhere between peak and residual strengths illustrated in Figure 13.12.

Back-analysis of *first-time* failures of cuttings in London Clay and Upper Lias Clay by Chandler and Skempton (1974) showed that in these materials the 'best fit' parameters were: London Clay, $c' = 1.0\,\text{kN/m}^2$, $\phi' = 20°$; Lias Clay, $c' = 1.0\,\text{kN/m}^2$, $\phi = 23°$ (brecciated), $c' = 1.5\,\text{kN/m}^2$, $\phi' = 25°$ (weathered and fissured). In a similar exercise involving coastal landslips in London Clay, Bromhead (1978) did not obtain a cohesion intercept. Obviously, analyses of this type are very sensitive to input parameters. Nevertheless, Chandler and Skempton (1974) make the point that if a small finite value of cohesion is not included (that is, if a $c' = 0$ assumption is made), the limiting slope angle of a cutting could be construed as being independent of its depth. This would be contrary to experience.

Figure 13.13 illustrates the difference in shear strength between hard unweathered blue London Clay taken from 34 m depth and more typical (weathered) brown London Clay. For these small samples (38 mm diameter) the cohesion intercepts shown may be exaggerated, compared with field conditions. Marsland (1971), for example, demonstrated that the intercept was reduced by about 75% in larger, nominally 100 mm diameter specimens, which may be expected to contain more discontinuities.

Skempton (1970b) suggested that the reduced shear strengths determined in first time failures might be attributed to the clay achieving full dilatancy, with an accompanying water content increase in the shear zone. This would be in accordance with *critical state* principles referred to earlier. Further details on 'critical state soil mechanics' can be found in Atkinson and Bransby (1978). For practical purposes this lower bound shear strength is represented in Figure 13.13 by the envelope pertaining to remoulded (fully softened) specimens. It is equivalent to the peak shear strength of the clay in its normally consolidated state as shown in Figure 13.12 (point B). The fully weathered mudrock in Figure 13.11 is an aggregated clayey silt. A further small drop in shear strength might be expected if it were tested again after being remoulded and disaggregated at the same water content.

The influence of discontinuities, anisotropy, loading rate, and creep all need to be considered in assessing the shear strength parameters for the design of long-term cutting slopes (see for example, the review by Simons and Menzies, 1978).

High horizontal stresses were discussed in section 13.3 in relation to the additional fractures and minor shears which may result during weathering. More directly concerned with design, Duncan and Dunlop (1969) used finite element analysis to model an excavated slope in both over-consolidated ($K_0 = 1.60$) and normally-consolidated ($K_0 = 0.81$) clays. The results indicated that the maximum shear stresses concentrated in the toe area of the cutting were about a factor of ten greater for a cutting made in the over-consolidated clay. It is pertinent to mention that progressive failures are likely to be initiated in the toe area of a slope.

Bjerrum (1967) concluded that over-consolidated clays and mudrocks with *strong* diagenetic bonds (section 13.3) would be the most dangerous type in terms of progressive failure because the latent strain energy they contained would only be released very slowly. However, swelling (discussed in section 13.4.1) is also regarded by many authors as another manifestation of strain energy release, following stress reduction by excavation or erosion. In this context important field evidence on the relationship between pore pressure equilibration and the delayed failure of cut slopes was obtained in the early 1970s.

The effect of equilibration on drained shear strength can be demonstrated by means of the Coulomb equation, in which the shear stress at failure (i.e. the shear strength of the material) is expressed as follows:

$$\tau = c' + (\sigma - u) \tan \phi' \tag{13.3}$$

where: τ is the shear stress at failure
c' is the cohesion with respect to effective stresses
ϕ' is the angle of internal friction, with respect to effective stresses
σ is the total applied normal stress
and u is the pore-water pressure.

If u is reduced and becomes negative when the clay is unloaded the effect will be to *increase* the second term in the equation. Swelling occurs as pore pressures equilibriate, which is commensurate with pore-water pressures that are increasing. Consequently, the drained (effective) shear strength will then decrease in accordance with the rate of swelling. Vaughan and Walbancke (1973) have produced field evidence which indicates that the main reason for delayed failures in London Clay is the slow rate of pore pressure increase (equilibration time). According to Chandler and Skempton (1974) the delay may be up to about 50 years in cuttings of 7–10 m in depth, and even in shallow cuttings of less than 4.5 m the time factor may be more than 10 years. Hence, it can be deduced that for early failures depicted in Figure 13.10 reduced pore pressures would have been operative. In the mudrocks of the Panama Canal, Banks *et al.* (1975) recorded reduced pore-water pressures in the *in situ* strata below canal water level some 55 years after it was opened to shipping.

First time slides involving small strains and pore-water pressure equilibration in man-made slopes are of relevance in understanding the complex interaction of mechanisms likely to be involved in slope deformation. However, it has already been mentioned that some natural slopes are much flatter than design slopes. This is because they contain both major and minor shear surfaces controlled by the residual shear strength or something approaching it (section 13.5.1). Under *landslide* conditions a large displacement of about 1 m or so is probably required before a continuous *principal slip surface* is induced, possibly by the coalescence of separately growing shear surfaces and zones. Given sufficient displacement the shearing resistance will drop from the *average* level close to the fully softened value (i.e. point B in Figure 13.12), to the residual value (point C).

The magnitude of the residual shear strength is influenced by a number of factors including the ratio of plate-shaped (clay minerals) to rotund particles. A high percentage of plate-shaped particles is required for a low residual strength, and to form a continuous shear surface (Skempton, 1964; Lupini *et al.*, 1981).

Residual strength has the character of pure friction and it was Kenney (1967) who first showed that pore fluid composition and clay mineral type are controlling factors. Smectites (montmorillonites) have ϕ_r' values ranging from 4° (Na) to 10° (Ca), illites 9° to 15°, kaolinites 14° to 22°, and hydrous micas *c.*16° to 24°. These should be compared with 'less brittle' (quartzitic) silts and sands for which ϕ_r' values are in the general range 28° to 35°. It is of interest that diagenetic clay mineral changes (Figure 13.1) should therefore be reflected to some extent in the topography of a region dominated by mudrocks. In general terms the flattest stable slopes should be expected in the *younger* geological formations with higher smectite contents, particularly if blanketing has retarded the weathering and destruction of expandable layers.

13.6 SUMMARY AND CONCLUSIONS

Although the majority of (pedological) soil types can develop on almost any mudrock or clay, slope formation entails a complex interaction of weathering and deformational processes. These processes can best be illustrated by weathering profiles in which an argillaceous character is retained throughout. Similarly, failure mechanisms of man-made cuttings in clay help in understanding the evolution of natural slopes.

The following factors would seem to be relevant to the relationship between weathering and slope formation:

1. Diagenetic changes which occur as a consequence of depth of burial generally impart an overall resistance to weathering. Nevertheless, deeply buried, heavily over-consolidated

clays and indurated mudrocks should be treated with great caution in the context of slope stability. This is because the stored strain energy and the acquired diagenetic bonds are believed to be released very slowly on unloading these materials.

2. Weathering is the converse of diagenesis. It is initiated by natural processes of uplift and erosion or in excavations made by man. Unloading leads to strain energy release and may give rise to high horizontal stresses close to the ground surface (expressed as $K_o = \sigma'_h/\sigma'_v$). Additional extensional fractures recorded in weathering profiles are attributed to this source and there is some theoretical evidence which favours the formation of slickensided shear surfaces as well. Importantly, the presence of existing discontinuities and the creation of additional ones is conducive to the ingress of water, and weathering, in the near-surface zone.
3. Physical disintegration (weathering) is more rapid than chemical weathering. It increases the surface area available for chemical reactions and consequently acts as a control on the latter. A low water table promotes both sets of weathering processes in the overlying partly-saturated zone.
4. Indurated mudrocks undergo a far greater degree of disintegration, to produce a wider range of particle sizes, than more weakly bonded clays. The following controls on the breakdown of indurated mudrocks are considered to be important: the incidence of sedimentary structures and discontinuities, slaking (compressed air breakage), expandable clay mineral content, particularly when Na is an interlayer cation.
5. The oxidation of pyrite, FeS_2 (when present), is probably of greater significance than the dissolution of calcite cement in chemical weathering. Other carbonates in mudrocks are relatively more resistant to chemical breakdown than calcite. The principal secondary sulphates formed as a consequence of pyrite oxidation are gypsum (from calcite) and jarosite (mainly from ions in clay minerals). All these chemical reactions involve volume changes in the parent material.
6. Apart from deposits which have been thoroughly reworked by surface erosion, solifluction, or possibly permafrost, weathering profiles in the UK indicate only minor formation of further (secondary) expandable clay minerals by acid weathering. Indeed, expandable minerals and chlorite can be depleted by weathering at the surface given the right conditions.
7. Reduced pore pressures and swelling are further effects of unloading. Pore pressure equilibration time is a function of the rate of swelling. The incidence of delayed failures (e.g. up to 50 years in London Clay cuttings) can be attributed to this process. Swelling on the scale of contained clay minerals is dominantly of an *intra*particle nature in expandable clays, while in some non-expandable species it is of an *inter*particle character.
8. Weathering schemes are used for comparative purposes in mudrocks. Significant weathering is above the Zone II/III boundary. Evidence suggests that the depth to this boundary may increase with the content of (expandable) smectite. Water content increases and shear strength decreases gradually upwards towards the surface in over-consolidated clays. In mudrocks, sharper changes are recorded. For natural slope investigations the effective stress shear strength parameters (c', ϕ') are used. From non-weathered to fully weathered strata there is a significant drop in cohesion, c', with a smaller decrease in ϕ'. Here again, a very large fall in cohesion is recorded in indurated mudrocks.

9. A trend of decreasing slope angle with time can be recognized in the formations for which data are collated in Figure 13.10. The simple concept of 'threshold slopes' is useful for comparing slope angles in different areas and different formations. Even so the concept is highly dependent on limit equilibrium stability analysis. Back-analysis of first-time failures in cuttings indicates that the *average* shear stress along the shear surface is much less than the *peak* shear strength of the material as measured in the laboratory. The shear strength parameters compatible with the average stress are not greatly different from those of a mudrock or clay in its fully softened state. Being brittle materials they are likely to be subject to *progressive failure* when overstressed. There is some evidence that K_0 may contribute to the concentration of high shear stresses at the toes of slopes where progressive failure is likely to be initiated.
10. Progressive failure is a complex load-shedding mechanism. Discontinuities may be exploited to act as stress concentrators, as well as helping to reduce the strength of the mass, and more locally, in causing softening as part of general weathering processes. Weathering alone would appear to reduce the shear strength of *indurated* mudrocks to a level which is probably close to the fully softened value. However, there are few *modern* natural slope failures for comparative purposes. Strain energy release with the dissolution of diagenetic bonds (para. 1 above) might be another factor in the progressive failure mechanism.
11. The flattest slopes are likely to be controlled by the minimum shear strength, namely, the *residual* strength. It is important to emphasize that residual strength can only be fully attained when a principal slip surface has been developed. Mass movements of about a metre would typically be necessary. Residual strength obtains on any pre-existing shear surface and has the nature of pure friction. Its magnitude is controlled by the clay content and clay mineralogy. In the latter respect it reflects the mineralogy of the mudrocks concerned.
12. Clearly the role of mass instability in slope development depends on the rates at which both slope steepening processes and weathering-induced softening operate, as well as other factors including piezometric pressures. In terms of instability style, a rapidly formed steep slope will probably be subject to rotational failure whereas slower rates of steepening may be associated with translational failures in which stability is controlled by the depth of weathering and the strength of the weathered material (see also the introductory section of Chapter 10). Back-analysis of first-time failures would be expected to yield values of average shear strength near to the fully softened value. Subsequent stability of the slope would then be controlled by the residual shear strength, which operates in all cases when a continuous shear surface is formed.

REFERENCES

Al-Shaikh-Ali, M. M. J. (1979). 'The measurement of K_0 in weathered Keuper Marl by the pressuremeter.' *7th Eur. Conf. Soil Mech. Fndtn. Eng.*, Brighton, **4**, 79–80.

Andersland, O. B., and Douglas, A. G. (1973). 'Soil deformation rates and activation energies.' *Géotechnique*, **20**, 1–16.

Anon. (1970). 'Working party report on the logging of rock cores for engineering purposes.' *Q. J. Eng. Geol., London*, **3**, 1–24.

Anon. (1977). 'Working party report on the description of rock masses for engineering purposes.' *Q. J. Eng. Geol., London*, **10**, 355–88.

Atkinson, J. H., and Bransby, P. L. (1978). *The Mechanics of Soils. An Introduction to Critical State Soil Mechanics.* McGraw-Hill, London.

Aylmore, L. A. G., and Quirk, J. P. (1959). 'Swelling of clay–water systems.' *Nature*, **183**, 1752–3.

Badger, C. W., Cummings, A. D., and Whitmore, R. L. (1956). 'The disintegration of shales in water.' *J. Inst. Fuel*, **29**, 417–23.

Banks, D. C., Strohm, W. E., De Angulo, M., and Lutton, R. J. (1975). *Study of clay shale slopes along the Panama Canal.* Report 3. Engineering analyses of slides and strength properties of clay shales along the Gaillard Cut. *Technical Report* S – 70 – 9. US Army Engineer Waterways Experiment Stations, Vicksburg., Miss. 114 pp + Appendices A–F.

Beckett, P. J., Cummings, A. D., and Whitmore, R. L. (1958). 'Interfacial properties of coal measure shales in water.' *Proc. Conf. on Science in the use of Coal, Sheffield University.* Paper 12, 14–19.

Berkovitch, I., Manackerman, M., and Potter, N. M. (1959). 'The shale breakdown problem in coal washing. Part 1—assessing the breakdown of shales in water.' *J. Inst. Fuel*, **32**, 579–89.

Bishop, A. W., Webb, D. L., and Lewin, P. I. (1965). 'Undisturbed samples of London Clay from the Ashford Common shaft: strength–effective stress relationships.' *Géotechnique*, **15**, 1–31.

Bjerrum, L. (1967). 'Progressive failure in slopes of overconsolidated clay and clay shales.' *Proc. J. Soil Mech. Fndtn Div., Am. Soc. Civil. Engrs.*, **93**, 1–49.

Blackmore, A. V., and Miller, R. D. (1961). 'Tactoid size and osmotic swelling in calcium montmorillonite.' *Proc. Soil. Sci. Soc. Am.*, **25**, 169–73.

Bolt, G. H. (1956). 'Physico-chemical analysis of the compressibility of pure clays.' *Géotechnique*, **6**, 86–93.

Bromhead, E. N. (1978). 'Large landslides in London Clay at Herne Bay, Kent.' *Q. J. Eng. Geol., London*, **11**, 291–304.

Bromhead, E. N. (1984). 'Slopes and embankments.' In *Ground Movements and their Effects on Structures.* Attewell, P. B., and Taylor, R. K. (eds.), Surrey University Press, Blackie Group, Glasgow, pp. 46–75.

Brooker, E. W. (1967). 'Strain energy and behaviour of overconsolidated soils.' *Can. Geotech. J.*, **4**, 326–33.

Burland, J. B., Longworth, T. I., and Moore, J. F. A., 1978. 'A study of ground movement and progressive failure caused by a deep excavation in Oxford Clay.' *Building Research Establishment Current Paper 33/78*, 36pp.

Burland, J. B., and Maswoswe, J. (1982). '*In situ* measurement of horizontal stress in overconsolidated clay using push-in spade shaped pressure cells.' *Géotechnique*, **32**, 285–6.

Carroll, D. (1970). *Rock Weathering.* Plenum Press, New York, 203pp.

Carson, M. A. (1969). 'Models of hillside development under mass failure.' *Geog. Anal.*, **1**, 76–100.

Carson, M. A. (1975). 'Threshold and characteristic angles of straight slopes.' In *Mass Wasting.* Yatsu, E., Ward, A. J., and Adams, F. (eds.), *Proc. 4th Guelph Symp. Geomorphology*, Guelph, 19–34.

Carson, M. A. (1976). 'Mass-wasting, slope development and climate.' In *Geomorpholgy and Climate.* Derbyshire, E. (ed.), John Wiley, London, pp. 101–36.

Carson, M. A., and Petley, D. J. (1970). 'The existence of threshold hillslopes in the denudation of the landscape.' *Trans. Inst. Brit. Geogrs.*, **49**, 71–96.

Chandler, R. J. (1969). 'The effect of weathering on the shear strength properties of Keuper Marl.' *Géotechnique*, **19**, 321–34.

Chandler, R. J. (1972). 'Lias Clay: weathering processes and their effect on shear strength.' *Géotechnique*, **22**, 403–31.

Chandler, R. J. (1974). 'Lias Clay: the long-term stability of cutting slopes.' *Géotechnique*, **24**, 21–38.

Chandler, R. J., and Skempton, A. W. (1974). 'The design of permanent cutting slopes in stiff fissured clays.' *Géotechnique*, **24**, 457–66.

Chenevert, M. E. (1970). 'Adsorptive pore pressures of argillaceous rocks.' In *Rock Mechanics Theory and Practice. Proc. 11th Symp. AIME.*

Cripps, J. C., and Taylor, R. K. (1981). 'The engineering properties of mudrocks.' *Q. J. Eng. Geol., London*, **14**, 325–46.

Curtis, C. D. (1980). 'Diagenetic alteration in black shales.' *J. Geol. Soc., London*, **137**, 189–94.

Deo, P. (1972). 'Shales as embankment materials.' *Report No. 45, Joint Highway Research Project*. Purdue University and Indiana Highway Commission, West Lafayette, December 1972.

Dewey, J. F. (1984). Personal communication.

Driscoll, R. (1984). 'The effects of clay soil volume changes on low-rise buildings.' In *Ground Movements and their Effects on Structures,* Attewell, P. B., and Taylor, R. K. (eds.), Surrey University Press. Blackie Group, Glasgow, pp. 303–20.

Duncan, J. M., and Dunlop, P. (1969). 'Slopes in stiff-fissured clays and shales.' *J. Soil Mech. Fndtn Eng. Div., ASCE*, **96**, 467–92.

Evans, L. J., and Adams, W. A. (1975). 'Chlorite and illite in some Lower Palaeozoic mudstones of mid-Wales.' *Clay Mins.*, **10**, 387–97.

Ferguson, H. F. (1967). 'Valley stress release in the Allegheny Plateau.' *Eng. Geol., Amsterdam*, **4**, 63–71.

Fleming, R. W., Spencer, G. S., and Banks, D. C. (1970). 'Empirical behaviour of clay shale slopes.' *US Army Eng. Nuclear Cratering Group (NCG), Tech. Rep.*, 15, **1**, 93pp and **2**, 304pp.

Harper, T. R., Appel, G., Pendleton, M. W., Szymanski, J. S., and Taylor, R. K. (1979). 'Swelling strain development in sedimentary rock in northern New York.' *Int. J. Rock Mech. Mining Sci. Geomech. Abstr.*, **16**, 271–92.

Harper, T. R., and Szymanski, J. S. (1983). 'Geological processes and the mechanical aspects of rock squeeze.' *Tectonophysics*, **91**, 119–35.

Hedberg, H. D. (1936). 'Gravitational compaction of clays and shales.' *Am. J. Sci.,* **31**, 241–87.

Hoek, E., and Brown, E. T. (1980). *Underground Excavations in Rock*. The Institution of Mining and Metallurgy, London, 527pp.

Hofmann, H. J. (1966). 'Deformational structures near Cincinnati, Ohio.' *Bull. Geol. Soc. Am.*, **77**, 533–48.

Horton, A. E., Manackerman, M., and Raybould, W. E. (1964). 'The shale breakdown problem in coal washing. Part 2—some causes of shale breakdown and means for its control.' *J. Inst. Fuel*, **37**, 52–8.

Hutchinson, J. N. (1967). 'The free degradation of London Clay cliffs.' *Proc. Geotech. Conf., Oslo*, **1**, 113–18.

Hutchinson, J. N. (1970). 'A coastal mudflow on the London Clay cliffs at Beltinge, North Kent.' *Géotechnique*, **20**, 412–38.

Kennard, M. F., Knill, J. L., and Vaughan, P. R. (1967). 'The geotechnical properties and behaviour of Carboniferous shale at the Balderhead Dam.' *Q. J. Eng. Geol., London*, **1**, 3–24.

Kenney, T. C. (1967). 'The influence of mineral composition on the residual strength of natural soils.' *Proc. Geotech. Conf., Oslo.* **1**, 123–30.

Jenny, H. (1951). 'Contact phenomena and their significance in plant nutrition.' In *Mineral Nutrition of Plants*. Truog, E. (ed.), University of Wisconsin Press, Madison, Wis., pp. 107–32.

Leach, B. A., Medland, J. W., and Sutherland, H. B. (1976). 'The ultimate bearing capacity of bored piles in weathered Keuper Marl.' *Proc. 6th Eur. Conf. Soil Mech. Fndtn. Eng., Vienna*, **1**(2), 507–14.

Little, A. J. (1972). *Slope stability in the Kimmeridge Clay*. Unpublished. M.Sc. Advanced Course in Engineering Geology Dissertation, University of Durham, 163pp.

Lupini, J. F., Skinner, A. E., and Vaughan, P. R. (1981). 'The drained residual strength of cohesive soils.' *Géotechnique*, **31**, 181–214.

Madsen, E. T. (1979). 'Determination of the swelling pressure of claystones and marlstones using mineralogical data.' *Proc. 4th Cong. Int. Soc. Rock Mech.*, **1**, 237–43.

Marsland, A. (1971). 'The shear strength of fissured clays.' *Stress–strain behaviour of soils. Roscoe Memorial Symposium*. Henley-on-Thames, Foulis, 59–68.

Marsland, A. (1977). '*In situ* measurement of the large-scale properties of Keuper Marl.' *Proc. International Conference on the Geotechnics of Structurally Complex Formations, Associatione geotechnica Italiana, Capri*, 1, 335–44.

Marsland, A., and Randolph, M. F. (1978). 'Comparison of the results from pressuremeter tests and large *in situ* plate tests in London Clay.' Building Research Establishment, Current paper 78/10, 26pp.

Matheson, D. S., and Thomson, S. (1973). 'Geological implications of valley rebound.' *Can. J. Earth Sci.*, **10**, 961–78.

Morgenstern, N. R., and Eigenbrod, K. D. (1974). 'Classification of argillaceous soils and rocks.' *J. Geotech. Div., ASCE,* **100**, 1137–56.
Nichelson, N. R., and Hough, V. N. D. (1967). 'Jointing in the Appalachian Plateau of Pennsylvania.' *Bull. Geol. Soc. Am.*, **78**, 609–29.
Nichols, T. C. (1980). 'Rebound, its nature and effect on engineering works.' *Q. J. Eng. Geol., London*, **13**, 133–52.
Olsen, R. E., and Mesri, G. (1970). 'Mechanisms controlling the compressibility of clays.' *J. Soil Mech. Fndtn. Div., ASCE*, **96**, 1863–78.
Parry, R. G. H. (1972). 'Some properties of heavily overconsolidated Oxford Clay at a site near Bedford.' *Géotechnique*, **22**, 485–507.
Penner, E., Eden, W. J., and Gillott, J. E. (1973). 'Floor heave due to biochemical weathering of shale.' *Proc. 8th Int. Conf. Soil Mech. Fndtn Eng.*, Moscow, **2**, 151–8.
Perrin, R. M. (1971). *The Clay Mineralogy of British Sediments*. Min. Soc. (Clay Minerals Group), London, 247pp.
Peterson, R. (1954). 'Studies of Bearpaw Shale at a damsite in Saskatchewan.' *Proc. Am. Soc. Civ. Engrs.*, **80**. *Proc. Sep.* No. 476, 28pp.
Peterson, R. (1958). 'Rebound in the Bearpaw Shale in Western Canada.' *Bull. Geol. Soc. Am.*, **69**, 1113–23.
Pettijohn, F. J. (1975). *Sedimentary Rocks*. Harper and Bros., New York, 618pp.
Potter, P. E., Maynard, J. B., and Pryor, W. A. (1980). *Sedimentology of Shale*. Springer-Verlag, New York, 306pp.
Prior, D. B. (1978). 'Some recent progress and problems in the study of mass movement in Britain.' In *Geomorphology: Present Problems and Future Prospects*. Embleton, C., Brunsden, D., and Jones, D. K. C. (eds.), Oxford Univ. Press, Oxford, pp. 84–106.
Quigley, R. M. (1975). 'Weathering and changes in strength of glacial till.' In *Mass Wasting* Yatsu, E., Ward, A. J., and Adams, F. (eds), *Proc. 4th Guelph Symp. Geomorphology*, 117–131.
Ringheim, A. S. (1964). 'Experiences with the Bearpaw Shales at the South Saskatchewan River Dam.' *Trans. 8th Int. Cong. on Large Dams, Edinburgh*, **1**, 529–50.
Rosenqvist, I.Th. (1968). *Mechanical Properties of Soils from a Mineralogical-physical-chemical viewpoint*. Report, Institut Geologi, Universitetet i Oslo.
Russell, D. J., and Parker, A. (1979). 'Geotechnical mineralogical and chemical inter-relationships in weathering profiles in an overconsolidated clay.' *Q. J. Eng. Geol., London*, **12**, 107–16.
Samuels, S. G. (1975). *Some properties of the Gault Clay from the Ely-Ouse Essex water tunnel.* Building Research Establishment Current Paper 3/75, 19pp.
Scott, J. S., and Brooker, E. W. (1978). 'Geological and engineering aspects of Upper Cretaceous shales in western Canada.' Geol. Survey of Canada. Paper 66–37, 75pp.
Shaw, D. B., and Weaver, C. E. (1965). 'The mineralogical composition of shales.' *J. Sedim. Petrol.*, **35**, 213–22.
Shaw, H. F. (1981). 'Mineralogy and petrology of the argillaceous sedimentary rocks of the UK'. *Q. J. Eng. Geol., London*, **14**, 277–90.
Simons, N. E., and Menzies, B. K. (1978). 'The long-term stability of cuttings and natural clay slopes.' In *Developments in Soil mechanics—1*. Scott, C. R. (ed.), Applied Science Publishers, London, pp. 347–91.
Skempton, A. W. (1948). 'The rate of softening in stiff fissured clays, with special reference to London Clay.' *Proc. 2nd Int. Conf. Soil. Mech. Fndtn. Eng.*, Rotterdam, **2**, 50–3.
Skempton, A. W. (1961). 'Horizontal stresses in overconsolidated London Clay.' *Proc. 5th Int. Conf. Soil Mech. Fndtn. Eng.*, Paris, **1**, 351–7.
Skempton, A. W. (1964). 'Long-term stability of clay slopes.' *Géotechnique*, **14**, 77–101.
Skempton, A. W. (1970a). 'The consolidation of clays by gravitational compaction.' *Q. J. Geol. Soc., London*, **125**, 373–411.
Skempton, A. W. (1970b). First-time slides in overconsolidated clays. *Géotechnique*, **20**, 320–324.
Skempton, A. W. (1977). 'Slope stability of cuttings in brown London Clay.' *Proc. 9th Int. Conf. Soil Mech. Fndtn. Eng., Tokyo*, **3**, 261–70.
Skempton, A. W., and Hutchinson, J. N. (1969). 'Stability of natural slopes and embankment sections'. *Proc. 7th Int. Conf. Soil Mech. Fndtn. Eng., Mexico*, State of Art, 291–340.

Skempton, A. W., and La Rochelle, P. (1965). 'The Bradwell slip, a short term failure in London Clay.' *Géotechnique*, **15**, 221–42.

Skempton, A. W., and Petley, D. J. (1967). 'The strength along structural discontinuities'. *Proc. Geotech. Conf., Oslo*, **2**, 29–46.

Smith, T. J. (1978). *Consolidation and other geotechnical properties of shales with respect to age and composition*. Unpublished Ph.D. Thesis, University of Durham, 452pp.

Spears, D. A. (1969). 'A laminated marine shale of Carboniferous age from Yorkshire, England.' *J. Sedim. Petrol.*, **39**, 106–12.

Spears, D. A. (1980). 'Towards a classification of shales.' *J. Geol. Soc., London*, **137**, 125–9.

Spears, D. A., and Taylor, R. K. (1972). 'Influence of weathering on the composition and engineering properties of *in situ* Coal Measures rocks.' *Int. J. Rock Mechn. Mining Sci.*, **9**, 729–56.

Sridharan, A., and Venktappa Rao (1973). 'Mechanisms controlling volume change of saturated clays and the role of effective stress concept.' *Géotechnique*, **23**, 359–82.

Taylor, R. K. (1984). *Composition and Engineering Properties of British Colliery Discards*. National Coal Board, London, 244pp.

Taylor, R. K., and Cripps, J. C. (1984). 'Mineralogical controls on volume change.' In *Ground Movements and their Effects on Structures*. Attewell, P. B., and Taylor, R. K. (eds.), Surrey University Press, Blackie Group, Glasgow, pp. 268–302.

Taylor, R. K., and Spears, D. A. (1970). 'The breakdown of British coal measure rocks.' *Int. J. Rock Mech. Mining Sci.*, **7**, 481–501.

Taylor, R. K., and Spears, D. A. (1981). 'Laboratory investigation of mudrocks.' *Q. J. Eng. Geol. London*, **14**, 291–309.

Tedd, P., and Charles, J. A. (1981). '*In situ* measurements of horizontal stress in overconsolidated clay using push-in spade shaped pressure cells.' *Géotechnique*, **31**, 554–8.

Terzaghi, K. (1961). 'Discussion of A. W. Skempton's paper "Horizontal stresses in overconsolidated London Clay".' *Proc. 5th Int. Conf. Soil Mech. Fndtn. Eng.*, Paris, **3**, 144–5.

Terzaghi, K., and Peck, R. B. (1948). *Soil Mechanics in Engineering Practice*. New York, John Wiley, 566pp.

Tourtelot, H. A. (1962). 'Preliminary investigation of the geologic setting and chemical composition of the Pierre Shale.' US Geological Survey, Prof. Paper 390.

Valeton, I. (1983). 'Palaeoenvironment of lateritic bauxites with vertical and lateral differentiation.' In *Residual Deposits: Surface Related Weathering Processes and Materials*. Wilson, R. C. L. (ed.), Blackwell Scientific Publications, Oxford, pp. 77–90.

Van Eeckhout, L. (1976). 'The mechanisms of strength reduction due to moisture in coal mine shales.' *Int. J. Rock Mech. Mining Sci.*, **13**, 61–7.

Vaughan, P. R., and Walbancke, H.-J. (1973). 'Pore pressure changes and delayed failure of cutting slopes in over consolidated clay.' *Geotechnique*, **23**, 531–9.

Verwey, E. J. W., and Overbeck, J.Th.G (1948). *Theory of Stability of Lyophobic Colloids*. Elsevier, Amsterdam.

Ward, W. H., Marsland, A., and Samuels, S. E. (1965). 'Properties of the London Clay at the Ashford Common shaft: *In situ* and undrained strength tests.' *Géotechnique*, **15**, 321–44.

Weaver, C. E. (1967). 'Potasium, illite and the ocean.' *Geochim. Cosmochim. Acta*, **31**, 2181–96.

Weeks, A. G. (1969). 'The stability of natural slopes in south-east England as affected by periglacial activity.' *Q. J. Eng. Geol., London*, **2**, 49–62.

Whitten, D. G. A., and Brooks, J. R. V. (1972). *A Dictionary of Geology*. Penguin Books, 493pp + Appendices.

Windle, D., and Wroth, C. P. (1977). '*In situ* measurement of the properties of stiff clays.' *Proc. 9th Int. Conf. Soil Mech. Fndtn. Eng., Tokyo*, **1**, 347–52.

Winmill, T. F. (1915–16). 'Atmospheric oxidation of iron pyrites. Pt IV' *Trans. Inst. Mining Engrs.*, **51**, 500.

Yaalon, D. H. (1962). 'Mineral composition of average shale.' *Clay Mins. Bull.*, **5**, 31–6.

Yong, R. N., and Warkentin, B. P. (1975). *Soil Properties and Behaviour*. (Developments in Geotechnical Engineering 5) Elsevier Scientific Publishing Company, Amsterdam. 449pp.

Young, A. (1961). 'Characteristic and limiting slope angles.' *Zeitchrift fur Geomorphologie*, **5**, 126–31.

Young, A. (1975). Longman, London, 288pp.

Slope Stability
Edited by M. G. Anderson and K. S. Richards

Chapter 14

Quick Clays

J. KENNETH TORRANCE
Geotechnical Science Laboratories, Department of Geography, Carleton University, Ottawa, Ontario, Canada, K1S 5B6

INTRODUCTION

Quick clays are defined in terms of their sensitivity (the ratio of the undisturbed to the remoulded shear strength) and their remoulded shear strength. Since the term 'quick clay' is a direct translation of the Norwegian term *kvikkleira*, and, given the long tradition of the Norwegians in investigating the properties of these materials, it is appropriate to adopt the current criterion used in Norway. We thus define as quick clays those soils exhibiting a sensitivity greater than 30 and having a remoulded shear strength less than 0.5 kPa (Norsk Geoteknisk Forening, 1974). Other sensitivity limits for quick clay designation have been proposed (see Torrance, 1983) but the Norwegian criterion is preferred by the author because it is the only one that requires the remoulded soil to behave as a fluid. It is important to note that quick clays are defined by their behaviour and not by their location, depositional environment or some other factor of origin or material.

The high sensitivity and, in particular, the fluidity of the remoulded (or failed) soil material make the quick clays particularly susceptible to landslide activity and dramatic geotechnical failures (Figure 14.1). Carson and Lajoie (1981) studied the morphology of landslides in sensitive marine clays and categorized them (Types I to V, Table 14.1) according to the nature of their penetration into the valley slope. The relative retrogression, R/H, was defined as the ratio of R, the distance of retrogression of the landslide from the position of the original slope, to H, the depth from the original land surface to the failure plane. For completeness, Type 0 has been added by the author to include in the typology simple rotational landslides which do not retrogress. The limits on the sensitivity scale (upper and lower) within which the various landslide types can occur have not been defined, and probably cannot be. Few deposits are sufficiently uniform to be characterized by a single value of sensitivity. Even the most sensitive deposits have a surface zone which has experienced weathering with its accompanying decrease in sensitivity.

Figure 14.1 Aerial view, looking upstream, of the 28 hectare landslide which occurred 16 May 1971 on the South Nation River, Ontario, Canada. Note the debris that has been forced upstream in the river channel. (*From Eden* et al., *1971*, Can. Geotech. J., **8**. *Reproduced by permission of National Research Council of Canada.*)

While all the aforementioned landslide types can probably occur in deposits which have the characteristics required for designation as quick clays, some are only possible if the post-failure strength is so low that the failed soil exhibits fluid behaviour. The nature and extent of retrogression are more controlled by the fluidity of the failed soil material than by the absolute value of the sensitivity. This is the case because sensitivity is defined as a ratio and hence soil materials with identical sensitivities can exhibit greatly different undisturbed and remoulded strengths. They will be stable under different stress conditions and exhibit different post-failure behaviour. It is essential, when considering stability conditions and landslide occurrence in quick clays, to recognize the implications of the dual requirements of a minimum sensitivity of 30 and a maximum remoulded shear strength of 0.5 kPa, with the latter implying fluid behaviour upon disturbance.

The most extensive areas of quick clay occur in Scandinavia and eastern Canada. They have also been reported in Alaska, Japan, the Soviet Union, and New Zealand. It seems unlikely that any large undiscovered areas of quick clay exist. It is probable that discoveries of additional small areas will continue to be made.

It is necessary, before proceeding further, to clarify the nomenclature that will be used for the Canadian marine clays since the names Leda clay, Champlain clay, and sensitive

Table 14.1 Landslide Typology for Sensitive Deposits (Mainly summarized from Carson and Lajoie, 1981)

Landslide Type		Relative Retrogression (R/H)	Limitation to Retrogression
0	—simple rotational failure	A single failure with low R/H value.	Debris remains at base of backscarp.
I	—two-dimensional spreading failure	Extreme of $1.5<R/H<46$ but commonly $4<R/H<27$. Range determined by valley width and depth.	The volume of the valley receiving the debris determines the volume of the failed area and hence the retrogression. This seems to be the most common failure type.
II	—aborted retrogression	Lower than would be predicted for Type I (as determined by valley characteristics).	Spatial variation in the nature of the sediment or some other constraint stops the retrogression before the constraint for Type I becomes applicable.
III	—excess retrogression	Greater than would be predicted for Type I (as determined by valley characteristics).	There is either minimal topographic limitation to down valley flow of debris or debris liquefaction and flow prevents debris accumulation. Retrogression stops for other than topographic reasons.
IV	—multidirectional retrogression	Comparable to Type III but with retrogression in several directions, often producing a 'bottlenecked' crater	Liquefaction of the debris presents no limitation to the retrogression. Retrogression stops for other than topographic reasons.
V	—flakeslide	Variable, but in the Type I range.	The whole area fails at once along a failure plane. The landslide size depends on topography and sediment characteristics.

clay have all been applied. They were originally called 'Leda clay' after a fossil commonly found in them. This fossil was misidentified but the name 'Leda clay' has persisted in common use to refer to all the sensitive marine clays of the basins flowing into the Gulf of St Lawrence. The term 'Champlain clay' has been used in recent years, synonymously with Leda clay, by many people. Strictly speaking, 'Champlain clay' should refer only to the sediments deposited in the Champlain Sea, which extended inland from Quebec City. Sometimes the even more general term 'sensitive clay' is used. This has value in that it implies no geographic restrictions and would include the sensitive marine clays of the James and Hudson Bay Lowlands, as well as fresh water deposits whose sensitivity exceeds some predetermined value. In this paper the term 'Leda clay' will be used when referring specifically to the deposits of the Gulf of St Lawrence drainage basins and 'Canadian or eastern Canadian sensitive marine clay' when the reference is more general. It should be emphasized that none of the above terms implies that the criteria for quick clay behaviour are met. The materials so named exhibit a wide range of sensitivity and remoulded strength.

A large number of scientists and engineers have been responsible for elucidating and developing an understanding of factors that can lead to quick clay behaviour. Torrance

Table 14.2 General Modal for Quick Clay Development. (*From Torrance, 1983*, Sedimentology, **30**. *Reproduced by permission of Blackwell's, Oxford.*)

Factors Producing a High Undisturbed Strengh	
Depositional	Post-Depositional
Flocculation*†	Cementation bonds
—salinity	—rapidly developed
—divalent cation adsorption	—slowly developed
—high suspension concentration	Slow load increase
	—time for cementation
	—'thixotropic' processes

Factors Producing a Low Remoulded Strength	
Material properties	Minimal consolidation
—Low activity minerals dominate*†	
	Salt removal (by leaching and/or diffusion)*
	—decrease in liquid limit greater than decrease in water content
	Dispersants†
	—decrease in liquid limit greater than decrease in water content

*Essential in marine clays.
†Essential in fresh water clays.

(1983) has recently summarized these factors into a 'general model for quick clay development' (Table 14.2). The model is formulated in terms of the influences of various material properties and natural processes on the sensitivity and remoulded strength. These are separated into depositional requirements and post-depositional factors which increase the undisturbed or decrease the remoulded strengths. This model will serve as the basis for discussion of the requirements for quick clay development.

The sensitivity of soil materials is commonly greater than 1, but the high sensitivity, >30, required for designation as quick clay is seldom, if ever, met by sedimentary materials at the time of their deposition because they are too weak in the undisturbed state, too strong in the remoulded state, or both. The development of quick clay appears to be a post-depositional response of these materials to change. The sensitivity of a soil material can obviously be increased by any influence which: (a) increases the undisturbed strength by a greater proportion than the remoulded strength; (b) decreases the remoulded strength by a greater proportion than the undisturbed strength; or (c) increases the undisturbed strength and decreases the remoulded strength. The questions that must be answered are: what makes a soil susceptible to an increase in sensitivity, and what are the factors which cause the change?

14.2 DEPOSITIONAL REQUIREMENTS FOR QUICK CLAY DEVELOPMENT

14.2.1 Flocculated Structure

For quick clay behaviour to develop, the sediment must have a flocculated structure (Rosenqvist, 1977; Quigley, 1980). This structure has edge-to-edge and edge-to-face linkages between particles and little, if any, preferred orientation of particles within aggregates or indeed in the structure as a whole. The effect of the flocculated structure is to provide a framework with an inherent strength which, after it has formed and stabilized, maintains the sediment at a constant void ratio, or at least at a higher void ratio than could otherwise be maintained, while other changes are occurring. If the soil is disturbed, collapse of the structure is the major factor in strength loss. The breakdown of structure is probably never complete and the degree of breakdown undoubtedly depends on the effort expended to cause breakdown. The nature and degree of interaction between the particles and fragments of the original structure after disturbance, which is determined by water content and chemical conditions, determines the post-failure or remoulded strength of the soil.

Flocculated structure develops at the time of deposition as a consequence of one or more of: high salinity, dominance of divalent cations on cation adsorption sites, and high suspension concentration. The marine clays of eastern Canada and Scandinavia consist of glacially ground rock material, i.e. rock flour, which was either discharged directly into the marine environment or, more commonly, was transported by fresh water streams and rivers to the marine environment. Under fresh water conditions the negatively charged mineral particles of fine silt and clay size, because of their negative charges, repel one another and act as individuals with little tendency to cluster into aggregates. The distance from the particle surfaces at which the interparticle repulsion is strong decreases when the salinity of the water in which the particle is suspended increases. Thus the mutual repulsion between particles decreases greatly when they enter the marine environment and it becomes possible for flocculation of particles to occur. A small net positive charge which may be present on the edges of the plate-shaped clay mineral particles also aids flocculation. The attraction of edges to faces and edges to edges along with the attraction of the edge of clay minerals to the surfaces of the other particles leads to the development of aggregates in which there is essentially random orientation of particles. Electron micrographs of undisturbed Leda clay suggest that these aggregates are connected by interaggregate linkages of fine particles (Delage and Lefebvre, 1984).

Flocculation to form aggregates occurs relatively rapidly upon entry into the marine water and the aggregates settle out of suspension to form a sediment in which silt- and clay-sized particles have not segregated according to particle size. The sediment appears uniform and has a relatively open, high-water-content structure. Flocculation may also occur, even under low salinity conditions, when divalent cations dominate in solution and on the cation exchange sites, since divalent cations are more strongly attracted to the negatively-charged mineral particles and decrease the distance from the particle surface to which repulsive forces dominate. Very high suspension concentrations also increase particle interaction and encourage flocculation. Despite these various possibilities for obtaining flocculated sediments, deposition under marine or brackish water conditions is the most common origin of the sediments in which quick clay behaviour has developed.

14.2.2 Material Properties

The material properties requirement for the development of quick clays is that low-activity minerals must dominate in the clay-size fraction. Activity was defined by Skempton (1953) as the ratio of the plasticity index of a soil to the percentage of clay-sized particles. The activity of non-swelling clay minerals and the primary silicates is generally < 1, whereas the activity of the swelling clay minerals is generally >1. The clay-sized fraction of quick clays of Canada and Scandinavia typically includes the primary minerals (such as quartz, feldspar, and amphibole) and the clay minerals (such as hydrous mica and chlorite), all of which are low-activity minerals. Only trace amounts of the higher activity, swelling clay minerals have been reported (see Brydon and Patry, 1961; Roaldset, 1972; Bentley and Smalley, 1979, Gillott, 1979; Quigley *et al.*, 1983). The quick clays reported from Japan contain smectite as 33–42% of the clay-sized fraction. For some reason, probably their high iron content, these smectites are non-swelling and of low activity (Egashira and Ohtsubo, 1983) and the response of the sediments to decrease of salinity is similar to that of the post-glacial marine clays of Canada (Ohtsubo *et al.*, 1982; Torrance, 1984).

In sediments of marine origin, leaching by fresh water with the accompanying decrease in salinity is a requirement for the development of quick clay behaviour. The liquid limit of low-activity minerals decreases when salinity decreases, whereas the liquid limit of high-activity minerals increases (Moum and Rosenqvist, 1961). This difference has fundamental implications for the response of the sediments to post-depositional change. Torrance (1970) demonstrated that a relatively small amount of high-swelling smectite added to a leached marine clay from Norway had dramatic behavioural implications through its effect of increasing the remoulded shear strength. The percentage of high-activity clay required to prevent quick clay development has never been investigated but undoubtedly depends on the character of both the low-activity and the high-activity minerals and probably other factors as well.

Smalley and his coworkers (Smalley, 1971; Cabrera and Smalley, 1973; Bentley and Smalley, 1978; Smalley *et al.*, 1984) have proposed an inactive-particle, short-range-bond hypothesis for high sensitivity. They stress the importance of clay-sized primary silicates, particularly quartz and feldspar, in the development of high sensitivities. Torrance (1983) argued that the clay-sized particles of primary silicates, which bear a net negative charge at the pH found in the sensitive marine clays and have a comparable surface charge density to the clay minerals (Yariv and Cross, 1979), are similar to the low-activity clay minerals in their response to salinity change. The Smalley hypothesis is essentially the same as the low-activity mineral requirement.

Experiments to investigate the influence of salinity on the yield stress of remoulded marine clays which have been enriched in primary minerals, by selective destruction of clay minerals, have recently been completed. These indicate that the primary-mineral-rich materials remain susceptible to decrease in the liquid limit and the remoulded strength

Table 14.3 Atterberg limits, at 20, 5, and 2 g/litre pore-water salinity and Na-saturation, of a natural Leda clay from the South Nation River landslide site and the same clay after enrichment in primary minerals by selective extraction of oxide minerals and layer silicates.

	Liquid Limit		Plastic Limit	
Salinity (g/litre)	Natural	Primary Mineral Enriched	Natural	Primary Mineral Enriched
20	48.4	32.6	20.7	17.9
5	44.7	24.8	21.2	17.7
2	39.5	—	17.9	—

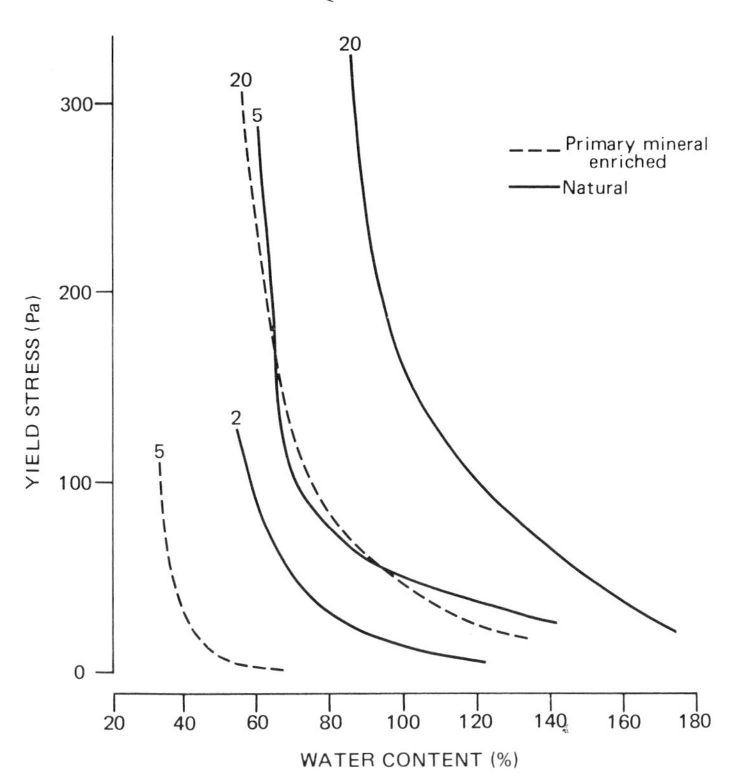

Figure 14.2 Yield stress–water content relationships of a Na-saturated Leda clay from the South Nation River landslide site, and the same Leda clay after primary mineral enrichment by selective extraction of oxide minerals and layer silicates. The numbers on the curves indicate the pore-water salinity in g/litre. The behaviour of both materials is affected by pore-water salinity, but the primary-mineral-enriched soil has the lower yield stress when salinity and water content are identical

(yield stress) at constant water content as the salinity decreases (Table 14.3 and Figure 14.2). At 2 g/litre salinity it was not possible to determine the liquid and plastic limits of the primary-mineral-enriched material. The Leda clay material used in these experiments, with its original clay mineral content, was capable of becoming a quick clay. Because the remoulded soil exhibited a similar response to changing salinity after enrichment with primary minerals, it would appear that quick clay could still be developed. The high clay mineral contents of the quick clays from Japan indicate that a relatively high content of low-activity clay minerals is consistent with quick clay development. This provides further evidence that it is mineral activity, not clay mineral content, that determines the susceptibility to sensitivity increase upon salt removal. The more general statement, namely that the minerals must be of low activity, seems preferable to accenting the importance of the clay-sized primary-mineral particles.

The role of iron and aluminium oxides and amorphous materials has received considerable attention. Most of the debate has centred around their possible role as cementing agents (discussed in the later section on cementation) but some investigation has also occurred into their effect on the behaviour of the remoulded soil. Hendershot and Carson (1978), using selective extraction techniques to remove the oxides and amorphous materials, found that they contributed to the plasticity of Leda clay. Their removal decreased the liquid limit and the plasticity. Dixon (1982) and Torrance (1984), working with several Leda clay materials and a Japanese quick clay, found that removal of oxides and amorphous materials decreased the yield stress, as determined by a coaxial viscometer, at constant water content and salinity. These materials acted as thickeners, presumably by increasing the degree of attractive interaction between the mineral particles.

Electrostatic repulsion between the negatively charged primary mineral and phyllosilicate particles is the principal reason for lack of attraction between particles in quick clays. The oxides and amorphous materials are generally present as coatings on other particles, and with isoelectric points somewhere between pH 6 and 9 (depending on their composition) have at most only a small negative charge at the pH in the Canadian and Scandinavian quick clays and a small positive charge at the pH in the Japanese quick clay. Their presence would be expected to decrease the repulsive forces between particles and increase the remoulded strength. The higher remoulded strength observed when oxides and amorphous materials are present suggests that their presence leads to lower sensitivities and a lower probability of quick clay developing.

14.3 POST-DEPOSITIONAL REQUIREMENTS FOR QUICK CLAY DEVELOPMENT

While the existence of a flocculated structure in a sediment dominated by low-activity minerals represents the composite, material-properties prerequisite for quick clay development, post-depositional factors must operate before quick clay behaviour is observed. It is rare, if indeed it occurs, that a sediment exhibits the required combination of high sensitivity and low remoulded strength unless some factor or factors have caused a post-depositional increase in the undisturbed strength, a post-depositional decrease in the remoulded strength, or both. Decrease of the remoulded strength to <0.5 kPa, while the undisturbed strength remains essentially constant, is the most common scenario. The flocculated structure provides the framework which maintains the soil at a constant water content, and maintains its undisturbed strength, while other factors decrease the remoulded strength.

Under the conditions of sedimentation, the water content of sediments, unless very deeply buried, is commonly observed to approximate the liquid limit for the material. Soils at their liquid limit have a remoulded strength of approximately 2.5 kPa. Since the sediment is saturated and its water content cannot be increased, the liquid limit must be decreased (and the liquidity index thereby increased) if the soil, after disturbance, is to exhibit the low remoulded strength required for designation as quick clay. Removal of salts and introduction of dispersing agents, both of which can occur without affecting the *in situ* water content, are possible factors that can cause the liquid limit, and hence the remoulded strength to decrease.

14.3.1 Salt Removal

Rosenqvist (1946, 1953) noted that a common feature of the quick clays of Norway was the low pore-water salinity of these originally marine deposits. After salt addition experiments, which confirmed that the remoulded strength increased when the salt was reintroduced, he proposed that leaching of salts from the flocculated marine sediment was the cause of the low remoulded strength and the high sensitivity. Penner (1965) likewise showed that in the marine clays of eastern Canada, very high sensitivity was only observed in soils which had low salinities. The major difference between their findings was that in Norway a consistent relationship was found between sensitivity and salinity, with essentially all the leached soils exhibiting quick clay behaviour, whereas in Canada no consistent relationship was found. Torrance (1983) reviewed sensitivity–salinity relationships as presented in the literature (Tables 14.4 and 14.5). He found that marine clays that retained a pore-water salinity above 2.4 g/litre were not quick. For marine clays in which the pore-water salinity had been decreased below 2 g/litre, quick clay behaviour was commonly,

Table 14.4 Sensitivity and Strengths for High and Moderate Salinity Marine Clays. (*Reproduced by permission of Blackwell's, Oxford.*)

	Undisturbed shear strength (kPa)	Remoulded shear strength (kPa)	Sensitivity S_t	Salinity (g/l)	Reference
NORWAY					
Djuprenna Basin, Oslofjorden	41.0–0.5	6.0–0.2	2–15	30–38	Richards (1976)
North Rauöyrenra, Oslofjorden	31.0–2.4	7.8–0.2	2–21	32–39	Richards (1976)
South Dramsfjorden	14.0–0.6	1.0–0.1	6–29	23–36	Richards (1976)
Sletter St (3.9 m) depth	12.5	0.6	21	38	Richards (1976)
(4.85 m) depth	19.0	0.6	31	34	Richards (1976)
Jelöya (0.75 m) depth	14.0	0.5	28	33	Richards (1976)
Slagentangen					
(0.75 m) depth	14.5	0.7	21	34	Richards (1976)
Dramsfjorden					
(2.7 m) depth	4.9	0.19	25	29	Richards (1976)
(4.1 m) depth	8.1	0.27	30	31	Richards (1976)
(4.9 m) depth	8.6	0.4	21	33	Richards (1976)
Resedimented Norwegian Clay	31	4.8	6.4	39	Bjerrum (1954)
Drammen	43	9.2	4.7	4.0	Torrance (1974)
	47	9.8	4.8	7.8	Torrance (1974)
	26.5	4.0	6.6	27	Torrance (1974)
CANADA					
Treadwell	150	12.2	12	18	Torrance (1979)
Bourget	162	7.9	20.5	10.4	Torrance (unpublished)
Plaisance	250	12.5	20	37	Torrance (unpublished)
Fassett	160	10	16	16	Torrance (unpublished)
Hawkesbury			20	≈6	Quigley (1980)
Hawkesbury			<10	>10	Quigley (1980)
St Valliers de Bellechase			20	4	Lefebvre and LaRochelle (1974)

Table 14.5 Sensitivities and Strengths for Leached Marine Clays

	Undisturbed Shear Strength (kPa)	Remoulded Shear Strength (kPa)	Sensitivity S_t	Salinity (g/l)	Reference
NORWAY					
Drammen	27.5	0.3	92	0.25	Torrance (1974)
	24.5	0.2	122	0.2	Torrance (1974)
	18.0	0.1	180	0.45	Torrance (1974)
Resedimented Norwegian Clay	8.6	0.1	≈90	1.0	Bjerrum (1954)
CANADA					
Touraine	10.6	0.37	286	1	Torrance (1979)
Chelsea			∞(vane)	0.23	Torrance (1979)
Angers	67	0.18	370	1.5	Torrance (1979)
Beach	21	1.2	17	1.1	Torrance (unpublished)
Chapeau	162	2.1	77	3.2	Torrance (unpublished)
Touraine	82	0.12	680	1.4	Torrance (unpublished)
Casselman	111	0.92	120	3.7	Torrance (unpublished)
Quyon	162	2.5	65	2.4	Torrance (unpublished)
Quyon	83	1.8	46	1.1	Torrance (unpublished)
Rockliffe, Ottawa			35	1	Mitchell and Markell (1974)
Pineview, Ottawa			18+	≈1.6	Mitchell and Markell (1974)
Sewerline, Ottawa			50–150	≈1.4	Mitchell and Markell (1974)
Notre Dame de la Salette			10–30	≈1.6	Mitchell and Markell (1974)
South Nation River			10–100	≈1.2	Mitchell and Markell (1974)
St Louis de Bonsecours			50	0.4	Lefebvre and LaRochelle (1974)
St Jean-Vianney			200–∞	<1	LaRochelle (1974)

(Reproduced by permission of Blackwell's, Oxford)

but not always, observed. Lebuis *et al.* (1983) examined over 90 earthflows in Leda clay, for which the morphology of the crater indicated that the remoulded soil must have exhibited fluid behaviour, and found that the pore-water salinity never exceeded 3 g/litre and rarely exceeded 2 g/litre. In agreement with the observations of Penner and Torrance, high sensitivity was only observed with leached marine clays but not all the leached soils exhibited high sensitivity or failed as earthflows. Thus for the Canadian marine clays it is apparent that only leached soils exhibit quick clay behaviour but leaching does not consistently result in quick clay behaviour.

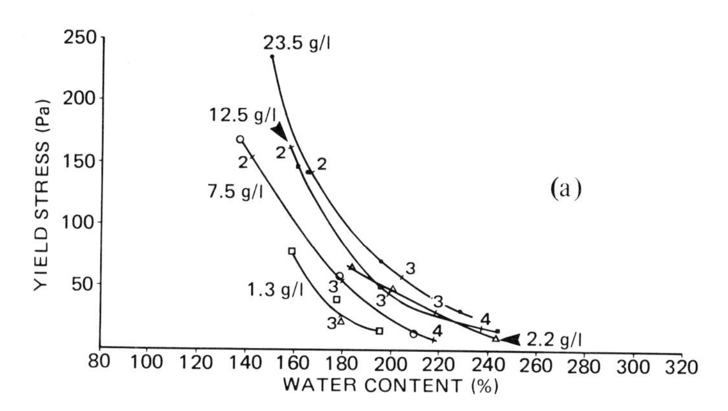

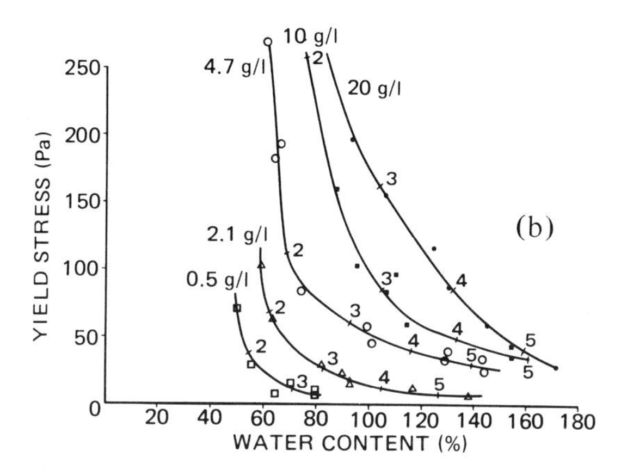

Figure 14.3 Yield stress–water content relationships for: (a) a low-activity smectite-containing, Na-saturated, marine clay from Ariake Bay, Japan; and (b) a Na-saturated Leda clay from the South Nation River landslide site. For both soils the yield stress (remoulded strength) decreases as the salinity of the pore water decreases. The numbers along each curve indicate the liquidity indices of the soils. (*From Torrance, 1984,* Soils and Foundations, **24**. *Reproduced by permission of the Japanese Society of Soil Mechanics and Foundation Engineering.*)

Salt removal also appears to be a requirement for the development of the quick clays reported from Ariake Bay, Japan. Egashira and Ohtsubo (1982) and Torrance (1984) have demonstrated that the liquid limits and flow behaviour of this material respond to salinity change in a similar manner to the sensitive Canadian marine clays.

Investigations of the influence of pore-water salinity on the yield stress of thoroughly remoulded samples of the Canadian and Japanese quick clays (Figure 14.3) have revealed a gradual decrease in the yield stress as these soils are leached from high salinities (Torrance, 1984). The field evidence, however, suggests that it is not until salinities below 2 g/litre that the repulsive forces between particles predominate and quick clay behaviour is exhibited.

Salt removal has been discussed in terms of the absolute salinity without consideration as to which ions are present. The natural material is commonly dominated by the sodium ion and most experiments have been done with sodium-saturated or sodium-dominated material. It has been demonstrated, however, that the various cations have different degrees

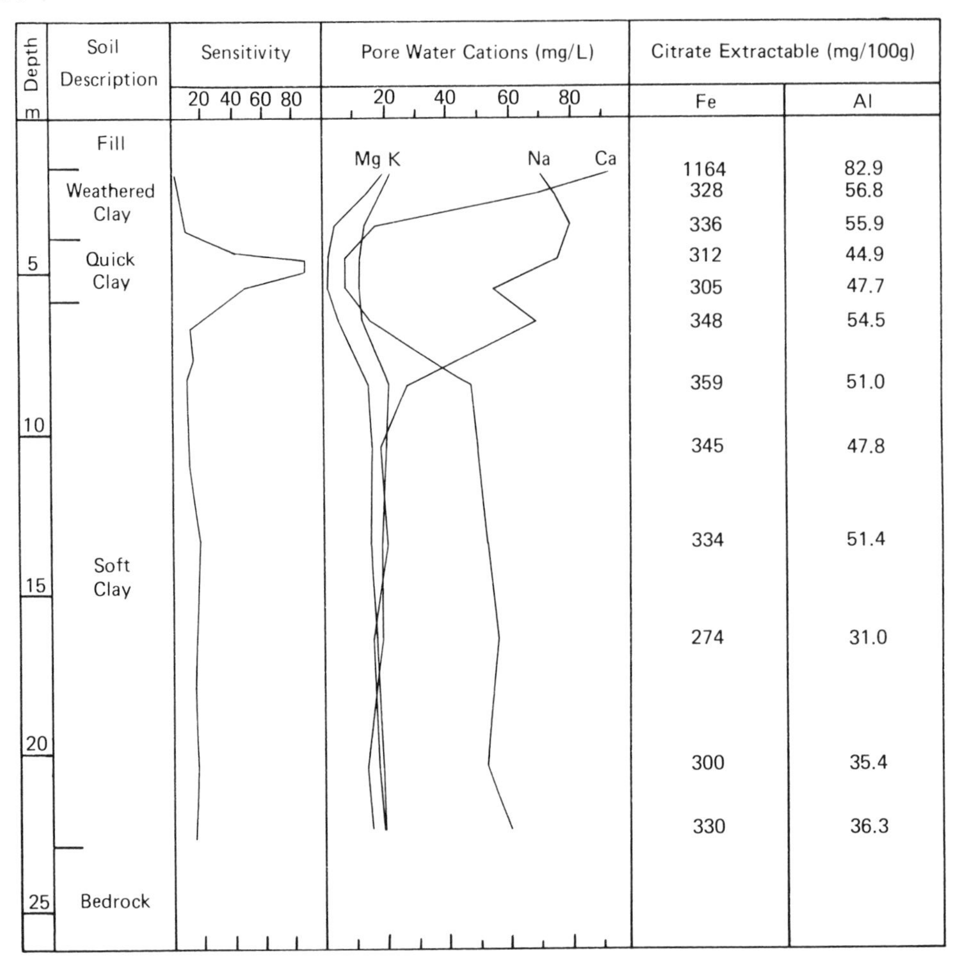

Figure 14.4 Sensitivity, pore-water chemistry, and citrate-extractable iron and aluminium in a marine clay from Drammen, Norway. (*Adapted from Moum* et al., *1971*, Geotechnique, **21**. *Reproduced by permission of the Institution of Civil Engineers.*)

of effectiveness in controlling soil behaviour. Penner (1965) recognized that quick clay behaviour did not develop in soils dominated by divalent cations and Rosenqvist (1955) recognized that potassium ion dominance led to higher remoulded strengths than if sodium dominated. Löken (1970) found that at constant water content and ionic strength the remoulded strength increased with changing ion saturation in the order $Na \leq Fe^{2+} \leq Mg \leq Ca < Fe^{3+} < K < Al$ for a Norwegian marine clay. Torrance (1975) found that the liquid limit of a Leda clay at constant salinity increased in the order $Na = Ca < K < Al$. The sensitivity exhibited by the soil is also influenced by ion saturation or pore-water composition. At Drammen, Norway (Moum *et al.*, 1971, 1972) and at Touraine, Quebec (Torrance, 1979) sensitivity reductions in part of the leached clay, between the entry point of the leaching water and the zone of maximum sensitivity, were attributed to increased pore-water concentrations of K, Ca, and Mg (Figures 14.4 and 14.5). It was suggested that these elements had been released to solution by mineral weathering.

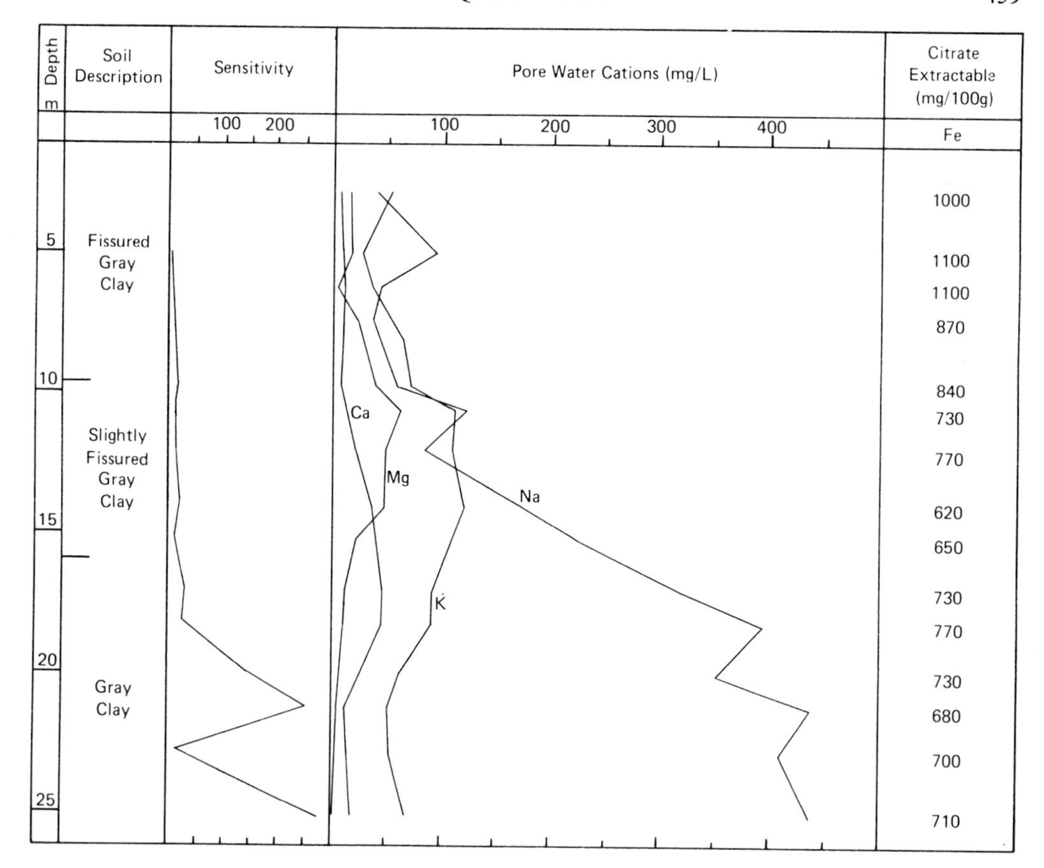

Figure 14.5 Sensitivity, pore-water chemistry and citrate-extractable iron in a Leda clay from Touraine, Quebec

Leaching of the saline pore water by the flow of fresh water through the sediment is generally viewed as the means of salt removal. The direction of leaching may be either from the surface downwards or, if artesian pressures are present, from the base of the sediment upward. It is also possible, but definitely not as common, for the salt concentration to be decreased by a diffusion process. This possibility was first brought to the author's attention in 1968 by the late J. Moum of the Norwegian Geotechnical Institute, Oslo, Norway. He attributed the salt concentration–depth profile at a site near sea level in Oslo to the diffusion of salts towards the surface where they were continuously removed from the site by lateral water flow in the near-surface zone. Torrance (1979) and Quigley *et al.* (1983) have also invoked this mechanism to explain some aspects of salt concentration profiles in the Ottawa Valley area of Ontario, Canada. Diffusion of salts will decrease the salinity much more slowly than commonly occurs with leaching. In light of the probable importance of diffusion, it is appropriate to use the generic term 'salt removal' in the general model rather than to use the more specific term 'leaching' which implies that water flow through the sediment must occur. 'Leaching' should still be used in specific cases when indeed it has been the mechanism of salt removal.

In summary, the accumulated evidence indicates that salt removal is necessary for the development of quick clay behaviour in a marine clay deposit, but that salt removal alone is not necessarily sufficient. Presumably there are factors which can block the development of the necessary high sensitivities.

14.3.2 Dispersing Agents

Dispersing agents are chemicals which decrease the attractive interactions between particles and hence decrease the tendency for soil particles to flocculate. The dispersing agents, either inorganic or organic chemicals, become associated with the particle surfaces and edges and either neutralize attractive forces or increase repulsive forces between particles.

The addition of dispersing agents to soils dominated by low activity minerals decreases the liquid limit. For undisturbed soils, in which the structure of the soil prevents volume change, this leads to a decrease in the remoulded strength and an increase in sensitivity. Penner (1965) observed that the remoulded strength of a Leda clay decreased upon addition of relatively small amounts of sodium hexametaphosphate (a well-known dispersing agent commonly used to prevent flocculation of soil particles during particle size analysis by sedimentation). In addition, Torrance (1975) has shown that low concentrations of sodium hexametaphosphate decrease the liquid limit of leached Leda clay but that higher concentrations encourage flocculation and increase the liquid limit in the same manner as other salts.

A wide range of organic chemicals present in leachate from bogs have a dispersing effect on the low-activity soil materials and are of importance to the development of high sensitivities and quick clay behaviour in some Swedish soils, particularly in some fresh water sediments (Söderblom, 1966). It is interesting that agricultural soil scientists look upon organic matter favourably, because of the better aggregation and more stable structure of the surface soil when organic matter is present, whereas soil engineers tend to look on it unfavourably, because of the decreased stability and poorer geotechnical behaviour of soil containing organic materials. Organic matter would appear to be an aggregating agent in the aerobic situations normally of concern to agriculture but a dispersing agent under the anaerobic conditions normally of concern to the engineer. Presumably the organic breakdown products formed under aerobic and anaerobic conditions have different characteristics.

Dispersing agents introduced to high salinity soils have no effect since they simply serve to increase the effective salinity and maintain flocculation. The dispersing agents also have little or no effect on the undisturbed strength of the marine clays since this is determined by the flocculated soil structure which is not broken down simply by the addition of a dispersant. If mechanical disturbance of a leached marine clay occurs, the presence of dispersing agents decreases the attractive forces between soil particles, to a greater degree than does salt removal alone, and the remoulded strength is lower and the sensitivity greater than in their absence. Thus, dispersing agents have little or no effect on marine clays that retain sufficient of their original pore-water salinity to regain the strongly flocculated state if disturbed, but may aggravate the problem of low remoulded strength and high sensitivity in the low-activity marine clays that have been leached.

The author believes that dispersants are not generally necessary for the development of quick clay behaviour in a marine clay deposit. Occasionally the decreased remoulded strength and increased sensitivity their presence provides may augment the effect of salt removal in 'borderline' cases to produce quick clay behaviour. Dispersants, however, are probably the main agents causing the development of quick clay in fresh water sediments.

14.4 OTHER FACTORS INFLUENCING SENSITIVITY

The factors discussed to this point are the ones believed to be essential to the development of quick clay behaviour in some or all cases. There exist other factors which affect either the undisturbed strength or the remoulded strength and have a bearing on the behaviour of the soil or on the possibility of quick clay development.

14.4.1 Cementation

Cementation in soil materials leads to an increase of the undisturbed strength and to the tendency for the soil to exhibit brittle behaviour. In contrast, cementation has little if any, direct effect on the remoulded strength and hence its presence would lead to an increase in the sensitivity. The magnitude of any undisturbed strength increase is presumably related to the type and amount of the cementing agent, the size, shape and mineralogy of the soil particles and the mode of action or method of emplacement of the cementing agent. That cementation is present in the quick clays is inferred from the mechanical behaviour of some of these soils.

Among the behavioural attributes that would be expected to result from cementation are: (a) an increased resistance to both shear failure and consolidation failure; and (b) a tendency for failure to occur suddenly after only small strains have occurred. The fluid behaviour of the quick clays after failure serves to increase the spectacular nature of failure in these soils and to accent the loss of strength.

Let us consider the consolidation behaviour of these soils. A sediment which has never carried a load greater than its present load is referred to as being normally-consolidated. Briefly, this implies that the soil has sufficient strength to retain its present void ratio while carrying its present load. If the load on a normally-consolidated soil is increased, by even a small amount, consolidation will commence through loss of water and increased interaction between particles until the soil gains sufficient strength to support the increased load. There is thus a relationship between the load being carried and the void ratio of the sediment (Figure 14.6(a)) which is called the virgin consolidation curve. A sediment which is at its equilibrium void ratio for the load being carried will plot on the virgin consolidation curve and has no ability to carry additional load without experiencing settlement or consolidation.

A soil material which can carry a load greater than its present load without experiencing consolidation is termed over-consolidated. Bjerrum (1967) identified four possible factors which can lead to a soil material responding in this manner: (a) unloading of the soil by erosion or by human activity, which would lead to the soil having previously carried a load greater than its present load (Figure 14.6(a)); (b) cementation, which would increase the strength of the soil after the present load has been achieved (Figure 14.6(b)); (c) delayed consolidation, which is a very slow process of consolidation which occurs under constant load conditions and makes the soil appear to have a resistance to consolidation in the short term, whereas the sediment really has a series of virgin consolidation curves depending

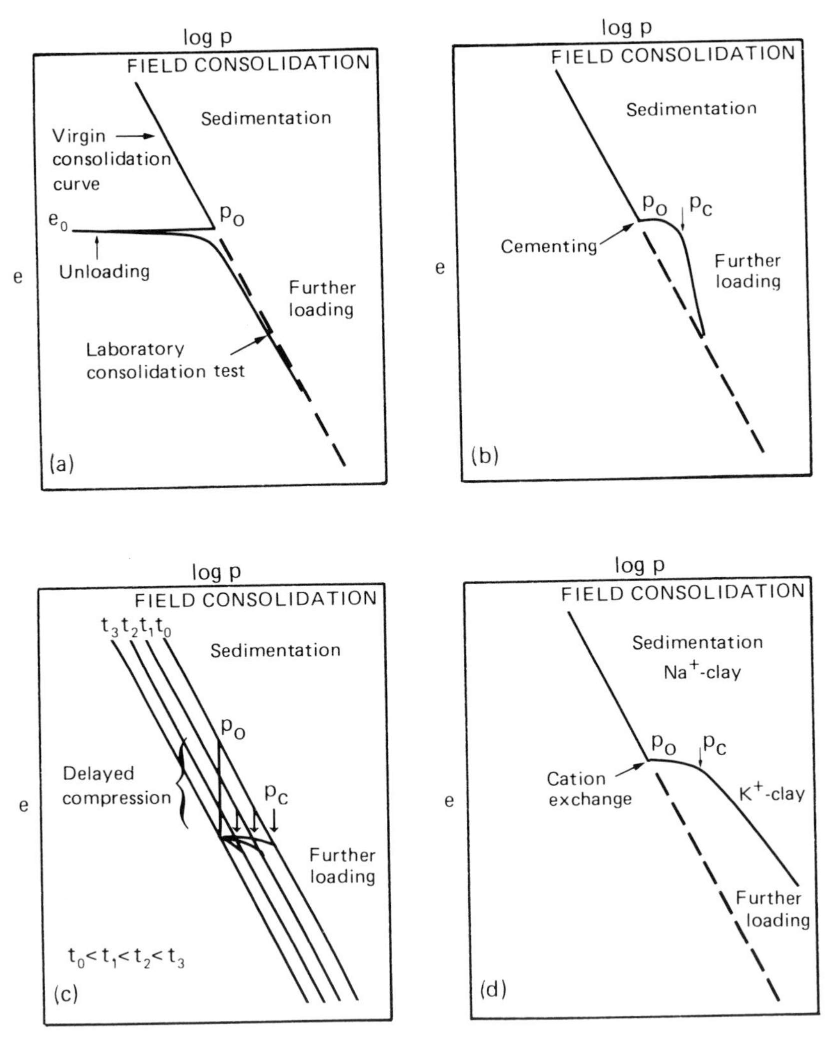

Figure 14.6 Development of pre-consolidation pressure, p_c (the ability to carry a load greater than the present load) by: (a) unloading from p_o, the *in situ* load, to some lower value. This soil can now carry a load equal to the load removed before experiencing further consolidation; (b) by cementation; (c) by delayed consolidation; and (d) by chemical change, such as weathering which may release K and lead to a K-dominated, rather than the original Na-dominated, clay. e_o and p_o represent the initial void ratio and overburden pressure, respectively. p_c represents the load that can be carried after the change has occurred. (*From Brown, 1969*, The Bolkesjö Symposium. *Reproduced by permission of the Norwegian Geotechnical Institute.*)

on the rate of load application (Figure 14.6(c)); and (d) chemical change, which can shift the virgin consolidation curve to allow greater loads to be supported (Figure 14.6(d)). This latter factor can be considered as a chemical weathering reaction.

The discussion has diverged from being solely a consideration of cementation, to encompass other factors which could lead to similar behavioural manifestations. This diversion has been to demonstrate that one must exercise caution in making inferences regarding the existence of cementation from the mechanical behaviour of soil.

A range of possible cementing agents has been proposed. These include carbonates (Townsend, 1965), hydrous and anhydrous oxides of iron and aluminium (Quigley, 1968), oxides of manganese (Loring and Nota, 1968), and amorphous materials (Yong *et al.*, 1979). The mechanisms proposed for cementation are precipitation of the cementing agent at interparticle contacts and growth in mineralogical continuity at the edges of mineral particles (Quigley, 1968). The existence of cementation in a material dominated by silt- and clay-sized particles is difficult to prove. Even the presence of material at particle contacts in electromicrographs is not conclusive proof, since the zones of particle contact would be the last places to retain water if the soil is prepared for electron microscopy by air-drying, freeze-drying or fluid displacement. The zones of particle contact would be the most probable locations for the soluble materials in the pore water to accumulate during preparation. Nevertheless, the probability of a cementing action is good, if the material accumulated at the contacts is dominated by a sparingly soluble material, such as the oxides proposed as cementing agents. The extraction of carbonates and iron compounds from undisturbed marine clay from Labrador using EDTA and comparison of before and after behaviour of otherwise similar samples suggested that cementation was present (Kenney *et al.*, 1967).

The mere presence of the minerals suggested as cementing agents should not be taken as confirmation that they are acting as cementing agents. All of these materials are commonly present in Leda clays but, as an example, much of the carbonate is associated with the $>2\,\mu m$ fraction (Bentley and Smalley, 1978) which suggests that it has limited cementing action. Investigations of the role of iron and aluminium oxides and amorphous materials have commonly involved their extraction with citrate–bicarbonate–dithionite (Mehra and Jackson, 1960) or the Segalen procedure (Segalen, 1968). Both of these procedures are sufficiently harsh to damage the layer silicates. The Segalen procedure, which was designed to dissolve the iron and aluminium oxides in oxide-rich tropical soils by alternately using strongly acidic and strongly basic extractants, is particularly aggressive and in addition to extracting the oxides and amorphous materials, decreased the content of hydrous mica and virtually eliminated chlorite from some Canadian quick clays (Figure 14.7).

It is not possible, with the current state of knowledge regarding cementation, to identify conclusively the materials involved but evidence is building that the oxides and amorphous materials are the most important. It is also not possible to estimate the proportion of the oxides and amorphous material extracted by the various procedures that are actively involved in cementation. The Norwegian quick clays, which are at most only weakly cemented (as evidenced by their exhibiting little or no apparent over-consolidation) contain these materials. In addition, no clear relationship between the amounts of oxides and amorphous materials extracted and the apparent degree of cementation has been demonstrated in the Canadian marine clays.

Cementation influences the sensitivity through increasing the undisturbed strength, but cementation acting alone cannot produce quick clay behaviour because it has little direct effect on the remoulded strength. The nature and timing of emplacement of the cementing agent is probably the important factor in determining if it plays a role in the development of quick clay. Iron and aluminium oxides in temperate region soils are normally present as surface coatings or partial surface coatings on other particles and only rarely as discrete particles (Jones and Uehara, 1973; El-Swaify and Emerson, 1974; Schwertmann and Taylor, 1977).

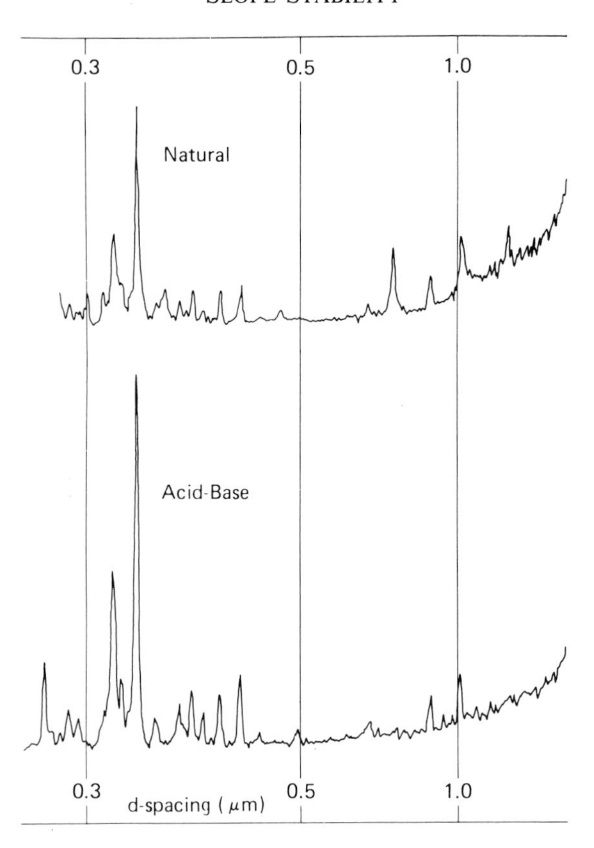

Figure 14.7 X-ray diffraction patterns for Leda clay from the South Nation River landslide site, in the natural state and after extraction by the alternating strong acid and strong base used in the Segalen procedure. Note that the acid–base extraction destroyed chlorite, as indicated by the loss of the peak at 1.4 μm, and decreased the amount of hydrous mica, as indicated by the decrease in the 1.0 μm peak relative to the other peaks

Discrete oxide particles should have little effect on quick clay behaviour, particularly if present in small amounts, but surface coatings could greatly influence the interaction of mineral particles. McKyes *et al.* (1974) report amorphous coatings on particles in sensitive clay soils. The pH of the Canadian and Scandinavian quick clays (>8), approximates the zero point of charge for many of the iron and aluminium oxides. The oxide surface coatings should bear only a small net charge (positive or negative depending on the pH and precise zero point of charge) which would definitely be less than that of the mineral it is coating. The effect would be to decrease the interparticle repulsion and increase the remoulded strength. At the low pH (4 to 5) in the Japanese quick clays, the oxides bear a net positive charge and presence of both discrete particles of oxides and surface coatings would increase the remoulded strength, with surface coatings having the greater effect.

It has been suggested (Sangrey, 1972; Bentley *et al.*, 1980) that the amorphous iron oxides are derived by the *in situ* weathering of hornblende. If this is the case, the oxide mineral content at the time of sedimentation should be low and cementation would build up slowly as the hornblende is weathered. The cementing compounds would thus have little influence on the void ratio of the sediment since the equilibrium void ratio for the overburden carried would be attained quickly in comparison to the rate of generation and emplacement of the oxides. Furthermore, the generation of oxide cements in a marine sediment should be very slow since weathering is inhibited until the pore-water salinity has been reduced to low concentrations (Shainberg, 1973).

Cementation should be most effective if the cementing agent becomes associated with the particles at the time of sedimentation. Oxide coatings on the particles at the time of flocculation would increase interparticle attraction, but would cement only weakly unless additional oxide accumulated or the existing films were redistributed. The latter possibility seems unlikely to occur quickly. None the less, if there exists some mechanism by which cementation can develop rapidly, during or shortly after sedimentation, to the point where the sediment can carry its weight, and that of additional sediment, without further consolidation the final void ratio will be higher than for a normally-consolidated uncemented sediment and the sensitivity would also be higher, both before and after leaching. Rapid cementation would increase the probability of a quick clay being developed upon leaching since it would lead to higher water contents and reduced remoulded strength.

The geotechnical behaviour of the Canadian quick clays strongly suggests that cementing agents are present and that the amount of cementing agent varies from very small quantities to quantities that have a major effect on the undisturbed strength of the soil. The degree of cementation appears to be the major difference between the Norwegian marine clays and the Leda clays and has led some authors to claim that the reasons for high sensitivity are different in the two materials. While this claim may be logically correct, it is perhaps more correct and useful to claim that the reasons for quick clay development appear to be the same but that the relative importance of the various factors is different. Cementation generally plays a more important role in the Leda clays. Since the remoulded strength for designation as a quick clay must not exceed 0.5 kPa, it seems unlikely that cementation alone can lead to quick clay behaviour in marine clays. Only in a few cases, where the undisturbed shear strength might otherwise be too low for the sensitivity to exceed 30 after salt removal, might cementation be considered essential to quick clay development, but even in these cases it plays a secondary role to salt removal.

The role of cementation, its very existence, the agents involved and how they act in the quick clays are being increasingly investigated by the geotechnical community, but it is still too early in the research for definitive statements to be made. Clearly cementation is important when present, but how and when it occurs and what agents are involved must still be considered areas of uncertainty.

14.4.2 Weathering

In the broadest context (see Chapter 13), weathering can be considered as any change that occurs in the soil material after its deposition. In this context, leaching and introduction of organic dispersing agents, both of which increase sensitivity by decreasing the remoulded strength, can be considered processes of weathering. It is normal, when considering quick clays, to separate these two processes from the rest of the weathering processes—including desiccation, oxidation, and mineral weathering—since the available evidence suggests that these other processes generally lead to an increase in the remoulded strength and a decreased sensitivity.

Figure 14.8 'Nodular' structure in the weathering crust of Leda clay from Gatineau, Quebec. The spatula blade is 2.5 cm across

The greatest degree of weathering has occurred in the surface zones. Here, desiccation and contact with the atmosphere and oxygen-bearing groundwaters have led to the development of a 'weathering crust' which has higher undisturbed and remoulded strengths and a lower sensitivity than the underlying sediment. Geotechnical profiles and chemical profiles for Drammen, Norway (Moum *et al.*, 1971) and Touraine, Quebec (Torrance, 1979) were presented in Figures 14.4 and 14.5. These soils have been subjected to desiccation in the near surface, to increased pore-water suction to some greater depth, and to various reactions involved in soil profile development including oxidation, hydrolysis, and the other reactions of chemical weathering (Buol *et al.*, 1980). A soil profile with A, B, and C horizons has developed in the upper metre but the influence of agents introduced at the surface is felt to greater depths. These influences are apparent from the changes in structure, citrate–dithionite-extractable cations, and pore-water cations.

The structural change is manifested in the development of a 'nodular' structure from the original massive sediment (Figure 14.8). The term 'nodular' was introduced by Eden and Mitchell (1970) to describe the small, roughly equidimensional blocks which are produced by the development of a relatively closely spaced fissure system in the near surface zone. The depth to which nodularity has developed varies from one to several metres, with the greater depths being observed near stream valleys. The presence of the valley leads to a lower water table, greater pore-water suctions, and weathering to greater depths in the immediately adjacent uplands. From the author's admittedly limited observations, nodularity is more strongly developed in the Canadian than in the Scandinavian marine clays. When slope failure occurs in a marine clay with nodular structure, the blocks remain intact and the failure plane follows the fissures. Some relatively small failures on the long-exposed terraces of the pre-Ottawa River, near Ottawa, Canada have occurred entirely within material with nodular structure.

The quantities of citrate–dithionite-extractable iron and aluminium are greatest in the near-surface zones (Figures 14.4 and 14.5). At Touraine the quantity of citrate–dithionite-extractable iron is greatest in the upper 10–15 metres where the intensity of development of fissuring and nodularity is greatest. The citrate-extractable iron and aluminium are also greatest in the surface zone at Drammen. Those oxides which were not present in the original sediment must be produced by the weathering of iron-bearing minerals and aluminosilicates. Along with desiccation, the oxides account for the higher undisturbed and remoulded strengths and the lower sensitivity in this weathered surface zone. In addition, as discussed in the section on material properties, the citrate–dithionite-extractable materials also contribute to higher liquid and plastic limits for these soils dominated by low-activity minerals.

The pore-water chemistry at both the Drammen and Touraine sites reveals other evidence of weathering, some of which has geotechnical implications. The hydrological evidence, namely an artesian pressure, indicates that water flow has been upward in the Drammen soil, except in the surface crust, whereas at Touraine groundwater flow has been downward. The Na concentration in the pore-water, which is residual from the original near-marine pore water and decreases with depth below 5 m at Drammen but increases with depth at Touraine, is consistent with this water flow direction. The concentrations of K, Mg, and Ca, for which there are weatherable mineral sources (micas, feldspars, chlorites, and carbonates) follow different patterns. Below about 7 m at Drammen and somewhere between 10 and 20 m depth at Touraine (depending on the element) the pore-water concentrations of these elements exceed the concentrations in the most leached zones. Since water flows from these zones of higher concentration towards the most leached zones, this concentration pattern is interpreted to indicate that the pore-water concentrations of these elements, at all depths between the entry point of the leaching water and the depth of current lowest concentration, have at some time in the past been reduced to values comparable to the lowest currently found and have subsequently been increased by weathering of K-, Mg-, and Ca-bearing minerals (Moum *et al.*, 1971; Torrance, 1979). The lowest remoulded strengths and highest sensitivities occur in the zones of lowest concentration of K-, Mg, and Ca. Where these concentrations have been increased by weathering their combined concentration is sufficient to affect the behaviour of the remoulded soil. Note that these results confirm that in the natural setting, the Na ion is less effective in increasing the remoulded strength than are K and the divalent cations.

Weathering also leads to changes in the clay mineralogy. Small quantities of swelling clay minerals including interlayered illite–vermiculite, interlayered chlorite–vermiculite, vermiculite and smectite have been reported in the Canadian marine clays (Brydon and Patry, 1961; Gillott, 1979; Bentley and Smalley, 1979; Quigley *et al.*, 1983). Small quantities of these swelling clay minerals may have been present at the time of sedimentation, but most investigators suggest that they have been produced by post-depositional weathering. Since these swelling clays are present in soils which exhibit quick clay behaviour, the amounts present are clearly below the threshold required to prevent quick clay behaviour (Torrance, 1970; Yong *et al.*, 1979). The presence of large amounts (37 to 50%) of a low-activity, non-swelling, high-iron content smectite as the dominant clay mineral in the marine clays reported from Ariake Bay, Japan (Egashira and Ohtsubo, 1983) indicates that not all smectites are inimical to the development of high sensitivity and quick clay behaviour.

Weathering of mineral materials is suppressed under high salinity conditions (Shainberg, 1973), although such reactions as bacterial reduction of iron compounds under anaerobic conditions to produce the iron sulphide reported by Haynes and Quigley (1978) can occur. It seems probable that relatively little weathering of the aluminosilicates in the sensitive marine clays occurs until the salinity has been reduced to low values. As already discussed this has important implications for the post-depositional production of cementing agents.

In summary, the weathering of the marine clays prevents the development of quick clay behaviour in the weathering crust and decreases the sensitivity in zones where leaching has been sufficient to produce quick clay. In terms of geotechnical behaviour, particularly with regard to decreasing the tendency for large flow slides, weathering would appear to be a positive factor.

14.4.3 Minimal Consolidation and Slow Load Increase

The role of minimal consolidation in influencing the remoulded strength and hence the sensitivity is obvious. The less post-depositional consolidation that occurs the higher will be the water content and the greater the probability that, after salt removal, the water content will sufficiently exceed the liquid limit for fluid behaviour to be exhibited upon remoulding. The flocculated structure acts to resist consolidation, and both relatively shallow burial (which leads to low overburden pressure with commensurate consolidation) and rapid development of cementation after sedimentation (but before the full depth of sediment or complete consolidation in response to the load has occurred) would decrease the remoulded strength and augment the sensitivity compared with a more heavily loaded or uncemented sediment.

The water content of the sediment is observed to decrease with depth below the surface, unless it has been rapidly and strongly cemented. There presumably exists some depth at which the overburden pressure causes sufficient consolidation that the water content no longer exceeds the liquid limit of the soil after salt removal by a sufficient amount for the remoulded strength requirement of quick clay (<0.5 kPa) to be met. The maximum depth of burial consistent with this requirement can be expected to vary from site to site depending on local conditions—including sediment composition, sedimentation conditions, extent of cementation and other factors which affect the undisturbed strength and hence the final water content.

Slow load increase, which implies a slow rate of sediment accumulation, minimizes the amount of consolidation at the final overburden pressure by increasing the time, at any void ratio, for cementation to occur. In addition, any thixotropic processes, which can be defined as any process causing a time dependent increase in strength, will have greater time to act. The importance of thixotropic processes is unknown, but if a quick clay is remoulded these processes lead to only a relatively small amount of strength regain

(Mitchell, 1960). It seems probable that, if cementation is excluded, they also have a relatively small effect on the undisturbed strength and hence on the final water content and sensitivity. Regardless, with slower load increase the final water content will be higher and the probability of quick clay behaviour after salt removal will be enhanced.

14.5 SCENARIO AND IMPLICATIONS

Quick clays are formed in fine-grained marine deposits dominated by low activity minerals through the action of salt removal (generally by leaching). The three factors—low-activity minerals, flocculated structure, and salt removal—appear to be essential to the development of quick clay behaviour with its characteristic high sensitivity and fluid behaviour of the remoulded material. None of the other factors which increase the sensitivity are essential in all cases, although they may be important in some 'borderline' situations.

The common scenario for quick clay development in a marine clay deposit appears to be roughly as follows (after Bjerrum *et al.*, 1969). Glacially ground rock material is carried by meltwater streams into a marine basin where the silt and clay particles flocculate and accumulate to form a high-water content sediment. Isostatic rebound of the earth's crust, in response to the decreased load following glacial retreat, raises the surface of these sediments above sea level. Fresh water may now enter the sediment and leaching of the salts from the sediment commences. When the pore-water salt concentration has been lowered to the order of 2 g/litre, the sensitivity and remoulded strength requirements for designation as quick clay may be met. A surface drainage network is established, simultaneously with leaching, as streams erode into the old sea bottom surface. Weathering of the near-surface zone also commences and increases its strength. None the less, as the streams erode their channels more deeply into the sediment the chances of slope failure increase (see Chapter 10).

The most common cause of slope failure in the marine clays is the undercutting of the slope by the river to the degree that conditions are favourable for a simple rotational landslide. Many of these occur every year. Most of these rotational landslides are small and many do not involve even the full height of the slope. The undisturbed strength determines the stability of the original slope. Occasionally one of these small slides serves as the initial or 'triggering' event for a major quick clay landslide. If the remoulded strength of the failed soil is so low that the debris of the initial slide does not remain in the landslide scar to provide stability to the backscarp (or is inadequate for this purpose), the backscarp in turn is unstable and fails. Thus begins a retrogressive landslide which may become very large.

Inasmuch as the stability of the original slope is determined by the undisturbed strength, the presence of cementation will be a factor. The angle and height at which slopes are stable increase with increased cementation. The remoulded strength of the initial slide debris determines whether support is provided to the backscarp. If the initial slide occurs entirely within the weathering crust, the debris has a nodular character and normally remains at the base of the slope to provide support. On the other hand, if the slide extends sufficiently deeply into the slope to involve quick clay, the quick clay with its fluid character after failure, can serve as a lubricant or agent which transports the debris away from the backscarp and leaves an unprotected slope. If the exposed backscarp contains quick clay, and fails in turn, the stage is set for a major retrogressive landslide in which the sequence of failure, liquefaction, and flow of the debris, followed by another failure and flow, continues until a situation is reached where the backslope is stable. This sequence can lead to very large landslides, such as the 26 hectare landslide at St Jean Vianney (Tavenas *et al.*, 1971), the 28 hectare landslide on the South Nation River (Figure 14.1, Eden *et al.*, 1971) or the approximately 20 square kilometre landslide that occurred about 500 years ago at the St Jean Vianney site (Lasalle and Chagnon, 1968).

In addition to the destruction of the landslide area, there may be damage both downstream and upstream from the site. The debris which flows downstream may endanger or destroy structures, such as the bridge destroyed by the St Jean Vianney debris (La Rochelle *et al.*, 1971). Upstream damage is generally caused by flooding, with the most severe recorded case being a landslide, in 1795, at Tesen, Nes in Norway, which dammed the Vorma River for 111 days and caused an 8 metre rise in the level of Lake Myosa (Löken *et al.*, 1970).

Quick clay is undoubtedly involved in all Type III and IV landslides (Table 14.1). At least a thin zone of quick clay is probably present in most Type V events in marine clays, with the failure plane occurring within the quick clay zone and the 'flake' riding on top of this fluid material. Most Type I landslides probably involve quick clay, with exceptions for some of those with low R/H values. The Type II landslides also generally involve quick clay, with the aborted retrogression occurring either because the weathered zone on the valley sides is thick and the debris from this weathered zone inhibits (but does not completely prevent) subsequent sliding, or because only a small zone of quick clay is involved. The Type 0 failures do not involve quick clay and most are restricted to the weathering crust on the valley sides.

The current state of engineering knowledge is such that the stability of slopes in the sensitive marine clays of Canada and Scandinavia can be reliably predicted. Geotechnical investigations are required, with appropriate remedial measures being taken when necessary, before construction activity is undertaken near slopes in areas of marine clay. Unfortunately, the number of potentially unstable slopes is great and it is not feasible to conduct detailed slope stability investigations throughout the marine clay areas. In the regional context, general guidelines with regard to the slope angles and slope heights that represent potential risks can be developed. These will vary with location but must, in all cases, err on the side of safety. A major unsolved problem is the prediction of the extent of retrogression to be expected if a landslide occurs in a quick clay. Past landslide activity in the nearby area, as assessed from aerial photographs or field survey, currently provides the best evidence.

The presence of quick clay has important implications for slope stability and landslide geomorphology. Knowledge of quick clay behaviour is critical to sensible development and land use in areas where it occurs. Land use regulations must take its existence into account and while it might be desirable to avoid all construction on quick clays, the large areas involved and their location make this impossible. We must continue to improve our ability to predict problems in these soils.

REFERENCES

Bentley, S. P., Clark, N. J., and Smalley, I. J. (1980). 'Mineralogy of a Norwegian postglacial clay and some geotechnical implications.' *Can. Mineralogist*, **18**, 535–47.

Bentley, S. P., and Smalley, I. J. (1978). 'Inter-particle cementation in Canadian post-glacial clays and the problem of high sensitivity ($S_t > 50$).' *Sedimentology*, **25**, 297–302.
Bentley, S. P., and Smalley, I. J. (1979). 'Mineralogy of a Leda/Champlain clay from Gloucester (Ottawa, Ontario).' *Eng. Geol.*, **14**, 209–17.
Bjerrum, L. (1954). 'Geotechnical properties of Norwegian marine clays.' *Géotechnique*, **4**, 46–69.
Bjerrum, L. (1967). 'Engineering geology of Norwegian normally-consolidated marine clays as related to settlement of buildings.' *Géotechnique*, **17**, 81–118.
Bjerrum, L., Löken, T., Heiberg, S., and Foster, R. (1969). 'A field study of factors responsible for quick clay slides.' *Int. Conf. Soil Mech. Fndtn Engng.*, Mexico City, **1**, 531–40.
Brown, J. D. (1969). 'Description of normally consolidated marine clays.' *The Bolkesjö Symposium on Shear Strength and Consolidation of Normally Consolidated Clays*, Norwegian Geotechnical Institute, Oslo, 20–2.
Brydon, J. E., and Patry, L. M. (1961). 'Mineralogy of Champlain Sea sediments and a Rideau Clay soil profile.' *Can. J. Soil Sci.*, **41**, 169–81.
Buol, S. W., Hole, F. D., and McCracken, R. J. (1980). *Soil Genesis and Classification* (2nd edn). Iowa State University Press, Ames, 406pp.
Cabrera, J. G., and Smalley, I. J. (1973). 'Quick clays as products of glacial action: a new approach to their nature, geology, distribution and geotechnical properties.' *Eng. Geol.*, **7**, 115–33.
Carson, M. A., and Lajoie, G. (1981). 'Some constraints on the severity of landslide penetration in sensitive deposits.' *Geog. Phys. et Quat.*, **XXXV**, 301–16.
Delage, P., and Lefebvre, G. (1984). 'Study of the structure of a sensitive Champlain clay and of its evolution during consolidation.' *Can. Geotech. J.*, **21**, 21–35.
Dixon, D. A. (1982). *Geochemical influences on the rheological properties of four post-glacial marine clays.* M. A. Thesis, Carleton University, Ottawa, 141pp.
Eden, W. J., Fletcher, E. B., and Mitchell, R. J. (1971). 'South Nation River landslide, 16 May 1971.' *Can. Geotech. J.*, **8**, 446–51.
Eden, W. J., and Mitchell, R. J. (1970). 'The mechanics of landslides in Leda clay.' *Can. Geotech. J.*, **7**, 285–96.
Egashira, K., and Ohtsubo, M. (1982). 'Smectite in marine quick-clays of Japan.' *Clays Clay Min.*, **30**, 275–80.
Egashira, K., and Ohtsubo, M. (1983). 'Swelling and mineralogy of smectites in paddy soils derived from marine alluvium, Japan.' *Geoderma*, **29**, 119–27.
El-Swaify, S. A., and Emerson, W. W. (1974). 'Changes in the physical properties of soil clays due to precipitated aluminium and iron hydroxides: I. Swelling and aggregate stability after drying.' *Soil Sci. Soc. Am. Proc.*, **39**, 1056–63.
Gillott, J. E. (1979). 'Fabric, composition and properties of sensitive soils in Canada, Alaska and Norway.' *Eng. Geol.*, **14**, 149–72.
Haynes, J. E., and Quigley, R. M. (1978). 'Framboids in Champlain Sea sediments.' *Can. J. Earth Sci.*, **15**, 464–5.
Hendershot, W. H., and Carson, M. A. (1978). 'Changes in the plasticity of a sample of Champlain clay after selective chemical dissolution of amorphous material.' *Can. Geotech. J.*, **15**, 609–16.
Jones, R. C., and Uehara, G. (1973). 'Amorphous coatings on mineral surfaces.' *Soil Sci. Soc. Am. Proc.*, **37**, 792–8.
Kenney, T. C., Moum, J., and Berre, T. (1967). 'An experimental study of bonds in natural clay.' *Proc. Geotech. Conf.*, **1**, Oslo, 65–9.
LaRochelle, P. (1974). *Rapport de synthèse des études de la coulée d'argile de Saint-Jean-Vianney.* Ministère des Richesses Naturelles, Gouvernment de Québec.
Lasalle, P., and Chagnon, J.-Y. (1968). 'An ancient landslide along the Saguenay River, Quebec.' *Can. J. Earth Sci.*, **5**, 548–9.
Lebuis, J., Robert, R. M., and Rissman, P. (1983). 'Regional mapping of landslide hazard in Quebec.' *Symp. Slopes on Soft Clays*, Report No. 17, Swedish Geotechnical Institute, 205–62.
Lefebvre, G., and LaRochelle, P. (1974). 'The analysis of two slope failures in cemented Champlain clays.' *Can. Geotech. J.*, **11**, 89–108.
Löken, T. (1970). 'Recent research at the Norwegian Geotechnical Institute concerning the influence of geochemical additions on quick clay.' *Geol. Foren. i Stockholm Forhand.*, **92**(2), 133–47.

Löken, T., Jorstad, F., and Heiberg, S. (1970). 'Gamle leirskred paa Romerike.' *Saeravtrykk av Romerike Historielags Aarbok VII 1970.*

Loring, D. H., and Nota, D. J. G. (1968). 'Occurrence and significance of iron, manganese and titanium in glacial marine sediments from the estuary of the St Lawrence River.' *J. Fish. Res. Board Can.*, **25**, 2327–47.

McKyes, E., Sethi, A. J., and Yong, R. N. (1974). 'Amorphous coatings on particles of sensitive clay soils.' *Clays Clay Min.*, **22**, 427–33.

Mehra, O. P., and Jackson, M. L. (1960). 'Iron oxide removal from soils and clays by a dithionite-citrate system buffered by sodium bicarbonate.' *Proc. Nat. Conf. Clays Clay Min.*, Washington, DC, 317–27.

Mitchell, J. K. (1960). 'Fundamental aspects of thixotropy in soils.' *Soil Mech. Fndtn Engng. Div., Am. Soc. Civil Eng.*, **86**, (SM3), 19–52.

Mitchell, R. J., and Markell, A. R. (1974). 'Flowsliding in sensitive soils.' *Can. Geotech. J.,* **11**, 11–31.

Moum, J., Löken, T., and Torrance, J. K. (1971). 'A geochemical investigation of the sensitivity of a normally consolidated clay from Drammen, Norway.' *Géotechnique*, **21**, 329–40.

Moum, J., Löken, T., and Torrance, J. K. (1972). 'A geochemical investigation of a normally consolidated clay from Drammen, Norway.' *Géotechnique*, **22**, 675–6.

Moum, J., and Rosenqvist, I.Th. (1961). 'The mechanical properties of montmorillonitic and illitic clays related to the electrolytes of the pore water.' *Proc. 5th Int. Conf. Soil Mech. Fndtn Engng.*, **I**, 263–7.

Norsk Geoteknisk Forening (1974). *Retningslinjer for presentasjon av geotekniske undersökelser*, Oslo, 16pp.

Ohtsubo, M., Takayama, M., and Egashira, K. (1982). 'Marine quick clays from Ariake Bay, Japan.' *Soils and Fndtns*, **22**, 71–80.

Penner, E. (1965). 'A study of sensitivity in Leda clay.' *Can. J. Earth Sci.*, **2**, 425–41.

Quigley, R. M. (1968). 'Discussion of landslide on the Toulnustouc River, Quebec.' *Can. Geotech. J.*, **5**, 175–7.

Quigley, R. M. (1980). 'Geology, mineralogy and geochemistry of Canadian soft soils: a geotechnical perspective.' *Can. Geotech. J.*, **17**, 261–85.

Quigley, R. M., Gwyn, Q. H. J., White, O. L., Rowe, R. K., Haynes, J. E., and Bohdanowiscz, A. (1983). 'Leda clay from deep boreholes at Hawkesbury, Ontario. Part I: Geology and Geotechnique.' *Can. Geotech. J.,* **20**, 288–98.

Richards, A. F. (1976). 'Marine geotechnics of the Oslofjorden region.' *Laurits Bjerrum Memorial Volume*, Norwegian Geotechnical Institute, Oslo.

Roaldset, E. (1972). 'Mineralogy and geochemistry of Quaternary clays in the Numedal area, Southern Norway.' *Norges Geol. Tidskr.*, **52**, 335–69.

Rosenqvist, I.Th. (1946). 'Om leires kvikkaktighet.' *Statens vegvesen Veglaboratoriet Medd.*, **4**, 5–12.

Rosenqvist, I.Th. (1953). 'Considerations on the sensitivity of Norwegian quick-clays.' *Géotechnique*, **3**, 195–200.

Rosenqvist, I.Th. (1955). *Investigations into the Clay–Electrolyte–Water System.* Norwegian Geotechnical Institute Publication No. 9, 125pp.

Rosenqvist, I.Th. (1977). 'A general theory for quick clay properties.' *Proc. 3rd Eur. Clay Conf.*, Oslo, 215–28.

Sangrey, D. A. (1972). 'Naturally cemented sensitive soils.' *Géotechnique*, **22**, 139–52.

Schwertmann, U., and Taylor, R. M. (1977). 'Iron oxides.' In *Minerals in Soil Environments.* Dixon, J. B., and Weed, S. B. (eds.), *Soil Sci. Soc. Am.*, pp. 145–80.

Segalen, P. (1968). 'Note sur une méthode de détermination des produits minéraux amorphes dans certaine sols a hydroxides tropicaux.' *Cah. ORSTOM, Ser. Pedol.*, **6**, 105–26.

Smalley, I. J. (1971). 'Nature of quick clays.' *Nature*, **231**, 310.

Smalley, I. J., Fordham, C. J., and Callendar, P. F. (1984). 'Discussion.' *Sedimentology*, **31**, 595–6.

Shainberg, I. (1973). 'Rate and mechanism of Na-montmorillonite hydrolysis in suspensions.' *Soil Sci. Soc. Am. Proc.*, **37**, 689–94.

Skempton, A. W. (1953). 'The colloidal activity of clay.' *Proc. 3rd Int. Conf. Soil Mech. Fndtn Engng.*, **1**, 57–61.

Soderblom, R. (1966). 'Chemical aspects of quick clay formation.' *Eng. Geol.*, **1**, 415–31.

Tavenas, F., Chagnon, J.-Y., and LaRochelle, P. (1971). 'The Saint-Jean-Vianney landslide: observations and eyewitness accounts.' *Can. Geotech. J.*, **8**, 463–78.

Torrance, J. K. (1970). 'Discussion.' *Eng. Geol.*, **4**, 353–8.

Torrance, J. K. (1974). 'A laboratory investigation of the effect of leaching on the compressibility and shear strength of Norwegian marine clays.' *Géotechnique*, **24**, 155–73.

Torrance, J. K. (1975). 'On the role of chemistry in the development and behaviour of the sensitive marine clays of Canada and Scandinavian.' *Can. Geotech. J.*, **12**, 326–35.

Torrance, J. K. (1979). 'Post-depositional changes in the pore water chemistry of the sensitive marine clays of the Ottawa area, Eastern Canada.' *Eng. Geol.*, **14**, 135–47.

Torrance, J. K. (1983). 'Towards a general model of quick clay development.' *Sedimentology*, **30**, 547–55.

Torrance, J. K. (1984). 'A comparison of marine clays from Ariake Bay, Japan and the South Nation River landslide site, Canada.' *Soils and Fndtns*, **24**, 75–81.

Townsend, D. L. (1965). 'Discussion on geotechnical properties of three Ontario clays.' *Can. Geotech. J.*, **2**, 190–3.

Yariv, S., and Cross, H. (1979). *Geochemistry of Colloid Systems*. Springer-Verlag, Berlin, 450pp.

Yong, R., Sethi, A. J., and LaRochelle, P. (1979). 'Significance of amorphous material relative to sensitivity in some Champlain clays.' *Can. Geotech. J.*, **16**, 511–20.

Slope Stability
Edited by M. G. Anderson and K. S. Richards

Chapter 15

Rock Slopes

M. J. SELBY
Department of Earth Sciences
University of Waikato, Hamilton, New Zealand

15.1 INTRODUCTION

In the introductory chapter to his book *Rock Control in Geomorphology*, Eiju Yatsu (1966, p. 9) remarked that many students of landforms use the phrase 'rock control' as a magic cloak, simply explaining landforms in relation to the differences in rocks: . . . 'this part remains not eroded because of hardness, that part is hard because it remains protrusive . . .'. The circular reasoning he was criticizing is evident, but in the intervening 20 years little attention has been paid to rock control by geomorphologists, and their concern has remained at the level of descriptive analysis of rock structures and the influence of structure on landforms.

Some part of the failure to study rock control is, no doubt, a tacit recognition of the inappropriateness of conventional geological descriptions and classifications as a basis for understanding rock strength. The geologist is largely concerned with mineralogical and petrographic factors as a basis of classification. Rock strength, however, is concerned primarily with the hardness of minerals, porosity and density of the rock, the bond adhesive strength between particles and across pores, the size of particles, the degree of anisotropy of mineral and grain orientations, the degree of interlocking of particles, flaws within the rock and its water content. Water content has often been ignored by geomorphologists but there is good evidence that in addition to pore pressures, the effect of pore fluid may reduce the uniaxial compressive strength of an oven-dry sample (Colback and Wiid, 1965; Broch, 1974).

An approach to understanding rock strength and other properties is to use the knowledge which has become increasingly formulated and systematized within the disciplines of rock and soil mechanics, known collectively as geomechanics. Few geomorphologists have, as yet, adopted this approach, but the recognition that it is the strength of rock masses which largely controls the form of weathering-limited slopes is inevitably drawing a few earth scientists into geomechanics.

As an introduction it is essential to make the distinction between intact rock and a rock mass. Intact rock is that body of rock material which has no major continuous fissures through it. The properties that influence its geomorphic behaviour under stress are those described in the second paragraph section 15.1. A rock mass is a body of rock containing continuous partings such as joints, faults, and planes of schistosity: the strength of rock masses is largely controlled by the shear strength available along the partings.

Recognition that it is the *major* discontinuities in rock masses which largely control their mechanical behaviour developed slowly during the 1950s and was forced upon the attention of civil engineers and engineering geologists by the failures of the Malpasset (1959) and Vaiont (1963) dams. The year 1966 may be regarded as the date of the formal foundation of rock mechanics with the first congress of the International Society of Rock Mechanics.

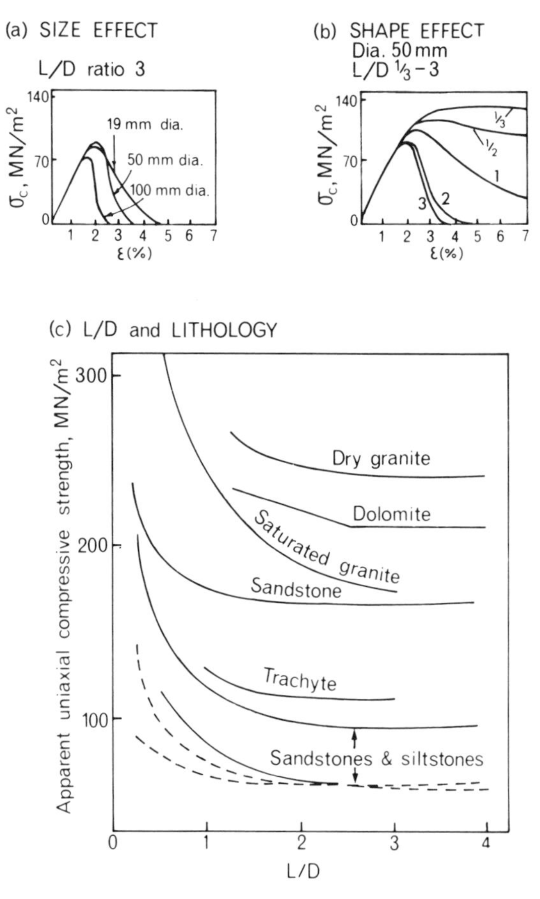

Figure 15.1 (a)(b) The influence of specimen size and shape on the deformation of a cylindrical specimen of marble (σ_c is stress and ϵ is strain). (*After Hudson* et al., *1971*, Proc. 13th US Rock Mech. Symp. *Reproduced by permission of the American Society of Civil Engineers*.) (c) The relationship between length/diameter of samples under uniaxial compressive stress for various lithologies. (*After Hawkes and Mellor, 1970*, Eng. Geol., **4**. *Reproduced by permission of Elsevier Science Publishers*.)

15.2 MECHANICAL PROPERTIES OF INTACT ROCK

In the middle to late nineteenth century, Cauchy introduced the concept of stress, and Navier that of strain, to form the basis of the current theory of elasticity which is widely used in geomechanics. In this theory, material is assumed to be a continuum made up of an aggregate of uniform elements. In the early twentieth century the atomic theory of matter developed, with the advent of X-ray technology by Roentgen and von Laue, with the theoretical work of Planck and Einstein, and the supporting evidence provided by Rutherford and Born. By the 1930s it was widely accepted that inorganic matter could be understood by representing it as being composed of quasi-spherical atoms arranged in regular geometrical units. Experiments on crystalline materials, however, showed that their tensile strength is several orders of magnitude less than the ideal strength predicted from atomic theory. It was concluded by Griffith (1921, 1925) that the real strength of crystalline materials is governed by the presence of imperfections such as cracks, grain boundaries and dislocations, and that the propagation of cracks forms the basis of an understanding of failure of intact materials such as glass.

In geomechanics the strength of intact rock is usually specified in terms of its uniaxial compressive strength. Users of such data must be aware that the properties of rocks determined in laboratory tests are functions of the true nature of the test material, the dimensions of the specimen and its representativeness, and also functions of the testing method and testing machine. The effects of specimen size and testing machine characteristics are well recognized. Stress–strain curves are similar in form for specimens larger than 50 mm in diameter for most intact rocks, but markedly different for sizes below that diameter (Figure 15.1). For samples with length/diameter ratios less than 2.0 there are large differences in the shape of the curves and a ratio of 2.5–3.0 is recommended (Brown, 1981) if experimental errors are to be avoided. In both loading and unloading the specimen the testing machine should have a stiffness which exceeds that of the specimen. In the grossly exaggerated example of Figure 15.2, it is evident that the testing machine's frame is distorting and storing elastic strain energy. At specimen failure this energy is released explosively; the platens of the machine move rapidly towards one another and the sample is shattered.

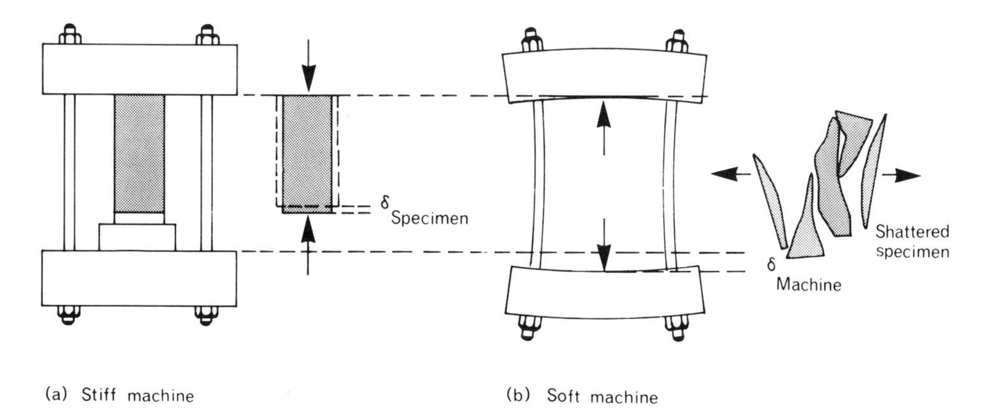

Figure 15.2 Cartoon of the results of uniaxially compressing a rock cylinder in a soft and in a stiff machine. (*In part, after Bock and Wallace, 1978, in* An Introduction to Rock Mechanics. *Reproduced by permission of Dr H. Bock and James Cook University, Australia.*)

Such catastrophic failure is rare in nature and simulation is not achieved in the test; furthermore, the information which could be derived from a post-failure stress–strain curve is lost. 'Stiff' testing machines have thick frames, or have feedback servo-control which prevents acceleration of the compression plates, by reducing the fluid pressure in the hydraulic jacks of the load mechanism, as failure commences. Even so, many data in current use are derived from tests on 'soft' machines, and full specifications of testing conditions are seldom quoted. It may be noted also that load applications in the 5–15 minutes of a common compression test do not simulate the duration of geological time that permits time-dependent behaviour, which is often of plastic rather than brittle deformation.

Similar reservations about uncritical use of laboratory test data apply to most attempts to define physical properties of rock.

15.3 ROCK MASS STABILITY

If laboratory results are chosen to represent a rock mass, they commonly over-represent stiffness by factors of 5–20 (Stagg, 1968, p. 125), and moduli of elasticity by factors of 10–20 (Muller, 1974, p. 19). In shear tests of rock joints the strength in one direction may, for example, be only 1% of that in other directions (Müller, 1974, p. 17). The scale of an experiment, consequently, has a major effect upon the results of that experiment and, as a general rule, the larger the scale the closer is the approximation to reality. To be realistic, testing conditions must represent the heterogeneous, anisotropic and discontinuous structural and hydrological properties of the rock mass. Such requirements can seldom be achieved in experiments, except for weakly consolidated Cenozoic rocks which behave as continuous materials, and which can be treated by the methods of soil mechanics. Tests on hard rock are best recognized as providing index information which can be used to compare one rock body with another, provided that standard methods of testing are used and quoted with the test data.

Quantitative studies of rock mass strength fall into two classes: those concerned with the analysis of rock slope stability against sliding, and those which study the strength of the whole rock mass as an indication of its resistance to deformation or erosional processes.

Analyses of rock slope stability usually involve estimates of the shear strength of pre-existing fractures on which movement is likely to occur. With few exceptions potential failure planes are partings which are continuous, or nearly so, and hence have a resistance to sliding which is dependent upon the frictional contact between the opposite walls of the critical parting.

15.3.1 Theory of Sliding Along a Parting

Views on the nature of friction between two rock surfaces can be divided into four groups (Lama, 1972).

1. Work in the late seventeenth and early eighteenth centuries assumed that frictional resistance is the result of microasperities lifting over each other (Kragelskii, 1965): a view which relates resistance to what we now call dilation.
2. The concept of the molecular nature of solids led to the assumption that friction is the result of overcoming the forces of molecular attraction between solids (Bowden and

Tabor, 1967). The laws of friction for unlubricated, or dry, surfaces are:

(a) the frictional resistance between two bodies is directly related to the normal force acting over the contact area between them, the resulting constant coefficient (μ) being known as the coefficient of friction;

(b) the frictional resistance and coefficient of friction are independent of the surface area of contact for constant normal forces;

(c) frictional resistance is independent of rubbing velocity;

(d) the magnitude of μ is determined partly by the mechanical properties of the minerals composing the two bodies, and partly by the asperities on the mineral surfaces. These laws hold for low velocities of shearing and for normal forces too low to shear or fracture the asperities: for lubricated surfaces the frictional resistance is independent of the normal force.

3. Work on metals, which indicates that during sliding of one body against another there is a penetration of asperities of one solid into the surface of another, has led to the concept of 'ploughing friction' in which a wave of deformation moves ahead of the penetrating asperity, so that resistance is due to displacement of the material surrounding the asperity.
4. A composite theory is implied in many modern analyses: it includes Coulomb's theory (equation (15.2) below), and represents friction as the result of interlocking of surface roughness and the lifting of microasperities over each other. In rocks, but not necessarily in metals, molecular attraction and the plastic deformation at low stress levels implied in ploughing friction are likely to be absent.

Newland and Allely (1957) were probably the first to recognize that shear resistance is not an intrinsic property of a material but depends upon the average angle of deviation of particle displacements from the direction of the applied stress. The major experimental work which is the basis for modern field and laboratory studies of frictional resistance along rock joints is that of Patton (1966a,b). Patton used models composed of kaolinite and gypsum plaster with completely interlocking asperities of controlled inclination. The result of his work, combined with that of Einstein *et al.* (1970), suggests that a realistic shear test of natural rock will produce a failure envelope with the general form indicated in Figure 15.3. At low stresses the failure line has a steep angle representing the angle of sliding resistance along a plane surface, which will approximate the angle of residual shearing resistance of the material plus the angle that the asperity surface makes with the horizontal. At low normal stresses the asperities slide over one another and shearing resistance can be represented by:

$$\tau_f = \sigma_n \tan(\phi_\mu + i) \qquad (15.1)$$

where τ_f is the shear stress at the onset of displacement over an asperity
σ_n is the normal stress
ϕ_μ is the angle of frictional sliding resistance along a plane surface
i is the inclination of the asperity to a common tangent to the base of the asperities.

After interlocked asperities have suffered a certain measure of dilation the strength of the asperities may be exceeded and they will shear through. This stage can be represented by:

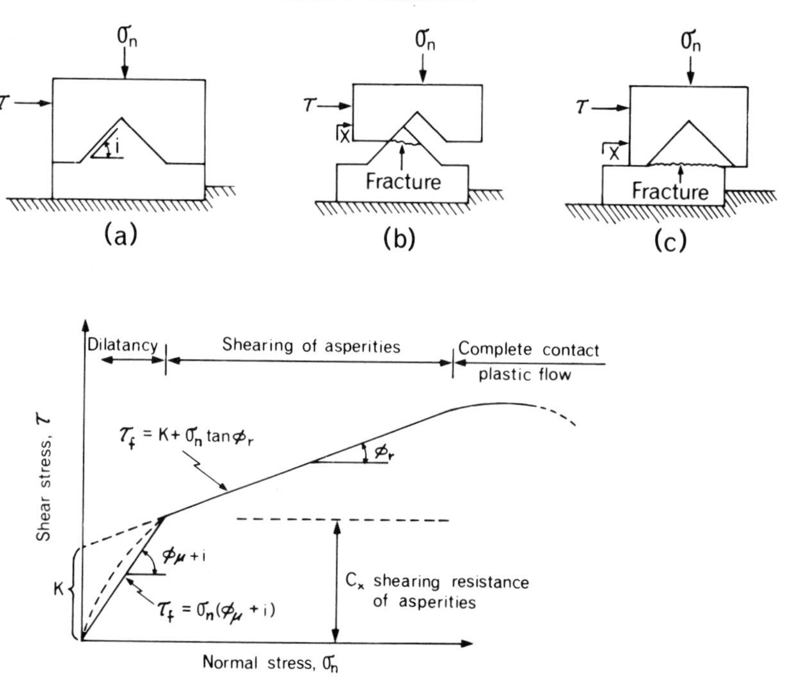

Figure 15.3 Failure of asperities and the resulting failure line

$$\tau_f = K + \sigma_n \tan \phi_r \tag{15.2}$$

where ϕ_r is the angle of residual shearing resistance
K is the constant, equal to the ordinate of the intersection with the shear stress axis of the straight line which can be used to approximate the τ–σ_n curve at relatively high normal stresses (the value of K is sometimes spoken of as an apparent cohesion in reference to a Coulomb failure criterion)

The relationship described above is wholly valid only where the asperities are entirely interlocking. This is rarely the case and it probably represents fresh tensile fractures only. The conditions in which asperities shear at low stresses along their bases represent a very weak material. Many experimental failure envelopes deviate from the bilinear model, and lie below those of the simple model involving 'pure' materials and totally interlocking surfaces, because interlocking is lost before failure as a result of displacements which are required before frictional contact is achieved along asperity faces, and because of the non-uniform stress distributions on the surface of asperities which partially fracture before maximum strength is reached. The failure line representing $(\phi_n + i)$ may be best represented by a curve (shown as a dashed line in Figure 15.3) indicating progressive failure of asperities. It is evident from these comments that strength tests of rock partings are likely to produce results with variable values depending upon the degree of interlocking, the form of the asperities, the magnitude of the shear displacement and the extent to which debris from fractured asperities accumulates along the partings. Added to this variability will be that induced by the testing machine and testing method, the inherent variability in the rock properties and its moisture content. A constant result from repeated testing of samples from the same formation is highly improbable. These considerations are discussed at length by Lama and Vutukuri (1978, Vol. IV).

In studies of rock joints it has to be recognized that joint infill material may have a large influence in joint strength. Whether the infill is washed-in soil, weathered rock, fault gouge or debris from crushed and sheared asperities, it will separate the asperities of joint walls, and as the infill becomes thicker it will progressively reduce contact between the walls. Where contact is wholly prevented the strength of the joint will be reduced to that of the infill.

15.3.2 Measurement of Joint Strength

There are no wholly satisfactory ways of measuring the frictional resistance of partings in rock. Ideally a large field shear box should be used which would simulate the stresses which are likely to induce failure in the field, and which would test the available strength along the whole of the joint, or a representative section of it. Large *in situ* tests have been carried out as part of the investigation for major foundation engineering projects, but the high cost of such tests—commonly several thousand dollars each—makes them impossible for most slope stability investigations. The commoner substitutes are described below.

A tilt test, Figure 15.4(a), requires that two blocks, which represent the upper and lower mated walls of the parting which is likely to become a failure plane, be tilted to an angle at which sliding of the upper block over the lower may just begin. This angle is the angle of plane sliding friction for that parting. The tilt test has a number of major disadvantages:

1. It must sample a potential slide plane in the direction of sliding and over the length of surface contact. Selecting and obtaining adequate samples may be impossible on a steep slope as the blocks must be large enough to be representative of roughness along the failure plane—a Herculean requirement.
2. There is increasing evidence that the value derived for $(\phi_\mu + i)$ is strongly dependent upon the length of the sample chosen for the test (Barton, 1981). This is due to the mobilization of larger, but less steeply inclined, asperities, as the sample size is increased. There is a reduction in both the dilation effect and asperity failure effect, as the sample becomes longer, with a corresponding decrease in the value of i so that, for a very long joint, the total frictional resistance is likely to be that of the residual angle of friction of the rock, ϕ_r, plus two or three degrees.
3. Single large asperities on the walls of one block may interlock with a mated depression in the opposite surface, especially where the slide plane has a 'stepped' profile. The range of angles achieved in tilt tests of several samples from one potential failure plane may thus vary from about 30°–75° depending upon the presence of such steep asperities and their frequency.
4. A tilt test does not measure the effect of the characteristic normal load acting across the parting.
5. Measuring a tilt angle to the nearest 5° is often the best that can be achieved in difficult field conditions involving large blocks.

A pull test, Figures 15.4(b) and 15.5, involves pulling a loose block of rock, with a joint surface forming its base, along an exposed joint plane. The force just required to

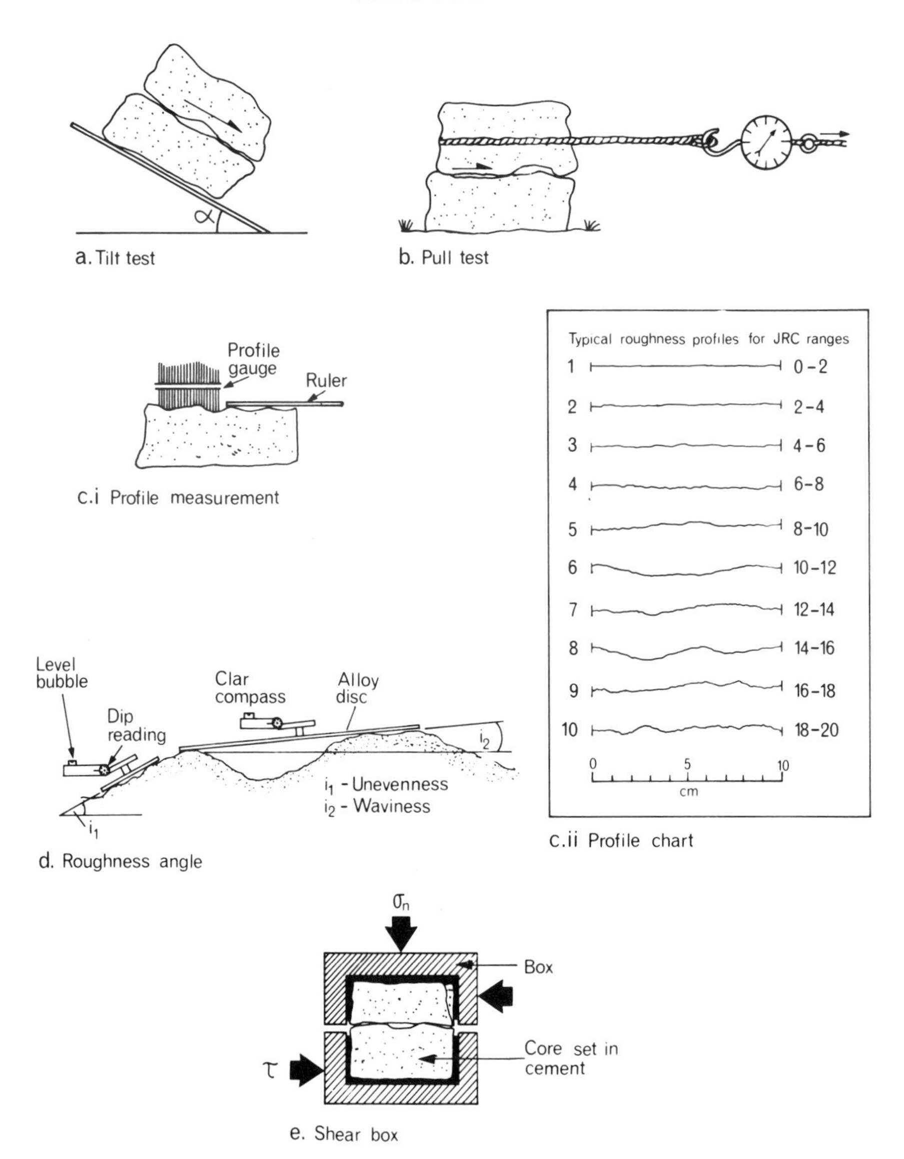

Figure 15.4 Methods of measuring joint frictional strength (c.ii, after Barton and Choubey, 1977; d, after Brown, 1981, Rock Characterization, Testing, and Monitoring. *Reproduced by permission of Pergamon Press.*)

move the block is recorded by a spring balance or similar instrument. The spring balance may also be suspended from a pole and used to measure the mass of the block. The frictional resistance along the joint is determined from

$$W.\cos\alpha \tan\phi_j = F + W.\sin\alpha$$
$$\tan\phi_j = (F + W.\sin\alpha)/W.\cos\alpha \qquad (15.3)$$

Figure 15.5 A pull test in progress

where $\tan\phi_j = (\phi_\mu + i)$
F is the spring balance reading
W is the mass of the block
α is the angle of inclination of the joint with the horizontal.

As with a tilt test the experiment should be repeated for a number of blocks and joint surfaces, and a mean value derived for ϕ_j. The limitations of pull tests are similar to those of tilt tests.

Joint roughness may be measured directly with a contour gauge, a ruler, or a plate of known dimensions (Figures 15.4(ci), (d) and 15.6–15.7). As indicated in Figure 15.4(d), joint roughness may be divided into two components: a first-order waviness which results from the large asperities, and a second-order roughness which results from the small asperities. The profile gauge is most reliable when it is used to measure first-order waviness. The gauge can record the roughness along many lines drawn on the surface of the joint and each record can be drawn from the gauge on to a field data sheet. The drawn profile may be used to derive directly measurements of the angle i or be compared with a published comparison chart (Figure 15.4(c.ii); Barton and Choubey, 1977). Alternatively, a ruler may be used to give a surface against which roughness may be assessed for comparison with the published chart (this is seldom a reliable practice), or by direct measurement of the depth of each depression between a pair of asperities (experience indicates that this method also is very difficult to apply with reliability).

Figure 15.6 A profile gauge on the joint surface of a sandstone with strong waviness of its asperities

Figure 15.7 Asperities of first and second orders on a fresh surface of dolerite subjected to tensile fracture. The scale is 25 cm long

The plates shown in Figure 15.4(d) are circular discs of aluminium alloy, with diameters of 5, 10, 20, and 40 cm. These are fitted in turn to an inclinometer. The Clar compass, manufactured by Breithaupt, is recommended as it incorporates a levelling bubble and a lid which is used to record dip directly. The first-order roughness angles are obtained by using the largest disc held against the joint surface in at least 25 positions, and recording dip angles and directions for each position. Up to 100 positions may be needed to record small asperity angles with the smallest disc (Brown, 1981).

A laboratory shear box is the only readily available device for measuring joint frictional strength under known stresses. Most boxes are so small that repeated testing is required if the whole joint surface is to be represented in the tests. In tests on models and specimens of rock it has been found that large asperities have roughness angles which are relatively low ($<20°$), and a second-order unevenness due to small asperities with roughness angles which may be as high as 60°, but more commonly fall in the range 40°–50°. Which of these components of roughness is significant depends upon the strength of the asperities and the magnitude of the normal and shear stresses. It has been shown by Selby (1982a, Figure 15.4) that, where critical joints for failure dip steeply out of a hillslope, the normal stresses due to the overburden are not high enough to crush asperities of most common minerals until they are weakened by prior shearing or weathering. At very low normal stresses then, the asperities remain intact, but as the normal stresses increase in magnitude the second-order, and then the first-order, asperities fail. Eventually the effective roughness angle is reduced to zero. An ultimate condition may occur in which the joint may become polished or slickensided by progressive failure so that a residual strength condition is reached in which the joint residual angle of friction (ϕ_{jr}) equals ϕ_r of the intact rock.

It has been stated already that the peak shearing resistance (ϕ_p) is related to a measure of joint roughness and the resistance of the joint wall asperities to crushing or shearing by an applied stress, as well as to ϕ_r. These components have been incorporated into an empirical formula by Barton and Choubey (1977):

$$\phi_p = JRC.\log_{10}(JCS/\sigma_n) + \phi_r \qquad (15.4)$$

where *JRC* is a joint roughness coefficient, and
JCS is the joint wall compressive strength.

For an estimation of the value of *JRC* two methods are available:

1. The roughness of the natural joint surface may be evaluated by pulling a natural joint block from a rock face and comparing the roughness of the joint with a chart of characteristic joint surface profiles provided by Barton and Choubey (1977, p. 19). This chart indicates the typical range of *JRC* values associated with the classes of roughness profile. Users of the chart are recommended to ensure that the chart is at true scale before comparisons are made. It has been pointed out by Tse and Cruden (1979) that rather small errors in estimating *JRC* could result in serious errors in the evaluation of peak shear strength and Barton and Choubey have recommended that tilt tests be used.
2. In a tilt test:

$$JRC = \frac{\alpha° - \phi°_{jr}}{\log_{10}(JCS/\sigma_{no})} \qquad (15.5)$$

where α is the tilt of the joint at failure, ϕ_{jr} is the residual friction angle of the joint, and σ_{no} is the normal stress induced by self-weight of the sliding block.

The tilt angle (α) is readily measured with an inclinometer. The residual friction angle of the joint may be derived from the relationship:

$$\phi_{jr} = (\phi_{\mu} - 20°) + 20(r/R) \tag{15.6}$$

where ϕ_{μ} is the basic friction angle of the intact wall rock determined either by tilt tests on dry unweathered sawn surfaces of the particular rock, or estimated from the table of values given by Barton and Choubey (1977, p. 6), R is the uniaxial compressive strength indicated by Schmidt hammer tests on smooth or sawn, dry, unweathered, rock surfaces, and r is the uniaxial compressive strength indicated by Schmidt hammer tests on wet, weathered, joint surfaces. (Note: A correlation chart for expressing L-type Schmidt hammer readings as compressive strength (MPa) is given by Barton and Choubey (1977, p. 10) after Deere and Miller (1966)).

A typical rock with $\phi_{\mu} = 30°$ and limited weathering ($r/R = 30/40$) has a theoretical minimum ϕ_{jr} value of 25°. If joint weathering has been more severe ($r/R = 20/40$), ϕ_{jr} is 20°. This method gives ϕ_{jr} values within 1° of the value derived by large scale laboratory tests.

The normal stress induced by self weight of the sliding block is given by:

$$\sigma_{no} = \gamma.h.\cos^2\alpha \tag{15.7}$$

where γ is the unit weight of the rock (kN/m^3), and h is the thickness of the upper block (metres). The *JCS* value required in equation (15.5) can be measured with a Schmidt hammer on the dry joint surfaces used in the tilt test. Three tilt tests are performed on each joint and the mean value is used for estimating *JRC*. Because of the very low stresses across the joint there is no visible damage and the test can be repeated on the same samples. In the determination of peak shear strength we are concerned with the minimum strength for stability so the *JCS* value used to evaluate equation (15.4) is that determined by a Schmidt hammer test on a saturated joint.

Barton's empirical relationship has been tested against experimental data and the failure criteria proposed by Hoek and Brown (1980) found to be an excellent fit to Mohr envelopes derived by the alternative methods (Hoek, 1983).

The derivation of ϕ_{μ} is a topic in need of further investigation. It is usually quoted as the angle of sliding friction for a smooth sawn surface. Experience has shown, however, that sawn surfaces are by no means uniform. The method recommended by Stimpson (1981), using three rock cores arranged as an elongated pyramid, provides two narrow surfaces of contact for a tilt test and, provided that the surfaces are uniformly cored to smooth, unridged, faces it is consistent in its results (W. Doolin, pers. comm.).

The importance of joint shear strength is emphasized by the amount of current research on the topic. Recent papers on the effect of joint size upon strength include those by Barton (1981), Bandis *et al.* (1981), and Barton and Bandis (1982). Joint stiffness has been studied by Swan (1983), complex behaviour by Bandis *et al.* (1983), and peak strength of joints by Fecker (1977). The dependence of test results upon the method of testing is, however, a matter of primary concern, for it emphasizes that strength parameters from any test cannot be considered true constant values: this has been amply demonstrated for residual strength of joints by Weissbach and Kutter (1978).

15.3.3 Hillslopes with Angles Critical for Stability

Stability analyses are dependent upon a knowledge of joint shear strength. It has been shown that there are several methods available for derivation of values of strength. Many of the methods lack precision and none is well founded in theory. Slope stability analyses, however, are carried out for practical reasons rather than 'pure' research. They may be used to analyse conditions in failures which have occurred in the past and have comparable value for earth scientists to that of post-mortem examination of a corpse for the medical profession. Studies of potential failures are most useful in hazard avoidance or control. These practical considerations justify empirical and imprecise, but effective, methods. Details of stability analyses are given in texts such as that by Hoek and Bray (1977) and by Hoek (1983). Discussions of rock slope failures are given in Brunsden and Prior (1984) and Voight (1978), and examples of practical methods of analysis of the effects on stability of joints and structures are considered in Chapter 5.

Theoretical analyses of joint shear strength in relation to rock slope forms and stability owe much to Terzaghi (1962). Subsequently Selby (1982a) has used the evidence that asperities fail under progressively increasing applied stresses to show that very high cliffs, with critical joints for stability which dip out of the slope, should have concave profiles. This arises because near the top of a cliff the joint is in a low stress regime and asperities survive. With increasing height of the cliff the asperities on joints at lower levels fail until at the base of the cliff $\phi_j = \phi_r$. As the cliff retreats critical joints are unbuttressed and fail, leaving a cliff face inclined at the angle of ϕ_j characteristic of that part of the cliff. The face of the Drakensberg in Natal provides one example of such a high cliff.

15.4 ROCK MASS STRENGTH

Engineers who are concerned with the strength of rock masses have, as their primary aim, the design of safe structures, whether these are dams or tunnels. It is not part of their aim to understand the significance of the physical characteristics of the rock except in so far as it permits prediction of behaviour under applied loads. Thus Bieniawski (1974) and Barton *et al.* (1974) were attempting to classify rock masses in respect to the design of tunnel supports. Hoek and Brown (1980) and Hoek (1983) presented a failure criterion for jointed rock which permits description of the strength of rock masses in numerical form, but the derivation of their empirical criterion was one of trial and error: 'apart from the conceptual starting point provided by Griffith theory (on the propagation of cracks), there is no fundamental relationship between the empirical constants included in the criterion and any physical characteristics of the rock (p. 192)'.

For geomorphic purposes a classification of mass strength need not be concerned with applied or residual stresses because at the surface of an outcrop these are either negligible in magnitude, or dissipated by the opening of joints. It does require to provide a basis for understanding the features of rock masses which give them resistance to processes of weathering and erosion, and it should be universally applicable so that it can provide a common basis of measurement. Such a classification has been proposed by Selby (1980).

Table 15.1 Maximum numerical ratings for classification parameters

Intact rock strength	20	18	14	10	5
Weathering	10	9	7	5	3
Spacing of joints	30	28	21	15	8
Joint orientations	20	18	14	9	5
Width of joints	7	6	5	4	2
Continuity of joints	7	6	5	4	1
Outflow of groundwater	6	5	4	3	1
	Very strong	Strong	Moderate	Weak	Very weak
Total rating	100–91	90–71	70–51	50–26	<26

Note: Joint infill is included in the 'continuity' parameter.

Its methodology and principles have been described in detail in the original paper and in Selby (1982b), and also briefly in Dackombe and Gardiner (1983). The method has been independently validated by Moon (1983). Details of the classification will not be repeated here, but to clarify subsequent discussion its main elements are described below.

The classification is based on a semiquantitative assessment of eight features of any rock unit or slope unit formed upon exposed rock. The measurements made of each of the eight features are weighted according to the significance of each feature, which has been elucidated by statistical analyses of data from over 300 rock units with a variety of lithologies and in several climatic environments. The eight features are: strength of intact rock as measured with a Schmidt hammer; state of weathering of the rock; the spacing of joints; orientation of the joints; width of the joints; continuity of the joints; infill of the joints and movement of water out of the rock mass. Each of these features is scored on a five-class scale using stated criteria, and the score for each rock unit, or hillslope segment, is totalled (Table 15.1).

Hillslopes with inclinations which have evolved so that their form is in equilibrium with the mass strength of their rocks are necessarily those which are not controlled by tectonic, structural, erosional or depositional processes. Equilibrium slopes also require considerable periods of time (usually >10 ky) for this evolution and adjustment to occur. By studying slopes in areas for which there is geomorphological evidence for an absence of these processes as controls on slope form and also for environmental stability over long periods, it has been possible to identify slopes controlled by the mass strength of their rocks. When the data points for slopes which are in equilibrium are plotted, with rock mass strength rating against the average slope angle supported by each rock unit, these data points all fall within a strength equilibrium envelope. The data points for slopes which are out of equilibrium fall above or below this envelope, and this type of plot is, therefore, useful for identifying the existence of structural controls, and as an approximate indicator of the maximum angle of slope a stable rock mass can support.

The use of five classes for scoring the rock features follows engineering practice and is convenient for applying the criteria to all of the eight features, but both intact rock strength and joint spacing can be measured on a continuous scale and a finer division of the data for these two features gives greater precision. Moon (1984) has provided a table with continuous scale ratings for these two parameters.

In the original work on strength equilibrium envelopes, data were available only from sites in Antarctica and New Zealand and it seemed inappropriate to specify the bounds

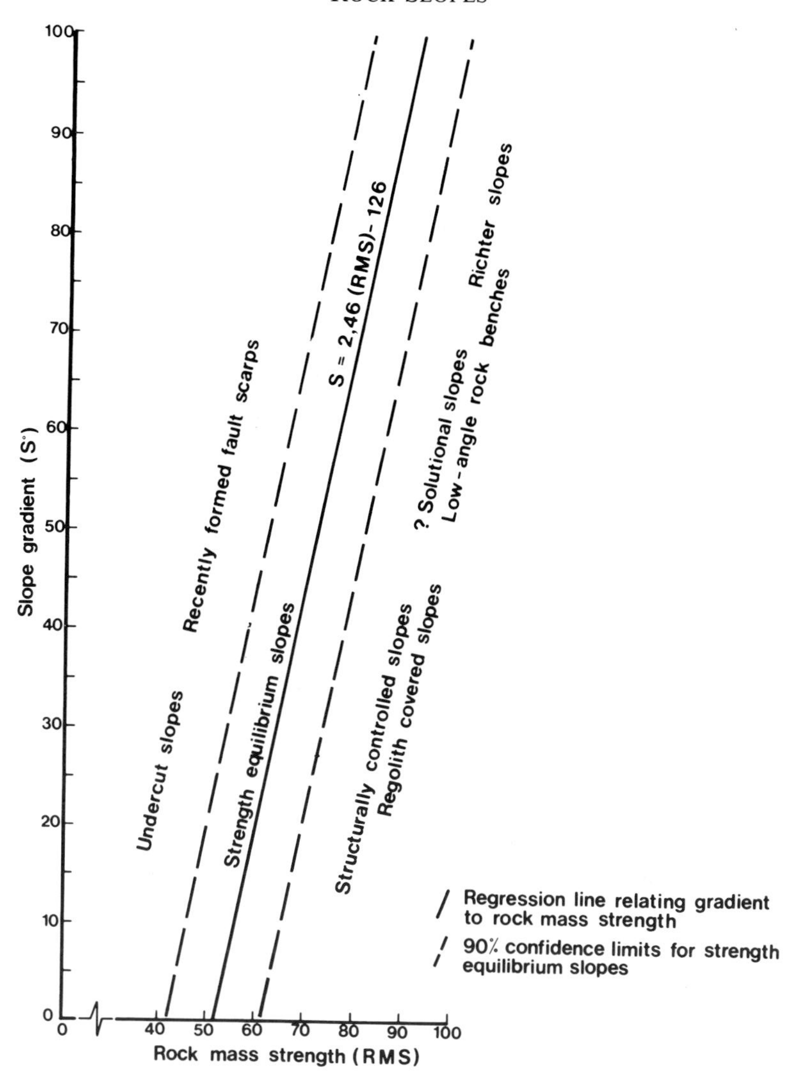

Figure 15.8 Revised plotting diagram for slope gradient and rock mass strength, with the strength equilibrium envelope. (*After Moon, 1984*, Earth Surf. Proc. Landforms, **9**.)

of strength envelopes until the method had been verified and applied in a greater variety of environments. Using both the original data and additional data from a variety of rock slopes in Africa, Moon (1984) has now specified the use of 90% confidence limits derived from the standard error of the regression of slope angle against rock mass strength, and provided a plotting diagram (Figure 15.8).

15.4.1 Applications of the Classification

The rock mass strength classification has now been used in many arid and semi-arid areas and in several mountain systems. It is evident from the results that slopes which have an inclination which is adjusted to the mass strength of their rocks are widespread, and that over-steepened slopes which are undercut by an erosional process, and some recently formed fault scarps, can be readily recognized. Slopes with inclinations which are less than those of strength equilibrium conditions may be recognized as: (a) exhibiting structural controls; (b) being Richter denudation slopes which have developed by cliff retreat above rock slopes with angles of inclination (usually 28°–36°) which are those of the angle of rest of talus on them, and which rock slopes may subsequently be laid bare as the talus material is weathered and removed by erosion; (c) being low-angle cut benches; or (d) being the result of solution processes. Regolith-covered slopes are best described as transport-limited and hence they fall outside the general class of rock slopes, but work in semi-arid zones has shown that many slopes with a thin veneer of slope deposits and soils may still have profiles which reflect the strength of the underlying rocks: in the field they are commonly recognizable by persistent outcrops of resistant strata (Selby, 1982c).

The widespread occurrence of strength equilibrium slopes, with over 300 rock slope units being in equilibrium, out of 350 so far tested, suggests that many rock slopes must retreat so that strength equilibrium is preserved. Consequently, as long as a rock unit has relatively constant mass strength properties, the slope will retreat parallel to itself. A steepening or decline of slope angle during retreat may result where mass strength increases or decreases, respectively, into the rock mass on which the slope is formed.

The validity of this argument seems to be generally confirmed by a study of the inselbergs in the Namib Desert (Selby, 1982c) and of the scarp forms around the margins of the central plateaux of southern Africa (Moon and Selby, 1983). The mass strength of the exposed rock units in the field study sites varies from 50 to 91 mass strength units and the slope angles on bare rock range from 45° to 90°. As many of the Namib Desert inselbergs are formed on steeply dipping sedimentary and metasedimentary and metamorphic rocks it is evident that slope retreat does not involve a simple form of retreat in which a constant slope angle is maintained. Rather there is a constant process of slope angle adjustment towards the strength equilibrium angle. This conclusion may be one of the most useful contributions of an application of the rock mass strength classification to resolution of a long-standing controversy in geomorphology. It both denies the possibility that landform evolution around the margins of southern Africa is accompanied by a flattening of slope angles and also modifies the contention of King (1953) and Fair (1948) that parallel retreat occurs because in unit time a superficial layer of rock is loosened and removed from the cliff face.

A second contribution to an old controversy involves an application of the classification to rock slopes which bear a veneer of boulders. Bryan (1922, p. 43) described slopes formed on hard rock with a veneer of boulders, in the Arizona deserts, as boulder-controlled. Subsequently, measurements have shown (Melton, 1965) that such slopes exist over the range of angles up to about 37° and that the existence of a boulder cover may be due to a number of causes. Melton suggested that steep angles of the debris (34–37°) may occur at the angle of static friction of that debris and that angles of around 26° may be related to its angle of sliding friction. Where the boulders are core stones left by removal of fine-grained saprolite of a former regolith, the boulders may lie at any angle up to 37° and they do not exert any control on the bedrock slope which is essentially a relict weathering front from the base of the regolith (Oberlander, 1972). Focusing attention on the boulders has obscured the possibility that the weathering and extension of joints in some granites produces a rock mass with rather low strength and hence a low strength equilibrium angle. In the Namib Desert the Amichab bornhardt has some boulder veneered units with mass strength ratings as low as 54 units and equilibrium angles of 22°. In Arizona, where a core-stone rich regolith has been stripped to leave a bornhardt, boulders may lie on rock surfaces at angles well below those required for strength equilibrium, but the available evidence indicates that all of the Namib inselbergs evolve towards strength equilibrium forms and hence the boulder veneers lie upon rock slopes which are adjusted to strength. The presence of the boulders is only an indication that they are stable on relatively low-angle slopes formed by backwearing (Selby, 1982d).

15.4.2 Evolution of Rock Slopes Towards Equilibrium Forms

Many of the early accounts of rock slope geomorphology mention briefly the evolution of rock slopes without specifying details of processes or forms. The object of this section is to discuss some of the processes involved in the progressive change of slopes in horizontally bedded sandstones and in massive granites. Examples of the slopes on sandstone are taken from the Olympus Range, and those in granite from the Taylor Valley, both in the arid polar environment of the McMurdo Dry Valleys of Antarctica (77°S, 162°E). Details of the geology of this area are given in Selby (1974).

The ridges forming the crests of the Olympus Range are cut by a major joint set which passes through the ridges at an angle of about 45° to the ridge faces. The joints are traceable across the floors of the intervening cirques and glacial valleys and are presumably of tectonic origin. Many of these joints occur in pairs which may both dip at about 70° ± 5° in either the same direction or in opposite directions, thus producing either inclined columns or triangular wedges. A few of the sets change their dip from cross-cutting to a columnar style at depth (Figure 15.9(a)). At 80° ± 8° to the first major set of joints is another set which results in each mountain ridge being divided into rectangular blocks. At the margins of each ridge, however, the slopes are inherited from an original glacial erosional form which cuts across the two joint directions. The glacial valley walls have been subsequently modified by snowpatch, wind and Richter-denudation slope processes, but the intersection of the tectonic joint pattern by a glacial valley wall has produced a plan-form margin to each ridge which has triangular or rectangular re-entrants which have been formed by large wedge failures or toppling failures respectively (Figures 15.9(a) and (b)). The cliff faces between re-entrants are developing unloading joints which are aligned parallel, or within 6° of parallelism, to the glacial valley axes. The unloading blocks are 1–5 m thick. They all begin at the foot of the cliff and terminate vertically in an overhang, behind which the unloading joint has little or no continuity. It is rare for pinnacles to form along ridge crests by joints propagating downwards.

At the end of each ridge large transection joints have opened. Where these joints dip steeply into the slope, as a result of being part of the major continuous tectonic set first described above, they may give rise to large (≃2 m) outwards displacements at the ridge crest before toppling or fracturing along the base of the column begins even though the column may be 50 m high (Figure 15.9(c)).

Forming the walls for central Taylor Valley are cliffs formed in granite (Figure 15.10). Two general classes of joint are recognizable: one is a closely spaced set of tectonic origin, and the other is a set of unloading joints nearly parallel with the valley walls. The joints of

(a)

(b)

Figure 15.9 (a) Slope on sandstone with a large wedge failure to the left and, to the right, joints tending to produce both columns and wedges. Olympus Range, Antarctica. (b) A large column has toppled from the rectangular cavity in the centre. Weathering is producing minor joint propagation in the face to the right, and major joints extend through the face to the left. (c) Column separation along major joints, with toppling, at the end of a large spur

tectonic origin control irregular slope forms on which ridges follow zones of more widely spaced joints, and intervening gullies are zones of closely spaced joints. As with the Olympus Range sandstones, unloading joints develop by upward propagation from the base of the cliff or upwards from a large sub-horizontal joint.

On both the sandstone and granite, weathering produces spalls of plate-like rock slabs which are unrelated to continuous joints and which terminate by either meeting a joint or fracturing to leave a sharp-edged depression on the cliff.

The significance of the examples given above is that they can only be understood by application of geomechanics principles.

Wedge failures developed from cross-joints have a stability which results from the shear strength of the joints. As the critical joints are usually opened by tensile fracturing through unweathered rock their surfaces are wavy and have a high roughness angle. The value of ϕ_j, therefore, commonly falls in the range of 65°–80°, and sliding may be prevented until weathering reduces joint strength to $\phi_j = 60°–65°$, which is a lower angle than that of the critical joint. Wedge failures are major features of crest lines of many mountains formed of steeply dipping sedimentary or metasedimentary rocks such as schists and gneisses. In many areas they may dominate the forms of the entire mountain face (Figure 15.11).

Topples and slab failures result from propagation of stress release joints. The removal of overburden and confining rock allows a partial stress relief to take place. The stress component normal to the cliff face will be zero at that face and progressively higher stresses will occur towards the interior of the rock mass. The mass will, therefore, tend to expand in the direction of the free face, giving rise to a slight arching and buckling-type of instability. This results in the splitting off of slabs parallel to the surface. In theory, the thickness

Figure 15.10 (a) Irregular topography formed on granite with variable joint directions and spacing, Taylor Valley, Antarctica. (b) Stress release on a granite cliff

Figure 15.11 Large wedge and planar failures have given a peaked and serrated form to this ridge in gneiss, Khumbu Himal, Nepal

of the spalling slabs could be calculated from a knowledge of the elastic modulus of the rock on confinement at the stress level corresponding to the depth of burial or tectonic compression. As this is seldom known, and *in situ* stress measurements provide data only on stresses residual in the rock at the time of measurement, little advance is to be expected in that direction. Ridge-end spreading is an unconfined type of stress release phenomenon which may occur along existing joints or it may involve propagation of stress-release fractures.

Where failures have occurred, weathering may cause minor fracturing and spalling. If the rate of weathering exceeds that of large-scale sliding or toppling, the cliff may remain at an angle controlled by the mass wasting process. If the rate of weathering is dominant, then the cliff will evolve towards strength equilibrium as spalling modifies cliff faces and permits rotation, falling, and sliding of small weathered blocks. Ultimately, the release of all residual stresses will permit weathering processes to turn all rock slopes into equilibrium forms. This occurs most readily where joints are closely spaced, and well-jointed rock masses are commonly found, therefore, to be in strength equilibrium.

15.4.3 Slopes Not In Strength Equilibrium

The rock mass strength classification can be used to distinguish the controls on slope form and inclination. Over-steepened slopes are essentially short term features of a landscape on a geological time-scale, and usually result either from undercutting by an erosional

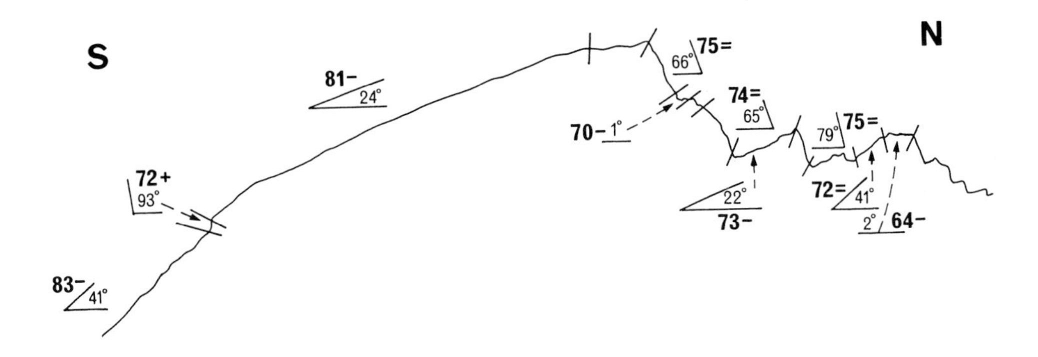

Figure 15.12 A profile of the limb of a fold in gneiss, showing the mass strength and inclination of major slope units. The site is about 10 km south of Cho Oyu in the Khumbu Himal, Nepal

process or from a recent structural process such as high-angle faulting. Over-steepend slopes start to evolve towards strength equilibrium angles as soon as joints begin to open, and joint blocks can rotate, slide or fall and so change the slope profile. Slopes with inclinations lower than those at which they would be in strength equilibrium may be the product of one or more processes, of which structural controls, planation, and the formation of a regolith in a transport-limited regime are the most obvious. The expression of a structural control is most evident in an area of active deformation, as on the flank of a growing anticline, but it is most commonly observed where erosion has stripped cover beds and left a resistant body of rock forming a surface; this situation commonly exists in areas of pre-Quaternary folded terrain or plutonic intrusions.

It is not always clear from field observation alone, whether a surface upon a rock mass is in, or out, of strength equilibrium. It is essential to recognize that a slope unit may have an inclination which places it in the strength equilibrium class by accident, while it may actually be, for example, structurally controlled. Yet it is an essential part of many geomorphological exercises to determine the controls on slope inclinations. In the example shown in Figure 15.12 a profile, near to the skyline of the ridge formed on a body of folded gneiss, has been studied and the mass strength numbers show that all dip slope units, except one which is the crown slope of a planar slide and over-steepened, have inclinations below those which could be supported by their rocks and are structurally controlled. All scarp slope units are in equilibrium. This example is taken from an area of active Late Quaternary glaciation and, like the examples already published from Antarctica (Selby, 1980), indicates that glacially over-steepened slopes are rapidly changing into strength equilibrium forms. Because of the wide spacing of the joints, and the hardness of the gneiss (N-type Schmidt hammer readings of 55 ± 6 units), slopes in the Khumbu region of the High Himalayas are commonly in strength equilibrium, even when they have inclinations in the range of 65° to 85°. Slope failure occurs because steeply inclined critical joints dip out of many glacially undercut slopes, and physical weathering and joint propagation through stress release, permit joints to open and large-scale rockfalls, planar slides and mixed rock and snow avalanches to occur.

15.4.4 Mapping Rock Mass Strength

In Figure 15.12 over-steepened, in equilibrium, and below equilibrium angles are indicated by +, =, and − signs. This procedure can be extended to maps of geomorphic features and either combined with other geomorphic mapping symbols or treated as a separate study. In Figure 5.13 a letter system has been added to the mathematical symbols to distinguish the controls on slope form. Most of these symbols have an obvious criterion for use, but judgement may have to be exercised in discriminating between structural and eroded forms. Where the rock surface has been stripped of cover beds to leave a clear joint plane exposed, then the symbol **S** is appropriate: where erosion has left an irregular surface, as is usually the case on exposed weak rocks, then the erosional form may be regarded as dominant and the symbol **E** used. All depositional or regolith-covered slope units have been classed together.

The mapping method may be useful for identifying landscape units which are undergoing evolution at several classes of rates or for identifying the extent of a given control.

15.4.5 Specific Modifications to a Mass Strength Classification

The original rock mass strength classification for geomorphic purposes (Selby, 1980) was related wholly to slope profiles. For some purposes, however, the inclination of joints with respect to angles critical for slope stability is irrelevant. In studies of susceptibility of

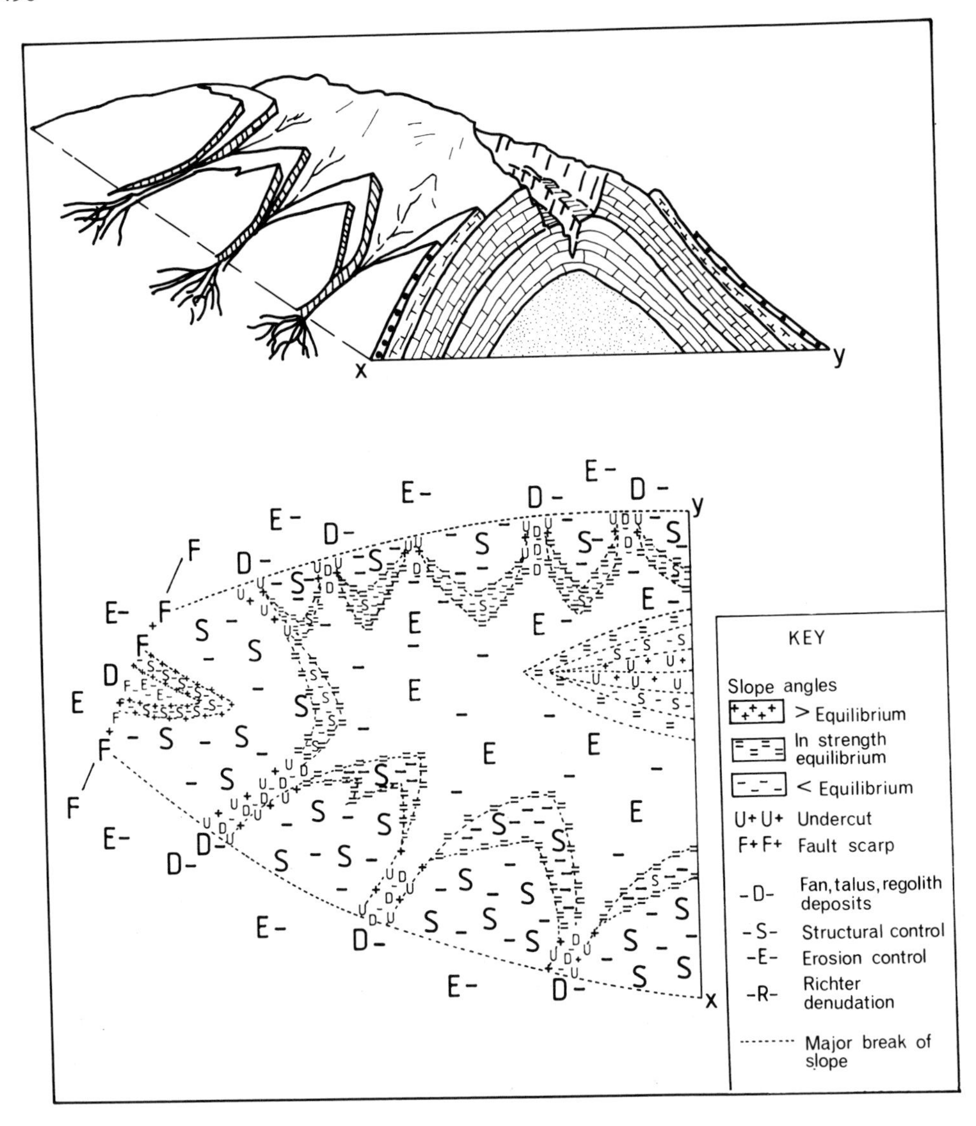

Figure 15.13 A schematic landscape, and map of the mass strength of its slope units

different types of rock mass to glacial plucking, for example, the critical inclination of a joint may be with respect to the angle it makes with the sole of the glacier. The inclination parameter may thus be redefined for particular purposes.

15.5 ROCK ELASTICITY AND STRENGTH AS INFLUENCES UPON SLOPE ANGLES AND DRAINAGE DENSITY

Engineers have found it necessary to take into account in their designs of structures in rock, the resistance of rock bodies to compression and distortion when those bodies are placed under stress, and the degree of recovery of original form (i.e. the elasticity) when the stress is removed. Geomorphologists have, as yet, not recognized the significance of rock response to stress, or stress reduction, but it may be speculated that erosional removal of weathered rock and soil from bedrock may permit that bedrock to swell, for joints in it to open, and for weathering processes to attack the rock. If this speculation is valid then it may be that the elastic response of bedrock to progressive erosion has an influence upon rates of weathering, cliff development, and on the development of drainage patterns in catchments in which the influences of joints in bedrock can reach the ground surface.

15.5.1 Elastic Moduli

The ability of a body to resist forces which tend to induce compressive, tensional, shear or volumetric deformations is determined by its elasticity. The theory of elasticity of solids involves an initial assumption that the solids will behave as perfect, or ideal, bodies with: (a) induced strains which are linear functions of applied stresses; (b) the same magnitude of strain resulting from a given stress; and (d) complete recovery of strain resulting from the removal of stress. Real bodies, however, behave in an ideal fashion only over a limited range of stresses, and it is essential to specify exactly the boundaries of the description of elastic behaviours.

The constant factors relating stress to resulting changes in dimensions of a body are known as moduli of elasticity. Their dimensions are given in units of stress, since strain is itself a ratio and hence dimensionless.

The moduli of elasticity for static conditions are:

Young's modulus, $$E = \frac{\text{longitudinal applied force/cross-sectional area}}{\text{change in length/original length}}$$

Shear (rigidity or torsion) modulus, $$G = \frac{\text{shear force/unit area}}{\text{angular deformation}}$$

Bulk modulus, $$K = \frac{\text{applied force/unit area}}{\text{change in volume/unit volume}}$$

Young's modulus is a measure of the ability of a body to resist elongation or shortening in the direction of the applied stress. The shear modulus is a measure of the ability of a body to resist a change in shape, and the bulk modulus is a measure of resistance to a change in volume. Both Young's and shear moduli may be measured in compression or tension, but it is common to do so in a uniaxial compressive test in which changes in length and diameter of a specimen are recorded for various stress levels (Figure 15.14). Most rocks have a stress–strain curve which is initially concave upwards: at low stress levels this curve reflects the reduction in porosity and from it is derived a measure of initial elasticity (E_i). Increases in stress may induce a nearly perfect elastic response with a linear relation between stress and strain, and this linear relationship may be followed by increasing deformation as internal cracking leads to final rupture: there is thus a convex segment of the stress–strain curve preceding rock yielding and dislocation. Values of Young's modulus are, therefore, constantly changing except in the linear part of the curve, and it is necessary

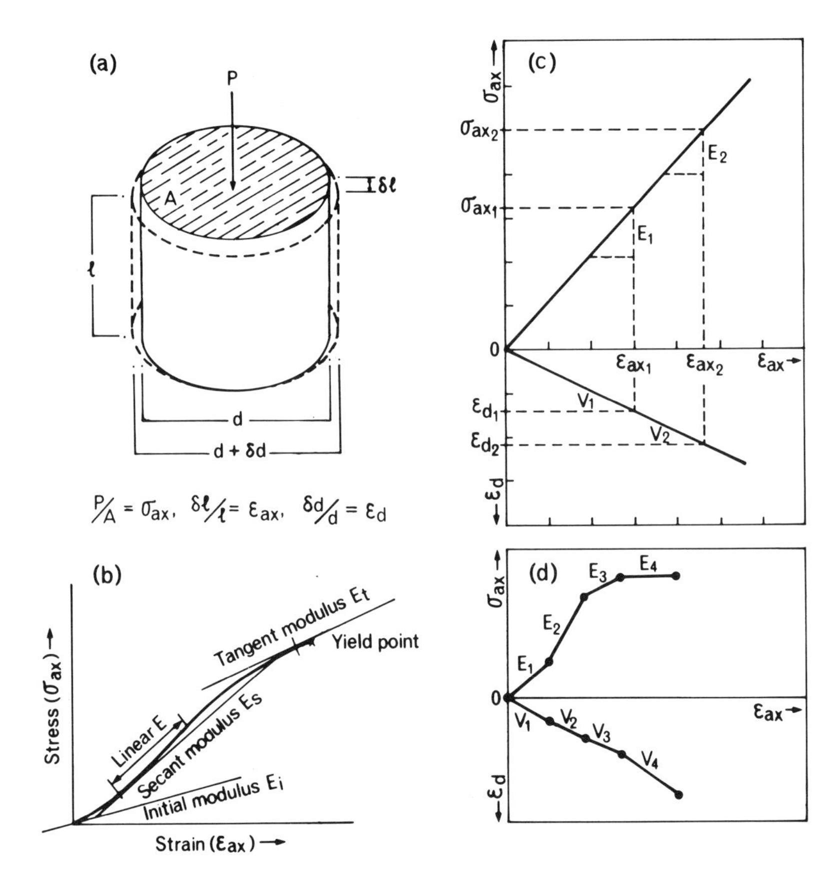

Figure 15.14 (a) Axial and diametral strain of a rock cylinder in uniaxial compression. (b) A generalized stress–strain curve for rock, up to the yield point, with definitions of the types of Young's modulus. (c) An idealization of a linear elasticity body showing constant values for Young's modulus and Poisson's ratio. (d) Variation in the values of Young's modulus and Poisson's ratio, for a rock, showing the strain response to increasing stress

to specify precisely the point or zone over which the modulus magnitude is being determined. It is common to take a tangent modulus at 50% of the ultimate strength, which is usually within the linear zone, or a secant value between two stated stress levels.

The relationship between longitudinal (axial) and lateral (diametral) strain of a specimen is a constant for that material, known as Poisson's ratio.

$$\text{Poisson's ratio, } \nu = \frac{\text{change in width/original width}}{\text{change in length/original length}}$$

Although Poisson's ratio is usually regarded as a constant this is not a reliable statement, for near the ultimate strength the stress–strain curves for most rocks indicate that the lateral strain increases more than the axial strain, resulting in an increased value of ν. Most published values of Poisson's ratio refer to values at, or below, 50% ultimate strength. For rocks, these values usually lie between 0.2 and 0.35. The stronger the rock the smaller will be its value of ν.

From a knowledge of E and ν it is possible to derive values for the other moduli:

$$G = \frac{E}{2(1+\nu)} \tag{15.8}$$

$$K = \frac{E}{3(1-2\nu)} \tag{15.9}$$

Rocks suffer transient loading by earthquakes (see Chapter 9) and some construction activities. Under such conditions the dynamic elastic constants are relevant and may be calculated from elastic wave velocities. Resonance and ultrasonic pulse methods are used in the laboratory and seismic propagation methods in the field. Knowledge of the wave velocities permits direct calculation of values of E and ν.

15.5.2 Elasticity, Strength, and Landforms

Measurements of rock strength and elasticity have been used only rarely in geomorphic studies. A notable effort is that of Cooks (1983) who collected core specimens from five major rock types—sandstone, tillite, granite, granite-gneiss, and gneiss—in South Africa and in the USA. Strength was determined of unweathered rock in uniaxial compression, and indirect tension (the Brazilian test involving splitting a sample along its longitudinal axis by compressing it over a rod). Elastic moduli were determined in the laboratory by ultrasonic velocity methods. Eighty samples of each of the five rock types were tested: and for fifty drainage basins, on each rock type, drainage density, valley side slope, and distribution of mass (hypsometry) were determined from maps at scales of 1:18,000 and 1:24,000 with contours at 15 ft (4.57 m) and 40 ft (12.20 m) intervals respectively.

Each of the dynamic constants E, G, and ν, and each of the strength measurements, uniaxial compressive, tensile and shear, were correlated with the morphometric measurements drainage density, angle of slope, and hypsometric integral. All of the relationships tested, except Poisson's ratio, were found to be significant at a very high confidence level.

The results indicate that, in general, the development of drainage basins can be correlated directly with the strength of the bedrock underlying them. More channels, deeper incision, and more extensive removal of material occur on the lower strength rocks and, *vice versa*, fewer channels, shallower incision, and more resistant masses are found on high strength material. The dynamic elastic moduli have related trends, with those rocks having the greatest resistance to deformation showing the same trends as the high strength rocks.

The work of Cooks (1983) is of considerable interest but it raises at least as many questions as it provides answers. The study areas were stated to be in fluvial environments, but no details were provided of their location, soil cover, vegetation or climate. All areas had mean slope angles lying between 14.1° and 10.4° and so were, presumably, areas of continuous soil cover. Drainage densities similarly have a very narrow range of means—from 12.3 to 15.3—and hypsometric integrals of 48.3 to 55.8. The results seem to imply that the material which rivers normally remove by erosion—weathered regolith and soil—has either little relationship to valley morphology or is controlled in its erodibility by rock strength. We are left in ignorance of what effects climate may have, which is a strange situation when the conclusions of numerous studies of the relationships between rainfall, runoff, and drainage density are recalled (see Richards, 1982, pp. 37–9 for a review). The relationship between drainage density and precipitation effectiveness seems to be close and well established.

It is also not clear what it is that is being measured in the dynamic elastic moduli studies of Cooks. Dynamic moduli methods usually involve low stresses and so they are usually regarded as being comparable only to initial tangent moduli determined by static methods. The parameters affecting propagation velocities of waves in rocks, and hence dynamic elastic constants, are: mineralogy, texture, density, porosity, anisotropy, water content, and temperature. But greater in importance, as far as morphometry is concerned, may be the degree of weathering, joint spacing, and joint infill of a rock mass: none of these was measured.

It is evident, therefore, that this study needs much replication and better definition of the study conditions before too many conclusions are drawn. It does, however, open many possibilities for future research.

15.6 CONCLUSIONS

For both the design engineer and geomorphologist traditional geological methods of classifying rocks are inappropriate and unhelpful, perhaps even misleading. Geological methods are designed to classify according to mineralogy and to provide an indication of the genesis of a rock. For the geomorphologist a method is required which classifies rock according to its capacity to resist geomorphic processes and to express its characteristics in landforms.

Some progress has been made in determining the features of rock masses which control their strength, and particularly their control on slope profiles were there is little or no soil cover. Research is now underway to elucidate the parameters which control the strength and elastic properties of intact rock. Related studies are being carried out to seek the relationship between geochemical changes and strength properties in weathered rock. Only when we have a deeper understanding of rock properties will it be possible to prepare a comprehensive study which can achieve the promise implied in the title of Yatsu's *Rock Control in Geomorphology*.

ACKNOWLEDGEMENTS

I am indebted to the Antarctic Division, New Zealand Department of Scientific and Industrial Research for support in the field. I am also indebted to John Wiley and Sons Ltd for permission to reproduce the material lying between equations (15.4) and (15.7) in this text which is taken from my paper published in *Earth Surface Processes and Landforms* (Selby, 1982a).

REFERENCES

Bandis, S., Lumsden, A. C., and Barton, N. R. (1981). 'Experimental studies of scale effects on the shear behaviour of rock joints.' *Int J. Rock Mech. Mining Sci. Geomech. Abstr.*, **8**, 1–21.

Bandis, S. C., Lumsden, A. C., and Barton, N. R. (1983). 'Fundamentals of rock joint deformation.' *Int. J. Rock Mech. Mining Sci. Geomech. Abstr.*, **20**, 249–68.

Barton, N. (1981). 'Some size dependent properties of joints and faults.' *Geophys. Res. Letters*, **8**, 667–70.

Barton, N., and Bandis, S. (1982). 'Effects of block size on the shear behaviour of jointed rock.' In *Issues in Rock Mechanics* Goodman, R. E. and Heuze, F. E. (eds.), Society of Mining Engineers of the American Institute of Mining, Metallurgical and Petroleum Engineers, New York, pp. 739–57.

Barton, N., and Choubey, V. (1977). 'The shear strength of rock joints in theory and practice.' *Rock Mech.*, **10**, 1–54.

Barton, N., Lien, R., and Lunde, J. (1974). 'Engineering classification of rock masses for the design of tunnel support.' *Rock Mech.*, **6**, 189–236.

Bieniawski, Z. T. (1974). 'Geomechanics classification of rock masses and its application in tunnelling.' *Proc. 3rd Int. Congr. Soc. Rock Mech., Denver,* **2**, Part A, 27–32.

Bock, H., and Wallace, K. (1978). 'Testing intact rock specimen'. In *An Introduction to Rock Mechanics*. H. Bock (ed.), James Cook University of North Queensland, Townsville, pp. 131–53.

Bowden, R. P., and Tabor, D. (1967). *Friction and Lubrication*. Methuen, London.

Broch, E. (1974). 'The influence of water on some rock properties'. *Proc. 3rd Int. Congr. Soc. Rock Mech.*, Denver, **2**, Part A, 33–8.

Brown, E. T. (ed.) (1981). *Rock Characterization, Testing and Monitoring*. Pergamon, Oxford.

Brunsden, D., and Prior, D. B. (eds.) (1984). *Slope Instability*. Wiley, Chichester.

Bryan, K. (1922). 'Erosion and sedimentation in the Papago country, Arizona, with a sketch of the geology.' *US Geol. Surv. Bull.*, **730-B**, 19–90.

Colback, P. S. B., and Wiid, B. L. (1965). 'The influence of moisture content on the compressive strength of rocks.' *Proc. 3rd Can. Rock Mech. Symp., Toronto*, 65–83.

Cooks, J. (1983). 'Geomorphic response to rock strength and elasticity.' *Zeitschrift fur Geomorphologie, N. F.*, **27**, 483–93.

Dackombe, R. V., and Gardiner, V. (1983). *Geomorphological Field Manual*. Allen and Unwin, London.

Deere, D. U., and Miller, R. P. (1966). 'Engineering classification and index properties for intact rock.' *Tech. Rep. No.* AFNL-TR-65-116, Air Force Weapons Laboratory. New Mexico.

Einstein, H. H., Bruhn, R. W., and Hirschfield, R. C. (1970). 'Mechanics of jointed rock. Experimental and theoretical studies.' *Department of Civil Engineering Report*, MIT, Cambridge, 115.

Fair, T. J. (1948). 'Hillslopes and pediments of the semi-arid Karoo.' *South African Geogr. J.*, **30**, 71–9.

Fecker, E. (1977). 'A new method for determining peak friction of rock joints.' *Bull. Int. Assoc. Eng. Geol.*, **16**, 198–200.

Griffith, A. A. (1921). 'The phenomena of rupture and flow in solids.' *Phil. Trans. Roy. Soc., London*, **A221**, 163–98.

Griffith, A. A. (1924). 'Theory or rupture', *Proc. 1st Congr. Appl. Mech.*, Delft, 1925, 55–63. Technische Bockhandel en Drukkerij, Delft.

Hawkes, I., and Mellor, M. (1970). 'Uniaxial testing in rock mechanics laboratories.' *Eng. Geol.*, **4**, 177–285.

Hoek, E. (1983). 'Strength of jointed rock masses.' *Géotechnique*, **33**, 187–223.

Hoek, E., and Bray, J. W. (1977). *Rock Slope Engineering* (2nd edn). The Institution of Mining and Metallurgy, London.

Hoek, E., and Brown, E. T. (1980). 'Empirical strength criterion for rock masses.' *J. Geotech. Eng. Div. Am. Soc. Civil Engrs.*, **106** (GT9), 1013–35.

Hudson, J. A., Brown, E. T., and Fairhurst, C. (1971). 'Shape of the complete stress–strain curve for rock.' *Proc. 13th US Rock Mech. Symp.*, University of Illinois, Urbana, 773–95.

King, L. C. (1953). 'Canons of landscape evolution.' *Bull. Geol. Soc. Am.*, **64**, 721–51.

Kragelskii, I. V. (1965). *Friction and Wear*. Butterworth, London.

Lama, R. D. (1972). 'Mechanical behaviour of jointed rock mass.' *Report* K-126, *Institute of Soil Mechanics and Rock Mechanics*, University of Karlsruhe, Karlsruhe.

Lama, R. D., and Vutukuri, V. S. (1978). *Handbook on Mechanical Properties of Rocks*, vols II–IV, Trans Tech, Clausthal.

Melton, M. A. (1965). 'Debris-covered hillslopes of the southern Arizona Desert—consideration of their stability and sediment contribution.' *J. Geology*, **73**, 715–29.

Moon, B. P. (1983). *Rock mass strength and the morphology of rock slopes in the Cape mountains.* Ph.D Thesis, University of the Witwatersrand, Johannesburg.

Moon, B. P. (1984). 'Refinement of a technique for determining rock mass strength for geomorphological purposes.' *Earth Surf. Proc. Landforms*, **9**, 189–93.

Moon, B. P., and Selby, M. J. (1983). 'Rock mass strength and scarp forms in southern Africa.' *Geografiska Annaler*, **65A**, 135–45.

Müller, L. (1974). 'Technical parameters of rock and rock masses.' In *Rock Mechanics*. Müller, L. (ed.), Springer, Vienna, pp. 16–34.

Newland, P. L., and Allely, B. H. (1957). 'Volume changes in drained triaxial tests on granular materials.' *Géotechnique*, **7**, 17–34.

Oberlander, T. M. (1972). 'Morphogenesis of granitic boulder slopes in the Mohave desert, California.' *J. Geology*, **80**, 1–20.

Patton, F. D. (1966a). *Multiple modes of shear failure in rock and related materials.* Ph.D thesis, University of Illinois, Urbana.

Patton, F. D. (1966b). 'Multiple modes of shear failure in rock.' *Proc. 1st Congr. Int. Soc. Rock Mech.*, **1**, 509–13.

Richards, K. (1982). *Rivers: Form and Process in Alluvial Channels*, Methuen, London.

Selby, M. J. (1974). 'Slope evolution in an Antarctic oasis.' *New Zealand Geogr.*, **30**, 18–34.

Selby, M. J. (1980). 'A rock mass strength classification for geomorphic purposes: with tests from Antarctica and New Zealand.' *Zeitschrift für Geomorphologie, N. F.*, **24**, 31–51.

Selby, M. J. (1982a). 'Controls on the stability and inclinations of hillslopes formed on hard rock.' *Earth Surf. Proc. Landforms*, **7**, 449–67.

Selby, M. J. (1982b). *Hillslope Materials and Processes*. Oxford University Press, Oxford.

Selby, M. J. (1982c). 'Rock mass strength and the form of some inselbergs in the central Namib Desert.' *Earth Surf. Proc. Landforms*, **7**, 489–97.

Selby, M. J. (1982d). 'Form and origin of some bornhardts of the Namib Desert.' *Zeitschrift für Geomorphologie N.F.*, **26**, 1–15.

Stagg, K. G. (1968). '*In situ* tests on the rock mass.' In *Rock Mechanics in Engineering Practice*. Stagg, K. G., and Zienkiewicz, O. C. (eds), Wiley, London, pp. 125–56.

Stimpson, B. (1981). 'A suggested technique for determining the basic friction angle of rock surfaces using core.' *Int. J. Rock Mech. Mining Sci. Geomech. Abstr.*, **18**, 63–5.

Swan, G. (1983). 'Determination of stiffness and other joint properties from roughness measurements.' *Rock Mechan. Rock Eng.*, **16**, 19–38.

Terzaghi, K. (1962). 'Stability of steep slopes on hard unweathered rock.' *Géotechnique*, **12**, 251–70.

Tse, R., and Cruden, D. M. (1979). 'Estimating joint roughness coefficients.' *Int. J. Rock Mech. Mining Sci. Geomech. Abstr.*, **16**, 303–7.

Voight, B. (ed.) (1978). *Rock Slides and Avalanches*, Elsevier, Amsterdam.

Weissbach, G., and Kutter, H. K. (1978). 'The influence of stress and strain history on the shear strength of rock joints.' *Proc. Int. Assoc. Eng. Geol. 3rd Int. Congr., Madrid*, Section II, **2**, 88–92.

Yatsu, E. (1966). *Rock Control in Geomorphology*. Sozosha, Tokyo.

Slope Stability
Edited by M. G. Anderson and K. S. Richards

Chapter 16

Mass Movement in Semi-Arid Environments and the Morphology of Alluvial Fans

ROGER LE B. HOOKE
Department of Geology and Geophysics
University of Minnesota, Minneapolis, Minnesota 55455, USA

16.1 INTRODUCTION

Although this volume is primarily concerned with mass movement, as opposed to fluvial processes, it is impractical to consider one in the absence of the other when discussing constructional landforms in semi-arid regions. Debris flows are the principal form of mass movement in such regions, although they are not restricted to them. Debris flows often evolve into fluvial events, and the fluvial part of the event generally results in important modifications of the debris-flow deposits. Furthermore the distinction between fluvial events and debris-flow events is not always clear as there seems to be a continuous gradation between the deposits formed by the two processes. Finally, the surficial characteristics of these constructional landforms are dominated by fluvial features.

A third process meriting discussion is sieve deposition, which combines elements of fluvial deposition and a variety of mass movement called grain flow (Middleton and Hampton, 1976). In grain flows, individual sediment particles are supported by each other, rather than by turbulent eddies in a transporting fluid. The two types of movement are distinct end members of what seems to be a continuum.

Both debris flows and grain flows are varieties of a class of flows called sediment gravity flows, in which the downslope movement of material is a result of gravity acting directly on the particles, rather than on a fluid which then entrains the particles. Hampton (1979) argues that these two types of flow are different because particles are supported by each other in grain flows, whereas they are supported by the matrix in debris flows. Takahashi (1980), on the other hand, suggests that this distinction is not significant because particles are supported by each other in both types of flow. In either case, the interstitial fluid, which can be air in the case of grain flows, is dragged along by the moving grains.

Because the fluvial end members of these processes are so important, and because there have been some exciting breakthroughs in the fields of fluvial mechanics and sediment transport in the past decade that have contributed significantly to our understanding of the morphology of depositional landforms in semi-arid and arid regions, we will begin this chapter with a discussion of fluvial processes. It would be a disservice to the imagination of the scientists who have contributed to these discoveries, as well as to the reader who may not be familiar with them, to do otherwise. In particular, we will begin by addressing the question of why channels on alluvial fans are braided.

16.2 FLUVIAL PROCESSES

Channels on alluvial fans are normally wide and shallow, or at least the flows in these channels are. There are probably two basic reasons for this. First, the banks of the channels are too weak to withstand the forces exerted on them by the flows. The banks thus erode rapidly when flows of any depth impinge upon them. The banks are weak, firstly, because there is little or no vegetation to stabilize them, and, secondly, because channels are normally steep so velocities, and hence bank shear stresses, are high. Alluvial fans typically have slopes of 0.05 to 0.20 in contrast to perennial rivers in more humid regions where slopes are one to three orders of magnitude less.

The second reason that channels are wide and shallow is that the transverse slope of the channel bottom towards its centreline is very low (Figure 16.1). This may seem circular, but it is an important point. When a transverse slope exists, particles entrained by the flow near the banks will have a tendency to migrate toward the centreline as they are swept downstream near the bed. The steeper the transverse slope, the greater this transverse flux of particles toward the centreline. In a steady-state situation there must be a process that balances this flux by moving particles outward towards the banks to replace those that are moved inward. Without such a process the channel would deepen near the banks, leading to increased bank erosion and hence widening.

In sand-bed streams, the balancing process is probably outward diffusion of sediment carried in suspension, and subsequent deposition of some of this sediment to replace that which was entrained at the bed and moved inward towards the centreline (Parker, 1978a; Pizzuto, 1984). In gravel-bed streams on alluvial fans, however, suspension will be small or negligible. Thus the channel must adjust itself so that the inward flux of sediment along the bed is also negligible. This adjustment occurs through erosion of the bed near the channel boundaries, and consequent widening of the channel until transverse bed slopes are too low to support a significant transverse flux of sediment. Width to depth ratios are thus large in such channels. Parker (1978b) shows that, under such conditions, there may be a redistribution of stress near the banks that reduces entrainment of sediment.

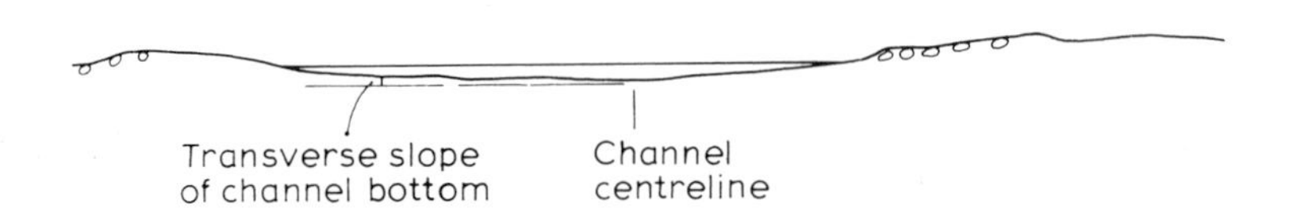

Figure 16.1 Sketch showing typical cross-section of channels on alluvial fans

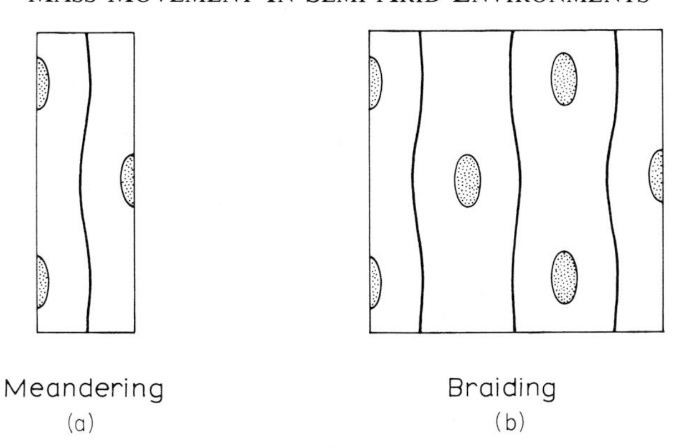

Figure 16.2 Relation between the number of bars that will form through growth of instabilities in a uniform flow over a movable bed and the width–depth ratio of the flow. Both flows are of the same depth. In the wider one bars will develop in the centre of the channel as well as along the banks. (*Modified from Parker, 1976, Fig. 3,* J. Fluid Mech., **76**. *Reproduced by permission of Cambridge University Press.*)

It might be worth noting that this conceptual model for explaining cross-sectional shapes of alluvial channels with large width to depth ratios is as applicable to steep, gravel-bedded channels on alluvial fans as it is to much gentler sand-bedded channels, so long as the flows in the latter are too slow to result in significant suspension, and so long as the sand is not stabilized by vegetation. An example is provided by channels on laboratory stream tables.

The processes that lead to the large width–depth ratios typical of channels on alluvial fans are important in fan morphology because of a fundamental instability that is a characteristic of all channels in alluvial materials, and that results in braiding of channels with large width–depth ratios. This instability, apparently first suggested by Callandar (1969), was investigated rigorously by Engelund and Skovgaard (1973) and Parker (1976).

Through elegant mathematical analysis, Engelund and Skovgaard and Parker have shown that there is an instability in water flowing in a straight uniform channel over a movable bed. The instability manifests itself through formation of alternate bars. An infinitesimal perturbation of the bed in one place will grow (into a bar) and give rise to simultaneous perturbations at predictable intervals along the channel. In a narrow deep stream ($w/d < 200$) the bars will be on opposite sides of the channel (Figure 16.2(a)) resulting in a sinuosity in the flow, which then develops into the well-known meandering geometry of such rivers. In channels with larger width–depth ratios there will also be central bars in the channel (Figure 16.2(b)). The number of such bars in a cross-section increases with increasing width–depth ratio. This is the fundamental cause of braiding in such rivers, and explains, in large part, why active surfaces of alluvial fans are characterized by braided channels.

16.3 MASS MOVEMENT PROCESSES: (i) DEBRIS FLOWS

Debris flows consist of a mixture of mud and rocks, along with whatever additional debris may be present on valley side slopes and in channel bottoms, including trees, and in modern times the occasional car or road grader (Figure 16.3), perhaps abandoned in haste by an unfortunate driver. To appreciate the power of debris flows, it is convenient to start by contrasting them with the fluvial processes we have been discussing.

Figure 16.3 Part of what has been reputed to be a road grader caught in a debris flow on Surprise Fan in Panamint Valley, California

Natural debris flows may have viscosities of over 1,000 poise and densities of 2.0 to 2.4 g cm^{-3} (Sharp and Nobles, 1953, pp. 552–3; Costa, 1984, p. 272). In contrast, the viscosity of water is ~ 0.01 poise, and the density of a stream carrying suspended sediment is only slightly greater than unity. Some authors (e.g. Costa, 1984, pp. 287–9) identify an intermediate type of flow, called a 'hyperconcentrated flow', with densities between 1.3 and 1.8 g cm^{-3}. Herein this distinction is not made, and flows of this type are simply considered to be a more fluid variety of debris flow.

The response of water to an applied shear stress, τ, can be represented by:

$$\tau = (\mu + \eta) \frac{du}{dy} \tag{16.1}$$

where du/dy is the rate of shear in turbulent flow and μ and η are the molecular and eddy viscosities, respectively. The eddy viscosity generally increases with rate of shear, but the molecular viscosity is constant (Rouse, 1950, p. 88). In constrast, debris flows behave as thixotropic substances, and their rate of deformation under stress is represented approximately by:

$$(\tau - \tau_c) = \mu(\tau) \frac{du}{dy}, \qquad \tau \geqslant \tau_c \tag{16.2a}$$

(Hooke, 1967; Johnson, 1965) where:

$$\tau_c = c' + (\sigma_n - u)\tan\phi' \qquad (16.2b)$$

(Johnson, 1965; Johnson and Rodine, 1984). In equation (16.2a), τ_c is the yield strength, and $\mu(\tau)$ indicates that the apparent viscosity, μ, is a function of the applied stress. No flow occurs until the yield strength is exceeded, but then the viscosity decreases gradually with increasing applied stress. The form of the function $\mu(\tau)$ differs among flows, owing to differences in flow density and in mineralogy of the silt–clay fraction. A constant value may be reached at high enough stress.

In equation (16.2b), which is based on the Mohr–Coulomb equation of soil mechanics (Terzaghi, 1950), c is the cohesive strength of the material, σ_n is the normal stress on the shear plane, u is the pressure in the pore fluid, and ϕ is the angle of internal friction. In most soil mechanics applications, the pore fluid is water, but in the case of debris flows the pore fluid is a water–clay–silt–sand–. . . mixture (Rodine and Johnson, 1976). This fluid, in fact, consists of the water plus all particles smaller than the ones that, for any purpose, are considered to form the 'clasts' in the flow. c and ϕ again vary with water content, mineralogy, and sorting of the granular material. In studies of material extracted from natural debris flows, for example, Johnson and Rodine (1984) found that c and ϕ decreased exponentially with increasing water content; c varied between about 40 and 1000 Pa and ϕ between about 0.5° and 25° with variations in water content of only 10% by weight. The sensitivity of these parameters, particularly ϕ, is due to the fact that in a poorly-sorted material larger particles cannot settle down into hollows among particles of similar size because these hollows are filled with smaller particles. Interlocking of particles is, therefore, decreased in importance, and the dilation that would normally be necessary to shear such a material is significantly reduced. Thus as long as there is ample fluid to slightly more than fill remaining pore space, settling of particles into interlocking positions, once they are dilated, is inhibited both by the buoyancy of the fluid and by the lack of spaces into which to settle (Rodine and Johnson, 1976). Consequently the cohesion of the material is reduced to that of the matrix, and the angle of internal friction approaches 0. Rodine (1974) and Johnson and Rodine (1984, pp. 290–310) discuss techniques for measuring these parameters.

Other possible constitutive relations between shear stress, shear strength, and velocity gradient are discussed by Johnson and Rodine (1984, pp. 277–90).

The shear stress at any level in a flow is given approximately by:

$$\tau = \varrho g d S \qquad (16.3)$$

where ϱ is the bulk density of the flow, g is the acceleration due to gravity, d is the depth below the surface, and S is the hydraulic gradient which is approximately equal to the surface slope. From consideration of equations (16.2) and (16.3) it will be seen that if either d or S is too low in a given flow, τ will not exceed τ_c and there will be no deformation. This is the case, for example, near the surface of a flow. Thus there should be a semi-rigid layer that rides along on the underlying viscously-deforming material. If the thickness of the rigid layer is greater than the thickness of the flow, the flow will stop. As expected, Johnson (1965) found that the thickness of the rigid layer was sensitive to the surface slope. The thickness of the rigid layer is also sensitive to variations in water content or grading of the material through the influence of these two parameters on ϕ.

Takahashi (1980) has also studied the conditions under which debris flows will stop and reached rather similar conclusions. His equations differ, however, because he ignores cohesion and assumes that the normal pressure at a given depth should be proportional to the water pressure at that depth plus the weight of overlying solids, rather than simply to the bulk density of the overlying material.

Owing to the high density of debris flows, forces on rocks in them are substantially different from those on comparable rocks in a stream. First, the submerged weight of a rock is reduced relative to its submerged weight in water. Because the bulk density of the displaced 'pore fluid' should be used in calculating this buoyancy effect, all but a few per cent of the weight of very large boulders may be supported by it (Rodine and Johnson, 1976, pp. 218–19, 233). Additional buoyancy may be provided by a pore-pressure increase associated with transfer of the weight of coarse grains to water in the fine matrix (Hampton, 1979). Secondly, as a result of the higher density, bed shear forces and drag forces are also higher than in a stream of comparable depth. Thus gravitational forces tending to prevent motion are reduced, and drag forces are increased.

The ability of a stream to transport sediment in suspension is closely governed by a balance between the settling velocity of the sediment and net upward transport of material by turbulent eddies (Vanoni, 1963). With the reduced settling velocity in debris flows, coarser material can be maintained in suspension. Dispersive pressure forces (Bagnold, 1954, 1955) may also keep material in suspension in debris flows with high concentrations of clasts.

These considerations suggest another fundamental difference between water flows and debris flows. Whereas streams vary their sediment load readily by deposition or erosion and will continue to flow as long as a slope exists, debris flows cannot selectively deposit any but possibly the coarsest fragments. This means that a debris flow cannot turn into a stream by deposition. Conversely evidence seems to be accumulating to the effect that water flows rarely if ever evolve into debris flows by moving over and entraining loose sediment.

Because of the behaviour described by equations (16.2), debris flows normally have steep rounded fronts, thicker parts move more rapidly than thinner parts, and surges in which the thickness is increased relative to that nearer the front move faster than the front and may overtake it (Jahns, 1949, pp. 12–13). Once the flow stops, such surges may remobilize it (Sharp and Nobles, 1953, p. 551; Johnson, 1965), or if they are too small to remobilize the whole flow, they may cause small pressure ridges to develop parallel to the flow borders (Jahns, 1949, p. 13). The main body of a debris flow will normally follow the main channel on a fan, but if the flow thickness exceeds the height of the channel banks, small lobes may form diverging from the channel. Wider lobes form further down-fan where the main channel, which normally has a slope lower than that of the adjacent fan surface, merges with this surface (Beaty, 1963). Debris flows are normally followed by water flows of substantially longer duration that erode the debris flow and re-excavate the main channel (Pack, 1923, p. 355; Blackwelder, 1928, p. 470; Beaty, 1963, pp. 520–1).

16.3.1 Generation of Debris Flows

As implied by the above discussion, the details of the processes by which debris flows are generated are not obvious. We know that they can result either from intense rainstorms

(Jahns, 1949), or from rapid snowmelt (Sharp and Nobles, 1953), but how does the concentration of sediment in the flow become so large that particles can no longer settle out, in contrast to the normal behavior of streams?

Johnson and Rodine (1984, pp. 310–33) have recently reviewed the status of observational evidence on this subject. They conclude that most debris flows are initiated by some form of sliding or slumping of material that has become saturated with water to the point where its shear strength is reduced below the applied shear stress. Once the material begins to slide, the movement results in tensile stresses that exceed the local tensile strength widely throughout the mass. Dilation occurs and cracks open admitting more water. This addition of water occurs at almost precisely the water content at which c and ϕ are most sensitive to changes in water content. τ_c thus quickly decreases and flow begins.

This explanation is satisfactory in situations in which there is reason to believe that the soil may have been saturated prior to initiation of the flow. Two examples are the debris flows initiated in soil over permafrost described by Sharp (1942), and the flows initiated by snowmelt near Wrightwood, Ca (Sharp and Nobles, 1953; Johnson, 1970, pp. 337–43).

Debris flows initiated by intense rainfall pose a slightly different problem. Four possibilities seem to exist in this case:

1. The rainfall, coupled with whatever less intense rainfall preceded it, suffices to saturate the soil to the point where failure occurs. This may happen when more water infiltrates into the regolith through the ground surface than is removed by percolation into deeper soil or rock units (Kessli, 1943, p. 347). The continuing rain then adds the small amount of additional water necessary to turn the slide into a flow.
2. The rainfall results in rivulets that, as they course down a steep slope, erode the soil and leave banks that are over-steepened. These banks fail due to their high moisture content (Johnson and Rodine, 1984, pp. 321–4). The failed banks, possibly only a few centimetres high, provide the slurry of mud, stones, and water that form the flow.
3. Streams formed by the cloudburst impact on debris at a break in slope, as at the base of a waterfall, and churn it up to such an extent that it mixes completely with the water and flows downslope as a debris flow.
4. The rainfall of itself, through splash erosion, churns up enough soil to form a debris flow.

Possibility 2 is supported by Johnson and Rodine's (1984, pp. 321, 326) observations of rills heading in arcuate scars in the source areas of some of the flows they studied. Possibility 3 is Johnson and Rodine's 'firehose' effect, and possibility 4, for which no field documentation exists, is similar to the 'firehose' effect, but on a smaller scale. Caine (1980), incidentally, found that debris flows rarely occur if the rainfall intensity in millimetres per hour is less than about 15 times the duration, in hours, raised to the -0.39 power. The physical significance such a limit might have, however, was not discussed. Perhaps there is some connection with infiltration rate, as mentioned in possibility 1.

An interesting combination of these processes is described by Means (cited by Costa, 1984, p. 271). Debris that slid or was washed down gullies in a steep valley side slope clogged the main channel at the base of the slope. A subsequent flow in that channel was dammed by this debris, infiltrated into it, and thus mobilized it into a debris flow. Similar combinations of events may be responsible for generation of debris flows in other situations.

Field observations thus seem to suggest that some special conditions are necessary to achieve the high sediment concentrations characteristic of debris flows. Rivulets or streams flowing over soil do not appear to be able to entrain enough material to form a debris flow. They may, however, undermine banks that, upon collapse, result in debris flows. (Johnson and Rodine, 1984, p. 331, cite one field observation and some laboratory studies in which a slurry comprised primarily of clay and water eroded enough additional debris to '. . . become a typical debris flow'. This, however, may be a case of a subspecies of debris flow, called a mudflow due to its lack of coarser material, evolving into a 'typical' debris flow by entraining coarser particles.)

Conversely debris flows do not appear to be able to evolve into stream flows by selective deposition of particles. However, addition of water to a debris flow, perhaps from tributary channels in which debris flows did not form, may dilute a flow to the point where it can deposit material selectively and, in the limit, become a stream flow.

In summary, while a continuum apparently exists between stream flows and debris flows, the transformation of one type of flow into the other apparently requires addition of either water or sediment through some process external to the flow itself.

16.3.2 Field Criteria for Recognition of Debris-Flow Deposits

Debris-flow deposits consist of cobbles and boulders embedded in a matrix of fine material. Individual beds are generally poorly sorted and unstratified. Denser flows commonly result in matrix-supported beds, whereas more fluid flows may leave clast-supported strata. Recent

Figure 16.4 Recent debris flow in Death Valley, California. Note older flows on fan surface in background. (Photo by N. Potter.)

debris flows have flat tops, steep sides, and a lobate form (Figure 16.4). The base makes an abrupt contact with underlying material. Rain and rill erosion, creep, and weathering eventually modify this distinctive morphology and internal character. Cobble and boulder accumulations with low relief are commonly the only remaining evidence of the flow, but their positions as levées along channels or as lobes diverging from channels usually leaves little doubt as to their origin.

In vertical or near-vertical exposures the only characteristic that unequivocally indicates a debris-flow origin is the matrix-supported character of the deposit. More fluid flows that result in clast-supported deposits often cannot be readily distinguished from stream gravels with a secondary matrix of fine material. The presence of large cobbles or boulders, or of unusually poor sorting is indicative of a debris-flow origin (Bluck, 1964; Bull, 1972) if it can be demonstrated that the coarse particles were actually transported by the flow, rather than, for example, dropped into the channel from an eroding bank, and if the poor sorting is clearly primary rather than being a result of infilling of pore space by later flows carrying finer material (Wasson, 1974, p. 88). Conversely, open pore space, torrential cross-bedding, scour and fill structures, or other distinctively fluvial features would be indicative of a fluvial origin. Some authors have suggested that tabular stones in a debris-flow deposit have a preferred orientation, with long axes either normal to the flow direction if the flow is particularly viscous (Bull, 1972, p. 70; 1977, p. 237), or parallel to the flow boundaries (Johnson and Rodine, 1984, p. 266) if the viscosity is lower. Coarse-grained fluvial deposits, on the other hand, may have an imbricate structure (Picard and High, 1973, pp. 98–100).

In summary, there appears to be a continuum between deposits of demonstrably fluvial origin and those of demonstrably debris-flow origin. This, of course, makes it difficult for those who would establish sedimentological criteria for distinguishing the two processes. To establish such criteria, one must have a suite of samples from deposits of both types, and must know *a priori* which process produced the deposit.

16.3.3 Effects of Debris Flows on Fan Morphology

One characteristic feature of many alluvial fans is a large number of cobbles and boulders. The occurrence of this material in levees and lobes and its association with material of recognizable debris-flow origin suggest that much of it is transported to the fan by debris flows.

Coarse material frequently accumulates at the front of a debris flow and is shoved aside by the advancing snout, forming levées that confine the remainder of the flow (Sharp, 1942, p. 225). The bouldery, sharp-crested levées common on many fans were probably formed in this way. Levées also form when a debris flow at peak discharge overflows channel banks. Such levées are generally wider and have more rounded crests than those described by Sharp.

A distinct sorting of stones, with pebbles on the inside (towards the flow) and coarser material on the outside, is commonly observed in natural debris-flow levées (Hooke, 1967; Johnson, 1965). This sorting probably does not result from selective deposition of coarser material by slower flow because the size difference is not large and the coarser material is probably not heavy enough to be dropped independently of the main body of the flow. Instead, coarser material selectively migrates to the surface (Jahns, 1949). Either the higher surface velocity or the sliding and rolling processes described by Sharp (1942, p. 225) then move it to the front and edges of the flow, where it is deposited when the edges stop moving.

As noted, debris flows stop when the shear stress on the bed, τ_b, no longer exceeds the yield strength of the mud, or $\tau_b < \tau_c$. This condition may result from a loss of water to the underlying dry fan material, thus increasing the yield strength, or from a decrease in either fan slope or flow depth as the flow moves down-fan and spreads out. Thus the areal extent of a debris flow is limited by its volume and yield strength, and by the slope of the fan surface. Furthermore, its down-fan extent is determined in large part by the degree to which existing channels prevent lateral spreading at the fanhead.

Consequently, most deposition near the toes of fans is probably by running water rather than by debris flows, although the material deposited may have been eroded from debris-flow deposits higher on the fan. In contrast, much of the deposition near the fanhead is probably caused by debris flows overtopping the channel banks (Figure 16.4). Thus the stratigraphy of fans on which debris-flow deposition has been important is inhomogeneous; debris-flow deposits nearer the fanhead interfinger in the midfan region with water-sorted material deposited nearer the toe.

Debris flows may also affect the difference in slope between the flanks of fans and the axis. This slope difference, which we will refer to as $\Delta\phi$, is principally a result of the fact that larger discharges have greater momentum, and are thus less readily diverted toward the flanks. Deposition on the flanks is consequently by smaller discharges that require steeper slopes to transport their sediment load (Hooke and Rohrer, 1979). Debris flows that spill out of the main channel near the fanhead and flow down the flanks may increase $\Delta\phi$, however, because they often do not reach the toes of the fans. They thus tend to deposit more material near the fanhead, increasing the slope. Larger values of $\Delta\phi$ were found on laboratory fans constructed, in part, by debris flows, and qualitative field observations suggest that this may also be true of natural fans.

16.3.4 Lithologic and Climatic Control of Debris Flows

As Blackwelder (1928, pp. 473–4) observed, the proportion of recognizable debris-flow material in and on fans varies widely. Some fans have practically no deposits of demonstrably debris-flow origin, but others consist predominantly of debris-flow material. These differences appear to be, in part, lithologically controlled. For example, the source areas of two fans in Deep Springs Valley, California (Hooke, 1967), are underlain by resistant carbonate or quartzitic rocks, and valley-side slopes are steep. Thus material is quickly transported to the fans with little pretransport weathering. As a result there is little fine material in the source area, and debris-flow formation is inhibited. In contrast, exposures of readily weathered sandy dolomite in the source area of an adjacent fan contribute substantial amounts of fine material and appear to be responsible for debris-flow deposition on this fan. Climatic and topographic factors do not vary significantly between these fans and hence are unlikely to be responsible for the differences in mode of deposition.

Another example of lithologic control is found in Death Valley, California. The youngest time stratigraphic units on Starvation Fan contain an usually high percentage of debris-flow material derived from a small quartz-monzonite plug, the Little Chief prophyry (Hooke, 1972). Some of these flows were enormous; Hunt and Mabey (1966) estimate

Figure 16.5 Sequence of three debris flow deposits, in this case averaging about 2 m in thickness, interbedded with thin fluvial units. Starvation Fan, Death Valley, California

volumes of between 6 and 20 million cubic metres, and one relatively recent flow topped a bank of the main channel which is now about 15 m high. In vertical exposures of some of the older gravels (Figure 16.5) debris flows averaging 1 m in thickness commonly comprise more than 75% of the gravels present. Other nearby fans that do not drain the area underlain by the quartz monzonite have appreciably less material of demonstrably debris-flow origin.

The proportion of debris-flow deposits on Starvation Fan also appears to vary with climate. Exposures of a time-stratigraphic unit older than those described above, and probably deposited during a pluvial period associated with the Illinoisan glaciation (Hooke and Lively, 1983), generally contain less than 50% debris-flow deposits. As lake levels in the valley during deposition of this time-stratigraphic unit were higher than at any time since, we infer that the climate was wetter and that there was more vegetation stabilizing the slopes.

16.3.5 Erosion by Debris Flows

There has been some discussion in the past regarding the ability of debris flows to erode. Lustig (1965), for example, believed that debris flows were responsible for eroding the deep channels that are so often found at the heads of fans. He and also Costa (1984, p. 274) argue that the large thickness and density of debris flows would result in high shear stresses on the bed, and hence in erosion. This, however, overlooks another important prerequisite for erosion: turbulent eddies must have the strength to break through any laminar sublayer that may exist near the bed. Otherwise this sublayer tends to protect particles from erosion. A laminar sublayer will exist if the boundary Reynolds number is less than about 3 (Vanoni, 1975, p. 231), which will often be the case in debris flows due to their high viscosity. This high viscosity will also tend to damp turbulence and increase the thickness of the laminar sublayer (Rouse, 1950, p. 101), thus inhibiting erosion.

Observational evidence for erosion by debris flows is ambiguous. In conversations with mechanical equipment operators who have cleared highways covered by debris flows, I have heard it said that pavements under flows were undisturbed, in contrast to the situation after water flows. In the absence of a highway pavement, some erosion may occur, of course. Johnson and Rodine (1984, p. 273), for example, mention channels being '. . . swept clean of debris . . .', presumably meaning that particles that projected well up into or through the laminar boundary layer were removed. They (1984, p. 321) also note, however, that in the source areas for debris flows there is frequently a transitional zone of no erosion between the slide scar whence the material came and the area where debris-flow levées indicate deposition.

Authors who view deeply scoured channels after the passage of a debris flow and write that 'most of the scour . . . was clearly due to' erosion by the debris flow, without presenting details of the observational evidence leading to this conclusion do not help resolve this problem. Among other things, such statements ignore the fact that in the absence of a subsequent water flow, a deposit of debris-flow material equal in thickness to the thickness of the rigid layer at the surface of the flow, should be found in the channel. Thus water must have eroded at least this much material from the channel bottom. How can they prove that this water was not responsible for more of the erosion?

On balance, the present evidence seems to be against significant erosion by debris flows, but the question is still open.

16.3.6 Relative Importance of Debris Flows and Water Flows

Volumetric estimates of the role of debris flows in transporting material to a fan are difficult to make. Among the problems are first the fact that the process by which material was transported to a fan is not necessarily the process by which that material was last reworked, and second the previously mentioned lack of adequate criteria for unambiguous recognition of debris-flow deposits. The characteristically heterogeneous matrix-supported deposits left by more viscous debris flows are often sorted and stratified by subsequent water flows, and deposits of less viscous debris flows may not be readily distinguishable from fluvial deposits. These factors may be partially responsible for the lower percentage of debris-flow material recognized by Blissenbach (1954, p. 179) on fans in areas of higher precipitation. Conversely, however, the best exposures of fan stratigraphy are usually in the main channel near the fanhead. Because debris-flow deposition is likely to be more common here, estimates may be biased in its favour.

Fans on which deposition has been dominated by viscous debris flows tend to have more irregular surfaces due to the persistence of debris-flow lobes. Surfaces of fans of predominantly fluvial origin, in contrast, will be smoother because braiding of fluvial flows requires less lateral variation in relief than splitting of debris flows, and fluvial events thus result in more uniform deposition. Phrased differently, if profiles are surveyed along a circular arc with its origin at the fanhead, fans composed of a large quantity of debris-flow material will tend to have many flat-topped positive relief features—the debris flow lobes—whereas similar profiles on fans dominated by fluvial deposition will have a predominance of flat-bottomed negative features—the braided channels.

In conclusion, based on several seasons of field work on fans in arid regions I would guess that the fan on which more than 50% of the material was *demonstrably* last deposited by debris flows would be rare, and that fans with less than 20% debris flow material would be most common. A contrary view is expressed by Beaty (1963).

16.4 MASS MOVEMENT PROCESSES: (ii) SIEVE DEPOSITION

Earlier students of alluvial fans have suggested that infiltration reduces the sediment-carrying capacity of fluvial flows and instigates deposition (Trowbridge, 1911, p. 738; Eckis, 1928, p. 237; Blissenbach, 1954, p. 178; Bull, 1964a, p. 17). However, Bull (1964b, p. 104) has shown that water on some fans composed of fine material does not percolate below the root zone of vegetation.

Laboratory observations (Hooke, 1967) suggest that permanent features attributable to infiltration will not be found on most fans. Surficial wetting of dry fan material may cause deposition during small flows or in the initial stages of large flows, but the amount of such deposition is small, and any features produced will be minor and will be destroyed by subsequent higher discharges.

If the fan material is sufficiently coarse and permeable, however, the entire flow may infiltrate before reaching the toe of the fan. Under these conditions a lobe of debris is deposited at the point where water is unable to effect further transport. Because water passes through rather than over such deposits, they act as strainers or sieves by permitting water to pass while holding back the coarse material in transport. The lobate masses thus formed are called 'sieve lobes' or 'sieve deposits', and the mode of formation is sieve deposition (Hooke, 1967). Extensive deposits inferred to have been formed in this way have been found on several fans.

To study this process, small alluvial fans were built in a 1.5 m square box in the laboratory by running water through a channel filled with coarse sand and granules, and thence into the box. Several 'episodes' were required to build each fan; each episode involved packing the inlet channel with sand and then running water over the sand for a fixed length of time.

On these laboratory fans, sieve deposition was initiated by deposition of granules which formed an initial debris barrier. The channel slope immediately upstream from this barrier was reduced by backfilling (Figure 16.6), and further deposition ensued. Deposition did not actually occur in layers as represented in Figure 16.6, but schematically this provides an easy way to visualize the process. When regrading of the channel above one barrier is complete, water is again able to transport material to the front of the lobe, and a new barrier is formed just up fan from the preceding one.

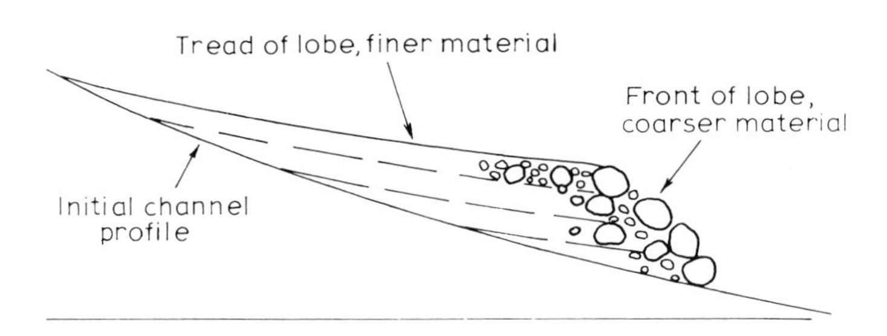

Figure 16.6 Schematic sketch of growth of a sieve lobe. (*From Hooke, 1967, Fig. 9,* J. Geol., **75**. *Reproduced by permission of the University of Chicago Press.*)

Deposition of coarser material along the lateral edges of a lobe, where competence is reduced, may confine the flow temporarily. However, lateral shifting of the flow, a decrease in infiltration as fine material is deposited upstream, or a slight increase in discharge can result in diversion of the flow over the lobe's flanks or front. In such cases rapid erosion ensues, and a fluidized debris mass, or grain flow, shoots down-fan a few centimetres and stops. Backfilling behind this deposit proceeds until this part of the lobe is built up to the elevation of the remainder.

The radial position of sieve lobes on laboratory fans was controlled by a balance between discharge and infiltration rate. For instance, with decreasing discharge and increasing fan thickness (increasing infiltration rate) the deposits were found progressively nearer the fanhead.

Thus, sieve deposition on laboratory fans is a consequence of the coarseness of the material and the resulting high infiltration rate. It may be initiated either by complete loss of discharge through infiltration, as is the case for sieve deposition near midfan, or by a break in slope, as at the toe of laboratory fans. Sieve deposition on natural fans is a result of the same factors, as the subsequent discussion of two examples will demonstrate.

16.4.1 Sieve Deposition on Natural Fans

The drainage basin of Shadow Rock Fan (Figure 16.7) is underlain predominantly by quartzite of the Lower Cambrian or Pre-Cambrian Campito Formation (Nelson, 1966). This rock is resistant to weathering, as evidenced by a dark desert varnish on rocks on older parts of the fan, and by the paucity of fine material in both recent and older deposits. Consequently infiltration rates are high in the older deposits as well as in recent sieve lobes, and sieve deposition can occur anywhere on the fan (Figure 16.7). The location of a given sieve lobe is determined by the size of the discharge that formed it; larger discharges continue further down-fan before loss of water is sufficient to cause deposition. Once the flow has infiltrated it apparently does not return to the fan surface. Hence freshly abraded gravel does not extend to the toe of the fan.

In this situation sieve lobes formed by low discharges are commonly modified by subsequent higher flows. Modification involves dissection of the original deposit, which is usually as wide as the channel that formed it, leaving several small disconnected lobes. For instance, deposits 5 and 6 on Shadow Rock Fan (Figure 16.7) apparently were dissected during construction of deposit 7, whereas lobes 1–4 are relatively unaltered and thus

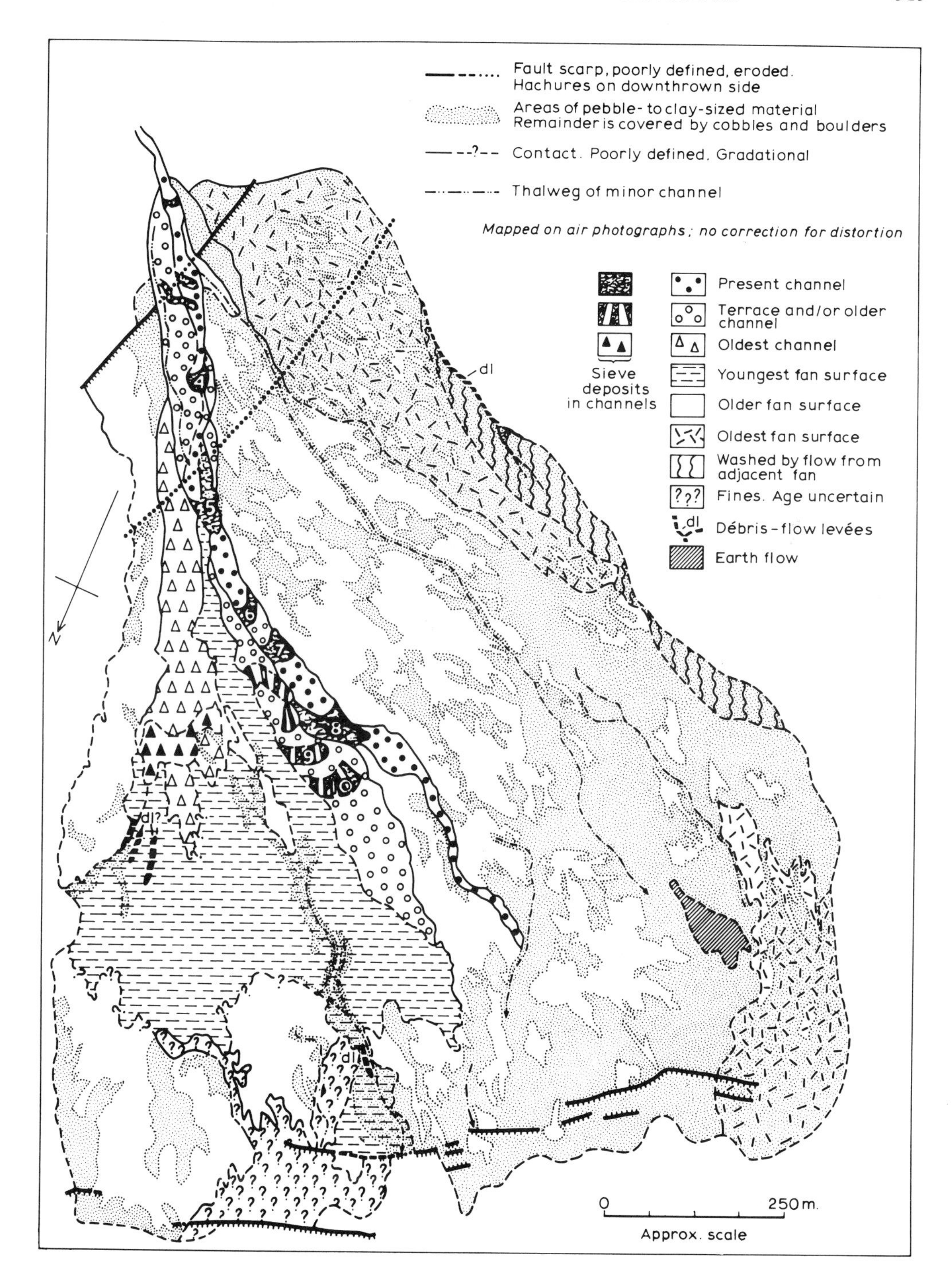

Figure 16.7 Geomorphic map of Shadow Rock Fan. (*From Hooke, 1967, Fig. 3,* J. Geol., **75**. *Reproduced by permission of the University of Chicago Press.*)

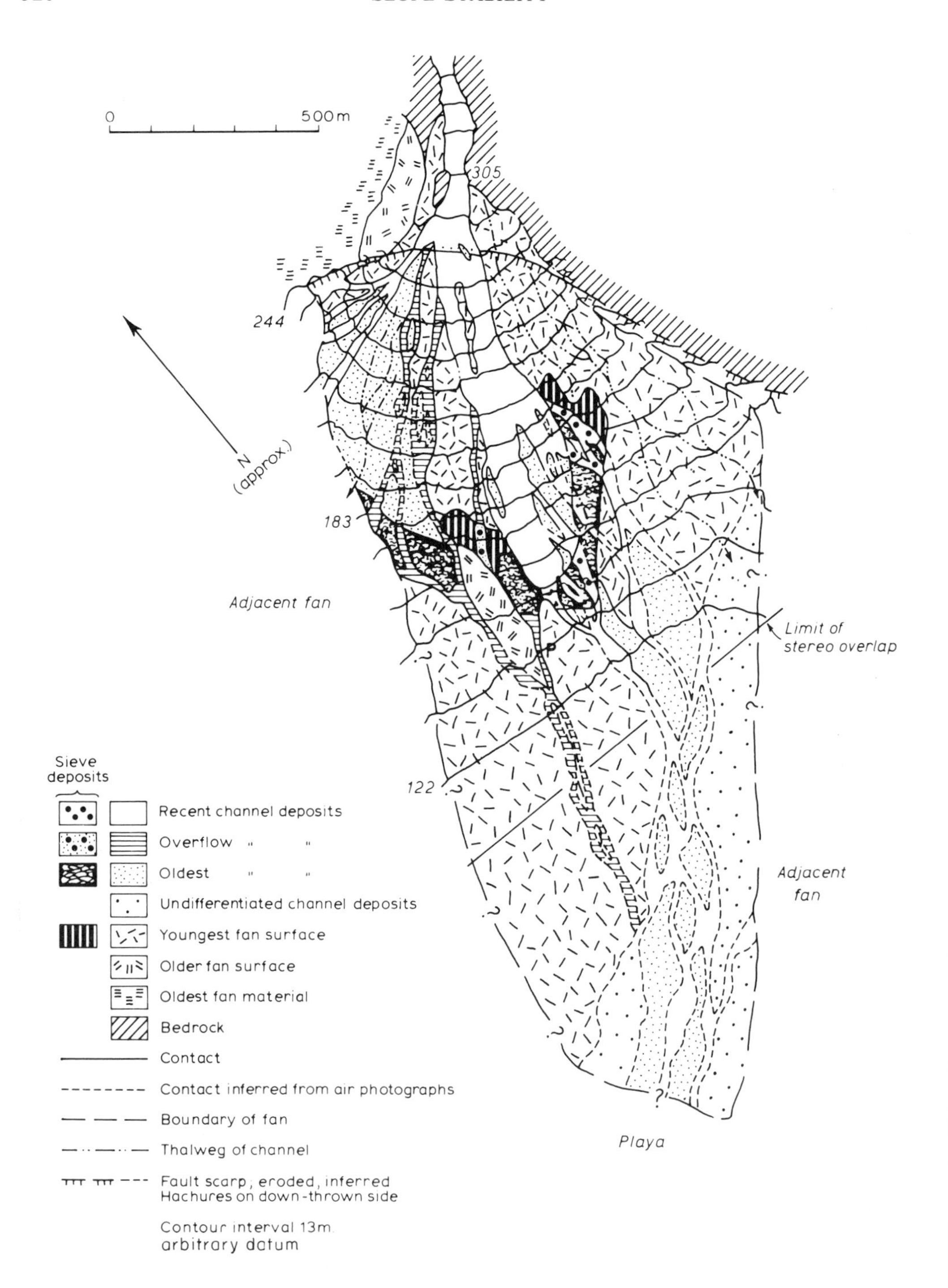

Figure 16.8 Geomorphic map of Gorak Shep Fan mapped in the field on hazy, high-altitude air photographs. (*From Hooke, 1967, Fig. 2.* J. Geol., **75**. *Reproduced by permission of the University of Chicago Press.*)

postdate 7. Modification of lobes is an important process, as unmodified deposits are rare on older parts of the fan surface.

Sieve deposition on Gorak Shep Fan (Figure 16.8) occurs at a break in slope at the down-fan limit of the upper part of the fan. The decrease in competence at the break in slope between the two segments presumably causes deposition and is believed to be responsible for the existing sieve deposits.

The source area of Gorak Shep Fan is underlain by a thick section of resistant early Palaeozoic (?) carbonate rocks, primarily dolomite. Slopes are steep, have virtually no soil cover, and supply the fan with pebble- and cobble-sized debris, with a minimum of fines. On the other hand, older deposits on the fan have a matrix of sand- to clay-sized particles produced by post-depositional weathering.

Because recent channel deposits rest on older material in which the void space is filled, infiltration alone is apparently not sufficient to instigate sieve deposition. A break in slope, as at the toe of laboratory fans, is also necessary. Water passing through sieve deposits on Gorak Shep Fan emerges below them and continues down-fan as surface runoff, as the extensive channel development on the lower segment indicates (Figure 16.8). Thus sieve deposition on Gorak Shep Fan occurs in a restricted zone, in contrast to the ubiquitous occurrence of such deposits on Shadow Rock Fan. This contrast reflects the coarser debris and much greater resistance to weathering of the quartzite on Shadow Rock Fan, where older deposits still have open pore space.

Fans on which sieve deposition has predominated may also have a larger than usual variation in slope with azimuth, $\Delta\phi$. This is because the coarser material of which such fans are constructed requires higher discharges for transport, and as noted, these higher discharges are less readily diverted to the flanks.

While sieve deposition is basically a fluvial process, the grain flows that periodically cascade down the fronts of developing sieve lobes are a form of mass movement.

16.5 DISTINGUISHING BETWEEN DEBRIS-FLOW AND SIEVE DEPOSITS

Both sieve and debris-flow deposition may take place on the same fan. However, for debris flows to form, substantial amounts of fine material must be present in the source area. Conversely sieve deposition cannot occur if too much fine material is available. Therefore, one process usually dominates, and conclusive evidence that one is significant usually implies that the other is much less important. For instance, Shadow Rock Fan is predominantly composed of sieve deposits, but four linear ridges (Figure 16.7) are probably debris-flow levées, possibly built by flows originating on the fan. A fifth levée, the long, branched one on the south side of the fan near the apex, was apparently built by a debris flow originating in the dolomitic terrane between Shadow Rock Fan and the fan to the southwest.

Distinguishing between materials deposited by these two processes after they have been modified by erosion and downslope movements is not easy. The following criteria were developed through comparison of deposits on various fans:

1. Recent sieve deposits are composed of pebble-sized to boulder-sized material without fines and thus have a high infiltration rate. In contrast, recent debris-flow deposits have a matrix of fine material.

2. Sieve deposits rarely contain the especially large boulders (>1 m diameter) found in many debris flows.
3. Relatively unmodified debris flows are typically 0.5–3 m thick and several times as wide. In contrast, the fronts of sieve deposits on Shadow Rock Fan are commonly 3 to 10 m high, but, owing to dissection, their treads are usually narrow and rounded in cross-section.
4. Contacts between debris flows and the underlying material are usually sharp and well defined, and flows appear to overlap the older deposits. Contacts between sieve deposits and underlying material are generally gradational and rarely give the impression of an overlapping relationship.
5. Debris flows can generally be traced some distance up-fan and have a slope approximately equal to the fan-surface slope, whereas sieve lobes typically have short treads with slopes less than that of the fan (Figures 16.6 and 16.7).
6. Debris-flow levées are distinctive and are indicative of debris-flow action.
7. Fresh sieve deposits are clearly related to a channel from the fanhead, but fresh debris flows may not be related to any visible channel.

16.6 ENTRENCHMENT OF FANS

The main channel on alluvial fans normally has a slope slightly lower than the slope of the adjacent fan surface. It thus merges with this surface at a point, commonly near midfan, called the intersection point (Figure 16.9) (Hooke, 1967). Material deposited here is generally coarser than the average material found in the channel, and in some instances forms a small secondary fan on the surface below the intersection point. The degree of entrenchment at the fanhead is commonly so great that water flows capable of filling the channel to overflowing are likely to be rare, if not virtually inconceivable.

The fundamental reason for the lower slope of the main channel near the fanhead is that the most recent flows have been able to transport the sediment load supplied on a slope lower than that of the local fan surface. In many cases this can be attributed to tectonic steepening of the surface (Bull, 1964b; Hooke, 1972). In other cases alternation between debris flows, requiring a steeper slope or at least capable of filling an incised channel and depositing material on the fan surface, and the fluvial events that commonly follow a debris flow may be responsible for incision (Hooke, 1967). In still other cases the latest fluvial event may have been large, or for some other reason able to transport the sediment load supplied on a slope that was lower than the average slope at the fanhead. Such flows would thus be capable of erosion. This is particularly likely in the waning stages of an event when sediment concentrations are low, a process often observed on laboratory fans (Hooke, 1967; Hooke and Rohrer, 1979). The lower sediment concentrations in the later stages may be attributed either to early removal of the most readily erodible material from the source area, or to the fact that the last water reaching the channel percolated through the soil on the valley-side slopes rather than running over it. Small changes in climate can also result in changes in vegetation and runoff, and hence in changes in sediment concentration in the flows. Still another possibility, proposed by Eckis (1928), is that gradual denudation of the source area results in lower valley-side slopes and hence lower sediment concentrations in the flows. Note that according to this explanation all fans not undergoing periodic rejuvenation by uplift of their source areas should be incised. In short, there are so many

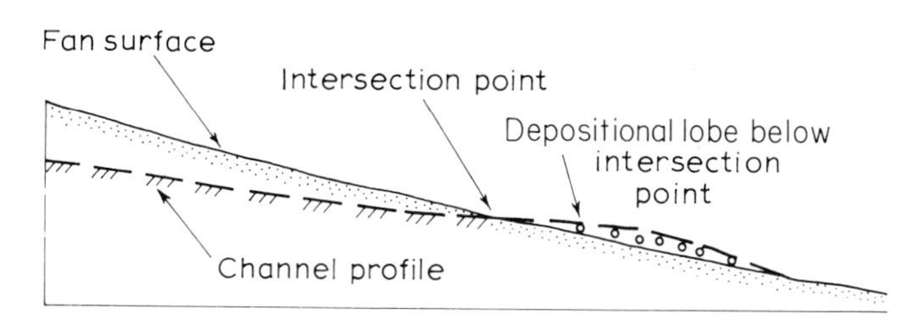

Figure 16.9 Idealized sketch of intersection-point relationships. (*From Hooke, 1967, Fig. 8,* J. Geol., **75**. *Reproduced by permission of the University of Chicago Press.*)

reasons for fanheads being at least moderately entrenched that it is not surprising to find in the field that some degree of entrenchment is the rule, rather than the exception.

16.7 SHIFTING OF CHANNELS

One process by which channels on alluvial fans change position involves a combination of lateral migration and aggradation of the channel bed at the intersection point. The aggradation results in gradual up-fan migration of the intersection point as low banks of the main channel are buried. Subsequent higher flows may then migrate laterally and erode a new channel, offset from the previous course, down the flank of the intersection-point deposit (Bull, 1964a, pp. 27–8; Eckis, 1928, pp. 234–36; Hooke, 1967, p. 457).

When such diversions take place, deposition resumes in an area of the fan that has not received sediment for some period of time. Such areas are generally topographically lower than laterally adjacent areas, and thus tend to 'capture' the flow. These areas have been referred to as 'low' areas, and the adjacent areas which had been built to elevations above the mean surface are called 'high' areas (Hooke and Rohrer, 1979).

Channels can also change position if, due to a rise in stage, the water overflows from the channel at a low point in the banks. Because most sediment in flows on alluvial fans is transported near the bed, such overbank flows are relatively sediment free, and thus can erode a new channel rapidly. On laboratory fans this erosion may result in a grain flow, but these do not appear to leave any permanent record in the stratigraphy. As the new channel deepens it captures more of the main flow. If the slope of this new channel is slightly greater than that of the old, capture is more likely.

Because of the complexity of these processes, we will consider only the broad patterns of deposition over long time periods rather than the details of a single diversion event. Stochastic methods are thus applicable. We will focus on the probability of a diversion from a high area, and treat the process as a Markov process. In a Markov process, a system or part of a system is considered to be in one of several 'states' and the probability of transition from one state to another is studied. The state of a particular area on an alluvial fan is determined by the height of the high area or depth of the low area.

Laboratory data have been used to determine the probabilities of transition from any given state to any other state (Hooke and Rohrer, 1979) and histograms have been prepared showing the frequency with which any given state is occupied (Figure 16.10). The state of any given area is measured in terms of the deviation of that area from a mathematically smooth surface. The mathematical function chosen to fit the surface assumes that the fan

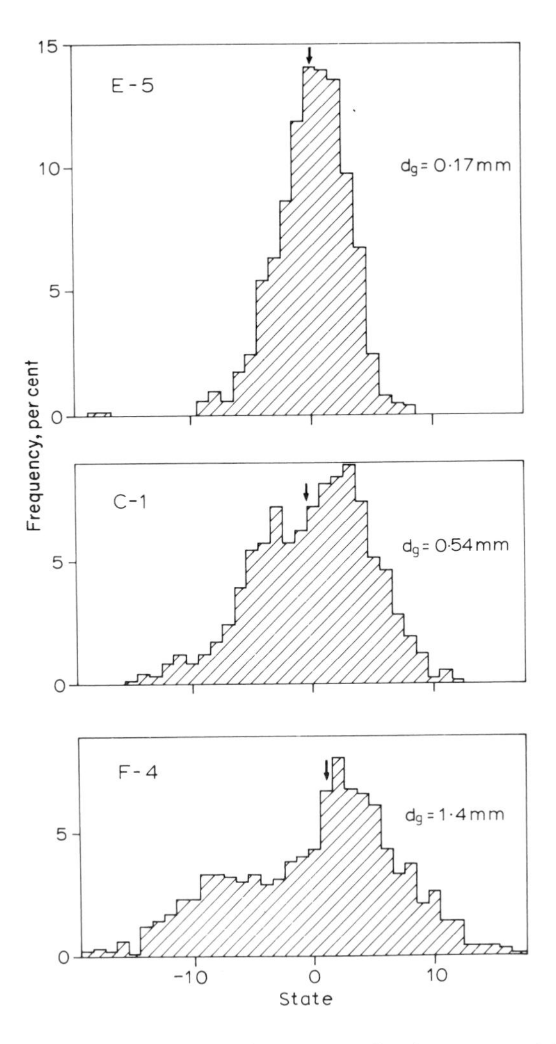

Figure 16.10 Histograms showing frequency of occurrence of various states on laboratory fans. Median values shown by arrow, dg = mean grain size. (*From Hooke and Rohrer, 1979, Fig. 5*, Earth Surf. Proc., **4**.)

is symmetrical about an axis that is a continuation of the main channel just above the fanhead. However, the fan slope may vary with azimuth and with distance from the fanhead. Areas that are within 0.5 mm of the smooth surface are assigned to state 0. The interval from −1.5 to −0.5 mm is state −1 and so forth. The transitions between states are then tabulated and a tally matrix is constructed showing the number of times any given change in state occurred.

The histograms in Figure 16.10 have two interesting features. Firstly, there is a pronounced broadening with increase in grain size. This implies that highs get higher and lows get deeper on fans built of coarser sand, and probably reflects the fact that higher velocities are required to transport the coarser material. Thus on fans built of coarser material lateral slopes into lows must be higher before diversion can occur. The second

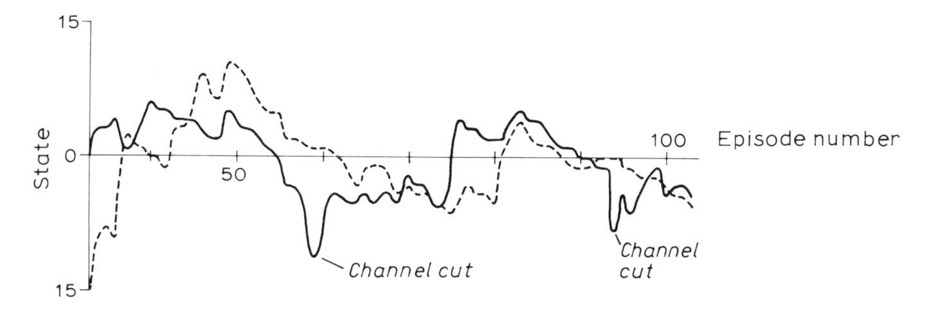

Figure 16.11 Change in state through time for two points on the same flank of a laboratory fan. Solid line is for point 50 cm from fan apex and 27° from axis. Dashed line is also for point 50 cm from fan apex, but 54° from axis. (*From Hooke and Rohrer, 1979, Fig. 6*, Earth Surf. Proc., **4**.)

interesting feature is the persistent asymmetry of the histograms. This is probably a result of channel cutting; high areas can develop only through deposition, but lows can be a result of either non-deposition or erosion. Thus low areas can be expected to be more pronounced.

More detailed study of the transitions suggests that this Markov process has memory. That is, transitions between states typically occurred in declining sequences, such as $3 \rightarrow 2 \rightarrow 1 \rightarrow 0 \rightarrow -1$ . . . This reflects the fact that if flow does not reach one area of the fan for a long time, other parts will be built higher. Thus the area of non-deposition will get progressively lower relative to the average surface. The progressive decrease in state reflects this tendency. Eventually flow is diverted into the resulting low and an abrupt increase in state results. Figure 16.11 shows the change in state through time for two positions on one laboratory fan; gradual decreases in state followed by abrupt increases can be seen in the figure, as can episodes of channel cutting in the position near the axis (solid line). The similarity in the pattern for these two positions 23 cm apart, is consistent with field data which indicate that low areas are often quite broad (Hooke and Rohrer, 1979).

Diversion of channels can also be caused by plugging of a channel with debris-flow material. Although, as noted previously, the most common situation is probably for the channel to be re-excavated in the same location by subsequent water flows, it is possible that the water flows will be diverted and excavate a new channel. Abandoned channels which disappear beneath debris-flow deposits when traced up-fan testify to the importance of this process (Beaty, 1963, pp. 527–8; Hooke, 1967, p. 451).

16.8 MAGNITUDE AND FREQUENCY OF DEPOSITIONAL EVENTS ON FANS

Wolman and Miller (1960) were among the first to discuss the importance of the frequency distribution of events in geomorphic studies. They found that the morphology of such diverse features as river channels, sand dunes, and beaches appeared to be controlled by events that had a recurrence interval of less than a year or two. In other words these features were not products of catastrophic events, but were instead adjusted to forces (discharges, wave heights, wind speeds, etc.) which could be expected to be equalled or exceeded every year or two. Such forces are considered to be of moderate magnitude.

In the case of alluvial fans, Hooke (1968) carried out laboratory studies designed to determine the frequency of the discharge which, if it occurred alone, would produce fans with the same slope as those produced by a distribution of discharges. He referred to this discharge as the dominant discharge, borrowing a term from Inglis (1949) and later Blench (1951, §6.16). Hooke and Rohrer (1979) found that this dominant discharge was equalled or exceeded by 25% to 35% of the discharges that occurred on the fans.

Other studies have suggested that the frequency of the 'dominant' event on several other landforms, chiefly rivers, may be that which is equalled or exceeded only 5% to 25% of the time (see Hooke and Rohrer, 1979, for a summary). Thus Hooke's experiments suggest that the relative magnitude of the dominant discharge, defined in terms of fan slope, is rather lower on fans than on at least some other landforms.

On most other landforms that have been studied, however, events that equal or exceed the dominant event occur once or twice a year, whereas on natural alluvial fans, years may pass without any depositional events of any size. The annual number of depositional events on a fan will vary with climate, of course, but for an order-of-magnitude figure we can assume a frequency of about one per year. For example, in something over 300 days of field work on fans spread over 10 years, I have witnessed only one such event, but I have seen evidence for several that occurred between visits to a particular area. On this basis it appears that flows equalling or exceeding the 'dominant' magnitude may occur only once in a half century to a century on alluvial fans in semi-arid regions.

Some authors (e.g. Beaty, 1974) think of such events as being catastrophic, and from a human point of view that assessment may have some validity. Certainly a large depositional event on an alluvial fan is spectacular, and if human works are affected it may be a catastrophe. However, from the point of view of fan morphology, a flood of roughly the dominant size, whether or not associated with a debris flow, is no more catastrophic than the mean annual flood in a perennial river. Phrased differently, it does not make sense to include periods of no flow in the time series used to study the frequency of events of different size on alluvial fans. To do so would mean that two 'identical' fans, one in the driest desert in the world and one in a semi-arid or even subhumid area, would have dominant discharges of the same magnitude but of very different frequency. This would be inconsistent with the basic intent of Wolman and Miller's (1960) paper; to establish a relation between morphology and the frequency distribution of geomorphic events.

16.9 SUMMARY

Although there are a number of constructional landforms in semi-arid regions, alluvial fans and associated features are the most important, volumetrically, in most such environments. Furthermore, with the possible exception of the grain flows that characterize the slip-faces of sand dunes, alluvial fans are the only constructional landform on which mass movements play an important role.

Of the various mass movement processes that occur on fans, debris flows are the most significant. Their importance, however, varies with climate and lithology. They tend to be more common, relative to fluvial events, in drier watersheds with less vegetation and in watersheds underlain by rock types that produce more fine material upon weathering. Many debris-flow deposits, however, are subsequently modified by fluvial processes so their origin is difficult to establish. Thus beds that are demonstrably of debris-flow origin usually comprise less than a quarter of the material in fans, and the dominant surface features on fans are of fluvial origin. If the process by which any given particle was transported to a fan could be determined, however, debris flows would be found to be more important.

Two other mass-movement processes that operate on fans are grain flows associated with sieve deposition and creep which leads to development of desert pavements. Neither is especially important in terms of the volume of material moved, but the degree of creep smoothing of a fan surface provides one criterion upon which relative age assignments can be based when mapping fans (Hooke, 1972, pp. 2075–8).

ACKNOWLEDGEMENTS

The critical comments of K. S. Richards and W. B. Bull resulted in significant improvements in this chapter. Their efforts are very sincerely appreciated. The University of Chicago Press kindly gave permission to reproduce large parts of an earlier paper with minor modification. Part of the cost of manuscript preparation was borne by the Department of Geology and Geophysics at the University of Minnesota. This is publication number 1088 of The School of Earth Sciences, Department of Geology and Geophysics, University of Minnesota, Minneapolis, MN 55455.

REFERENCES

Bagnold, R. A. (1954). 'Experiments on a gravity-free dispersion of large solid spheres in a Newtonian fluid under shear.' *Proc. Royal Soc. London*, Ser. **A225**, 49–63.
Bagnold, R. A. (1955). 'Some flume experiments on large grains but little denser than the transporting fluid, and their implications.' *Inst. Civil Engrs. Proc.*, pt. 3, **4**, 174–205.
Beaty, C. B. (1963). 'Origin of alluvial fans, White Mountains, California and Nevada.' *Assoc. Am. Geogrs. Annals*, **53**, 516–35.
Beaty, C. B. (1974). 'Debris flows, alluvial fans, and a revitalized catastrophism.' *Zeitschrift fur Geomorphologie Suppl.* Bd. **21**, 39–51.
Blackwelder, E. (1928). 'Mudflow as a geologic agent in semi-arid mountains.' *Geol. Soc. Am. Bull.*, **39**, 465–80.
Blench, T. (1951). *Hydraulics of Sediment-bearing Canals and Rivers*. Vancouver, Evans Industries.
Blissenbach, E. (1954). 'Geology of alluvial fans in semi-arid regions.' *Geol. Soc. Am. Bull.*, **65**, 175–89.
Bluck, B. J. (1964). 'Sedimentation of an alluvial fan in southern Nevada.' *J. Sedim. Petrol.*, **34**, 395–400.
Bull, W. B. (1964a). 'Alluvial fans and near-surface subsidence in western Fresno County, California.' US Geol. Survey Prof. Paper 437-A, 1–71.
Bull, W. B. (1964b). 'Geomorphology of segmented alluvial fans in western Fresno County, California.' US Geol. Survey Prof. Paper 352-E, 79–129.
Bull, W. B. (1972). 'Recognition of alluvial fan deposits in the stratigraphic record.' In *Recognition of Ancient Sedimentary Environments*. Rigby, J. K., and Hamblin, W. K. (eds.), Soc. Econ. Paleontologists and Mineralogists, Spec. Publ. No. 16, pp. 63–83.
Bull, W. B. (1977). 'The alluvial fan environment.' *Prog. Phys. Geog.*, **1**, 222–70.
Caine, N. (1980). 'The rainfall intensity–duration control of shallow landslides and debris flows.' *Geografiska Annaler*, **62A**, 23–7.
Callandar, R. A. (1969). 'Instability and river channels.' *J. Fluid Mech.*, **36**, 465–80.
Costa, J. E. (1984). 'Physical geomorphology of debris flows.' In *Developments and Applications of Geomorphology*. Costa, J. E., and Fleisher, P. J. (eds.), Springer-Verlag, NY, 268–317.
Eckis, R. (1928). 'Alluvial fans of the Cucamonga District, southern California.' *J. Geol.*, **36**, 224–47.
Engelund, F., and Skovgaard, O. (1973). 'On the origin of meandering and braiding in alluvial streams.' *J. Fluid. Mech.*, **57**, 289–302.
Hampton, M. A. (1979). 'Buoyancy in debris flows.' *J. Sedim. Petrol.*, **49**, 753–8.

Hooke, R. LeB. (1967). 'Processes on arid-region alluvial fans.' *J. Geol.*, **75**, 438–60.
Hooke, R. LeB. (1968). 'Steady-state relationships on arid-region alluvial fans in closed basins.' *Am. J. Sci.*, **266**, 609–29.
Hooke, R. LeB. (1972). 'Geomorphic evidence for Late-Wisconsin and Holocene tectonic deformation, Death Valley, California.' *Geol. Soc. Am. Bull.*, **83**, 2073–98.
Hooke, R. LeB., and Lively, R. L. (1983). 'Dating of late Quaternary deposits and associated tectonic events by U/Th methods, Death Valley, California.' *Final Report for NSF Grant* EAR-7919999.
Hooke, R. LeB., and Rohrer, W. L. (1979). 'Geometry of alluvial fans: Effect of discharge and sediment size.' *Earth Surf. Proc.*, **4**, 147–66.
Hunt, C. B., and Mabey, D. R. (1966). 'Stratigraphy and structure Death Valley, California.' US Geol. Survey Prof. Paper 494-A, 162.
Inglis, C. G. (1949). 'The behaviour and control of rivers and canals (with the aid of models), Part 1.' *Central Water Power Irrigation Research Station, Poona, India, Research Publications* No. 13.
Jahns, R. H. (1949). 'Desert floods.' *Eng. and Sci. J.*, **23**, 10–14.
Johnson, A. (1965). *A model for debris flow.* Ph.D. Thesis, Pennsylvania State University, 232pp.
Johnson, A. M. (1970). *Physical Processes in Geology*. Freeman, Cooper, and Co., San Francisco, 577pp.
Johnson, A. M., and Rodine, J. R. (1984). 'Debris flow.' In *Slope Instability*. Brunsden, D., and Prior, D. B. (eds.), pp. 257–361.
Kessli, J. E. (1943). 'Disintegrating soil slips of the coast ranges of central California.' *J. Geol.*, **51**, 342–52.
Lustig, L. K. (1965). 'Clastic sedimentation in Deep Springs Valley, California.' US Geol. Survey Prof. Paper 352-F, 131–92.
Middleton, G. V., and Hampton, M. A. (1976). 'Subaqueous sediment transport and deposition by sediment gravity flows.' In *Marine Sediment Transport and Environmental Management*. Stanley, D. J., and Swift, D. J. P. (eds.), John Wiley, NY, pp. 197–218.
Nelson, C. A. (1966). 'Geologic map of the Blanco Mountain Quadrangle, Inyo and Mono counties, California.' *US Geol. Survey Map* CQ-529.
Pack, F. J. (1923). 'Torrential potential of desert waters.' *Pan-American Geologist*, **40**, 349–56.
Parker, G. (1976). 'On the cause and characteristic scales of meandering and braiding in rivers.' *J. Fluid. Mech.*, **76**, 457–80.
Parker, G. (1978a). 'Self-formed rivers with equilibrium banks and mobile bed. Part 1. The sand-silt river.' *J. Fluid. Mech.*, **89**, 109–25.
Parker, G. (1978b). 'Self-formed rivers with equilibrium banks and mobile bed. Part 2, The gravel river.' *J. Fluid Mech.*, **89**, 127–46.
Picard, M. D., and High, L. R., Jr (1973). 'Sedimentary structures of ephemeral streams.' *Developments in Sedimentology*, **17**, Elsevier, 223pp.
Pizzuto, J. E. (1984). 'Equilibrium bank geometry and width of shallow sandbed streams.' *Earth Surf. Proc. Landforms*, **9**, 199–207.
Rodine, J. D. (1974). *Analysis of the mobilization of debris flows.* Unpublished Ph.D. Dissertation, Stanford University, Stanford, CA, 226pp.
Rodine, J. D., and Johnson, A. M. (1976). 'The ability of debris, heavily freighted with coarse clastic materials, to flow on low slopes.' *Sedimentology*, **23**, 213–34.
Rouse, H. (ed.) (1950). '*Engineering Hydraulics*'. John Wiley, New York, 1,039pp.
Sharp, R. P. (1942). 'Mudflow levees.' *J. Geomorph.*, **5**, 222–7.
Sharp, R. P., and Nobles, L. H. (1953). 'Mudflow of 1941 at Wrightwood, southern California.' *Geol. Soc. Am. Bull.*, **64**, 547–60.
Takahashi, T. (1980). 'Debris flow in prismatic open channel,' *Am. Soc. Civil Engrs. J. Hydr. Div.*, **106**, 381–96.
Terzaghi, K. (1950). 'Mechanism of landslides.' In *Application of Geology to Engineering Practice*. Berkey Volume, Geological Society of America, 83–123.
Trowbridge, A. C. (1911). 'The terrestrial deposits of Owens Valley, California.' *J. Geol.*, **19**, 706–47.
Vanoni, V. A. (1963). 'Sediment transportation mechanics: suspension of sediment.' *Proc. Am. Soc. Civil Engrs. J. Hydr. Div.*, HY9, 45–76.

Vanoni, V. A. (ed.) (1975). *Sedimentation Engineering*. Am. Soc. Civil Engrs., NY, 745pp.
Wasson, R. J. (1974). 'Intersection point deposition on alluvial fans: An Australian example.' *Geografiska Annaler*, **56A**, 83–92.
Wolman, M. G., and Miller, J. P. (1960). 'Magnitude and frequency of forces in geomorphic processes.' *J. Geol.*, **68**, 54–74.

Slope Stability
Edited by M. G. Anderson and K. S. Richards

Chapter 17

Mechanisms of Mass Movement in Periglacial Environments

CHARLES HARRIS
Geography Section
Department of Geology, University College Cardiff

17.1 INTRODUCTION

Since Andersson's early work on Bear Island (Andersson, 1906), the importance of mass movement in periglacial environments has become firmly established by numerous field studies. A distinctive suite of periglacial mass movement processes may be defined, since all are characteristically associated with the thawing of frozen ground. Initiation of instability and subsequent movement velocities depend more on the ice content of the thawing soil and its rate of thaw than on slope gradients. Since ice contents of frozen sediments are often extremely high, thawing may lead to the release of a greater volume of water than can be normally accommodated by the soil pore space. Under these circumstances slope failures are likely even on very gentle slopes.

A genetic classification of periglacial mass movements is probably the most useful to engineers and geomorphologists, and the mechanism and rate of displacement are major elements in such a classification. McRoberts and Morgenstern (1974a) sub-divided landslides in Arctic Canada into fall-, flow- and slide-dominated mass movements. They considered that flow-dominated failures were most common and these could be classified according to speed of movement and surface form into three major categories; solifluction, skinflows, and bimodal flows.

McRoberts (1978) considered that solifluction is the characteristic deforming mode of many naturally occurring active layers and is typically a slow flow affecting the whole or part of the active layer. Brown and Kupsch (1974) defined solifluction as 'slow, downslope movement of saturated, nonfrozen earth material behaving apparently as a viscous mass, over a surface of frozen material' and quoted typical rates of movement of between 0.5 cm and 10 cm per year. It is generally accepted that solifluction involves two closely related

Figure 17.1 Turf-banked solifluction lobe, Okstindan, northern Norway. Surface gradient approximately 4°, altitude 710 m, parent material silty sand till. Note rucksack for scale

though distinct processes; frost creep and gelifluction (e.g. Washburn, 1979, pp. 198–201). Frost creep was defined by Washburn (1967) as the ratchetlike downslope movement of particles as a result of frost heaving of the ground and subsequent settling on thawing, and the term gelifluction was coined by Baulig (1956, 1967) to describe downslope flow of saturated soil associated with the thawing of frozen ground. McRoberts and Morgenstern (1974a) stressed the difficulty of distinguishing between the two processes in the field and French (1976) used the term solifluction in a general sense to include downslope movements due to both frost creep and gelifluction. Harris (1981) pointed out the value of this general term in describing topographic features such as lobes and terraces (Figure 17.1) where the relative importance of frost creep and gelifluction is unknown.

Skinflows, active layer glides (Mackay and Mathews, 1973), or active layer detachments (Stangl *et al.*, 1982), are more rapid failures involving the detachment of a thin veneer of vegetation and soil which flows or slides downslope over a frozen subsoil (Figure 17.2). Skinflows are usually ribbon-like in plan and result from a sudden disturbance of thermal conditions in the active layer due to such events as forest fires, heavy rainfalls or unusually high air temperatures.

The third common mode of slope failure in permafrost areas, bimodal flows, result from rapid thawing of ice-rich permafrost. Basal erosion of valley sides or coastal cliffs frequently initiates such failures when exposed ground ice forms a scarp and ablates back, releasing sediment as it retreats. The sediment flows and slides to the base of the scarp where it flows away in a mudflow apron or lobe of much lower gradient (Figure 17.3).

Figure 17.2 Skinflow or active layer glide, Mackenzie Valley, North-West Territories, Canada. Failure resulted from active layer thickening following a forest fire. (*Photo courtesy of J. R. Mackay.*)

Figure 17.3 Bimodal flow caused by fluvial steepening of valley side, Banks Island, Western Canadian Arctic. (*Photo courtesy of H. M. French.*)

Alternative terms include retrogressive thaw flow slides (Rampton and Mackay, 1971; Hughes, 1972; Brown and Kupsch, 1974), ground ice slumps (French and Egginton, 1973; French, 1974b), thaw slumps (Washburn, 1979) and thermo-erosion cirques (Czudek and Demek, 1970).

The presence of frozen subsoil results in the mass movements described above being largely translational rather than rotational in character. However, deep seated rotational slides do occur through frozen ground, though the bases of such failures generally pass through underlying unfrozen clays (McRoberts and Morgenstern, 1974b).

17.2 ICE SEGREGATION

17.2.1 Theoretical Considerations

Both Taber (1929, 1930) and Beskow (1935) inferred the presence of films of unfrozen 'adsorbed' water in frozen soils, separating soil ice from soil particles, and enabling particle-free ice lenses to develop. The theory of 'primary frost heaving' considers ice lenses to grow in the larger pores at the freezing front where adsorption pressures are lowest. Water migration from the unfrozen soil below maintains ice lens growth and, through phase change, heat flux to the surface. When the upward heat flux required by the thermal boundary conditions can no longer be maintained by the release of latent heat at the freezing interface, the interface descends to a new plane where ice lens growth begins again. The suction force responsible for drawing water to the growing ice lenses is explained either by the need to maintain an equilibrium thickness of the adsorbed water films at the freezing front (e.g. Takagi, 1978, 1979), or by the thermodynamics of the curved boundaries between ice and water at the freezing front (Jackson and Chalmers, 1958; Penner, 1959; Everett, 1961; Everett and Haynes, 1965; Williams, 1968).

The theory of 'secondary frost heaving' has been developed by Miller (1972, 1978), Miller *et al.* (1975), and Konrad and Morgenstern (1980, 1981). Laboratory experiments have shown that ice lens growth does not take place at the freezing front, but some distance behind it (Loch and Kay, 1978). Between the growing ice lens and the 0 °C isotherm is a frozen fringe in which larger pores contain some ice, separated from mineral grains by unfrozen water. According to Miller, water entering the frozen fringe from below is transferred to the growing ice lens via the unfrozen adsorbed water and *via* pore ice, which moves upwards in the direction of decreasing temperature. Motion of the pore ice is sustained by phase change at the ice–water interfaces at which latent heat is either released or absorbed. Loch and Miller (1975) reported heaving pressures recorded during laboratory experiments exceeding those predicted by the primary heaving model by factors as large as 3 to 6, and took this as support for the concept of secondary heaving.

Konrad and Morgenstern (1980, 1981, 1982) emphasized the role of permeability in the frozen fringe in regulating ice lens initiation and growth. They showed that during freezing of a fine grained soil, the Gibb's free energy of the unfrozen water films is lower at lower negative temperatures, so that water migration across the frozen fringe to the base of the growing ice lens occurs in response to the resulting suction gradient. The temperature at the base of the frozen fringe is defined as the '*in situ* freezing temperature' which is higher than the 'segregation freezing temperature' at the base of the growing ice lens (the upper boundary of the frozen fringe). Since frozen soil permeability falls more or less exponentially with falling temperature (Burt and Williams, 1976), there is a rapid fall in permeability across the frozen fringe.

According to Konrad and Morgenstern, ice segregation will continue as long as permeability in the frozen fringe is sufficient to allow the passage of water to the ice lens, but falling temperatures in the fringe will eventually reduce permeability sufficiently to stop this water flow. A new ice lens will then begin to grow lower in the frozen fringe where temperature, permeability, and water supply are higher. The importance of water flow into the frozen fringe in controlling rates of heave was confirmed in laboratory experiments by Mageau and Morgenstern (1980). Experiments by Konrad and Morgenstern (1981) showed that under thermal steady-state conditions the rate of frost heaving was linearly related to the temperature gradient across the frozen fringe and they defined a segregation potential (*SP*) for fine grained soils such that:

$$v = SP \text{ grad } T \qquad (17.1)$$

where v is the velocity of arriving pore water (mm s^{-1}), grad T is the thermal gradient across the frozen fringe, and *SP* has units of $mm^2\ s^{-1}{}^\circ C$.

The frost heave increment Δh_s that occurs in a time interval t as a result of moisture migration across the frozen fringe may be obtained from:

$$\Delta h_s = 1.09 v \Delta t \qquad \text{(Nixon, 1982)} \qquad (17.2)$$

To this may be added the amount of heave due to *in situ* freezing of pore water, given by:

$$\Delta h_i = 0.09 n \Delta X \qquad (17.3)$$

where n is the porosity, ΔX is the increase in depth of the frozen fringe in time Δt, and 0.09 is the volumetric expansion during phase change. Nixon (1982) has presented field data from Calgary which shows good correspondence with calculated frost heave using the Konrad and Morgenstern model.

17.2.2 Factors Influencing Ice Segregation

Four factors may be cited as being of particular significance in affecting the amount of ice segregation during soil freezing; the pore size of the soil, the moisture supply, the rate of heat extraction, and the confining pressure. Theory and observation indicate that the susceptibility of a soil to ice segregation increases as pore size decreases. However, the low permeability of heavy clays may restrict water migration sufficiently to prevent significant ice segregation (Penner, 1968). In view of the close correspondence between pore size and grain size, and the relative ease with which the latter may be measured, frost susceptibility criteria based on soil textures are frequently used, such as those of Beskow (1935) and Casegrande (1931). Such factors as variation in grain size distribution, soil density, clay mineralogy, and soil fabric (Lovell, 1983) result in a considerable range in observed frost heave for soils of similar granulometry (Figure 17.4).

The amount of segregation ice formed during freezing of a frost susceptible soil must clearly depend in part on water supply. McGaw (1972) showed experimentally that the rate

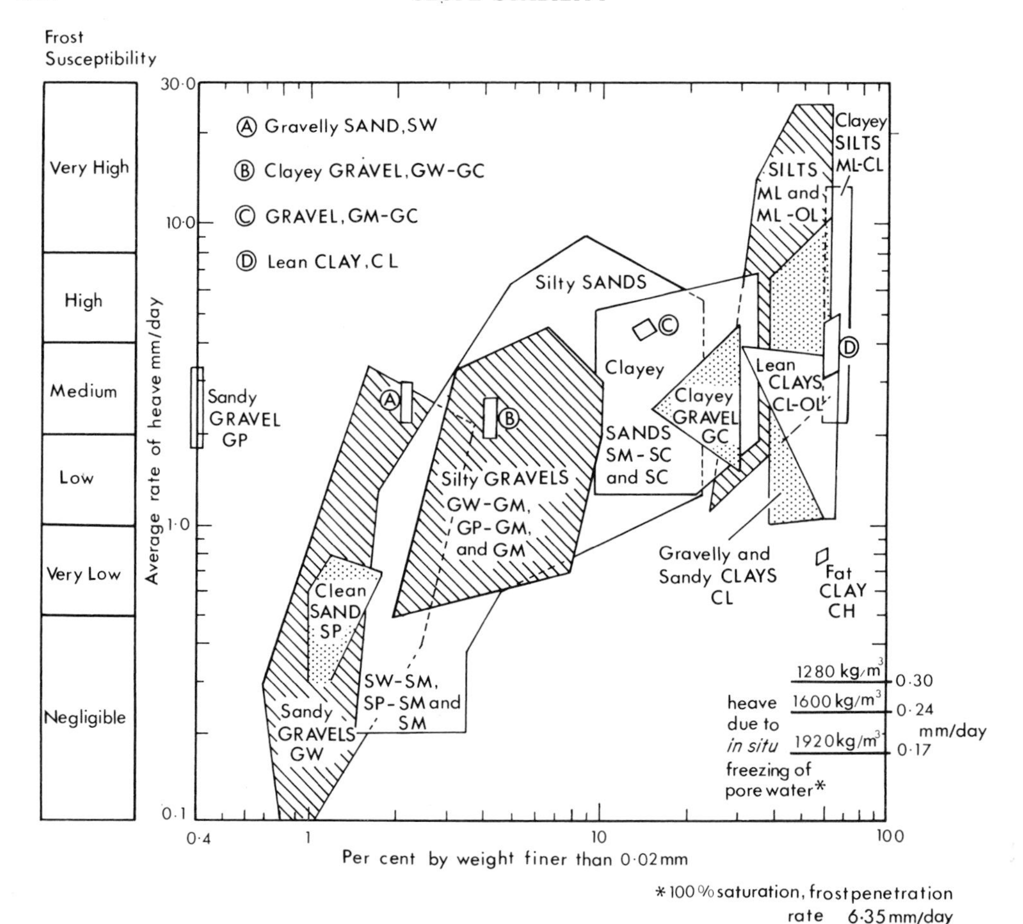

Figure 17.4 Average rate of heave plotted against percentage finer than 0.02 mm from laboratory tests of a range of natural soils. (*From Kaplar, 1974,* CRREL Report, **250**. *Reproduced by permission of US Army Cold Regions Research and Engineering Laboratory.*)

of heave fell rapidly as depth to water table increased from 0 m to 2.5 m, the rate of reduction being greatest at higher rates of freezing. The influence of moisture content as expressed by soil moisture suction at the freezing front was illustrated by Konrad and Morgenstern (1981, 1982). Increased suction reduced ice segregation by lowering the ice segregation potential.

The rate of heat extraction is a major factor in controlling the rate of penetration of the freezing front. This is in turn directly related to the rate of frost heave (Anderson *et al.*, 1978) and inversely related to the heaving ratio (heave per unit depth). Kaplar (1970), Penner (1972), and Chen *et al.* (1983) showed that high freezing rates reduce the time available for water migration to the freezing front so that the heaving ratio is reduced, eventually to that resulting simply from freezing of *in situ* pore water. For Devon silt Konrad and Morgenstern (1982) suggested a limiting rate of cooling of the frozen fringe of approximately 2.5 °C/hour, above which no migration of water occurred.

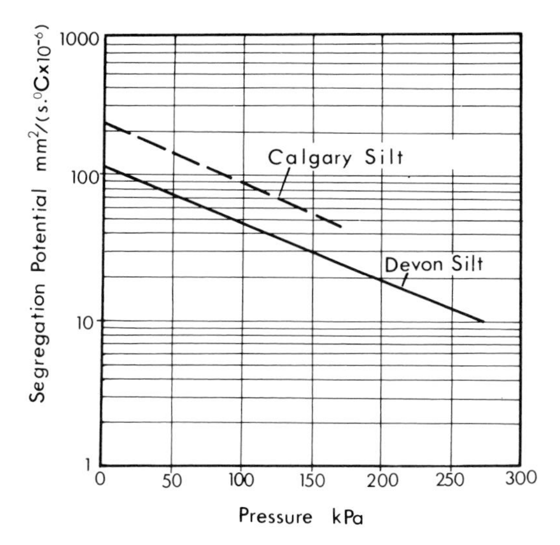

Figure 17.5 Ice segregation potential plotted against overburden pressure. (*Nixon, 1982*, Can. Geotech. J., **19**. *Reproduced by permission of the National Research Council of Canada.*)

The reduction in frost heave resulting from increasing confining pressure is well documented (e.g. Kaplar, 1970) and is elegantly demonstrated by the relationship between pressure and ice segregation potential (Figure 17.5) (Nixon, 1982). Williams (1968) suggested that the reduction in segregation ice observed in permafrost in the Mackenzie Delta with increasing depth was due to increasing overburden pressures.

17.3 GROUND ICE DISTRIBUTION IN ARCTIC PERMAFROST

Mackay (1972) classified ground ice other than pore ice into three major groups according to the mode of ice accumulation; ice-wedge ice, segregation ice, and injection ice. Mackay estimated that in the Northern Hemisphere, more than 2.5 million square kilometres of tundra and boreal forest are underlain by ice-wedge ice, and in these areas up to 50% of the surface 3 m may consist of ice. Segregation ice was sub-divided into epigenetic and aggradational, the former developing in sediments deposited before permafrost formation, the latter resulting from surface sedimentation above permafrost leading to the upward movement of the permafrost table. Mackay (1983) showed that water migration from the base of the active layer into the top of the permafrost may significantly increase the amount of segregation ice in that zone. Injection ice results from expulsion of pore water from saturated sands and gravels during growth of permafrost (McRoberts and Morgenstern, 1975), and occurs as extensive horizontal sheets. Massive ground ice bodies mainly consisting of injection ice, in Alaska and Arctic Canada, may be up to 1 km^2 in area and 30 m in thickness. The tops of most such ground ice masses are within 30 m of the ground surface (Mackay and Black, 1973) so that thawing of the permafrost results in considerable subsidence and instability.

French and Harry (1983) stressed the importance of granulometry in affecting the excess ice content of near-surface permafrost in Banks Island, North West Territories, Canada. Silty sands contained 20% to 75% excess ice by volume while medium well-sorted sands contained less than 10% excess ice. Excess ice contents and cumulative potential thaw subsidence were estimated for the Sachs River Lowlands, Banks Island, by French and Harry (1983). They showed that thawing of 8 m of permafrost in this area would result in over 3 m of surface subsidence. Using data from Lawrence and Proudfoot (1977), Pollard and French (1980) have estimated the volume of ground ice on Richards Island, Mackenzie Delta, Canada. They concluded that in the upper 10 m of permafrost a maximum of 47.5% of the total volume consists of ground ice, of which over 85% is pore ice and segregation ice. Excess ice content in the upper 10 m was shown to average approximately 14.3%, giving potential thaw subsidence of 1.4 m for 10 m of thawing. These figures are similar to those reported by Brown (1967) for the arctic slope of Alaska.

Crampton (1981) analysed factors influencing ground ice contents to produce a terrain sensitivity map of the Mackenzie River Valley. He reported highest ice contents approximately 5.5 m below the surface where fine silts lay immediately below coarser sands and gravels. Ice contents increased northwards, the most rapid increase corresponding to the transition from discontinuous to continuous permafrost.

17.4 SLOPE STABILITY DURING THAW OF ICE RICH SOILS

The susceptibility of thawing ice-rich soils to instability, even on very gentle slopes, has long been acknowledged by geotechnical engineers and geomorphologists. The development of the thaw consolidation theory (Morgenstern and Nixon, 1971) to explain this phenomenon followed observations by Taber (1943) who described the loss in strength of ice-rich soils subjected to rapid thawing and slow drainage. Impedance of drainage arises from the effectively impermeable nature of the underlying permafrost (Woo and Steer, 1982, 1983). In areas with seasonal soil freezing, impedance of drainage during thawing downwards from the surface is terminated with final clearance of frozen ground, and slopes may then become freely drained (Harris, 1972, 1977, 1981).

The one-dimensional thaw consolidation theory (Morgenstern and Nixon, 1971) is based on Terzaghi's theory of consolidation and the Neuman equation for thaw penetration through frozen soil. The time-dependent factors, thaw rate and rate of consolidation, are combined in the thaw consolidation ratio R, such that:

$$R = \frac{1}{2}\left(\frac{\alpha}{\sqrt{C_v}}\right) \tag{17.4}$$

where C_v is the coefficient of consolidation and α is the thaw parameter in the Neuman equation:

$$X_t = \alpha\sqrt{t} \tag{17.5}$$

X_t being the depth of thaw and t time

McRoberts (1975) used the Stephan solution of the Neuman equation to estimate the value of α as follows:

$$\alpha = \sqrt{\frac{2K_u T_s}{L}} \tag{17.6}$$

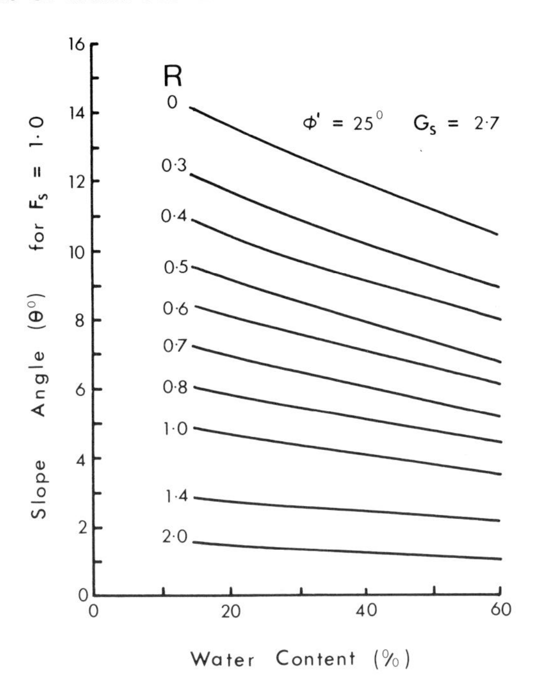

Figure 17.6 Solution to infinite slope analysis for a thawing slope, in terms of the thaw-consolidation ratio R. (*From McRoberts and Morgenstern, 1974a,* Can. Geotech. J., **11**. *Reproduced by permission of the National Research Council of Canada.*)

where K_u is the thermal conductivity of the thawed soil; T_s the stepwise change in temperature causing thaw; and L the volumetric latent heat of the frozen soil. He showed that α is relatively insensitive to variations in thaw rate and ice content, ranging from 0.015 cm $s^{-0.5}$ for extremely slow thaw (assumed step temperature change 2 °C, soil ice content 70%) to 0.095 cm $s^{-0.5}$ for extremely rapid thaw (step temperature change 20 °C, ice content 5%).

Thaw consolidation theory has been applied to the prediction of slope stability using the infinite slope model (McRoberts, 1978; Morgenstern, 1981). The slope is assumed to thaw-consolidate under self-weight, with a permafrost table parallel to the surface. For a cohesionless soil the factor of safety is given by:

$$F_s = \frac{\gamma'}{\gamma} \frac{1}{1+2R^2} \frac{\tan \phi'}{\tan \beta} \tag{17.7}$$

where γ' is the submerged unit weight of soil, γ is the unit weight of saturated soil, and β is the slope angle. This equation can be solved in terms of R and the water content for a typical set of soil properties, and McRoberts and Morgenstern (1974a) provided a graphical solution for a soil with $\phi' = 25°$, and specific gravity of particles 2.7 (Figure 17.6). It is apparent that for R values greater than 1 (rapid thawing) failure can occur on very gentle slopes.

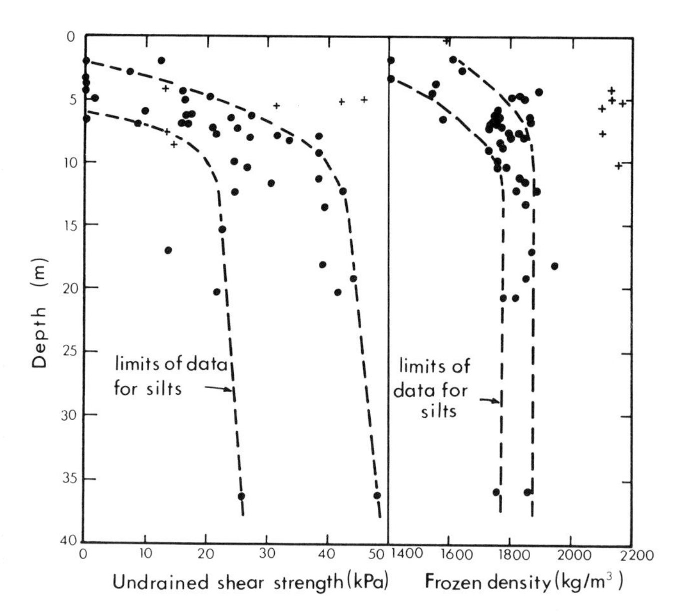

Figure 17.7 Profiles of undrained strength and density at Niglintgak, Mackenzie Delta, North-West Territories, Canada. (*From Nixon and Hanna, 1979,* Can. Geotech. J., **16**. *Reproduced by permission of the National Research Council of Canada.*)

Excess pore pressures are, therefore, predicted by the thaw consolidation theory when R values are high. McRoberts *et al.* (1978) reported field measurements of pore pressures at two sites in the Mackenzie River Valley where disturbance of vegetation led to slope failures on modest slopes. Excess pore pressures consistent with those predicted by the thaw consolidation theory were recorded, the highest during the first year of permafrost disturbance when thawing was most rapid.

If R values are high, effective stress levels approach zero (McRoberts and Morgenstern, 1974a), and the soil is effectively undrained. Nixon and Hanna (1979) have determined the strength of a large number of thawed undrained permafrost samples from the Mackenzie Delta, NWT, Canada. Most samples were deltaic silts, and undrained shear strength tests showed that unfrozen strength decreased as frozen density decreased, reaching zero for frozen densities of less than about 1,670 kg m^{-3}. This corresponded to water contents greater than 35% to 42% (Figure 17.7). Low frozen densities, corresponding to high ice contents, were encountered within 5 m of the surface, with thawed strength at or close to zero. Below 5 m the unfrozen strength started to increase and became roughly constant below 10 m, in the range of 23 kPa to 43 kPa, corresponding to frozen densities of between 1,780 kg m^{-3} and 1,870 kg m^{-3}.

17.5 PROCESSES AND MECHANISMS OF PERIGLACIAL MASS MOVEMENT

17.5.1 Solifluction

The two components of solifluction, frost creep and gelifluction, both result from thaw consolidation of ice-rich soils, so that quantifying their relative importance in the field is generally difficult if not impossible. However, maximum potential frost creep may be computed by assuming vertical settlement of the soil during thaw from:

$$C = h \tan \beta \tag{17.8}$$

where C is downslope displacement at the soil surface, h is heave normal to the surface, and β is the slope angle.

Washburn (1967) estimated ratios of creep to gelifluction ranging from 0.36:1 to 5.3:1 in silty diamictons (poorly-sorted sediments containing gravel, sand, and finer grain sizes) at Mesters Vig, Greenland; the lower ratio corresponding to gelifluction-dominated wet sites, the higher ratio to creep-dominated drier sites. Where vegetation is sparse relatively rapid frost creep in the surface few centimetres of soil may result from ground freezing and thawing (Higashi and Corte, 1971; Washburn, 1979, p. 92). For instance Walton and Heilbronn (1983), on slopes with sorted stripes in south Georgia, observed displacements in the uppermost 1 cm–2 cm of soil of up to 7 cm per year, associated with displacements of no more than 2 mm in the subjacent layers. This high rate of surface movement was attributed to the action of needle ice.

Gelifluction is generally considered to involve slow saturated flow of the active layer when frictional strength is reduced by excess pore pressure generated during thaw consolidation. The role of a frozen subsoil in preventing free drainage during ground thaw was stressed by Harris (1977) and Gamper (1983) who concluded that following clearance of frozen ground in non-permafrost areas soils drain rapidly and become stable. In such non-permafrost areas, therefore, solifluction is likely only during spring and early summer. In Kärkevagge, Sweden, Rapp (1962) observed that saturation of the active layer in spring over a layer of frozen subsoil caused solifluction but no slides; saturation in autumn in the absence of a frozen subsoil caused slides but no solifluction.

Field measurements of soil movement generally utilize buried columns or flexible plastic tubes to monitor variations with depth, and painted stones or pegs to monitor surface rates. A review by Harris (1981) showed that movement generally decreases with depth and dies out at between 0.2 m and 2 m below the surface. Profiles of movement range from concave to convex downslope (see for instance Jahn, 1978), even within a single solifluction lobe (Figure 17.8). Major factors influencing subsurface patterns of movement include the distribution of segregation ice within the profile, the thickness of the surface vegetation, and the frequency of freezing and thawing of the surface soil layers causing frost creep. Rein and Burrous (1980) in their large-scale laboratory experiments showed that most soil movement during thaw occurred in soil layers with ice content in excess of 150% by weight, so that uneven distribution of segregation ice will lead to zones of greater shearing within the profile. Surface movement may be restricted by a thick vegetation mat, giving profiles of movement which are strongly convex downslope (Rudberg, 1964; Benedict, 1970; Harris, 1972; Price, 1973; Jahn, 1978). Particularly active frost creep in the near-surface soil layers may, on the other hand, generate profiles of movement which are strongly concave downslope (Jahn, 1978).

In an important paper on solifluction in the permafrost environment of Garry Island, NWT Canada, Mackay (1981) stressed the contrasting thermal regimes of active layers in permafrost and non-permafrost areas. In the former, the active layer freezes both downwards from the surface and upwards from the permafrost table, while in the latter

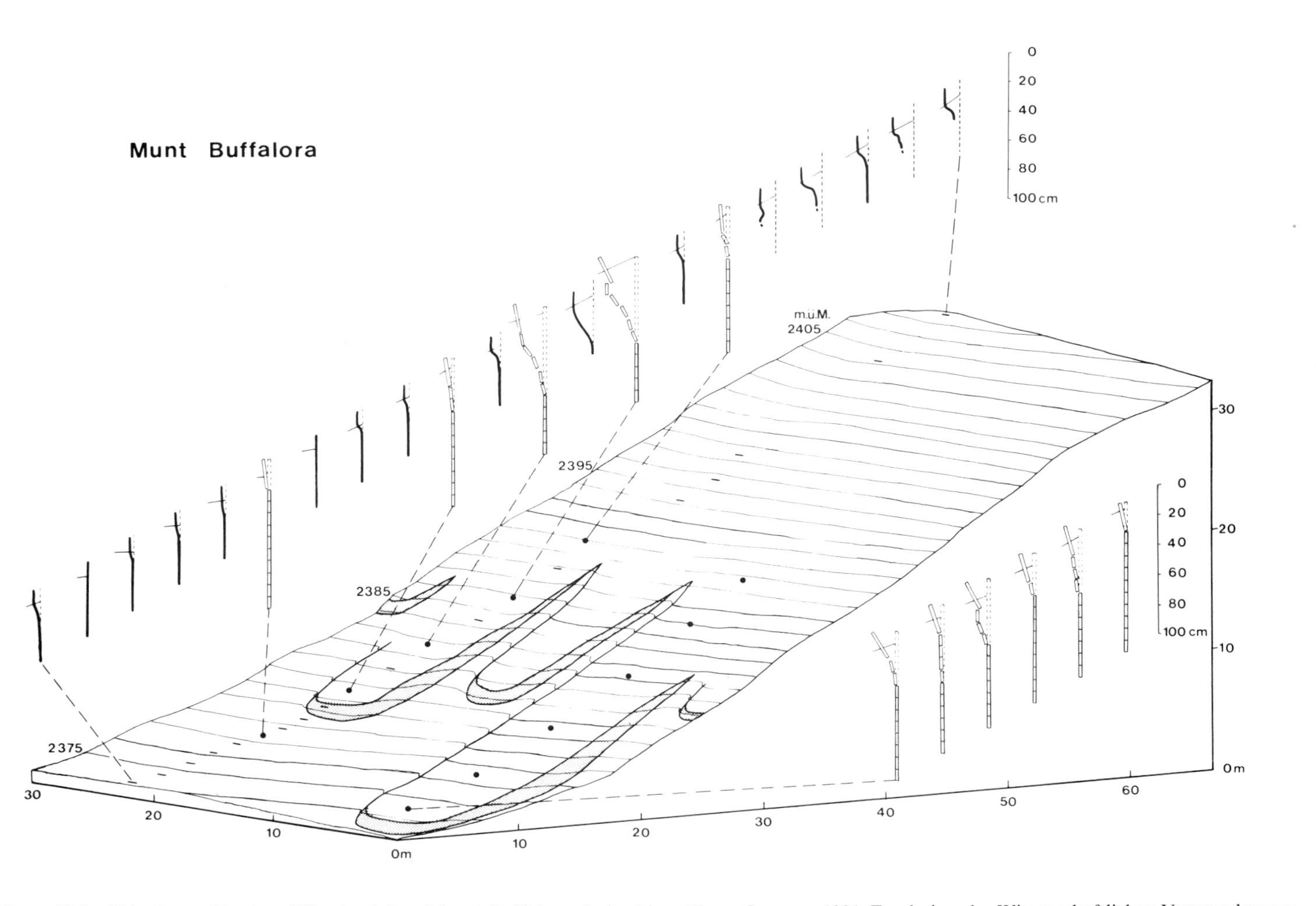

Figure 17.8 Velocity profiles in solifluction lobes, Mount Buffalora, Swiss Alps. (*From Gamper, 1981*, Ergebnisse der Wissenschaftlichen Untersuchungen im Schweizerischen Nationalpark, **15**. *Reproduced by permission of the author.*)

freezing from the surface only is possible. In Garry Island Mackay reported that upward freezing generated a zone of high ice content at the base of the active layer. Ice contents in this zone were further increased by downward percolation and refreezing of water during summer thaw. Mackay recorded 14.4 mm of heave in the lower part of the active layer during the summer thaw period due to this mechanism. Above the ice-rich basal zone the active layer showed little excess ice so that thawing caused negligible consolidation or downslope movement. However, when thaw penetrated to the basal ice-rich zone in late summer the release of excess water led to downslope movement of the active layer in a plug-like fashion, the relatively undisturbed active layer sliding over the thawing ice-rich basal zone. In contrast to alpine areas with no permafrost, therefore, solifluction in Garry Island took place mainly in late summer at the time of maximum thaw penetration.

This late summer plug-like movement of the active layer may possibly provide a modern analogue for the low-angled mud slides observed in clays in southern Britain (Weeks, 1969; Skempton and Weeks, 1976; Chandler, 1970a,b; Chandler *et al.*, 1976). Skempton and Weeks (1976) showed that the most recent slides in the Weald date from the Late Glacial and they concluded that the high pore pressures necessary to initiate movement ($r_u = 0.83$–0.93) were due to thaw consolidation in an active layer over permafrost.

In the Brooks Range, Alaska, Reanier and Ugolini (1983) described solifluction lobes above the tree line where permafrost was present and stressed the importance of basal sliding. They considered that ice lenses provided planes of potential sliding as the active layer thawed, and described soil peds with slickensided surfaces due to small-scale shears. Van Vliet-Lanoë *et al.* (1984) also reported lateral displacements along small slip planes associated with ice lenses during the thaw of the active layer in Spitsbergen. Micromorphological analysis of field and laboratory samples showed that ice lens development during soil freezing caused compression of the soil between lenses and the formation of a platy microfabric. Partings between plates formed the loci for ice lens growth during subsequent freezing cycles and for localized slip during thaw (Figure 17.9). Reviews of soil micromorphological phenomena resulting from soil freezing and related processes, including solifluction, are given by Van Vliet-Lanoë (1982) and Harris (1985).

In his review of soil movement rates due to solifluction, Harris (1981) reported mean surface rates in non-permafrost alpine areas ranging from 2 mm per year (Colorado Front Range, Benedict, 1970) to 213 mm per year (Tasmania, Caine, 1968). The latter very high value may have resulted partly from needle ice action beneath the painted stones used to monitor surface displacements. Typical mean rates were between 10 mm per year and 20 mm per year. Recently published measurements of surface movement rates confirm these general levels of activity; in the eastern Swiss Alps, for instance, Gamper (1983) reported a maximum annual movement of 48.2 mm and average annual rates of 0.2 mm to 11.1 mm, the higher values being associated with vegetation-free areas. Gamper measured year to year variation in movement rates as high as a factor of 8, with high values in years with little snow, deep freezing of the soil, early clearance of snow in spring, and subsequent thawing from the surface. Low movement rates resulted from high snowfall, shallow freezing, and late clearance of snow resulting in thawing of the frozen soil largely from the bottom upwards.

Most studies of alpine solifluction relate to surface movement rates on slopes with lobes and terraces (Figure 17.10), and all show large spatial variations (Figure 17.8). Generally rates are highest on the treads of lobes and terraces in axial locations (e.g. Benedict, 1970) and decrease towards the sides and fronts of the features.

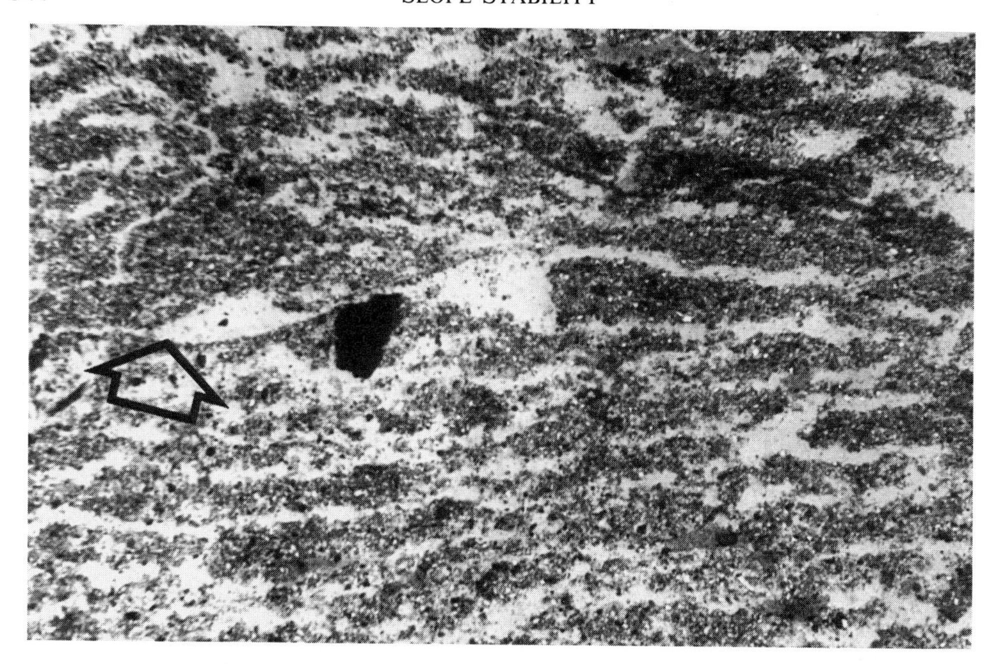

Figure 17.9 Shear plane associated with thawing ice lenses produced experimentally after six cycles of alternate freezing and thawing (*Van Vliet-Lanoë* et al., *1984*). Frame width 26 mm. (Photomicrograph courtesy of B. Van Vliet-Lanoë)

Figure 17.10 Measurement of surface soil movement rates, Okstindan, Norway. The line of pegs traverses the upper part of the lobe shown in Figure 17.1

Figure 17.11 Instrumented solifluction slope, Garry Island, North-West Territories, Canada. The hummocks on the slope above the frontal bank are moving downslope *en masse*. (Photo courtesy of J. R. Mackay.)

In high latitude permafrost areas field studies show mean rates of movement to be similar or slightly higher than in the alpine zone, ranging from 9 mm per year to 40 mm per year, with a maximum value of 150 mm per year reported by Holgate *et al.* (1967) for Signy Island. Rates of movement appear much more uniform across a given slope in the permafrost zone than in alpine areas, resulting generally in smooth solifluction sheets rather than lobes or terraces (Figure 17.11, and see for instance French, 1974a, Figures 2 and 3, pp. 73 and 74, and Stangl *et al.*, 1982, Figures 8 and 9, pp. 142 and 143). Mackay (1981) in Garry Island, NWT described turf hummocks developed on solifluction slopes which moved downslope with little distortion, maintaining their identities for hundreds if not thousands of years. Mackay measured average surface movement rates of between 0.2 mm per year and 10 mm per year on slopes of between 1° and 7°.

Volumetric transport rates were calculated by Harris (1981) from the data of Rudberg (1964), for northern Sweden (mean 39 $cm^3\ cm^{-1}\ yr^{-1}$), Benedict (1970) for the Colorado Front Range (mean 59 $cm^3\ cm^{-1}\ yr^{-1}$), Price (1973) for the Ruby Range, Yukon (mean 32 $cm^3\ cm^{-1}\ yr^{-1}$), Harris (1977) for Okstindan, Norway (26 $cm^3\ cm^{-1}\ yr^{-1}$) and Williams (1966) for Schefferville, Quebec (114 $cm^3\ cm^{-1}\ yr^{-1}$). Mackay (1981) estimated for a hill-slope as a whole in Garry Island a mean volumetric transport rate of 15 $cm^3\ cm^{-1}\ yr^{-1}$ due to solifluction. In comparison, Young (1972) suggested average rates of soil creep in humid temperate areas of between 0.5 $cm^3\ cm^{-1}\ yr^{-1}$ and 3 $cm^3\ cm^{-1}\ yr^{-1}$, so that on this basis transport rates due to slow mass movements in the periglacial zone would appear to exceed those in the temperate zone by at least one order of magnitude. However, radiocarbon dating of palaeosols buried by the advance of solifluction lobes in alpine regions indicates

Figure 17.12 Unvegetated stone-banked solifluction lobe, Okstindan, northern Norway. Altitude approximately 1400 m

average rates of frontal advance of only between 0.6 mm and 3.5 mm per year (Benedict, 1976), though higher rates have occurred in the past at most sites (e.g. Alexander and Price, 1980; Gamper and Suter, 1982; Gamper, 1983; Reanier and Ugolini, 1983). Taking the average thickness of lobes as between 0.5 m and 1 m, such rates of frontal advance suggest volumetric transport of sediment of between 0.3 $cm^3\ cm^{-1}\ yr^{-1}$ and 35 $cm^3\ cm^{-1}\ yr^{-1}$. On alpine slopes at least, therefore, sediment transport by solifluction may vary considerably both spatially and temporally.

An assessment of the relative importance of surface wash and solifluction in Kärkevagge, Swedish Lappland has been made by Strömqvist (1983). In this non-permafrost area where slopes are well vegetated, the study showed that erosion by running water was localized and sediment transfers to the fluvial system were mainly associated with catastrophic events such as mudflows and avalanches. Compared with solifluction transport, the amount of material carried by surface wash was found to be diminutive. In the permafrost environment of Banks Island, NWT, French and Leikowicz (1981) also compared slope wash with solifluction, in this case on two low-angled slopes (2° to 4°). They estimated solifluctional transport of between 22 $cm^3\ cm^{-1}\ yr^{-1}$ and 30 $cm^3\ cm^{-1}\ yr^{-1}$, which represented rates at least two orders of magnitude greater than the measured rates of slopewash.

Gelifluction and frost creep may affect a wide range of sediments, ranging from alluvial silts and clays through diamictons of glacial or periglacial origin, to screes. Talus shift involves creep movements of individual scree fragments due to disturbance by such agencies as rock fall, needle ice, localized sliding, etc. (Gardner 1969, 1979), and affects slopes close to their angle of rest. On gentler periglacial slopes residual soils and till often suffer frost

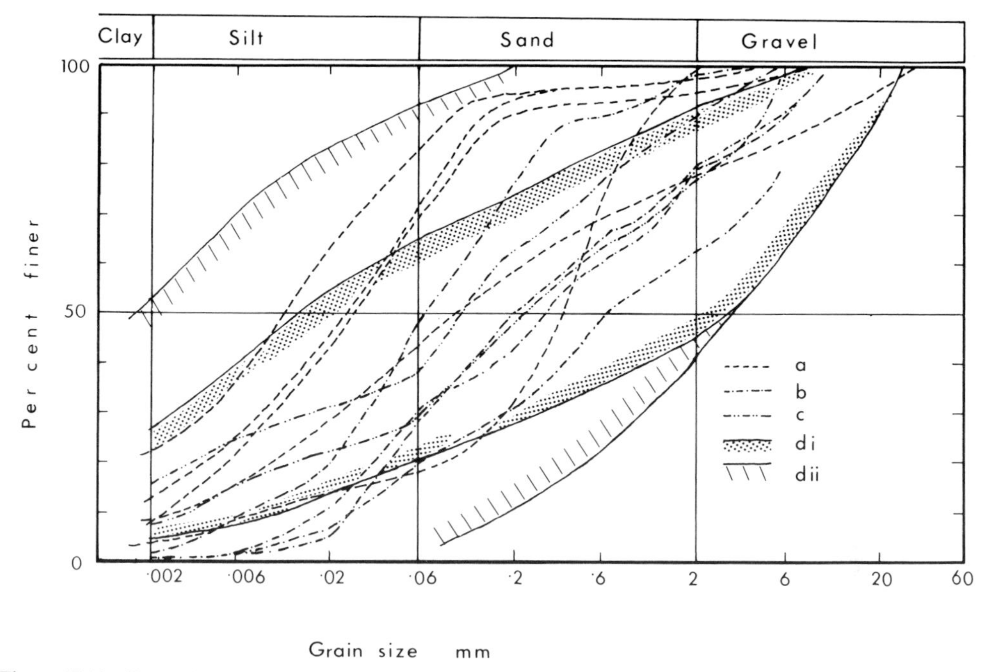

Figure 17.13 Examples of grain-size distributions of soliflucted sediments: (a) Vestspitsbergen (*Chandler, 1972*). (b) Sköddefjellet, Spitsbergen (*Jania, 1977*). (c) Okstindan, Norway (*Harris, 1977*). (d) Envelopes of soils developed from Palaeozoic siltstones, eastern Melville Island, North-West Territories, Canada, (i) major concentration, (ii) minor concentration (*Stangl* et al., *1982*)

sorting which brings coarser material to the surface, leaving fines at depth (e.g. Rudberg, 1964). This fine grained substrate may well be frost susceptible, leading to frost heaving in winter and solifluction during the summer thaw. Sorted stripes, stone banked lobes (Figure 17.12), and on steeper slopes, shallow lobate slides, represent the morphological expression of such downslope movement. In Spitsbergen these features have been described in detail by Jania (1977) and typical grain size distributions are shown in Figure 17.13.

Where till forms the substrate in which solifluction occurs, the sediment is generally matrix supported, the matrix commonly consisting of silt or silty sand (Figure 17.13). In Melville Island, NWT, Canada, Stangl *et al.* (1982) reported solifluction associated particularly with silty soils produced by weathering of fine sandstone or siltstone facies (Figure 17.13). Such sandy and silty soils generally have low liquid limits and plasticity indices. In the Unified Soil Classification System (US Bureau of Reclamation, 1963), solifluction soils are grouped at the lower end of the A-line in the CL and ML categories (Figure 17.14).

However, soils with significant clay contents have also been reported in studies of solifluction, for instance, the reworked tills of Garry Island contain greater than 25% clay (Mackay, 1981), and some of Washburn's sites in Mesters Vig, Greenland, show clay contents of up to 30%. The frequency of sliding over distinct shear planes rather than saturated flow, with shearing distributed throughout the moving soil, is likely to increase in heavy clay soils, as is suggested by the fossil periglacial low-angled shears in clays in southern England. These clays show significantly higher plasticity and liquid limits than the typical solifluction soils (Figure 17.14).

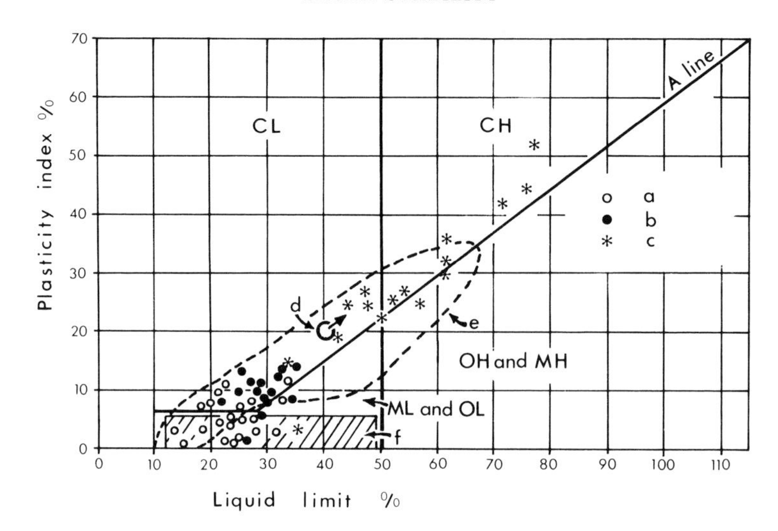

Figure 17.14 Plasticity Chart. (a) Active solifluction sediments, Banks Islands (*French, 1974*); Greenland (*Everett, 1967*; *Washburn, 1967*); and Norway (*Harris, 1977*). (b) Fossil solifluction sediments, England (*Mottershead, 1971; Harris and Wright, 1980; Wilson, 1981*). (c) Fossil planar slides in clay, England (*Chandler, 1970a,b*; *Chandler* et al., *1976*). (d) Minimum plasticity and liquid limit from Garry Island, North-West Territories, Canada, solifluction study (*Mackay, 1981*). (e) Range of values reported for near-surface sediments, District of Keewatin, North-West Territories, Canada (*Shilts, 1974*). (f) Range of values reported for solifluction lobes in the Swiss Alps (*Furrer* et al., *1971*)

Where shear strength parameters are reported, they generally reflect the non-cohesive silty or sandy nature of most solifluction soils, with ϕ' values varying from 23° in the Mackenzie Valley, NWT Canada (McRoberts and Morgenstern, 1974a) to 35° in north Norway (Harris, 1977), and 36° in Spitsbergen (Chandler, 1972).

Finally, one of the most characteristic features of soliflucted diamictons is that the clasts tend to become oriented during soil movement (Harris, 1981). Indeed, the organization, though less strongly developed, extends down to the sand-sized particles (Harris and Ellis, 1980). Clasts tend to show low-angled dips and strong preferred orientations parallel to the slope direction. Cailleaux and Taylor (1954) suggested that between 63% and 95% of elongate clasts in solifluction deposits lie parallel to the slope direction. From studies in Quebec and Baffin Island Brochu (1978) reported between 63% and 93% of elongate clasts with long axis orientations parallel to the slope azimuth, compared with median values of 36% for gravity slope debris and 33% for fluviatile or fluvioglacial deposits.

17.5.2 Failures of the Skinflow Type

While solifluction is the characteristic mass movement in many active layers, occurring annually under average conditions (McRoberts, 1978), sudden failures involving rapid localized movements generally result from an unusual event such as heavy rainfall, fire, high temperature, or rapid slope erosion by a river or the sea. In high latitude permafrost areas such events generate unusually rapid and deep thawing of the permafrost. Ice contents in permafrost immediately below the active layer are often high, so that thawing is likely to cause subsidence, and on slopes, instability, due to release of excess water.

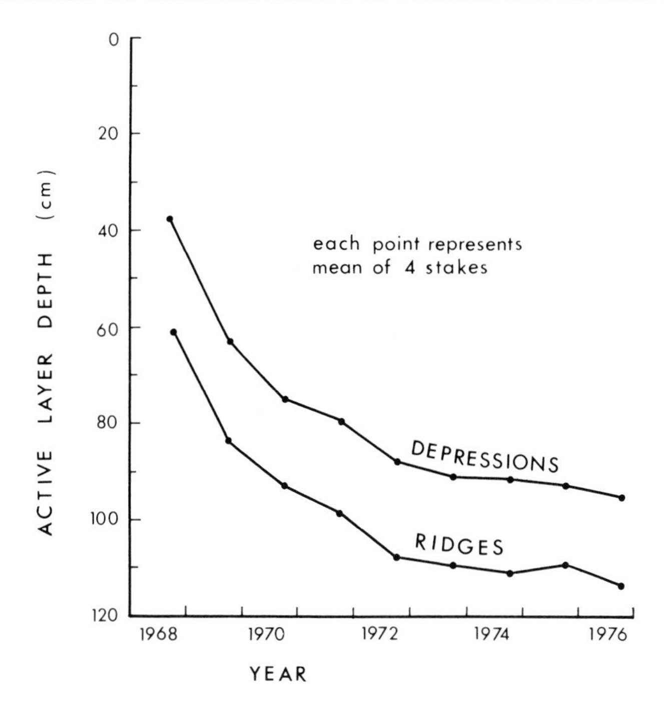

Figure 17.15 Observed thickening of the active layer near Inuvik, Mackenzie Delta, Canada, following fire in 1968. (*From Mackay, 1977,* Rept Activities, Pt B. Geol. Surv. Can. Paper 77–1B. *Reproduced by permission of Geological Survey of Canada.*)

The rate of thickening of the active layer at Inuvik, Mackenzie Delta, Canada, following a forest fire in August 1968 was most rapid during the first season (Figure 17.15), and decreased roughly exponentially over the next eight years (Mackay, 1977). McRoberts *et al.* (1978) modelled degradation rates in the Mackenzie Valley for high (40%) and low (20%) ice contents and showed most rapid thawing in the first season following thermal disturbance, with fastest degradation in the low ice content case.

The term skinflow was used by McRoberts and Morgenstern (1974a) to describe the detachment of a thin veneer of soil and vegetation and its subsequent displacement downslope over a planar surface, generally the permafrost table. They suggest that flow is the dominant process of movement (Figure 17.16). Skinflows are generally shallow ribbon-like flows with a small arcuate escarpment at the head and arcuate lobe at the toe. They are common in the arctic tundra (French, 1976) and subarctic boreal forest (McRoberts, 1978).

In the Mackenzie River Valley, Canada, McRoberts and Morgenstern (1974a) described skinflows in colluvial silts developed on slopes of between 6° and 25°. Also in the Mackenzie Valley, Mackay and Mathews (1973) reported shallow skinflow-type failures which they referred to as active layer glides. These authors emphasized the role of planar sliding over the permafrost table, with the vegetation mat preventing much internal disturbance of the displaced mass.

Shallow planar landslides in the active layer were also described by Hardy and Morrison (1972) and Hughes (1972) in northwestern Canada and by Stangl *et al.* (1982) in eastern

Figure 17.16 Skinflow, Horton River, North-West Territories, Canada. Note erect tree moving along flow channel, and two people for scale. (Photo courtesy of J. R. Mackay.)

Melville Island. Stangl *et al.* (1982) referred to these failures as active layer detachments. They concluded that slope inclination and moisture content were the two most important factors controlling stability at a site, and that active layer detachments resulted from accelerated thawing triggered by heavy rainfall events. In Alaska Holmes and Lewis (1965) observed a skinflow which was initiated during rapid thickening of the active layer in a very warm spell during a milder than normal summer. Movement in this case lasted for three days.

The terms mudflows and earthflows are also used to describe active layer failures of the skinflow type. In the hilly terrain around Umiat, Alaska, earthflows on slopes up to 20° were described by Carter and Galloway (1981) developed on Cretaceous silts and shales. Flows were associated with areas of runoff concentration, ranged from 10 m to 50 m wide, and were up to 450 m long and 1.5 m deep. They consisted of a flat-floored, often striated upper scar bounded by nearly vertical scarps, and a lobate or hummocky lower lobe. Carter and Galloway concluded that the presence of readily-hydrated bentonite clays made the active layer particularly prone to these landslides, and they occurred during warmer than average summers when the active layer was thicker. Intense summer rainstorms were considered to trigger individual failures.

Such shallow planar active layer failures are of sufficient frequency to concern engineers working on transport and routes over permafrost terrain. Stabilization procedures have been proposed by Pufahl and Morgenstern (1979). The technique they described requires a cover of insulative material to slow the rate of melting which in turn reduces the release of excess soil water and hence generation of positive pore pressures. In addition the slide was given a surcharge load of free-draining sand or gravel, which calculations showed increased the normal effective stress disproportionately to any increase in shearing stress.

In mountainous terrain, both in permafrost and non-permafrost areas, slopes are steep, and shallow active layer failures generally rapid. Such failures are morphologically similar to skinflows and are referred to as mudflows or debris flows. Recent reviews by Innes (1983) and Costa (1984) indicate that mudflows and debris flows form part of a continuum, with mudflows being drier and slower, and debris flows wetter and faster. In both cases failures are generally initiated by unusually heavy rainfall (Caine, 1980). However, neither mudflows nor debris flows are restricted to the periglacial zone (see for example, Chapter 16).

In the Longyear Valley, Spitsbergen, eye witness reports described mudflows and debris flows triggered by a rainstorm of 30.8 mm in 12 hours in July 1972 (Larsson, 1982). The active layer first became oversaturated on the plateau surfaces above the valley and lateral seepage initiated rapid ribbon-like failures. The active layer moved over the permafrost table initially as slides before flowing downslope to deposit debris flow lobes at the base of the slopes. The flow tracks were marked by ribbons of debris having rounded lobate fronts and flanked by levées up to several metres wide. Boulders were oriented parallel to the levées, but became transversely oriented at the lobe fronts. This debris flow/mudflow morphology corresponds closely to descriptions given by Rapp (1960) and Rapp and Nyberg (1981) in northern Scandinavia, Sharp (1942) in the St Elias Range, Yukon, Holmes and Lewis (1965) in northern Alaska, and Jahn (1976), also in Spitsbergen.

17.5.3 Bimodal Flows

These failures differ significantly from solifluction and skin flows in that they are not restricted to the active layer, but result from rapid and deep thawing of ice-rich permafrost. They develop on river valley sides, coastal slopes, and around thermokarst depressions, where ground ice is exposed to accelerated lateral melting. The term bimodal flow refers to two morphological sectors (McRoberts and Morgenstern, 1974a; McRoberts, 1978), an upper headscarp where ablation releases sediment which slides, flows or falls down to a gently inclined mudflow lobe, where it flows away (Figure 17.17).

Bimodal flows have been described in the Canadian High Arctic (Victoria Island, Washburn, 1947; Ellef Ringes Island and Axel Heiberg Island, Lamothe and St Onge, 1961; Garry Island, Kerfoot, 1969; Kerfoot and Mackay, 1972; Banks Island, French and Egginton, 1973; French, 1974b; Harry *et al.*, 1983; Melville Island, Stangl *et al.*, 1982); the Mackenzie Valley and Delta (Mackay, 1966; Hughes, 1972; Mackay and Mathews, 1973; McRoberts and Morgenstern, 1974a; Pufahl and Morgenstern, 1979), and east Siberia (Czudek and Demek, 1973). A cycle of development and decay has been proposed by Lamothe and St Onge (1961) whereby lateral erosion by a river rapidly steepens a valley side, exposing ice-rich permafrost. Rapid ablation leads to retreat of an ice-rich scarp at the top of the slope, forming a semicircular hollow opening towards the river (Figure 17.18). Silt released by the melting scarp flows out of the hollow as a gently inclined mudflow tongue. Lamothe and St Onge suggested that eventually the scarp becomes buried by silt, further ablation is prevented, and the bimodal flow stabilizes, this occurring at diameters of around 40 m on Ellef Ringes Island.

Figure 17.17 Headscarp of bimodal flow, exposing glacially deformed ice-rich silts, Garry Island, North-West Territories, Canada. (Photo courtesy of J. R. Mackay.)

On Banks Island, French (1976) reported diameters of up to 300 m, and calculated that bimodal flows there stabilize approximately 30 to 50 years after initiation. Mackay (1966) noted that polycycle bimodal flows are common in the Mackenzie Delta. In these examples earlier mudflow lobes become stabilized by the termination of thawing of the headscarps, and the toes subsequently suffer further erosion, exposing undisturbed ground ice to a new cycle of ablation. A new scarp then retreats across the floor of the original flow, releasing further silty sediment. Stangl *et al.* (1982) also reported reactivation of stabilized bimodal flows on Melville Island.

Mechanisms of headscarp retreat were discussed in detail by Mackay (1966), Kerfoot (1969), and McRoberts and Morgenstern (1974a). Most common is simple retreat of the ice-rich scarp due to ablation, which leaves the active layer above unsupported. This eventually collapses and slides down to become incorporated in the mudflow (Figure 17.17).

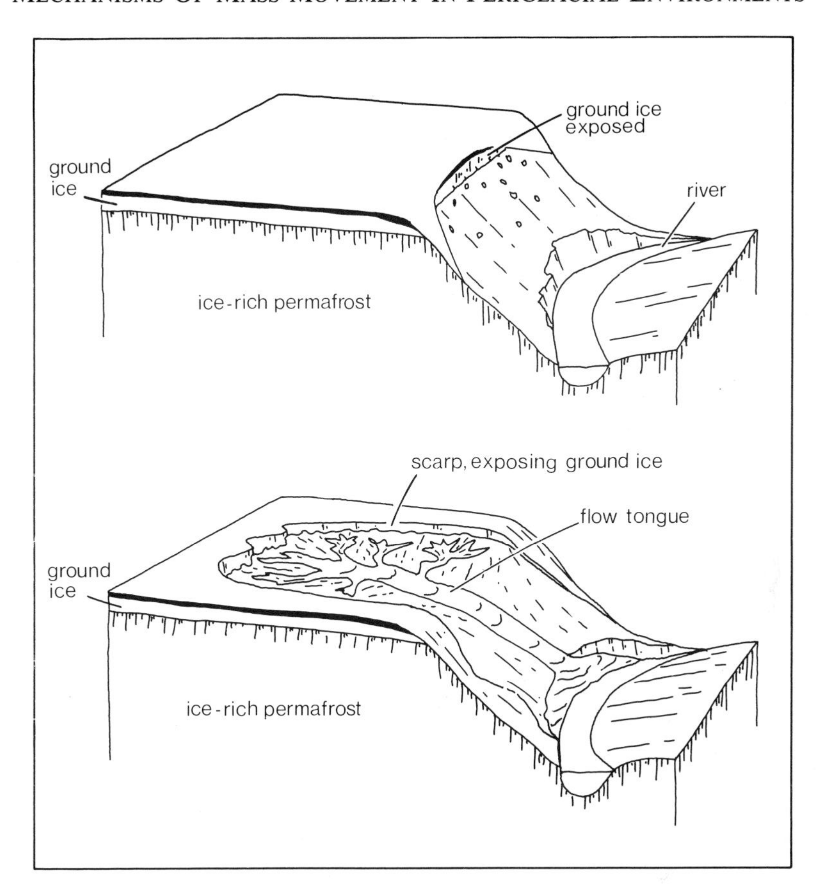

Figure 17.18 Block diagram showing development of a bimodal flow. (*Modified from Lamothe and St Onge, 1961.*)

In some cases, mineral and organic material is released more gradually by scarp retreat, and flows in pulses down over the ice face to the mudflow lobe. Where the scarp is formed in less ice-rich permafrost, thawing may proceed back from the face to some critical depth, when rapid failure occurs, and the thawed layer slides down to the mudflow exposing still frozen permafrost to further ablation.

Rates of scarp retreat were reviewed by McRoberts and Morgenstern (1974a), who showed rates ranging from 1 cm per day to 20 cm per day during the thaw season. More recent measurements by Hegginbottom (1977) in eastern Melville Island fall within this range, with values of 19 cm per day and 14 cm per day in mid-July in successive seasons.

The energetics of the ablating headscarp were discussed by Pufahl and Morgenstern (1979), who showed that the magnitude of heat flux may be estimated from the formula;

$$q = vL + vCT \tag{17.9}$$

where v is the rate of retreat (m day^{-1}), L is the volumetric latent heat of fusion of the soil (MJ m^{-3}), C is the volumetric heat capacity of the soil (kJ m^{-3}), and T is the ground temperature in degrees C below zero. If the ground temperature approaches zero, this reduces to;

$$q = vL \tag{17.10}$$

The authors showed that the three significant terms in the total heat flux were net radiation, latent heat transfer, and sensible heat transfer. Field measurements in the Mackenzie Valley showed approximate contributions of 43% by net radiation, 35% by sensible heat transfer, and 24% by latent heat transfer.

17.6 CONCLUSION

As this chapter has shown, the most important aspect of periglacial slopes formed in unconsolidated soils is the presence of ice-rich frozen ground. Thawing of such frozen ground is liable to release excess water and promote significant loss of strength in the newly thawed soil as normal stresses are transferred to the pore water during consolidation. Mass movements on very gentle slopes may result. Solifluction represents the annual, widespread, slow, mass movement of the active layer during thaw. Sporadic localized failures producing rapid flows and slides are generally superimposed on slopes suffering some degree of solifluction. They usually result from accelerated thawing initiated by such events as abnormally high temperatures, high rainfall, or fire.

The susceptibility of periglacial slopes to thaw-induced instability is of particular concern to engineers, especially where engineering works may disturb the thermal equilibrium of ice-rich permafrost. Careful terrain analyses such as those described by Krieg and Reger (1976), Berg *et al.* (1978), and Linell and Tedrow (1981) are therefore necessary in advance of more detailed site investigations. It is in this field of terrain evaluation that the most obvious practical interaction between engineers and geomorphologists takes place.

REFERENCES

Alexander, C. S., and Price, L. W. (1980). 'Radiocarbon dating of the rate of movement of two solifluction lobes in the Ruby Range, Yukon Territory.' *Quaternary Res.*, **13**, 365–79.

Anderson, D. M., Pusch, R., and Penner, E. (1978). 'Physical and thermal properties of frozen ground.' In *Geotechnical Engineering for Cold Regions*. Andersland, O. B., and Anderson, D. M. (eds.), pp. 37–102.

Andersson, R. G. (1906). 'Solifluction, a component of subaerial denudation.' *J. Geol.*, **14**, 91–112.

Baulig, H. (1956). 'Peneplaines et pediplaines.' *Soc. belge études geographie*, **25**, 25–58.

Baulig, H. (1957). 'Peneplains and pediplains.' *Bull. Geol. Soc. Am.*, **68**, 913–30.

Benedict, J. B. (1970). 'Downslope soil movement in a Colorado alpine region: rates, processes and climatic significance.' *Arctic and Alpine Res.*, **6**, 55–76.

Benedict, J. B. (1976). 'Frost creep and gelifluction: a review.' *Quaternary Res.*, **6**, 55–76.

Berg, R., Brown, J., and Haugen, R. K. (1978). 'Thaw penetration and permafrost conditions associated with the Livengood to Prudoe Bay Road, Alaska.' *Proc. 3rd Int. Permafrost Conf.*, Edmonton, **1**, 615–21.

Beskow, G. (1935). 'Tjäbildningen och tjällyftningen med särskild hänsyn till Vägar och jarnvägar.' *Sveriges Geol. Unders.*, Arsbok 26, No. 375.

Brochu, M. (1978). 'Disposition des fragments rocheux dans les dépôts de solifluxion, dans les éboulis de gravité et dans les dépôts fluviatiles: mesures dans l'est de L'Arctique Nord Americain et comparison avec d'autres régions du Globe.' *Biul. Peryglac.*, **27**, 35–51.

Brown, J. (1967). 'An estimation of volume of ground ice, coastal plain, northern Alaska.' *US Army Corps of Engineers, CRREL*, Hanover, N.H., *Manuscript*, 22pp.

Brown, R. J. E., and Kupsch, W. O. (1974). *Permafrost Terminology*. Nat. Res. Council, Canada, 14272, 62pp.

Burt, T. P., and Williams, P. J. (1976). 'Hydraulic conductivity in frozen soils.' *Earth Surf. Proc.*, **1**, 349–60.

Cailleux, A., and Taylor, G. (1954). 'Cryopedologie, étude des sols gelés.' *Expeditions Polaires Francaises Missions Paul-Emil Victor IV Paris, Hermann et Gie, Act. Sci. Ind.*, 1203, 218pp.

Caine, T. N. (1968). 'The log-normal distribution and rates of soil movement: an example.' *Rev. Geomorph. Dyn.*, **18**, 1–7.

Caine, T. N. (1980). 'The rainfall intensity–duration control of shallow landslides and debris flows.' *Geografiska Annaler*, **62A**, 23–7.

Carter, L. D., and Galloway, J. P. (1981). 'Earth flows along the Henry Creek, Northern Alaska.' *Arctic*, **34**, 325–8.

Casagrande, A. (1931). 'Discussion of "A new theory of frost heaving" by A. C. Benkelman and F. R. Olmstead.' *Highway Res. Bd. Proc.*, **11**, 168–72.

Chandler, R. J. (1970a). 'The degradation of Lias Clay slopes in an area of the East Midlands.' *Q. J. Eng. Geol.*, **2**, 161–81.

Chandler, R. J. (1970b). 'A shallow slab slide in the Lias Clay near Uppingham, Rutland,' *Géotechnique*, **20**, 253–60.

Chandler, R. J. (1972). 'Periglacial mudslides in Vestspitzbergen and their bearing on the origin of fossil "solifluction" shears in low angled clay slopes.' *Q. J. Eng. Geol.*, **5**, 223–41.

Chandler, R. J., Kellaway, G. A., Skempton, A. W., and Wyatt, R. J. (1976). 'Valley slope sections in Jurassic strata near Bath, Somerset.' *Phil. Trans. Roy. Soc., London*, **A283**, 527–55.

Chen, X., Wang, Y. and Jiang, P. (1983). 'Influence of penetration rate, surcharge stress, and groundwater table on frost heave', *Proc. 4th Int. Permafrost Conf.*, Alaska, 131–5.

Costa, J. E. (1984). 'Physical geomorphology of debris flows.' In *Developments and Applications of Geomorphology.*' Costa, J. E., and Fleisher, P. J. (eds.), Springer-Verlag, Berlin, pp. 268–317.

Crampton, C. B. (1981). 'Analysis of synergistic systems for evaluating terrain sensitivity to disturbance of icy permafrost in the Mackenzie River Valley, Canada.' *Biul. Peryglac.*, **22**, 15–31.

Czudek, T., and Demek, J. (1970). 'Thermokarst in Siberia and its influence on the development of lowland relief.' *Quaternary Research*, **1**, 103–20.

Czudek, T., and Demek, J. (1973). 'The valley cryopediments in eastern Siberia.' *Biul. Peryglac.*, **22**, 117–30.

Everett, D. H. (1961). 'The thermodynamics of frost damage to porous solids.' *Trans. Faraday Soc.* **57**, 1541–51.

Everett, D. H., and Haynes, J. M. (1965). 'Capillary properties of some model pore systems with special reference to frost damage.' *Int. des Laboratories d'Essoir et de Reserches sur les Materieux et Constructions Bulletin*, **27**, 31–8.

Everett, K. R. (1967). 'Mass-wasting in the Taseriaq area, West Greenland.' *Medd. om Grønland*, **165**, 1–32.

French, H. M. (1974a). 'Mass-wasting at Sachs Harbour, Banks Island, NWT, Canada.' *Arctic and Alpine Res.*, **6**, 71–8.

French, H. M. (1974b). 'Active thermokarst processes, eastern Banks Island, Western Canadian Arctic.' *Can. J. Earth Sci.*, **11**, 785–94.

French, H. M. (1976). *The Periglacial Environment*. Longman, London.

French, H. M., and Egginton, P. (1973). 'Thermokarst development, Banks Island, Western Canadian Arctic.' *N. American Contribution, 2nd Int. Permafrost Conf.*, Yakutsk, 203–12.

French, H. M., and Harry, D. G. (1983). 'Ground ice conditions and thaw lakes, Sachs Harbour Lowlands, Banks Island, Canada.' In *Mesoformes des Reliefs in heutigen Periglazialraum*. Schunke, H., and Poser, H., (eds.), Abh. Akad. Wissenchaften Göttingen, Math–Phys. Kl. 35, 70–81.

French, H. M., and Leikowicz, A. G. (1981). 'Periglacial slopewash investigation, Banks Island, Western Arctic.' *Biul. Peryglac.*, **28**, 33–45.

Furrer, G., Bachmann, F., and Fitze, P. (1971). 'Erdströme als Formelemente von Soliflukhousdecken im Raum Munt Chevagi/Munt Buffalora.' *Ergebnisse der Wissenschaften Untersuchungen im Schwizerischen Nationalpark*, **11**, 188–269.

Gamper, M. (1981). 'Heutige Solifluktionsbeträge von Erdströmen und klimamorphologische Interpretation fossiler Böden.' *Ergebnisse der Wissenschaftlichen Untersuchungen im Schweizerischen Nationalpark* Bnd. **15**, 443pp.

Gamper, M. (1983). 'Controls and rates of movement of solifluction lobes in the Eastern Swiss Alps.' *Proc. 4th Int. Permafrost Conf.*, Alaska, 328–33.

Gamper, M., and Suter, J. (1982). 'Postglaziale klimageschichte der Schweizer Alpen.' *Geographica Helvetica*, **37**, 105–14.

Gardner, J. (1969). 'Observations of surficial talus movement.' *Z. Geomorph.* **13**, 317–23.

Gardner, J. (1979). 'The movement of material on debris slopes in the Canadian Rocky Mountains.' *Z. Geomorph.*, **23**, 45–57.

Hardy, R. M., and Morrison, H. A. (1972). 'Slope stability and drainage considerations for arctic pipelines.' *Proc. Canadian Northern Pipeline Res. Conf., Ottawa, Nat. Res. Council Canada, Tech. Mem.*, **104**, 249–66.

Harris, C. (1972). 'Processes of soil movement in turf-banked solifluction lobes, Okstindan, Northern Norway.' In *Polar Geomorphology*. Price, R. J., and Sugden, D. E. (eds.), Inst. Brit. Geogr. Spec. Publ. 4, 155–74.

Harris, C. (1977). 'Engineering properties, groundwater conditions, and the nature of soil movement on a solifluction slope in North Norway.' *Q. J. Eng. Geol.*, **10**, 27–43.

Harris, C. (1981). *Periglacial Mass-Wasting: A Review of Research*. B.G.R.G. Research Monograph 4, Geo Abstracts, Norwich.

Harris, C. (1985). 'Geomorphological applications of soil micromorphology with particular reference to periglacial sediments and processes.' In *Geomorphology and Soils*. Richards, K. S., Arnett, R. R., and Ellis, S. (eds.). George Allen & Unwin, London, pp. 219–32.

Harris, C., and Ellis, S. (1980). 'Micromorphology of soils in solifluction materials, Okstindan, Northern Norway.' *Geoderma*, **23**, 11–29.

Harris, C., and Wright, M. D. (1980). 'Some last glaciation drift deposits near Pontypridd, South Wales.' *Geol. J.*, **15**, 7–20.

Harry, D. G., French, H. M., and Clarke, M. J. (1983). 'Coastal conditions and processes, Sachs Harbour, Banks Island, Western Canadian Arctic.' *Z. Geomorph.*, **47**, 1–26.

Hegginbottom, J. A. (1977). 'An active retrogressive thaw flow slide on eastern Melville Island, District of Franklin.' *Geol. Surv. Can., Paper* 78-1A, 525–6.

Higashi, A., and Corte, A. E. (1971). 'Solifluction: a model experiment.' *Science*, **171**, 480–2.

Holgate, M. W., Allen, S. E., and Chambers, M. J. G. (1967). 'A preliminary investigation of the soils of Signy Island, South Orkney Islands.' *Brit. Antarctic Survey Bull.*, **12**, 53–71.

Holmes, G. W., and Lewis, C. R. (1965). 'Quaternary geology of the Mount Chamberlin area, Brooks Range, Alaska.' *US Geol. Survey Bull.*, 1201-B.

Hughes, O. L. (1972). 'Surficial geology and land classification, Mackenzie Valley Transportation Corridor.' *Proc. Northern Pipeline Res. Conf.*, National Research Council, Canada.

Innes, J. L. (1983). 'Debris flows.' *Progr. Phys. Geog.*, **7**, 469–501.

Jackson, K. A., and Chalmers, B. (1958). 'Freezing of liquids in porous media with special reference to frost heaving in soils.' *J. Appl. Phys.*, **29**, 1178–81.

Jahn, A. (1975). 'Solifluction.' Chapter 14 in *Problems of the Periglacial Zone*. P.W.N. Polish Scientific Publishers, Warsaw, 221pp.

Jahn, A. (1976). 'Contemporaneous geomorphological processes in Longyeardalen, Vestspitzbergen (Svalbard).' *Biul. Peryglac.*, **26**, 253–68.

Jahn, A. (1978). 'Mass washing in permafrost and non-permafrost environments.' *Proc. 3rd Int. Permafrost Conf.*, Edmonton, 295–300.

Jania, J. (1977). 'Debris forms on the Sköddefjellet slope.' *Results of investigations of the Polish Scientific Spitsbergen Expeditions, 1970–1974 Vol. II, Acta. Univ. Wratislaviensis*, 387, 91–117.

Kaplar, C. W. (1970). 'Phenomenon and mechanism of frost heaving.' *Highway Res. Record*, **304**, 1–13.

Kerfoot, D. E. (1969). *The geomorphology and permafrost conditions of Garry Island, NWT.* Unpubl. PhD Thesis, Univ. British Columbia, Vancouver B.C., 308pp.

Kerfoot, D. E., and Mackay, J. R. (1972). 'Geomorphological process studies, Garry Island, NWT.' In *Mackenzie Delta Area Monograph.* Kerfoot, D. E. (ed.), 22nd Int. Geogr. Congr. Montreal, 115–130.

Konrad, J.-M., and Morgenstern, N. R. (1980). 'A mechanistic theory of ice lens formation in fine-grained soils.' *Can. Geotech. J.*, **17**, 473–86.

Konrad, J.-M., and Morgenstern, N. R. (1981). 'The segregation potential of a freezing soil.' *Can. Geotech. J.*, **18**, 482–91.

Konrad, J.-M., and Morgenstern, N. R. (1982). 'Prediction of frost heave in the laboratory during transient freezing.' *Can. Geotech. J.*, **19**, 250–9.

Krieg, R. A., and Reger, R. D. (1976). 'Preconstruction terrain evaluation for the trans-Alaska pipeline project.' In *Geomorphology and Engineering.* Coates, D. R. (ed.), Dowden, Hutchinson and Ross, Stroudsburg, Pa., pp. 55–76.

Lamothe, L., and St Onge, D. (1961). 'A note on periglacial erosion processes in the Isachsen Area, NWT.' *Geog. Bull.*, **16**, 104–13.

Larsson, S. (1982). 'Geomorphological effects on the slopes of Longyear Valley, Spitzbergen, after a heavy rainstorm in July 1972.' *Geog. Ann.*, **64A**, 105–25.

Lawrence, D. E., and Proudfoot, D. A. (1977). 'Mackenzie Valley geotechnical data bank.' *Geological Survey of Canada, Ottawa, Open Files* 421–5, p. 24.

Linell, K. A., and Tedrow, J. C. F. (1981). *Soil and Permafrost Surveys in the Arctic.* Clarendon Press, Oxford.

Loch, J. P. G., and Kay, B. D. (1978). 'Water redistribution in partially frozen saturated silt under several temperature gradients and overburden loads.' *Soil Sci. Soc. Am. J.*, **42**, 400–6.

Loch, J. P. G., and Miller, R. D. (1975). 'Tests of the concept of secondary frost heaving.' *Soil Sci. Soc. Am. Proc.*, **39**, 1036–41.

Lovell, C. W. (1983). 'Frost susceptibility of soils.' *Proc. 4th Int. Permafrost Conf.*, Alaska, 735–9.

Mackay, J. R. (1966). 'Segregated epigenetic ice and slumps in permafrost, Mackenzie Delta, NWT.' *Geog. Bull.*, **8**, 59–80.

Mackay, J. R. (1972). 'The world of underground ice.' *Ann. Ass. Am. Geog.*, **62**, 1–22.

Mackay, J. R. (1977). 'Changes in the active layer from 1968 to 1976 as a result of the Inuvik fire.' *Rept. of Activities, Pt. B, Geol. Surv. Can. Paper* 77-1B, 273–5.

Mackay, J. R. (1981). 'Active layer slope movement in a continuous permafrost environment Garry Island, Northwest Territories, Canada.' *Can. J. Earth Sci.*, **18**, 1666–680.

Mackay, J. R. (1983). 'Downward water movement into frozen ground, western arctic coast, Canada.' *Can. J. Earth Sci.*, **20**, 120–34.

Mackay, J. R., and Black, R. (1973). 'Origin, composition and structure of perennially frozen ground and ground ice: a review.' *N. American Contribution to 2nd Int. Permafrost Conf., Yakutsk*, 185–92.

Mackay, J. R., and Mathews, W. H. (1973), 'Geomorphology and Quaternary history of the Mackenzie River Valley near Fort Good Hope, NWT, Canada.' *Can. J. Earth Sci.*, **10**, 26–41.

McGaw, R. (1972). Frost heaving versus depth to water table. *Highway Research Board*, **395**, 45–55.

Mageau, D. W., and Morgenstern, N. R. (1980). 'Observations on moisture migration in frozen soils.' *Can. Geotech. J.*, **17**, 54–60.

McRoberts, E. C. (1975). 'Field observation of thawing in soils.' *Can. Geotech. J.*, **12**, 126–30.

McRoberts, E. C. (1978). 'Slope stability in cold regions.' In *Geotechnical Engineering for cold Regions.* Andersland, O. B., and Anderson, D. M. (eds.), McGraw-Hill, New York.

McRoberts, E. C., Fletcher, E. B., and Nixon, J. F. (1978). 'Thaw consolidation effects in degrading permafrost.' *Proc. 3rd Int. Permafrost Conf.*, Edmonton, 694–9.

McRoberts, E. C., and Morgenstern, N. R. (1974a). 'The stability of thawing slopes.' *Can. Geotech. J.*, **11**, 447–69.

McRoberts, E. C., and Morgenstern, N. R. (1974b). 'The stability of slopes in frozen soil, Mackenzie Valley, NWT.' *Can. Geotech. J.*, **11**, 554–73.

McRoberts, E. C., and Morgenstern, N. R. (1975). 'Pore water expulsion during freezing.' *Can. Geotech. J.*, **12**, 130–8.

Miller, R. D. (1972). 'Freezing and heaving of saturated and unsaturated soils.' *Highway Res. Record*, **393**, 1-11.

Miller, R. D. (1978). 'Frost heaving in non-colloidal soils.' *Proc. 3rd Int. Permafrost Conf.*, Edmonton, 708-13.

Miller, R. D., Loch, J. P. G., and Bresler, E. (1975). 'Transport of water and heat in a frozen permeameter.' *Soil Soc. Am. Proc.*, **39**, 1029-41.

Morgenstern, N. R. (1981). 'Geotechnical engineering and frontier resource development.' *Géotechnique*, **31**, 305-65.

Morgenstern, N. R., and Nixon, J. F. (1971). 'One dimensional consolidation of thawing soils.' *Can. Geotech. J.*, **8**, 558-65.

Mottershead, D. N. (1971). 'Coastal head deposits between Start Point and Hope Cove, Devon,' *Field Studies*, **5**, 433-53.

Nixon, J. F. (1982). 'Field frost heave predictions using the segregation potential concept.' *Can. Geotech. J.*, **19**, 526-9.

Nixon, J. F., and Hanna, A. J. (1979). 'The undrained strength of some thawed permafrost soils.' *Can. Geotech. J.*, **16**, 420-7.

Penner, E. (1959). 'The mechanics of frost heaving in soils.' *Highway Res. Bd. Bull.*, **225**, 1-22.

Penner, E. (1968). 'Particle size as a basis for predicting frost action in soils.' *Soils and Fdtns*, **8**, 21-9.

Penner, E. (1972). The influence of freezing rate on frost heaving.' *Highway Res. Record*, **393**, 56-64.

Pollard, W. H., and French, H. M. (1980). 'First approximation of the volume of ground ice, Richards Island, Pleistocene Mackenzie Delta, North West Territories, Canada.' *Can. Geotech. J.*, **17**, 509-16.

Price, L. W. (1973). 'Rates of mass-wasting in the Ruby Range, Yukon Territory'. *North American Contribution, 2nd Int. Permafrost Conf.*, Yakutsk, 235-45.

Pufahl, D. E., and Morgenstern, N. R. (1979). 'Stabilization of planar landslides in permafrost.' *Can. Geotech. J.*, **16**, 734-47.

Rampton, V. N., and Mackay, J. R. (1971). 'Massive ice and icy sediments throughout the Tuktoyaktuk Peninsula, Richards Island, and nearby areas, District of Mackenzie.' Geol. Survey Can., Paper 71-21.

Rapp, A. (1960). 'Recent development of mountain slopes in Kärkevagge and surroundings, Northern Sweden.' *Geog. Ann.*, **42A**, 71-200.

Rapp, A. (1962). 'Kärkevagge, some recordings of mass movements in the Northern Scandinavian Mountains.' *Biul. Peryglac.*, **11**, 287-309.

Rapp, A. and Nyberg, R. (1981). 'Alpine debris flows in Northern Scandinavia. Morphology and dating by lichenometry.' *Geog. Ann.*, **63A**, 183-96.

Reanier, R. E., and Ugolini, F. C. (1983). 'Gelifluction deposits as sources of palaeoenvironmental information.' *Proc. 4th Int. Permafrost Conf.*, Alaska, 1042-7.

Rein, R. G., and Burrous, C. M. (1980). 'Laboratory measurements of subsurface displacements during thaw of low-angle slopes of a frost susceptible soil.' *Arctic and Alpine Res.*, **12**, 349-58.

Rudberg, S. (1964). 'Slow mass movement processes and slope development in the Norra Storfjall area, Southern Swedish Lappland.' *Z. Geomorph.*, **4**, Suppl. 5, 192-203.

Sharp, R. P. (1942). 'Soil structures in the St Elias Range, Yukon Territory.' *J. Geomorph.*, **5**, 274-301.

Shilts, W. N. (1974). 'Physical and chemical properties of unconsolidated sediments in permanently frozen terrain district of Keewatin.' *Geol. Survey Can., Paper* 74, 229-35.

Sigafoos, R. S., and Hopkins, D. M. (1952). 'Soil stability on slopes in regions of perennially frozen ground.' *Highway Res. Bd. Spec. Rept.*, **2**, 176-92.

Skempton, A. W., and Weeks, A. G. (1976). 'The Quaternary history of the Lower Greensand escarpment and Weald clay vale near Sevenoaks, Kent.' *Phil. Trans. Roy. Soc., London,* **A283**, 493-526.

Stangl, K. O., Roggensack, W. D., and Hayley, D. W. (1982). 'Engineering geology of surficial soils, eastern Melville Island.' *Proc. 4th Can. Permafrost Conf.*, 136-47.

Strömqvist, L. (1983). 'Gelifluction and surface wash, their importance and interaction on a perglacial slope.' *Geog. Ann.*, **65A**, 245-54.

Taber, S. (1929). 'Frost heaving.' *J. Geol.*, **37**, 428–61.
Taber, S. (1930). 'The mechanics of frost heaving.' *J. Geol.*, **38**, 303–17.
Taber, S. (1943). 'Perennially frozen ground in Alaska: its origin and history.' *Bull. Geol. Soc. Am.*, **54**, 1433–548.
Takagi, S. (1978). 'Segregation freezing as the cause of suction force for ice lens formation.' *C.R.R.E.L. Rept.*, 78, Hanover, New Hampshire.
Takagi, S. (1979). 'Segregation freezing as the cause of suction force for ice lens formation.' *Eng. Geol.*, **13**, 93–100.
US Bureau of Reclamation (1963). *Earth Manual.* Washington, DC.
Van Vliet-Lanoë, B. (1982). 'Structures et microstructures associées à la formation de glace de ségrégation: leurs consequences.' In R. J. E. Brown Memorial volume, *Proc. 4th Canadian Permafrost Conference*, Calgary, 116–122.
Van Vliet-Lanoë, B., Coutard, J-P., and Pissart, A. (1984). 'Structures caused by repeated freezing and thawing in various loamy sediments. A comparison of active, fossil and experimental data.' *Earth Surf. Proc. Landforms*, **9**, 553–66.
Walton, D. W. H., and Heilbronn, T. D. (1983). 'Periglacial activity on the subAntarctic island of South Georgia.' *Proc. 4th Int. Permafrost Conf.*, Alaska, 1356–61.
Washburn, A. L. (1947). 'Reconnaisance geology of portions of Victoria Island and adjacent regions, Arctic Canada.' *Geol. Soc. Am. Mem.*, **22**, 142pp.
Washburn, A. L. (1967). 'Instrumental observations of mass-wasting in the Mesters Vig district, NE Greenland.' *Medd. om Grønland.*, **166**, 1–297.
Washburn, A. L. (1979). *Geocryology.* Edward Arnold, London.
Weeks, A. G. (1969). 'The stability of slopes in south-east England as affected by periglacial activity.' *Q. J. Eng. Geol.*, **5**, 223–41.
Wilson, P. (1981). 'Periglacial valley-fill sediment at Edale, North Derbyshire.' *E. Midland Geog.*, **7**, 263–71.
Williams, P. J. (1966). 'Downslope soil movement at a Sub-Arctic location with regard to variation in depth.' *Can. Geotech. J.*, **3**, 191–203.
Williams, P. J. (1968). 'Ice distribution in permafrost profiles.' *Can. J. Earth Sci.*, **5**, 1381–6.
Woo, M. K., and Steer, P. (1982). 'Occurrence of surface flow on arctic slopes, Southwestern Cornwallis Island.' *Can. J. Earth Sci.*, **19**, 2368–72.
Woo, M. K., and Steer, P. (1983). 'Slope hydrology as influenced by thawing of the active layer, Resolute, NWT.' *Can. J. Earth Sci.*, **20**, 978–86.
Young, A. (1972). *Slopes.* Oliver and Boyd, Edinburgh.

Slope Stability
Edited by M. G. Anderson and K. S. Richards

Chapter 18

Dating of Ancient, Deep-Seated Landslides in Temperate Regions

R. H. JOHNSON
Formerly, School of Geography, University of Manchester, Manchester M13 9PL

18.1 INTRODUCTION

This chapter deals with large landslides that occurred hundreds if not thousands of years ago in the present temperate zone, and is therefore concerned with very long-term slope stability/instability conditions. The landslides involved were generally very large, and their failure surfaces were located deep within the hillslopes. Indeed, such slides are only preserved because of their size for there has been insufficient time to destroy such large landforms (Erskine, 1973). As Voigt and Pariseau (1978) have aptly commented 'the larger the mass movement (usually) the further back in time the event occurred, and in consequence, the more descriptive and less quantitative is our knowledge of the specifics of the event'. Direct and indirect methods of dating such landslide events, and their reliability, are examined first, and it is shown that while few landslides can be dated absolutely, relative dating is often possible. Stabilized or fossil landslides are perhaps too often considered as a finite condition which occurs in a short time interval, but in many instances, landsliding results from slope material deterioration over long periods of time when the original stresses were initiated by valley-floor bulging, cambering or topple flexuring. These deformation structures were often created at times which anticipated the landslip movement by several thousand years, but they had a profound influence upon the way in which the subsequent event took place.

Temperate regions are characterized by stabilized landslide features although the conditions which facilitated these movements are no longer present. In such circumstances, inferences have to be made concerning the morphology of the landslides, the former condition of the soil materials, and the stress conditions at the time of failure. Many of these inferences are untestable and in Voigt's words rest on 'somewhat fragile foundations'. There is, therefore, much uncertainty and speculation regarding the modes of slope failure as models conceived for use in dealing with recent, small slip movements are imprecise or even misleading when applied to large, ancient slope movements. Nevertheless, some models have proved useful for evaluating potential slope hazard conditions and for studying changing slope forms through long time periods. One such model is presented in the last section of this chapter. This is based upon the analysis of slope forms in one south Pennine catchment where the rocks vary greatly in their resistance to denudation, and where the valley-side crests are formed (usually) by a sandstone outcrop. It shows that larger mass movements actually form a distinct stage in the evolution of the slopes. Examples of the problems of dating landslides, particularly those preceded by the development of deformation structures, are also drawn from the south Pennines.

18.2 THE DATING OF FOSSIL LANDSLIDES

18.2.1 Absolute Methods

Opportunities for determining the absolute age of a landslide do not occur frequently, but sometimes vegetation is either overridden or incorporated into a landslip. Should it then be discovered, radio-carbon assays may be used to determine when life in the organism ceased, although the difficulties in obtaining reliable dates for the failure from such materials may not always be appreciated. First, the wood, plant or peat residues are most commonly found either in or beneath the slump-toe where they are most likely to be contaminated by groundwater seepage. Second, the cause of death in the organism may be due to causes other than the landslide event. Finally, it is not always possible to determine whether the organic remains were overridden during the main slide movement, or were incorporated and transported at a later time.

Despite these difficulties, it has been possible to establish the age of some large-scale mass movements using radio-carbon techniques in conjunction with other dating techniques. At Szymbark and other sites in Poland (Gil *et al.*, 1974), dateable organic materials were found in a number of slump hollows which had become infilled with peat and mineral deposits. As the peat pollen stratigraphy could be correlated with the regional pollen stratigraphy, it was possible to use the carbon-14 dates to identify the times when the first mass movements took place (*c.* 8,210 $\pm$ 150 years BP).

A less accurate but useful chronology was also established at Mam Tor, in north Derbyshire (Figure 18.1). On the summit of the hill overlooking the slide there is a prehistoric encampment whose ramparts were partly destroyed by the landslide. The site was first occupied some 3,000 years ago (Jones and Thompson, 1965; Coombs, 1971), but the defences were not constructed until much later. The main scar probably did not disrupt the outer camp defences which were built on the then stable slopes upslope of the landslide, but since that time the main scar has retreated and its lower face is now buried by talus scree. The back of the scar is now well within the encampment and is currently still very unstable: evidence of this is indicated by a small wedge slide which disrupts the line of the scar crest, and there are tension cracks to the rear of the scar which suggest that movements will continue. On the slump itself peat began to form in surface depressions some time after the beginning of Pollen Chronozone VIII, i.e. *c.* 2,500–2,000 years BP (J. H. Tallis, personal communication). This marks an upper time limit for the main slide movement which at its base overrides head deposits of Late glacial age. According to

Figure 18.1 Air photograph of the Mam Tor landslide, north Derbyshire. The damage to the Iron Age fort defences caused by the main slide movement is clearly visible. Several landslide slumps in the Edale valley can be seen to the north and east of the hill fort

Chandler (personal communication) the toe of the slide has been displaced some 300 m in the last 3,000 years. Archaeological and botanical evidence therefore indicates that the main landslide failure must have taken place at least 2,000 years ago, but during the past 180 years many slump movements had been reported at Mam Tor as a trunk road was constructed across the lower parts of the landslide (Figure 18.1). The route was only maintained with great difficulty and detailed investigations at the site have shown that both the initial and the subsequent failures can be attributed to the presence of weak pyritic shales within the failure surface zone whose decomposition was caused by the build-up of high groundwater levels within the slope (Vears and Curtis, 1981).

Secondary displacements similar to those at Mam Tor must have occurred at many other landslide sites in Britain. In most instances such events have not been dated, but at Lawrence Edge, in Longdendale, some wood fragments of a birch tree were recovered from the forward edge of a mudslide which had formed when part of an older slump collapsed. A radio-carbon age determination (Birm. 481) for the birchwood gave a date of 1,970 ± 100 years BP (Johnson and Walthall, 1979). At Hadleigh Castle in Essex, Hutchinson and Gostelow (1976) were also able to date (using radio-carbon) several slip movements which had taken place on an old sea cliff in the past 10,000 years (Figure 18.2). The study is of particular interest as it demonstrates that a reliable interpretation can only be made if the organic materials are placed in their correct stratigraphical and spatial positions within the slump deposits. Figure 18.2 shows that a simple reconstruction based on the position of the organic debris in the succession would be misleading as younger beds have been buried by older regolith materials transported in subsequent mudslides. The major failure occurred in prehistoric times, some time before 5,843 ± 120 years BP, when colluvial slide debris buried organic materials previously deposited in an adjacent coastal marsh. Other organic material provided radio-carbon dates that could be related to successive mudslide events and to the stratigraphy of the other slope materials.

This method of establishing the timing of a slope movement was also used by Christiansen (1983) for a massive landslip near Denholm, in Saskatchewan, Canada. Here, the movement on the slope first started when glacial meltwaters caused the North Saskatchewan River to erode its channel down to the level of the potential shear surface at the base of the slope. The subsequent failure surface affected not only the slope, but also the alluvium of the channel bed, which has been disturbed by the shear surface passing upwards through it. By radio-carbon dating the organic material found in the alluvium, Christiansen was able to determine both the rate of sedimentation in the valley (2.4 mm yr^{-1}), and the rate at which the shearing had progressed through the valley floor infill (35 mm yr^{-1}). He calculated that the sliding had commenced some 11,000 years ago, at the time when the river ceased to be a glacial spillway, then continued intermittently for about 4,000 years.

One other method of direct dating was used by Terasmae (1975), who showed that tree ring growth is affected by slumping and that dendrochronological techniques could be used to record landslide events in the last 300 years. Of course, the length of time for which such a chronology can be used is restricted, but Terasmae argued that it could be used where other dating methods are not feasible.

18.2.2 Indirect (Relative) Dating Methods

Given the meagre possibilities for absolute dating most investigators try to establish a relative time-scale to which a particular slope failure can be referred. This may involve compiling a

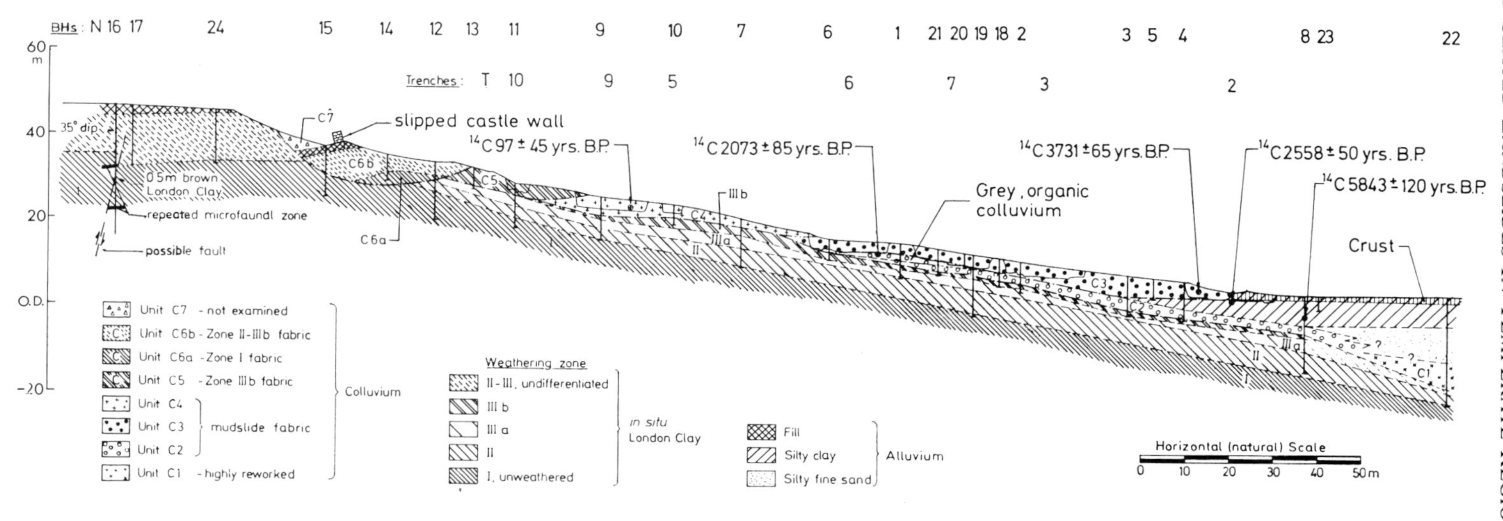

Figure 18.2 Overall cross-section showing engineering geology of Hadleigh Cliff, Essex. (*From Hutchinson and Gostelow, 1976*, Phil. Trans. R. Soc. Lond., **A283**. *Reproduced by permission of The Royal Society*.)

stratigraphic record of slope deposits, or reconstructing a denudational chronology in which changes in the environment are identified but not dated exactly. Examples of both approaches are presented here.

In some locations it has been possible to find the relative age of a landslide by using pollen-analytical techniques to study the stratigraphy of peats accumulated in slump hollows (Sidaway, 1963; Morariu, 1964; Franks and Johnson, 1964; Gil *et al.*, 1974; Tallis and Johnson, 1980). Such stratigraphies often preserve a record of past vegetation and climatic changes which have taken place in the vicinity of the slope failure and these can be correlated with other palynological records from nearby upland sites. It is then possible to establish the approximate time when slide movements ceased. Sometimes quite local slope changes can be monitored as hillwash is frequently inter-bedded with the peat and any local rockfalls or secondary slumping will be detected in the pollen record of those plants which are sensitive to changes in their habitat (Johnson, 1965; Gil *et al.*, 1974).

There are, however, certain limitations to this technique. Firstly, because many floristic changes did not occur synchronously throughout Britain but varied from region to region, only local pollen-analytical studies should be used for correlating landslide events with specific pollen zones. Secondly, the total pollen spectrum through all the preserved peat layers must be examined. This is because 'the pollen spectra for the basal biogenic deposits of the various slide areas show considerable differences which appear to be a reflection of local vegetation differentiation in relation to topography . . .' 'There is no reason to suppose that accumulation did not follow closely upon the cessation of slide movements . . .' but the evidence of local water conditions, of dominance of *Salix*, and of periodic soligenous episodes, all suggest the importance of local hydrology rather than of regional climate in determining peat accumulation in the slide depressions (Tallis and Johnson, 1980). Correlation errors then, will almost certainly occur if the investigation is confined to the basal deposits.

Thirdly, it must be recognized that peat growth only takes place when conditions at the site are favourable, which will be some time after movement on the slope has ceased. The peat stratigraphy, therefore, does not record when failure occurred, but rather the time when the slope became stabilized. Finally, it is often difficult to ascertain where the greatest thickness of peat in a hollow is located, because the floor of the depression is often uneven or covered with boulders or rock fragments. On an irregular slope, the peat cover forms first in the deeper slope hollows which are usually waterlogged, and then extends both up and down slope from them (Tallis, 1964). Younger peat can, therefore, form upslope of that found in deeper depressions on the lower slopes, and there is little chance of obtaining reliable samples to be used for radio-carbon age determinations at such sites. For as Chambers (1982) has stated, 'the resolution of a peat initiation date could be impaired by the current requirement for a not inconsiderable weight (and hence vertical thickness) of peat if each sample is to be fractionated and radio-carbon-dated by conventional counters. For example, beneath blanket bogs, the raw humus layer can contain a condensed sequence . . . equivalent to a thousand years of deposition.' Hopefully, new dating techniques may resolve this problem, but despite these difficulties there have been a number of successful datings of landslides using this method.

In 1966, Starkel argued that the timing of Flandrian landslide failures had not been wholly random, but had been climatically controlled. He postulated that most failures occurred at times either when permafrost was generally waning (i.e. between 11,000 and 9,000 years

BP), or when climate was generally more humid and warmer (*c*. 7,000–5,000 and 1,500–500 years BP). Starkel's hypothesis would appear to have some support from more recent studies (Hutchinson, 1969; Hutchinson and Gostelow, 1976; Brunsden and Jones, 1972; Pitts, 1979; Selby, 1979; Johnson and Vaughan, 1983). However, in their south Pennine Longdendale studies, Tallis and Johnson (1980) established that although several mass movements had taken place at similar times in the early Postglacial, there was no strong climatic factor which had determined when failures took place.

The fact that an individual slope failure takes place only after a long history of denudation and many climatic vicissitudes cannot be over-emphasized. All slope materials attain their threshold states independently of local climate change, and indeed Whalley *et al.* (1983) have demonstrated that the most favoured climatic periods for sliding are not necessarily the times at which sliding has occurred. Their data, from the glaciated parts of Iceland, show that climatic influences were not a particularly important control on either the magnitude or the frequency of rockslide events, and that the 'triggering' of slides was essentially a response to mechanical conditions induced by the effects of high 'cleft' water pressures.

For this reason geomorphologists are now less concerned with dating specific events, and are studying more closely the spatial and temporal relationships of mass movements with other phenomena. Correlations are, therefore, sought between landslides and other landforms such as terraces or moraines. Sometimes, associated dating opportunities arise. For example, the thickness of weathering rinds and the growth of lichen on landslide rock debris have been measured and compared with that found on boulders deposited in moraines whose age is known (Birkeland, 1973; Dowdeswell, 1982).

Relationships with other landforms vary in their complexity. For example, at Empingham in Leicestershire, there is a terrace known to contain Ipswichian (last interglacial) faunal remains. This terrace is overlain by regolith deposits containing periglacial head interbedded with debris from local landslides. The slides, therefore, are of Devensian age (Chandler, 1976). In the Swainswick and Avon valleys near Bath, past slope failures have been correlated with specific periods of river downcutting and aggradation (Chandler *et al.*, 1976). Whereas some small landslides in these valleys are Flandrian, the larger landslides are older and probably failed when the streams were downcutting to a base level some 27 m below the present flood plain. This erosion took place in an early phase of the Main Devensian Glaciation when permafrost became established in the valleys. The slopes then became unstable because of high cleft pore-water pressures developed above the frozen groundwater level. The slumps extended onto the valley floor, but subsequently aggradation took place within the valleys and the slide debris was trimmed back. A river terrace was formed in the Late Devensian period and the valley-sides became covered with younger head materials. These much later became highly unstable and liable to mudsliding.

Sometimes landslide features can be correlated with the known limits of former ice sheets. This association has been commented upon by Whalley (1974) and Sissons (1976) and has recently been studied extensively in Scotland by Holmes (1985). According to Whalley, many rockslides and rockfalls occur at ice margins because of abundant meltwater available from the ice and adjacent snow banks. This water percolates into the ground or forms ice wedges in joints or rock crevices thereby creating and sustaining high pore-water pressures, but both Whalley and Holmes note that these conditions did not result in immediate slope failure in either Iceland or Scotland. Furthermore, they suggest that although ice over-steepened the slopes it rarely made them unstable. In a number of instances, Holmes has shown that rock mass creep was a major factor in causing slope instability where deep-seated failures occurred. He believes that the sustained high pore-water pressures necessary for the creep action were induced by the presence of glacial drift or soliflucted regolith on the footslopes.

Rock mass creep also appears to have been a factor in the development of the south Pennine rotational landslides. Many of these are situated on slopes that were either formerly covered by the Devensian ice sheet, or were in near proximity to it. Others were formed on slopes that were subjected to strong periglacial influences. Deep-seated creep took place where permafrost was most severe, but as in Scotland, large-scale mass movements were delayed until some time later. Rockfalls and rockslides took place on the upper slopes, but their dates are unknown. In some localities, the rockfalls appear to predate other landslip movements as their debris is now found beyond the immediate fall area having been transported in later slump movements (Johnson and Walthall, 1979). If this interpretation is correct then the rockfalls are probably of Devensian age as the main Pennine slip movements occurred during the Flandrian (Tallis and Johnson, 1980). They, therefore, form part of the long history of slope instability which began with the formation of valley-floor bulges and other deep-seated structures, to which some reference must now be made.

18.3 VALLEY-SIDE CAMBERING AND VALLEY-FLOOR BULGING

The association between low-angled, deep-seated failure surfaces and superficial structures such as cambers and bulges has long been recognized. Cambering and bulging of strata have both been found in landslide areas since they were first observed in the late nineteenth century by engineers constructing dams in the southern Pennines (Bateman, 1884). Found mostly in deep valleys which underwent several phases of intense downcutting and erosion during the Pleistocene, the structures have since been encountered in almost every formation of Carboniferous or later age.

Lapworth (1911) first drew attention to two of their most important characteristics. He reported on the effects of creep deformation observed by Bateman at Woodhead in Longdendale, where it extended 100 m into the valley sides and to depths of 37 m beneath the valley floor. Lapworth also noted very high artesian water pressures in one of the bulge structures in the Langsett valley near Sheffield. Similar high water pressures were later discovered in other Derbyshire bulge structures at Fernilee (Morton, 1932) and Ladybower (Hill, 1949) where the strata were deformed to depths of 30 m and 60 m respectively. From the study of Bateman's section (Figure 18.3), Lapworth concluded that the underlying beds had been squeezed out from under the valley sides when they became saturated and could no longer bear the weight of the overlying rocks, but it was not appreciated until much later that these structures could vary enormously in size. Whereas most of the Pennine bulge structures measure only a few metres in height, depth and breadth, those in the Colorado Valley are up to 35 km long and cause major unfolding and deformation of the strata in the valley sides to at least 600 m above the valley-floor. Such a structure cannot be attributed to simple valley-side load stresses and as a consequence other theories have been put forward to account for their formation (Potter and McGill, 1978).

The first generally accepted explanation for valley bulging came from Hollingworth *et al.* (1944), who investigated slope conditions in the Northamptonshire iron ore field. Here,

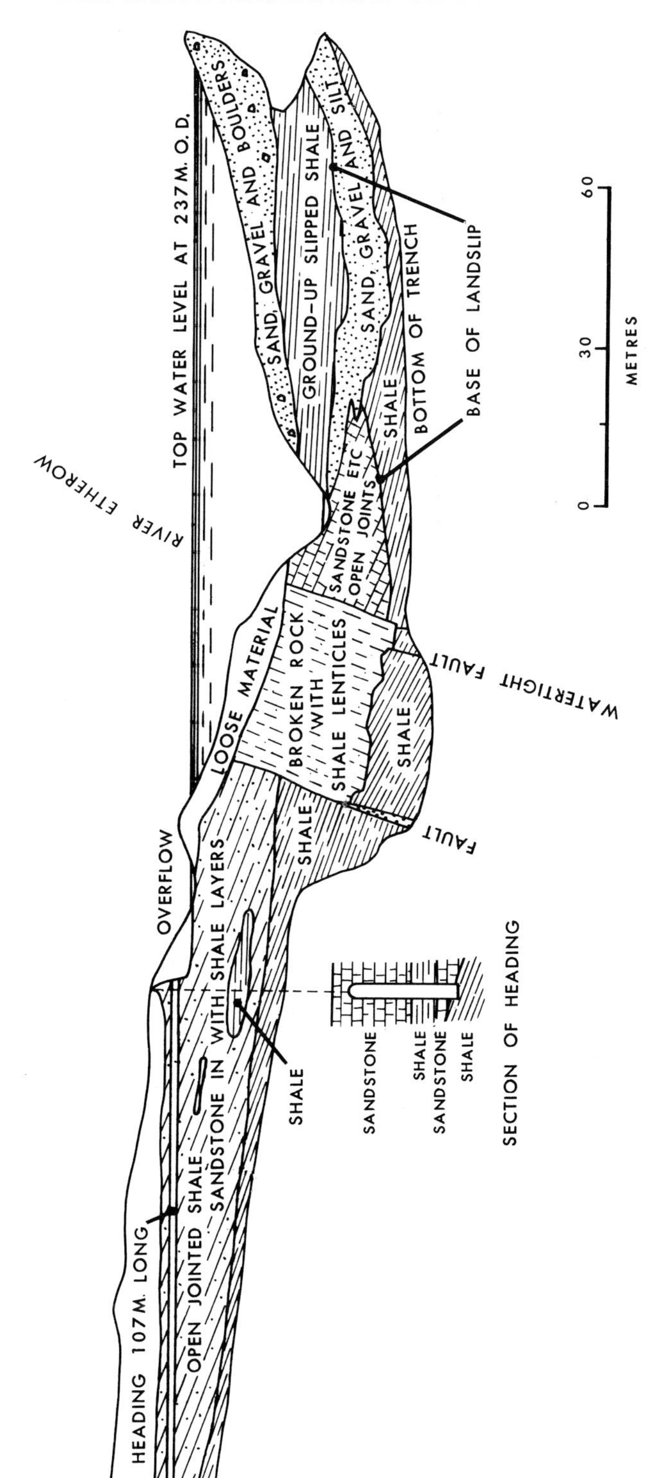

Figure 18.3 Ancient landslip (Lawrence Edge) and deformed valley-floor structures at Woodhead Dam, Longdendale (re-drawn from Bateman, 1884). The Shale Grit strata to the north of the dam have been deformed by progressive creep within the shales causing cambering or more probably topple flexuring on the footslope

they found cambered slopes extending upslope for considerable distances away from the incised sections of the valleys in which bulge structures had been formed. The cambered rocks were strongly-fractured with each block being separated by deep wedge-shaped cracks—gulls—infilled with tills of Mid-Pleistocene (Anglian?/Wolstonian?) age. At first, the upward tilting and arching of the strata in the valley-floor was attributed to ice loading of the slope. This, it was thought, caused the squeezing out of the more ductile rocks that overlie or are inter-bedded with more rigid rocks. Further studies by Horswill and Horten (1976) have shown, however, that the weakened clays have been over-consolidated at some time in the past and that deformation took place both before, and after, the period of ice transgression.

From a detailed geotechnical study, Vaughan concluded that it was not ice, but permafrost which had caused the over-consolidation. This had developed in the later Devensian (last) glaciation, when the clay beds crept and deformed towards the valley centre from beneath still frozen, but permeable and stronger strata. Such attentuation of the beds near the valley-floor undoubtedly stretched, thinned and cambered the upper slopes whose rocks then became fissured in the process. The stresses were directed horizontally, but no major discontinuities developed until after the permafrost thawed. When this occurred, the thrust forces were then deflected upwards from under the valley-floor causing brecciation and disturbance within the clay beds. The development of such structures no doubt help to weaken the slope materials for 'the thinning of the ice inclusion causes a sudden rise in moisture content, induces excess hydrostatic pressure in the shattered rock masses, and leads to flowage or instability along bedding planes and joints' (Higginbottom and Fookes, 1971). They also greatly facilitated the movement and storage of groundwater generated by the release of snow- and ice-meltwaters. This caused high pore-water pressures to develop wherever solifluction debris sealed permeable strata (Denness, 1972).

The classic studies in the East Midlands led to their recognition in other parts of the world (Ter-Stepanian, 1977a; Radbruch-Hall, 1978), but cambering of slopes on such an extensive scale as that observed in Northamptonshire has rarely been observed elsewhere. Also, it is now well known that past cold climate conditions are not a necessary factor for the formation of bulge structures. Providing there is a quick removal of the overburden with consequent valley-floor rebound, the strata may be deformed at any time following deep valley incision. Under such conditions, shear strength is reduced by the subsequent swelling and softening of clays or silts and this aggravates any instability due to over-steepening in the near-surface zone. Cambering does not necessarily occur, however, even if there is an increased lateral transfer of material downslope. In fact the tensional fracture of strata and the outward movement of rigid rocks towards unloaded surfaces can be achieved in a number of ways as the deformational creep of the soft rocks frequently leads to topples and other displacements which may ultimately result in wedge, slab or rotational modes of hillslope failure. For example, in very steep, high slopes where very weak strata outcrop, the rocks will deform by bending, folding or flowing plastically prior to toppling or otherwise failing (Caine, 1982). Alternatively, where there is inter-bedding of competent and incompetent rocks, such movements may take place within a thick zone of uniform rock such as mudstone, or creep will be outwards and downdip in weakened basal beds overlying the harder rocks: this will occur without any buckling of the strata taking place. Mass rock creep can occur incrementally in rocks such as schists or progressively where foliation planes permit a continuing decrease of frictional strength towards a residual condition (Mahr and Nemčok, 1977). 'Whatever the causes and there may be several, mass rock creep will take place long before failure occurs and before the material ruptures or is subject to visco-plastic deformation or a combination of both' (Radbruch-Hall, 1978). Because the creep takes place at depth it is rarely exposed and almost impossible to detect. Its action is largely unpredictable, but some surface features are now widely recognized as indicating that deformation has taken place within a slope. Selby (1982) has noted that where creep has taken place 'the more coherent rocks become rafted, tilted, rotated, depressed or raised in a chaotic manner and the weaker rocks are affected, depressed or raised in a chaotic manner and the weaker rocks are affected by intense surface rippling'. Such features may be found within some landslide slumps, as for example, at Alport Castles in North Derbyshire (Johnson and Vaughan, 1983), where the deformation induced other modes of slope failure which took place some time later.

18.4 A SOUTH PENNINE CASE STUDY

18.4.1 Introduction

This review of the dating of stabilized landslides has shown that there are many difficulties to be overcome if reliable dates are to be obtained for particular mass movement events, and that the problem of detecting any periodicity in such movements is likely to be complicated. These complexities can best be evaluated for a region with a relatively uniform geology, relief and denudational history, where several slopes failed at similar times and with common elements controlling the shear-stress conditions which caused periodic failure occurrences.

The particular area selected is the Peak District National Park in the southern Pennines, and includes that part of north Derbyshire known as 'The Dark Peak'. Its relief is formed of limestones, sandstones, gritstones, shales and other mudrocks and in one of its valleys, Longdendale (Figure 18.4), there are several large landslides including one known as The Rollick Stones. This landslide has not been described in detail before but it typifies extremely well such landforms in this region.

18.4.2 The Southern Pennines: Geology and Relief

The Dark Peak is an upland area whose relief has been eroded, for the most part, in sandstone and mudrock formations of Carboniferous (Namurian) age (Figure 18.5). It is a region of tabular hills, low plateaus, and deeply incised valleys that form the headwater catchments of the Rivers Trent and Mersey. The shale/grit terrain extends southwards to enclose the limestone plateau area of the Peak District and the broad structural pattern of the whole upland area is shown in Figure 18.6, which is based on the work of Fearnsides (1932) and the Geological Survey (Bromehead *et al.*, 1933; Stevenson and Gaunt, 1971). Prolonged Tertiary denudation has exposed the Carboniferous Limestone in a broad anticlinal structure. These rocks form a terrain consisting of extensive tracts of undulating plateau and low hills dissected by steep-sided dry valleys. In these 'dales' the valley sides are mostly scree-covered with terracettes present on the deeper regoliths. Extensive slope movements are rare, being limited to slopes in which volcanic rocks outcrop. These provide an impermeable base on which sliding has sometimes taken place (Burek, 1977).

Figure 18.4 An aerial view of the south side of Longdendale, north Derbyshire. (Fairey Air photograph: 1. 548 69112.) Many landslides are shown in this view and their morphological features have been described by Johnson and Walthall (1979). The Rollick Stones landslide is in that part of the valley overlooking the Torside reservoir. The Bradwell Sitch–Lawrence Edge landslide immediately to the east of the Woodhead reservoir dam whose trench section is shown in Figure 18.3 provides a good example of a multi-storey landslide, but was not classified as such in the author's earlier papers

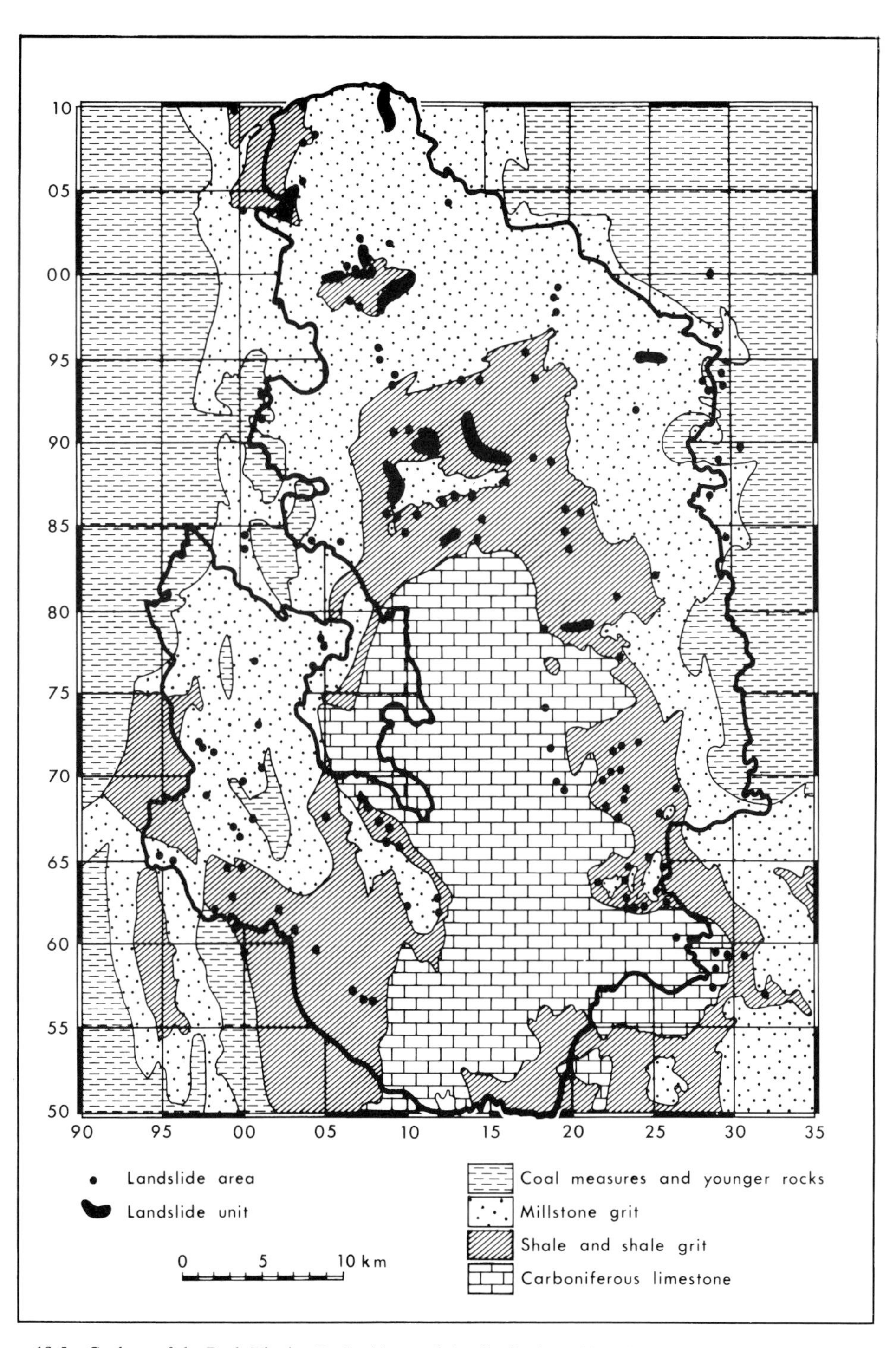

Figure 18.5 Geology of the Peak District, Derbyshire, and the distribution of individual landslides and landslide areas

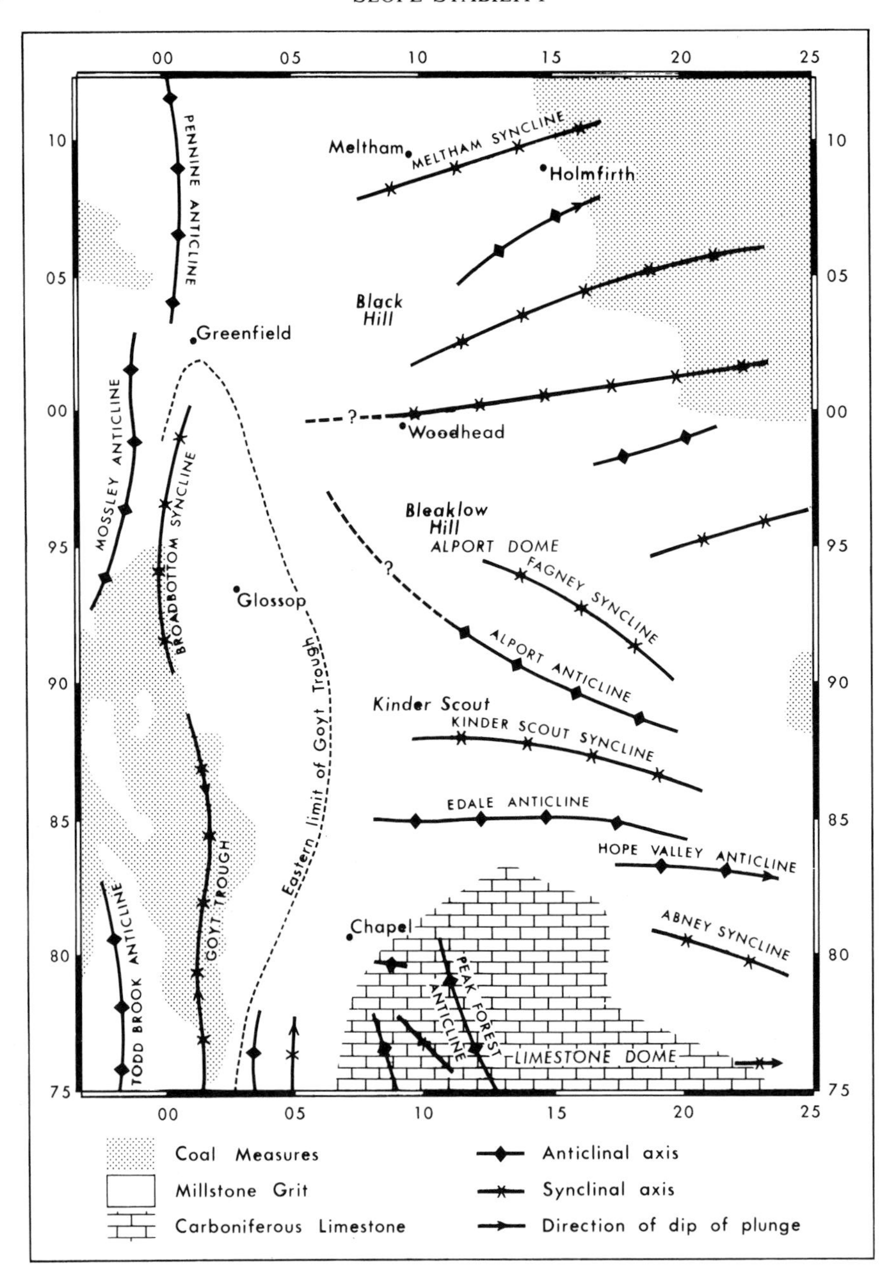

Figure 18.6 Structure of the northern part of the Peak District. (*After Fearnsides, 1932 and the Geological Survey memoirs, 1933; 1971.*)

The influence of the 'dome' structure extends northwards to cause an upwelling of the strata near Alport and Woodhead (Figure 18.6), where parts the Shale Grit succession are now exposed on the lower valley slopes. Further north, the broad dome is replaced by the main Pennine Anticline, a broad asymmetrical fold whose core is exposed in the Tame

Valley, near Greenfield. On the flanks of these two anticlines there are a number of other fold flexures that are Late Carboniferous in age. These, although strongly faulted and displaced, exert a strong influence upon the local relief pattern of scarps, cuestas, and shale vales. According to Clayton and Straw (1976), differences between the cuesta/vale topographies of the three areas result from three factors. First, the upper rock formations are normally thinner and less coarse than those which occur in the lower parts of the succession; second, the pattern of pitching folds varies from area to area; and third, the adjustment of drainage is more complete in western areas than elsewhere.

In the west, the Pennine strata have been strongly folded into a series of north–south pitching folds arranged *en echelon*. In the central area, the folding is very shallow with the axial trends being west to east. Here, the Kinderscout Grit formation contains massive sandstones which provide high escarpments or 'Edges' wherever they outcrop. Tabular hills such as Black Hill, Bleaklow, and Kinderscout stand above extensive plateaux whose edges are formed by the massive sandstones and overlook deeply incised valleys. Most of the Millstone Grit succession is exposed in the sides of these valleys and consists of great thicknesses of shale, mudstone, and siltstone. The superimposition of competent over incompetent strata is thus a prime factor leading to slope failure in this region. In the east the rock dips are dominantly to the east, but are less steep than those which influence the escarpments in the southwest Pennines and few of the folds are totally inverted. The cuestas in consequence are larger, rise to higher elevations, and are often capped by tors or rock crags: the landslides are also larger than those found in the western areas.

During the Quaternary, valley incision was characterized by alternate phases of rapid downcutting and periods when the slopes became stabilized and the valley-floors aggraded. The whole region was covered by ice more than once, but during the last (Devensian) glaciation the ice sheet was confined to the lower western hillslopes. As a consequence, slopes elsewhere were subject to intensive periglacial erosion and to the removal of their regolith by solifluction (Waters and Johnson, 1958; Johnson, 1968; McArthur, 1977, 1981; Wilson 1981). As the permafrost first became discontinuous and then disappeared from the slopes, many different forms of mass movement took place, further degrading the slopes. The removal of material by solution and other weathering processes was no doubt quite considerable but mass movements such as falls, topples, flows, and slides have been commonplace agents of hillslope denudation (Johnson, 1980). The most effective agents for transporting material downslope, however, were the rockslides which were rotational with deep-seated failure surfaces. The distribution of the south Pennine landslide areas is shown in Figure 18.5, which was compiled from vertical air photography and from map sheets of the British Geological Survey. Altogether some 230 separate landslide units have been identified within the National Park area of which 67 were simple slide forms including earthflows and the rest compound, multiple, or complex, rotational slides. All appear to be largely stabilized, but there have been instances of secondary movements within slump areas where roads have been constructed across the toe of the slides (e.g. at Mam Tor in the Derwent catchment and at Cowms Moor in the Woodlands Valley). All the slides studied occur in areas where the precipitation is greater than 1,150 mm and are at altitudes above 300 m OD. The average slope angle on the failed sites is never less than 20°, but the aspects of the slopes vary, with 30% of the slides facing west or east; 43% north or south; and 27% having either a southeasterly or northwesterly aspect. Where slopes have failed the adjacent slopes are often dissected by strong rill and gully erosion

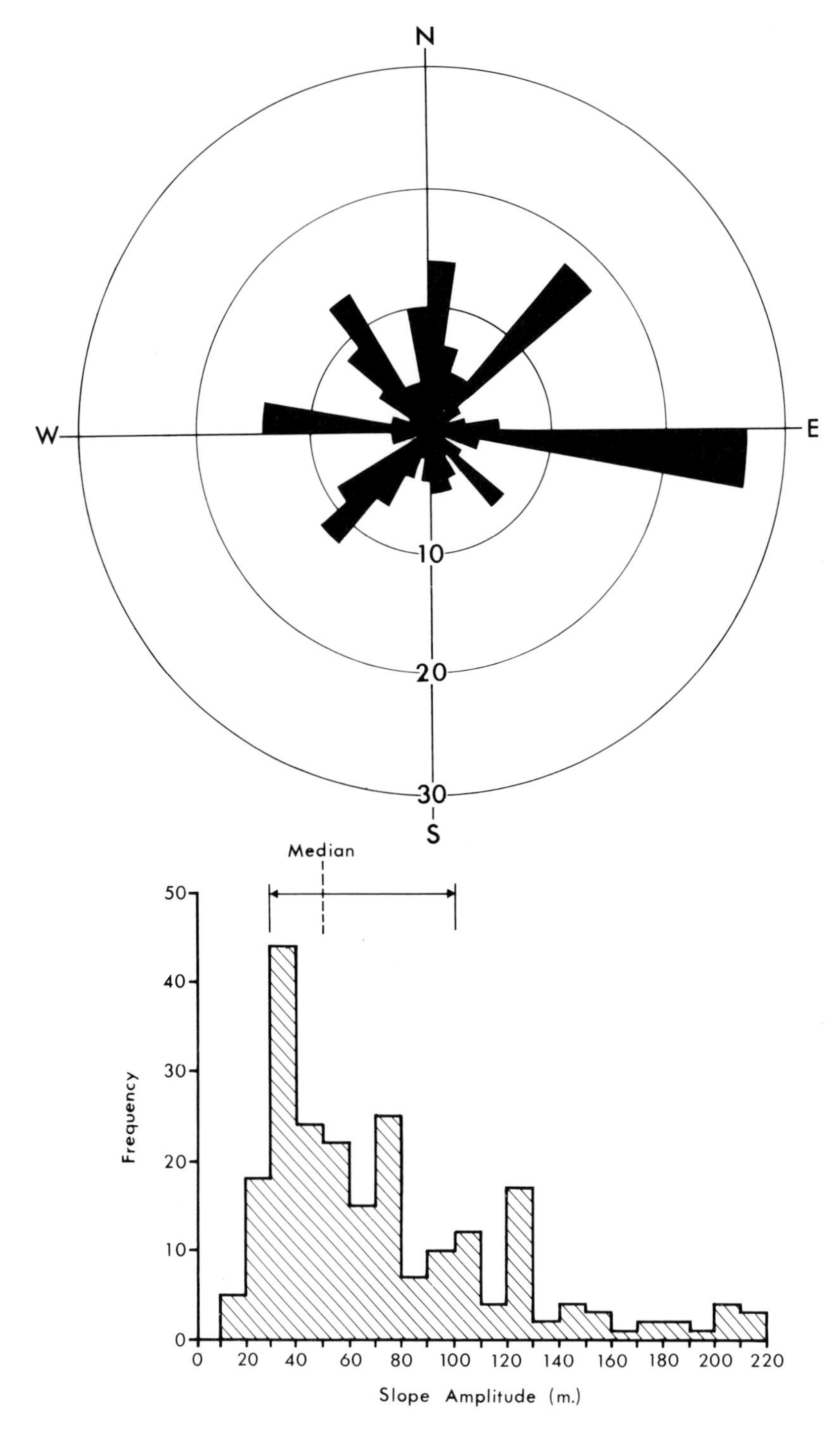

Figure 18.7 Rose diagram and frequency histogram showing distribution of landslides with reference to slope aspect and height

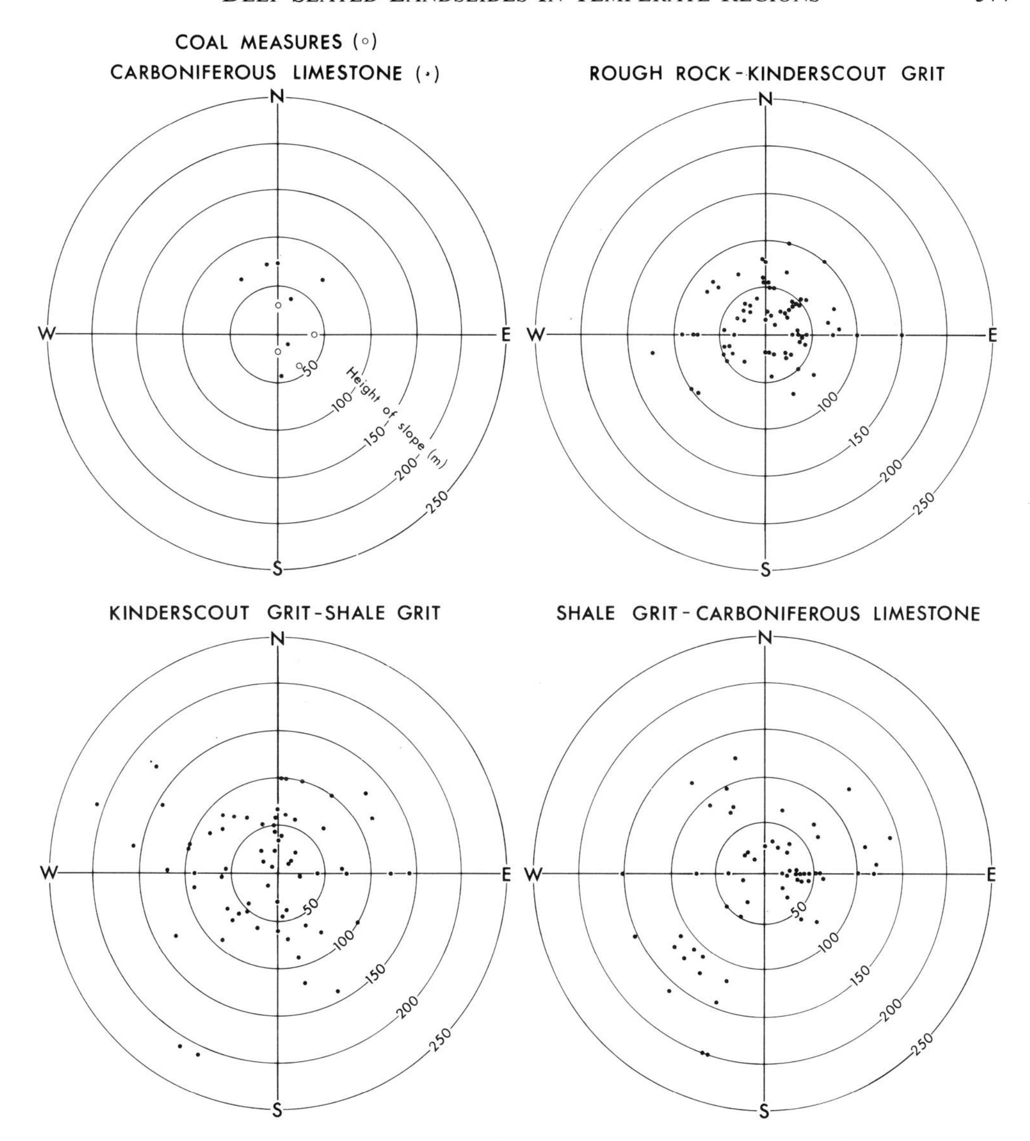

Figure 18.8 Landslide aspect and height (scar crest to toe) correlated with appropriate rock formations. (NB Coal Measures and Carboniferous Limestone data put for convenience on one diagram.)

which has removed the regolith and allowed free egress of groundwater from the water-table springs.

Figure 18.7 shows the distribution of the landslides in terms of their aspect and slide height: the histogram shows their frequency in different classes of slide height, which is defined as the height difference between the toe of the slump and the scar crest. The smaller landslides are to be found mostly in the tributary valleys in the upper catchment areas and where the slopes are broadly convex in profile. Undercutting by streams is a major factor accounting for the slope failures and most of the smaller slides occur on the footslopes. This factor does not appear to be so important when the larger landslides are considered. These are found in the larger valleys on slopes where the profiles are convex–concave–convex and much longer and higher in terms of their local relief; the landslide height usually exceeds 120 metres.

The spatial distribution can also be examined in terms of geological factors (Figure 18.8). Rockslides in either the Carboniferous Limestone or Coal Measures rocks are too few to consider, but the diagram shows that many of the smaller slides have taken place within strata that occur in the higher parts of the Carboniferous succession. These are, therefore, found in the southwest or southeast Pennines. Where the Carboniferous Limestone–Shales Series outcrop to the west of the main limestone massif there are also a number of small landslides. These occur on the flanks of the small isolated hills in the Dove and Manifold Valleys where the slopes are rather more rectilinear in form. The larger landslides are found in the rocks of the central Dark Peak area and on slopes which face either north or south

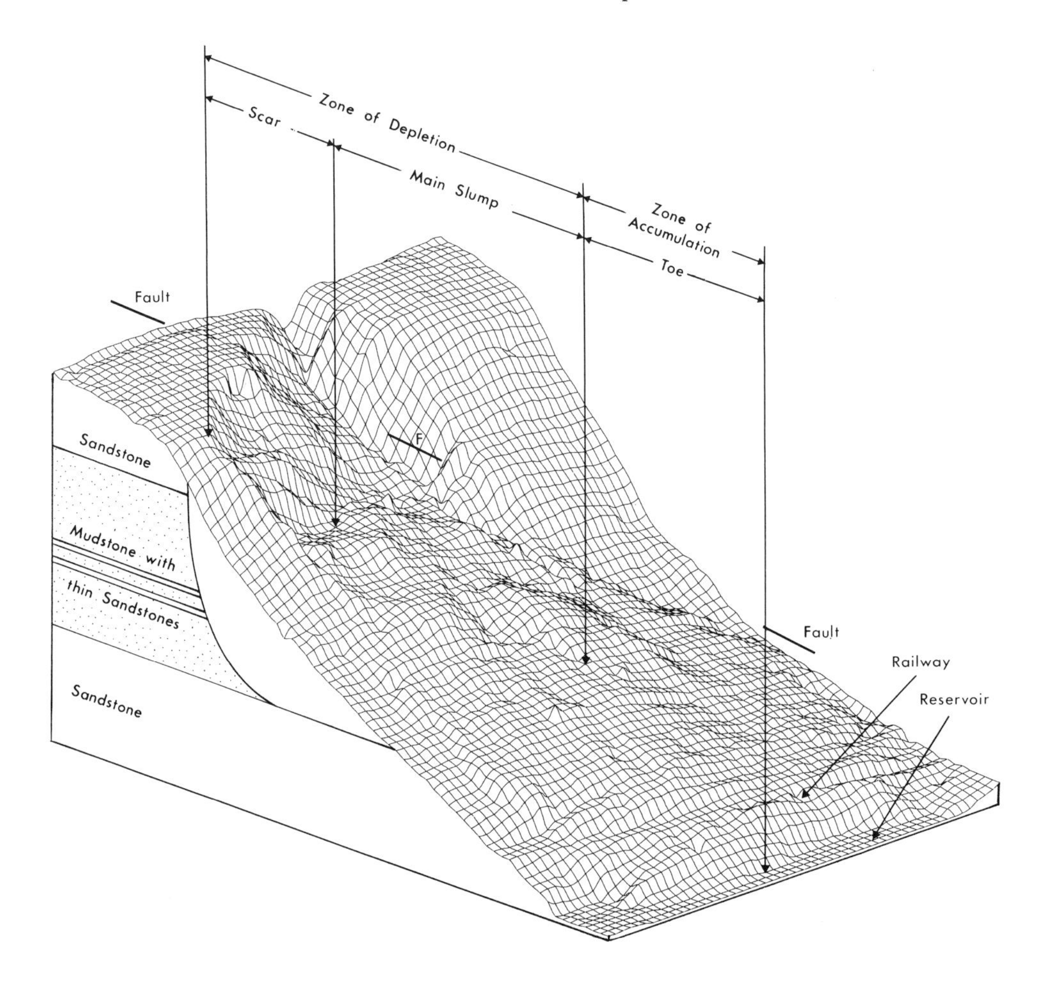

Figure 18.9 Block diagram of Rollick Stones landslide drawn using a SYMAP computer program

or northeast or southwest. As elsewhere their orientation is dependent upon the local fold structure alignments which determine where the rock formations most susceptible to landsliding outcrop. Geological factors do not, however, explain why varied types of rotational landslides have taken place in this region, nor do they account for the large number of landslides taking place in a short time and in such a limited area. Examination of one landslide helps to indicate the factors that led to its instability and conditioned the type of landslide involved.

18.4.3 The Rollick Stones Landslide

The main features of this landslide are shown in Figures 18.9 to 18.13. In terms of Varnes' classification (1978) it is a Slump-Earthflow, for sliding occurred on a much weakened shale/clay slip-surface, but more properly it should be classified as a Rockslump-Earthflow as the slope movement involved a large mass of coherent rock and only at its base was the material reduced to a weak clay consistency.

The slope movement was rotational with some backward tilting of the slump blocks in the upper part of the slide, but the following features characterize the landslide form (Figure 18.9). Its zone of depletion is divided into the 'scar' and 'main slump' with the present scar standing some 100 m above the back of the slump. Capped by a sandstone

Figure 18.10 The Rollick Stones — the main slump ridges and scar. Note the topple structures at the forward edge of the scar and the detached sandstone pillar in the photo left centre. This has moved downslope from its parent slump ridge probably as a result of topple flexuring

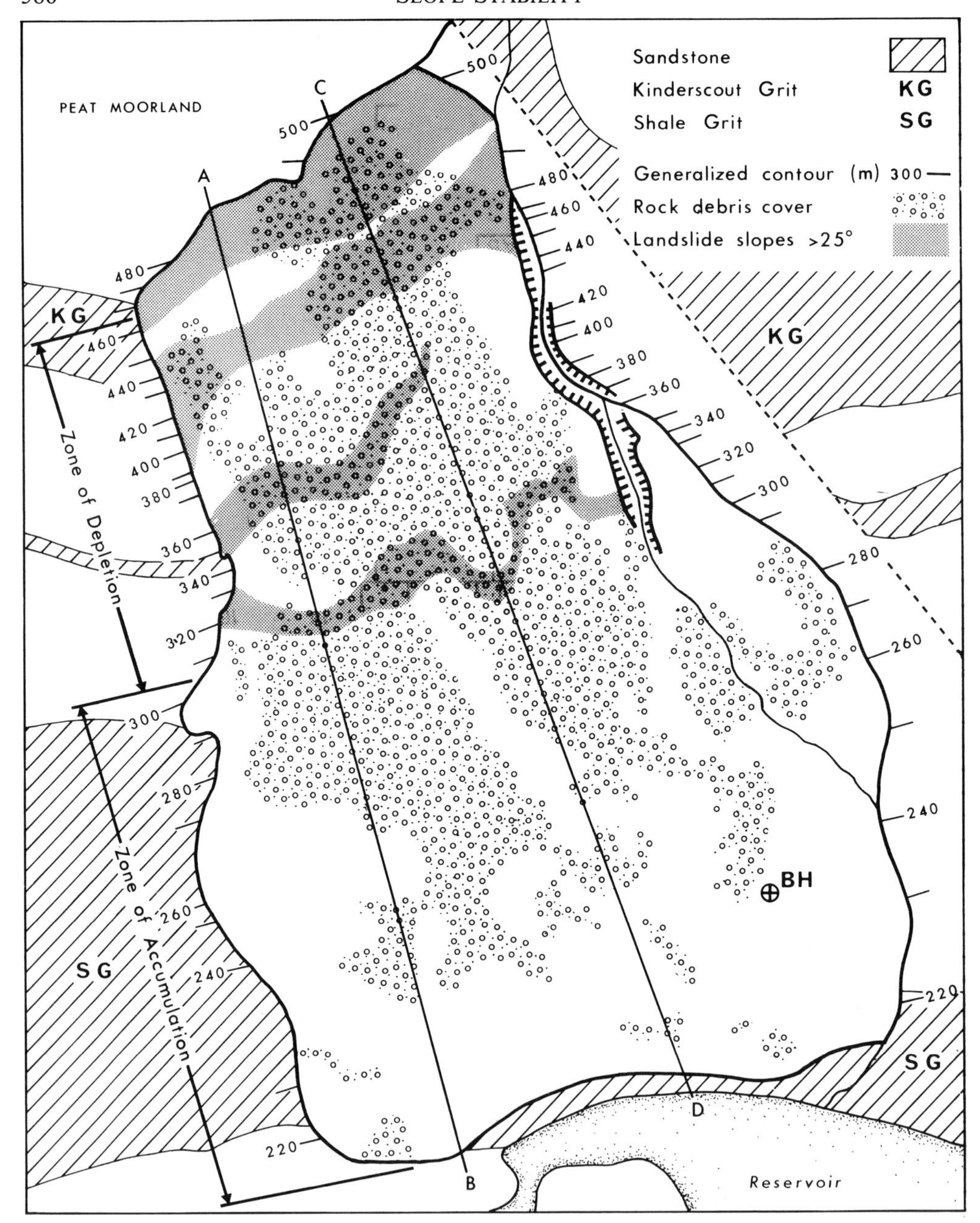

Figure 18.11 The Rollick Stones: geology, boulder fields, and general configuration

cliff 10 m high, this cliff has been affected by topple movements and flexuring and this has increased the inclination of the strata valleywards (Figure 18.10).

The larger slump units are formed of coherent sandstones that were displaced from the upper parts of the slope and became broken into large slump-block units in the process (Figure 18.9). Their removal left much of the scar unsupported and this caused more blocks

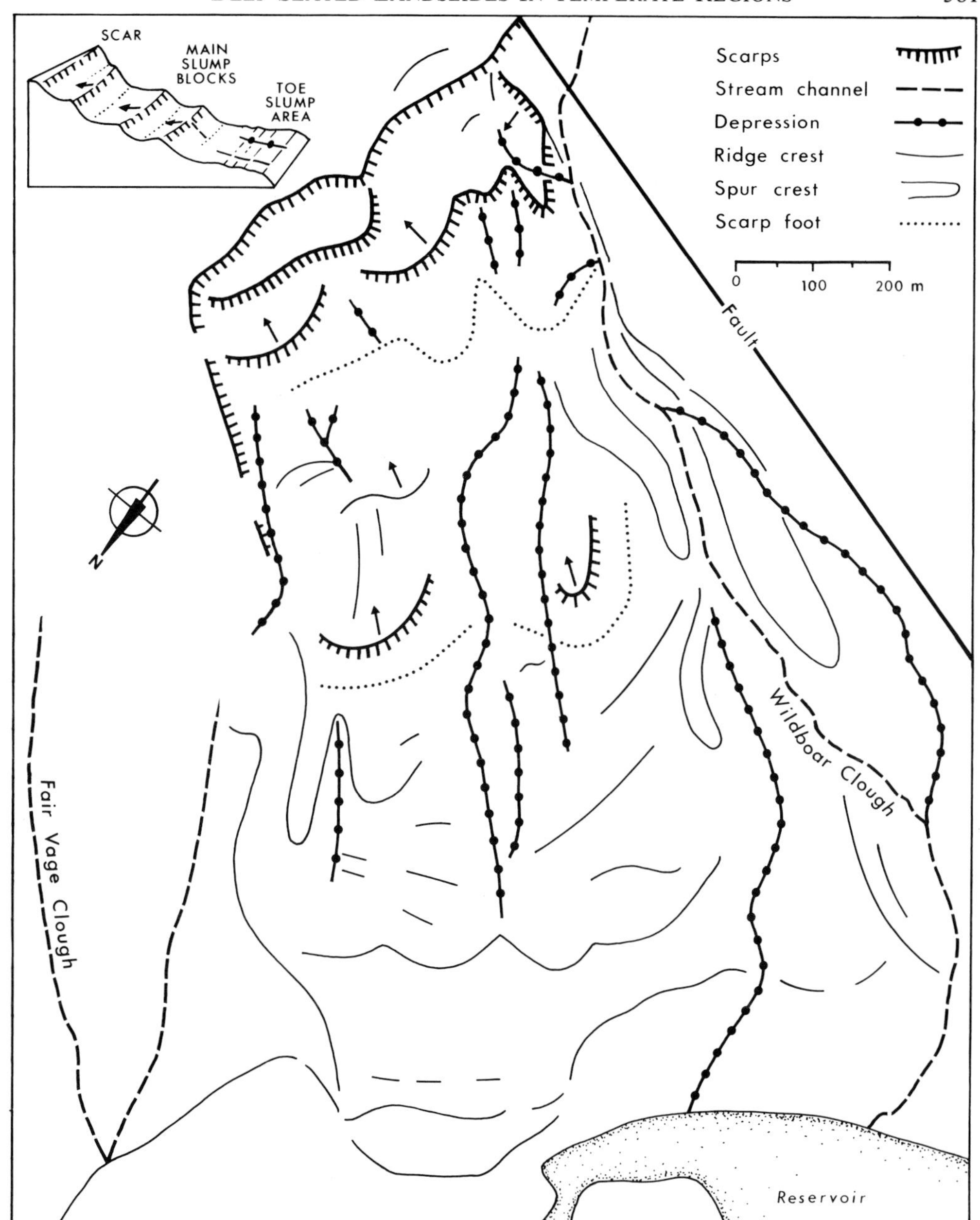

Figure 18.12 The Rollick Stones: main morphological features

to break from the scar and slip down into a graben structure at the rear and southern side of the main slump mass. The tilted blocks in this part of the slump form a series of steps with strong riser faces on which sandstone strata, still largely intact, are exposed: some of these have undergone some flexuring following their transportation and Figure 18.10 shows that, in one locality, a pillar of rock has moved some 10 metres forward from the slump ridge without falling.

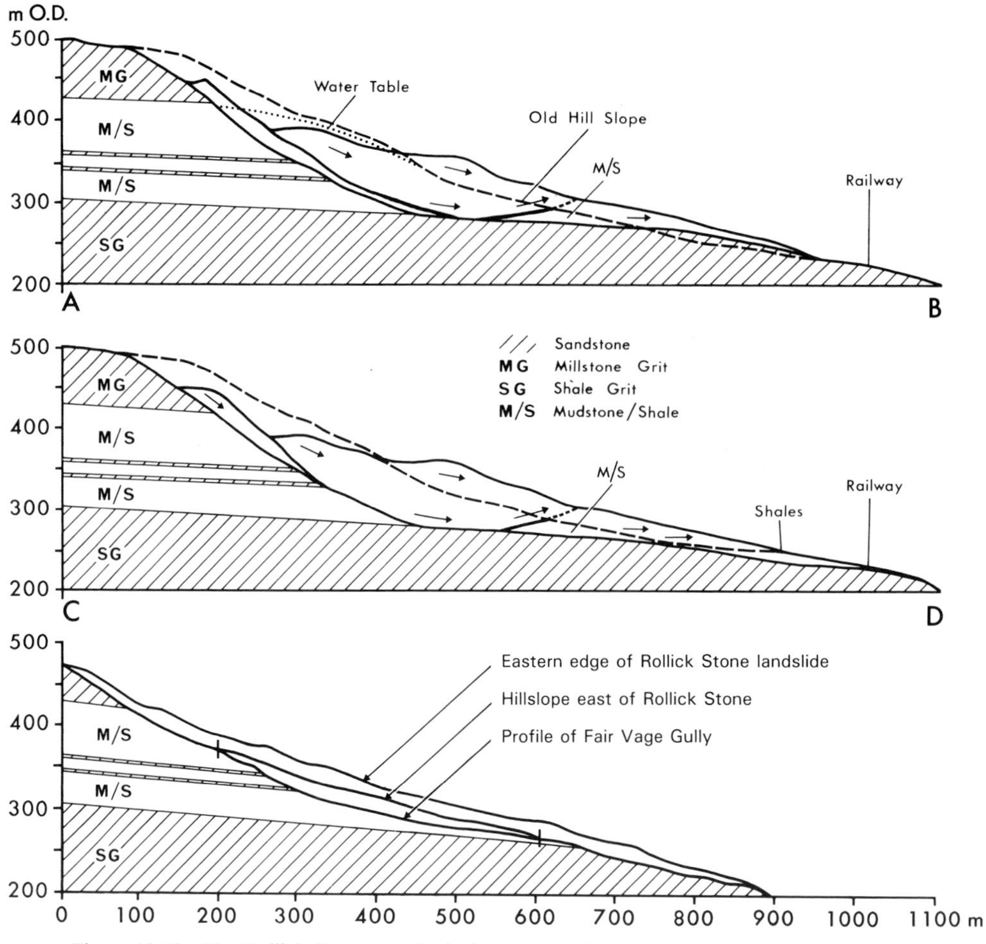

Figure 18.13 The Rollick Stones: geological sections with slump ridges shown schematically

In the zone of accumulation, the geometry of the slide is characterized by the considerable displacement and dilation of the slump. The mobile material in the toe has moved downslope far beyond the place where the rupture surface would have intersected the pre-landslide slope profile, spreading as it moved. During its displacement, the lower slump became deformed with several transverse ridges and radial cracks resulting from the tensional stresses incurred. These are readily detected from the air photograph (Figure 18.4) from which they can be mapped (Figure 18.12).

The toe debris is between 13 and 17 metres thick: at the base of a borehole (see Figure 18.11), slip surfaces were observed in the mudrocks overlying the thick Shale Grit sandstone. This sandstone increases in thickness and rises upslope beneath the landslide. It was not directly affected by the slope movements, but its upper surface has broken and fractured by flexuring or cambering, making it highly permeable. The shales lying immediately above were also weakened by the tensional stresses and their loss of cohesion was probably due to small fractures across the bedding planes, and possibly to pre-existing failure surfaces.

The residual angle of internal friction in the slip surface shales is 21° which is well below the mean slope angle inferred from the generalized contours shown in Figure 18.11. The geology of the slide is shown schematically in Figure 18.13 and is based on the assumption that the slide was non-circular with considerable lateral displacement occurring along the bedding planes.

There is some evidence to show that the slope was first affected by progressive (rock) mass creep at the base of the slide and that this induced failure in the shales and resulted in the rotation and displacement of the more coherent, massive strata further upslope. Two factors make such a hypothesis appear tenable. First, a substantial thickness of permeable and impermeable strata overlie the weak shale mudrock in the slope and this loading stress would have fluctuated according to the groundwater levels present. Second, the two thin sandstone members present in the slope influence its form in two ways. Interbedded with shales and mudstones, they are not only more resistant to brittle fracture, but they also provide aquifers within the slope. These facilitate shale decomposition in the sandstone/mudrock contact zones and such weathering can be seen in stream sections along the bed of the Fair Vage Clough immediately to the east of The Rollick Stones. Here, in the same rocks in which sliding took place, hydration has caused the mudstones to weather spheroidally and to be highly decomposed along the joint surfaces. Prolonged subsurface weathering would have reduced these rocks to a stiff kaolinite clay incapable of sustaining the imposed load and subject to creep deformation before its shear strength was lowered to failure level.

Several other factors were conducive to landsliding on this slope. Valley-floor bulging had previously taken place in the vicinity, and its effect on the Shale Grit Sandstone has already been noted. Another contributory factor was the part played by solifluction in sealing the aquifers, when at the same time other frost-associated processes were attacking the higher parts of the slope and opening up joints that became the location of high 'cleft' pore-water pressures. The long-term modification of the slope resulting from these factors did much to reduce its stability, but they did not in themselves provide the trigger mechanism for the landslide event itself. This took place when the Wildboar Clough (Figure 18.12) undercut the base of the spur on which the slide took place. The main slump was directed outwards toward the Longdendale valley-floor, but its movement also resulted in the diversion of the clough (stream) causing it to erode a new channel along the flank and within the toe section of the slump (Figures 18.11 and 18.12).

The Rollick Stones contains one very distinctive feature from which it derives its name. Covering the whole of the slump surface there is an extensive boulder field in which many of the rocks measure more than 1 m^3 in size (Figure 18.14). The boulders were either released in rock falls from the original scarp face or else they were toppled from the forward edge of the scar and from the forward edges of the slump ridges at a much later time, possibly during, or after the formation of the rockslump-earthflow. If the criteria used by Caine (1982) to differentiate between cambering and toppling are applied to this slope, then there is some evidence that topple flexuring took place with rock mass creep operating in a manner very similar to that which occurred in slope movements in the Slánske Mountains (Malgot, 1977). Thus, the boulders and rock slabs were transported on a deformed layer of mudrocks, becoming increasingly fragmented in the process. At the Rollick Stones, on the lower part of the slump, the boulders are concentrated into lobes, but were carried downslope as far as the valley-floor having moved over the Shale Grit

Figure 18.14 The Rollick Stones — the boulder field overlying the lower parts of the slump

footslope (Figure 18.11). The reason for this concentration is not known but was probably due to rapid surface flow when the slump dilated. Boulder spreads and other morphological features similar to those of this slide have been observed in the Tione di Trento region of the Italian Alps, and at one site near the Vedretta d'Ambiez glacier, the landslide occurred within an eight-hour period in 1957. Its form is similar to that found at Rollick Stones but it was generated on slopes of 35° to 40° (J. C. Flageollet, personal communication).

Other examples of rockslump-earthflows occur in Longdendale at Millstone Rocks, Birchen Bank and Tintwistle Knarr, and some of these have been dated using palynological techniques (Tallis and Johnson, 1980). As in the case of the Rollick Stones, they are Postglacial with most movement having taken place sometime before 6,000 years BP. Other similar types occur in the Tame Valley near Greenfield (i.e. the Stable Stones), and in the upper Derwent catchment areas of Edale, Alport Dale, and Hope Dale. At all of these sites the strata dip outwards, towards the valley-floor, but many other Pennine slides have developed in similar situations, by rotating into simple, multiple, compound, multi-storey or complex forms during their displacement (Ter-Stepanian, 1977b). Not all have been studied in detail, but it would appear that progressive creep may well have taken place in many of them, especially where rapid downcutting in the Late Devensian considerably modified the slope profiles. These long-term mass movement processes will now be examined and considered as an integral part of slope evolution for the south Pennine region as a whole.

18.4.4 Landslide Formation and Slope Development in the South Pennines

The history of the Rollick Stones Landslide provides a further illustration of slope movements occurring when climatic conditions and denudational processes were less

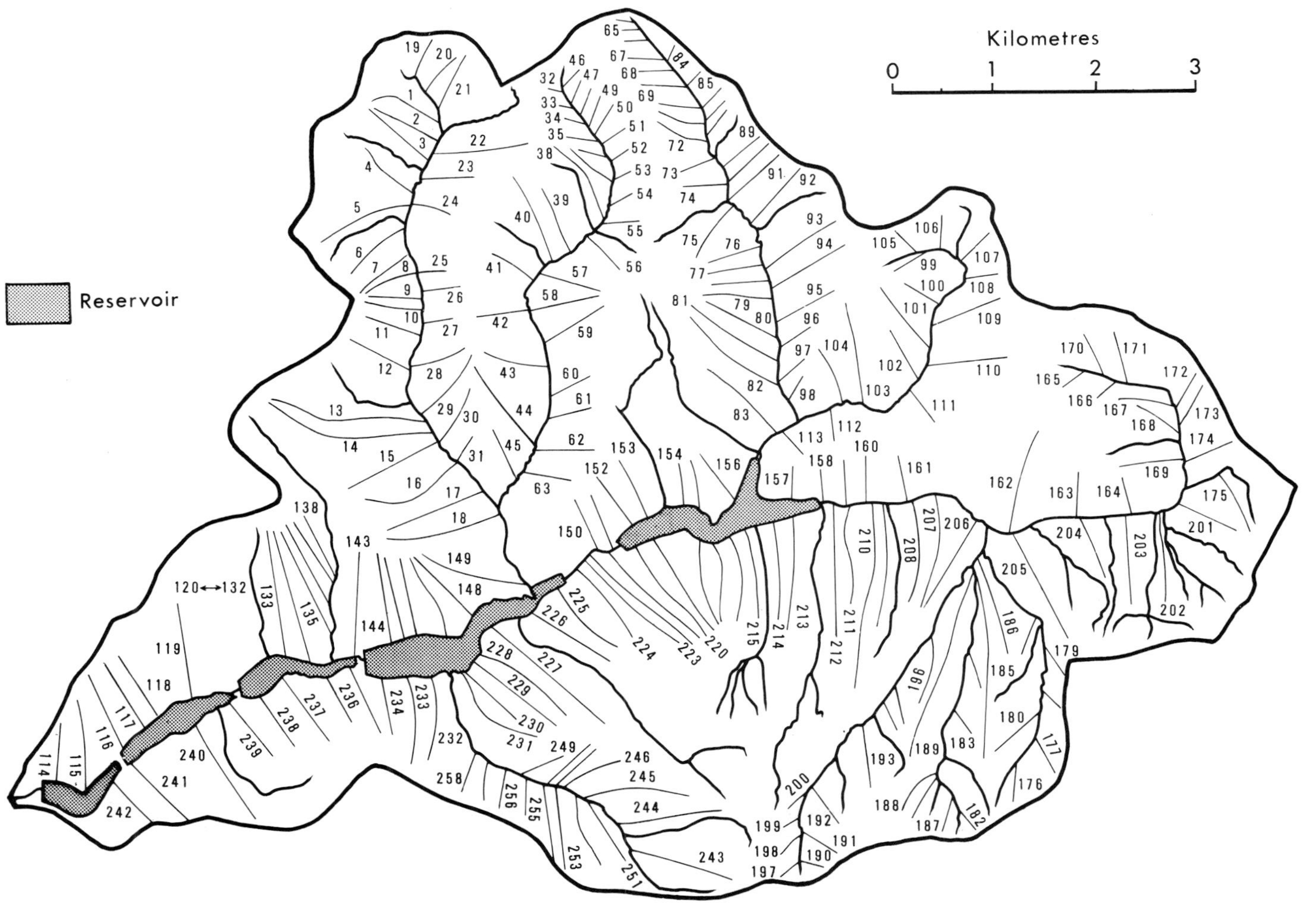

Figure 8.15 Longdendale: distribution of slope profiles used to provide slope profile data

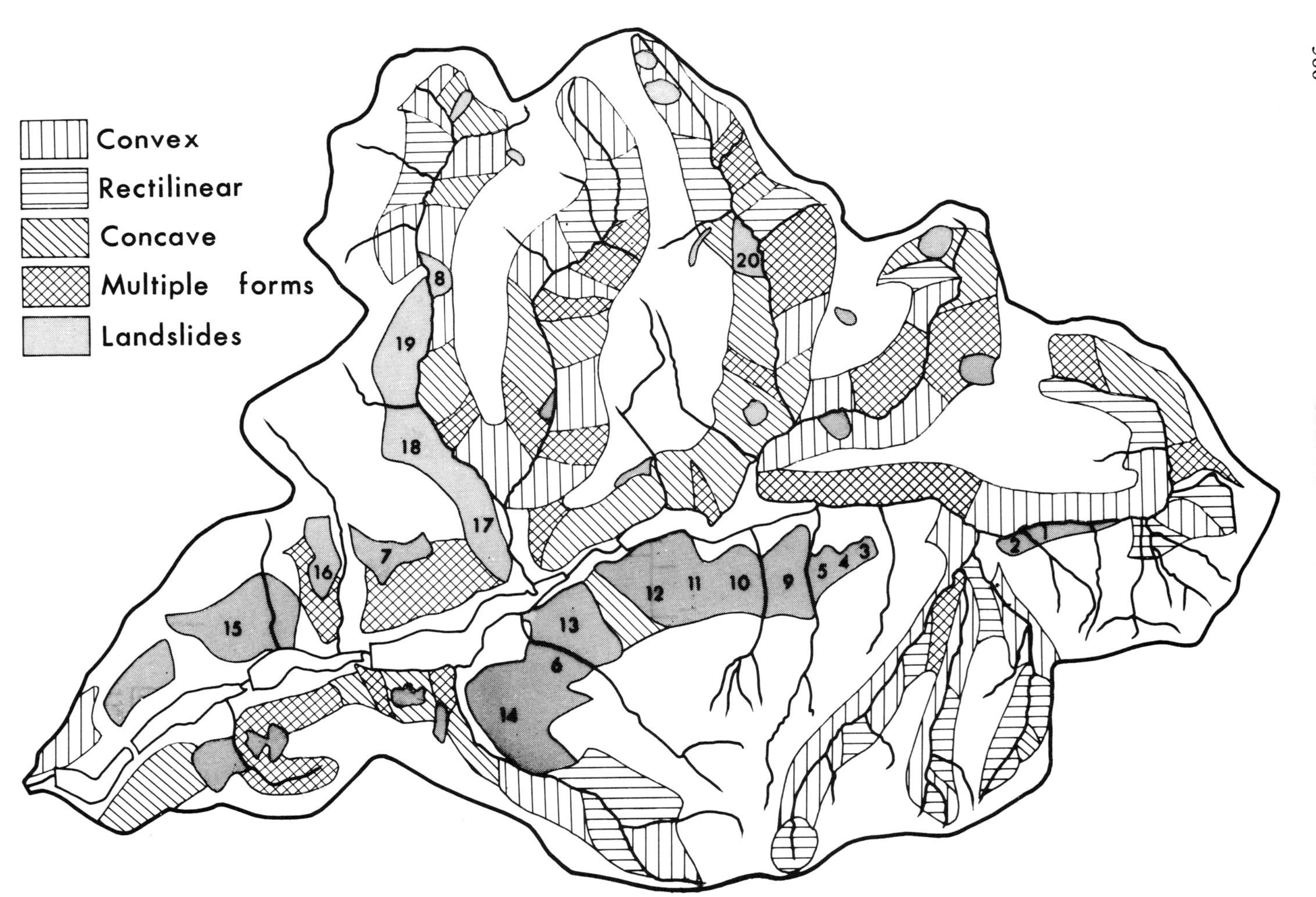

Figure 18.16 Longdendale: distribution of slope types using Young's (1972) method of slope analysis

vigorous than they had been. The slope must have experienced prolonged degradation before its materials became weakened and progressive creep led first to valley-floor bulging, cambering or toppling. These affected only parts of the slope, however, and antedated the main slide event by some hundreds of years. It is, therefore, useful to consider first, the role that these processes and the later deep-seated mass movements play in the evolution of the slopes in Longdendale; and second, whether a slope model representing long-term slope changes can be formulated given these particular geological and topographical conditions.

To investigate slope evolution in Longdendale, a cartographic study was undertaken which involved drawing 254 slope profiles using map sheets at a scale of 1:2,500 with a 2.5 m contour interval: the maps were compiled from a photogrammetric survey and the constructed profiles were checked in the field to establish that they provided an accurate representation of the relief (Figure 18.15).

The slopes were first classified into rectilinear, convex, concave, compound or complex profiles using the criteria suggested by Young (1972), and a distribution map was then drawn (Figure 18.16). Comparisons were made between this map and others (Johnson, 1981) showing landform morphology, regolith materials, soils and potential landslide hazard. It was found that the concave slopes are most common in the larger valleys, where they are thickly mantled with regolith or slump debris and have strongly gleyed soils. As expected, the greatest landslide hazard occurs on these slopes, especially where groundwater levels have remained high in slopes capped by massive sandstones that overlie thick mudrocks. In contrast, the straight and convex slopes are more characteristic of the tributary valleys, where they are often scree-covered, better drained, and have thin acidic soils. Stream undercutting is the greatest threat to stability in these valleys and the landslips there are generally on a much smaller scale.

To study the profiles in more detail, a series of statistical parameters was obtained from the slope data collected for drawing the profiles. Many of these were first used by Crozier (1973) and Parsons (1978, 1982) in their work on landslide and slope morphology, but only four of the twenty parameters measured are important and account for 80% of the variance. They are the slope curvature and roughness, the maximum slope angle, and the position of any scarp forming outcrops present in the hillslope. All the measured parameters were used in clustering procedures and in discriminant analysis, as a means of classifying the profiles. At first, the data from all 254 profiles were used as the sample for clustering, but this did little more than relate slopes in terms of their size. The sample was, therefore, divided subjectively into two sets containing: (a) slopes where no deep-seated mass movements had taken place; and (b) slopes which had failed and were now stabilized. The sub-division was checked using discriminant analysis and only seven profiles were found to be incorrectly classified. Cluster analysis was then undertaken for each group of profiles and the dendrograms are shown in Figure 18.17.

Thirteen clusters were identified and their members and characteristics are listed in the Appendix. Eight of these are in the first dendrogram, being differentiated by a dissimilarity coefficient of 0.206 but of these, three (clusters 3, 4, and 5) can be combined to form a fairly homogeneous group (one of these clusters contains less than ten profiles all differing from each other and from those in the other clusters). Figure 18.18 shows the spatial distribution of the profiles in clusters 1–8, and it would appear that for these slopes, slope amplitude is the controlling factor in the distribution. The configuration of the slopes varies,

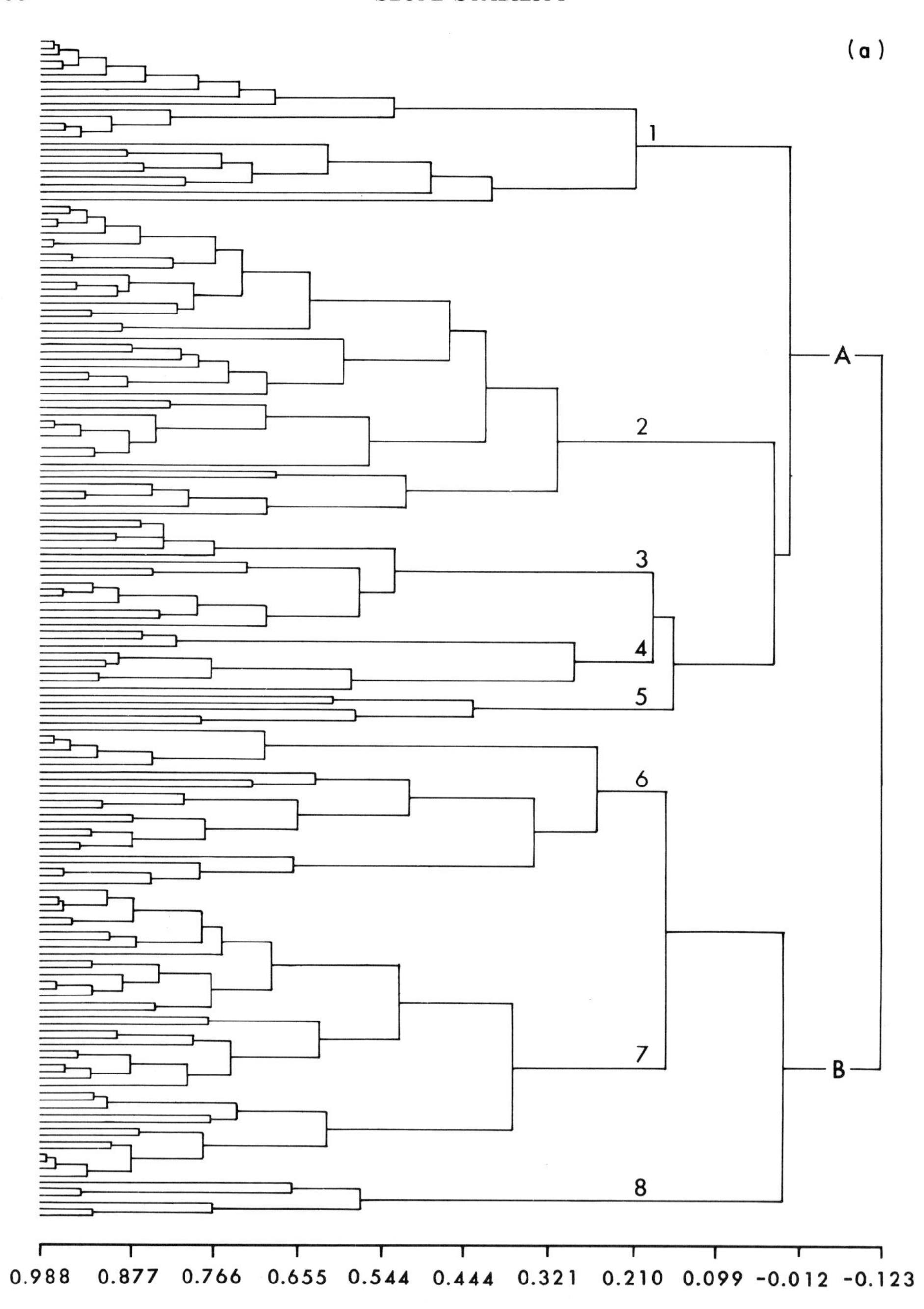

but similar profiles are often found on opposite sides of the valleys even when the geology would favour an asymmetrical cross-valley profile.

The second dendrogram has four separate clusters at a dissimilarity coefficient of 1.5 and two larger clusters if the coefficient is increased to 2.164: slide height/length, slope curvature and maximum slope angle are the parameters which are the most important for

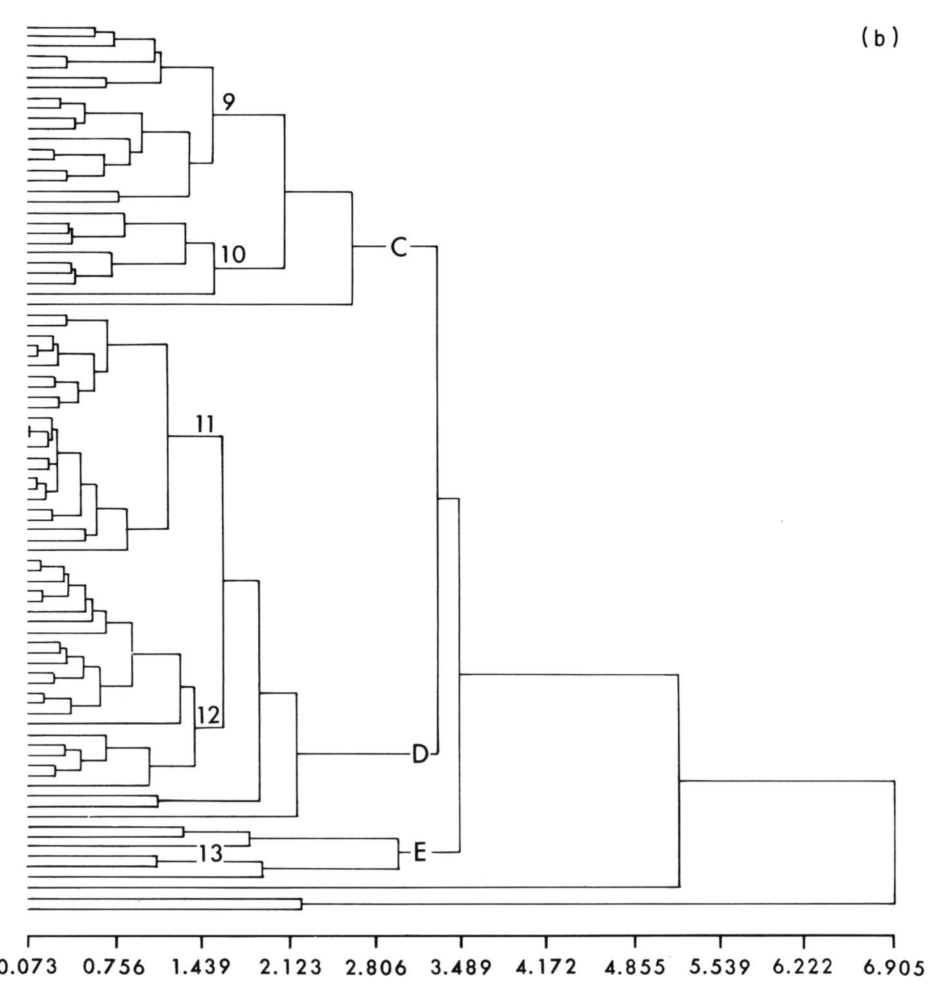

Figure 18.17 Longdendale: cluster dendrograms (for explanation see text and appendix)

distinguishing between the landslide profile types, but one cluster is of little importance as it contains only a small number of diverse profile forms.

It is, therefore, possible to reduce the number of clusters to seven and to identify a slope profile type for each. The main characteristics for each type are given in Table 18.1.

Figure 18.19 is a temporal model covering thousands of years of slope evolution, and is therefore untestable. It is based on spatially-distributed profile forms identifiable in the various parts of the catchment, and on the assumption that the upper catchment slopes will ultimately progress towards profiles comparable to those in the lower part of the present valley. The model assumes that initially the slopes were formed by downcutting through a succession of alternating arenaceous and argillaceous strata of variable thickness, and that the irregularities formed by the thinner sandstones were quickly removed or covered by regolith. Additional constraints include the threshold slope angles for Pennine slopes of 35°, 20°, 14°, and 10° (Carson and Petley, 1970). It is also assumed that each slope

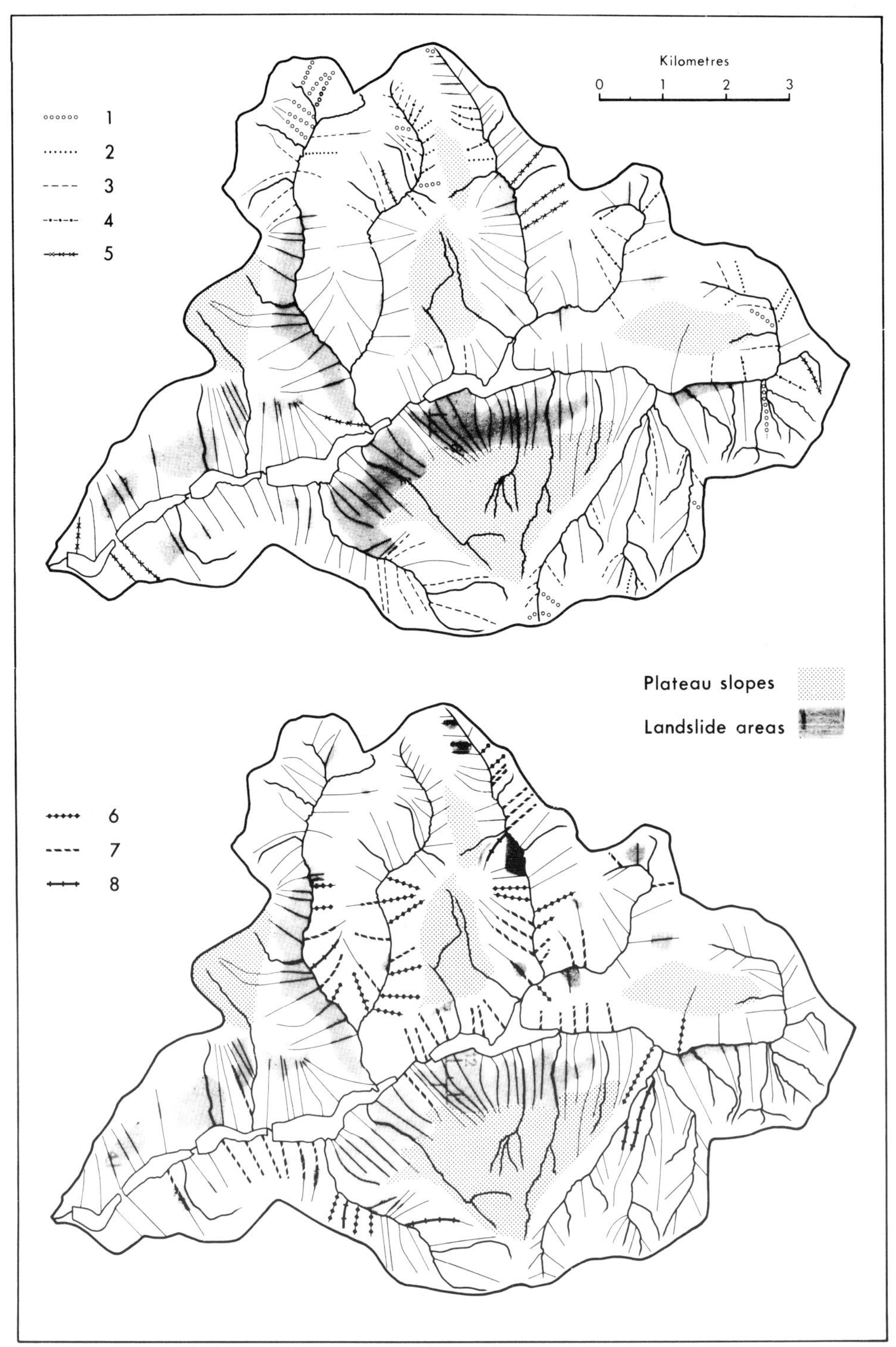

Figure 18.18 Longdendale: distribution of slope types based on cluster analysis

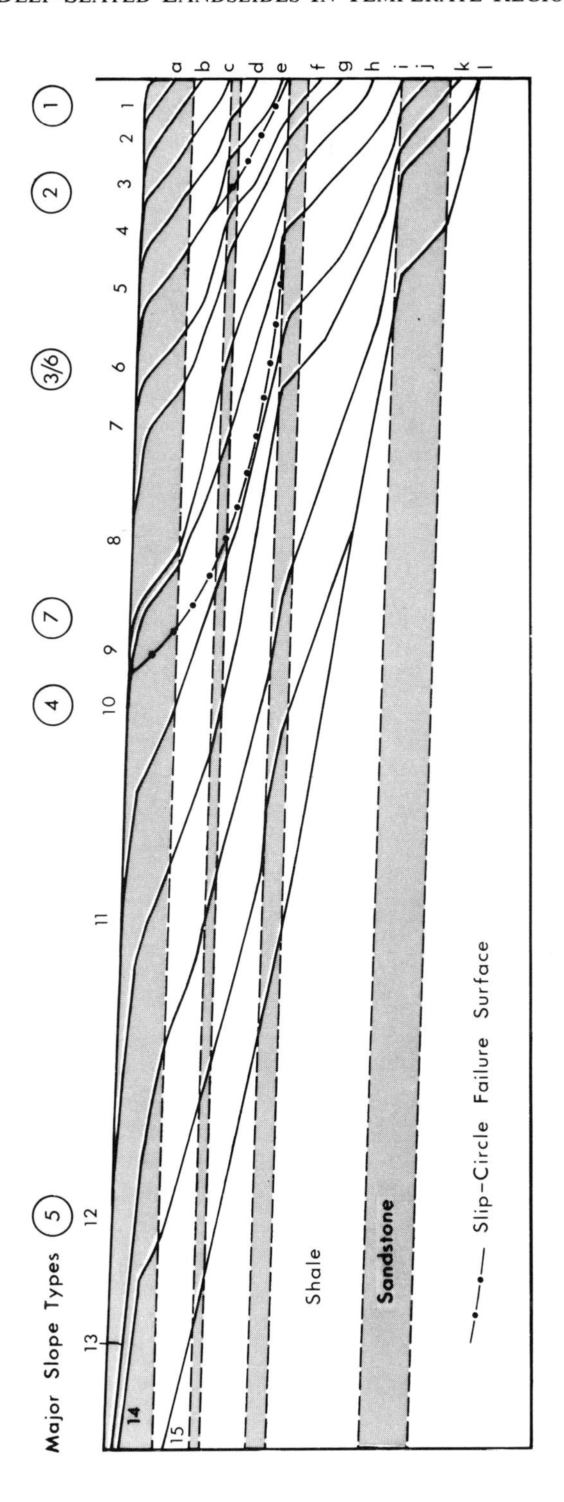

Figure 18.19 The Longdendale slope evolution model

Table 18.1 Slope Profile Types in Longdendale

1. Long gently sloping valley sides with a convex footslope, usually found in upper catchment areas.
2. Slopes broadly convex in form, but with a rectilinear slope element forming the footslope.
3. Broadly convex profiles that are more complex in form having one or more structural benches forming a distinct slope element.
4. Slopes that are concave overall, but have a structural bench of some width at their base.
5. Slopes which are broadly concave, but are longer and more irregular in profile than those in type four.
6. Broadly convex slopes, but with the lower slopes affected by past mass movements.
7. Major landslide areas developed on the higher and middle sections of the slope profile.

element doubled its length during each phase of downcutting, slope back wasting continued at times when the base level did not change and that the sandstones were uniformly resistant and degraded at a rate ten times slower than the mudrocks. In the slope model, the influence of any hard outcrops depends upon their thickness and the length of time their free faces remain exposed.

In the early stages, the slopes had long rectilinear elements in their profiles whose gradients were determined by either the dip of the strata or by the thickness of the regolith overburden. With continued downcutting, each slope element increased in length at the expense of the elements lying further upslope, the regolith became comminuted and new thresholds determining slope gradients were attained. Thirteen stages of slope development have been calculated and include the seven profile types noted in Table 18.1. Such an evolutionary sequence, however, would not have been without interruptions, as a variation of stability through time implies that backwearing or downwasting was not continuous but proceeded only when the regolith was unstable, and ceased when it became stabilized. For the initial stages, the profiles have been drawn to conform to the mathematical curves computed by Hirano (1975), as these slopes would have been moulded by hillwash and rill erosion. They would have had an erosion rate eight times faster than that achieved by soil creep, but one-third slower than the rate at which most slump deformations occur (Leopold *et al.*, 1964). The first stages are ones where convex elements are produced, but as the slope amplitude increases, rectilinear slopes are formed. These are maintained so long as regolith is removed from the base of the slope and will increase in length, according to Summerfield (1976), as long as downcutting continues. Ultimately, there is a progression towards a profile with convex–straight–concave elements with the gradient in each element reflecting an equilibrium between the processes which remove, transfer or deposit soils on the slope.

A critical condition in the model is reached between stages 5 and 6 when the sandstone outcrops form the local base levels for regolith transport on the upper slope, but are themselves being undercut by stream erosion. The undermining of the mudrocks in the footslope results in failure and the displacement of the buttress rock. At this stage, the failed Longdendale slopes are about 120 m in height, the profiles are still broadly convex, but with some irregularity in their form. Many of the slopes probably survive this first phase without experiencing slope failure and it can be assumed that the slope would then evolve progressively to at least stage 9 when the thickness of the regolith on the slope probably increased sufficiently to impede groundwater movement to the surface. Stress levels must also have increased progressively with further downcutting.

With continued development the upper slope, itself now more than 100 m high, becomes increasingly susceptible to slope failure. The basal mudrocks become weakened through a loss of cohesion resulting from seepage weathering and erosion and, during cold climates, the creation of high cleft or other pore-water pressures. Progressive creep followed by rotational slides along non-circular slip surfaces are then the most likely outcomes in the evolution of such slopes. Such slope movements at stages 5 and 9 are essentially mechanisms of slope decline and contrast with rock falls, rock slides, and avalanches which are agents of slope replacement or slope retreat (Carson, 1976). Ultimately, all create rectilinear elements inclined at their limiting angles of stability as these are determined by the condition of the fractured material: eventually the slumps become replaced by taluvium-mantled slopes inclined at their threshold angles.

Not all slopes have failed, however, and some have evolved to forms similar to those in stages 11 and 12 of the model. It is unlikely that the later stages of the slope model will ever be reached as long as caprocks outcrop on the upper slopes in the valleys. But where a caprock is removed or overlain by a thick regolith, stages 14 and 15 would be possible outcomes given long periods of geologic time. They may well represent conditions which once prevailed in pre-Quaternary periods. If so they represent the *de initio* slope forms from which the present valley side slopes have evolved.

In Longdendale, estimates of the maximum rate of scarp retreat in the main valley can be calculated. These range from 0.76 to 0.95 mm yr^{-1}, based on McArthur's (1977) estimate of 1 to 1.25 million years for the incision of the Dark Peak valleys. The slope model thus suggests that the first critical phase would be reached in the main valley some time between 0.66 and 0.75 million years ago since the second phase is known to have taken place in the last 10,000 years. Obviously the exact duration of this time period is impossible to calculate with any accuracy, but if the slope model is valid, then it clearly emphasizes the importance of Brunsden's (1979) observation concerning the role of the caprock formations in landslide areas. These are always an inhibiting factor in the degradation of any slope, but where a slump protects a scar face the stresses on the slope will only increase very slowly. Long intervals of time must, therefore, elapse between failure periods as debris has to be removed or become degraded before high stress conditions can return to the slope.

18.5 CONCLUSIONS

Three important issues are raised by this study. The first identifies the inadequacies of the existing dating methods and the paucity of data useful for investigating periodicities in past landslide activity. The second emphasizes the importance of progressive rock mass creep for inducing failure conditions, and the third is concerned with the role of deep-seated mass movements as part of a slope evolutionary sequence. As yet there is insufficient evidence available for a proper testing of Starkel's (1966) hypothesis which correlates high frequencies of past landsliding with periods of severe cold or high *average* precipitation, although the occurrence of the 'Little Ice Age' in Scandinavia and elsewhere did result in an increase in the number of rockfalls and avalanches (Grove, 1972). In western Europe, in historic times, there have been many instances of rockfalls or slides occurring during heavy rain, and in slump areas, secondary movements have been frequently observed following periods of heavy rainfall (Hutchinson *et al.*, 1980). The current data suggest, however, that major rockslides have not occurred with any marked periodicity. Even in areas close to former ice sheet margins where slope processes were particularly active, the over-steepening was rarely sufficient to induce immediate landslipping although subsequently rockslides, rockfalls, avalanches, and landslides have all taken place. This indicates (a) that the rocks retained shear strengths at levels well above the critical factor of safety, only failing when the climate substantially improved, and (b) that 'triggering' rainfall does not have to be associated with secular wet periods.

In areas of variable lithology and rapid downcutting during the Quaternary, it is not unusual to find deformational structures present both beneath the floors and on the sides of the larger valleys. These features show the importance of progressive rock mass creep which operated in conditions quite different from those which prevail. At times long before failure, a slope would be weakened by the effects of bulging, cambering, and toppling and would undergo increased tensional stress affecting particularly the strata having the least shear strengths. These strata outcrop not only on the footslope, but also in mid-slope positions. Here, they are most susceptible to topple flexuring or other rock deforming processes which increase shear stresses on rock surfaces separating competent from incompetent strata. The resulting progressive deterioration has been a prime factor in causing the failure at the Rollick Stones and elsewhere, but the 'trigger' mechanism in this particular case was provided by stream erosion.

The third issue is one which refers to deep-seated mass movements and their functional role in the evolution of the Longdendale valley-sides. The model, while admittedly based on 'fragile' foundations, suggests that these hillslopes have evolved over long periods of time, with or without deep-seated mass movements taking place. The individual stages were not of equal length being dependent upon the extent to which caprock scarps controlled the backwearing of the slopes, but most of the middle valley slopes have become vulnerable to mass rock creep and failure more than once during their history. The potential failure hazard was undoubtedly increased if the slope attained a critical condition during a severe climate regime when degradation rates were high, but Voight and Pariseau's comment (1978) on the age of landslides remains valid for modern hillslopes. The single rotational landslide type is a feature of an early stage in a slope's history and as such its debris could be removed long before the slope attains an amplitude where shear stresses again become critical. If failure then occurs it will take place in the rocks that form the mid-slope element of the profile and a much larger mass movement will be formed. Protected at their base by hard rock outcrops, the larger Longdendale slumps have become stabilized and will persist for a long time given present denudational conditions. Their ancient quality is thus a characteristic of a relatively recent modification to the hillslopes but their inherent instability is derived from factors that have operated since the downcutting first began.

ACKNOWLEDGEMENTS

Thanks are due to Mr Graham Bowden for preparing the text figures; Air Views (Manchester) for permission to reproduce Figure 18.1; Clyde Surveys Ltd for permission to reproduce Figure 18.4; The Director, The North-West Road Reconstruction Unit, Preston, for access to unpublished information on borehole and resistivity data from the Rollick Stones Landslide; Professor J. N. Hutchinson and The Royal Society for permission to reproduce Figure 18.2; Dr G. Holmes and Professor J. C. Flageollet for permission to cite unpublished material.

Appendix

The Longdendale Slope Profiles

Diagnostic features and Profile numbers associated with each cluster group: (Figures 18.15, 18.17 and 18.18).

STABLE SLOPE PROFILES

Group A (Figure 18.18A)
Cluster 1
Dendrogram Profile sequence (DPS). 1, 35, 178, 2, 3, 55, 20, 21, 202, 168, 190, 191, 197, 19, 64, 23, 51, 187, 74, 53, 171, 173, 166, 172. $n=24$.

Gently sloping profiles in areas with small relief amplitude: low slope integrals (slope height/slope length) and slope curvature indices: slope gradients regular with 45% of slope segments* at less than 5° inclination.

Cluster 2
DPS 4, 183, 36, 37, 5, 170, 34, 184, 167, 252, 188, 195, 165, 22, 251, 46, 192, 40, 109, 7, 32, 163, 100, 246, 243, 48, 244, 185, 6, 47, 111, 169, 65, 199, 198, 179, 24, 41, 33, 247, 181, 253, 254, 102, 200, 194. $n=46$.

Slope gradients steeper than those in cluster 1 with extensive rectilinear elements on the profiles. Found mostly in the upper catchment, they are characteristic dip slope profiles. A significant proportion of the slope segments have slopes <9°, but slope integrals are often high (55% or more) with steepest slopes occurring in footslope or valley-side crest areas.

Cluster 3
DPS 70, 72, 182, 201, 56, 49, 73, 164, 71, 101, 177, 99, 54, 107, 52, 50. $n=16$.

The most important cluster in Group A. Profiles have moderate declivity (5°–18°) with mid-slope elements being either rectilinear or concave. Steepest slope elements at valley-side crest, relief amplitude 40–90 m. These profiles occur chiefly in upper valley areas.

*For this present cartographic analysis, a slope *segment* is defined as having a length of 25 m as measured from the map sheets. A slope *element* measures more than 100 m (ground distance) and has a uniform curvature overall.

Cluster 4

DPS 114, 242, 241, 149, 93, 92, 39, 175, 94. $n=9$.

Mid-valley profiles with relief amplitude >90 m. Profiles are not very steep, but are irregular, with structural benches a characteristic of this type of profile.

Cluster 5

DPS 58, 186, 176, 193, 180. $n=5$.

A cluster set which is transitional between Groups A and B. Slopes are long and irregular with 80% of profile segments at less than 9°.

Group B (Figure 18.18A)

Cluster 6

DPS 45, 26, 255, 27, 256, 258, 67, 96, 58, 206, 95, 162, 85, 78, 86, 57, 79, 59, 113, 31, 61, 62, 63. $n=23$.

Slopes have a marked upper and lower slope elements with structural benches often forming the footslope. Mid-slope concavities less marked than those found in cluster 12. 62% of slope elements between 10° and 30°. Curvature index is high *c.* 27%, but slopes have moderate amplitude between 110 and 150 m.

Cluster 7

DPS 60, 81, 28, 30, 98, 158, 87, 42, 88, 158, 80, 97, 83, 151, 152, 156, 160, 89, 104, 103, 43, 108, 105, 153, 159, 154, 161, 82, 44, 91, 90, 135, 155, 224, 150, 236, 232, 240, 233, 235, 234, 237. $n=43$.

Long slopes exceeding 715 m. Slope segments generally of moderate declivity 18°–30°. Index of curvature 25%, slope integral *c.* 47%. Maximum slope angles found in upper part of the profile.

Cluster 8

DPS 75, 257, 29, 189, 196, 245. $n=6$.

A small group of profiles that are irregular in form with maximum slope angles exceeding 30°.

LANDSLIDE SLOPE PROFILES

Group C (Figure 18.18B)

Cluster 9

DPS 13, 76, 11, 249, 250, 25, 84, 112, 9, 77, 137, 116, 18, 208, 138, 209, 8, 10. $n=18$.

Profiles of moderate amplitude (>130 m), characterized by single slip failures. Slopes moderately steep with 65% of slope segments >10° and most segments between 18° and 30°. Curvature index 24%; slope integral 54%. Maximum steepness found in mid-slopes. Slumps forms only a small part of profiles and the general profile irregularity is therefore small.

Cluster 10

DPS 106, 110, 203, 204, 205, 68, 69, 248, 207. $n=9$.

The single landslips form a larger component of the profile than they do in cluster 9. Slopes have an amplitude of only 100 m and most are in the upper catchment areas. 89% of slope segments have slopes between 5° and 18°.

Group D (Figure 18.18)
Cluster 11
DPS 66, 117, 118, 133, 145, 147, 134, 143, 144, 212, 213, 146, 126, 128, 228, 119, 131, 127, 132, 139, 216, 223, 136, 141, 211. $n=25$.

Multiple failures have occurred on these slopes in mid-valley locations. Marked irregularity is the major characteristic of these profiles: slope amplitude is *c*. 250 m, slope length exceeds 1,000 m. The steepest segments occur on the landslide scars: 60% of segments are between 5° and 18°.

Cluster 12
DPS 148, 124, 125, 122, 123, 130, 238, 129, 121, 230, 218, 219, 221, 222, 227, 226, 220, 120, 214, 215, 217, 225, 142. $n=23$.

Greater slope amplitude (>260 m) and generally lower slope segment declivities distinguish this group of landslide profiles from those in cluster 11.

Group E (Figure 18.18B)
Cluster 13
DPS 239, 231, 229, 12, 17, 140, 15, 16, 210, 14, 115, 174. $n=12$.

Minor cluster of profiles having multiple slump landslides covering more than 60% of the slope length. The profiles differ from those in clusters 11 and 12 by occurring up-valley and in places where the amplitude is about 200 m and the slope irregularity is less.

REFERENCES

Bateman, J. F. L. (1884). *A History and Description of the Manchester Waterworks*, VII. Manchester, 292pp.
Birkeland, P. W. (1973). 'Use of relative dating methods in a stratigraphic study of rock glacier deposits, Mt Sopris, Colorado.' *Arctic and Alpine Res.*, **5**, 401–16.
Bromehead, C. E. N., Edwards, W., Wray, D. A., and Stephens, J. V. (1933). *Geology of the Country around Glossop and Holmfirth*, Mem. geol. surv. GB, HMSO, London, 209pp.
Brunsden, D. (1979). 'Mass movements.' In *Process in Geomorphology*. Embleton, C., and Thornes, J. (eds.). Arnold, London, pp. 130–86.
Brunsden, D., and Jones, D. K. C. (1972). 'The morphology of degraded landslide slopes in south-west Dorset.' *Q. J. Eng. Geol.*, **5**, 1–18.
Burek, C. (1977). 'The Pleistocene Ice Age and after.' In *Limestone and Caves of the Peak District*. Ford, T. D. (ed.), GeoAbstracts, Norwich, pp. 87–128.
Caine, N. (1982). 'Toppling failures from alpine cliffs on Ben Lomond, Tasmania.' *Earth Surf. Proc. Landforms*, **7**, 133–52.
Carson, M. A. (1976). 'Mass wasting, slope development and climate.' In *Geomorphology and Climate*. Derbyshire, E. (ed.). Wiley, London, pp. 101–36.
Carson, M. A., and Petley, D. J. (1970). 'The evidence of threshold hillslopes in the denudation of the landscape.' *Trans. Inst. Brit. Geog.*, **49**, 71–95.
Chambers, F. M. (1982). 'Blanket peat initiation—a comment.' *Q. Newsletter*, **36**, 37–9.
Chandler, R. J. (1976). 'The history and stability of two Lias clay slopes in the Upper Gwash valley, Rutland.' *Phil. Trans. Roy. Soc., London*, **A283**, 463–91.
Chandler, R. J., Kellaway, G. A., Skempton, A. W., and Wyatt, R. J. (1976). 'Valley slope sections in Jurassic strata near Bath, Somerset.' *Phil. Trans. Roy. Soc., London*, **A283**, 527–56.
Christiansen, E. A. (1983). 'The Denholm Landslide, Saskatchewan. Part 1: Geology.' *Can. Geotech. J.*, **20**, 197–207.
Clayton, K. M., and Straw, A. (1976). *Eastern and Central England*. Methuen, London, 247pp.
Coombs, D. (1971). 'Mam Tor: a Bronze Age hillfort?' *Current Arch.*, **3**, 100–2.

Crozier, M. J. (1973). 'Techniques for the morphometric analysis of landslips.' *Zeitschrift für Geomorph.*, **12**, 60–76.
Denness, B. (1972). 'The reservoir principle of mass movement.' *Inst. Geol. Sci. Rept.*, 72/7, 1–13, HMSO, London.
Dowdeswell, J. A. (1982). 'Relative dating of Late Quaternary deposits using cluster and discriminant analysis, Audobon Cirque, Mt Audobon, Colorado Front Range.' *Boreas*, **2**, 151–61.
Erskine, C. F. (1973). 'Landslides in the vicinity of the Fort Randall Reservoir, South Dakota.' *U.S.G.S. Prof. Paper*, 675, 65pp.
Fearnsides, W. G. (1932). 'Geology of the eastern part of the Peak District.' *Proc. Geol. Assoc.*, **43**, 152–91.
Franks, J. W., and Johnson, R. H. (1964). 'Pollen analytical dating of a Derbyshire landslide.' *New Phytol.*, **63**, 209–16.
Gil, E., Gilot, E. G., Kotarba, A., Starkel, L., and Szczepanek, K. (1974). 'An early Holocene landslide in the Neski Beskid and its significance for palaeogeographical reconstructions.' *Geom. Carpatho.*, **8**, 69–83.
Grove, J. M. (1972). 'The incidence of landslides, avalanches and floods in western Norway during the Little Ice Age'. *Arct. Alpine Res.*, **4**, 131–8.
Higginbottom, I. E., and Fookes, P. G. (1971). 'Engineering aspects of periglacial features in Britain.' *Q. J. Eng. Geol.*, **3**, 85–117.
Hill, H. P. (1949). 'The Ladybower Reservoir.' *J. Inst. Water Engrs.*, **35**, 414–25.
Hirano, M. (1975). 'Simulation of developmental processes on interfluvial slopes with reference to graded form.' *J. Geol.*, **83**, 113–23.
Hollingworth, S. E., Taylor, J. H., and Kellaway, G. A. (1944). 'Large-scale superficial structures in the Northamptonshire Ironstone Field.' *Q. J. Geol. Soc., London*, **100**, 1–44.
Holmes, G. (1985). *Rock-slope failure in parts of the Scottish Highlands.* Unpublished Ph.D. Thesis, Univ. of Edinburgh.
Horswill, P., and Horton, A. (1976). 'Cambering and valley bulging in the Gwash valley at Empingham.' *Phil. Trans. Roy. Soc., London*, **A283**, 427–51. See also Appendix by Vaughan, P. R., 'The deformation of the Empingham valley slope.' 452–62.
Hutchinson, J. N. (1969). 'A reconsideration of the coastal landslides at Folkestone Warren, Kent.' *Geotechnique*, **19**(1), 6–38.
Hutchinson, J. N., and Gostelow, T. P. (1976). 'The development of an abandoned cliff in London Clay at Hadleigh, Essex.' *Phil. Trans. Roy. Soc., London*, **A283**, 557–604.
Hutchinson, J. M., Bromhead, E. N., and Lupini, J. F. (1980). 'Additional observations on the Folkestone Warren landslides.' *Q. J. Eng. Geol.*, London, *13*, 1–31.
Johnson, R. H. (1965). 'A study of the Charlesworth landslide near Glossop, North Derbyshire.' *Trans. Inst. Brit. Geogrs.*, **37**, 111–25.
Johnson, R. H. (1968). 'Four temporary exposures of solifluction deposits on Pennine hillslopes in north-east Cheshire.' *Merc. Geol.*, **2**, 379–87.
Johnson, R. H. (1980). 'Hillslope stability and landslide hazard—a case study from Longdendale, north Derbyshire, England.' *Proc. Geol. Assoc.*, **91**(4), 315–22.
Johnson, R. H. (1981). 'Four maps for Longdendale—a geomorphological contribution to environmental management in an upland valley.' *Manchester Geog.*, **2**(2), 6–24.
Johnson, R. H., and Vaughan, R. D. (1983). 'The Alport Castles, Derbyshire: A south Pennine slope and its geomorphic history.' *E. Midland Geog.*, **59**(3), 79–88.
Johnson, R. H., and Walthall, S. (1979). 'The Longdendale Landslides.' *Geol. J.*, **14**(2), 135–158.
Jones, G. D. B., and Thompson, F. H. (1965). 'Excavations at Mam Tor and Brough-en-Noe.' *Derby. Archaeol. J.*, **85**, 89–93.
Lapworth, H. (1911). 'The geology of dam trenches.' *Trans. Inst. Water Engrs.*, **16**, 25–66.
Leopold, L. B., Wolman, M. G., and Miller, J. G. (1964). *Fluvial Processes in Geomorphology.* Freeman, San Francisco, pp. 92–4.
McArthur, J. L. (1977). Quaternary erosion in the upper Derwent Basin and its bearing on the age of surface features in the Southern Pennines.' *Trans. Inst. Br. Geog.*, **2**(4), 490–7.
McArthur, J. L. (1981). 'Periglacial slope planations in the Southern Pennines.' *Biul. Peryglac.*, **28**, 85–97.

Mahr, T., and Nemčok, A. (1977). 'Deep-seated creep deformations in the crystalline cores of the Tatry Mountains.' *Bull. Assoc. Eng. Geol.*, **16**, 104–6.

Malgot, J. (1977). 'Deep-seated slope deformations in neovolcanic mountain ranges of Slovakia', *Bull. Assoc. Eng. Geol.*, **16**, 106–9.

Morariu, T. (1964). 'Age of landsliding in the Transylvanian tableland.' *Rev. Roum. Geol. Geoph. Geogr. ser. Geog.*, **3**, 149–57.

Morton, E. (1932). 'Geology of the Goyt valley,' *Trans. Inst. Water Engrs.*, **37**, 244–254.

Parsons, A. J. (1978). 'A technique for the classification of hillslope forms.' *Trans. Inst. Brit. Geogrs.*, **3**(4), 432–43.

Parsons, A. J. (1982). 'Slope profile variability in first-order drainage basins.' *Earth Surf. Proc. Landforms*, **7**, 71–8.

Pitts, J. (1983). 'The temporal and spatial development of landslides in the Axmouth–Lyme Regis undercliffs, National Nature Reserve, Devon. *Earth Surf. Proc. Landforms*, **8**, 589–603.

Potter, D. B., and McGill, E. (1978). 'Valley anticlines of the Needles District, Canyon Lands National Park, Utah.' *Bull. Geol. Soc. Am.*, **89**(1), 952–60.

Radbruch-Hall, D. H. (1978). 'Gravitational creep of rock masses on slopes.' In *Rockslides and Avalanches, 1. Natural Phenomena.* Voigt, B. (ed.), Ch. 17, Elsevier, Amsterdam, pp. 605–65.

Selby, M. J. (1979). 'Slopes.' In *Man and Environmental Processes*, Gregory, K. J., and Walling, D. E. (eds.), Dawson, Folkestone, pp. 105–20.

Selby, M. J. (1982). *Hillslope Materials and Processes.* University Press, Oxford, pp. 264.

Sidaway, R. (1963). 'Buried peat deposits at Litton Cheney.' *Proc. Dorset Nat. Hist. and Archaeol. Soc.*, **85**, 78–86.

Sissons, J. B. (1976). *Scotland.* Methuen, London, pp. 150.

Starkel, L. (1966). 'The palaeogeography of mid- and eastern Europe during the last cold stage and West European comparisons.' *Phil. Trans. Roy. Soc., London*, **B280**, 351–72.

Stevenson, I. P., and Gaunt, G. D. (1971). *Geology of the Country around Chapel en le Frith.* Mem. geol. surv. GB, HMSO, London, pp. 444.

Summerfield, M. A. (1976). 'Slope form and basal stream relationships: a case study in the Derwent Basin of the Southern Pennines, England.' *Earth Surf. Proc. Landforms*, **1**, 89–96.

Tallis, J. H. (1964). 'Studies on Southern Pennine Peats: II, The pattern of erosion.' *J. Ecol.*, **52**, 333–44.

Tallis, J. H., and Johnson, R. H. (1980). 'The dating of landslides in Longdendale, north Derbyshire, using pollen-analytical techniques.' In *Timescales in Geomorphology.* Cullingford, R. A., Davidson, D. A., and Lewin, J. (eds.), Wiley, London, pp. 189–207.

Terasmae, J. (1975). 'Dating of landslides in the Ottawa River Valley by dendrochonology—a brief comment.' In *Mass Wasting.* Yatsu, E., Ward, A. J., and Adams, F. (eds.), 4th Guelph Symposium on Geomorphology, GeoAbstracts, Norwich, pp. 153–8.

Ter-Stepanian, G. I. (1977a). 'Deep-reaching gravitational deformation of mountain slopes.' *Bull. Assoc. Eng. Geol.*, **16**, 87–94.

Ter-Stepanian, G. I. (1977b). 'Types of compound and complex landslides.' *Bull. Assoc. Eng. Geol.*, **16**, 72–4.

Varnes, D. J. (1978). 'Slope movements, types and processes.' In *Landslides, Analysis and Control.* Schuster, R. L., and Krizek, R. J. (eds.), Nat. Acad. of Sci. Spec. Rept. 76, Washington, USA, pp. 12–33.

Vears, A., and Curtis, C. (1981). 'A quantitative evaluation of pyrite weathering.' *Earth Surf. Proc. Landforms*, **6**, 191–8.

Voigt, B., and Pariseau, W. G. (1978). 'Rockslides and avalanches: an introduction.' In *Rockslides and Avalanches, 1. Natural Phenomena.* Voight, B. (ed.), Elsevier, Amsterdam, pp. 1–67.

Waters, R. S., and Johnson, R. H. (1958). 'The terraces of the Derbyshire Derwent.' *E. Midland Geog.*, **2**, 3–15.

Whalley, W. B. (1974). 'The mechanics of high-magnitude, low-frequency rock failure and its importance in a mountainous area.' *Geog. Paper 27*, Univ. of Reading, Department of Geography, 48pp.

Whalley, W. B., Douglas, G. R., and Jonsson, A. (1983). 'The magnitude and frequency of large rockslides in Iceland in the Post-Glacial.' *Geog. Ann.*, **65**, A(1–2), 99–110.
Wilson, P. (1981). 'Periglacial valley-fill sediments at Edale, North Derbyshire.' *E. Midland Geog.*, **7**(7), 263–71.
Young, A. (1972). *Slopes*. Oliver and Boyd, Edinburgh, 288pp.

Slope Stability
Edited by M. G. Anderson and K. S. Richards

Chapter 19

Slope Development Through the Threshold Slope Concept

S. C. FRANCIS
Merlin Profilers (Velpro) Ltd, Duke House, Woking GU21 5BA

19.1 INTRODUCTION: THE CONCEPT OF THRESHOLD SLOPE ANGLES

The introduction of soil mechanics to geomorphology has improved understanding of the rapid mass movement processes common on relatively steep slopes. Many attempts have now been made to use this knowledge in interpreting slope form measurements because an understanding of the processes which control slope form enables present, future, and to some extent, past slope forms to be modelled with the aim of furthering an understanding of landscape evolution. A theme central to studies which have attempted to interpret slope characteristics in terms of slope stability analyses is that at some critical inclination under given conditions a slope soil will become unstable and undergo landsliding, which acts to restore slope stability by reducing the slope angle. In the simplest hypothetical case, given 'uniform' steep slope soil conditions and a relative absence of other slope degradation processes, the steepest slope angle would be limited to the inclination at which the slope soil would become unstable and landsliding would occur. Lower angled slopes would have a reduced limit of steepness at which landslide debris would come to rest or regain stability under the most adverse of ground conditions. Therefore, if a distinct style of landsliding has been widespread in a particular area then many slopes may be inclined at angles which are similar to those which could be predicted from stability analyses of the slope soils (see Figure 19.1). Determination of the two limiting slope angles in this simple system is not only of interest to geomorphologists studying environments in which mass movement may be the dominant control over slope characteristics. The lower angle may represent an upper bound for the response of slopes to processes which are not directly controlled by mass movement, while the steepest angle will be the lowest angle at which rock outcrop will occur when a substantial soil cover is unable to develop due to excessive slope steepness.

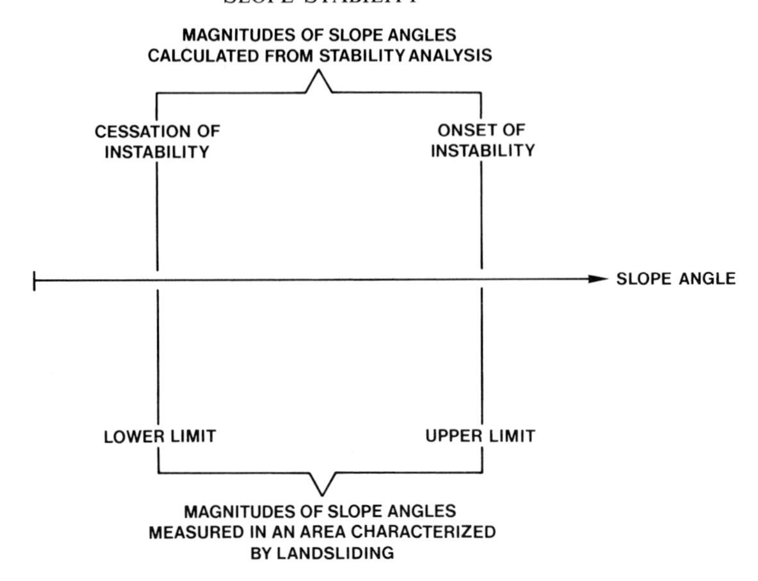

Figure 19.1 The relationship between landsliding and slope angle

Despite the simplicity of this concept, it has been applied to real slopes with varying degrees of success (for geomorphological examples, see Davies, 1963; Carson, 1976; Rouse and Farhan, 1976; Van Asch, 1983). Carson and Petley (1970) have outlined the potential importance of mass movement in slope evolution, and subsequently Carson has reviewed the concept and its applications (Carson and Kirkby, 1972; Carson, 1975). Through these studies, the term 'threshold slope' has become widely used in association with interpretations of slope form using slope stability analyses. However, a single definition of a threshold slope does not exist. For instance, Carson (1975, p. 19) states that:

> 'The concept of the threshold slope angle is a simple one: that, for a natural slope cut in earth material, there is a single angle of ground slope above which rapid mass movement will occur from time to time, and below which the slope material is stable with respect to rapid mass-wasting processes. . . .'

From this, the threshold slope angle could be taken as a parameter which may be used to distinguish between sections of slopes which are dominated by different evolutionary processes. However, this usage can be restrictive as any single slope could during its evolution have been characterized by a series of threshold angles as implied by Carson (1969), or any series of slopes at a particular time may be characterized by a range of threshold angles, as stated by Carson (1976). To prevent the term threshold slope being applied in an unnecessarily restrictive manner, it should refer to simple specified critical or limiting conditions which significantly influence slope angle, as discussed above, and as used, for example by Anderson and Richards (1981) and Chandler (1982).

The concept of threshold slope angles has undoubtedly led to a greater understanding of slope form, and has shown that the steepness of slopes can be controlled by mass movement. This chapter evaluates some of the implications of the concept and its application, together with an investigation of a number of other little-discussed factors which are relevant to long-term mass movement behaviour. Some of the issues raised illustrate the value of better integration between geomorphology and engineering geology in the study of slope form, slope stability, and slope development.

19.2 STABILITY MODELLING IN STUDIES OF SLOPE EVOLUTION

19.2.1 The Application of Simple Stability Analyses

Undemanding slope stability analyses are intuitively more suitable for generalized applications, and this attribute may offset their lack of precision. Implicit in the use of a simple method such as the 'infinite slope' analysis (see Chapter 2), is that the validity of results may be limited because of the assumptions made during its use (Chandler, 1982). Chandler's critical viewpoint concerns the explicit and implicit assumptions made when applying this stability model to cohesive soils, and he expresses concern about its unquestioned and familiar usage by many geomorphologists. While accepting these criticisms in the context of cohesive soil studies and supporting a more cautious approach to stability modelling in general, some geomorphological applications of the infinite slope analysis have been shown to be useful (for example, see Carson and Petley, 1970; Rouse, 1975; Richards and Anderson, 1978; Van Asch, 1983). The source of Chandler's apprehension is clearly illustrated through examination of the invariably simplistic discussions of slope stability analyses in most introductory works on soil mechanics, such as Craig (1978) and Smith (1968), which have become even further simplified in geomorphological texts (for example, see Carson and Kirkby, 1972; Statham, 1977; Selby, 1982; and Williams, 1982). It is evident that geomorphologists have been attracted to the theoretical rigour of soil mechanics methods in spite of the obvious complexity of landforming processes and the difficulties of model calibration. The main application of soil mechanics is, however, to provide design solutions in the face of economic constraints and, in the past, empirical methods have often been adequate for construction purposes. While many geomorphological applications have exploited the simpler soil mechanics methods, Chapters 4, 6, and 7 illustrate the increasing interaction between geomorphologists and engineering geologists in the development of more process-based models.

19.2.2 Refinement of Stability Models

Refinement of elementary slope stability analyses has taken place to allow more precise modelling of slopes for engineering purposes, and this has resulted in a demand for input data which is often difficult to satisfy in practice. In general, the more rigorous and accurate that a slope stability analysis becomes, the more restricted that analysis is to specific slopes. Consequently the geomorphologist is faced with insurmountable problems of data acquisition when attempting to model mass movements over large areas and long time scales (see Chapter 12).

An alternative approach to stability modelling of hillslope form is to use a refined slope stability analysis which is rigorously applied to at least one slope for which high quality soil, hydrological and geometrical data are available. Undertaking this analysis will substantially enhance appreciation of the slope characteristics in general, highlight the fact that no slope stability analysis can be thorough, and indicate that a degree of latent

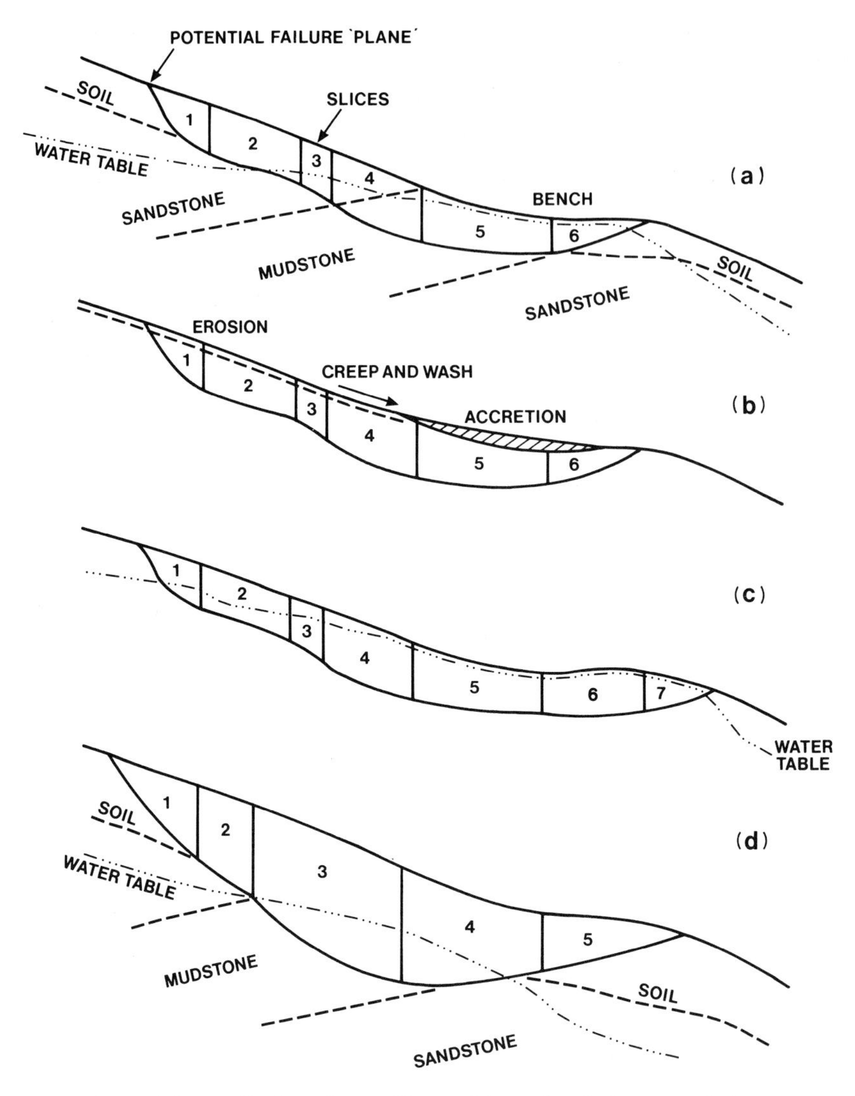

Figure 19.2 Application of a stability model to a benched section of slope

compromise must always be present because of both subjective and objective analytical uncertainties. Armed with this experience, ranges of soil, water, and geometric parameters can be inserted into the stability model which can then be run for a great variety of conditions to simulate effects of mass movement upon slope evolution. Modelling is not only confined to using ranges of parameters which may be presently typical of the slopes under study, but also allows definition of soil and water conditions to simulate long-term stability behaviour. At this point it may be worth while to use a probabilistic approach to assess factors of safety, because of the increased uncertainty concerning changes in material properties with weathering. Probabilistic stability assessments are becoming increasingly popular in assessing the reliability of slope designs (see, for example, Alonso, 1976;

Priest and Brown, 1983; St George, 1984) and appear to be ideally suited to geomorphological use in the future.

An example of the use of a conventional stability analysis in slope evolution modelling is given in Figure 19.2. Here, the evolution is considered of a benched slope whose current form is geologically controlled, with the present slope section shown in Figure 19.2(a). In Figure 19.2(b) soil creep and wash processes have redistributed surface soil, and assuming that all other stability parameters remain constant for a specified slip circle, there is an increase in the factor of safety because the soil mass redistribution has changed the moments about the potential centre of rotation. In Figure 19.2(c) pedogenic processes have formed an illuviated horizon which has increased water-table levels in slices 1–4 through perching, and has impeded basal drainage from slice 6 where the water table has risen sufficiently to necessitate the inclusion of a further slice in the stability analysis to extend the toe. These circumstances would reduce the factor of safety against failure. In Figure 19.2(d) soil thickness has increased following weathering, and the potential failure 'plane' has moved downwards to the deeper soil–rock interface. In this case the slices and centres of rotation must be redefined, the shear strength properties of the soil may need to be changed (this is discussed in greater detail later in the chapter) and the water-table position may be modified by changes in the bulk properties of the soil. Under these circumstances the factor of safety may increase or decrease in value. The operation can be repeated for several 'type' slopes in a region. Countless variations on this theme exist, and are amenable to computer processing provided that realistic estimates of magnitudes and limits of the parameters are available. It must be emphasized that this approach is also a compromise, but acknowledges that intensive site-specific experiences should form the foundation for evolutionary modelling.

19.3 SOME PROBLEMS IN EVALUATING THE CONTROL OF MASS MOVEMENT OVER SLOPE FORM DEVELOPMENT

19.3.1 Introduction: The Nature of Threshold Slopes

An explicit assumption made when interpreting slope development through the threshold slope concept is that mass movement dominates this development. To determine the extent to which threshold slopes are controlled by mass movement processes, it is essential to have some knowledge of the history of the slope or slopes under study. Even when present slope attributes are exclusively used to define the starting point of a predictive slope evolution exercise, this information is still extremely useful. Two examples of different slope histories serve to highlight the importance of assessing whether or not mass movement processes have controlled slope form.

Firstly, Wu *et al.* (1984) presented a series of essentially rectilinear slope profiles from which the slope angle histograms shown in Figure 19.3 have been derived, with virtually all of the slopes measured during their photogrammetric survey being inclined at less than 20°. These distinctive characteristics could invite interpretation in terms of the maximum slope angle being controlled by mass movement, however, the data were obtained from the Olympus Mons basaltic shield volcano on Mars, a planet whose topography has been distinctively controlled by volcanic activity. Each of the slope profiles shown in Figure 19.4 are between 260 and 460 km long, the slope crests being approximately 26 km above the Mars datum, with slope angles measured over sections of slope 5 km long.

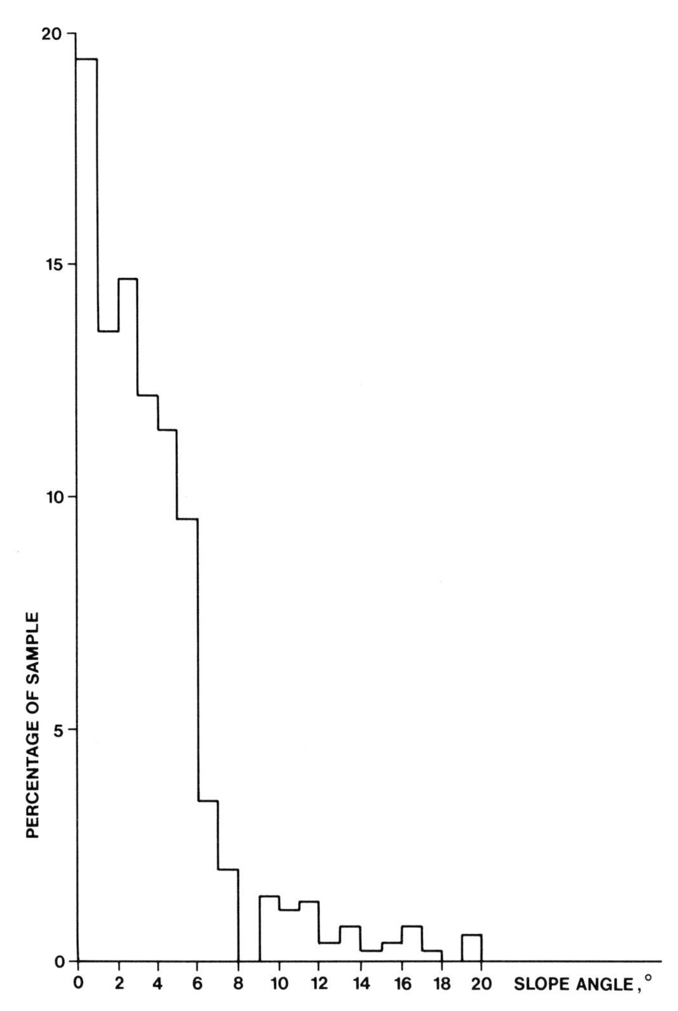

Figure 19.3 Slope angle histogram, Olympus Mons

Secondly, Francis (1984) has shown that rectilinear slopes of similar steepness have developed in valley systems of different character on sandstones in south east Wales. Figure 19.5 shows valley cross-sections, one from a typically 'U' shaped glaciated valley and the other from a typically 'V' shaped valley of fluvial origin. Slope development in the latter case would be weathering limited, but in the former case slope basal erosion appears to be absent. Interpretation of these slopes is further complicated by the presence of different styles of instability, with both shallow landsliding controlled by soil shear strength properties and large-scale deep seated landsliding controlled by weak rock shear strength properties being common (Davies, 1963; Rouse, 1969; Farhan, 1976; Gostelow, 1977). Whether or not these different mass movement processes give rise to different slope forms in the long term has not been thoroughly assessed, though the important works of Savigear (1952) and Hutchinson (1967) do suggest that different instability processes resulting from different erosional conditions result in different slope form development. Chapter 18 illustrates the

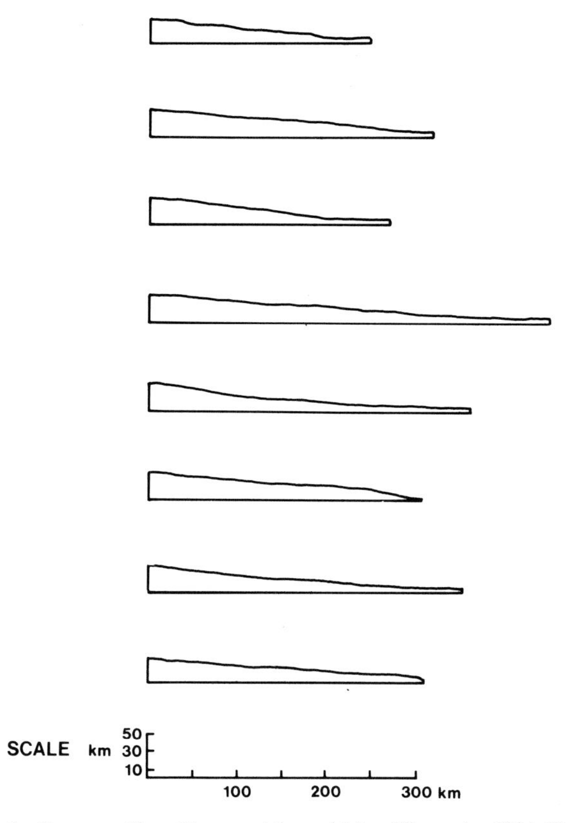

Figure 19.4 Slope profiles, Olympus Mons. (*After Wu* et al., *1984*, Nature, **309**.)

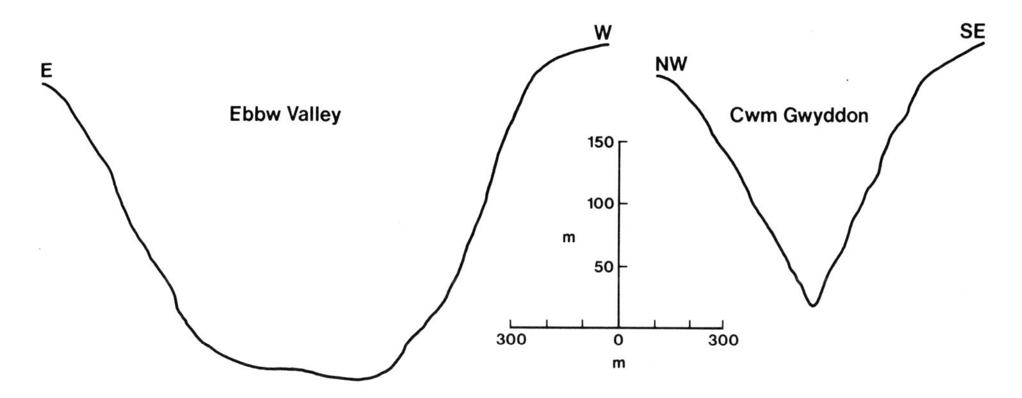

Figure 19.5 Valley cross-sections, south east Wales

complex interaction of a range of instability processes in producing, over late Quaternary time-scales, the slope morphology of an area in the English North Midlands. Other examples where a good knowledge of slope history has assisted slope form interpretation include the work of Wallace (1977) and Dunkerley (1980) concerning slopes developing on fault scarps and raised coral reefs respectively: here geological knowledge has enabled time-scales of slope evolution to be established.

In summary, it appears that some slopes may seem to be threshold slopes on the basis of their form, but knowledge of the history of a slope is required to establish whether or not mass movement processes have influenced its development. Evaluation of the long-term response of slope form to mass movement processes will always be incomplete because of the transient nature of measurement methods. Three approaches to evaluation exist, however, provided that sites can be located where evolutionary changes can be unambiguously identified. The first ideal setting occurs where events can be correlated with suitably preserved organic material which may provide useful radio-carbon dates (Hutchinson and Gostelow, 1976), the second occurs where evolutionary changes are rapid enough to allow a sequence of measurements to be made (Brunsden and Kesel, 1973), and the third is where reliable space/time substitutions can be made which enable a sequence of evolutionary changes to be deduced, usually following the cessation of basal erosion (Savigear, 1952; Hutchinson, 1967). In all cases the availability of historical documentation of landsliding events is particularly useful.

19.3.2 Slope Attributes

It is often convenient to assume that slopes have only a few simple attributes, but both casual observation and the detailed engineering geomorphological mapping now routinely used during landslide assessment (see Brunsden *et al.*, 1975) indicate the true complexities of slope morphology. Parsons (1979) states that three-dimensional slope form is rarely considered in slope studies, while his data for Chalk slopes show good correlations between slope angle and contour curvature. His work represents a strong criticism of the routine presentation of slope angle histograms to characterize slope steepness irrespective of the distribution of hillslope spurs and hollows. Further, in areas apparently dominated by steep slopes, low-angled slopes are always present and often cover a large proportion of catchment areas (see Gregory and Brown, 1966; see also Figure 19.6). Simple attempts to enhance the substance of histogram data have been made by Farhan (1976) and Rouse and Farhan (1976) who correlate slope angles with lithology and slope length. Also, Carson and Petley (1970) and Carson (1971, 1975) have attempted to associate specific slope angle ranges with slopes underlain by particular soil types, together with the degree of soil weathering and basal erosion conditions. All of these works have correlated their slope angle distributions with the results of soil stability analyses, but only Farhan (1976) has presented a crucial detailed study of local slope morphology. Unfortunately his engineering geomorphological mapping only supplemented and was not incorporated with the discussion of slope stability analysis.

Another factor which influences the assessment of slope attributes is the scale of measurement. In threshold slope studies reference is continually made to the importance of rectilinear slopes (see Carson, 1969) but the apparent uniformity of slope angles usually reflects the detail of the slope profiling survey and the consistency of slope angle suggested,

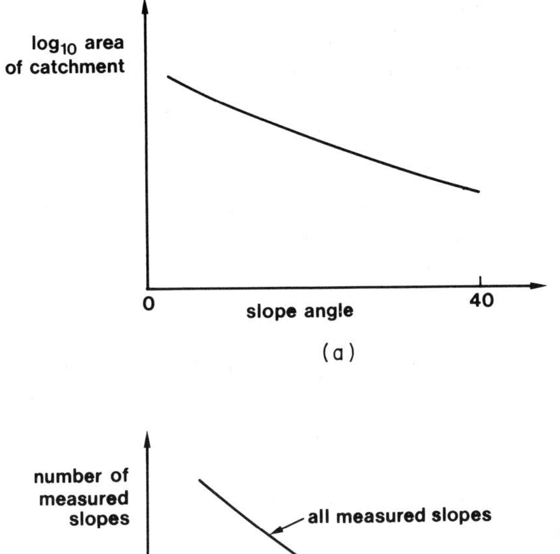

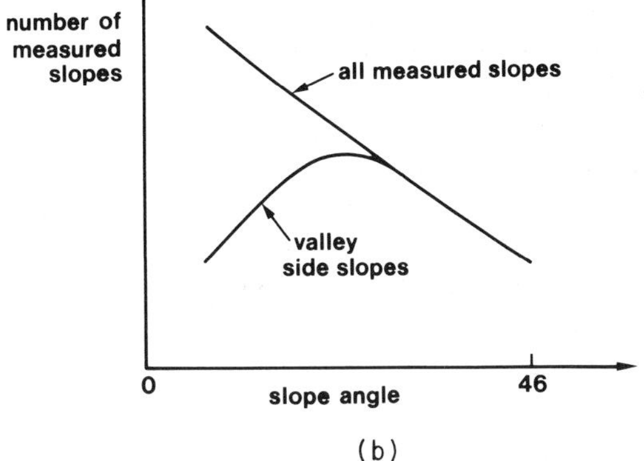

Figure 19.6 Relationships between slope angle, catchment area, and slope population. (a) (*After Gregory and Brown, 1966*, Z. Geomorph., **10**.) (b) (*After Rouse and Farhan, 1976*, Q. J. Eng. Geol., **9**.)

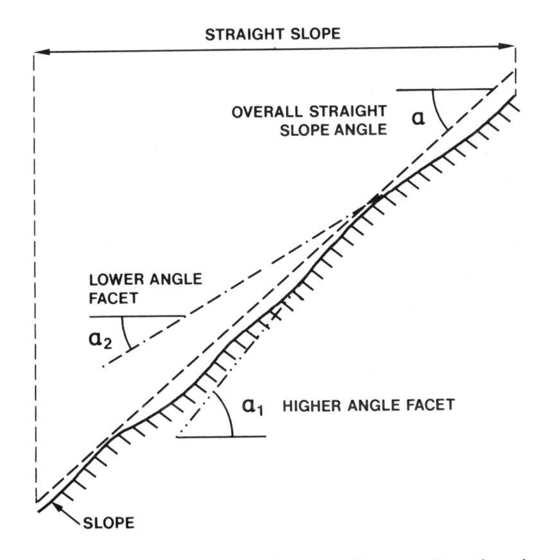

Figure 19.7 The effect of scale upon slope angle estimation

for instance, by 1:10,000 map contour intervals is rarely reflected by pantometer profiles (see Figure 19.7). It appears that the scale of slope angle measurement to be used when assessing the effects of mass movement upon form should be related to the size of landslides and the slopes themselves. As the ratio of the latter variables (each in itself a possible summary statistic of the style of instability) is obviously an important factor to be considered in slope evolution, perhaps it could be used as a guideline when designing slope form measurement programmes. There are other constraints on the choice of measurement scale, particularly the size of a study area. In site-specific studies detailed slope surveys are a prerequisite for stability analysis, but when attempting a regional assessment of slope evolution the logistical requirements for detailed measurements are formidable and it may be necessary to turn to remotely sensed data, which effectively includes large-scale maps. Further, many areas of steep slope cannot be effectively surveyed because of vegetation cover: it is virtually impossible to use a pantometer on slopes covered with thick heather and extremely difficult to use conventional surveying equipment on bracken-covered slopes during the summer period.

Many steep slopes have areas of rock outcrop downslope from which are found extremely steep sections of slope (Davies, 1963). As such slopes are often degraded or stabilized screes the slope form may be a response to rockfall processes with a steep rectilinear section adjacent to the rock outcrop or headwall. The slope angle may decline downslope to give a basal concavity (Statham, 1973). Scree weathering will eventually result in instability if drainage becomes impeded due to particle size reduction. Consequently the occurrence and position of outcrop on a slope should always be considered in assessments of slope evolution, particularly in upland areas where scree degradation has taken place obscuring evidence of the genetic slope processes.

19.3.3 Slope Evolution And Soil Attributes

A factor which should be considered in long-term slope evolution studies is the influence of weathering upon the soil and rock properties which are relevant to slope stability. Based upon examinations of vertical soil and rock profiles it has been clearly shown that weathering affects the shear strength, bulk and hence permeability properties of rock and its derived soil cover (Lumb, 1962; Chandler, 1969, 1972; Gumperayarnnont, 1976; Baynes, 1977; Irfan, 1978; Francis, 1984). This aspect has been discussed in the context of slope stability by Francis (1984), Carson and Petley (1970), Kirkby (1973), and Statham (1974, 1977) who have argued that if weathering modifies those attributes which control landsliding, then weathering must control slope evolution (see Chapter 13). Francis (1984) regards weathering processes as slope processes, and identifies soil properties which are both modified through weathering, and control soil shear strength. These properties include soil particle size, size distribution, particle form and roundness, consequent soil void ratio, particle strengths and interparticle 'friction' behaviour during shear. Further, these properties also effectively control pore and pore neck sizes, and hence soil hydraulic conductivity which affects the development of pore-water pressures. To evaluate the influence of the weathering of sandstone soils upon slope stability a weathering model was proposed in terms of granular soil mechanics, where both the parent rock and derived soil were considered as particulate masses. Treating rock as a large-scale particulate system with blocks of rock separated by discontinuities, it can be regarded as a coarse grained soil with effectively complete

Table 19.1 Changes in shear strength with weathering

	ROCK AND SOIL PARTICLE ASSEMBLAGE PROPERTIES RELEVANT TO SLOPE EVOLUTION				
	OF MOST RELEVANCE TO ROCK SLOPE/SCREE EVOLUTION		OF MOST RELEVANCE TO SOIL COVERED SLOPE EVOLUTION		
MATERIAL CHARACTERISTIC, CONTROLLED BY WEATHERING STATE	Unweathered rock, totally interlocked particles, directionally dependent maximum density granular soil/'rockfill'	Slightly weathered rock, open discontinuities, typical granular soil/'rockfill'	Transitional granular/cohesive soil*	Highly weathered normally consolidated cohesive soil	Completely weathered, normally consolidated residual clay cohesive soil
MAIN CONTROL OF SHEAR STRENGTH BEHAVIOUR	Void ratio and discontinuity orientations	Void ratio and particle mineralogy	Clay/granular particle proportions	Clay content, sensitivity, and mineralogy	
DETERMINATION AND MAGNITUDE OF SHEAR STRENGTH PROPERTIES	*In situ*/model testing, ϕ' between 70° and ϕ' of 35°, directionally dependent	*In situ*/model testing, large-scale laboratory testing, ϕ' between 50° and ϕ'_{cv} of 30°	Laboratory testing, ϕ' between 45° and ϕ'_r of 5°	Laboratory testing, *in situ* testing, ϕ' between 30° and ϕ'_r of 5°	Laboratory testing, *in situ* testing, ϕ' between 30° and ϕ'_r of 5°
LITERATURE REVIEW SOURCE	Barton (1973)	Marsal (1973), Koerner (1970)	Lupini *et al.* (1981), Kenney (1977).		Kenney (1967)

NB Clay defined by size, $<2\,\mu m$.
*Note decomposition effects; see Baynes and Dearman (1978).

interlocking between angular particles. Considerable dilation is required to mobilize a failure zone in a direction inclined to the discontinuity sets which results in a potentially large difference between ϕ' and ϕ'_{cv} (see Table 19.1). When failure is mobilized parallel to a discontinuity set then ϕ' may tend towards ϕ'_{cv} with dilation limited to the amplitude of asperities on the discontinuity surface. The magnitude of the dilation component of ϕ' under low confining stresses close to the ground surface may be highly variable from one locality to another because it is also controlled by the compressive, tensile, and shear strength of asperities. Weathering characteristically develops away from discontinuity surfaces, and even though the bulk of the rock may be intact as relatively unweathered 'corestones' the altered asperities may degrade under relatively low stresses in the same way as conventionally assumed when discussing the failure of high strength asperities under high confining stresses (Barton, 1973). This situation is analogous to particle degradation not only in strong rockfill tested under high stress conditions, but also in most weathered regoliths subject to a low stress regime (Francis, 1984). Overall, when considering rock as an engineering soil the only special characteristics to consider are the strength anisotropy, this property perhaps having a similar effect to soil fabric, and the tight interlocking. This latter characteristic can result in a ϕ' value potentially much higher than occurs in conventional soils, being up to 70°, as opposed to the values of about 50° which can be attained by extremely dense granular soils.

A particularly interesting state of weathering is found around the 'rock/soil interface'. This interface may vary in character from an abrupt contact typical of most steep slope soils to a complex zone containing corestones within a grus matrix. The latter type of interface has been well documented in studies of granitic rock weathering, and within the engineering literature the works of Moye (1955) and Ruxton and Berry (1957) have become particularly popular. Despite the geotechnical predilection with defining weathering grades for purely practical purposes two useful by-products of this type of work, particularly in Hong Kong, have been the recognition that the soil/rock interface when defined where the corestone/grus contents are equal (Dearman, 1974, 1976) may be more abrupt than is commonly envisaged, and that the concept of an interface defined in this way can have mechanical significance despite its arbitrary origins. The fact that soil landslides on steep tropical slopes are basally limited by this interface was detailed by the perceptive work of Wentworth (1943).

As can be deduced from Table 19.1, weathering of rock and soil will cause changes in shear strength with time. Further, the establishment of a soil catena over a hillslope will cause downslope variation in soil properties due to water movement. On slopes subject to landsliding soils are often eroded down to relatively unweathered bedrock which must result in distinctively episodic rather than continual changes in soil properties along particular slope profiles, reflecting the interplay of mass movement and weathering processes (Kirkby, 1973). Associated with soil stripping are vegetation cover changes, perhaps similar to the effects of deforestation which has been shown on the basis of stream solute and suspended sediment load measurements to change hillslope weathering rates (Bormann *et al.*, 1974). This may also result in changes in the soil void ratio, hence in the unit weight, infiltration rates, and pore-water pressure distributions, and cause increased soil erosion through rainsplash and overland flow. It follows that the interactions between hillslope mechanical, hydrological, pedological, and weathering processes must also be considered in slope evolution studies (Kirkby, 1977), but unfortunately little work has been conducted to this end perhaps highlighting the youthfulness of slope evolution studies. However, it is possible to make predictions of the changes in shear strength that may take place as a consequence of extended periods of weathering in two ways. Firstly, the considerable data base of experimental work on the shear strength of soils, particularly the results of studies on the effects of the variation of specific soil particle parameters upon shear strength can be applied to slope evolution problems; Table 19.1 represents a simple example of this approach. Secondly, a programme of strength testing can be conducted, using various mixtures of real and artificial soil particles, which is intended to simulate soil behaviour at some stage of future weathering.

Following the work of Kirkby (1973) and Statham (1974), Francis (1984) considered in detail the above approach as applied to sandstone soils. Emphasis was placed upon the study of ϕ'_{cv} because at that time this soil property was considered to be both important to long-term slope evolution, and a material constant over the range of stresses encountered on natural hillslopes. Further, its magnitude is dependent upon the specific soil particle characteristics which are incorporated in the weathering model mentioned earlier. In contrast, most earlier work had considered the effects of soil particle characteristics upon peak shear strength ϕ' (i.e. Holtz and Gibbs, 1956; Kolbuszewski and Frederick, 1963; Koerner, 1970) and little is known about the effects of soil particle characteristics upon ϕ'_{cv}, although it is generally assumed that they are the same.

A particular problem experienced in the study of changes in soil characteristics with weathering concerns the large-sized particles found in most real soils. For instance, soils developed on the Pennant Sandstone, South Wales, contain 26% cobbles and boulders and 43% gravel: conventional testing apparatus is incapable of holding soils which contain such large particles, but as changes in particle size with weathering may have significant effects upon shear strength (Kirkby, 1973) it is necessary to consider their influence. The experience gained in using the apparatus required for assessment of rockfill (very coarse-grained aggregate) dam construction (Marachi *et al.*, 1972; Marsal, 1973) is of limited applicability to slope studies because the stress ranges considered over which the machines were designed to operate are too high to allow meaningful comparison with the relatively insignificant stresses relevant to thin hillslope soil stability. Also, when attempting to compare the behaviour of highly dilatant soils which invariably degrade during straining over a large stress range, Coulomb's law becomes inapplicable due to substantial envelope curvature which is primarily caused by interparticle deformation mechanisms whose influences are ignored in this empirical relationship. In order to assess the effect of large particles upon slope soil shear strength, large-scale shear tests at low normal loads and high void ratios are required. As there is presently insufficient experimental evidence available to distinguish between the effects of large particles upon shear strength and the effects of changes in void ratio upon shear strength which accompany selective inclusion or complete exclusion from soils (scalping) of large particles (of usually unspecified form attributes) the development of suitable testing equipment should be a research priority. Limited results obtained from large-scale angle-of-repose measurements conducted by the author suggest that gravel and cobble sized particles give ϕ'_{cv} values in the range of 38° to 40°, while sand sized particles give lower values between 32° and 36° (see Figure 19.8). If a comparison with cohesive soils is acceptable, then the opposite scale dependence of strength is seen: cohesive soil shear strength parameters reduce in value with increase in scale due to the inclusion of macrofabric discontinuity sets in the test sample (Bishop and Little, 1967).

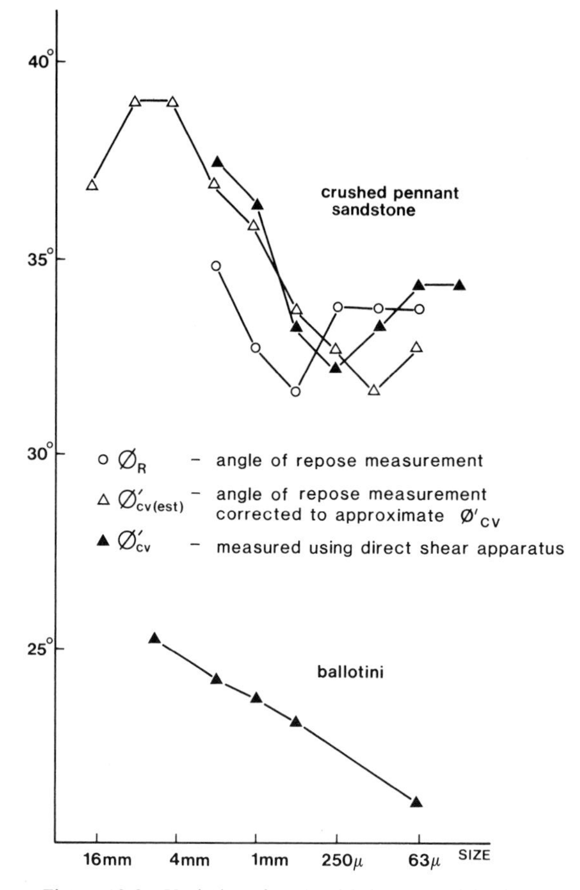

Figure 19.8 Variations in ϕ'_{cv} with large particle sizes

Back-analysis of closely studied landslides allows the estimation of shear strength at true scale (Chandler, 1976), but the need for detailed groundwater data to make a reliable back-analysis restricts this approach to cohesive soils. With granular soils relatively free drainage reduces the reliability of estimates of pore-water pressures during failure.

Assuming that a scale-dependent upper bound to ϕ'_{cv} for granular soils is approximately 40°, it is useful to consider the results of shear strength testing of various properties on real and artificial soils which can be imagined as simulated granular soils at progressive stages of weathering. An extensive shear testing programme conducted by Francis (1984) using sandstone-derived soils and crushed rock showed that only small changes in ϕ'_{cv} occur for a wide range of particle sizes, roundness, form, and gradings, all results being within the range of 32°–40°. Particle angularity was shown to have the greatest effect upon ϕ'_{cv}, but even highly rounded millet-seed sand grains gave ϕ'_{cv} values above 32°. The extremely low ϕ'_{cv} values in the 21°–25° range typical of ballotini are certainly not characteristic of real soils. For the soils used (particularly those derived from Pennant Sandstone) both particle form and grading varied substantially, but had negligible influence upon ϕ'_{cv}. The importance of grading as a control over ϕ'_{cv}. The importance of grading as a control over ϕ'_{cv} was emphasized by Statham (1974), but a series of tests conducted using similar gradings of crushed rock and ballotini could not reproduce the variability reported in his work.

There is only one grading condition when significant changes in ϕ'_{cv} of a real soil may occur as a consequence of weathering. This arises when the clay mineral content of a soil has increased to the point where soil particle movement during shear changes from the dilatory sliding and rolling movements typical of granular soils (Skinner, 1969) to a style of motion dominated by sliding between clay micelles which become progressively aligned parallel to the direction of shear with increased straining. This leads to the definition of relatively narrow 'shear planes' along which displacement takes place, rather than the 'shear zones' typical of granular soils. Coincident with the change in deformation style is a reduction in strength due to the minimal dilatory movement required to allow straining, together with a change in interparticle friction regime. By further increasing the clay content of a soil through weathering, shear behaviour becomes totally dominated by clay mineralogy, when at large strains the soil residual strength, ϕ'_r, is attained (see Table 19.1). In a detailed study of residual strength Lupini (1980), following the work of Kenney (1977), has shown that the transition between ϕ'_{cv} and ϕ'_r as a consequence of increasing the clay content of simple soils (mainly bimodal gradings) occurs within a narrow range of soil gradings referred to as the 'transitional zone'. Whether or not this type of behaviour will occur with well-graded soils is uncertain. However, it is certain that, as weatherable soil constituents degrade to clay minerals, their distribution within the soil will depend upon pedological processes with weathering rims, illuviation, and eluviation leading to relatively clay-rich and clay-deficient zones occurring within the soil. It should also be noted that a possibly comparable strength transition occurs at peak strength conditions (ϕ'_p), where by increasing the clay content of normally consolidated artificial soil a dramatic strength decrease is observed (Holtz and Ellis, 1961; Kawakami and Abe, 1970).

In the above discussion it is implicitly assumed that a slope soil has been derived from rocks which contain constituents which will weather to produce clay minerals, such as feldspathic and ferromagnesian rocks, heavily over-consolidated mudrocks, and rocks containing argillaceous rock clasts and matrix. Eventually the clay-rich soil may modify its mineralogy as a response to a change in the weathering environment; further small changes in ϕ'_r may then occur, since as Kenney (1967) has shown, ϕ'_r is dependent on clay mineralogy (see Table 19.1).

Another factor which must be considered during slope soil weathering is the effect of the soil normal stress (σ), being the product of overburden thickness and unit weight above a plane of interest; this has two important consequences for slope stability. Firstly, a granular soil strength envelope is effectively linear at very low confining stresses (Ponce and Bell, 1971; Francis, 1984), although it becomes significantly curved at high stresses. Consequently it is reasonable to assume that ϕ' and ϕ'_{cv} are independent of σ at shallow depths. However, the residual strength failure envelope for a cohesive soil is found to be convex at low confining stresses when determining ϕ'_r from ring shear tests (Bishop *et al.*, 1971) while the peak strength envelope is practically linear. Therefore, variations in slope soil normal stress may affect the stability of clay soil slopes and it would be unreasonable to expect both unit weight and depth of overburden to remain indefinitely constant. One situation where residual strength would certainly be reduced through increasing normal stresses within the soil would occur downslope of a translational landslide, where the deposition of mobilized soil would increase the overburden and possibly double the normal load on a potential failure plane and hence reduce the factor of safety of the slope.

Secondly, a real granular soil will always experience significant particle degradation with strain even under low confining stresses, resulting in substantial regrading. It follows that feedback may occur, since as grading changes during straining both before and after failure, so does deformation style. The most important practical consequence of this for natural slopes may be the reduction in soil permeability along a shear zone rather than a change in the angle of internal shearing resistance.

While considering normal stresses in a soil, it is useful to examine the unit weight parameter in more detail. The relationship

$$\gamma_s = \gamma_w \left(\frac{SG + e}{1 + e} \right)$$

where γ_s is the saturated unit weight of soil, γ_w is the unit weight of water, e is the void ratio, and SG the specific gravity of the soil solids, indicates that γ_s is fairly insensitive to changes in e. For this reason alone precise estimates of γ_s are rarely required in practice, but approximations are often necessary simply due to measurement problems, particularly with granular soils containing large particles. Nevertheless, soil unit weight is an important parameter to be considered in virtually all stability analyses and its relevance to steep slope evolution must be considered in terms of the variations in soil void ratio and specific gravity which occur with time. As soil structure, which not only controls total soil void space but also void space size distribution, is dependent to a large extent upon low density soil organic matter, then void ratio and specific gravity are interdependent. Nevertheless it is convenient to speculate on their contribution to long-term slope evolution separately, especially as the role of void ratio in controlling shear strength can be relatively easily visualized (as for instance illustrated by Bishop and Green, 1965).

The higher the organic content of a soil, the lower its unit weight; thus, since the organic content of a soil declines with depth, there is an increase in unit weight with soil depth. The dependence of unit weight upon organic content of the soil indicates that land use may be particularly important to slope stability. Soils underlying bracken covered areas tend to have low unit weights compared with areas of grassland, and the impact on slope stability of substantial deforestation through changes in both the organic content and its actions upon the soil, needs clarification. On a longer time-scale soil horizon development is a consequence of the soil organic content which may not change overall soil unit weight significantly, provided that soil translocation processes act only in a vertical sense. Over such a period it can be argued that climate could modify soil unit weight through vegetation control. In Figure 19.9, typical values of unit weight, void ratio, and specific gravity are given for a leached brown earth soil which are characteristic of steep slope soils in temperature areas. Unit weight can be seen to vary little within this profile until the soil–rock interface is reached. Here a marked reduction in void ratio occurs at the base of highly discontinuous rock which often lies between the C horizon of the soil and true 'rock'. As a rough guide, organic-rich soils may attain saturated unit weights of 14–16 kN m^{-3} but compact, virtually organic-free soils may return saturated unit weights as high as 18–20 kN m^{-3}. Higher values of up to 22 kN m^{-3} can be achieved but such soils are either virtually mudrocks or granular soils with void ratios tending towards e_{min} (Kolbuszewski,

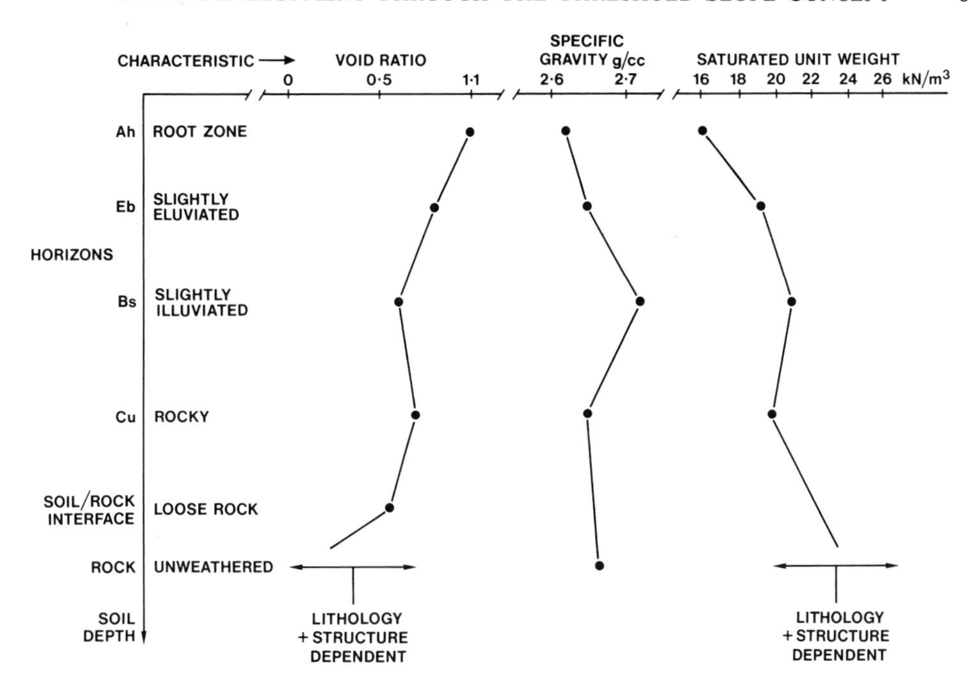

Figure 19.9 Variations in void ratio, specific gravity, and unit weight with soil depth

Table 19.2 Results of stability analyses using various combinations of ϕ' and γ_s

		γ_s (kN m^{-3})				
		14.7	17.5	18.2	19.5	20.2
ϕ' (°)	35	13	17	—	—	—
	38	15	19	20	21	22
	43	—	—	—	25	26
	53	—	—	—	—	34

Maximum stable slope angles (°)

1948) which require minimally weathered rock or intense compaction, to give a void ratio of around 0.5. Many areas of steep slope are uplands and if dense tills, particularly lodgement tills, are present then such values may be attained.

Unfortunately it is unrealistic to consider soil unit weight as an independent variable in slope stability analysis. This is because low void ratios result in high values of both unit weight and ϕ', further complicated by increases in unit weight enhancing the factor of safety when the water table lies above the potential basal shear zone of a slope soil. This is illustrated by reference to Table 19.2 which was compiled by using the infinite slope analysis for saturated soil with water flow parallel to the slope and shows the maximum stable slope angles for various soil strengths and unit weights.

The measurements were made upon granular soils and are discussed in detail by Francis (1984). Essentially, it is only realistic to consider high values of ϕ' when void ratios are low and, therefore, unit weights are high. For instance, if saturated unit weights were as high as 19.5–20.2 kN m^{-3}, ϕ' would be greater than 38° because of the low void ratio indicated by the high unit weight: in this case a value of 43° is more realistic. The unit weight of 20.2 kN m^{-3} corresponds to a ϕ' value of 53°, which is extremely high and perhaps only representative of heavily consolidated tills. It should also be noted that by excluding the results in Table 19.2 which refer to the low unit weight of 14.7 kN m^{-3}, any single soil sample having undergone suitable compaction treatment could give rise to the entire range of stable slope angles shown. From the foregoing discussion it is evident that the major influence of unit weight upon slope stability in the long term is indirect, through its association with void ratio and hence ϕ'. In addition, vegetation colonization of previously barren areas such as following climatic change or in fresh landslip scars may cause reduction in soil unit weight. The effects of vegetation on slope stability have been discussed in Chapter 6.

When a shallow landslide occurs in a granular soil most of the displaced soil is disturbed and consequently the post-failure unit weight would be expected to differ from the original undisturbed unit weight; this difference not being confined to the basal shear zone. At the onset of failure a high ϕ' value together with a high unit weight value (if the slip surface were subject to positive pore-water pressures) would be relevant. However, at the cessation of instability the important parameters controlling stabilization are ϕ'_{cv} and a low unit weight, as during failure the void rato will tend towards e_{cv} which is only slightly less than e_{max}, the maximum void ratio attainable by the soil. This is illustrated by reference to Table 19.2 where if at failure $\phi' = 43°$ and $\gamma_s = 20.2$ kN m^{-3} movement would reduce ϕ'_{cv}, in this case to 38°. If we assume the same unit weight to be applicable then stability is regained with a slope angle of 22°. It would be more realistic to use a reduced post-failure unit weight, perhaps 17.5 kN m^{-3}, which would correspond to a stable angle of 19°. This approach neglects the other consequence of decreasing the unit weight of the soil through increasing the void ratio. For example, if the volume of water in a landsliding soil either remains constant or reduces in value then the pore-water pressure variable m must decrease. This will act to increase the stability of the moving soil mass and bring it to rest, not 'at the worst possible conditions' where $m = 1$ but at a higher angle with $m < 1$. This would be the case even if the soil mass did not drain during failure, which would be highly unlikely considering the soil disruption and the increases in void ratio which would occur, possibly even creating significant negative pore-water pressures.

As discussed earlier, on steep slopes a failed soil mass may disintegrate, and discussion of water-table levels within the soil mass becomes spurious. Given conditions when pore-water pressures reduce during landsliding, as internal deformation enhances void space and drainage, then the entire concept of threshold slope angles reflecting landsliding with saturated conditions collapses. It is perhaps surprising that the effects of drainage during failure have not previously been considered in this context, considering the steep backscar and toe lobe angles commonly seen to occur with landslides, together with the levées observed at the margins of debris flows.

It is appropriate to consider the main problem revealed by research which has attempted to correlate slope angles with the results of slope stability analyses. This research shows that virtually all steep slope angle populations tend to be normally distributed about slope angles in the 20°–30° range, typically between 22° and 28° (see the histograms in the works of Carson, 1975; Rouse and Farhan, 1976; Van Asch, 1983; Francis, 1984). In environments where the infinite slope analysis is appropriate these slope angles cannot be convincingly explained in terms of mass movement for saturated ($m = 1$) or dry ($m = 0$) conditions. This is contrary to Carson's (1975, 1976) suggestion that it is only relevant to consider these extremes of groundwater conditions because they are likely to be experienced in the long term. While finding this suggestion convincing, the available data indicate that within the limits of the stability models available, steep slope angle populations are distributed about slope angles which can only be explained by using a value of m intermediate between the two extreme groundwater conditions. Incorporating slope angle measurements with soil property data and conducting back-analyses enables values of m for limiting stability to be estimated. m values corresponding to mean observed slope angles have been estimated to lie in the ranges of 0.37–0.69 (Van Asch, 1983) and 0.28–0.89 (Francis, 1984). These findings do not preclude soil saturation having taken place in the past provided that drainage occurred during failure.

19.3.4 Slope Evolution and Hydrology

The conclusion of the preceding section suggests that, while discussing relatively discriminative uses of slope stability models in slope evolution studies, it is worth considering simple modifications to the hydrological assumptions used in assessing the influence of slope stability on form. Hillslope hydrology studies in the last 20 years have led to an important understanding of groundwater movement on slopes. As it is commonly acknowledged that the effects of groundwater upon slope stability are difficult to assess because of measurement problems, an input of hydrological expertise to slope stability evaluation is urgently needed (see Anderson and Howes, 1985). Two aspects of hillslope hydrology which are particularly relevant to stability assessment are the effects of convergent flow into hollows, and the upslope extension of the saturated wedge adjacent to drainage channels. For the moment, however, only two simple modifications of popular groundwater assumptions will be considered.

Firstly, it is common practice to assume that the worst possible groundwater flow condition for stability occurs when the ground is saturated and flow is parallel to the slope surface. The importance of this condition has been heavily emphasized by advocates of threshold slopes (Carson, 1975; Rouse, 1975) with the former defining the maximum slope angle possible under these conditions as the 'semi-frictional' type of threshold slope angle as predicted by the infinite slope analysis (Carson, 1975, 1976). However flow lines are not always parallel to the slope, with horizontal flow lines at valley side spring lines forming an important example (Knox, 1927; Fookes *et al.*, 1972; Francis 1984). Taking this flow condition, the pore-water pressure on a hypothetical shear plane would approximate to $mz\,\gamma_w$ (where m is the ratio of water-table height above shear plane to the depth z of the shear plane below the surface, with γ_w being the unit weight of water), which is greater than under conditions of parallel flow when pore-water pressure would equal $mz\,\gamma_w \cos\alpha$ (where α is the slope angle). Restating the infinite slope analysis for conditions of horizontal water flow at failure gives

$$\frac{m\gamma_w}{\gamma_s} \tan\phi'(\cos^2\alpha)^{-1} + \tan\alpha - \tan\phi' = 0$$

(where γ_s is the saturated unit weight of soil and ϕ' is the angle of internal shearing resistance).

From this relationship α can be found using Newton's method. The difference in results obtained using the two flow conditions can be significant. For example, with $m=0.5$, $\gamma_s = 18\ \mathrm{kN\,m^{-3}}$ and $\phi' = 42°$, then for flow parallel to the slope $\alpha = 33°$ and for horizontal flow $\alpha = 30°$. A situation in which this difference is important occurs when localized horizontal flow about a spring acts as a trigger mechanism, with associated instability reducing toe support for an otherwise stable uphill section of slope, resulting in more widespread retrogressive failure through strain softening. Even less favourable stability conditions will result if groundwater flow is upwards, but such occurrences would require confinement. This can be provided by low permeability iron pans and illuviated horizons capping gravel-rich subsoil which is above low permeability bedrock, such as seen on Exmoor slopes.

The second reconsideration of groundwater assumptions concerns the hydrological characteristics of the soil mass which has actually failed. Implicit in the use of the $m=1$ condition to predict the lowest angle of slope controlled by mass movement is the assumption that a failed soil mass comes to rest at an angle controlled by the same groundwater conditions that occurred at failure. However, if the hydrological characteristics of the soil change as a consequence of failure then the section of slope from which soil has been eroded, the 'backscar', will stand at a different angle from the translated, failed soil mass. With weathered clays of low hydraulic conductivity and good coherence original hydrological characteristics may be maintained by the failed mass, but Chandler's (1982) findings indicate that changes may take place. He reported that the lowest-angled Lias Clay slopes studied in Central England have stabilized for conditions of $m<1$, and that if $m=1$, then the slopes would become unstable again. Chandler's reasoning seems to imply that these slopes became unstable when m was less than unity, but it is the author's contention that the slopes may have mobilized at $m=1$ and drained rapidly enough in the shear zone to reduce m to less than one, in that stability was gained partly because of reducing pore-water pressure. This argument is proposed because of recent experience with granular soils (Francis, 1984), where rapid pore drainage can take place once landsliding is initiated. Further, granular soil drainage rates are enhanced because the void ratio of the soil increases about the basal shear zone and throughout the entire failed soil mass. This is particularly characteristic of shallow failures resulting in the void ratio tending towards its maximum value. On steep slopes the failed soil mass may entirely disintegrate and its downslope depositional behaviour may be controlled by individual soil particle movement, perhaps in a fashion similar to that seen on scree slopes but with retardation of particle movement being essentially controlled by vegetation. It is unlikely that this type of behaviour can be modelled by conventional stability analysis methods. At present it appears that more caution should be exercised when assuming pore-water pressure conditions which affect landslide restabilization, and in many cases the assumption of saturation which may be valid at the onset of instability may be unsound when applied to the failed soil mass.

The potential changes in soil hydraulic conductivity with soil weathering should also be considered in the context of mass movement as drainage conditions will affect the magnitude of pore-water pressures. There appear to be several factors which control the hydraulic conductivity of slope soils during weathering. Firstly, weathering reduces the particle sizes of a soil. Assuming for the moment that other soil particle characteristics remain constant a reduction in particle sizes causes a reduction in hydraulic conductivity; also the greater the range of particle sizes in a soil, the lower the hydraulic conductivity

(Graton and Fraser, 1935). As the cross-sectional area of pore necks is the main control over water movement (ignoring wetted surface area, capillarity, tortuosity, etc.) and provided that a soil is relatively free of fine-grained particles, then void ratio can be used as a soil-specific indicator of soil hydraulic conductivity when considering changes in particle size distribution with weathering (Fraser, 1935; Beard and Weyl, 1973; Statham, 1974).

If it is not assumed that soil particle characteristics are independent of size then particle shape, roundness, and mineralogy (especially with finer soils) also have to be considered as these parameters are size dependent, varying considerably with the state of weathering (Francis, 1984) and affecting the void ratio and hence hydraulic conductivity (Fraser, 1935; Gaither, 1953; Rogers and Head, 1961; Beard and Weyl, 1973). A further important control over soil hydraulic conductivity is the soil organic material, which promotes soil structure with the development of a secondary soil hydraulic conductivity, and differentiation into soil horizons of different hydrological properties.

19.4 SUMMARY

This chapter has not attempted to provide answers to the problem of assessing the influence of mass movement upon slope evolution. Instead it has highlighted a series of unresolved problems which are relevant to stability modelling. It is evident that the state of the art is insufficiently precise to allow sweeping evolutionary statements to be made (Anderson *et al.*, 1980). In the short term, site-specific studies indicate that mass movement can control the form of steep slopes, but inevitably evolutionary interpretations become more speculative over longer time-scales. This is a particular problem as most studies of slope development have been made in temperate regions where palaeoclimates have been variable: perhaps this variability can be minimized by looking in more detail at slopes which have evolved under more constant environmental conditions.

REFERENCES

Alonso, E. E. (1976). 'Risk analysis of slopes and its application to slopes in Canadian sensitive clays.' *Geotechnique*, **26**, no. 3.

Anderson, M. G., and Howes, S. (1985). 'Development and application of a combined soil water–slope stability model.' *Q. J. Eng. Geol.*, **18**, 225–36.

Anderson, M. G., and Richards, K. S. (1981). 'Geomorphological aspects of slopes in mudrocks of the United Kingdom.' *Q. J. Eng. Geol.*, **14**, 363–72.

Anderson, M. G., Richards, K. S., and Kneale, P. E. (1980). 'The role of stability analysis in the interpretation of the evolution of threshold slopes.' *Trans. Inst. Brit. Geogrs.*, **5**, 100–12.

Barton, N. (1973). 'Review of a new shear-strength criterion for rock joints.' *Eng. Geol.*, **7**, 287–332.

Baynes, F. J. (1977). *Engineering characterisation of weathered rock*. Ph.D. Thesis. Univ. Newcastle upon Tyne.

Baynes, F. J., and Dearman, W. R. (1978). 'The relationship between the microfabric and the engineering properties of weathered granite.' *Bull. Int. Assoc. Eng. Geol.*, **18**, 191–7.

Beard, D. C., and Weyl, P. K. (1973). 'Influence of texture on porosity and permeability of unconsolidated sand.' *Bull. Am. Assoc. Petrol. Geol.*, **57**, 349–69.

Bishop, A. W., and Green, G. E. (1965). 'The influence of end restraint on the compressive strength of a cohesionless soil.' *Geotechnique*, **15**, 243–66.

Bishop, A. W., Green, G. E., Garga, V. K., Andresen, A., and Brown, J. D. (1971). 'A new ring shear apparatus and its applications to the measurement of residual shear strength.' *Geotechnique*, **21**, 273–328.

Bishop, A. W., and Little, A. L. (1967). 'The influence of size and orientation of the sample on the apparent strength of the London Clay at Maldon, Essex.' *Proc. Geotech. Conf.*, Oslo, **1**, 89-96.
Bormann, F. H., Likens, G. E., Siccama, T. G., Pierce, R. S., and Eaton, J. S. (1974). 'The export of nutrients and recovery of stable conditions following deforestation at Hubbard Brook.' *Ecol. Monogr.*, **44**, 255-77.
Brunsden, D., Fookes, P. G., Jones, D. K. C., and Kelly, J. M. H. (1975). 'Large scale geomorphological mapping and highway engineering design.' *Q. J. Eng. Geol.*, **8**, 227-53.
Brunsden, D., and Kesel, R. H. (1973). 'Slope development on a Mississippi River bluff in historic time.' *J. Geol.*, **81**, 576-97.
Carson, M. A. (1969). 'Models of hillslope development under mass failure.' *Geog. Anal.*, **1**, 76-100.
Carson, M. A. (1971). 'Application of the concept of threshold slopes to the Laramide Mts, Wyoming. I.B.G. Special Publication No. 3.' *Slopes: Form and Process*. Brunsden, D. (ed.), pp. 31-49.
Carson, M. A. (1975). 'Threshold and characteristic angles of straight slopes.' In *Mass Wasting*. Yatsu, E. *et al.* (eds), *Proc. 4th Guelph Symp. Geomorph*, 19-34.
Carson, M. A. (1976). 'Mass wasting, slope development and climate.' In *Geomorphology and climate*. Derbyshire, E. (ed.), Wiley, pp. 101-36.
Carson, M. A., and Kirkby, M. J. (1972). *Hillslope Form and Process*. Cambridge. pp. 475.
Carson, M. A., and Petley, D. J. (1970). 'The existence of threshold slopes in the denudation of the landscape.' *Trans. Inst. Br. Geog.*, **49**, 71-92.
Chandler, R. J. (1969). 'The effect of weathering on the shear strength of Keuper Marl.' *Geotechnique*, **19**, 321-34.
Chandler, R. J. (1972). 'Lias Clay: weathering processes and their effect on shear strength.' *Geotechnique*, **22**, 403-31.
Chandler, R. J. (1976). 'The history and stability of two Lias clay slopes in the Upper Gwash Valley, Rutland.' *Phil. Trans. Roy. Soc. London*, **A283**, 463-91.
Chandler, R. J. (1982). 'Lias Clay slope sections and their implications for the prediction of limiting or threshold slope angles.' *Earth Surf. Proc. Landforms*, **7**, 427-38.
Craig, R. F. (1978). *Soil Mechanics* (2nd edn). Van Nostrand Reinhold.
Davies, P. (1963). *A study of some aspects of landsliding and slope development in an area of South Wales*. Ph.D Thesis, Univ. Wales, Swansea.
Dearman, W. R. (1974). 'Weathering classification in the characterisation of rocks for engineering purposes in British practice.' *Bull. Int. Ass. Eng. Geol.*, **9**, 33-42.
Dearman, W. R. (1976). 'Weathering classification in the characterization of rock, a revision.' *Bull. Int. Ass. Eng. Geol.*, **13**, 123-7.
Dunkerley, D. L. (1980). 'The study of the evolution of slope form over long periods of time: a review of methodologies and some observational data from Papua New Guinea.' *Zeit. für Geomorph.*, **24**, 52-67.
Farhan, Y. I. (1976). *A geomorphological engineering approach to terrain classification*. Ph.D. Thesis, Univ. Wales, Swansea.
Fookes, P. G., Hinch, L. W., and Dixon, J. C. (1972). 'Geotechnical considerations of the site investigation for stage IV of the Taff Vale trunk road in South Wales.' *Proc. 2nd Brit. Reg. Conf. P.I.A.R.C.* Cardiff, 1-23.
Francis, S. C. (1984). *The geotechnical properties of weathering sandstone regoliths*. Ph.D. Thesis, Univ. London.
Fraser, H. J. (1935). 'Experimental study of the porosity and permeability of clastic sediments.' *J. Geol.*, **43**, 910-1010.
Gaither, A. (1953). 'A study of porosity and grain relationships in experimental sands.' *J. Sedim. Petrol.*, **23**, 180-91.
Gostelow, T. P. (1977). 'The development of complex landslides in the Upper Coal Measures at Blaina, South Wales.' *Int. Symp. Geotech. Structurally Complex Formations*, Capri. Ital. Geotech. Ass. 255-68.
Gratton, L. C., and Fraser, H. J. (1935). 'Systemic packing of spheres, with particular relation to porosity and permeability.' *J. Geol.*, **43**, 785-909.
Gregory, K. J., and Brown, E. H. (1966). 'Data processing and the study of landform.' *Zeit. für Geomorphol.*, **10**, 237-63.

Gumperayarnnont, H. (1976). *Engineering geological aspects of weathered sandstones from South-West England*. Ph.D. Thesis, Univ. Newcastle upon Tyne.
Holtz, W. G., and Ellis, W. (1961). 'Triaxial shear characteristics of clayey gravel soils.' *US Dept. Interior Bureau Reclam.* Rept. No. EM577.
Holtz, W. G., and Gibbs, H. J. (1956). 'Triaxial shear tests on pervious gravelly soils.' *Proc. Am. Soc. Civ. Eng. J. Soil Mech. Fndtn Div.*, **82**, 1–22.
Hutchinson, J. N. (1967). 'The free degradation of London Clay cliffs.' *Proc. Geotech. Conf.*, Oslo, **1**, 113–18.
Hutchinson, J. N., and Gostelow, T. P. (1976). 'The development of an abandoned cliff in London Clay at Hadleigh, Essex.' *Phil. Trans. Roy. Soc., London*, **283A**, 557–604.
Irfan, T. Y. (1977). *Engineering properties of weathered granite*. Ph.D. Thesis, Univ. Newcastle upon Tyne.
Kawakami, H., and Abe, H. (1970). 'Shear characteristics of saturated gravelly clays.' *Trans. Jpn. Soc. Civ. Eng.*, **2**, 295–8.
Kenney, T. C. (1967). 'The influence of mineral composition on the residual strength of natural soils.' *Proc. Geotech. Conf.*, Oslo, 123–9.
Kenney, T. C. (1977). 'Residual strengths of mineral mixtures.' *Proc. 9th Int. Conf. Soil Mech.*, **1**, 155–60.
Kirkby, M. J. (1973). 'Landslides and weathering rates.' *Geol. Appl. Idrogeologia* Bari. 8/1, 171–83.
Kirkby, M. J. (1977). 'Soil development models as a component of slope models.' *Earth Surf. Proc. Landforms*, **2**, 203–30.
Knox, G. (1927). 'Landslides in South Wales valleys.' *Proc. Trans. South Wales Inst. Eng.*, **43**, 161–247, 257–90.
Koerner, R. M. (1970). 'Effect of particle characteristics on soil strength.' *Proc. Am. Soc. Civ. Eng. J. Soil Mech. Fndtn Div.*, **96**, 1221–34.
Kolbuszewski, J. (1948). 'An experimental study of the maximum and minimum porosities of sands.' *Proc. 2nd Int. Conf. Soil Mech. Fndtn. Eng.*, Rotterdam, 158–65.
Kolbuszewski, J., and Frederick, M. R. (1963). 'The significance of particle shape and size on the mechanical behaviour of granular materials.' *Proc. Eur. Conf. Soil Mech. Fndtn. Eng.*, Wiesbaden, **1**, 253–63.
Lumb, P. (1962). 'The properties of decomposed granite.' *Geotechnique*, **12**, 226–43.
Lupini, J. F. (1980). *The residual strength of soils*. Ph.D. Thesis, Univ. London.
Lupini, J. F., Skinner, A. E., and Vaughan, P. R. (1981). 'The drained residual strength of cohesive soils.' *Geotechnique*, **31**, 181–214.
Marachi, H. D., Chan, G. K., and Seed, H. B. (1972). 'Evaluation of properties of rockfill materials.' *Proc. Am. Soc. Civ. Eng., J. Soil Mech. Fndtn Div.*, **98**, 95–112.
Marsal, R. J. (1973). 'Mechanical properties of rockfill.' In *Embankment Dam Engineering*. Hirshfeld, R. C., and Poulos, S. J. (eds.), Wiley, pp. 109–200.
Moye, D. G. (1955). 'Engineering geology for the Snowy Mountains scheme.' *J. Inst. Eng. Aust.*, **27**, 281–99.
Parsons, A. J. (1979). 'Plan form and profile form of hillslopes.' *Earth Surf. Proc.*, **4**, 395–402.
Ponce, V. M., and Bell, J. M. (1971). 'Shear strength of sand at extremely low pressures.' *Proc. Am. Soc. Civ. Eng., J. Soil Mech. Fndtn Div.*, **97**, 625–38.
Priest, S. D., and Brown, E. T. (1983). 'Probabilistic stability analysis of variable rock slopes.' *Trans. Inst. Mining Metallury*, **92**, A, Jan.
Richards, K. S., and Anderson, M. G. (1978). 'Slope stability and valley formation in glacial outwash deposits, North Norfolk, *Earth Surf. Proc.*, **3**, 301–18.
Rogers, J. J., and Head, W. B. (1961). 'Relationships between porosity, median size, and sorting coefficients of synthetic sands' *J. Sedim. Petrol.*, **31**, 467–70.
Rouse, W. C. (1969). *An investigation of the stability and frequency distribution of slopes in selected areas of West Glamorgan*. Ph.D. Thesis, Univ. Wales, Swansea.
Rouse, W. C. (1975). 'Engineering properties and slope form in granular soils.' *Eng. Geol.*, **9**, 221–35.
Rouse, W. C., and Farhan, Y. I. (1976). 'Threshold slopes in South Wales.' *Q. J. Eng. Geol.*, **9**, 327–38.

Ruxton, B. P., and Berry, L. (1957). 'Weathering of granite and associated erosional features in Hong Kong.' *Bull. Geol. Soc. Am.*, **68**, 1263–92.

Savigear, R. A. G. (1952), 'Some observations on slope development in South Wales.' *Trans. Inst. Brit. Geog.*, **18**, 31–51.

Selby, M. J. (1982). *Hillslope Materials and Processes*. Oxford Univ. Press, Oxford, 264pp.

Skinner, A. E. (1969). 'A note on the influence of inter-particle friction on the shearing strength of a random assembly of spherical particles.' *Geotechnique*, **19**, 150–7.

Smith, G. N. (1968). *Elements of Soil Mechanics for Civil and Mining Engineers*. Crosby Lockwood, London, 341pp.

St George, J. D. (1984). 'A probabilistic approach to slope stability assessment in surface mining operations.' *Dept. of Min. Mag. Univ. Nottingham*, **36**, 51–7.

Statham, I. (1973). *Process-form relationships in a scree system developing under rockfall*. Ph.D. Thesis, Univ. Bristol.

Statham, I. (1974). 'The relationship of porosity and angle of repose to mixture proportions in assemblages of different sized materials.' *Sedimentology*, **21**, 149–62.

Statham, I. (1977). *Earth Surface Sediment Transport*. Clarendon Press, Oxford, 184pp.

Van Asch, Th. W. J. (1983). 'The stability of slopes in the Ardennes region.' *Geol. Mijnbouw*, **62**, 683–8.

Wallace, R. E. (1977). 'Profiles and ages of young fault scarps, north-central Nevada.' *Bull. Geol. Soc. Am.*, **88**, 1267–81.

Wentworth, C. K. (1943). 'Soil avalanches on Oahu, Mauaii.' *Bull. Geol. Soc. Am.*, **54**, 53–64.

Williams, P. J. (1982). *The Surface of the Earth: An Introduction to Geotechnical Science*. Longman.

Wu, S. S. C., Garcia, P. A., Jordan, R., Schafer, F. J., and Skiff, B. A. (1984). 'Topography of the shield volcano, Olympus Mons on Mars.' *Nature*, **309**, 432–5.

Author Index

Subject Index